Developments in Volcanology 2

ARC VOLCANISM

DEVELOPMENTS IN VOLCANOLOGY

Further titles in this series

1 *H. TAZIEFF and J.C. SABROUX*
FORECASTING VOLCANIC EVENTS

DEVELOPMENTS IN VOLCANOLOGY 2

ARC VOLCANISM

Selected papers from the international symposium on "Arc Volcanism" held in Tokyo and Hakone and sponsored by the Volcanological Society of Japan and the International Association of Volcanology and Chemistry of the Earth's Interior, August 31–September 5, 1981

edited by

S. ARAMAKI

Earthquake Research Institute, University of Tokyo, Bunkyo-ku, Tokyo 113, Japan

and

I. KUSHIRO

Geological Institute, University of Tokyo, Bunkyo-ku, Tokyo 113, Japan

reprinted from *Journal of Volcanology and Geothermal Research* Volume 18

ELSEVIER, Amsterdam – Oxford – New York – Tokyo 1983

ELSEVIER SCIENCE PUBLISHERS B.V.
Molenwerf 1
P.O. Box 211, 1000 AE Amsterdam, The Netherlands

Distributors for the United States and Canada:

ELSEVIER SCIENCE PUBLISHING COMPANY INC.
52, Vanderbilt Avenue
New York, NY 10017

Library of Congress Cataloging in Publication Data
Main entry under title:

Arc volcanism.

(Developments in volcanology ; 2)
"Reprinted from Journal of volcanology and geothermal research, volume 18."
Includes bibliographies.
1. Island arcs--Congresses. 2. Volcanism--Congresses.
I. Aramaki, S. (Shigeo), 1930- . II. Kushiro, I. (Ikuo), 1934- . III. Nihon Kazan Gakkai.
IV. International Association of Volcanology and Chemistry of the Earth's Interior. V. Journal of volcanology and geothermal research. VI. Series.

QE511.2.A7 1983 551.2'1 83-14115
ISBN 0-444-42234-X

ISBN 0-444-42234-X (Vol. 2)
ISBN-0-444-42235-8 (Series)

Printed in The Netherlands

CONTENTS

INTRODUCTION

In the later summer, 1981, the international symposium on "Arc Volcanism" was held in Tokyo and Hakone under the sponsorship of the Volcanological Society of Japan and the International Association of Volcanology and Chemistry of the Earth's Interior (IAVCEI). More than 400 scientists attended the symposium and about 220 papers were read during the main session from August 31 through September 5. The topics covered a wide range: large-scale pyroclastic eruptions, calderas, mechanics and dynamics of ascent and eruption of magmas, narrative and description of recent eruptions including many papers on Mount St. Helens (1980—1981) and Usu (1977—1981) eruptions, monitoring of various geophysical and geochemical parameters before, during and after the eruptions, hazard assessment and study on mitigation of volcanic activity, petrological and experimental approach on magma genesis related to arc-type magma series, across- and along-the-arc variations of chemical features of magmas and local descriptions of geology, geophysics and geochemistry of the volcanic bodies and terrains.

In this special issue, we have attempted to collect some of the representative papers given at the symposium in the fields of petrology, geochemistry and field volcanology. Although individual papers presented can very well stand on their own, we believe that a collection in a single volume will greatly help the readers to grasp a general knowledge of the status quo of the science of arc volcanism related to the above fields.

SHIGEO ARAMAKI
IKUO KUSHIRO
(Editors)

Journal of Volcanology and Geothermal Research, 18 (1983) 1—56
Elsevier Science Publishers B.V., Amsterdam — Printed in The Netherlands

THE COMPOSITIONALLY ZONED ERUPTION OF 1912 IN THE VALLEY OF TEN THOUSAND SMOKES, KATMAI NATIONAL PARK, ALASKA

WES HILDRETH

U.S. Geological Survey, 345 Middlefield Road, Menlo Park, CA 94025 (U.S.A.)

(Received October 20, 1982)

ABSTRACT

Hildreth, W., 1983. The compositionally zoned eruption of 1912 in the Valley of Ten Thousand Smokes, Katmai National Park, Alaska. In: S. Aramaki and I. Kushiro (Editors), Arc Volcanism. J. Volcanol. Geotherm. Res., 18: 1—56.

On June 6—8, 1912, ~15 km³ of magma erupted from the Novarupta caldera at the head of the Valley of Ten Thousand Smokes (VTTS), producing ~20 km³ of air-fall tephra and 11—15 km³ of ash-flow tuff within ~60 hours. Three discrete periods of ash-fall at Kodiak correlate, respectively, with Plinian tephra layers designated A, C-D, and F-G by Curtis (1968) in the VTTS. The ash-flow sequence overlapped with but outlasted pumice fall A, terminating within 20 hours of the initial outbreak and prior to pumice fall C. Layers E and H consist mostly of vitric dust that settled during lulls, and Layer B is the feather edge of the ash flow. The fall units filled and obscured the caldera, but arcuate and radial fissures outline a 6-km² depression. The Novarupta lava dome and its ejecta ring were emplaced later within the depression. At Mt. Katmai, 10 km east of the 1912 vent, a 600-m-deep caldera of similar area also collapsed at about this time, probably owing to hydraulic connection with the venting magma system; but all known ejecta are thought to have erupted at Novarupta. Mingling of three distinctive magmas during the eruption produced an abundance of banded pumice, and mechanical mixing of chilled ejecta resulted in deposits with a wide range of bulk composition. Pumice in the initial fall unit (A) is 100% rhyolite, but fall units atop the ash flow are >98% dacite; black andesitic scoria is common only in the ash flows and in near-vent air-fall tephra. Pumice counts show the first half of the ash-flow deposit to be 91—98% rhyolite, but progressive increases of dacite and andesite eventually reduced the rhyolitic component to <2%. The later, rhyolite-poor flows were hotter, less mobile, and widely produced partially welded tuff and vapor-indurated sillar.

The main ash flow was too deflated and sluggish 16 km from the vent to surmount a 25-m-high moraine in its path but was diverted around it and continued 5 km down-valley, engulfing and charring trees but not toppling all of them. Thin ash-flow veneers feather 30—40 m up the enclosing valley walls but only where a constriction in the central VTTS locally raised the flow level. In the upper VTTS, the "high sand mark" is not a veneer but a marginal bench formed in thick tuff by differential compaction. Flooding from adjacent glaciers led to phreatic explosions that ejected blocks of tuff more welded than any yet exposed. A cluster of phreatic craters dammed a lake atop the tuff, the breaching of which caused a flood that scoured the ash-flow surface in the central VTTS, transported 50-cm blocks of welded tuff >20 km to the lowermost VTTS, and deposited 1—8 m of debris there.

0377-0273/83/$03.00

Rhyolitic ejecta contain only 1—2% phenocrysts but andesite and dacite have 30—45%. Quartz is present and augite absent only in the rhyolite, but all ejecta contain plagioclase, orthopyroxene, titanomagnetite, ilmenite, apatite, and pyrrhotite; rare olivine occurs in the andesite. The zoning ranges of phenocrysts in the rhyolitic and intermediate ejecta do not overlap. New chemical data show the bulk SiO_2 range to be: rhyolite 77 ± 0.6, dacite 66—64.5, and andesite 61.5—58.5%. The dacitic and andesitic ejecta contrast in color and density, and it is not certain whether they form a compositional continuum. Analyses reported by Fenner within the 66—76% SiO_2 range were of banded pumice and lava and of bulk tephra that mechanically fractionated and mixed during flight. Despite the gap of 10% SiO_2, Fe-Ti-oxide temperatures show a continuous range from rhyolite (805—850°C) through dacite (855—955°C) to andesite (955—990°C). Thermal continuity and isotopic and trace-element data suggest that all were derived from a single magmatic system, whether or not they were physically contiguous before eruption. If the rhyolitic liquid separated from dacitic magma, extraction was so efficient that no dacitic phenocrysts were retained and no bulk compositions in the range 66—76% SiO_2 were created; if it were a partial melt of roof rocks atop an intermediate magma body, then such rocks had no O- or Sr-isotopic contrast with the andesite-dacite magma and clearly did not include the Jurassic arkosic or granitic basement. The presence of Holocene domes of pre-1912 glassy dacite adjacent to the 1912 vent suggest that the 7 km^3 (or more) of high-silica rhyolitic magma (a composition rare in the Aleutian arc) was generated in less than a few thousand years. The 1912 vent is semi-encircled by several andesitic stratocones and is as close to Mageik, Trident, and Griggs volcanoes as it is to Mt. Katmai. The erupted magma probably occupied only shallow levels of an extensive system of injection and storage under a cluster of several stratovolcanoes. Although Quaternary basalt is not known to have erupted here, the intrusion of basaltic magma probably sustains the greater-VTTS magmatic system.

INTRODUCTION

Few historic eruptions have been as provocative for petrologists and volcanologists as that of 1912 in the Valley of Ten Thousand Smokes (VTTS) at the northwest foot of Mt. Katmai on the Alaska Peninsula (Figs. 1, 2). Among historic eruptions it is virtually unique in having generated a large volume of pumiceous pyroclastic flows that (unlike those of Krakatau and Santorini) came to rest on land. It was the century's most voluminous eruption by far, one of the three largest in recorded history, and one of the few historic eruptions to have produced welded tuff. An abundance of compositionally banded pumice evoked petrologic controversy, and the high-temperature fumaroles in the ash-flow sheet, which persisted vigorously for about 15 years, attracted attention to the vapor transport of metals (Allen and Zies, 1923; Zies, 1929).

Although several expeditions reached the VTTS between 1916 and 1923, the eruption and its products remained poorly understood, owing in part to remoteness and difficult field conditions but principally to a lack of prior attention to pyroclastic-flow deposits. Most of the early confusion can be attributed to a poor understanding of the relations between pyroclastic fall and flow deposits, to a lack of appreciation for the mobility of such flows, and to failure to identify the actual vent for the 1912 ejecta. An outstanding

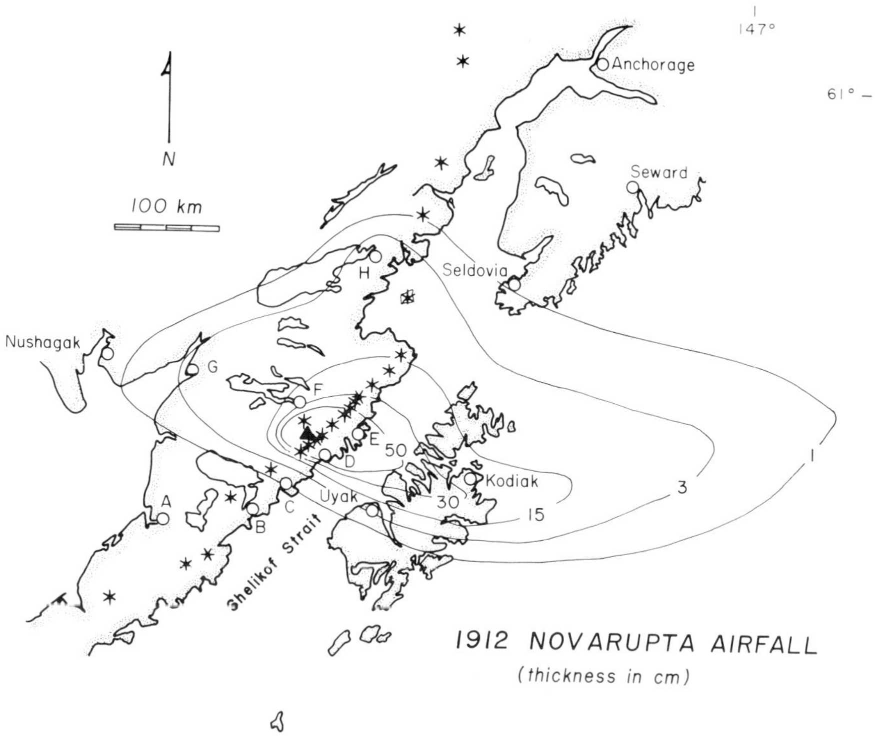

Fig. 1. Location map for the upper Alaska Peninsula, Kodiak Island, and Cook Inlet region. Stars are active or potentially active volcanic centers, mostly andesite-dacite stratocones, all with late Holocene activity. Triangle is the 1912 vent at Novarupta, with tephra isopachs adapted from Griggs (1922). Letters refer to settlements that, in addition to the named towns, figured in reports on the 1912 activity: *A* = Ugashik; *B* = Kanatak; *C* = site of Cold Bay (Puale Bay); *D* = site of Katmai village; *E* = Kaflia Bay; *F* = site of Savonoski; *G* = Naknek; and *H* = Iliamna.

stratigraphic and petrographic study by Curtis (1968) set most of the record straight. Curtis demonstrated that most or all of the 1912 ejecta vented within the 2-km-wide subsidence structure surrounding the Novarupta dome, proved that shallow assimilation had been unimportant, noted that most fumaroles were rooted in the ash-flow tuff itself, and, by means of comparative stream profiles, provided a convincing estimate of the thickness of the tuff in the upper VTTS. The present effort, for which this is a progress report prompted by the IAVCEI symposium, emphasizes the chemical petrology and the emplacement mechanisms of the 1912 ejecta.

GEOLOGIC SETTING

The 1912 vent at Novarupta (Fig. 3) is semi-encircled by five andesitic to dacitic composite volcanoes (Fig. 2), all of which exhibit active fumaroles

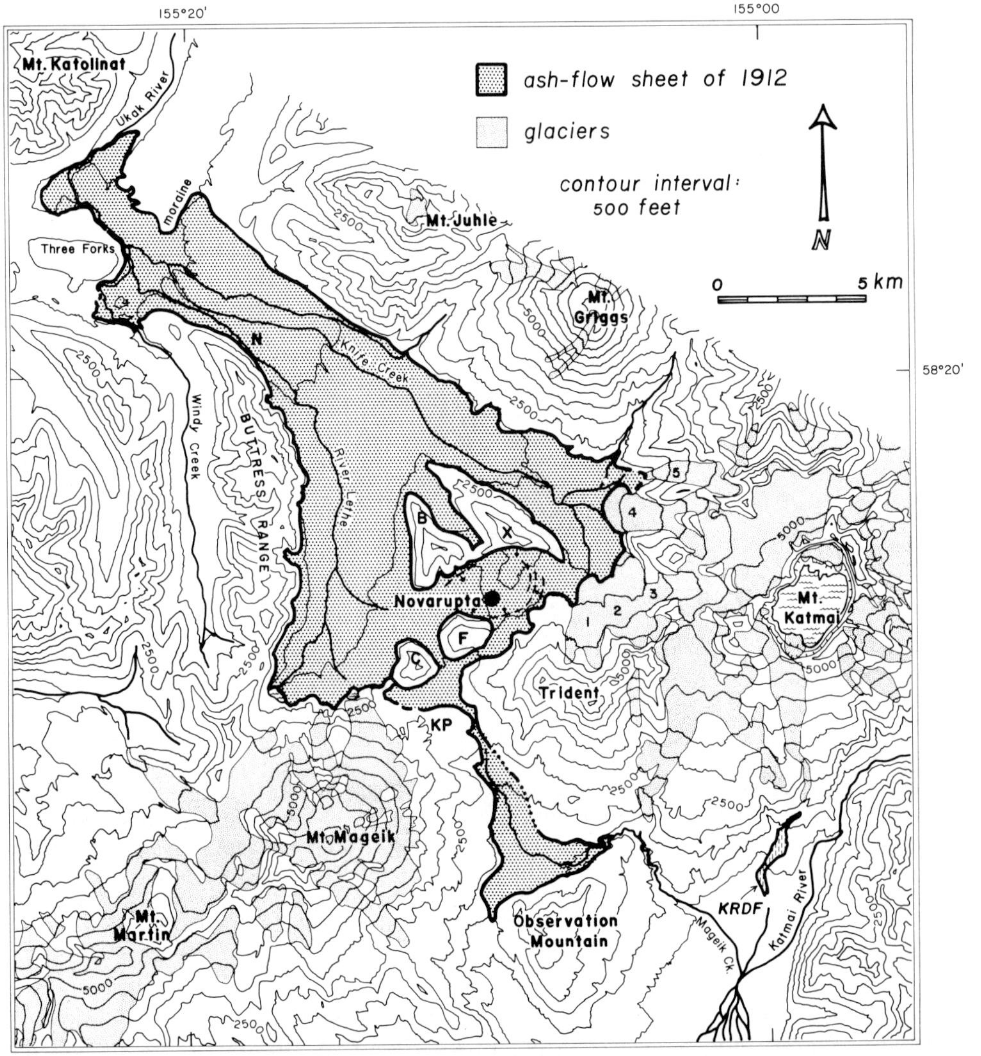

Fig. 2. Location map for the Valley of Ten Thousand Smokes, showing the 120-km^2 extent of the 1912 ash-flow sheet and the five composite volcanoes nearby. Mts. Griggs, Katmai, Trident, Mageik, and Martin all have active fumaroles and have undergone Holocene eruptions. One ash flow penetrated Katmai Pass (*KP*) and ponded up to 25 m thick in Mageik Creek on the Pacific slope, but 99% of the ash-flow volume moved northwestward to form the VTTS. Its thickness is not known in the upper VTTS but may be as great as 250 m (Fig. 14); it thins to ~30 m just above the moraine that partially blocked its path near Three Forks (Fig. 15) and to <10 m at its now-eroded snout on the Ukak River. Mt. Cerberus (*C*) and Falling Mtn. (*F*) are Holocene dacite domes adjacent to the 1912 vent at Novarupta (Fig. 3). Baked Mtn. (*B*) and Broken Mtn. (*X*) consist of Jurassic sedimentary rocks, which, like Falling Mtn., were truncated by the eruption and subsidence in the Novarupta caldera (Fig. 13). *KRDF* = Katmai River debris flow, a deposit of remobilized ejecta (Fig. 7) that debouched from a single narrow gorge fed from the glaciated south slope of Mt. Katmai, which was in the sector of heaviest tephra fall. *N* is the VTTS constriction where the ash flows left marginal veneers and constructional hills of tuff. Part of the Mageik Creek ash-flow lobe is now covered by the 1953–1960 lava flows from Trident. The Knife Creek glaciers are numbered for reference.

Fig. 3. View southwestward across Novarupta dome, approximately 65 m high and 380 m in diameter, encircled by its ejecta ring and centered within the 2-km-diameter basin of subsidence that was the source of the 1912 ejecta. The sheared-off face of the Holocene dacite dome, Falling Mountain, was the site of vigorous fumarolic activity, some of which persisted into the 1930's (Fenner, 1938). Another Holocene dacite dome, Mt. Cerberus, lies on the right flank of Falling Mountain, and beyond them rises snow-mantled Mt. Mageik, the summit of which is 10 km from Novarupta and contains a hot crater lake with abundant jet-like fumaroles. At the far right, beyond the ash-flow plain in the southwestern corner of the VTTS, is a glacially dissected volcanic center thought to be of Pleistocene age, in the core of which is exposed a hornblende-diorite porphyry and a thick dike complex of 75%-SiO_2 aplite. (Photo by T.P. Miller.)

and have undergone (undated) Holocene eruptions. The Quaternary volcanic piles here are generally <1500 m thick and mantle a ruggedly glaciated terrain of gently dipping sandstones and shaly siltstones of Late Jurassic age (Keller and Reiser, 1959; Curtis, 1968). Deeply dissected andesitic-dioritic centers are also exposed (Fig. 3), but it is not yet known if these are Quaternary or Tertiary in age.

The setting of the VTTS is not particularly unusual for the predominantly andesitic-dacitic Alaska—Aleutian arc, except in the clustering of five discrete composite volcanoes (and a few lesser centers) within 25 km of each other and in the location of a major center (Mt. Griggs) 10 km behind the axis of the main chain (Figs. 1, 2). Pumiceous pyroclastic-flow deposits of Holocene and late Pleistocene age are fairly common along the arc (Miller and Smith,

1977), and a few exceed the 1912 sheet in volume. Lavas and ejecta that contain >66% SiO_2 are uncommon but do occur at six or eight centers; along the entire 2500-km-long arc, however, high-silica rhyolite (75—77% SiO_2) is known to have erupted only on Umnak Island (Byers, 1961) and at Novarupta in 1912.

There were no pre-1912 eruptions of the five Katmai-group volcanoes in the memory or traditions of the native inhabitants, but layers of andesitic and dacitic tephra in the Brooks River area (48 km NW of Novarupta) record at least nine other explosive events during the last 7400 ^{14}C years (Dumond, 1979). Published compilations notwithstanding, there are no credible reports of historic eruptions in the Katmai group except at Novarupta in 1912 and at Trident in 1953—1968 (Ray, 1967). In particular, a pumice shower said by a fisherman to have issued from Mt. Mageik in August 1927 (Jaggar, 1927) cannot be corroborated by detectable deposits in the volcano's crater or on its flanks.

CHRONOLOGY OF THE 1912 ERUPTION

The record of tephra falls in surrounding inhabited areas restricts the main eruptive activity in 1912 to a period of ~60 hours between the afternoon of 6 June and the early morning of 9 June. Principal uncertainties concerning the eruptive sequence have been the times of ash-flow emplacement relative to the series of tephra falls (Figs. 1, 4) and the time of collapse of Katmai caldera, 10 km east of the 1912 vent (Fig. 2). Because there were no geophysical instruments in Alaska, no scientific observers in the district, and no eyewitnesses with a view of the VTTS, most reports of seismic, acoustic, and eruptive phenomena were not necessarily very accurate, chronologically or descriptively. The most reliable data are the visual observations from aboard the steamer *Dora* and the record of tephra falls at Kodiak (Figs. 1, 4; Martin, 1913; Griggs, 1922). Martin visited Kodiak, Uyak, Katmai village, and Cold Bay (Fig. 1) in August of 1912 and gathered what accounts he could from residents of the district, but he did not venture inland to the volcanoes nor was he able to interview anyone from the two bands of native hunter-fishermen said to have been within 30 and 40 km, respectively, north and south of Novarupta at the time of the initial outburst. Reconstructions of the eruptive sequence by Griggs (1922) and Fenner (1923, 1925, 1950) are heavily inferential, overinterpretive both of Martin's data and of latter-day interviews with people said to have been at Savonoski when the eruption began (see below), and suffer badly from the incorrect assumption that much of the tephra had come from Mt. Katmai. The following is a critical reappraisal of what appear to be the facts, as assembled largely by G.C. Martin (1913; and unpublished USGS field notes); all times cited are adjusted to Alaskan local time.

Earthquakes were reported to have been felt at Katmai village (30 km SSE of Novarupta) as early as the evening of 31 May, and severe shocks were felt

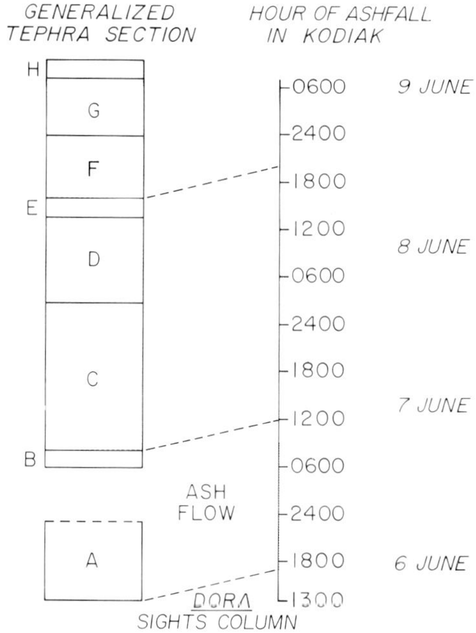

Fig. 4. Generalized 1912 eruptive sequence and proposed correlation of tephra layers defined by Curtis (1968) in the VTTS with the periods of ashfall at the town of Kodiak (Martin, 1913; Griggs, 1922). On 6 June, the steamer *Dora* was in Shelikof Strait north of Uyak (Fig. 1) from 0845 onward, the Katmai volcanoes in full view with a "strong westerly breeze and fine clear weather", but no one aboard saw the tephra column until 1300. The inclined dashed lines represent the 4-hour interval before fine ash began falling in Kodiak, 170 km east of Novarupta. The tripartite ash at Kodiak is today approximately 25 cm thick, whereas the equivalent fall units near Novarupta exceed 10 m. Layers E and H are phenocryst-depleted units that consist largely of fine-sand to dust-sized glass that settled during lulls in the eruption; Layer B is also in part such a deposit, but (as discussed in the text) it is principally the feather edge of the rhyolitic ash flow. The ash-flow sequence was largely contemporaneous with rhyolitic fall-unit A but outlasted it. Dacitic Layer C rests either upon B, upon the ash flow, or upon locally intercalated fluvial and debris-flow deposits atop the ash flow.

at Uyak, Kanatak, and Nushagak (200 km WNW) on 4 and 5 June. On the morning of 6 June, explosions were heard at Nushagak and at Seldovia (240 km ENE), but there were no reports of accompanying seismicity or ash clouds. The same morning the *Dora* left Uyak at 0845 and steamed northeastward up Shelikof Strait, the Katmai-group volcanoes in full view with a "strong westerly breeze and fine clear weather", but no one aboard noticed a tephra column until 1300 (when the vessel was ~88 km southeast of Novarupta). That column was clearly Plinian and, as shown below by stratigraphic data, it was predominantly rhyolitic; the cloud overtook the *Dora* by 1500 and began dropping ash at Kodiak (170 km ESE; Figs. 1, 4) by 1700.

Separate interviews with two individuals said to have been in or near Savonoski (30 km N) on the morning of 6 June were conducted in 1918 (Griggs, 1922, p. 17) and in 1923 (Fenner, 1925, p. 216); these have been interpreted to suggest that the ash flow in the VTTS began in the morning several hours prior to the first high tephra column. This is quite unlikely in view of: (1) the record of the *Dora*; (2) the fact that the basal tephra layer in the VTTS, the distinctively rhyolitic Layer A, does not occur atop the ash-flow deposit (Fig. 4); and (3) the common association of pumiceous pyroclastic flows with the collapse of vertical eruption columns. Preliminary eruptive activity at Novarupta on the morning of 6 June is quite reasonable and may be the best explanation of the noise and dust reported that morning at Savonoski and the explosions heard at Nushagak and Seldovia; but a major eruption column and a rhyolitic ash flow several km^3 in volume are certainly excluded prior to 1300.

Major shocks that punctuated virtually continuous seismic activity were reported locally at ~1300 and ~2300 on 6 June and ~2240 on 7 June. These times are not necessarily very accurate. The first instrumentally recorded teleseism (at Seattle) is reported to have arrived at 1241 on 6 June (Fenner, 1925). Harvard, Ottawa, and several Eurasian stations recorded many teleseisms originating in southwestern Alaska between 1805 on 6 June and ~2100 on 11 June. One of these near midnight on 6/7 June has been estimated at magnitude 6.4 and another at 0606 on 10 June at magnitude 7.0 (Coffman and von Hake, 1973). Earthquakes following the main phases of the eruption were felt at Cold Bay (60 km SSW) on 50 of the 70 days through mid-August (Fenner, 1925).

A great explosion, accompanied by an earthquake felt at Cold Bay ~1300 on 6 June (Martin, 1913), was audible throughout the region and close in time to the first recorded teleseism and to the first sighting of tephra by the *Dora*. Another blast at ~1500 on 6 June, for which no accompanying earthquake was noted, may have been the most severe outburst of all, being heard for hundreds of kilometers (Martin, 1913). Martin suggested, but did not cite any evidence for, explosions synchronous with the large earthquakes felt nearby at ~2300 on 6 June and 2240 on 7 June. There were, however, reports of loud noises between midnight and 0200 on 8 June, at Katanak (95 km SW) and at Cordova (590 km ENE) (Martin, 1913; Fenner, 1925). Explosions continued to be heard, some as far away as Juneau (1200 km) until 10 June, though how many of the noises may have been thunder is impossible to assess.

Major tephra eruptions, first noted by the *Dora* at 1300 on 6 June, continued to fall heavily on downwind settlements nearby until 9 June. The principal sector of dispersal was southeastward (Fig. 1), and the most distant positive record of ashfall was in Puget Sound (2400 km SE); atmospheric effects were worldwide (Griggs, 1922; Volz, 1975a). In the main downwind direction at Kodiak (170 km ESE) there were three discrete periods of ashfall (Fig. 4): (1) 1700 6 June until 0910 7 June; (2) ~1200 7 June until 1430 8 June; and (3) during the night of 8/9 June. The 9th of June dawned clear, and no further ash-falls were recorded on Kodiak Island. A major vapor plume and, possibly, sporadic ashfalls close to the source lasted all summer (Martin, 1913; Griggs, 1922; Fenner, 1925). Timing of the emplacement of the Novarupta dome is poorly known; its extrusion followed the last major tephra fall and was complete at the time of its discovery on 31 July 1916.

Details of correlation between audible explosions, tephra falls, and seismic events are not well known. Martin (1913) is the principal source, but his article has been misread and overinterpreted and ensuing errors propagated in successive publications. The seismicity does not correlate very well with eruptive events, and indeed there need be no correspondence (e.g., Filson et al., 1973; Nairn et al., 1976; Yokoyama et al., 1981). Much of the seismic activity seems more likely to have been related to fitful subsidence of Katmai and Novarupta calderas.

At the onset of eruption, no one is known to have had a view of the VTTS (much less the Novarupta area itself), and only the party of native fishermen then southwest of Katmai village en route to Cold Bay may have been in a position to see Mt. Katmai. The Savonoski interviews (above) smack of ex post facto embellishment, although Martin (1913, p. 147) accepted a third-hand report in 1912 that the villagers there may have seen Mt. Katmai after its collapse, but before 9 June. It is not clear where they might have had such a view, because, even notwithstanding the ash clouds, Mt. Katmai is not visible from Savonoski or along their escape route down the lake to Naknek (Fig. 1). Martin, Griggs, and Fenner accepted such reports to mean that Mt. Katmai had lost its top by the afternoon of 6 June. This may be true, but it is not clear how Mt. Katmai could have been seen at all during the eruptive interval at Novarupta or what meaning should be attached to such expressions as "blown off" or "blown up". No definitive geologic evidence has been found to fix the timing of Mt. Katmai's collapse, but it may be a reasonable inference that in addition to the syneruptive earthquakes, much of the summer-long seismicity can be attributed to such collapse, as well as to tectonic adjustments over a somewhat wider area, or even to dome emplacement at Novarupta.

STRATIGRAPHY OF THE EJECTA

In at least three main intervals of high-column activity on 6—9 June 1912, a total of ~15 km^3 of magma was ejected over a span of ~60 hours from the area now occupied by the 2-km-wide, pumice-filled depression surrounding the small lava dome, Novarupta (Figs. 2, 3, 14). This magma expanded to generate 20—25 km^3 of air-fall tephra (much of which fell at sea; Fig. 1; Nayadu, 1964; Federman and Scheidegger, 1981) and 11–15 km^3 of ash-flow deposits that came to rest in the VTTS, Katmai Pass, and Mageik Creek (Fig. 2). Griggs (1918) and Fenner (1920, 1923) realized early that the ash flows had originated somewhere in the VTTS rather than at Mt. Katmai, and

Shipley (1920) unequivocally attributed it to Novarupta. The early workers, however, were all so impressed with the summit caldera of Mt. Katmai that they assumed some large fraction of the tephra falls had vented there as well as at Novarupta. Not surprisingly, they gave no criteria by which to distinguish deposits from the two supposed sources. In 1953–1954 Curtis (1968) subdivided the tephra into nine fall units, measured many sections in the Mt. Katmai—Novarupta area, constructed isopach maps for each unit, and demonstrated that at least 99% of the 1912 ejecta had vented near Novarupta. The present study has examined the compositional and physical variability of the ash-flow deposits and of Curtis' nine fall units.

The rhyolite air-fall: Layer A

The juvenile ejecta of the initial Plinian outburst consisted entirely of high-silica-rhyolitic pumice, silky white and phenocryst-poor, which was accompanied by 5—10% of accidental lithic ejecta that becomes sparser upward in the deposit. Curtis (1968, pp. 167–175) described Layer A at length, noting its poor sorting relative to subsequent fall units, the pink to terra-cotta fine rhyolitic ash that pervades but does not fill the interstices between pumice lapilli, and the appearance near the very top of the unit of small amounts of dacite pumice, banded pumice, and accessory fragments of vitrophyre. Layer A is better sorted on the Pacific slope of the volcanic range (Figs. 1 and 2) than in the near-vent outcrops of the VTTS.

Because Layer A is the basal fall deposit adjacent to the ash-flow sheet (as elsewhere) but nowhere overlies it, its eruption must have preceded and/or coincided with some part of the ash-flow emplacement sequence. Where the base of the ash-flow sheet is exposed in the lower VTTS, it is commonly underlain by a discrete 1–3 cm layer of fine-grained, rhyolitic, air-fall ash. This could be the deposit of preliminary activity on the morning of 6 June, or it could be an upwind equivalent of the basal part of Layer A itself. In either case, compared to the proportion of Layer A (roughly one-third) in continuous air-fall sections remote from the ash-flow sheet, this subjacent ash layer is far too thin at any given distance from the vent to represent all of A. Hence, it seems required that fall unit A have been in large part synchronous with emplacement of the predominantly rhyolitic (Table I) early ash flows. Concurrent deposition from a high column and from a turbulent lateral eruption cloud, perhaps in conjunction with strong atmospheric turbulence and condensation, may account for the poor sorting, fine-ash entrapment, and stratigraphic relations of Layer A within ~8 km of the vent. In Mageik Creek, south of the range, Layer A is much better sorted, includes a little dacite and andesite pumice in the upper third of its 60-cm thickness, and directly underlies the 25-m-thick deposit of an ash flow that spilled through Katmai Pass toward the middle or end of the pyroclastic-flow sequence.

The ash-flow sequence

The first half or more of the ash-flow volume to be emplaced was also largely rhyolitic, and its dominant pumice type is compositionally, texturally, and mineralogically identical to that of Layer A. At no exposed level, however, is pumice in the ash-flow deposit exclusively rhyolitic like it is in the first 80—90% of Layer A. Dacitic, andesitic, and banded pumice joined the rhyolite in proportions that generally increased throughout the ash-flow emplacement sequence (Table I).

Griggs, Fenner, and Curtis all emphasized distinctions between the quartz-bearing, phenocryst-poor rhyolite and the intermediate ejecta (which they generally lumped as "pyroxene-bearing pumice"), but none drew sufficient attention to the bimodal character of the latter. On fresh surfaces, the intermediate pumice is either light grey or nearly black, and, though both are rich in phenocrysts, this simple macroscopic distinction is corroborated by bulk, glass, and mineral analyses (Figs. 8–12; Tables II, III). Medium-grey pumice in the range between the two populations with 58—60% and 64–66% SiO_2 is exceedingly uncommon. Table I gives the results of an effort to utilize this ready field distinction to quantify progressive changes in the proportions of the three simultaneously erupting magmas.

The down-valley distal portion of the ash-flow sheet (except for its top few meters) is predominantly rhyolitic, up to 98% so. Because no fall unit later than Layer A contains much rhyolite, either the 8–12 km^3 of rhyolite-rich ash flows were emplaced concurrently with Layer A or they followed it and had little accompanying tephra fall. The fact that banded and intermediate pumice constitute 2–20% of the principal rhyolitic ash-flow unit but are absent throughout most of Layer A shows that these components did not contribute to a high tephra column in the early stages of eruption; instead, they joined some fraction of the co-erupting rhyolite in collapsing directly into pyroclastic flows. A marked reduction of column height, nearly concurrent with the first appearance of intermediate ejecta after deposition of most of Layer A, could explain the absence of a coarse air-fall layer that is compositionally equivalent to the bulk of the ash-flow tuff.

Up-valley toward Novarupta (as well as in the top few meters distally, as near Three Forks), the ash-flow sequence contains a strikingly greater proportion of dacitic, andesitic, and banded pumice. Nevertheless, in the deep gorges of the central VTTS the dominantly rhyolitic tuff still constitutes the principal volume, though it underlies a sourceward-thickening wedge of tuff that is relatively poor in rhyolite (Table I). In the upper regions of Knife Creek and the River Lethe (Fig. 2) most ash-flow exposures have a dark- to medium-grey ashy matrix and contain abundant dacitic and andesitic pumice clasts. On gorge walls in the central VTTS the transition is seen to take place gradually over a height of >10 m *within* the single main emplacement unit. A few discrete rhyolite-poor flow units (Fig. 5), each only a few meters thick, cap the ash-flow sequence in the upper VTTS and

TABLE I

Representative pumice-count data for 1912 ejecta

Stratigraphic unit and location			Rhyolite	Dacite	Andesite	Banded	No. of stations
Layer A	Mt. Griggs		100	0	0	0	3
	Mageik Creek		100	0	0	0	3
Ash flows	Ukak lobe:	main unit	83—95	1— 8	1— 3	3—14	5
		top 2 m	77	11	2	10	1
	Three Forks area:	main unit	77—91	1— 5	0— 9	6—16	6
		upper dark unit	49—72	1— 8	15—35	8—17	4
		top 1m	13—36	21—41	18—45	5—21	4
	Central VTTS:	deep in gorges	55—86	0— 4	5—35	7—12	8
		upper dark unit	18—69	0—10	16—54	8—37	9
		topmost 2-m unit	2—35	36—68	8—28	4—12	6
		Juhle Fork, deep	90—95	1— 2	2— 4	2— 4	2
		Juhle Fork, top	36—48	2— 9	35—52	7— 9	3
	Upper Knife Creek:	main surficial unit	1— 9	50—61	35—44	1— 4	12
		margin at Glacier 3	41	3	55	1	1
		fluvially deposited pumice beds	2	83	11	4	1
		block-rich late pumice flow	0— 8	58—85	9—32	1—12	4
	Upper River Lethe:	SW margin, basal unit	69—73	0—2	13—18	9—16	2
		S margin, lowest exposed unit	22	76	0	2	1
		surficial unit around margins	18—39	10—42	35—50	5—25	6
		surficial unit, axial	1—17	54—75	5—25	3—10	7

	Katmai Pass	4	66	15	15	1
	Katmai Pass	50	48	1	1	1
	Mageik Creek lobe	35—52	46—61	0— 2	0— 2	6
	Mageik Creek lobe, basal	52	47	1	0	1
	Thin late flow units near Novarupta	0— 4	66—77	5—33	6—29	4
Layer B	Upper Knife Creek	80—92	8—16	0—11	0—3	4
Layer C	Upper Knife Creek	<1	>99	0	0	2
	Lower Mageik Creek	1	99	0	0	1
Layer D	Upper Knife Creek	2	97	1	0	1
Layer F	Upper Knife Creek	<1	>99	0	0	2
Layer G	Upper Knife Creek	0	>99	0	<1	2
	Lower Mageik Creek	1	99	0	0	1
Thick coarse tephra, 0.7 km NE of Novarupta, top		0	49	24	27	1
Total (rough estimate)		40—50	40—50	5		

Proportions were determined in the field by identifying and tallying all pumice blocks and lapilli larger than ~1 cm within a selected area of outcrop (typically several m^2). Usually 200—500 pumice lumps were counted at each station, and many were broken open to assure correct identification. Owing to compositional gaps, the four pumice types are readily distinguishable in the 1912 VTTS suite.

Fig. 5. View of successive ash-flow units of contrasting composition, on NW side of the lake in the extreme SW corner of the VTTS (Fig. 2). Lower unit is non-indurated, and pumice counts in it give 73% rhyolite, 18% andesite, 9% banded, and no dacite. Upper ash-flow unit is incipiently welded and, at its top, it is oxidized and vapor-phase indurated; its pumice proportions are 21% rhyolite, 50% andesite, 13% banded, and 16% dacite. The intercalated meter of plane-stratified tephra is compositionally like the lower unit. Dacite pumice appears near the base of the upper unit and increases upward. Partial welding begins abruptly with the sharp jump in the andesite/rhyolite ratio of both lapilli and matrix. Exposed thicknesses here are >15 m and 5—7 m, respectively, but both units wedge out toward the valley wall <100 m away; they dip ~15° toward the valley axis and thicken as exposure is lost toward the poorly incised valley floor. The upper unit is typical of most surface exposures of the upper-valley ash-flow tuff.

overlie the main rhyolite-rich unit as a thin sheet all the way to the Ukak River below Three Forks, as far as 18 km from Novarupta. Though intermediate-composition material is abundant in near-vent fall deposits, the andesite-rich tuff of these later flow units has no equivalent in stratified airfall deposits more than 4—5 km from Novarupta. This suggests that most of the andesitic ejecta were products of low fountaining, which contributed to the pyroclastic flows but not to the Plinian column.

Because deep levels of the tuff are not exposed in the upper half of the VTTS, there is no control on the proportion of rhyolite there, but, despite this uncertainty, rhyolite must constitute at least 60% of the total ash-flow volume. If a large amount of rhyolite-rich tuff is concealed by the largely intermediate sheets at the surface, as seems likely, then the proportion of rhyolite could be as high as 75% for the pyroclastic flows as a whole.

There was clearly a progressive increase in the total abundance of dacite and andesite during formation of the ash flows, but no systematic shift in the ratio of these two components is evident in the deposits. In a general way, the latest thin ash flows, along the axial portions of both arms of the upper VTTS, are dacite-dominant, whereas the intermediate component in the rhyolite-rich ash flows tends to be largely andesitic; but wide fluctuations in the dacite/andesite ratio took place during the emplacement sequence. Except at the very onset, however, all three magmas vented concurrently and remained available throughout the eruption. They mixed in varying proportions, both as liquids to produce the banded pumice and as solid ejecta within the pyroclastic flows and tephra falls.

Layer B: The feather edge

As defined by Curtis, this unit is found only along the northeast side of the Knife Creek arm of the upper VTTS, where it has been traced from the margin of the ash-flow sheet up to elevations 300—400 m higher. Its top is sharply defined by the conformable base of dacitic Layer C, but its contact with subjacent Layer A is gradational. Layer B is essentially a finer-grained continuation of A, as both are largely rhyolitic and poorly sorted in this near-vent downwind sector. It consists of pink to terra-cotta ash with 2—5% each of phenocrysts and small lithics, as well as scattered small pumice lapilli, which also occur locally in discontinuous lenses. Curtis (1968, p. 175) pointed out that the pumice and phenocrysts of Layer B are mixtures of rhyolitic and intermediate material; in this respect, B contrasts with most other tephra units, as the essential ejecta in A are almost all rhyolite and in C to H almost all dacite. Heavy-liquid separations of its ash matrix show Layer B to be 94—95% glass and dominantly rhyolite, but as much as 20% of its phenocryst fraction is derived from andesitic and dacitic pumice.

The glassy ash of Layer B includes a small fraction (<1%) of discrete or clustered, thin-walled glass spheres, 0.2 to 1 mm in diameter, which coexist with the dominant stretched-vesicle and angular pumiceous shards. These delicate spherical bubbles also occur in Layer A and in the rhyolitic component of downwind ash deposits (as at Kodiak), but in near-source tephra they are most prominent in the fine-grained Layer B.

Layer B is not a co-ignimbrite air-fall ash (Sparks and Walker, 1977), because it does not occur atop the ash flow nor adjacent to it at distal locations. Instead, I think that it was deposited by the feather edge of the main rhyolite-rich ash flow itself. This was not initially obvious, because in the upper VTTS the main mass of rhyolitic ash-flow tuff is everywhere covered by the later, more indurated, andesite-dacite flow units. From its first outcrops adjacent to the edge of these units along the valley-floor margin, Layer B thins rapidly from ~4 m to only a few cm at a distance of 2–3 km up the slopes of the surrounding peaks. Its internally lenticular structure may be attributable to outward surging of finer ash from the margin of the moving pyroclastic flows. Such an explanation may also suit parts of Layer A, which, adjacent to the ash-flow margin and only 5–6 km from its source, is also poorly sorted and irregularly to lenticularly stratified.

Deposits that consist in large part of pink to terra-cotta rhyolitic ash have been noted at the stratigraphic position of Layer B in three other settings:

(1) Immediately south of Mt. Katmai, where Layer A is a well-sorted coarse pumice fall ~1 m thick, it is overlain by ~2 cm of pinkish ash, which probably represents the settling of fines during a quiescent interval preceding the Plinian eruption of dacitic Layer C (Fig. 4).

(2) On Broken Mountain, only 4 km north of Novarupta and 100 to 200 m above the ash-flow surface, ~10 cm of sandy, terra-cotta, vitric ash (with a sprinkling of rhyolitic pumice granules) lie between Layer A and a stack of local tephra layers rich in both andesite and dacite pumice. Pumice-fall units that contain abundant andesite exist only in the near-vent area, and since these layers are capped by Layer C, they must be approximately time-equivalent to the uppermost (rhyolite-poor) ash flows.

(3) On the steep slope of the Buttress Range 10–13 km northwest of Novarupta, the edge of the ash-flow deposit is a wedge of fines-enriched, pinkish, vitric ash that feathers out upslope at elevations as high as 40 m above the general level of the adjacent valley-filling sheet. The ash is mostly rhyolite, overlies 1 cm of (upwind) Layer A, is capped by Layer C, and closely resembles Layer B in the up-valley areas described by Curtis. Where >50 cm thick it contains abundant pumice lapilli, but where it tapers laterally upslope to <20 cm only ash and dust remain. If this deposit were simply a veneer left by passage of an inflated ash flow, its position might suggest ~50% inflation of the flow as it passed this area, where the compacted axial thickness is now ~50—80 m. Alternatively, I favor the interpretation, more consistent with the depletion of coarse clasts and with evidence (given below) for a much lower degree of inflation, that the ash wedge was formed by marginal surging or jetting, perhaps from a somewhat turbulent upper zone of the moving flow.

The dacite air-fall: Layers C to H

Layers C, D, F, and G, as defined by Curtis (1968), are Plinian deposits that are each one to several meters thick in the upper VTTS and in which the essential pumice is almost exclusively quartz-free dacite (Table I). The rhyolitic, andesitic, and banded pumice so abundant in the earlier ash-flow sequence are virtually absent in these late fall units, which account for two-thirds of the air-fall and 30—40% of the total eruptive volume.

Layers E and H are thin beds of sandy to silty ash that settled during the lulls that followed major outbursts. As anticipated for fine-grained elutriates, they consist of 90—95% glass shards. Separations in heavy liquids show them to be almost devoid of rhyolitic quartz and of dark andesitic glass. Figure 6 demonstrates that, unlike the Plinian layers that thin rapidly away from the source, Layer E is the deposit of a regional dust cloud, which settled slowly during a break in the eruptive sequence. It is not a co-ignimbrite ash because its constituents are predominantly dacitic, though some fraction of the vitric dust elutriated from the pyroclastic flows could conceivably have remained aloft, even after flushing by fall units C and D.

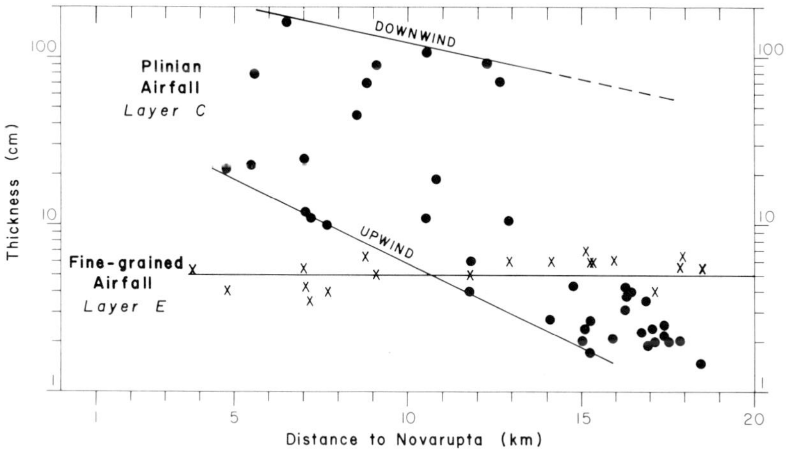

Fig. 6. Plot of thickness vs. distance to the source at Novarupta for tephra layers C and E (Fig. 4; Curtis, 1968), as measured at 40 locations between the Ukak and Katmai Rivers (Fig. 2). For these units, "downwind" was toward the SE quadrant, and "upwind" toward the NW, relative to Novarupta. Data are most complete and correlation most reliable for Layer C, which was produced by the first dacitic pumice fall after the ash flow; but the other Plinian units (Fig. 4) probably show a similar pattern. Layer E appears to represent an area-wide dust cloud that settled during an eruptive lull, probably on 8 June.

Layer H of Curtis also consists mostly of crystal-depleted vitric ash, which forms a conformable muddy capping upon uneroded tephra sections in the VTTS. It is composite, however, and includes not only the ash that settled after pumice-fall G but ash beds from a few small subsequent eruptions that left only a local record. In the upper VTTS, at least three such units are present, each 6—15 cm thick within 5 km of Novarupta. All are predominantly silt-sized glass but have basal 0.5—1.0 cm layers of sandy crystal-rich ash, each crystal-vitric couplet being the deposit of a late eruptive event. The

Layer H sequence widely contains accretionary lapilli, and its vitric matrix has an abundance of vesicle-like voids.

In his near-source study area, Curtis used thin (0.5—3 cm) interlaminae of similar vitric ash, along with changes in mean pumice size, to define his C—D and F—G boundaries. The ash interlaminae could represent the flushing of fines from condensation clouds that formed in rarefaction shadows behind the shock waves generated by major explosions. The two boundaries do not appear to have regional significance because, as noted by Curtis (1968, p. 178), they cannot be readily correlated with similar breaks in the dacite pumice sections on the Pacific slope. Detailed correlation of dacite fall units is of particular interest, however, because this would help tie the sequence of strata in the VTTS to the three discrete periods of ashfall at Kodiak, the only place where a fairly accurate ash-fall chronology was recorded (Fig. 4).

At the south base of Mt. Katmai near Katmai River, 12–14 km toward Kodiak from Novarupta, undisturbed tephra sections consist of a basal meter of rhyolite pumice and ~2.5 m of dacite pumice separated by a 20—60-cm interval in which these two pumice types coexist. Composition alone requires that the rhyolite be Layer A, that the mixed zone correlate roughly with the second half of the ash-flow sequence, and that the dacite correlate with Curtis' Layers C to H atop the ash flow. The dacite is divisible into a thick (1.0—1.8 m), coarse, normally graded, lower unit and an upper interval that is 0.9—1.4 m thick and laterally variable in the thicknesses and size-grading of its 7 or 8 constituent layers. These upper layers, however, include only two (normally graded) beds of coarse pumice, the rest consisting of granular to silty ash. The only logical correlation of these two upper coarse units is with Layers F and G in the VTTS. The lowest and thickest dacite unit in Katmai River sections must then be equivalent to both Layers C and D, in accord with the virtual continuity of these two units in the VTTS (see photos in Curtis, 1968, pp. 195, 197). Ongoing granulometric studies may refine these correlations, but my tentative conclusion is that Layer E accumulated during the only lull of significant duration (a few hours?) during deposition of dacitic pumice layers C through G (Fig. 4). At Kodiak, 170 km downwind, near-source complexities imposed by variable eruption rate and atmospheric turbulence had blurred into three main intervals of ashfall, equivalent to Layers A, C—D, and F—G.

On the basis of a wind shift deduced from his isopach maps, Curtis (1968, p. 183) suggested that the second and third ashfalls at Kodiak *both* correlated with his Layer C. I cannot exclude this, but I find it unlikely for the following reasons:

(1) There is no break within the Layer C air-fall section (and scarcely any between C and D), which could well have accumulated continuously within a few hours, failing then to explain the hiatus of several hours duration at Kodiak.

(2) Tephra C is a bit thinner than the sum of the other Plinian dacite layers (D, F, and G) in Curtis' measured sections, but the grain-size distribu-

tion and the maximum sizes of pumice and of lithic clasts are similar in all four units. This is also true in sections at Katmai River where the existence of three coarse dacite units shows that Layer C was not the only part of the C—G sequence to be transported a significant distance toward Kodiak. After the third ashfall at Kodiak, it dawned clear on 9 June (Griggs, 1922, p. 13) and no later ashfalls were recorded there. It seems improbable that the later dacite eruptions, apparently as powerful (if not individually as long-lasting) as that of C, could deposit comparably coarse tephra at Katmai River and not also at least dust at Kodiak.

(3) The relative thicknesses of the dacitic and rhyolitic fractions of the air-fall sequence are similar in the VTTS and 12—14 km toward Kodiak at Katmai River; the dacite layers comprise ~65–75% of the stratified tephra in both areas. Martin's field measurements in 1912 for locations in and near Kodiak show that the second and third layers there comprise 55—67% of the total ash thickness. This argues against a marked hindrance to eastward transport of ash by a change in wind direction during the main eruptive events. A difficulty with this line of reasoning, however, is that turbulence aloft resulted in some degree of mixing between components erupted at different stages, as shown by the small admixture of rhyolitic and andesitic components present in the largely dacitic second and third layers at Kodiak. It is hard to be quantitative about this yet, because the vitric ash from the dacite is itself rhyolitic.

Novarupta and its near-vent ejecta

The small dome of contaminated rhyolite, Novarupta (Fig. 3), apparently reached its final level of emplacement at the end of the eruptive sequence, possibly several months after the last major pumice eruption. Accompanying explosions that hurled dense blocks of vitrophyre and agglutinate >1 km from the extruding dome also dispersed any uplifted tephra cover from its roof.

The blocky lava dome itself has been described at length and illustrated by Fenner (1923, pp. 51, 59; 1950, pp. 715—721) and Curtis (1968, pp. 192—194). Its diameter is 380 m, its height ~65 m, and the intermediate (mostly dacite) lava interbanded with the rhyolite constitutes no more than 5% of the exposure; most of the conspicuous banding reflects textural variation in the rhyolite. Angular xenoliths, 0.5—10 cm across, are common in both rhyolitic and intermediate bands and are mostly finely crystalline andesite with few to abundant phenocrysts.

After emplacement of the ash flows and formation of the Novarupta depression, the depression was filled and its margins mantled by deposits of very coarse tephra, which extend to a distance of ~1.5 km from the dome. The deposits are in part near-vent facies of air-fall units C—H, but accumulation of the lithic-rich ejecta ring immediately around the dome (Figs. 3, 14) clearly outlasted the regional pumice falls. Maximum thickness of the ring ejecta is at least 100 m and could be much greater, but basal exposures are lacking. The ring deposits consist of a wide array of ejecta, dominated by crudely stratified dacite pumice but including an abundance of andesitic scoriae, dacitic and banded vitrophyre, finely vesicular breadcrusted blocks and lapilli, sedimentary rocks and pre-1912 intermediate lavas, fragments of hydrothermally altered lava and pumice, fused tuff, and densely to slightly agglutinated masses of recycled ejecta. Conspicuous for their scarcity around a rhyolite dome are rhyolitic ejecta, which constitute no more than 1% of the material exposed beyond the dome's talus apron. Near Novarupta, blocks of most lithologies are as large as 50 cm, but the extremes I noted are a 3-m breadcrusted block of densely agglutinated pumice 1 km east of the dome and a 5-m block of dense dacite vitrophyre 900 m south of it. Some of the most heterogeneous agglutinates revesiculated and breadcrusted after re-ejection, and equally remarkable are blocks of dense vitrophyre that contain angular inclusions of incompletely collapsed pumice.

Intercalated with the dacite air-fall deposits just beyond the ejecta ring are a few 2—4-m-thick pyroclastic flows, which outcrop in gullies 1 km south, 1—1.5 km west, and 2.5 km ENE of the dome. They include dacite pumice blocks up to 40—70 cm in diameter, as well as a much greater proportion of andesitic scoriae (up to 40%) than the enclosing pumice layers. Also within ~2.5 km of the dome are several gully-wall outcrops of crudely stratified lag-fall deposits, some slightly agglutinated, each 0.5—2 m thick, and all consisting of variable mixtures of coarse dacite pumice, andesitic scoriae, and lithic fragments. One of these grades up into a 2—3-m-thick pyroclastic flow, as indicated by loss of sorting and a great increase in the fine-grained matrix over a transition zone only several cm thick. Fall deposits that contain an abundance of andesitic scoriae do not occur >4 km from the dome. After the main ash-flow sequence was completed, it appears that subsequent outbursts of andesite were not energetic enough to do more than fall back locally, as stratified tephra or as minor near-vent pyroclastic flows.

Secondary deposits: Interaction of H_2O and hot ejecta

Ash-flow tuff was deposited against the margins of Glaciers 1, 2, 3 and 4 in upper Knife Creek (Fig. 2) and against two glaciers at the north foot of Mt. Mageik. Both Mageik glaciers have melted back to leave small lakes at the sites of their 1912 termini (Fig. 2), but positions of the Knife Creek ice fronts are relatively unchanged and some have even advanced several tens of meters out over the edge of the tuff. The interaction of ice, snowfields, rain, and running water with warm tephra and hot ash-flow tuff soon produced a variety of secondary deposits, including high-energy fluvial sands and gravels, debris-flow sheets, pond deposits of stratified ashy mud and pumiceous diamicton, phreatic craters with aprons of rippled surge beds, and a major flood deposit that was swept >20 km to the lowermost VTTS.

Fluvial sand and gravel

Curtis (1968, p. 179) illustrated such a deposit conformably intercalated between ash-flow deposits and air-fall Layer C. This is the only stratigraphic interval (and proglacial upper Knife Creek the only area) where deposits from running water are interstratified within the sequence of primary eruptive units. Stream, fan, and terrace alluvium younger than these are, of course, widespread in the VTTS. The intercalated fluvial unit is usually 1—3 m thick, but up to 7 m thick near the snouts of Glaciers 3 and 4. It rests upon the partly welded ash-flow tuff itself or upon <1 m of plane-laminated, poorly sorted tephra overlying it, accounting for incipient lithification of the fluvial material in several places by hot fluids permeating upward from the tuff. The sediments consist of alternating thin beds and laminae of phenocryst-rich coarse sand and rounded pumiceous gravel, commonly lenticular and channelled, washed free of fines, and parallel-, cross-, or dune-bedded. They contain a few percent rhyolite, in addition to andesitic, banded, and the dominant dacitic pumice; this contrasts with the superjacent, exclusively dacitic, air-fall layers C to H, within which fluvial sediments do not recur. Evidently, within a few hours of torrential flooding after emplacement of the final, lithologically heterogeneous ash flows, runoff from the glaciers was temporarily suppressed by the heavy fallout of dacitic tephra.

Block-rich pumice-flow deposit

On the banks of gorges incised along the snouts of Glaciers 1, 3, and 4 and for an additional 2 km downstream, discontinuous outcrops of an indurated pyroclastic flow, 1—3 m thick and unusually rich in dacite pumice blocks (up to 50 cm), occur atop or locally within the fluvial section just described. Especially near its base, parts of the deposit consist of framework-supported coarse pumice with little matrix, but commonly this grades into an upper half that is matrix-supported and rich in fines. Whether this material was emplaced as a hot debris flow or as one of several near-vent pumice flows I am not certain, but its incipient welding, its content of up to 85% dacite pumice (Table I), and its occurrence at the end of the ash-flow sequence make me favor the latter. Its presence at the northwest side of Glacier 1 virtually precludes a source on Mt. Katmai (Curtis, 1968, p. 201), on the slopes of which no pyroclastic-flow deposits have been found in any case. This outcrop is only 2 km from the Novarupta depression and strongly resembles other sintered, dacite-rich, blocky pumice flows exposed in gulches 1 km south, 1.5 km west, and 2.5 km ENE of the dome. At none of these localities do fluvial sediments accompany the block-rich unit as they do in proglacial upper Knife Creek.

Rhyolite-rich debris-flow deposit

Talus cones and sheets of air-fall tephra reworked from steep surrounding slopes today lap out onto the valley floor along most of its perimeter. In the easternmost corner of the VTTS north of Glacier 4, however, a pumiceous mudflow deposit 1—4 m thick is intercalated between the ash flow and the dacite tephra layers. It is massive, non-indurated, poorly sorted, and consists of ~50% rhyolitic ashy matrix, ~25% rhyolitic pumice lapilli, only 1—2% dacite or andesite pumice, and ~20% 1—20 mm lithic fragments of shale, andesite, and diorite. The lithic constituents represent the bedrock exposed on the steep adjacent slopes and help to demonstrate that the deposit formed by rapid remobilization of the earliest air-fall ejecta and some of its colluvial substrate. By contrast, similarly remobilized deposits (up to 12 m thick nearby and in Mageik Creek) that *overlie* the dacite air-fall contain an abundance of both dacitic and rhyolitic pumice, and they are poor in lithics.

The intercalated mudflow locally overlies 0.5—2 m of fluvially stratified pumiceous gravel and crystal sand and contains a few lenses of similar material up to 30 cm thick. It was important to verify the secondary nature of these deposits because they resemble rhyolite-rich pyroclastic flows, they are locally penetrated by alteration zones and degassing pipes from the subjacent tuff, and, if primary, they would have added a further complication to the eruptive compositional sequence. Rhyolite-rich eruptive units have nowhere been found to overlie the late, rhyolite-poor part of the ash-flow sequence.

Phreatic craters and surge deposits

Phreatic explosions within the ash-flow sheet produced an abundance of circular craters, mostly 20—60 m but up to 150 m wide, surrounded by aprons of cross-bedded ejecta that commonly include blocks of partly welded tuff. Several such craters lie within 250 m of Glaciers 3 and 4, and a zone containing more than 50 of them is preserved along the north side of Knife Creek for 6 km farther downstream. The distribution of their deposits shows that many more once occupied the present stream channel and probably helped to establish its incised course. At least 20 craters also formed near the ash-flow margin along the base of glaciated Mt. Mageik, others where major gulches in the Buttress Range debouch onto the tuff, and a large cluster where the Juhle fork of Knife Creek veers inward from the tuff margin. All are at places where water could pour onto the ash-flow surface or against its margins. Even near the Ukak terminus, a crater formed in non-welded tuff only 10 m thick where a side stream poured directly onto the ash-flow deposit that had dammed its course.

Every crater examined cuts the dacite air-fall layers and deposited part or all of its phreatic ejecta blanket on top of them. This indicates that the phreatic explosions generated within the ash-flow deposits persisted for at least the two additional days while Layers C to G accumulated. At least some of the explosions took place only after considerable fluvial erosion of the primary deposits, possibly days or weeks later, perhaps when compaction-related fractures permitted access of water to the interior of the tuff.

The explosion deposits thin radially away from constructional crater rims, where they are typically 2—5 m thick but >15 m thick around a cluster of craters in upper Knife Creek. They consist of crudely sorted thin beds that are alternately enriched in phenocryst sand or in pumice lapilli, both kinds having a crystal-vitric matrix of muddy ash. Parallel bedding dominates, but dunes, antidunes, festooned cross-sets, pinch-and-swell bedding, and extensively rippled bedding are all common. Truncated bedding sets give evidence of multiple outbursts and of mutual interference of deposits from adjacent craters. For an illustration, see Curtis (1968, p. 181).

The deposits contain abundant blocks of partly welded ash-flow tuff, commonly up to 0.5 m, but many are as large as 2 m; some show impact sags and/or thin zones of alteration of the enclosing tuff matrix. Many blocks are more welded (ρ = 1.8—1.9) than tuff at any level yet exposed by stream incision in the VTTS. Most of the phreatic ejecta are derived from the rhyolite-poor ash flows emplaced late in the eruptive sequence (and from the overlying dacite air-fall), but a few craters near Glacier 4 and along the northeast side of the VTTS expelled material that is up to 80% rhyolitic. Although this implies a somewhat deeper (or near-marginal) source for these particular explosions, in general it appears that most of the outbursts ejected material only from the upper and intermediate levels of the tuff.

Pond deposits

Deposition of the ash flows temporarily impounded streams, permitting local accumulation of pumice-rich ashy diamicton and fluvio-lacustrine strata of ashy mud, sand, and gravel. The largest of such areas were "Fissure Lake" (Griggs, 1922, p. 212), which extended 1 km eastward from the present lake in the southwest corner of the VTTS, and a 1-km-long lake where Windy Creek enters the tuff. Neither had fully breached its ash-flow dam by 1919, but both did so subsequently. Radially columnar-jointed blocks of partly welded tuff that are enclosed in chaotic ashy sediments at the site of "Fissure Lake" suggest explosive excavation of parts of the lake basin.

Between the southwest margin of Glacier 4 and the snout of Glacier 3, present stream cuts expose the deposits of one or more ice-marginal ponds where up to 10 m of coarse to fine fluvial and lacustrine sediment accumulated atop the block-rich pumice flow and against walls excavated by phreatic explosions in the dacite air-fall and the capping surge beds. The deposits consist of sand and pumiceous gravel, partly stratified but with up to 5 m of unstratified sand containing lithic dropstones (to 20 cm) and disoriented blocks of layered dacite tephra that were evidently frozen when they tumbled in.

The largest lake, ~1.5 km long, was impounded along Knife Creek just WNW of Glacier 4, not by primary deposits but by explosive disruption of the welded tuff and accumulation of more than 15 m of phreatic explosion debris around a cluster of ~20 phreatic craters; a disrupted area near the "damsite" is illustrated by Curtis (1968, p. 181). The basin in which 10—15 m of lake sediment then accumulated had been formed prior to its closure by means of differential axial compaction of the welded tuff, fluvial erosion of the overlying air-fall tephra, and shallow excavation of the tuff by earlier phreatic explosions. The lake sediments rest either directly upon disrupted tuff or upon a meter or so of intercalated stream alluvium. Their lower half consists of grey muddy diamicton in several 0.5—2 m unsorted units, rich in pumice lapilli and separated by 10—20 cm intervals of stratified sand and pumice. The upper half is laminated to thin-bedded mud with a few lenses and interbeds of pumiceous sandy alluvium. Scattered blocks of pumice and of coherent clots of tens to hundreds of pumice lapilli (frozen when deposited?) may have been emplaced in mudflows or as dropstones, tumble-ins, and snowbank slurries. Welded-tuff blocks up to 2 m wide probably slid into the lake from the fringing surge deposits, as impact sags have not been identified. At the upstream end of the lake-basin deposits there are additional outcrops of disrupted welded tuff, overlain by interfingering lake muds, alluvium, and pumiceous mudflow deposits, into which had calved rotated blocks, several meters in size, of the layered dacite tephra.

Flood deposit in the VTTS

In much of the lower VTTS, including the Windy Creek embayment and the terminal Ukak lobe, the ash flow and the stratified dacite air-fall above it are overlain by a debris-flow deposit, which is typically 1—2 m but locally up to 8 m thick. Its total volume is in the range 3—6 $\times$ 10^6 m^3. Most of the deposit is pumiceous diamicton with pumice blocks to 35 cm, a sprinkling of dioritic and andesitic clasts to 15 cm, and abundant small fragments of Jurassic shale in a muddy matrix of crystal-vitric ash. The deposit is generally unsorted and unstratified, and, though the thickest distal sections consist of a few conformable flow units, its emplacement was probably a single continuous event. It overran the ash-flow sheet, pouring beyond its terminus, engulfing upright uncharred trees, and feathering out into the woods.

The debris-flow deposit contains an abundance of welded-tuff blocks (as large as 50 cm), indicating that transport took place after some or all of the phreatic outbursts and requiring that the source and/or transport path included one of the areas with phreatic craters, 15—20 km up-valley. The River Lethe arm of the VTTS (Fig. 2) is mantled by an intact air-fall section and reveals no evidence for the passage of such a flow. The only likely site of origin for the debris flow is thus the temporary lake just described in upper Knife Creek. This is the only location satisfactory as a source for the exotic constituents, the welded-tuff blocks, and the abruptly released volume of water required for high-energy transport.

After breaching of the lake, the flood was confined to a broad channel, now occupied by Knife Creek, for 5—6 km before spreading across the central VTTS, where it widely stripped the air-fall and scoured the top meter or two of nonwelded tuff and vapor-indurated sillar. The scoured surface includes distinctive ash-flow tuff that is relatively poor in dacite but rich in both rhyolite and andesite (Table 1); such material is prominent in the debris-flow deposits in the broad area south and southeast of Three Forks. The northeast margin of the flood path can be tracked for several kilometers in the central VTTS as a 10-m-wide mudflow veneer strewn with welded-tuff blocks along the edge of the scoured axial zone.

Several gorges in the central VTTS were already incised 20—30 m into the tuff when first viewed in 1917 (Griggs, 1922, p. 254). Since these would effectively have intercepted and confined such a debris flow (as they have the annual runoff ever since), it is likely that the flood took place in the summer of 1912 or 1913.

Katmai River pumiceous debris flow

At the southern foot of Mt. Katmai (Fig. 2) is an unstratified sheet of poorly sorted 1912 ash and pumice lapilli (Fig. 7), which is up to 18 m thick near the center of its remaining 0.6 km^2 outcrop. Griggs (1918, pp. 124, 130) recognized it as a secondary deposit emplaced at low temperature atop the stratified dacite air-fall, but several subsequent workers have misinterpreted it as a hot pyroclastic-flow deposit. This is not surprising, because it closely resembles the nonwelded rhyolite-rich ash-flow sheet in the VTTS. Its nature is of more than peripheral interest because both Fenner (1950, p. 700) and Curtis (1968, p. 201) argued that it had erupted from the summit of Mt. Katmai, 6 km north of the deposit. If this were true, the sheet would represent the only 1912 ejecta known not to have erupted at Novarupta.

The evidence is compelling that deposition was secondary, by remobilization of Novarupta tephra down a single steep canyon from the snow-and-ice-covered southern slope of Mt. Katmai:

(1) The deposit contains uncharred twigs, branches, and a few rooted bushes, some with bark still intact.

(2) It lacks elutriation pipes or zones of fumarolic alteration, both common features of the ash-flow sheets in Mageik Creek and the VTTS.

(3) No air-fall material overlies it, but the sheet rests directly upon the intact or gently bevelled top of the stratified pumice section equivalent to Layers A through G. Its emplacement clearly postdated all of the regional tephra falls.

(4) None of its pumice lapilli are distinguishable from the dacite and rhyolite ejected at Novarupta. It is essentially devoid of andesitic and banded pumice, consistent with their general absence in fall deposits >4 km from the Novarupta vent. The ratio of dacite to rhyolite pumice lapilli is 7 : 3, unlike that of any exposed 1912 ash-flow unit (Table I) but virtually identical to that of the bulk air-fall tephra on Mt. Katmai. The top 2—3 m of the debris-flow unit are even more dacite-rich, up to 93% of the pumice being dacite.

(5) A marked scarcity of lithic fragments ($\ll$1%) and a local hint of reversed grading of pumice lapilli (Fig. 7) are consistent with (but do not require) secondary emplacement.

The surviving outcrop is an eroded remnant, but even had the original area been three times as large, the volume of the sheet could not have exceeded

Fig. 7. View of the Katmai River pumiceous debris-flow deposit, consisting entirely of remobilized 1912 tephra, redeposited at the mouth of the gorge draining the narrow glacial tongue due south of Katmai caldera (Fig. 2). It overlies the dacite tephra layers that cap the ash-flow sequence in the VTTS and along Mageik Creek; photographs of this relationship were published by Griggs (1918, p. 124) and Fenner (1950, Plate 1). Griggs called it the "Katmai Mud Flow", but several subsequent workers thought it an ash-flow deposit, which it superficially resembles. However, it contains uncharred twigs and bark, has no degassing pipes or fumarolically altered zones, and has a ratio of pumice-clast types unlike any 1912 ash flow (Table I) but similar to the bulk pumice fall on Mt. Katmai (rhyolite/dacite ~3/7 with no andesite).

0.01 km^3. The catchment area from which the tephra could have remobilized into the funnelling gorge (Fig. 2) was no less than 4 km^2, most of it a snow-mantled glacier with 11—14° slopes, upon which an average of >5 m of tephra surely fell. This would provide a minimum source volume of ~0.02 km^3, roughly double what was redeposited by the debris flow; these proportions seem to be satisfactory. The potential source area could have been much larger had the summit of Mt. Katmai not yet collapsed; or is the collapse what remobilized the tephra?

Two other scenarios warrant comment. (1) If a major tephra column above Novarupta were blown strongly eastward and collapsed from high elevation to generate ash flows over Mt. Katmai, this would still not explain the pumice ratios, low-temperature emplacement, or restricted distribution of the Katmai River deposit. (2) Alternatively, if the deposit were interpreted as a distal part of the Mageik Creek ash-flow lobe (Fig. 2) that had poured down to Katmai River and back-filled the area of present outcrop, this would be in conflict with the position of the debris flow on top of the dacite air-fall, the pumice-clast ratios, the lack of degassing features, and the downhill inclination of uncharred bushes overrun by the debris flow.

Study of this deposit has been quite instructive and emphasizes the caution necessary in interpreting rapidly remobilized pyroclastic material. In every respect except temperature and manner of initiation, this *was* a pumiceous pyroclastic flow, consisting of the same kinds of ejecta that produced primary ash flows only 12 km away a few days earlier.

Volume relations

Griggs (1922) calculated the total tephra fall to be ~20 km^3, and this remains as accurate an estimate as the regional data allow. On the basis of worldwide pyrheliometric measurements of solar spectral extinction, Volz (1975b) estimated the mass of aerosol still aloft in September 1912 to have been ~13 million metric tons or only ~0.005 km^3 dense rock equivalent (DRE).

The ash flow has a surface area of ~120 km^2, and Curtis (1968) meticulously and elegantly calculated its volume in the VTTS by means of comparative valley profiles. His estimate of 11 km^3 of ash-flow tuff is as good as the uncertainties warrant and is probably conservative. Addition of the Mageik Creek lobe and a few km of eroded distal extremities would raise the volume by only 0.1—0.2 km^3, but tuff concealed by the latest ejecta in the Novarupta caldera and in the ~3 km^2 area ENE of it (Figs. 2, 13) could add ~1 km^3 or so.

On the basis of pumice counts (Table I), estimated volumes of sub-units, and the assumption that components of the ashy matrix are roughly in the same proportion as the counted lapilli, I have calculated the volumes of essential rhyolite, dacite, and andesite ejecta to be as follows. The 20 km^3 of air-fall ejecta consist of ~28% rhyolite, 70% dacite, and no more than 2% andesite. If downwind ash is progressively enriched in the rhyolitic components, the fraction of rhyolitic tephra could exceed one-third. The ash-flow deposits consist of at least 60% rhyolite, with roughly 25% dacite and 15% andesite; if the wedge of rhyolite-rich tuff concealed in the upper VTTS is as thick as 200 m axially (Fig. 13), then these proportions could be as extreme as 75/16/9% (R/D/A) in the ash-flow fraction. For the entire eruptive volume, such estimates combine to give a range from 54/41/5% (R/D/A) for rhyolite-maximizing assumptions and an ejecta volume of 35 km^3 to 40/54/6%

for rhyolite-minimizing assumptions and a volume of 31 km^3. This amounts to 14—16 km^3 DRE. To the extent that Griggs' air-fall volume is an over-estimate, the proportion of rhyolite would be higher, as rhyolite is a larger component of the ash-flow than of the air-fall deposits.

PETROLOGY: A PRELIMINARY SUMMARY

The petrochemical study is not yet complete, but ample data are at hand (Figs. 8—12; Tables II—IV) to present a summary, which is provided here in order to support the volcanological discussion, to point out problems with published data from Katmai, and to draw attention to the wide compositional gap within the 1912 suite.

Bulk composition

In most respects the Novarupta ejecta form a typical Aleutian-arc calcalkaline suite, although the most mafic material erupted in 1912 was medium-K silicic andesite (Taylor, 1969) with an Mg-number of only ~55 and FeO^*/MgO ~1.8. Mt. Griggs and Mt. Katmai have erupted andesitic lavas a few percent less silicic than any 1912 ejecta (Fenner, 1926; Kosco, 1981), but no basaltic rocks have been found in place on any Katmai-group volcano. Basalt was evidently erupted earlier, however, as both Fenner and Kosco analyzed basaltic boulders (Fig. 9) collected from the caldera rim and floor of Mt. Katmai.

The 1912 suite is calcic (not calcalkalic) with an alkali-lime index of 63 (Fig. 8; Peacock, 1931); it falls well inside the calcalkaline field on a plot of FeO^*/MgO vs. SiO_2 (Miyashiro, 1974). On an AFM diagram (Fig. 9), the suite plots close to Kuno's (1968) hypersthenic magma series, to the Aleutian calcalkaline data of Coats (1952, 1962) and Byers (1961), and virtually on top of the main volcanic trend for Unalaska Island (Perfit et al., 1980).

The samples analyzed here are glassy pumice, scoria, and vitrophyre, each with unexsolved titanomagnetite and few groundmass microlites. All, however, are vesicular and have lost most of their magmatic H_2O and S, presumably CO_2, and probably some Cl (Table II). The rather coherent variation trends for Al_2O_3 and CaO (Fig. 8) and for Eu/Eu^* (Fig. 10; Hildreth, 1981, fig. 9), as well as the lack of any apparent correlation between these factors and bulk plagioclase content (Ewart, 1976), suggests that feldspar accumulation has not seriously affected the analyzed rocks.

Compositional gap and thermal continuity

Chemical analyses of the 1912 rocks (Table II, Fig. 8) give the following ranges in SiO_2 contents (all data recalculated to 100% anhydrous; see Fig. 8): Rhyolite 77 ± 0.6%, dacite 66—64.5%, and andesite 61.5—58.5%. The dacitic and andesitic ejecta contrast in color and vesiculated density, and it remains uncertain whether there is a complete compositional continuum between the light-grey pumice and the black scoriae. So far,

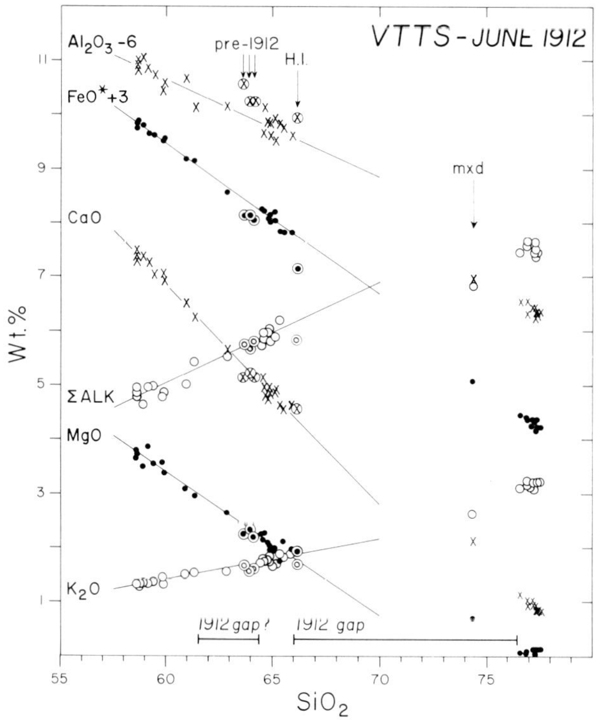

Fig. 8. Variation of major elements (in wt.%) for the June 1912 eruption in the Valley of Ten Thousand Smokes. New analyses by XRF, USGS laboratories at Menlo Park and Denver. For internal comparability, all data are normalized to 100%, excluding H_2O and CO_2 but including halogens and ~30 trace oxides. The pile-up of points at 77.1—77.6% SiO_2 represents 11 analyses of rhyolitic pumice. Most samples are 1912 pumice and scoria from air-fall and ash-flow emplacement units, except the following. *Mxd* = Fenner's (1923) analysis of rhyolitic lava slightly contaminated by co-eruptive dacite in the Novarupta dome; new analyses of the least contaminated bands in the dome are plotted here at 65.1 and 76.9% SiO_2. *H.I.* = Fenner's (1950) analysis of the horseshoe-island lava on the Katmai caldera floor. *Pre-1912* points are the precaldera lava domes, Falling Mtn. and Mt. Cerberus, and a fluvially stratified deposit of pumice lapilli in the lower VTTS; these plot respectively at 64.2, 64.0, and 63.7% SiO_2 and partially occupy a possible second compositional gap apparent in the 1912 zonation. Note that the linear-regression lines fit the 1912 andesite-dacite data rather well but do not project directly toward the rhyolitic data arrays.

only one analyzed 1912 sample, a medium-grey pumice block from the topmost dacite pumice fall, lies in the range 61.5—64.5% SiO_2 (Fig. 8), despite persistent search for such material. It can be seen in Figs. 8 and 9 that the two pre-1912 dacite domes, Falling Mountain and Mt. Cerberus, fall close to the 1912 trends and within the apparent break in continuity just mentioned.

The larger gap, between the intermediate suite and the rhyolite, is further illustrated by the data of Table III. Banded pumice of dacite and rhyolite that commingled during the eruption is indeed present, but I doubt that

TABLE II

Representative analyses of the Novarupta system[1]

	Pre-1912 Domes		1912 Ejecta					Novarupta Dome		
#	Falling Mtn. K-177	Mt. Cerberus K-164	Topmost ash flow K-123	Top of Layer G K-101	Top of Layer C K-45	Basal ash flow K-139	Base of Layer A K-42	Purest rhyolite K-16	Crystal-poor band K-22	Crystal-rich band K-18
SiO_2	64.2	64.0	58.6	62.9	64.6	77.3	77.4	76.9	76.6	65.1
TiO_2	0.62	0.63	0.71	0.72	0.71	0.17	0.17	0.17	0.23	0.71
Al_2O_3	16.2	16.2	16.9	16.2	15.65	12.35	12.4	12.5	12.55	15.5
Fe_2O_3	2.59	2.33	2.96	2.00	1.97	0.44	0.39	0.55	0.48	1.91
FeO	2.72	3.03	4.16	3.78	3.44	0.81	0.85	0.89	1.04	3.50
(FeO*)	5.05	5.13	6.83	5.58	5.21	1.21	1.20	1.39	1.47	5.22
MnO	0.10	0.10	0.14	0.13	0.12	0.05	0.05	0.05	0.06	0.12
MgO	2.20	2.34	3.80	2.65	2.14	0.09	0.11	0.02	0.08	1.95
CaO	5.15	5.22	7.31	5.67	4.97	0.93	0.87	0.95	1.13	4.85
Na_2O	4.19	4.13	3.54	4.01	4.20	4.21	4.25	4.36	4.36	4.21
K_2O	1.60	1.56	1.29	1.57	1.74	3.25	3.20	3.22	3.10	1.68
P_2O_5	0.13	0.14	0.13	0.13	0.135	0.03	0.03	0.02	0.04	0.15
F	—	—	400	—	400	500	500	500	—	—
Cl	20	<10	800	630	700	800	1375	1125	950	500
S	<10	<10	140	240	150	80	<10	60	10	180
Sr	303	320	327	240	237	47	40	52	58	215
Ba	595	575	410	530	600	1025	990	960	950	550
V	112	105	175	120	105	<2	<2	<2	4	102
Y	28	26	7	15	24	42	46	48	40	13
Zr	117	110	113	145	151	157	158	157	146	165

La	12	12	11	12	13.5	18.5			18.5	13
Ce	25	26	24	28	29	41.5			42	30
Nd	14	15	14	16	16	23.5			23	17
Sm	3.5	3.5	3.8	4.2	4.25	5.6			5.7	4.4
Eu	0.90	0.91	0.97	1.26	1.24	1.01			1.0	1.3
Tb	0.64	0.64	0.62	0.75	0.76	1.03			1.05	0.82
Yb	3.0	3.0	2.8	3.3	3.45	4.7			4.8	3.5
Lu	0.46	0.45	0.42	0.50	0.54	0.61			0.64	0.54
Rb	29	29	20	—	38	71			71	29
Cs	0.7	0.5	1.2	1.5	1.5	3.6			2.8	1.4
Ba	550	570	440	580	625	975			995	600
Th	2.8	2.8	2.9	3.35	3.25	5.65			5.6	3.2
U	1.2	1.3	1.3	1.5	1.4	2.5			2.55	1.5
Hf	3.4	3.4	3.2	4.5	4.5	5.3			5.2	4.5
Ta	0.30	0.29	0.3	0.34	0.33	0.52			0.49	0.34
Sb	0.6	0.3	0.5	0.6	0.6	1.1			0.9	0.6
Sc	15.5	16.6	25.0	17.6	18.3	7.8			8.7	19.5
Cr	5.1	7.0	27	9	8	2			1.6	6
Fe%	3.85	3.99	5.27	—	3.95	0.91			1.15	4.00
Co	11.5	12.4	18	—	—	—			1.6	—
Zn	65	62	85	72	73	37			30	86
(H_2O)	0.25	0.18	0.88	1.56	2.52	2.83	2.66	0.26	0.48	0.26

[1] All samples of 1912 ejecta were pumice blocks, the Novarupta samples microvesicular vitrophyres, and the two pre-caldera samples nonvesicular vitrophyres. Major elements and Cl through Zr were analyzed by XRF, USGS laboratories at Menlo Park; FeO by titration; F by specific-ion electrode; REE and lower list (Rb through Zn) by INAA at Lawrence Berkeley Laboratory by methods of Perlman and Asaro (1969). All data are normalized to 100%, H_2O-free; major elements in wt.%, others in ppm. Total H_2O contents by Penfield analysis are given at the bottom.

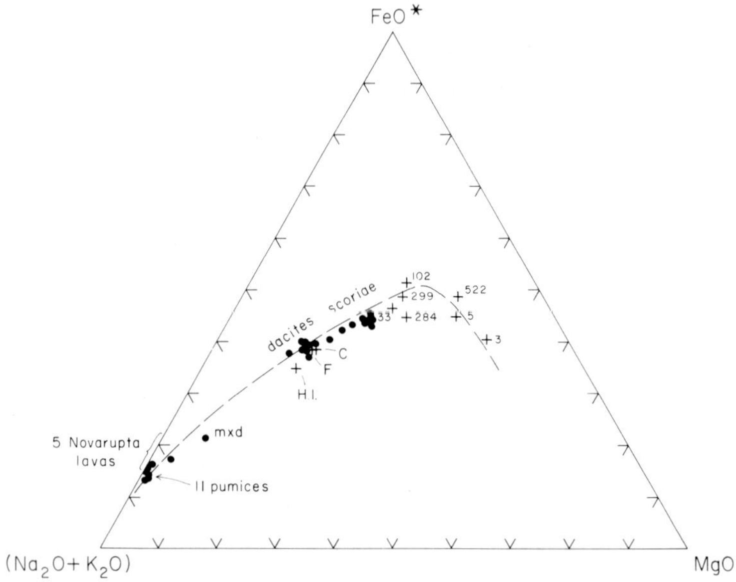

Fig. 9. AFM plot of the 1912 VTTS suite and some related or nearby lavas. Solid dots are 1912 ejecta and Novarupta dome lavas. As in Fig. 8, *mxd* is Fenner's (1923) Novarupta sample 153-C. Crosses are pre-1912 lavas, as follows: Precaldera domes (Table II), *F* = Falling Mtn. and *C* = Mt. Cerberus. *H.I.* = Fenner's (1950) sample of the "horseshoe island" lava on the floor of Katmai caldera. For comparison, a few analyses of rocks more mafic than any erupted in 1912 are plotted from Fenner (1926) and Kosco (1981): 3, 5, 102, 522 from Mt. Katmai, 299 from Mt. Katolinat; 284 from near Mt. Juhle; and 33 from Mt. Griggs. Basaltic samples 522, 5, and 3 occur as blocks on the rim and floor of Katmai caldera; no basalts have been found in place in the Katmai district. The dashed line is the trend for the volcanic rocks of Unalaska Island (Perfit et al., 1980).

homogeneous pumice in the range 66—76% SiO_2 has escaped my attention. Analyses reported by Fenner (1923, 1950) within this range were of banded pumice and lava and of bulk tephra that had mechanically fractionated and/ or mixed during flight (see Fenner, 1950, pp. 697—699). Fenner argued that his straight-line variation diagrams signified assimilation, but, though mixing of a sort indeed took place, most of the mixing that he documented was aeolian or in the conduit. The banded pumice illustrates commingling of all three distinctive magmas during the eruption, rhyolite-andesite being the commonest combination, but contacts between bands are sharp and there is little approach to homogenization. The 1912 rocks showing the most advanced contamination are the banded lava dome and the agglutinated recycled ejecta surrounding it; these may have been extruded many months after the June eruption.

The exchange between Bowen (1928) and Fenner (1926, 1931) debating the linearity or curvature of Katmai variation trends (Wilcox, 1979) did not really concern the 1912 suite; rather, the argument dealt with 18—19 analyses by Fenner of samples that included five intrusive rocks, one 1912

TABLE III

Contrasts and similarities across the compositional gap in the eruptive suite of June 1912

	Rhyolites	Intermediates
Fe/Mg	>15	3.4 — 2.2
K/Na	0.87— 0.82	0.53 — 0.40
Rb/Sr	1.5 — 1.2	0.13 0.06
Ti/Zr	6 — 8	26 — 38
Ba/Sr	25 — 18	2.7 — 1.6
Zn/Cu	8.5 — 5	3.3 — 1.5
Eu/Eu*	0.5 — 0.6	0.8 — 1.1
V/Ni	<1	5 — 10
Ba	1025 —950	600 —400
CaO	0.84— 0.96	4.6 — 7.4
K_2O	3.3 — 3.1	1.9 — 1.3
Na_2O	4.4 — 4.2	4.3 — 3.4
K/Rb	380 —430	420 —580
K/Ba	27 — 26	26 — 24
Ba/La	55 — 51	44 — 37
Ce/Yb	9.1 — 9.0	9.1 — 8.4
La/Sm	3.3 — 2.9	3.1 — 2.6
$(La/Sm)_N$	2.1 — 1.7	2.0 — 1.6
$(Tb/Yb)_N$	0.9 — 1.1	0.9 — 1.2
Th/U	2.1 — 2.3	2.2 — 2.3
Zr/Hf	28 — 32	32 — 39
Phenocrysts (wt.%)	0.3 — 2.2	30 — 45
$^{87}Sr/^{86}Sr$ ($\pm$0.0002)	0.7036	0.7034— 0.7036
$\delta^{18}O$ plag ($\pm$0.1)	6.4 — 6.1	6.1 — 5.8

pumice, and a variety of lavas from Mts. Griggs, Katmai, Mageik and Martin a few of them morainal boulders and others probably agglutinates. Fenner did not report his several 1912 tephra analyses until 1950.

In Fig. 8, the actual trends for 1912 are seen to be very linear in the 58–66% SiO_2 range, but extrapolation across the compositional gap to the rhyolite would require curvatures in the sense discussed by Bowen (1928 pp. 114—124), inconsistent with simple mixing. All of the present samples were homogeneous and were mostly analyzed together as a batch in the same laboratory. Hybridization, fractionation, and repeated internal mixing (Hildreth, 1981; Sakuyama, 1981; DePaolo, 1981) are probably part of the pre-eruptive history of any long-lived intermediate magmatic system, but the superficial syneruptive mixing so prominent in the 1912 ejecta was not responsible for the compositional gradients evident in the suite. It is not absolutely certain whether the andesite, dacite, and rhyolite were part of a single zoned reservoir or coexisted in two or more discrete bodies prior to eruption, but relative to most rock suites studied for their petrochemical variation and evolution, at least the eruptive age relations of the samples plotted in Figs. 8 to 12 are well known.

Despite the gap in magma composition of 10% SiO_2, titanomagnetite-ilmenite equilibration temperatures show a *continuous* thermal variation across the whole compositional range (Fig. 11): Rhyolite 805—850°, dacite 855—955°, and andesite 955—990°C. ^{18}O fractionations of 3.9—4.2 per mil between titanomagnetite and quartz support the rhyolitic temperature range, as do measurements of homogenization temperatures of vapor-glass inclusions in 1912 quartz phenocrysts by Clocchiatti (1972). The continuity in T and f_{O_2} for the entire sequence, the similarity of the rhyolitic and intermediate ejecta in $^{87}Sr/^{86}Sr$ (~0.7036), and the identity or continuity of the whole suite in many elemental ratios (Table III, lower half) each suggests that all the ejecta were derived from a contiguous magmatic system. In particular, the similarity of K/Rb, Th/U, $\delta^{18}O$, Na_2O, and REE patterns (Fig. 10) on both sides of the gap weighs against a separate origin for the rhyolite by melting of roof- or wall-rocks unrelated to the system. If the rhyolite were a partial melt of roof rocks atop an intermediate magma body, then such rocks had no Sr-isotopic contrast with the andesite-dacite magma and clearly did not include the Jurassic arkosic or granitic country rocks. If dioritic intrusives of late Tertiary or Quaternary age were partially melted to produce the rhyolite, an especially attractive hypothesis, the ^{18}O data show that they had not been hydrothermally altered.

On the whole, the evidence suggests that the entire 1912 suite was comagmatic in a broad sense, but this does not exclude internal remixing of variably fractionated magma batches or partial remelting of crystallized forerunners within a thermally long-lived injection system. The geochemical

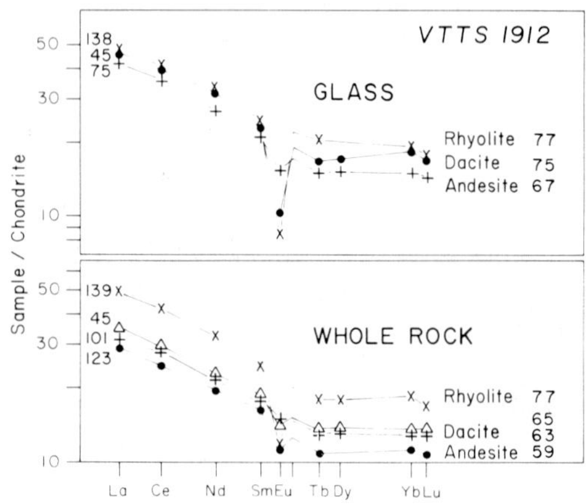

Fig. 10. Chondrite-normalized REE abundances for 1912 pumice and scoriae and for high-purity glass separates, plotted vs. ionic radii. Sample numbers at left; wt.% SiO_2 at right. Note parallelism of patterns (excluding Eu) from 58 through 77% SiO_2. Several more analyses not plotted are virtually identical; Novarupta rhyolitic and dacitic bands closely match plotted samples 139 and 45, respectively. Patterns for precaldera domes, Falling Mtn. and Mt. Cerberus (Table II), are closest to 1912 sample 123, which is 5.5% lower in SiO_2.

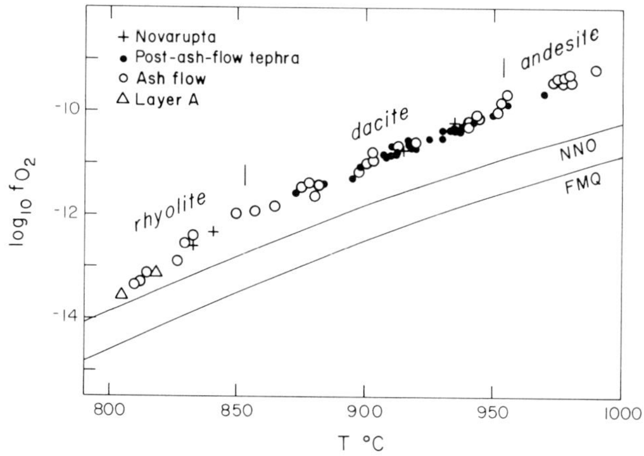

Fig. 11. Values of Fe-Ti-oxide equilibration temperature and oxygen fugacity, determined for titanomagnetite-ilmenite pairs from 65 glassy samples of 1912 ejecta and lava. NNO and QFM are experimental oxygen-buffer curves, for reference only. Symbols represent eruptive units, as shown. Rhyolites range from 805 to 850°C, dacites 855 to 955°C, and dark andesite scoriae 955 to 990°C. Values determined by the methods of Buddington and Lindsley (1964) and Carmichael (1967)

data do not require that the magma reservoir have been areally coextensive or vertically continuous; the strongest evidence for physical continuity is the Fe-Ti-oxide data of Fig. 11. An eruptive sequence in which initially predominant rhyolite was joined progressively by andesite and lesser dacite, then succeeded by Plinian outbursts that were almost exclusively homogeneous dacite, is difficult to reconcile with a simply stratified magma chamber. Discrete reservoirs (as suggested by Curtis, 1968) or a sill-like antechamber cannot be excluded (see below).

Mineralogy

In the course of separating minerals from ~120 samples, thirty were done with quantitative care, supplementing thin-section study to yield the following data. The rhyolites contain 0.5—2 wt.% phenocrysts and both the dacites and andesites 30–45%, with an apparent tendency for dacites to be more crystal-rich. All 1912 ejecta contain plagioclase, hypersthene, titanomagnetite, ilmenite, apatite, and pyrrhotite in that order of abundance. In addition, quartz is similar in abundance to plagioclase in the rhyolites and augite about half as abundant as hypersthene in the intermediate rocks. Traces of euhedral amphibole (as yet unanalyzed) occur in separates from the rhyolitic lavas of Novarupta, but they may be vapor-phase crystals. Zircon has not been found nor have unincluded crystals of apatite. In the andesites and dacites, a typical ratio is plag 5, opx 2, cpx 1, mt 1, whereas in the rhyolites it is quartz 5, plag 5, opx 2, mt 1; the other phases are at least an order of magnitude scarcer.

Because only the rhyolites contain quartz and lack clinopyroxene, and only the andesitic scoriae contain a trace of olivine (Fo_{74-82}), mineral separations of certain fine ashes can provide stratigraphic information. For instance, the virtual absence of quartz in Layer E (Fig. 6) shows that this largely vitric fallout was not elutriated from the ash flows but from the dacitic ejecta of Layers C and D. Similarly, nearly quartz-free separates from 1912 ash at Brooks Lake (50 km NW of Novarupta) confirm field evidence that fine-grained dacitic tephra were dispersed in the upwind direction to a much greater extent than the earlier-erupted rhyolite.

Curtis (1968) and Fenner (1923, 1926, 1950) provided petrographic descriptions and photomicrographs of the 1912 ejecta, so the following notes largely concern microprobe investigations, the preliminary results of which are summarized in Table IV and Fig. 12. Plagioclase is barely zoned (An_{25-29}) in the rhyolites but strongly so in the dacites and andesites, in which both normal and reverse zoning occur. The range of zoning found in dacites is An_{34-71}, but only An_{36-53} is common. Most feldspars in the andesitic scoriae zone widely between An_{35} and An_{83} (largely An_{47-79}), but a sparse subpopulation appears to be restricted to An_{73-83}, and a few discrete grains are virtually unzoned anywhere in the range An_{87-94}.

The pyroxenes are only moderately zoned in 1912 intermediate ejecta, and the rhyolitic hypersthene is hardly zoned at all. In addition to the sparse grains of anorthite and olivine, the andesites contain a minor subpopulation of magnesian clinopyroxenes with 5—8 wt.% Al_2O_3 (Fig. 12). Titanomagnetite is notably abundant, making up ~3 wt.% of all intermediate rocks, but ilmenite is about 100 times scarcer. Neither oxide is zoned, and both exhibit systematic variation in minor element contents (Table IV) across the range of temperature and bulk composition represented here.

It needs to be emphasized that, in addition to the differences in assemblage, the rhyolitic and intermediate rocks are also distinct in that there is no overlap (Fig. 12) between the compositional ranges of the phenocrysts they do have in common, plagioclase, orthopyroxene, and the oxide minerals. The dacite is clearly not a product of mixing between the rhyolitic and andesitic magmas that erupted in 1912.

Inclusions of glass, pyrrhotite, and apatite are common in the pyroxenes and oxides. In the feldspar, all three are present, but only glass is common. Titanomagnetite and ilmenite are commonly included by both pyroxenes in the intermediate rocks but rather uncommonly by the rhyolitic hypersthene. Glass inclusions are abundant in the splendidly euhedral quartz phenocrysts. Microprobe analyses of these inclusions exhibit summation deficiencies which have been interpreted to indicate a volatile content of 4—5% by weight in the entrapped glass (Clocchiatti, 1972). The inclusions that contained vapor bubbles gave homogenization temperatures of 850 ± 10°C (108 determinations; Clocchiatti, 1972), the upper end of the Fe-Ti-oxide temperature range I have found for quartz-bearing samples of 1912 ejecta (Fig. 11).

In contrast to the pumice of the main eruptive phases, Novarupta lava and

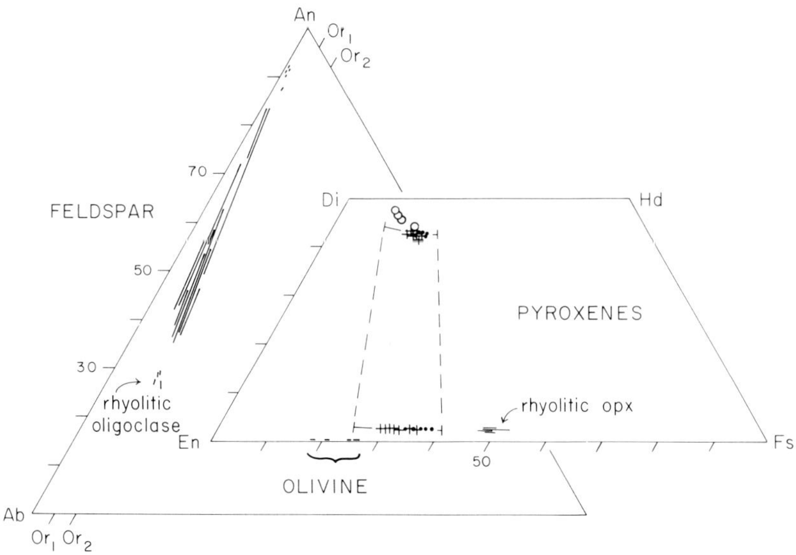

Fig. 12. Compositions of representative pyroxenes, feldspars, and olivines in the 1912 ejecta. Note that for clarity the Or scale has been exaggerated 4-fold. The extensive zoning of plagioclase (individual sample ranges plotted) in both andesite and dacite contrasts with the barely zoned rhyolitic oligoclase and with occasional grains of unzoned anorthite, which along with rare olivines have been found only in andesitic scoriae. Average compositions of zoned pyroxenes are shown for several representative samples; the bracketing lines and tie lines show the extremes of zoning yet analyzed in 1912 andesitic-dacitic pyroxenes. Crosses andesites; dots dacites; open circles, uncommon Al-Ti-rich clinopyroxenes in andesites. Note the lack of overlap between the ranges of phenocryst compositions in the rhyolitic and intermediate suites.

much of the vitrophyric and agglutinated ejecta in its ejecta ring are somewhat contaminated. In addition to a few percent of discrete dacitic bands and the angular lithic inclusions in the dome, even the rhyolite yields a small amount of clinopyroxene xenocrysts and comminuted lithic debris when separated in heavy liquids. This accounts for the apparently less evolved compositions of most analyzed Novarupta samples (Figs. 8, 9; Table II) relative to rhyolitic pumice in Layer A and in the ash flows.

Implications of the petrological data

The abundance of sharply banded pumice in the ash flows (Table I) demonstrates commingling of magmas from different levels or discrete reservoirs during the eruptive process. The mixed pumice does not support either the concept of an eruption triggering arrival of new mafic magma or the idea that the dacites are hybrid mixtures of andesite and rhyolite. The large volume (7—8 km^3), the phenocryst-rich nature (30—45% crystals), and the low temperature (855—990°C) of the andesite and dacite magma erupted in 1912 indicate that it was part of a long-established zoned reservoir, which was

TABLE IV

Representative electron-microprobe analyses of 1912 phenocrysts[1]

Unit	A		Early distal ash flow		C		G		Late proximal ash flow		Early distal ash flow		
#	K-42		K-139		K-45		K-101		K-75		K-140		
	C	R	C	R	C	R	C	R	C	R	C	R	Al-rich
Orthopyroxene													
SiO_2	51.0	50.0	51.1	50.7	52.5	52.3	52.7	52.0	54.2	52.8	53.0	51.5	51.5
Al_2O_3	0.34	0.28	0.42	0.32	0.64	0.66	2.60	0.40	1.35	1.04	1.46	1.52	2.5
TiO_2	0.13	0.12	0.12	0.09	0.21	0.23	0.23	0.10	0.12	0.18	0.25	0.19	0.32
FeO	29.3	32.0	28.5	30.2	21.85	23.7	16.4	24.4	16.0	22.5	18.1	23.8	18.0
MgO	17.0	15.4	17.0	16.4	21.9	20.5	25.8	20.8	26.8	21.7	25.4	21.5	25.0
MnO	1.46	1.48	1.46	1.53	0.72	0.88	0.32	0.94	0.31	0.65	0.32	0.61	0.32
CaO	1.13	1.13	1.19	1.23	1.43	1.25	1.59	1.33	1.53	1.42	1.33	1.40	1.64
Na_2O	0.01	0.01	0.02	0.01	0.01	0.03	0.01	0.01	0.01	0.04	0.03	0.04	0.04
Total	100.37	100.42	99.81	100.48	99.26	99.55	99.65	99.98	100.32	100.33	99.89	100.56	99.32
Clinopyroxene	absent		absent										
SiO_2					52.1	52.1	51.0	52.0	52.1	52.7	53.0	52.5	49.5
Al_2O_3					1.92	1.37	2.40	0.91	3.39	1.65	2.3	0.97	6.1
TiO_2					0.41	0.49	0.55	0.27	0.38	0.39	0.37	0.21	0.9
FeO					9.5	10.9	9.1	10.9	5.9	10.4	7.95	10.6	8.5
MgO					14.2	13.9	14.8	13.6	16.3	13.9	16.55	13.7	14.4
MnO					0.30	0.45	0.30	0.49	0.16	0.35	0.18	0.50	0.18
CaO					21.0	20.2	20.9	20.6	21.5	20.6	20.0	21.0	20.4
Na_2O					0.30	0.34	0.30	0.32	0.22	0.36	0.18	0.34	0.29
Total					99.73	99.75	99.35	99.1	99.95	100.35	100.53	99.82	100.27

Plagioclase													
Na_2O	8.12	8.20	8.00	8.12	6.34	7.16	4.90	3.46	1.95	5.75	4.15	7.22	0.89
K_2O	0.39	0.39	0.40	0.38	0.28	0.37	0.19	0.25	0.06	0.27	0.14	0.35	0.02
CaO	5.64	5.47	5.91	5.96	8.95	7.27	11.5	8.75	17.0	10.15	12.67	7.10	18.72
Al_2O_3	24.2	23.6	24.5	24.0	27.0	25.3	29.7	27.0	34.3	28.2	29.9	25.2	35.4
SiO_2	61.5	62.0	61.0	61.5	57.7	59.7	54.0	57.5	47.0	56.0	52.9	59.4	44.9
Total	99.85	99.66	99.81	99.96	100.27	99.8	100.29	99.96	100.31	100.37	99.76	99.27	99.93
	mt	ilm	mt	ilm	mt	ilm	mt	ilm	mt	ilm	mt	ilm	olivine
Fe-Ti oxides													
TiO_2	10.5	45.0	10.3	45.4	9.90	41.6	9.5	41.3	9.1	38.9	8.6	38.7	0.03
FeO[1]	81.5	51.0	81.9	50.1	80.3	51.7	79.9	52.4	79.0	52.5	78.9	53.0	17.2
MnO	0.78	1.26	0.76	1.20	0.56	0.58	0.43	0.68	0.44	0.33	0.35	0.40	0.22
MgO	0.78	1.46	0.72	1.45	1.60	2.59	2.20	2.62	2.85	3.73	3.02	3.25	42.8
ZnO	0.22	0.02	0.19	0.01	0.16	0.01	0.06	0.01	0.11	0.01	0.08	0.01	—
Al_2O_3	1.26	0.13	1.25	0.15	1.90	0.25	2.45	0.27	3.20	0.54	3.41	0.45	0.02
V_2O_3	0.16	0.08	0.18	0.11	0.52	0.24	0.62	0.23	0.69	0.38	0.68	0.35	—
Cr_2O_3	0.09	0.02	0.12	0.05	0.08	0.04	0.12	0.04	0.11	0.06	0.11	0.06	—
(FeO)	38.8	36.6	38.8	37.0	37.3	32.2	36.4	31.7	35.2	28.0	35.6	28.6	0.17 CaO
(Fe_2O_3)	47.4	16.0	47.9	14.55	47.8	21.7	48.3	23.0	48.7	27.2	49.3	27.1	39.7 SiO_2
Total	99.9	100.5	100.1	99.9	99.8	99.2	100.0	99.8	100.3	99.1	100.0	98.9	100.1

[1] The paired analyses give the extremes of zoning for each sample, cores (C) in the first column, rims (R) in the second. Although the microprobe study is incomplete, these analyses bracket the data now at hand. The first two samples are rhyolites (Table I), the next two dacites, and the last two andesites. The Al-rich opx, cpx, and anorthite are virtually unzoned and, along with slightly zoned olivine, occur sparsely in the andesitic scoriae. The Fe-Ti oxides are not zoned; their iron values are recalculated by the method of Carmichael (1967).

cooling, crystallizing, and differentiating in the crust. Phenocryst-poor, higher-temperature ejecta that might represent fresh injections have been looked for but not found. The lack of any real transition from banding to blending in 1912 pumice and the absence of mutual contamination of the rhyolitic and intermediate bands by their contrasting suites of phenocrysts (except in the lava dome emplaced later) imply that the obvious mixing was syneruptive and superficial.

A major question is the origin of the high-silica rhyolite and its pre-eruptive situation relative to the intermediate magmas. This occurrence of high-silica rhyolite is unique on the Alaska Peninsula, but the intermediate magmas are the common stuff of the surrounding volcanoes. That ~ 3 km^3 of rhyolitic magma were released before being joined by *any* of this intermediate magma suggests either that a rhyolitic cap (perhaps a discrete convection cell) had existed on top of a sharply zoned system or that the zoned dacite-andesite reservoir broke into and flowed through a quasi-discrete rhyolitic reservoir that was closer to the vent. The strongest evidence that the rhyolite was contiguous or close to the intermediate chamber is the complete continuity of all the T-f_{O_2} data (Fig. 11) and the need for a heat source to keep the rhyolite poor in phenocrysts.

No rhyolite is known to have erupted previously anywhere in the Katmai district, and it erupted in 1912 only at Novarupta. The Falling Mountain and Mt. Cerberus domes, however, had erupted adjacent to the 1912 vent (Figs. 2, 3) only a few thousands or hundreds of years earlier, and their compositions are close to that of 1912 dacite (Fig. 8; Table II). The Falling Mountain dome overlapped and was disrupted by the 1912 vent (Fig. 3); its position suggests that at least 7 km^3 of rhyolitic magma evolved within the short time after its eruption, and indicates that a well-evolved intermediate magma system had already penetrated to the surface at the very site of what became the 1912 vent.

If the rhyolitic liquid separated from dacitic magma by filter-pressing of interstitial melt or by ascent of buoyant boundary layers (Hildreth, 1979, 1981; Shaw, 1974; McBirney, 1980), then extraction was so efficient that no dacitic phenocrysts were retained (Fig. 12) and no bulk compositions between 66 and 76% SiO_2 were created. If, on the other hand, the rhyolite were a partial melt of roof rocks atop an intermediate magma body, then such rocks had no Sr- or O-isotopic contrasts with the andesite-dacite magma. Because the exposed country rocks are feldspar-bearing sedimentary rocks of Jurassic age, the only likely source rock consistent with the isotopic constraints in a partial-melting argument would be unaltered Tertiary or Quaternary igneous forerunners. The thermal and mechanical problems of extracting a 77% SiO_2 partial melt from dioritic or andesitic rocks are at least as difficult as cleanly removing such a liquid from a crystallizing dacite magma. Quantitative assessment of the feasibility of such removal awaits additional analytical data on glasses and crystal-liquid partitioning in this suite, but preliminary examination of the data in Tables II, III and IV reveals no major geochemical obstacle to a model of efficient crystal-liquid fractionation. The main problem is the mechanics of extraction.

NOVARUPTA CALDERA AND KATMAI CALDERA

Because the 1912 ejecta vented at Novarupta, contemporaneous collapse of a 600-m-deep caldera at the summit of Mt. Katmai (Figs. 2, 13) indicates that its conduit system was somehow hydraulically connected to the venting system 10 km away. The area of Katmai caldera is ~4.2 km^2 at lake level and ~7 km^2 at the rim, similar to the area of subsidence around Novarupta (Figs. 2, 14), and, perhaps fortuitously, the floor of Katmai caldera is at about the same elevation as the pre-eruptive level of the 1912 vent.

Subsidence at Mt. Katmai

The evidence that apparently led Griggs to infer that Katmai caldera had formed in 1912 was never made explicit, and, although the strong seismicity summarized above would alone seem to demand it, this could conceivably have all been related to eruption and subsidence at Novarupta. The pre-1912 configuration of Mt. Katmai was poorly known (Griggs, 1922, p. 275), no pre-1912 photographs have ever come to light, no one is known to have set foot on the mountain before 1916, and the eyewitness reports of the eruption were dramatized, at best ambiguous, and reached investigators second-hand. After a period of skepticism, nevertheless, the following evidence persuaded me that Mt. Katmai did collapse at the time of the Novarupta eruptions.

(1) On Griggs' first ascent to the caldera rim in 1916, there were no intra-caldera glaciers on the slump-block benches where ice is now thick.

(2) In 1916—1919 the extracaldera glaciers beheaded by caldera collapse still presented inward-facing ice cliffs up to 50 m high, flush with the rock precipices below. By 1953, however, the ice fringing the caldera was more subdued in aspect and, although one ice tongue was flowing back into the abyss, there had generally been recession of the ice cliffs from the brink itself. The slow ablation rate of the ice cliffs, attributable to the 1912 pumice mantle on the glaciers, itself argues for near-contemporaneity of collapse and eruption.

(3) Many major fumaroles issued from talus slopes peripheral to the caldera floor during 1916—1923 (Griggs, 1922; Fenner, 1930). Some orifices were fringed with yellow and white incrustations (never visited), presumably precipitated from gases rising along caldera faults. No fumaroles have been reported around the caldera lake between 1953 and the present, although yellowish-green patches locally discolor the lake water and sulfurous odors persist within the caldera.

(4) A ten-meter-deep lake occupied the caldera floor during all ascents in the 1916—1919 interval, but Fenner (1930) found it empty in 1923 (Fig. 13). This behavior may have reflected continuing structural instability follow-

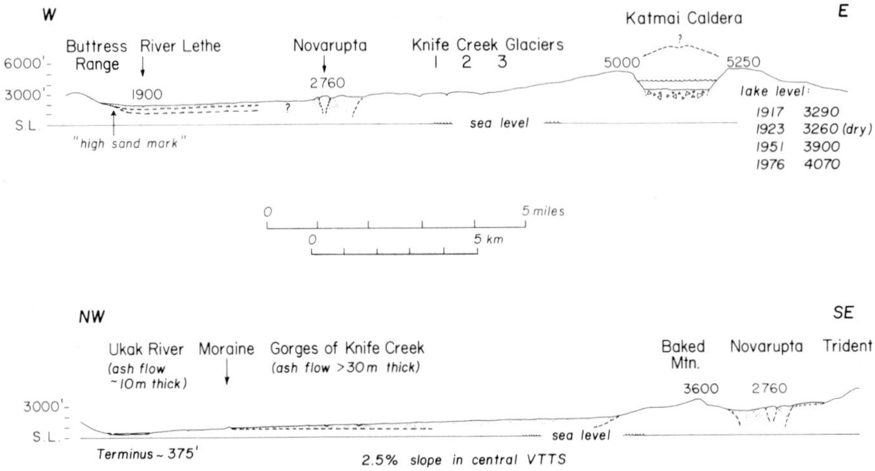

Fig. 13. Sections across the head of the VTTS and down its axis; no vertical exaggeration. Constructed from USGS 15′ topographic maps with elevations in feet (1 m = 3.28 feet). The base of the ash-flow tuff is not exposed in the upper VTTS, but Curtis (1968) estimated a thickness of ~250 m on the basis of the gradients of several nearby streams of comparable discharge, base level, and glacial history. In the upper cross-section, hypothetical thicknesses of 100 m and 250 m are shown for the tuff in the upper part of the River Lethe arm of the VTTS. The pre-collapse elevation of Mt. Katmai is poorly known but may have been ~7500 feet (Griggs, 1922, p. 270). Lake levels are from Griggs, 1922; Fenner, 1930; USGS 15′ Mt. Katmai (B-3) quadrangle map; and personal observation in 1976.

ing collapse or, alternatively, merely slow uptake by the fracture system of surplus meltwater formed from the glacial ice that collapsed into the caldera and temporarily overwhelmed the outlet capacity. There are *no* dry years on Mt. Katmai.

(5) In 1923 on the caldera floor Fenner photographed boiling pools, mudpots, and a mud geyser that erupted with great frequency to a height as great as 60 m (Fenner, 1930). In addition to the many fumaroles on the central floor and in the fringing talus, these features indicate that convective heat flow was far higher in 1923 than the 5°C lake water (measured by R. Motyka in July 1975) would suggest for the present.

Although the evidence thus seems to be compelling that its caldera formed at that time, there is no good evidence for any summit eruption of juvenile magma at Mt. Katmai in 1912. Fenner (1930, 1950) proposed that the "horseshoe island" (now covered by the caldera lake), which consisted of platy dacitic lava, was a final non-pyroclastic product of the 1912 activity. Roughly circular in outcrop and about 400 m in diameter, the unit was ropy in places but severely "broken up", it had visible relief of ~20 m, and it contained a sheer-walled central lagoon ~75 m wide and at least 20 m deep. If it were a 1912 lava flow or agglutinated spatter pile, it had been badly faulted or explosively reamed subsequent to eruption. Fenner (1950) observed that his analysis did not fit his linear variation diagrams for the 1912 ejecta, and this is true for most elements analyzed (Fig. 8). I am inclined to regard the horseshoe island a post-subsidence eruptive unit on the caldera floor, but the question remains unsatisfactorily resolved.

Curtis (1968, pp. 200—206) attributed >99% of the 1912 ejecta to the Novarupta vent, but he argued that four discrete units (which would amount to <0.1 km³ in total volume) had erupted from Mt. Katmai. Most or all of this material can be accounted for in other ways:

(1) The Katmai River debris flow, with which I have dealt above, consists of remobilized air-fall tephra from Novarupta.

(2) The block-rich pumice flow in upper Knife Creek, which is interbedded with fluvially reworked tephra beneath Layer C, as related above, is similar to several other outcrops of blocky, dacite-rich pyroclastic flows that are distributed radially around Novarupta and occupy roughly similar stratigraphic positions.

(3) A remnant of andesitic "welded tuff" just inside the west rim of Katmai caldera (Curtis, 1968, p. 204) turns out to be a poorly sorted fall unit that zones from nonwelded to a dense agglutinate and, as shown by glacial striae, is of pre-1912 age. All welded outcrops around the west and south rims of the caldera (Curtis, 1968, Plate 9) appear to be pre-1912 near-vent agglutinates.

(4) Finally, Curtis suggested that a thin capping deposit of crystal-rich ash, which he designated Layer I, was a widespread tephra layer erupted from Mt. Katmai. Correlations among the thin topmost layers in the VTTS and on Mt. Katmai are difficult to establish because of erosion and aeolian reworking, but my conclusion is that "Layer I" must be a composite of several things. The uppermost unreworked strata are various thin layers of lithic- and crystal-enriched ash that appear to be related to growth of the dome and its ejecta ring (Fig. 14), to phreatic outbursts from explosion craters in the ash-flow sheet, and to aeolian stripping of the vitric top member of the sharply graded crystal-vitric couplets of Layer H. Wind-winnowed lithic-crystal lags are also common, but these are generally distinguishable as such.

On the rim of Katmai caldera, all pumice lapilli appear to be of ordinary Novarupta air-fall origin, and no bombs or breadcrusted blocks have been found that might indicate 1912 activity in the crater itself. In addition to the pre-1912 agglutinates mentioned above, caldera-rim deposits include reworked ash of uncertain age, minor till, and periglacial ashy diamicton, which is probably a post-1912 residue of glacial ablation. Some of the lithic dust and grit on the rim may have been deposited from avalanche clouds sent up during piecemeal collapse of the caldera walls, similar to the deposits formed at Fernandina caldera during its 1968 collapse (Simkin and Howard, 1970). In summary, then, I know of no good evidence that any juvenile material erupted from Mt. Katmai proper, in or since 1912; the horseshoe island may be an exception.

The Novarupta depression

Figure 14 provides a stereo aerial view of the actual vent of 1912, as well as a sketch of the pattern of radial and scalloped arcuate fractures that mark the tephra-filled depression. This was an area of intense fumarolic activity, and it outlasted all the fumaroles that were rooted in the ash-flow sheet, as several of the faults seen in Fig. 14 still emit warm wispy vapors. Fenner (1923, 1938) pointed out that several fumaroles in this area issued from bedrock, from both the dacite of Falling Mountain and the Jurassic siltstone of Baked Mountain. Some were still mildly active along the *west* side of Baked Mountain in 1980, as far as 3 km from the dome and well outside the area of obvious subsidence, suggesting a somewhat wider hydrothermal, if not magmatic, system at depth.

Griggs and Fenner recognized that major parts of Falling, Baked, and Broken Mountains were removed by the eruption, and Curtis estimated ~250 m of subsidence around Novarupta. Cryptic subsidence is likely to have been much greater and a significant volume of ash-flow tuff is probably concealed by the stratified tephra. Truncation of Falling Mountain must have taken place on the first day of the eruption because the brow of its inward-facing scarp is mantled only by rhyolite-poor surge deposits and the dacite air-fall but by neither Layer A nor rhyolite-rich ash-flow material. In like manner, the dacite tephra rests directly upon Jurassic bedrock on the inboard side of the peculiar saddle between Baked and Broken Mountains that is truncated along the north-marginal fault system of Novarupta basin (Fig. 14). Such relationships appear to indicate that the 2-km-wide basin was created during or immediately after eruption of Layer A and the ash flows but was partly filled and obscured by dacite ejecta on the following days. Compactional adjustments in the thick pyroclastic fill may account for most of the subsidence-related faulting evident in Fig. 14, but the depression itself formed earlier.

So, is Novarupta basin a crater or a caldera? On the basis of size, shape, and at least some subsidence, the 1912 vent could be called a caldera. But the formation of what we may infer to be a flaring funnel-shaped conduit was probably initiated by explosive ejection of abundant lithic debris, by inward slumping of the walls, and by continued reaming of the widening orifice during the eruption. This is akin to what Yokoyama (1963, 1981) has proposed, on the basis of geophysical studies, for Krakatau and several Japanese calderas. If 5% of Layer A and of the ash flows were lithic material, this would amount to ~1 km^3 of ejecta, about half the volume of an inverted cone 2 km deep and 2 km in diameter. Considering the uncertainties in conduit geometry, volume of lithic ejecta, and the unexposed volume of intravent lithic fallback debris, this is consistent with a major component of explosive excavation during formation of the Novarupta basin, though probably in conjunction with some unknown amount of local (and possibly regional) flexural subsidence caused by the release of 15 km^3 of magma. Some regional adjustment seems required because the volume accounted for by passive enpp. 47--48

pp. 45—46

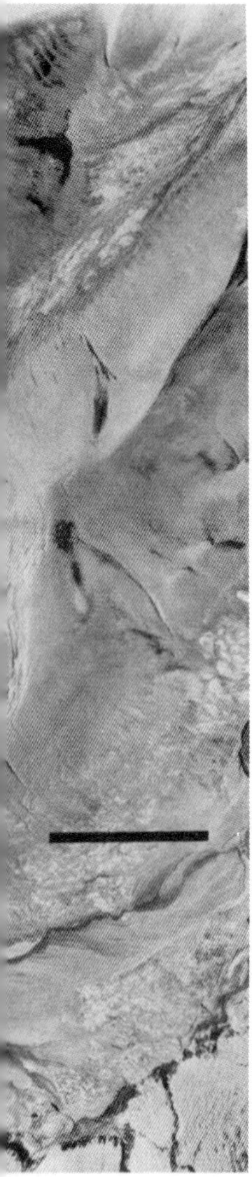

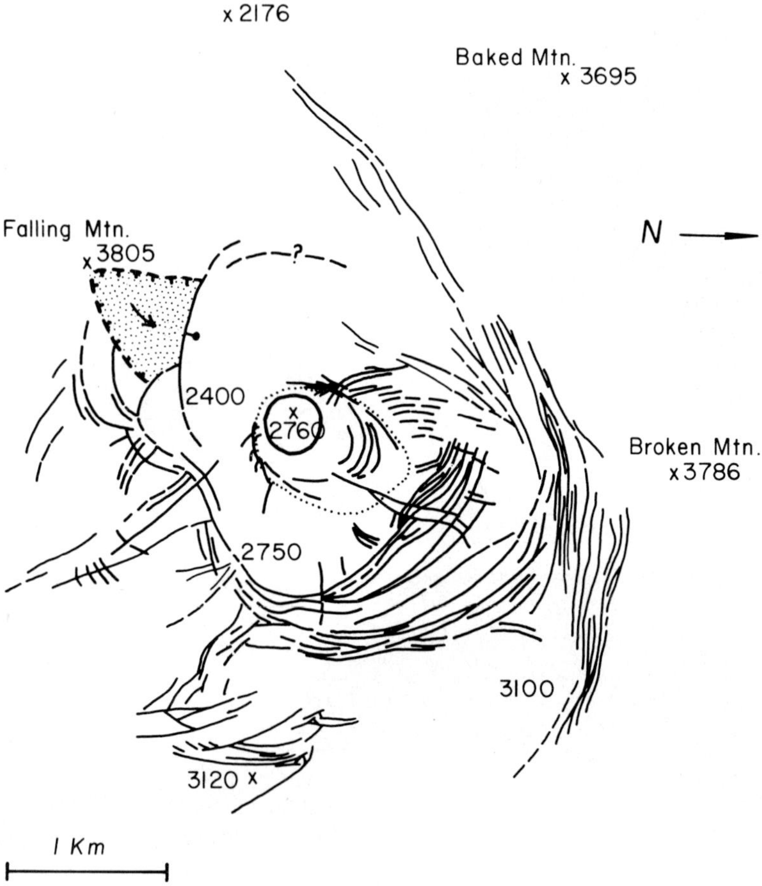

Elevations in feet (1m = 3.28 feet)

.... Crest of Novarupta tephra ring

⊥⊥⊥⊥ Margins of Falling Mtn. scarp

Heavy line outlines Novarupta

banked against Glacier 1. Most light-colored patches are fumarolically altered tuff. The severely faulted hill immediately NNE of Novarupta exposes only 1912 ejecta; in 1980 it still had a few wispy, almost odorless, fumaroles, as did Novarupta and the E-W fault zone along the right margin. Scale bar is 1 km.

b. Fracture pattern outlining the tephra-filled Novarupta caldera; same orientation as the photos in (a).

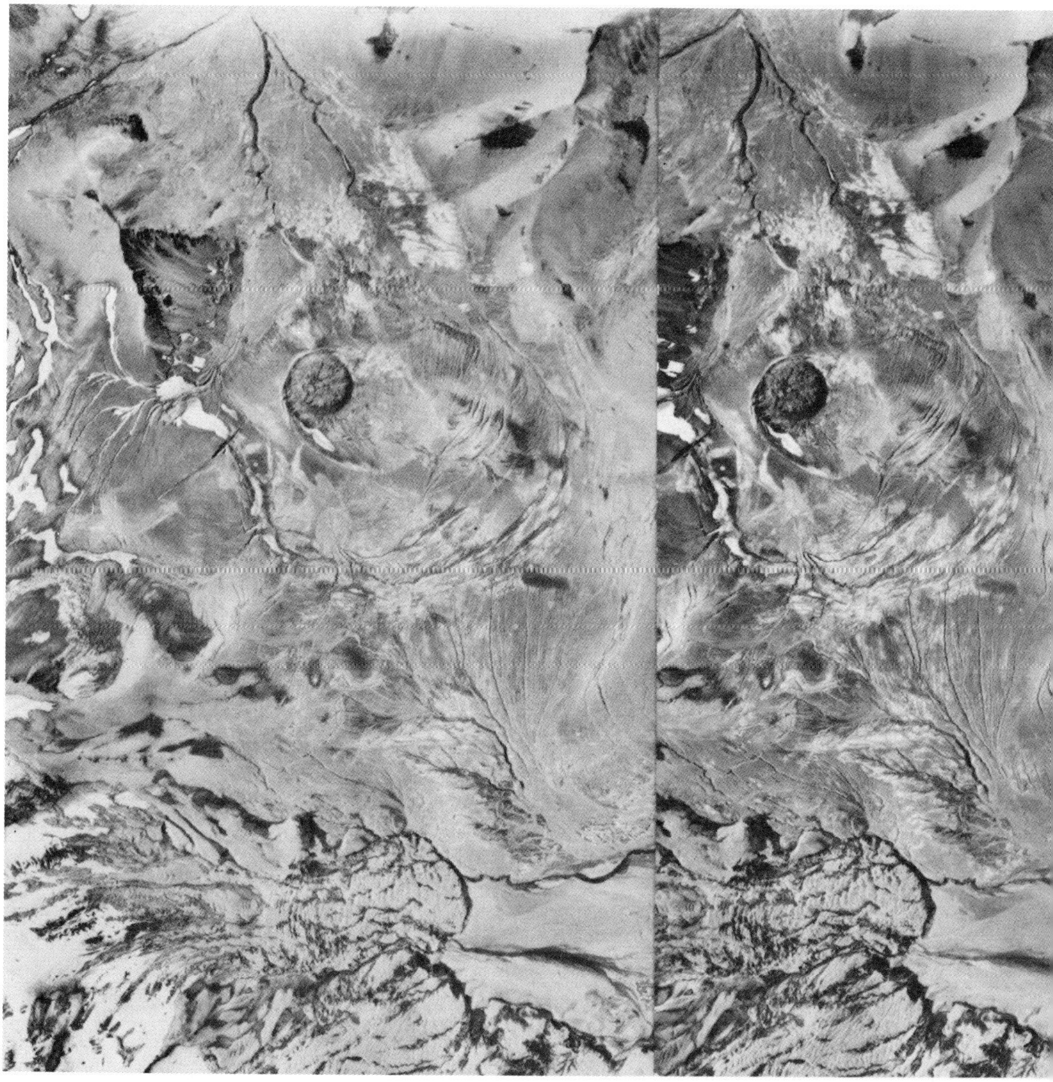

Fig. 14. a. Stereo pair of aerial photographs taken in 1951, showing the 1912 vent area around the 380-m-wide lava dome, Novarupta. North is to the right. Scarp at the left is the sheared-off face of Falling Mountain (Fig. 3), a Holocene dacite dome. Tephra-mantled bedrock at the upper right is flat lying Jurassic siltstone. All other exposure is 1912 air-fall tephra or reworked equivalents, except topmost flow units of the ash-flow tuff, which crop out in gorges at the top of the photos. At bottom, 1912 ejecta overlie and are

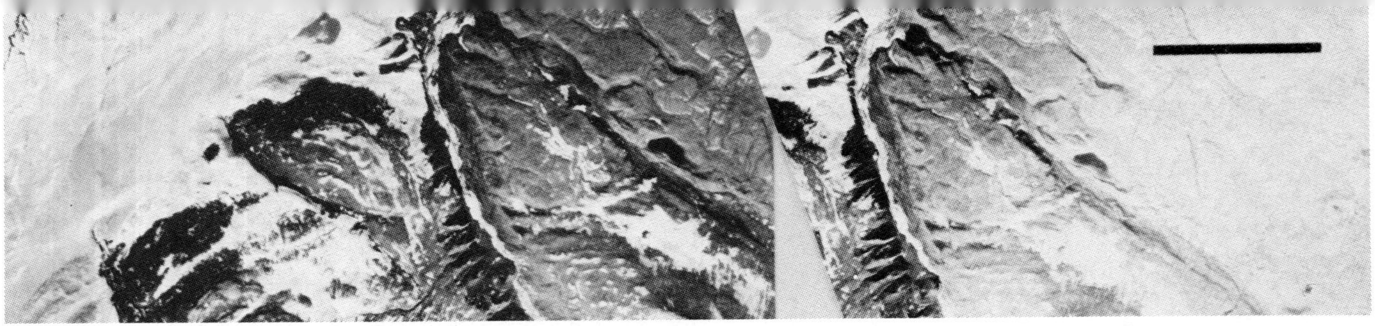

Fig. 15. Stereo pair of aerial photographs taken in 1951, showing the cross-valley moraine (apprximately 16 km from Novarupta) that partially blocked the 1912 ash flow and directed its further movement along pre-existing drainages. Here in the lower VTTS, tuff is nonwelded and largely rhyolitic. It is 20—30 m thick where it wedges out against the central part of the morainal barrier, but the patchy remnants immediately behind the moraine are only 1—4 m thick. Just upstream from prominent gorge junction at left center, tuff exceeds 33 m in thickness, but it thins downstream and does not generally exceed 10 m in either its terminal lobe or along its main connecting outcrop, where it filled the old channel of the Ukak River. Bedrock exposed in surrounding hills and along gorge bottom near Three Forks (top) is horizontally stratified Jurassic siltstone. North is to right. Scale bar is 1 km.

gulfment at Katmai caldera (6–8 km^3), if hydraulically transferred to Novarupta, still only amounts to half of the magma erupted.

The magma reservoir

The mechanics of magma transfer and the geometry of magma reservoirs beneath the volcanoes are among the intriguing questions raised by the 1912 eruption. A microearthquake study (Matumoto and Ward, 1967) showed S-wave shadowing and associated attenuation of high-frequency P-waves under much of the volcanic axis adjacent to the VTTS, leading Matumoto (1971) to infer discrete magma reservoirs at $\leqslant$10 km beneath the Martin-Mageik edifice, at $\leqslant$10 km under the whole Griggs-Katmai-Novarupta-Trident area, and a third at 20–30 km depth beneath the latter group. The cluster of several stratocones here may indeed be the expression of some such complex and voluminous magma storage system, perhaps along the lines portrayed in fig. 15D of Hildreth (1981). The 1912 vent is semi-encircled by a cluster of andesitic volcanoes and is as close to Mageik, Trident, and Griggs as it is to Mt. Katmai. It seems likely that such a level of magmatic focusing upon so limited an area must somehow be responsible for evolution of the large volume of high-silica rhyolite.

The rhyolitic reservoir that erupted in 1912 was large relative to many of the dacite magma bodies that commonly intrude or erupt through intermediate stratocones. If the 7 km^3 of rhyolitic magma that vented at Novarupta were imagined to be the tip of a vertically cylindrical zoned chamber 1 km in diameter, the rhyolite would have to have been at least 9 km thick. More realistically, if the chamber underlay all of Novarupta basin and were ~3 km wide, the rhyolitic layer only had to be 1 km thick. If, alternatively, the rhyolite were a sill-like sheet only 100 m thick, it could be thought of as a 10-km disc that extended all the way to Mt. Katmai; but a sill of such large aspect ratio that penetrated so far from the volcanic axis should have cooled to a greater degree than had the practically aphyric 1912 rhyolite.

Any satisfactory model of the configuration of the pre-eruptive magma reservoir must account for: (1) the phenocryst-poor nature of the rhyolite; (2) the thermal continuity of all the 1912 ejecta; (3) the young dacite domes adjacent to the 1912 vent; (4) the collapse of Mt. Katmai 10 km away; and (5) the complex sequence of eruption in which the pumice was at first entirely rhyolite, later mostly andesite and dacite, and finally almost exclusively dacite. In view of these constraints, Matumoto's seismic data, and the volume relations, I favor the idea of a complex injection zone of mafic dikes underlying the entire cluster of volcanoes here, a zone that locally grades upward into a few discrete reservoirs of andesite and dacite, two of which underlay Katmai and Novarupta in 1912. Compensation for half of the erupted magma by collapse of Mt. Katmai suggests that a hydraulic connection existed between the two subsystems, probably through a common root zone of dikes and sills subjacent to the zoned reservoirs. As much as 8 km^3 of magma from Mt. Katmai may have moved toward or into the Novarupta subsystem in response to the eruptive losses, but the horseshoe island probably represents the only Mt. Katmai magma that actually vented in 1912. It appears likely that all the magma erupted at the Novarupta vent belonged to the Novarupta subsystem. This is suggested by the petrologic affinities (Table III) and thermal continuity (Fig. 11) of all the ejecta and by the need for a major subjacent reservoir of hotter magma to maintain >7 km^3 of rhyolite in a nearly aphyric condition.

The 1912 outburst may have temporarily aborted development of the kind of large zoned magma chamber that sometimes produces 50-km^3 ash-flow sheets and large calderas that truncate *clusters* of stratovolcanoes, as, for example, at Aniakchak and Fisher calderas (Miller and Smith, 1977) and at Mt. Mazama (Bacon, 1983, this volume). The Captains Bay pluton in the Aleutian Islands (Perfit et al., 1980) appears to be the shallow plutonic equivalent of such a complex. Although hardly any Quaternary basalt is known to have erupted at Katmai, the intrusion of basaltic magma probably sustains the entire magmatic system (Poreda et al., 1981; Hildreth, 1981, p. 10182).

EMPLACEMENT OF THE ASH FLOWS

Griggs, (1918, p. 141) saw no evidence in the ash-flow deposits for rapid transport, unusual mobility, or momentum sufficient to ride very far up the enclosing valley walls; accordingly, he reasoned that the pyroclastic material had "welled quietly up" from fissures all over the valley floor and in Katmai Pass. He was wrong about widespread sources and he underestimated the mobility of the tuff slightly, but his observation of relatively sluggish emplacement was fundamentally sound.

The behavior of the pyroclastic flows in the VTTS contrasted markedly with recent reports of spectacularly mobile, highly fluidized, or extremely explosive emplacement of other circum-Pacific ash flows (Aramaki and Ui, 1966; Yokoyama, 1974; Miller and Smith, 1977; Walker et al., 1980). No 1912 ash flows penetrated passes that are only 560 m and 700 m higher than Novarupta and only 5.7 km and 5 km southeast of the dome in the main downwind direction, casting doubt upon the likelihood of column collapse from high elevation. Westward-moving ash flows failed to ride up the flanks of the Buttress Range (Fig. 2) higher than a few tens of meters and did not penetrate passes in that range that are about the same elevation as Novarupta, only 150—200 m above the present valley floor, and only 7.5 km on an unobstructed straight-line path from the source. One flow did traverse Katmai Pass into Mageik Creek on the Pacific slope (Fig. 2), and a late andesite-rich pulse feathered out in the pass itself; but Katmai Pass is actually lower in elevation than Novarupta, and ash-flow veneers left within the pass occur only up to a few tens of meters higher than the locally ponded level of the tuff.

Moraine barrier

After travelling 16 km down the VTTS from the vent, the main ash flow was too deflated and sluggish to surmount a 25-m-high moraine perpendicular to its path (Figs. 2, 15). Parts of the flow penetrated gaps in the moraine, but much of it was diverted around the low west end to continue another 5—6 km down-valley, where it engulfed and charred trees but did not topple all of them. The tuff is >30 m thick only 1 km south of the moraine and >20 m thick within 400 m of it; there are no basal exposures any closer, but no evidence has been found to suggest that the tuff is any thinner than 20 m where it begins to lap up against the barrier. Yet in several places where the moraine stands only 5 m higher than the present ponded level of the sheet, there is no ash-flow deposit beyond it, indicating that the slowly moving flow had neither the momentum nor the degree of inflation necessary to cross so trivial an obstacle. There is no obvious compaction-related deformation of the tuff, which is completely non-welded in the lower VTTS.

Blowdown of trees did occur behind the moraine, but neither surge nor unequivocal ash-cloud deposits have been identified there. Primary ash deposits up to 0.4 m thick do overlie parts of the moraine and the sheltered areas behind it (Fig. 15), but these are the same regional layers of dacite air-fall that overlie the adjacent ash-flow sheet. Both the dacite tephra and the ash flow are underlain by 1—3 cm of fine-grained rhyolitic ash of Layer A. An ash layer 1—3 cm thick immediately above Layer A behind the moraine is >90% glass and could conceivably be an ash-cloud vitric elutriate, but in contrast to the ash flow here, it appears to be largely dacitic.

VTTS narrows

Six kilometers from the moraine and 10 km down-valley from Novarupta, the floor of the VTTS narrows from as much as 7 km wide to only 3 km (Fig. 2). Along both margins of the valley in this constricted area, the outer few hundred meters of the ash flow came to rest in mounds and ridges that stand with up to 15 m relief above its elsewhere-smooth surface. Since the hills are draped conformably by the stratified dacite air-fall, they are clearly primary constructional features. The high level of internal friction suggested by such initial relief on the surface of a pyroclastic-flow deposit is additional evidence for its poorly fluidized outflow.

In this same area of valley constriction, ash flows left thin veneers of poorly sorted but fines-enriched tuff that feather up the valley walls to elevations as high as 30—40 m above the general level of the adjacent valley-filling sheet. As suggested above for Layer B, these fine-grained feather edges of the ash-flow deposit may have formed by marginal jetting from the moving flows, but the evidence against rapid or fluidized emplacement weighs against the likelihood of deposition from a pyroclastic flow sufficiently expanded to have subsequently deflated by 40 m. These veneers are not general around the VTTS but are prominent on both sides of the valley's narrowest reach; this suggests that the constriction itself caused a temporary local elevation of the flow level, which permitted deposition of the high marginal veneers during an interval that was terminated by lowering of that level as the rate of outflow eventually slackened.

High sand mark

In the upper VTTS where the tuff is thick (Fig. 13), the "high sand mark" (Griggs, 1922; Fenner, 1923) is not a veneer, and it was deposited neither by passage of an inflated flow nor by back-drainage of a temporarily high-riding flow. Rather, as correctly interpreted by Fenner (1923, p. 13), it is a marginal bench formed in thick welded tuff, largely by differential compaction. The feature fades out gradually down-valley as the tuff thins, but around the upper basin of the River Lethe, the 30—60 m difference between the valley floor and the fissured hingeline of the bench (Fenner, 1923, p. 40) suggests a thickness of 150—250 m for the tuff, in agreement with Curtis' (1968) estimate from comparative stream profiles. The absence of so pronounced a bench in upper Knife Creek suggests a lesser thickness of tuff in that area.

In the lower VTTS, the distally thinning rhyolite-rich tuff is wholly nonwelded, and its base is widely exposed. In the central VTTS, gorges floored in partly welded tuff have reached a maximum measured depth of 33 m; some of these had been incised to >20 m within 5 years after the eruption (Griggs, 1922). In the upper VTTS, however, the high-temperature rhyolite-poor flow units that were emplaced late in the ash-flow sequence produced partly welded tuff and sillar, which are widely exposed but not yet incised by more than several meters.

Compaction faults with centimeters to meters of displacement are common near the margins of the tuff and especially so at the hingeline of the "high sand mark" bench. Because many of these displace the overlying dacite air-fall as well as the ash-flow tuff itself, compaction clearly persisted for more than two days after deposition. Because some of the distinctively rhyolite-poor, late ash flows were channelled along broad axial swales atop the main volume of the tuff, however, it also seems clear that compaction was underway virtually immediately after emplacement.

Segregation units

In gorge-wall exposures in the central VTTS, most of the dominantly rhyolitic tuff can be seen to consist essentially of a single thick flow unit overlain by two, and locally three, thin flow units of contrasting composition. But both distally and marginally, such gorges also reveal a gradation within tens of meters laterally from unsorted and unstratified tuff to crudely sorted, thick or thin stratification, evidently representing a transition from plug to laminar flow. Most such segregation units are several meters thick, but in one gorge just up-valley from the moraine (Fig. 15) the tuff divides into one hundred or more laminae, many only 1—3 cm thick. Many such units, both thick and thin, have their own fine-grained basal layers and crude internal grading, normal and reversed for lithics and pumice respectively. It would be easy to mistake such features for discrete emplacement units, but all they appear to represent is internal shear related to internal velocity gradients as the main flow was slowing to a halt, rather like a sliding deck of cards.

ACKNOWLEDGEMENTS

I thank Garniss Curtis for providing samples that first kindled my interest in going to Katmai for a petrologic study and for not warning me what I was in for. My additions and amendments in no way diminish the importance of his classic study. To successive field companions, Larry Jager, Anita Grunder, and Judy Fierstein, I am grateful for putting up with typhoons, brown bears, the inescapable ashy grit, and my variable humor under such conditions. In 1979, we were accompanied by the late David Johnston, who measured many of the sections plotted in Figure 6 and displayed energy and enthusiasm I shall never forget. Dan Kosco, Terry Keith, and Peter Shearer joined us at various times, contributing welcome perspectives, cheer, and strong backs. Bruce Kaye and several other National Park Service personnel facilitated our transit and provided valuable logistical information. The National Geographic Society permitted access to its photographic archives of the Griggs expeditions. Fierstein and Anne Gartner meticulously prepared samples in the laboratory and drafted the diagrams, while Dorothy Johnson and Lynn Buchholz cheerfully typed the manuscript. I am grateful to I.S.E. Carmichael for microprobe facilities and 1976 field expenses under NSF Grant EAR-74-12782. Helpful reviews by Charles R. Bacon, Robert L. Christiansen, and A. Ewart are thoroughly appreciated.

REFERENCES

Allen, E.T. and Zies, E.G., 1923. A chemical study of the fumaroles of the Katmai region. National Geographic Society, Contributed Technical Papers, Katmai Series, 2: 75—155.

Aramaki, S. and Ui, T., 1966. Aira and Ata pyroclastic flows and related caldera depressions in southern Kyushu, Japan. Bull. Volcanol., 29: 29—48.

Bacon, C.R., 1983. Eruptive history of Mt. Mazama, Cascade Range, U.S.A. In: S. Aramaki and I. Kushiro (Editors), Arc Volcanism. J. Volcanol. Geotherm. Res., 18: 57—115.

Bowen, N.L., 1928. The Evolution of the Igneous Rocks. Princeton Univ. Press, Princeton, N.J., 332 pp.

Buddington, A.F. and Lindsley, D.H., 1964. Iron-titanium oxide minerals and synthetic equivalents. J. Petrol., 5: 310—357.

Byers, F.M., Jr., 1961. Petrology of three volcanic suites, Umnak and Bogoslof Islands, Aleutian Islands, Alaska. Geol. Soc. Am. Bull., 72: 93—128.

Carmichael, I.S.E., 1967. The iron-titanium oxides of salic volcanic rocks and their associated ferromagnesian silicates. Contrib. Mineral. Petrol., 14: 36—64.

Clocchiatti, R., 1972. Les cristaux de quartz des ponces de la Vallée des Dix Milles Fumées (Katmai, Alaska). C.R. Acad. Sci. Paris, 274: 3037—3040.

Coats, R.R., 1952. Magmatic differentiation in Tertiary and Quaternary volcanic rocks from Adak and Kanaga Islands, Aleutian Islands, Alaska. Geol. Soc. Am. Bull., 63: 485—514.

Coats, R.R., 1962. Magma type and the crustal structure in the Aleutian arc. In: G.A. Macdonald and H. Kuno (Editors), The Crust of the Pacific Basin. Am. Geophys. Union, Monogr., 6: 92—109.

Coffman, J.L., and von Hake, C.A., 1973. Earthquake history of the United States. U.S. Dept. Commerce, Natl. Oceanic and Atmospheric Administration, Publ. 41-1, 208 pp.

Curtis, G.H., 1968. The stratigraphy of the ejecta from the 1912 eruption of Mount Katmai and Novarupta, Alaska. In: R.R. Coats, R.L. Hay, and C.A. Anderson (Editors), Studies in Volcanology. Geol. Soc. Am., Mem., 116: 153—210.

DePaolo, D.J., 1981. Trace element and isotopic effects of combined wallrock assimilation and fractional crystallization. Earth Planet. Sci. Lett., 53: 189—202.

Dumond, D.E., 1979. People and pumice on the Alaska Peninsula. In: P.D. Sheets and D.K. Grayson (Editors), Volcanic Activity and Human Ecology. Academic Press, New York, N.Y., pp. 373—392.

Ewart, A., 1976. Mineralogy and chemistry of modern orogenic lavas — some statistics and implications. Earth Planet. Sci. Lett., 31: 417—432.

Federman, A.N. and Scheidegger, K.F., 1981. A comparison of abyssal and terrestrial deposits of tephra from the 1912 eruption of Katmai/Novarupta. EOS, Trans. Am. Geophys. Union, 62: 1062.

Fenner, C.N., 1920. The Katmai region, Alaska, and the great eruption of 1912. J. Geol., 28: 569—606.

Fenner, C.N., 1923. The origin and mode of emplacement of the great tuff deposit of the Valley of Ten Thousand Smokes. National Geographic Society, Contributed Technical Papers, Katmai Series, 1, 74 pp.

Fenner, C.N., 1925. Earth movements accompanying the Katmai eruption. J. Geol., 33: 116—139 and 193—223.

Fenner, C.N., 1926. The Katmai magmatic province. J. Geol., 34: 673—772.

Fenner, C.N., 1930. Mount Katmai and Mount Mageik. Z. Vulkanol., 13: 1—24.

Fenner, C.N., 1931. The residual liquids of crystallizing magmas. Mineral. Mag., 22: 539—560.

Fenner, C.N., 1938. The phenomena of Falling Mountain. Am. J. Sci., 35-A: 35—48.

Fenner, C.N. 1950. The chemical kinetics of the Katmai eruption. Am. J. Sci., 248: 593—627 and 697—725.

Filson, J., Simkin, R. and Lei-Kuang Leu, 1973. Seismicity of a caldera collapse: Galápagos Islands 1968. J. Geophys. Res., 78: 8591—8622.

Griggs, R.F., 1918. The great hot mud flow of the Valley of Ten Thousand Smokes. Ohio J. Sci., 19: 117—142.

Griggs, R.F., 1922. The Valley of Ten Thousand Smokes. National Geographic Society, Washington, D.C., 340 pp.

Hildreth, W., 1979. The Bishop Tuff: Evidence for the origin of compositional zonation in silicic magma chambers. Geol. Soc. Am., Spec. Pap., 180: 43—75.

Hildreth, W., 1981. Gradients in silicic magma chambers: Implications for lithospheric magmatism. J. Geophys. Res., 86: 10153—10192.

Jaggar, T.A., 1927. Eruption of Mageik in Alaska. The Volcano Letter, no. 147, Oct. 20.

Keller, A.S. and Reiser, H.N., 1959. Geology of the Mount Katmai area, Alaska. U.S. Geol. Surv. Bull., 1058-G: 261—298.

Kosco, D.G., 1981. Characteristics of Andesitic to Dacitic Volcanism at Katmai National Park, Alaska. Ph.D. Thesis, Univ. California, Berkeley, 249 pp.

Kuno, H., 1968. Differentiation of basalt magmas. In: H.H. Hess and A. Poldervaart (Editors), Basalts, vol. II. Wiley Interscience, New York, N.Y., pp. 624—688.

Martin, G.C., 1913. The recent eruption of Katmai volcano in Alaska. National Geographic Magazine, 24: 131—181.

Matumoto, T., 1971. Seismic body waves observed in the vicinity of Mt. Katmai, Alaska, and evidence for the existence of molten chambers. Geol. Soc. Am. Bull., 82: 2905—2920.

Matumoto, T. and Ward, P.L., 1967. Microearthquake study of Mt. Katmai and vicinity, Alaska. J. Geophys. Res., 72: 2557—2568.
McBirney, A.R., 1980. Mixing and unmixing of magmas. J. Volcanol. Geotherm. Res., 7: 357—371.
Miller, T.P. and Smith, R.L., 1977. Spectacular mobility of ash flows around Aniakchak and Fisher calderas, Alaska. Geology, 5: 173—176.
Miyashiro, A., 1974. Volcanic rock series in island arcs and active continental margins. Am. J. Sci., 274: 321—355.
Nairn, I.A., Hewson, C.A.Y., Latter, J.H. and Wood, C.P., 1976. Pyroclastic eruptions of Ngauruhoe volcano, central North Island, New Zealand, 1974. In: R.W. Johnson (Editor), Volcanism in Australasia, Elsevier, Amsterdam, pp. 385—405.
Nayadu, Y.R., 1964. Volcanic ash deposits in the Gulf of Alaska and problems of correlation of deep-sea ash deposits. Mar. Geol., 1: 194—212.
Peacock, M.A., 1931. Classification of igneous rock series. J. Geol., 39: 54—67.
Perfit, M.R., Brueckner, H., Lawrence, J.R. and Kay, R.W., 1980. Trace element and isotopic variations in a zoned pluton and associated volcanic rocks, Unalaska Island, Alaska: A model for fractionation in the Aleutian calcalkaline suite. Contrib. Mineral. Petrol., 73: 69—87.
Perlman, I. and Asaro, F., 1969. Pottery analysis by neutron activation. Archaeometry, 11: 21—52.
Poreda, R., Craig, H. and Motyka, R., 1981. Helium isotope variations along the Alaskan—Aleutian arc. EOS, Trans. Am. Geophys. Union, 62: 1092.
Ray, D.K., 1967. Geochemistry and Petrology of the Mt. Trident Andesites, Katmai National Monument, Alaska. Ph.D. Dissertation, Univ. Alaska, Fairbanks, 198 pp.
Sakuyama, M., 1981. Petrological study of the Myoko and Kurohime volcanoes, Japan: Crystallization sequence and evidence for magma mixing. J. Petrol., 22: 553—583.
Shaw, H.R., 1974. Diffusion of H_2O in granitic liquids. In: A.W. Hofmann et al. (Editors), Geochemical Transport and Kinetics. Carnegie Inst. Washington, Publ., 634: 139—170.
Shipley, J.W., 1920. Some chemical observations on the volcanic emanations and incrustations in the Valley of Ten Thousand Smokes, Katmai, Alaska. Am. J. Sci., 50: 141—153.
Simkin, T. and Howard, K.A., 1970. Caldera collapse in the Galápagos Islands, 1968. Science, 169: 429—437.
Sparks, R.S.J. and Walker, G.P.L., 1977. The significance of vitric-enriched air-fall ashes associated with crystal-enriched ignimbrites. J. Volcanol. Geotherm. Res., 2: 329—341.
Taylor, S.R., 1969. Trace element chemistry of andesites and associated calc-alkaline rocks. In: A.R. McBirney (Editor), Proceedings of the Andesite Conference. Oreg. Dept. Geol. Miner. Resour. Bull., 65: 43—64.
Volz, F.E., 1975a. Distribution of turbidity after the 1912 Katmai eruption in Alaska. J. Geophys. Res., 80: 2643—2648.
Volz, F.E., 1975b. Burden of volcanic dust and nuclear debris after injection into the stratosphere at 40°-58°N. J. Geophys. Res., 80: 2649—2652.
Walker, G.P.L., Heming, R.F. and Wilson, C.J.N., 1980. Low-aspect ratio ignimbrites. Nature, 283: 286—287.
Wilcox, R.E., 1979. The liquid line of descent and variation diagrams. In: H.S. Yoder, Jr. (Editor), The Evolution of the Igneous Rocks: Fiftieth Anniversary Perspectives. Princeton Univ. Press, Princeton, N.J., pp. 205—232.
Yokoyama, I., 1963. Structure of caldera and gravity anomaly. Bull. Volcanol., 26: 67—72.
Yokoyama, I., Yamashita, H., Watanabe, H. and Okada, H., 1981. Geophysical characteristics of dacite volcanism — the 1977—1978 eruption of Usu volcano. J. Volcanol. Geotherm. Res., 9: 335—358.

Yokoyama, S., 1974. Mode of movement and emplacement of Ito pyroclastic flow from Aira caldera, Japan. Tokyo Kyoiku Daigaku Chishitsugaku Kobutsugaku Kyoshitsu Kenkyu Hokoku, C, 12: 17—62.

Zies, E.G., 1929. The Valley of Ten Thousand Smokes: I. The fumarolic incrustations and the bearing ore deposition. II. The acid gases contributed to the sea during volcanic activity. National Geographic Society, Contributed Technical Papers, Katmai Series, 4: 1—79.

Journal of Volcanology and Geothermal Research, 18 (1983) 57—115
Elsevier Science Publishers B.V., Amsterdam — Printed in The Netherlands

ERUPTIVE HISTORY OF MOUNT MAZAMA AND CRATER LAKE CALDERA, CASCADE RANGE, U.S.A.

CHARLES R. BACON

U.S. Geological Survey, 345 Middlefield Road, Menlo Park, CA 94025 (U.S.A.)

(Received October 20, 1982)

ABSTRACT

Bacon, C.R., 1983. Eruptive history of Mount Mazama and Crater Lake caldera, Cascade Range, U.S.A. In: S. Aramaki and I. Kushiro (Editors), Arc Volcanism. J. Volcanol. Geotherm. Res., 18: 57—115.

New investigations of the geology of Crater Lake National Park necessitate a reinterpretation of the eruptive history of Mount Mazama and of the formation of Crater Lake caldera. Mount Mazama consisted of a glaciated complex of overlapping shields and stratovolcanoes, each of which was probably active for a comparatively short interval. All the Mazama magmas apparently evolved within thermally and compositionally zoned crustal magma reservoirs, which reached their maximum volume and degree of differentiation in the climactic magma chamber ~ 7000 yr B.P.

The history displayed in the caldera walls begins with construction of the andesitic Phantom Cone ~ 400,000 yr B.P. Subsequently, at least 6 major centers erupted combinations of mafic andesite, andesite, or dacite before initiation of the Wisconsin Glaciation ~ 75,000 yr B.P. Eruption of andesitic and dacitic lavas from 5 or more discrete centers, as well as an episode of dacitic pyroclastic activity, occurred until ~ 50,000 yr B.P.; by that time, intermediate lava had been erupted at several short-lived vents. Concurrently, and probably during much of the Pleistocene, basaltic to mafic andesitic monogenetic vents built cinder cones and erupted local lava flows low on the flanks of Mount Mazama. Basaltic magma from one of these vents, Forgotten Crater, intercepted the margin of the zoned intermediate to silicic magmatic system and caused eruption of commingled andesitic and dacitic lava along a radial trend sometime between ~ 22,000 and ~ 30,000 yr B.P. Dacitic deposits between 22,000 and 50,000 yr old appear to record emplacement of domes high on the south slope. A line of silicic domes that may be between 22,000 and 30,000 yr old, northeast of and radial to the caldera, and a single dome on the north wall were probably fed by the same developing magma chamber as the dacitic lavas of the Forgotten Crater complex. The dacitic Palisade flow on the northeast wall is ~ 25,000 yr old. These relatively silicic lavas commonly contain traces of hornblende and record early stages in the development of the climatic magma chamber.

Some 15,000 to 40,000 yr were apparently needed for development of the climactic magma chamber, which had begun to leak rhyodacitic magma by 7015 ± 45 yr B.P. Four rhyodacitic lava flows and associated tephras were emplaced from an arcuate array of vents north of the summit of Mount Mazama, during a period of ~ 200 yr before the climactic eruption. The climactic eruption began 6845 ± 50 yr B.P. with voluminous airfall deposition from a high column, perhaps because ejection of ~ 4—12 km^3 of magma to form the lava flows and tephras depressurized the top of the system to the point where vesiculation at depth could sustain a Plinian column. Ejecta of this phase issued

0377-0273/83/$03.00

from a single vent north of the main Mazama edifice but within the area in which the caldera later formed. The Wineglass Welded Tuff of Williams (1942) is the proximal featheredge of thicker ash-flow deposits downslope to the north, northeast, and east of Mount Mazama and was deposited during the single-vent phase, after collapse of the high column, by ash flows that followed topographic depressions. Approximately 30 km^3 of rhyodacitic magma were expelled before collapse of the roof of the magma chamber and inception of caldera formation ended the single-vent phase. Ash flows of the ensuing ring-vent phase erupted from multiple vents as the caldera collapsed. These ash flows surmounted virtually all topographic barriers, caused significant erosion, and produced voluminous deposits zoned from rhyodacite to mafic andesite. The entire climactic eruption and caldera formation were over before the youngest rhyodacitic lava flow had cooled completely, because all the climactic deposits are cut by fumaroles that originated within the underlying lava, and part of the flow oozed down the caldera wall.

A total of ~ 51—59 km^3 of magma was ejected in the precursory and climactic eruptions, and ~ 40—52 km^3 of Mount Mazama was lost by caldera formation. The spectacular compositional zonation shown by the climactic ejecta — rhyodacite followed by subordinate andesite and mafic andesite — reflects partial emptying of a zoned system, halted when the crystal-rich magma became too viscous for explosive fragmentation. This zonation was probably brought about by convective separation of low-density, evolved magma from underlying mafic magma. Confinement of postclimactic eruptive activity to the caldera attests to continuing existence of the Mazama magmatic system.

INTRODUCTION

Mount Mazama and its spectacular caldera containing Crater Lake have attracted the interest of scientists since the turn of the 20th century, when the area was designated a national park and J.S. Diller published his pioneering monograph (Diller and Patton, 1902). The caldera provides the only direct evidence for the recent existence of a large shallow magma chamber in the Quaternary High Cascades; it formed by collapse during a voluminous pyroclastic eruption, whose products were compositionally zoned from rhyodacite to mafic andesite, ~ 6800 yr B.P. Because of superb access and exposure and the recency of the caldera-forming eruption, Mount Mazama is ideal for study of the evolution of a magmatic system related to a convergent plate boundary. Furthermore, the history of Mount Mazama is of significance for the study of igneous-related geothermal systems and for the assessment of volcanic hazards.

Forty years after Diller, Williams (1942) published his masterpiece on the geology of Crater Lake National Park — a work that has stood virtually unmodified for four decades. Crater Lake also served as an important example in Williams' (1941) classic paper on calderas. Notable topical contributions have been made over the years by Nelson (1961), McBirney (1968), Noble et al. (1969), Lidstrom (1971), Ritchey (1979, 1980), and Williams and Von Herzen (1983). This report summarizes preliminary results of an ongoing study of the evolution of Mount Mazama, which includes detailed geologic mapping, K-Ar and radiocarbon dating, paleomagnetic investigations, mineral and whole-rock chemistry, and Pb and Sr isotopic studies.

Many of the ideas presented here naturally are parallel to those of previous workers, and no attempt is made to cite every instance of confirmation of earlier work. However, some important conclusions of previous workers are shown to be erroneous in the light of the present study, and so a major revision of the history of Mount Mazama is necessary. In particular, reevaluation of the climactic eruptive sequence sheds new light on the chronology of explosive eruptions and caldera collapse and suggests physical reasons for the different phases of the climactic eruption.

Williams (1942) presented a detailed description of the compositionally zoned ash-flow sheet from the caldera-forming eruption of Mount Mazama and was, perhaps, the first to specify that the observed zonation from rhyodacitic pumice to overlying andesitic scoria represents eruption from a magma chamber which was zoned from silicic magma at its top to mafic magma in its roots. The evolution of Mount Mazama is here treated as a record of the growth and development of a zoned magmatic system (cf. Smith, 1979; McBirney, 1980; Hildreth, 1981), in which the climactic eruption and its immediate precursors preserve, in effect, an instantaneous view of a vertically zoned chamber.

The present report was prepared after the International Association of Volcanology and Chemistry of the Earth's Interior 1981 Symposium on Arc Volcanism. Many of my conclusions are preliminary, and, as work progresses, some details will probably be changed. My revisions to the story of Mount Mazama should be viewed in the context of the broad scope and lasting significance of the work of Howel Williams, to whom this report is dedicated.

TERMINOLOGY

Two problems in nomenclature still persist in accounts of Quaternary history and "arc" volcanism: chronology of glacial stages and terminology of volcanic rock compositions. The present report relies on the work of Scott (1977) on glaciation in the central Oregon Cascades and probable correlations with the marine oxygen-isotope record (Shackleton and Opdyke, 1973) for a chronology of glacial culminations and interglacial periods. Correlation of the continental record of climatic variation with the marine oxygen-isotope record has been substantiated by recent palynological studies (Heusser and Shackleton, 1979; Adam et al., 1981). The term "Wisconsin" used herein refers to the Wisconsin Glaciation, occurring 75,000–12,000 yr B.P.; the Wisconsin Glaciation is further subdivided with reference to the marine oxygen-isotope record, as used by Colman and Pierce (1981, pp. 26–27). Pre-Wisconsin chronology is divided by oxygen-isotope stages.

The present classification of the volcanic rocks of Mount Mazama is arbitrary but not devoid of physical significance. I use SiO_2 content (anhydrous basis) as a convenient index: basalts contain <53% SiO_2; andesites, 53 to 63%; dacites, 63 to 68%; and rhyodacites, 68 to ~ 72%. Discrimination of a small group of rhyolites is not currently practical. Andesites are subdivided into mafic (53—57% SiO_2) and silicic (60—63% SiO_2) where analytical data indicate that such distinctions are meaningful.

Throughout this report, I refer to the Mazama magmatic system. The areal boundaries of this system, which probably varied as it evolved, are defined here as the inward limit of the most recently active mafic vents for any particular interval. The upper boundary would be the top of a contiguous reservoir of magma or of a swarm of closely spaced small intrusive bodies. The depth to the roof very likely decreased over time and reached a minimum value of perhaps a few km during the Holocene. No discrete lower boundary is envisioned; the crustal part of the system probably grades downward into a zone of intrusion of basaltic magma in the lower crust and upper mantle.

GEOLOGIC SETTING

Regional tectonic features

The chain of volcanoes of the High Cascades approximately parallels the plate boundary and appears to be related to subduction of the small Juan de Fuca and Gorda plates beneath the North American plate. Plate convergence is approximately normal to the trend of the volcanic chain in British Columbia and northern Washington, where Quaternary volcanism has been restricted to a few local eruptive centers. In southern Washington, Oregon, and northern California, the trend of the Quaternary volcanic chain is oblique to the plate-convergence direction, and discrete long-lived eruptive centers are superimposed on a broad belt of monogenetic basaltic volcanoes and small shields. The ultimate magma source for Cascade volcanism thus appears to be related to subduction, but the proportion of mafic magma leaked to the surface, being much greater in the zone of oblique convergence (White and McBirney, 1978), is probably controlled by the regional tectonic stress field. This regional field is more affected by Pacific plate—North American plate interaction than by subduction. Evidence from earthquake focal mechanisms (Rogers, 1979), as well as a pattern of approximately north-trending normal faults and northwest-trending right-lateral and northeast-trending left-lateral strike-slip faults within the range (cf. Lawrence, 1976), and of east-trending fold axes east of the Cascades (Campbell and Bentley, 1981), indicates approximately north—south maximum horizontal compressive stress in the upper crust throughout the Pacific Northwest.

South of the Brothers fault zone in central Oregon, the High Cascades effectively mark the west limit of the Basin and Range province (Lawrence, 1976). The High Cascades of central Oregon partly fill a graben (McBirney, 1978; Taylor, 1981), bounded on the west by a fault zone across which major changes in heat flow (Blackwell et al., 1982) and Bouguer gravity (Couch et al., 1982) occur. Long-lived volcanic centers may occur where profound northwest-trending crustal discontinuities intersect the axis of the High Cascades and the flux of mafic magma becomes focused. Mount Mazama lies at the north end of the Klamath graben, where this deep rift (Veen et al., 1981) bends northward into coincidence with the High Cascade trend. Faults with a major normal component of displacement immediately south and north of Mount Mazama trend slightly west of north; evidence for faulting west of Mount Mazama suggests north-northeastward strikes. No regional faults have been recognized in the walls of Crater Lake caldera.

Fig. 1. Physiography of Crater Lake caldera and vicinity, Oregon. Illustration by S.E. Mayfield.

Alignments of monogenetic basaltic vents

Fissures and alignments of coeval vents may delineate the intersections of dikes with the surface and, therefore, can be used to infer the azimuth of the axis of maximum horizontal compressive stress (Nakamura, 1977). In the Crater Lake region and throughout most of the Cascades, such alignments tend to be north—southward, as would be inferred from other indicators of tectonic stress. Other alignments commonly include vents of varying age and probably reflect preferential leakage of magma in structurally disturbed zones.

Immediately north and south of Mount Mazama, Quaternary monogenetic volcanoes erupted olivine-bearing basalt to mafic andesite, similar to rocks elsewhere in the High Cascades. Residence of such magma near the crust/mantle interface and its injection into the lower crust may explain the elevated regional heat flow in the Oregon Cascades, as reported by Blackwell et al. (1982). Most well-preserved mafic cinder cones and associated lava flows in the Crater Lake region are probably late Pleistocene (e.g. Red Cone, Fig. 1, which has been heavily glaciated); evidence for older monogenetic mafic volcanism is probably hidden beneath younger deposits.

Local stress field

The gradual decrease in displacement and the disappearance of approximately north—south-trending faults toward Mount Mazama from the south suggest that deformation near the caldera was dominated by the local upper-crustal stress field induced by the volcano. Numerous subradial dikes are evident on the caldera walls. The Forgotten Crater vent system on the west side of Mount Mazama and the Sharp Peak domes on the northeast (Figs. 1, 6A) define late Pleistocene radial fracture systems. A cluster of vents for four Holocene rhyodacitic lava flows (Fig. 6B), products of the last preclimactic eruptions, suggests the presence of a shallow magma chamber ~ 6800 yr B.P.

Foundations of Mount Mazama

Quaternary stratocones, such as those that made up Mount Mazama, lie on a pile of interspersed shields, stratocones, small monogenetic volcanoes, and valley-filling flows of basalt, mafic andesite, and andesite that compose the volcanic rocks of the High Cascades. Absence of regional tilting distinguishes these rocks from underlying rocks of the Western Cascades (Thayer, 1937). The following summary is based on recent work by Smith et al. (1982), including K-Ar dates listed by Fiebelkorn et al. (1982), between lat 42° and 43° N. In the area studied by Smith et al. (1982), the High Cascade flows range in age from Quaternary to ~ 7 m.y.; mafic andesite is the most voluminous rock type. The volcanic rocks of the Western Cascades are exposed ~ 20 km west of Crater Lake and also probably underlie Mount Mazama. The Western Cascade pile is $\gtrsim$ 5500 m thick in the area investigated by Smith et al. (1982), where it forms a generally homoclinal gently east-dipping sequence composed of subequal amounts of (1) basalt, mafic andesite, and andesitic lava flows and tuff breccia and (2) andesitic to rhyolitic ash-flow tuff, lava flows, and small intrusions. The Western Cascade rocks range in age from ~ 15 to 35 m.y.; no volcanism is documented between ~ 15 and ~ 7 m.y. B.P. Evidence for pre-Oligocene rocks beneath the High Cascades is inconclusive.

Volcanic rocks adjacent to Mount Mazama

Descriptions of the volcanoes near Crater Lake can be found in Williams (1942). Two groups of volcanic rocks deserve special mention because they may be important to development of the Mazama volcanic focus.

Cones and domes east of the caldera

Many silicic domes and associated lava flows occur within ~ 10 km of the eastern caldera wall (Williams, 1942, fig. 6). Rather uniform dacite or rhyodacite was erupted from at least 25 vents distributed over ~ 120 km^2. Two small basaltic cinder cones lie at the northeast margin of this silicic volcanic field. A mafic andesitic (?) lava flow from a vent southeast of Mount Scott flowed down the drainage of Scott Creek; the andesite overlies the silicic rocks. The andesitic to dacitic (?) stratocone of Mount Scott (Fig. 1) is the most prominent edifice in the area. Lava flows from Mount Scott overlie the silicic rocks. As yet, none of these rocks is well dated, but a K-Ar dated flow (~ 340,000 yr B.P.) exposed low on the caldera wall between Redcloud Cliff and Sentinel Rock (see *Danger Bay*) resembles the lavas of Mount Scott and may have come from that source. The apparent volume and uniformity of the silicic rocks indicate that the Crater Lake area has been a locus of silicic volcanism for considerably longer than implied by the geology of the immediate vicinity of the caldera. A K-Ar age of 5.5 ± 0.5 m.y. on plagioclase in rhyodacite lava from Pothole Butte (E.H. McKee, pers. commun., 1982) suggests that the silicic rocks east of the park may be considerably older than exposed lavas of Mount Mazama.

Southern rhyodacites

Rhyodacitic lava flows border the canyon of Sun Creek south of the andesitic stratocones of the south caldera walls. Advanced hydration of glass and extensive glacial erosion indicate that these flows are pre-Wisconsin. Although their relations to the andesite flows of Mount Mazama have not yet been clearly defined, these rhyodacitic flows are overlain by andesitic lavas that probably correlate with rocks on the caldera walls that have been dated at ~ 340,000 yr B.P. (unit *db*, Fig. 3B). Quaternary rhyodacitic volcanism near the axis of the Cascades occurred here but apparently not for at least 100 km to the north or south.

PETROGRAPHY

Petrographic descriptions of the Mount Mazama volcanic rocks were given by Diller and Patton (1902), Williams (1942), and Ritchey (1979). Nevertheless, a few general observations are useful here.

Lava flows and pyroclastic rocks

Andesites and dacites of Mount Mazama typically contain plagioclase as the major phenocryst phase, commonly in multiple textural populations. Both augite and hypersthene are present; hypersthene is more abundant, especially in dacites. Hornblende is present in some andesites of the Hillman Peak eruptive center, in andesite south of the summit of Garfield Peak, and in most of the Holocene rocks. Fe-Ti oxide microphenocrysts are most conspicuous in the more silicic rocks. Olivine phenocrysts are minor but nearly ubiquitous; they typically are resorbed, rimmed by pyroxene, or buried within glomeroporphyritic clots. Many textural features suggestive of reaction and partial dissolution of phenocrysts imply sudden changes in heat content only explicable by the mixing of magmas of contrasting temperatures (Hibbard, 1981; Sakuyama, 1981), although there is no reason to believe that mixing was solely responsible for all the observed compositional diversity.

Basalts erupted near Mount Mazama and mafic andesites, both in the Mazama pile and elsewhere nearby, contain higher proportions of olivine and may lack hypersthene. Textures suggestive of mixing occur in the mafic andesites, though less so than in the andesites and dacites.

Mafic inclusions

Spheroidal inclusions, from $\sim$ 1 mm to nearly 1 m in diameter and generally showing high porosity (cf. Eichelberger, 1980), are common in the intermediate and silicic Mazama lavas (Fig. 2). Porous inclusions are rare or absent in phenocryst-poor mafic andesite sheets, which may have been fed by lava fountains. These inclusions range in composition from mafic andesite to andesite and are typically poorer in SiO_2 by $\gtrsim$ 3% than their hosts. Olivine phenocrysts are common, as are crystal morphologies indicative of rapid crystallization in an undercooled environment (e.g., acicular pyroxene, and hopper and swallowtail plagioclase). Inclusions in the Holocene rhyodacitic Llao Rock flow contain hornblende, as well as olivine and plagioclase phenocrysts. Concentric zonation in crystal size or vesicularity is common in the larger inclusions (cf. Heiken and Eichelberger, 1980); the margins, which may be fine grained and vesicle poor, commonly show a crenulated surface, convex toward the host. Vesicular glass is generally present, but much of the porosity is due to void spaces within a network of crystals.

Chemical analyses and the abundance and proportions of phenocryst species suggest magmatic compositions. These mafic inclusions appear to represent blobs of mafic magma from somewhat deeper in a zoned system than their host. Upon incorporation, they cooled rapidly and froze, exsolving volatile constituents in the process. The resulting lowered bulk density inhibited mechanical separation of the inclusions from the host magmas. The mafic inclusions are evidence for incomplete mixing, perhaps as suggested by Huppert et al. (1982), in compositionally and thermally zoned reservoirs.

Fig. 2. Andesitic inclusion in obsidian of Llao Rock flow. Inclusion is vesicular porous hornblende- and olivine-phyric blob of magma that crystallized after being incorporated into ryodacitic magma. Note crenulated vesicle-poor chilled margin. Knife is 8 cm long.

Cognate inclusions, consisting of aggregates of euhedral phenocrysts within brown glass, are common in Mazama intermediate lavas. These inclusions are distinct from the porous mafic inclusions in their: (1) near absence of vesicles; (2) general excessive phenocryst content, commonly in nonmagmatic proportions, to represent blobs of liquid magma; and (3) sizes rarely larger than 2 cm in diameter. The cognate inclusions seem to be bits of highly viscous crystal-enriched (?) magma, derived from the crystallizing margins of reservoirs or conduits, in which crystals have been mechanically concentrated or from which melt has been extracted. Many small cognate inclusions are difficult to distinguish from the glomeroporphyritic clots that are very common in Mazama lavas. In contrast, a few angular xenoliths in the pyroclastic deposits and lava flows show partial-melting textures.

TABLE I

Summary of the eruptive history of Mount Mazama

Age or interval[1] (yr)	Age control[2]	Composition[3]	Unit on Figs. 3 and 4	Unit or event
<6845	P, S	A, R	—	Postcaldera volcanism: Merriam Cone, Wizard Island, and rhyodacite platform.
6845 ± 50	C, P	MA, A, R	*cl*	Climactic eruption: (1) Plinian eruption of air-fall tephra; (2) Wineglass Welded Tuff; (3) collapse of caldera concurrent with emplacement of ash-flow deposits and backflow of Cleetwood lava into caldera; (4) phreatic (?) ejection of intracaldera debris.
7015 ± 45	C, P	R	*lp*	Eruption of lower pumice fall from Llao Rock eruptive center.
6845—7200?	C, P, S	R	*rc, lr, cc*	Emplacement of four lava flows and associated tephra: Redcloud (rc), Llao Rock (lr), Grouse Hill (gh), Cleetwood (cc); rc is oldest; relative ages of lr and gh are not well established; cc is youngest.
>12,000?	BG, P, S	B, MA	—	Vents outside Mazama shadow zone. Ages vary and are poorly constrained; one may be Holocene.
22,000—26,000	C, AG, BG	D	*lwl*	Palisade flow erupted from vent south of Palisade Point.
22,000—30,000	AG, BG, P, S	B, MA, A, D	—	Forgotten Crater complex; eruption of basalt and mixed intermediate lavas from west-northwest-trending radial-fissure system.
22,000—30,000	BG, S	D?	—	Emplacement of dome east of Llao Rock.
22,000—30,000?	BG	D?	—	Sharp Peak domes; erupted on northeast-trending radial fractures.
22,000—50,000	AG, BG, P, S	D	*df, lpf*	Debris-flow and lithic-pyroclastic-flow deposits in Munson Valley, Sun Valley, and on west caldera rim; probably represent dome emplacement high on southwest flank.
30,000—50,000	AG, BG, K, S	A, D	*mwl*	Middle Wisconsin lava flows; present on southwest, north, and northeast wall.

50,000—60,000	AG, S	D	*pc*	Tephra and local lava flows of "Pumice Castle" and ash-flow deposits on southwest wall.
50,000—60 000	BG, AG, S	D	—	Lavas of Scott Bluffs; may belong to early part of "Pumice Castle" eruptive episode.
50,000—75,000	AG, BG, S	A, D	*ewl*	Early Wisconsin lava flows; some probably emplaced near boundary between oxygen isotope stages 3 and 4.
>50,000	S	MA	*cb*	Lava sheets of Cloudcap Bay.
60,000—75,000	AG, S	—	—	Early Wisconsin glacial deposits.
67,000—73,000?	BG, K, S	A	*hp*	Lavas and tephra of Hillman Peak.
75,000—110,000	BG, S	—	—	Hydrothermal alteration of most pre-Wisconsin units exposed on caldera walls; may have been multiple episodes, some of which could be older than 110,000 yr B.P.
110,000—125,000	AG, K, S	A, D	*mp*	Lava flows and domes and lithic-pyroclastic-flow deposits of Merriam Point.
110,000—185,000	BG, K	MA, A	*lb*	Lavas sheets of Llao Bay.
75,000—340,000	AG, BG, S	A, D?	*sr*	Lava flows of Sentinel Rock; extensive glacial erosion of underlying lavas (db) suggest sr flows are younger than ~ 250,000 yr (Shackleton and Opdyke, 1973).
190,000—340,000	BG, K, S	A, D	*dc*	Lavas of Dutton Cliff.
340,000	BG, K, S	A, D	*db*	Lavas of Danger Bay.
400,000	K, S	A?	*pcn*	Phantom Cone lavas and pyroclastic rocks.
>340,000	BG, S	R	—	Southern rhyodacitic lava flows; may or may not be related to Mazama magmatic system.
>75,000	BG	All?	—	Mount Scott and other volcanic rocks east of caldera; may or may not be related to Mazama magmatic system; mafic cones may be younger than 75,000 yr.

[1] Interval refers to bracketing dated units or events and does not indicate the duration of an event. Ages based on preliminary K-Ar dates are rounded.
[2] AG = lies on glaciated surface; BG = glaciated; C = radiocarbon age of associated soil or carbonized wood (Table II); K = preliminary K-Ar whole-rock age (M.A. Lanphere, written commun., 1981, 1982); P = paleomagnetic pole position (D.E. Champion, written commun., 1981, 1982); S = superposition. See text for explanation.
[3] B = basalt; MA = mafic andesite; A = andesite; D = dacite; R = rhyodacite. See section entitled "Terminology" in text for explanation.

A cluster of pre-Wisconsin polygenetic volcanoes

Geologic mapping of the Crater Lake caldera walls reveals the presence of at least five independent volcanoes of pre-Wisconsin age, described below in tentative chronological order (Table I).

Phantom Cone

The oldest rocks in the caldera walls are exposed low on Dutton Cliff at the southeast shore of Crater Lake (Williams, 1942). Because of steep initial dips of the fragmental deposits and lava sheets, and because a small associated island, composed in part of related dikes, was named "Phantom Ship", Williams (1942) referred to this neatly sectioned stratocone as Phantom Cone (unit *pcn*, Fig. 3A). Preliminary K-Ar results indicate an age of ~ 400,000 yr for the andesitic lava at lake level (M.A. Lanphere, written commun., 1981). Apparently conformable relations between the strata suggest that Phantom Cone grew by repeated eruption from a central vent over a comparatively short period.

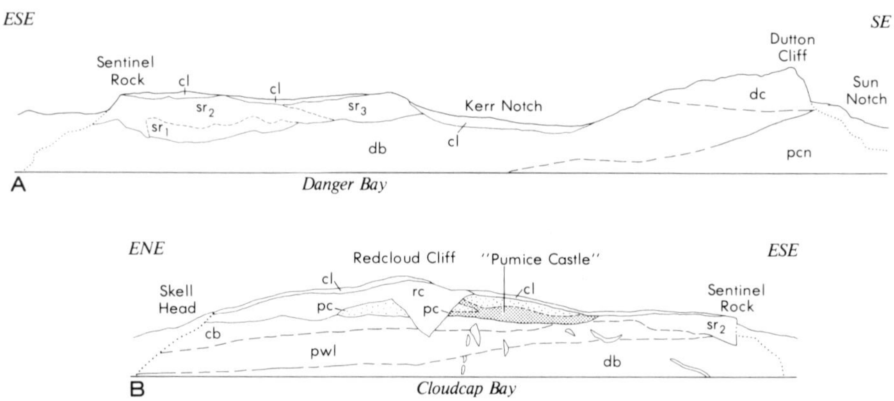

Fig. 3. Geologic sketch maps of caldera walls; bases from photographic panoramas and not to scale. Solid lines, skyline, shoreline, and contacts between geologic units; long-dashed lines, inferred or concealed contacts; short-dashed lines, significant intraunit contacts; dotted lines, edges of prominences on caldera walls.
A. Southeast wall. *pcn* = lavas of Phantom Cone; *db* = lavas of Danger Bay; *dc* = lavas of Dutton Cliff; *sr* = lava flows and clastic rocks of Sentinel Rock; *cl* = undifferentiated deposits of climactic eruption (includes lower pumice fall here). Units *pcn* and sr_2 include vent areas. Two thick lava flows, units *sr* and sr_2, partly fill glaciated valley cut in unit *db*. Kerr Notch is a beheaded glacial valley.
B. East wall. *pwl* = pre-Wisconsin intermediate lava flows and lithic-pyroclastic-flow (?) deposits; *cb* = lava sheets of Cloudcap Bay; *pc* = tephra (heavy stippling) and lava flows (light stippling) of "Pumice Castle"; *rc* = Redcloud flow. Various intermediate dikes are also shown (thicknesses exaggerated). Deposits of climactic eruption not shown where less than a few meters thick.

Danger Bay

Gently dipping thick lava flows lap onto the south and east flanks of Phantom Cone at Danger Bay and east of Sun Notch (unit *db*, Fig. 3A, B). This dominantly andesitic pile includes thick dacitic lava flows near its top, one of which, low on the wall between Redcloud Cliff and Sentinel Rock, gave a preliminary K-Ar age of ~ 340,000 yr (M.A. Lanphere, written commun., 1981). Several short eruptive episodes with intervening periods of quiescence may be represented by this extensive stack of lava flows. Typical of intermediate lava flows, most of the flows in this pile and elsewhere at Crater Lake have autobrecciated tops that can account for as much as half the thickness of a flow unit. A few debris-flow or lithic-pyroclastic-flow deposits are also present.

Dutton Cliff

The uppermost third of the caldera wall at Dutton Cliff consists of intermediate lava flows that may be separated from the underlying rocks of Danger Bay by a significant time break (unit *dc*, Fig. 3A; Williams, 1942, pl. 6). This sequence of flows extends westward to Applegate and Garfield Peaks (Fig. 4A), where chemical analyses show compositions ranging from andesite to dacite. The andesitic lava flows and lithic-pyroclastic-flow (?) deposits between Sentinel Rock and Skell Head (unit *pwl*, Fig. 3B), and andesitic lavas low on the southwest wall (unit *pwl*, Fig. 4A), may be of similar age.

The lava flows at the south caldera rim probably were erupted from the highest stratocone of the ancestral Mount Mazama cluster. This edifice was deeply sculpted by multiple glaciations, including the Wisconsin and earlier events probably corresponding to oxygen-isotope stages 6 and 8 (Shackleton and Opdyke, 1973). This suggests the Dutton Cliff sequence is $\gtrsim$ 190,000 yr old.

Sentinel Rock

Two thick superposed intracanyon intermediate lava flows lie unconformably on glaciated lavas of the Danger Bay sequence at Sentinel Rock (units sr_1 and sr_2, Fig. 3A, B). The apparent feeder for the upper flow (unit sr_2 = Sentinel Rock Flow of Williams, 1942) forms the cliff of Sentinel Rock. The U-shaped glacial valley that these flows filled probably dates from oxygen-isotope stage 6 or 8. The lava flows are overlain to the south by interbedded thinner flows and pyroclastic deposits (unit sr_3) with gentle initial dips. Whether these rocks are significantly younger than the two flows at Sentinel Rock or how either set may be related to the flows on Dutton Cliff is not yet known, although the younger set has been extensively eroded by Wisconsin glaciers that occupied the adjoining Kerr Notch (Fig. 3A). Impressive columnar jointing at the bases of flows of both sets, and structures within interbedded clastic rocks of the upper set suggesting deposition by pyroclastic-surge mechanisms, imply eruption during a comparatively wet, though ice-free (at this elevation), period.

Llao Bay

A broad shield of nearly flat-lying thin lava flows forms a distinctive band at the base of the caldera walls from just north of Hillman Peak to Pumice Point (units lb_1 and lb_2, Fig. 4B, C; Fig. 5). The shield is thickest below Llao Rock, and its center may lie northwest of the caldera. An apparent unconformity within the shield at Llao Bay led to the division into units lb_1 and lb_2 shown on Fig. 4B. Chemically, these flows are mafic andesite to andesite, as also suggested by comparatively thin rubbly flow tops and the broad lateral extent of many flows. Replicate K-Ar dates on a lava flow at Steel Bay gave a weighted mean age of 109,000 ± 10,000 yr (M.A. Lanphere, written commun., 1981). The absence of evidence for interaction with ice is consistent with the implication of the K-Ar ages that these flows were erupted during interglacial oxygen-isotope stage 5. Glacial striae on the top flow of the shield at its northeast extremity near Pumice Point

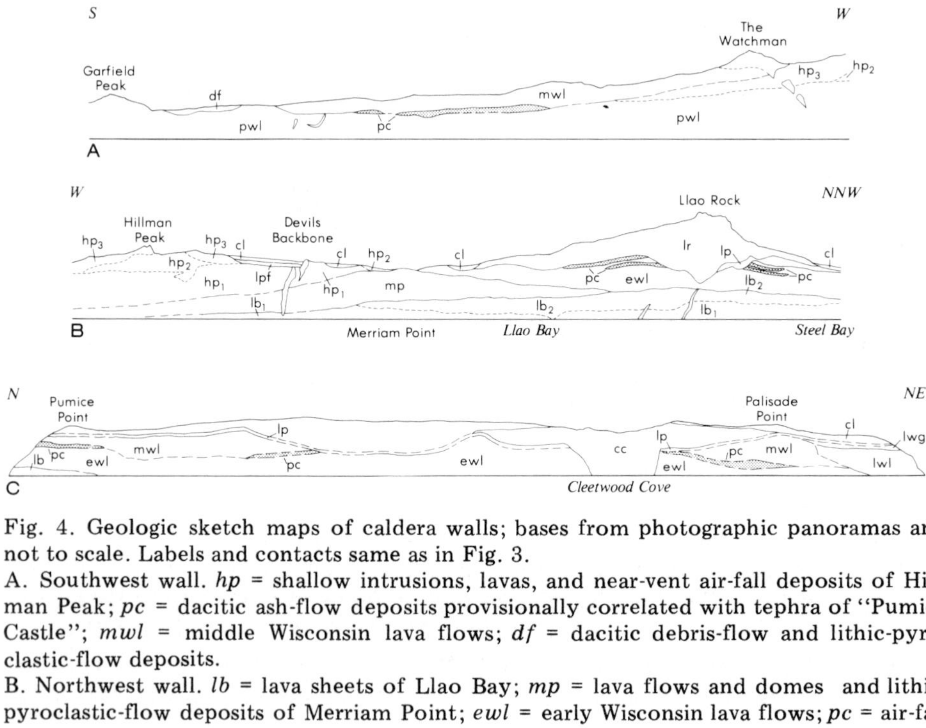

Fig. 4. Geologic sketch maps of caldera walls; bases from photographic panoramas and not to scale. Labels and contacts same as in Fig. 3.
A. Southwest wall. *hp* = shallow intrusions, lavas, and near-vent air-fall deposits of Hillman Peak; *pc* = dacitic ash-flow deposits provisionally correlated with tephra of "Pumice Castle"; *mwl* = middle Wisconsin lava flows; *df* = dacitic debris-flow and lithic-pyroclastic-flow deposits.
B. Northwest wall. *lb* = lava sheets of Llao Bay; *mp* = lava flows and domes and lithic-pyroclastic-flow deposits of Merriam Point; *ewl* = early Wisconsin lava flows; *pc* = air-fall tephra provisionally correlated with tephra of "Pumice Castle"; *lpf* = lithic-pyroclastic-flow deposits; *lp* = lower pumice fall; *lr* = Llao Rock flow; *cl* = undivided deposits of climactic eruption. Most climactic deposits on Llao Rock not visible from this angle. Two dikes in unit *lb* below Llao rock (thicknesses exaggerated) are hornblende rhyodacite derived from climactic magma chamber (see text and Fig. 7).
C. North wall. *pc* = air-fall tephra provisionally correlated with tephra of "Pumice Castle"; *lwl* = late Wisconsin lava (Palisade flow); *lwg* = late Wisconsin till; *lp* = lower pumice fall (thickness exaggerated for clarity); *cc* = Cleetwood flow. Note backflow at Cleetwood Cove (see Fig. 11). Near-vent tephra of Cleetwood eruptive center is included in unit *lb* beneath Cleetwood flow (see Fig. 8).

probably date from oxygen-isotope stage 4. At lake level at Llao Bay, a flow in unit lb_1 gave a preliminary K-Ar age of ~ 185,000 yr (M.A. Lanphere, written commun., 1982), but it remains to be confirmed that the Llao Bay lavas can be subdivided into two groups of significantly different age. If the latter is true, erosion of the older lavas took place during the second half of oxygen-isotope stage 6; glacial striae on the top flow of unit lb_1 south of Devil's Backbone may be of this age.

Fig. 5. Llao Rock. Lava sheets of Llao Bay form bottom third of wall, below vegetation. Great cliff is formed by rhyodacitic Llao Rock flow. Lower pumice fall makes up white band below Llao Rock flow and thick wedge of near-vent tephra at right. Note craggy top of Llao Rock flow and climactic pumice fall above. Pumice deposits have been eroded from flanks by climactic pyroclastic flows, which left dark-gray deposits at top of section. Compare with Fig. 4B.

Merriam Point

Most of the caldera wall above the Llao Bay lavas at Merriam Point is made up of a few thick intermediate lava flows or broad domes (unit *mp*, Fig. 4B). Bedded mudflow deposits that mark the northern edge of this pile are overlain by a thick lava flow that gave a provisional K-Ar age of 115,000 ± 5000 yr (M.A. Lanphere, written commun., 1981). A conspicuous fragmental (lithic-pyroclastic-flow?) deposit of light-gray dacite, containing prismatically jointed blocks, lies on a dacite lava flow which forms the south base of the Merriam Point sequence, just south of the Devil's Backbone dike. Atwood (1935, p. 162) reported glacial striae on the andesite lava flow (unit lb_1, Fig. 4B) underlying this dacite. The equilibrium-line altitudes of Scott (1977) suggest that only a major glacial advance could have affected this surface at its ~ 2000-m elevation. Because the underlying Llao Bay lavas are probably about 185,000 yr old and were glaciated prior to eruption of the Merriam Point sequence, which is overlain by flows with provisional K-Ar ages of ~ 115,000 yr and $\gtrsim$ 73,000 yr (see *Hillman Peak*), eruption of the Merriam Point sequence may have closely followed the latest glacial advance of oxygen-isotope stage 6.

Hydrothermal alteration

Most rocks of the caldera walls just described are hydrothermally altered. Permeability was a controlling variable in the alteration processes; flow-top breccia and clastic layers are more altered than dense flow interiors. Mafic minerals are commonly altered to green clay. Bleached areas low on the walls below Garfield and Applegate Peaks contain abundant disseminated pyrite and irregular veins and cavities of drusy quartz (T.E.C. Keith, written commun., 1981). Oxidation of pyrite causes the yellow coloration of interflow breccia zones.

The age of hydrothermal alteration is not known, although rocks stratigraphically above the ~ 110,000-yr-old rocks in Llao Bay appear unaffected. Several alteration events may have occurred during successive volcanic episodes. The alteration on the south wall resembles that in other heavily glaciated Cascade stratovolcanoes (e.g., Brokeoff Mountain in Lassen Volcanic National Park, California).

Implications for pre-Wisconsin topography

Exposures on the caldera walls indicate that Mount Mazama consisted of a large glacially sculpted andesitic to dacitic stratovolcano cluster centered 1—2 km south of the center of the lake. This complex was composed of at least four overlapping sequences, each probably constructed during a relatively short interval. The summit was never higher than ~ 3700 m, and may not have approached this elevation. A broad shield of mafic andesitic to andesitic lavas (max elev. ~ 2000 m) was centered ~ 6 km north-northwest of the focus of this stratovolcano cluster and lapped against its flanks. This shield was surmounted by the thick lava flows and domes of the Merriam Point sequence. Any intervening cones are now lost beneath the caldera floor.

To the east of Mount Mazama, many domes and lava flows formed a pile ~ 2300 m high, surmounted by the stratocone of Mt. Scott (present elev. 2721 m). Thick silicic lava flows south of Mount Mazama reached ~ 1900 m. Union Peak volcano (present elev. 2347 m; Williams, 1942) and other, unnamed mafic shields (~ 1700 m) flanked Mount Mazama on the southwest and northwest, respectively.

Hydrothermal alteration or the earliest Wisconsin glacial advance appears to have affected all these rock units. Accordingly, the remaining volcanic sequences are tentatively assigned to the period since ~ 75,000 yr B.P., the approximate date of the beginning of the earliest Wisconsin glaciation at the boundary between oxygen-isotope stages 5 and 4 (Colman and Pierce, 1981, p. 26).

Wisconsin volcanic rocks and glacial deposits

Eruption of andesite and dacite from several centers during the early Wisconsin resulted in rapid growth of stratocones and piles of thick lava flows or domes around all but what are now the south and southeast limits of the caldera.

Cloudcap Bay

Midway up the caldera walls at Cloudcap Bay, below Redcloud Cliff, are several gently dipping, relatively thin sheets of mafic andesite (unit *cb*, Fig. 3B). Varying oxidation and vesicularity, which impart a subtle irregular banding to these flows, suggest that they were fed by lava fountains. Many dikes are visible above the south end of Cloudcap Bay, some of which may be related to vents for the mafic andesite. The lavas resemble those of Llao Bay, but are not affected by green clay alteration that is virtually ubiquitous in the Llao Bay rocks. The mafic andesites at Cloudcap Bay lie on the surface of andesitic lavas (unit *pwl*, Fig. 3B) that may have been eroded by pre-Wisconsin glaciers; further work is needed to establish the ages of both sets of andesites.

Hillman Peak

Hillman Peak is the westernmost andesitic stratocone in the cluster of polygenetic volcanoes that made up Mount Mazama. The internal structure of the Hillman cone is neatly bisected by the caldera wall (units hp_{1-3}, Fig. 4A, B). Immediately north of Hillman Peak, this wall is composed of three readily separated units overlying the Merriam Point sequence: (1) thin fountain-fed flows of mafic andesite to andesite, derived from a nearby vent; (2) a thick cone of bedded pyroclastic-fall deposits and associated shallow intrusions of hornblende-bearing mafic andesite; and (3) capping andesite flows, many consisting of thoroughly agglutinated bombs. Unit hp_1 is cut by the spectacular Devil's Backbone segmented andesite dike north of the limits of continuous exposures of units hp_2 and hp_3. This dike may mark the end stage of the episode responsible for unit hp_1; Wisconsin glacial erosion has removed whatever edifice this dike fed. Lavas of unit hp_1 appear to extend southward to the southwest quadrant of the caldera (Fig. 4A, B). Andesitic lava flows overlying rocks of the Merriam Point sequence above Llao Bay may also be correlative with unit hp_1 (unit *ewl*, Fig. 4B). The age of unit hp_1 is not yet well constrained.

Flows of hornblende andesite sandwiched between two series of pyroxene andesite lavas west of Hillman Peak indicate that the hornblende-bearing lava was more voluminous than might be judged from the caldera wall exposures. Vent breccia and bedded tephra in the Hillman cone have been altered in places by fumarolic gases, and interflow rubble lower on the walls shows the effects of hydrothermal alteration, associated with small intrusive bodies. Hornblende andesite from unit hp_2 yielded a K-Ar age of 73,000 ± 6000 yr; the summit pyroxene andesite of Hillman Peak (unit hp_3) gave a K-Ar age of 67,000 ± 6000 yr (M.A. Lanphere, written commun., 1982).

Early Wisconsin glacial deposits

Local deposits of glacial drift and fluvial sediment near lake level between the Palisades and Grotto Cove are the thickest glacial deposits above water level. They were probably laid down during the retreat of ice that occupied, and extended north from, a broad amphitheater bounded by the Llao Bay—Merriam Point edifice, the older stratocone complex of the south wall, and the Sentinel Rock—Cloudcap Bay volcanic pile. Relations to underlying and overlying units, and the low elevation of the base of the deposits (< 1900 m) indicate an early Wisconsin age, probably 60,000 to 70,000 yr (Colman and Pierce, 1981).

Early Wisconsin lavas

Silicic andesitic to dacitic lava flows in the middle of the caldera wall above Steel Bay overlie the Merriam Point and Llao Bay sequences but underlie dacitic pyroclastic-flow deposits tentatively correlated with the tephra of "Pumice Castle" (Fig. 4B, C). Glacial striae on the top flow of the Llao Bay sequence near Pumice Point suggest that the overlying andesitic lava flow is younger than the earliest Wisconsin glacial advance. Although ice probably was absent at the time of eruption, some flows lie on lenses of fluvial (?) sediment and are strikingly columnar; they may have been emplaced at a time near the boundary between oxygen isotope stages 3 and 4, possibly ~ 60,000 yr B.P. Some andesitic lava flows exposed midway up the caldera walls between the Palisades and Grotto Cove may also be of this age. The thick silicic andesitic lava flow near the boat landing at the west end of Cleetwood Cove (Fig. 4C), which lies on monolithologic avalanche (?) deposits, is overlain by the tephra of "Pumice Castle" and is probably early Wisconsin in age. A conspicuous thick lava flow and nearby small dome at Steel Bay were fed by thick dikes that merge into the extrusions (Fig. 5). These features and other overlapping laterally restricted thick lava flows indicate that the eruptive style of Mount Mazama had changed by this time from (1) short spurts of dominantly andesitic lava from central vents to (2) a more diffuse pattern of vents that commonly erupted dacite.

Dacitic lavas of Scott Bluffs

Thick lobate dacitic lava flows, which issued from vents between Cloudcap and Mount Scott and flowed northeastward, form Scott Bluffs (Fig. 1). Remnants of marginal levees are present, although the flanks of the flows are heavily glaciated. Similar but less extensive dacitic flows were erupted at Cloudcap, where they are overlain by agglutinated tephra of "Pumice Castle". The lava flows of Scott Bluffs lie on a surface that probably was glaciated during oxygen-isotope stage 4. These relations suggest that the dacites of Scott Bluffs are not much older than the tephra of "Pumice Castle" and may belong to the same eruptive episode.

Tephra of "Pumice Castle"

Dacite pumice containing distinctive coarse plagioclase and pyroxene occurs locally in the caldera walls from the southwest quadrant clockwise to Cloudcap Bay (unit *pc*, Figs. 3B, 4A—C). The most prominent outcrops are above Cloudcap Bay at the "Pumice Castle" ("Cottage Rock" of Williams, 1942). "Pumice Castle" stands out as a tower of alternating welded and partially welded air-fall pumice, supported from below by a particularly resistant remnant of dacitic lava. A major vent was near "Pumice Castle", where thick air-fall deposits grade northward from barely coherent to welded vitrophyre over a distance of ~ 300 m. Rhythmic alternation of nonwelded and more densely welded layers implies pulsating flow at the vent; the nonwelded layers might represent greater eruption heights and, thus, increased exit velocities. A wedge of dacitic lava lies within the welded air-fall deposits at their north end, where the entire section has been truncated by the vent system of the rhyodacitic Redcloud flow (Fig. 3B). The uppermost layers of air-fall deposits have been fused by heat and pressure from the overlying dacitic lava flow. This flow was probably erupted as the relatively degassed tail end of the "Pumice Castle" eruption. A poorly developed soil is exposed locally beneath the pumice.

Air-fall pumice deposits sandwiched between andesitic lava flows between Palisade Point and Pumice Point are evidently correlative with those at "Pumice Castle" (unit *pc*, Fig. 4C). At Pumice Point, air-fall deposits lie on soil developed in thin drift on glaciated andesitic lava, as well as below andesite flows, the highest of which has a K-Ar age of 47,000 ± 15,000 yr (M.A. Lanphere, written commun., 1981). The pumice deposits were eroded before the overlying andesitic lava flows were emplaced. Similar air-fall deposits are present low in the wall below the Palisade flow (see *Palisade flow*).

Two layers of dacitic air-fall deposits occur below either side of the Llao Rock flow (unit *pc*, Fig. 4B; Fig. 5), interbedded with autobrecciated dacitic lava flows. The upper layer, which dips as much as 45° into the caldera wall, is partly to densely welded. The welding suggests that the pumice was erupted from a nearby vent. Similar deposits, although nonwelded, are present in the wall above Steel Bay. The lower layer appears to correlate with the dacite pumice at Pumice Point.

Dacitic ash-flow deposits lie within a broad paleovalley cut into andesitic lava flows in the southwest caldera wall (unit *pc*, Fig. 4A). Complex internal structures in nonwelded deposits suggest interaction with water or a damp substrate. An andesitic lava flow overlying the eroded ash-flow deposit yielded a K-Ar age of 52,000 ± 4000 yr (M.A. Lanphere, written commun., 1981).

If all these pyroclastic rocks indeed represent the same brief eruptive episode, they constitute a unique Pleistocene stratigraphic horizon on the caldera walls. All the rocks were apparently emplaced at comparatively high temperature. The distribution of welded air-fall deposits and interbedded lava flows implies source vents near Llao Rock and Redcloud Cliff. Airborne pumice probably blew northward and eastward, whereas pyroclastic flows traveled down a valley south of the Hillman Peak—Merriam Point edifice. At least one ash layer in Pleistocene lacustrine deposits near Summer Lake, ~ 110 km east of Crater Lake, can be correlated with tephra of "Pumice Castle" (J.O. Davis, written commun., 1982). Because the pyroclastic rocks commonly lie on glaciated surfaces or poor soils and because they are overlain by lava flows dated at ~ 50,000 yr, these eruptions of dacite probably took place during a brief interval between 60,000 and 50,000 yr ago, early in oxygen-isotope stage 3. Any minor collapse structure that may have been associated with the "Pumice Castle" eruptions would have been engulfed by the caldera that formed during the climactic eruption.

Middle Wisconsin lava flows

Andesitic and dacitic lava flows that were scoured by late Wisconsin ice occur above the tephra of "Pumice Castle" on the southwest and north caldera walls (unit *mwl*, Fig. 4A, C). In the southwest wall, middle Wisconsin lava flows fill the same paleovalley occupied by the underlying dacite ash-flow tuff. Dikes in the caldera wall suggest that some flows issued from vents near the south end of exposure; dips imply that other flows came from near The Watchman. The Watchman flow (Williams, 1942, pp. 44—46) contains ~ 67.5% SiO_2. Its feeder dike crops out amidst talus on the caldera wall. The sides of the Watchman flow have been glaciated. A preliminary K-Ar age on the flow of 50,000 ± 3000 yr (M.A. Lanphere, written commun., 1981) indicates that this glaciation occurred both during a middle Wisconsin advance in oxygen-isotope stage 3 time, between 50,000 and 30,000 yr ago (Colman and Pierce, 1981), and during the late Wisconsin, corresponding to oxygen-isotope stage 2. The K-Ar ages of the Watchman flow and the flow overlying the ash-flow deposits to the south imply that these and the intervening lavas were erupted over a short period, possibly < 1000 yr.

Other glaciated middle Wisconsin lava flows occur above the tephra of "Pumice Castle" at Steel Bay, Pumice Point, and Palisade Point. The ages of these flows are bracketed between ~ 22,000 and ~ 50,000 yr by the underlying tephra and the absence of evidence for interaction with late Wisconsin glaciers (Colman and Pierce, 1981).

Thick flows of silicic andesite lie on early Wisconsin glacial deposits or on thinner andesitic flows in the north caldera walls between the Palisades and Skell Head. The spectacular subhorizontal to radiating columnar joints in the flow at the Wineglass, and the glassiness of the lava, suggest interaction with ice. This flow, which is overlain by drift composed dominantly of similar andesite at its west margin, may be a subglacial lava flow. The bases of other thick silicic andesite flows near the top of the walls are commonly columnar and glassy where they lie on glacial deposits, such as at the Palisades. Three of these flows retain some of their original morphology and can be traced northeastward as much as 3 km from the caldera rim; they appear to rest on a surface that probably was glaciated in oxygen-isotope stage 4. This sequence of flows has been glaciated more extensively than lavas known to have been glaciated only in the late Wisconsin; because they are apparently younger than the tephra of "Pumice Castle", the thick silicic andesites are probably between ~ 50,000 and ~ 30,000 yr old.

Late Pleistocene dacitic rocks

Dacitic lava was vented from at least five eruptive centers in the late Pleistocene (Fig. 6A). The ages of three of these are well constrained to be between ~ 22,000 and ~ 30,000 yr B.P.; the other two are possibly also of this age. Hornblende, which commonly is abundant in Holocene silicic rocks, is present in most of these dacitic rocks; its presence is thought to link the dacite to the early development of a large silicic magma chamber which later became the source of the ejecta of the climactic eruption.

Dacite debris in Munson Valley and on the west caldera rim

Fragmental deposits at the head of Munson Valley (unit *df*, Fig. 4A) consist of a massive monolithologic debris-flow deposit, ~ 40 m thick, overlain by one or more comparatively thin heterolithologic lithic-pyroclastic-flow deposits. Clasts in the debris-flow deposit are nearly all of porphyritic dacite containing traces of hornblende and abundant porous andesitic inclusions. Paleomagnetic study shows that the material was emplaced below its Curie temperature (D.E. Champion, oral commun., 1980). However, bleaching and oxidation related to postemplacement fumarolic alteration locally is pervasive, perhaps owing to emplacement on the wet floor of Munson Valley. Similar deposits occur high on the west flank of Garfield Peak and locally on the west side of the valley of Sun Creek. The debris-flow deposits there and in Munson Valley may record the destruction of a cooling lava flow or dome that was extruded at an elevation of at least 2400 m somewhere north of Garfield Peak (Fig. 6A).

Lithic-pyroclastic-flow deposits, containing dacitic blocks similar to those in the underlying debris flow, as well as juvenile prismatically jointed blocks of gray dacite and rare pumiceous dacite and rhyodacite, overlie the thicker debris-flow material in Munson Valley. These deposits are ten-

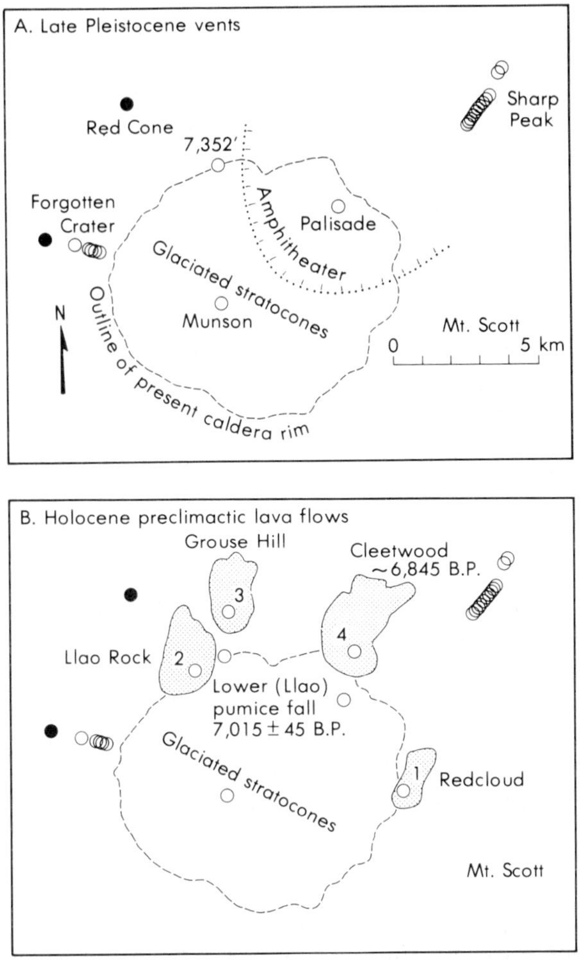

Fig. 6. Late Pleistocene and Holocene vent distribution. A. Late Pleistocene vents (~ 22,000—~ 30,000 ? yr B.P.); filled circles are basaltic vents, open circles are intermediate or dacitic vents. Locations of vents for Palisade flow (Palisade) and parent dome(s) for debris-flow and lithic-pyroclastic-flow deposits (Munson) inferred. B. Preclimactic rhyodacitic lava flows (shaded) and vents in order of eruption.

tatively interpreted as recording the eruption of fresh lava that was responsible for destruction of the dome now represented by the debris-flow deposit. Similar prismatically jointed blocks are present in lithic-pyroclastic-flow deposits south of The Watchman and near the top of the caldera wall south of Devil's Backbone (Fig. 4B). Paleomagnetic data on these prismatically jointed blocks strongly suggest that the deposits at the three localities are correlative because they all yield an unusual pole position (D.E. Champion, written commun., 1981).

The distribution of dacitic debris-flow and lithic-pyroclastic-flow deposits on the south and west rim of the caldera suggests a common source south-

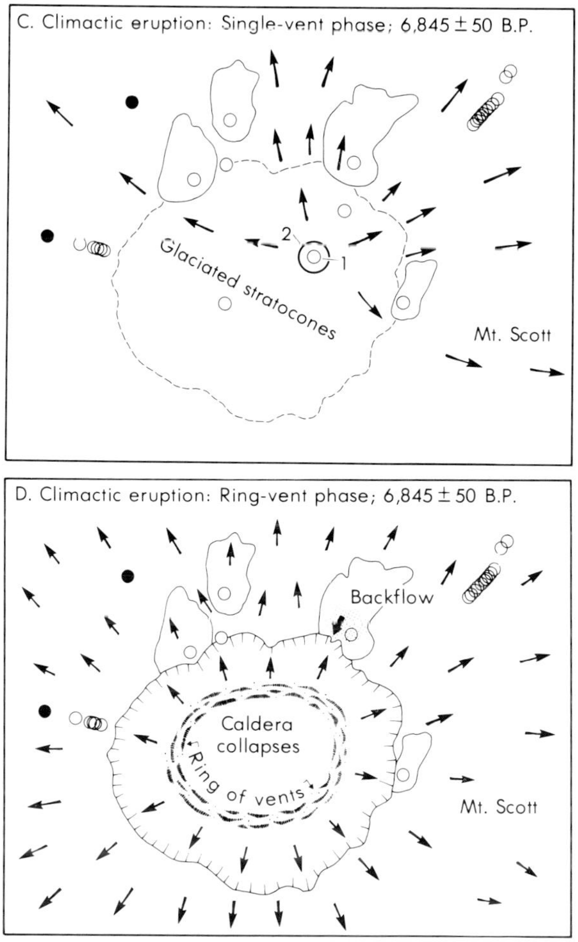

Fig. 6. Late Pleistocene and Holocene vent distribution. C. Singel-vent phase of climactic eruption. Location of vent inferred; *1* indicates vent for Plinian eruption, *2* indicates enlarged vent during eruption of ash flows that deposited the Wineglass Welded Tuff; relative sizes of vents schematic. Arrows show direction of movement of ash flows, which were confined to topographically low areas. D. Ring-vent phase of climactic eruption. Arrows show direction of movement of ash flows, which surmounted most topographic barriers. Blackflow of Cleetwood flow into caldera indicated by shading and arrow. Location of ring of vents approximate.

east of Wizard Island and north of Garfield Peak. These deposits rest on $\gtrsim$ 50,000-yr-old glaciated lava flows. All known exposures are either very thick or are remnants left in areas that would have been protected from vigorous glacial erosion. Azimuths of glacial striae on the caldera rim (Williams, 1942, fig. 31) and the thicknesses and locations of remnants of the deposits suggest that these deposits are older than the late Wisconsin glaciation, and are probably between 22,000 and 50,000 yr old. Such an age is consistent with the paleomagnetic pole positions obtained for the lithic-pyroclastic-flow deposits (D.E. Champion, written commun., 1982).

Sharp Peak domes

Ten dacitic (or rhyodacitic?) domes, the highest of which is Sharp Peak, form a northeast-trending line radial to the caldera. Two more domes continue the trend, but are offset ~ 0.7 km to the north (Fig. 6A). Hornblende phenocrysts are present in the domes closest to the caldera. All twelve domes probably were emplaced in the same eruptive episode, but the time of this event is the least well known of the late Pleistocene dacitic eruptions. The southeast side of the domes was little affected by glaciation; pumiceous blocks are present and large boulders on the slope of Sharp Peak appear to have disintegrated in place during cooling. However, steep cliffs and exposures of devitrified rock on the west and northwest sides of the higher domes suggest that late Wisconsin ice reached a higher level in the drainage to the west. Glacial striae and polish are locally poorly preserved above talus on the higher domes. Consequently, the domes are > 22,000, and possibly < 30,000 yr old.

Dome east of Llao Rock

The highest lava in the caldera wall between Llao Rock and Pumice Point is a hornblende-bearing dacitic or rhyodacitic dome (Hill 7352', Fig. 6A). This dome apparently has been glaciated, but lies on eroded andesitic lava which is younger than the tephra of "Pumice Castle". Erosion of the underlying andesite probably took place in the middle Wisconsin. A poorly-developed soil is present in the top of marginal talus deposits, which are overlain by Holocene tephra, exposed in the caldera wall at the east edge of the dome. Thus, the dome probably was emplaced between ~ 22,000 and ~ 30,000 yr B.P.

Forgotten Crater

The cinder cone of Forgotten Crater[1] and related lava flows (Williams, 1942, pp. 56—57) deserve special mention because of their position < 1 km west of the caldera rim, wide range in compositions, and apparent youth. The Forgotten Crater complex consists of: (1) olivine basaltic lava and a fissure vent system; (2) an olivine basalt cinder cone; and (3) one extensive flow and three small flows or domes of commingled andesitic to dacitic lava. Field relations and the preliminary results of paleomagnetic investigations (D.E. Champion, written commun., 1981) indicate that all these eruptive products are similar in age. The basalt flow (1) and tephra (2) underlie the mixed flows (3). The fissure vent, cinder cone, and the four vents for the mixed lavas are aligned N70°W, approximately radial to the caldera (Fig. 6A). The head of the basalt flow lies near the projection of

[1] The informal name "Forgotten Crater" does not appear on maps. An effort is underway to officially name the cinder cone after Howel Williams.

an inferred normal fault trending N10°E that is now buried by products of the climactic eruption. The Forgotten Crater eruptive sequence may have been initiated by an eruption of typical High Cascade basaltic magma (51.6% SiO_2). Because of its proximity to the Mazama magmatic system, this disturbance came under the influence of the local stress field and allowed the basaltic magma access to the margins of a compositionally zoned system by opening of radial fractures. Eruption of highly complex, varingly mixed and commingled magmas followed.

A few olivine basalt bombs in the cinder cone have cores of banded pumiceous andesite and dacite; commonly these cores are somewhat angular, presumably owing to rapid fracturing of viscous magma. All the intermediate lava in the Forgotten Crater complex contains poorly vesiculated spheroidal basaltic inclusions, ~ 1 cm to > 1 m in size, many of which are compositionally identical with the basalt of the cinder cone and lava flow; other inclusions have up to ~ 54% SiO_2, are also represented as very rare cores of basaltic bombs in the cinder cone, and are probably mixtures of the basaltic and andesitic magmas. The proportion and size of inclusions are greater in the more mafic andesitic hosts. These inclusions were apparently entrapped in the intermediate magmas as blobs of magma but had crystallized sufficiently to fracture by the time of the cinder cone eruption.

The commingled lava flow described by Williams (1942, p. 57) was erupted from a vent just west of the cinder cone. At its distal end, this flow is a nearly homogeneous dark-gray olivine-bearing andesite (~ 59—62% SiO_2), charged with inclusions. Bands of inclusion-bearing dacite (~ 64—67% SiO_2) as little as a few cm wide and containing minor hornblende become more abundant toward the source, until the silicic component dominates. Identical dacite containing mafic andesite bands and basaltic inclusions was also erupted from two vents within the cinder cone and from a third vent a short distance to the east, just west of the rim road. Ongoing study of the Forgotten Crater complex may elucidate the details of compositional zonation near the margin of the Mazama magmatic system.

No material suitable for radiocarbon dating has yet been recovered from beneath the tephra or lava flows of Forgotten Crater. Forgotten Crater basaltic tephra lie on the lithic-pyroclastic-flow deposits near Devil's Backbone. Mafic tephra beneath the Palisade flow may have been erupted from Forgotten Crater. The main banded lava flow retains much of its original form, including steep flow fronts and marginal levees. A few outcrops of lava show glacial striae and polish; the basalt flow displays similar evidence of minor glaciation to elevations at least as low as 2000 m. It seems highly unlikely that early Holocene alpine glaciers would have existed at elevations as low as 2000 m (Scott, 1977, fig. 11). Forgotten Crater itself is banked against a subdued arête, and the lavas lie on deeply glacially eroded andesite from the Hillman Peak eruptive center. The glaciation responsible for gently scouring the Forgotten Crater complex probably occurred during latest Wisconsin time. Forgotten Crater must be younger than the middle Wisconsin culminations, discussed by Colman and Pierce (1981), that resulted in significant erosion of the Hillman Peak edifice and the Watchman flow. Therefore, the Forgotten Crater complex is probably between 22,000 and 30,000 yr old.

Palisade flow

The dacitic (65.0% SiO_2) Palisade flow (unit *lwl*, Fig. 4C; Williams, 1942, p. 32) occupies a glaciated paleovalley cut into the northwest margin of the adjoining silicic andesite flow at The Palisades and the eastern edge of the silicic andesite flow that forms Palisade Point (unit *mwl*, Fig. 4C). The vent for the flow lay to the south of the present outcrops (Fig. 6A). The upper surface of the Palisade flow is craggy, unlike other north-wall intermediate lava flows, which have had their spiny tops and rubble carapaces entirely removed by glacial erosion; however, glacial till (unit *lwg*, Fig. 4C) is present on its top at a much lower elevation than the adjacent flow at The Palisades. Beneath the Palisade flow are related avalanche or lithic-pyroclastic-flow deposits resting on ~ 40 cm of mafic tephra, which lies on mudflow (?) and fluvial deposits overlying tephra provisionally correlated with that of "Pumice Castle". Poorly developed soil from the top of the mudflow (?) deposit yielded a radiocarbon age of 25,480 ± 580 yr (sample 7, Table II). Because the Palisade flow can have been glaciated only in the late Wisconsin and because there is no evidence for interaction with ice, the flow must be between ~ 22,000 and ~ 26,000 yr old.

Late-Wisconsin topography

The distribution of dacitic pyroclastic rocks and the azimuths of glacial striae on the caldera rim (Williams, 1942, fig. 31) provide clues to the form of Mount Mazama during the late Wisconsin. Mount Mazama apparently consisted of a cluster of overlapping eroded stratocones that formed an irregular asymmetric ridge from Hillman Peak to Mount Scott (Figs. 1, 6A); the Garfield Peak—Dutton Cliff complex, Hillman Peak, the Cloudcap Bay and Sentinel Rock volcanoes, and Mt. Scott were the major summits. North of these summits was a broad glacial amphitheater containing the source vents for several thick intermediate lava flows. One or more dacitic domes lay between Hillman and Garfield Peaks and remnants of an apron of related debris survived in favorable sites in the valleys to the west and south. This configuration of Mount Mazama apparently persisted until the preclimactic rhyodacitic lava flows were emplaced ~ 7000 yr B.P.

Extra-caldera monogenetic cones and small shields

The concept of a shadow zone (Smith and Shaw, 1975), inside of which mafic melts do not escape to the surface because of the presence of overlying less dense magma, appears to apply to the Mazama magmatic system at least as early as ~ 100,000 yr B.P. (post-Llao Bay shield). Forgotten Crater provides spectacular evidence for the late Pleistocene location of the west edge of the shadow zone.

Williams (1942) identified about 20 mafic cinder cones and patches of scoria outside the caldera in Crater Lake National Park, of which the Timber Crater shield, ~ 8 km north-northeast of the caldera, is the most voluminous. Ejecta from most of these vents consists of olivine-bearing basalt or mafic andesite that resembles the eruptive products of other Quaternary monogenetic volcanoes in the Oregon Cascades. Such lavas are interpreted as basaltic melts, somewhat modified by crystal fractionation and interaction with deep crustal rocks, that constitute a regional background flux of mantle-derived magma. These rocks may represent the magma(s) underlying Mount Mazama, and perhaps most of the High Cascades, at depth and providing a source of thermal energy and mass necessary to maintain the observed long-lived magmatic system. To refer to the cinder cones as "parasitic" (e.g., Williams, 1942, p. 55) is inappropriate, since this implies that they are uniquely related to the Mazama magmatic system (cf. Hildreth and Fierstein, 1983).

Williams (1942) believed that many of these mafic cones had not been glaciated. Red Cone, however, thought by Williams (1942, p. 56) to be one of the youngest, has been extensively glaciated on its north side. None of the extracaldera cones within the park has the symmetrical youthful form displayed by the Holocene cone of Wizard Island. A cinder cone ~ 0.8 km northeast of the confluence of Castle and Little Castle Creeks, however, may be early Holocene because it is apparently banked against the glaciated wall of the valley of Castle Creek. Moreover, exposed beneath ash-flow deposits of the climactic eruption in Castle Creek are thin basaltic lava flows, which probably came from the vent marked by the cinder cone, that have well preserved ropy surfaces; unless the present exposure is atypical, the upper flow was not glaciated. The lava is the most primitive basalt yet analyzed from the Crater Lake area (47.7% SiO_2, 0.10% K_2O). The non-differentiated composition of the basalt is consistent with the position of the vent west of the axis of the High Cascades, ~ 7 km west-southwest of the caldera, which indicates minimal opportunity for crustal residence (cf. Blackwell et al., 1982). A nearby mafic andesitic cinder cone, ~ 1 km northeast of the basaltic cone, is clearly much older. The ages of the mafic vents apparently vary considerably, and further work is necessary to establish a chronology.

PRECURSORS TO THE CLIMACTIC ERUPTION

Crater Lake caldera provides an ideal natural laboratory for reconstruction of the chronology of precursory volcanic eruptions, the climactic eruption itself, and collapse of the caldera. Although some aspects of this sequence are not fully known, many refinements can be made to the scenario outlined by Williams (1942). Critical to a new interpretation is recognition of a widespread lag breccia, or proximal facies of the zoned ash-flow deposits, preserved on the caldera rim and upper slopes. Williams (1942), who had tentatively interpreted this deposit as glacial drift, confused the near-vent stratigraphic relations of climactic ejecta and precursory pumice falls and lava flows, all of which are now known to have been erupted during the Hypsithermal interval of the Holocene, when no ice would have existed at present Mazama elevations. Paleomagnetic secular-variation studies and radiocarbon dating have been crucial to refining a chronology of events; however, as discussed below, the main conclusions regarding timing can be drawn on the basis of field relations alone.

Preclimactic rhyodacitic lava flows and tephra

Lava flows from four eruptive centers (Fig. 6B) and one widespread pumice fall preceded the climactic eruption by at most a few centuries. The precursory rhyodacitic lava flows and tephra are compositionally continuous with, and commonly slightly more differentiated than, early products of the climactic eruption (Fig. 7). All these rocks contain hornblende as a significant component of the phenocryst assemblage, unlike most Pleistocene Mazama eruptive products. By the time of these Holocene eruptions, Mount Mazama had apparently experienced no major cone building activity for possibly 40,000 yr, the interval since voluminous andesitic to dacitic eruptions within the area of what is now the north half of the caldera. A period of ~ 15,000 yr had elapsed since eruption of the last intermediate products that left a recognizable record. This period was apparently necessary for accumulation of the rhyodacitic part of the climactic magma chamber. A long period of relative quiescence, followed by eruption of silicic magma away from a central vent and controlled by a nontectonic stress field, signaled the onset of the climactic eruptive sequence.

Redcloud eruptive center

The vent for the Holocene Redcloud eruptive center (Figs. 3B, 6B) virtually coincides with the older "Pumice Castle" vent, just east of the caldera wall (Williams, 1942, p. 53). Emplacement of a thick (~ 100 m) rhyodacitic lava flow was preceded by the eruption of hornblende-bearing pumice; this pumice-fall deposit occurs below the lava flow in the caldera wall north of the vent and forms a wedge to the east of the margin of the lava flow along the road to Cloudcap. Restoration of the small fraction that was lost during collapse of the caldera gives a minimum estimated volume of 0.2 km^3 for the flow.

The age of the Redcloud flow (Williams, 1942, p. 53) is bracketed by the 7015 ± 45 yr radiocarbon age of the overlying tephra from the Llao Rock vent (see *Llao Rock eruptive center*) and the observation that the flow does

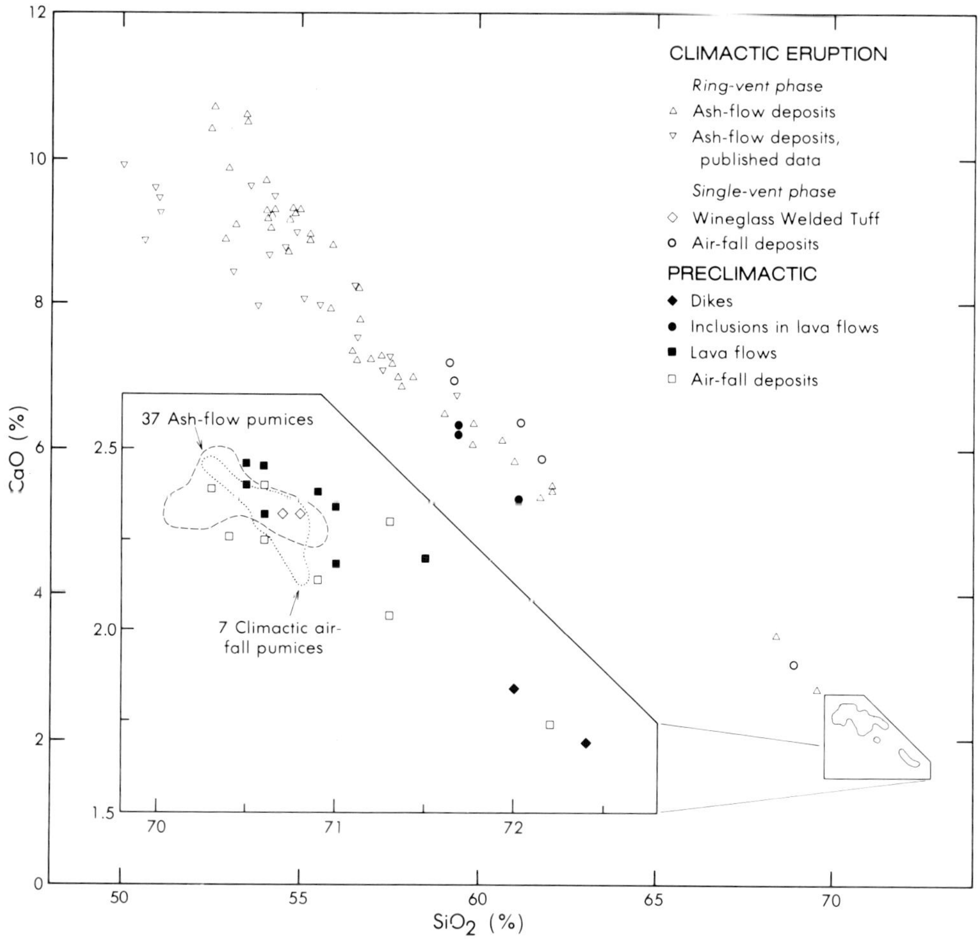

Fig. 7. CaO vs. SiO_2 contents for climactic ejecta and related Holocene rocks. Inset shows data in stippled fields of large plot. Lava flows are Redcloud, Llao Rock, Grouse Hill, and Cleetwood. Inclusions are from Llao Rock and Grouse Hill flows. Dike samples are glassy selvages of two dikes below Llao Rock (Fig. 4B). The most silicic air-fall pumice is from base of wedge of near-vent tephra from Llao Rock eruptive center (Figs. 4B, 5). All analyses recalculated volatile free, with $Fe^{2+} = 4Fe^{3+}$. All data from present study (X-ray fluorescence analyses performed at the U.S. Geological Survey laboratory in Denver, Colorado) except inverted triangles, which are for scoria from ash-flow deposits, analyzed by Ritchey (1979) and K. Aoki (in McBirney, 1968).

not appear to have been glaciated. The upper surface of the flow at the caldera rim is craggy, a smooth north margin was probably abraded by the climactic pyroclastic flows, and remnants of the east flow margin along the Cloudcap road evidently owe their dipping form to slumping of the flow interior rather than to glacial erosion. Preliminary results of paleomagnetic measurements show a steep inclination that is consistent with an age between ~ 7000 and ~ 7200 yr (D.E. Champion, oral commun., 1981). Eruption of the Redcloud flow and tephra sometime between ~ 7200 and ~ 7000 yr B.P. was probably the first in a series of volcanic precursors to the climactic eruption.

Llao Rock eruptive center

The great cliffs formed by the rhyodacitic Llao Rock flow (Figs. 4B, 5; Williams, 1942) are among the most spectacular features of Crater Lake National Park. The lava flow contains more abundant inclusions of porous andesite (Figs. 2, 7) than any other rhyodacite flow. The reconstructed flow (Fig. 6B) has a minimum volume of 0.5 km^3. Williams (1942, pp. 47—48) thought that the vent for this flow was fortuitously situated in a glacial valley, but I believe that the form of the base of the flow (Fig. 4B) can be attributed to emplacement within a crater excavated by related explosive eruptions.

The initial eruptions built a thick wedge of coarse fall deposits east of the vent (pumice to ~ 30 cm and lithic fragments to ~ 1 m, both decreasing in size eastward), which includes finer air-fall or surge layers extending east of Llao Rock for ~ 1 km. Pumice in the lowermost layers of the wedge commonly contains an unusually high proportion of hornblende. The wedge of near-vent deposits rests on a thin soil developed on underlying glacial (?) deposits and glaciated intermediate lava flows. The near-vent deposits comprise the proximal facies of widespread air-fall tephra; these deposits are here informally termed the "lower pumice fall", to distinguish them from deposits of the climactic eruption. The unit forms a conspicuous white band beneath the east half of cliffs of the rhyodacite of Llao Rock (Figs. 4B, 5); its upper part becomes progressively "fused" upward to the base of the lava flow beneath the high cliffs, and columnar joints within the flow extend downward into the "fused" tuff.

Deposits of the lower pumice fall are sparse west of Llao Rock but can be traced clockwise around the caldera at least as far as Garfield Peak; these deposits invariably contain hornblende-bearing mafic inclusions similar to those of the lava flow of Llao Rock and may contain as much as ~ 20% of lithic fragments. Air-fall tephra $\lesssim$ 1 km west of Garfield Peak and south of the caldera as far west as Munson Valley, which contains a high proportion of hornblende, may correlate with the earliest pumice erupted from the Llao Rock vent. The lower pumice fall is the only known voluminous pyroclastic deposit associated with eruption of the preclimactic rhyodacitic lava flows; that the products of the Llao Rock vent contain several orders of magnitude more mafic inclusions than those of the other eruptive centers may be relevant to this observation (see *Initiation of the climactic eruption*).

The age of the lower pumice fall is 7015 ± 45 yr, on the basis of radiocarbon dating of carbonized twig fragments from the upper ~ 1 cm of a 15-cm-thick soil developed on glacial drift beneath the deposit at the Wineglass (Table II, sample 6). The common occurrence of carbonized-twig-bearing material, some of which contains small pumice lapilli, just beneath the lower pumice fall suggests that the upper part of this soil may be composed of ash related to the first eruptions from the Llao Rock vent. Results of paleomagnetic measurements on pumice in the tephra wedge east of Llao Rock are consistent with the radiocarbon age (D.E. Champion, written commun., 1981).

Ash beds that are apparently correlative with the lower pumice fall are present in northern Washington (Mack et al., 1979), ~ 700 km from Crater Lake, in eastern Washington (Blinman et al., 1979), near Steens Mountain, southeastern Oregon (Blinman et al., 1979; P.J. Mehringer, oral commun., 1982), and in the Lake Lahontan area, western Nevada (Tsoyawata Bed of Davis, 1978; J.O. Davis, written commun., 1982). The lower pumice fall must have been deposited over a period of at least several days in order for sufficient variation in wind direction to occur to account for the wide dispersal. The volume of magma equivalent to the lower pumice fall deposits is probably ~ 2—10 km^3.

Two dikes below Llao Rock, each ~ 2 m thick, of hornblende-phyric rhyodacite, which have compositions similar to that of the earliest erupted pumice of the Llao Rock tephra wedge (Fig. 7), are slightly more differentiated than either the lava flow or most of the lower pumice fall. Preliminary results of paleomagnetic studies suggest that the dikes were emplaced before the eruptions at the Llao Rock vent (D.E. Champion, oral commun., 1982).

Grouse Hill lava flow and dome

The chemical composition, morphology, and stratigraphic position of the rhyodacitic lava flow and capping dome of Grouse Hill (Fig. 6B; Williams, 1942, pp. 49—50) indicate that they belong to the preclimactic rhyodacites. The lava of Grouse Hill is more coarsely porphyritic and contains fewer inclusions than that of Llao Rock. The combined volume of the lava flow and dome is at least 0.5 km^3. Tephra related to the Grouse Hill vent has not been recognized in the caldera-wall exposures.

The flow and dome of Grouse Hill are blanketed by air-fall and pyroclastic-flow deposits of the climactic eruption. This covering is discontinuous because much was removed by the climactic pyroclastic flows (see *Erosion by pyroclastic flows*). No lower-pumice-fall deposits have been found between the Grouse Hill lava and overlying climactic deposits; the base of the flow is not exposed owing to onlapping of ash-flow deposits. On the basis of these relations, the Grouse Hill eruptive center is older than the climactic eruption and should be younger than the lower pumice fall; preliminary results of paleomagnetic investigations by D.E. Champion (written commun., 1982) are consistent with this inference.

Cleetwood eruptive center

The rhyodacitic Cleetwood flow, which forms the cliffs above Cleetwood Cove (Figs. 4C, 6B; Williams, 1942, pp. 50—52), was the last preclimactic rhyodacite and had not completely cooled at the time of the climactic eruption. Its composition is virtually identical with that of pumice of the climactic eruption. Rare mafic inclusions are present. The reconstructed minimum volume of the lava flow is ~ 0.6 km^3.

The lava flow at Cleetwood Cove was preceded by eruption of a small volume of lithic fragments and pumice that apparently formed a ring of coarse, near-vent fall deposits overlying the lower pumice fall (Fig. 8) and now exposed at the base of the cliffs on either side of Cleetwood Cove and along the trail to the boat landing. Low hills northeast of the rim road may mark the location of the vent (Fig. 6B), although surface features of the lava flow are complicated by effects related to collapse of the caldera. The age of the Cleetwood flow is constrained by that of the overlying climactic pumice fall; the flow must be no more than a few months older than the pumice fall because: (1) air-fall pumice of the climactic eruption is oxidized and welded only on top of the southern Cleetwood flow; (2) the pumice and overlying proximal ash-flow deposits locally were altered by fumarolic gases that originated within the lava flow; and (3) where thickest, the lava oozed down the caldera wall (see *Timing of caldera collapse* and Fig. 13).

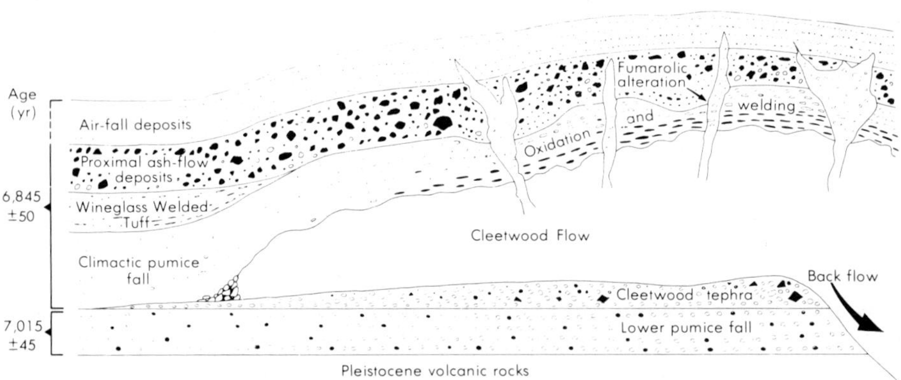

Fig. 8. Schematic diagram of stratigraphic relations at Cleetwood Cove (not to scale). Oxidation and welding of climactic pumice fall and fumarolic alteration of all units overlying lava flow indicate that: (1) lava was hot at time of climactic eruption; and (2) all events from emplacement of Cleetwood flow through ejection of intracaldera debris took place over a short interval. Cleetwood tephra consists of near-vent deposits of Cleetwood eruptive center.

Significance of preclimactic rhyodacites to the climactic eruption

Vent distribution

Precursory events that may forewarn of catastrophic pyroclastic eruptions are important for evaluating potential volcanic hazards. Crater Lake is the only collapse caldera in the late Quaternary Cascades, excluding the neighboring and significantly different Newberry and Medicine Lake Volcanoes. Although other volcanic centers, notably South Sister, have erupted silicic lava, the vents for these extrusions commonly define linear trends that suggest control by tectonic stress regimes. Mount Mazama alone erupted a series of late Pleistocene and Holocene silicic lava flows and pyroclastic rocks whose documented vents form a diffuse cluster or lie on radial trends. Furthermore, the compositions of the Holocene rhyodacitic rocks are very similar and evidently reflect venting of the uppermost part of a magma chamber. The four rhyodacitic vents (Fig. 6B) suggest an arc but the two thin rhyodacitic dikes below Llao Rock are more nearly radial to the caldera than parallel to the walls and the Pleistocene Forgotten Crater and Sharp Peak vents lie on radial trends (Fig. 6A). Conceivably, other Holocene rhyodacitic vents existed within the area of the caldera, and the arc is an artifact of eccentric collapse. Regardless of the existence of an arc of vents, the pattern of silicic vents reflects the local stress field, established in the roof rocks above an upper-crustal magma chamber several km in diameter, dominated by stresses related to that body of magma rather than by regional tectonics.

Williams (1941, 1942) emphasized the arcuate nature of the preclimactic vent pattern. His "northern arc of vents" includes the Llao Rock, Grouse Hill, Cleetwood, and Redcloud eruptive centers. However, Williams (1941, 1942) also chose to include the Watchman flow, Hillman Peak complex, and Sentinel Rock in this arc because the vent systems for these features are nearly bisected by the caldera walls. Because these vents are much older, their positions are irrelevant to the dimensions of the climactic magma chamber.

Initiation of the climactic eruption

Stratigraphic relations indicate that the climactic eruption began shortly after extrusion of the Cleetwood flow. Slow extrusion of preclimactic rhyodacitic magma from the top of the chamber may have taken place when internal pressure of the accumulating silicic magma overcame the sum of the minimum horizontal compressive stress and the strength of the roof. By the time it reached the surface, most of the preclimactic rhyodacitic magma was largely degassed because of sluggish flow in conduits, and this viscous magma solidified and plugged feeders that had initially erupted pumice. The eruption of the lower pumice fall may have lasted longer and produced a much larger volume of ejecta than the other preclimactic rhyodacitic eruptions because entrainment of mafic inclusions had decreased the viscosity of the rhyodacitic magma, either owing to heating or, possibly, to increase in water content caused by crystallization of the inclusions; the lowered viscosity would have allowed the relatively high flow rate in the conduit necessary for a sustained Plinian eruption.

Because of complete temporal, compositional (Fig. 7), and mineralogic continuity between the preclimactic rhyodacite and pumice of the climactic eruption, something other than composition must have triggered the catastrophic pyroclastic eruption. A sustained Plinian eruption was apparently made possible here by the extrusion of the ~ 4—12 km³ of magma represented by the preclimactic rhyodacites over a period as short as 100 to 200 yr. Presumably, this volume of magma had to be removed from the top of the system before pressure decrease and vesiculation could take place at sufficient depth to feed a voluminous Plinian eruption.

THE CLIMACTIC ERUPTION

The climactic, caldera-forming eruption of Mount Mazama can be divided into two phases: single vent and ring vent. Onset of collapse of the caldera probably caused the switch from the first to the second phase of eruption; the timing of this switch can be documented through stratigraphic relations.

Single-vent phase

Deposits of the single-vent phase consist of two units erupted successively from the same vent area: the climactic pumice fall and the Wineglass Welded Tuff of Williams (1942). The climactic pumice fall represents air-fall deposition from a high Plinian column; the Wineglass Welded Tuff was deposited by ash flows derived from a collapsed column (Fig. 9).

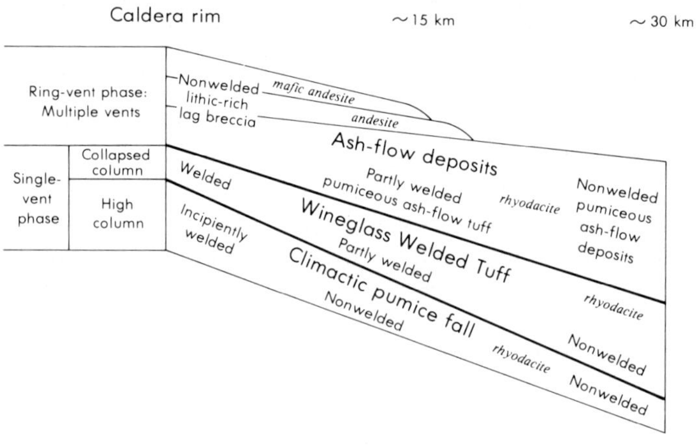

Fig. 9. Schematic diagram illustrating facies of deposits of climactic eruption. Thicknesses not to scale; relative thickness of Wineglass Welded Tuff inferred at right.

Climactic pumice fall

Deposits of the climactic pumice fall account for a large part of the material erupted during the single-vent phase. These deposits consist of bedded air-fall pumice lapilli and blocks up to 60 cm in diameter and lithic fragments to over 1 m in diameter at the caldera rim. Because of subsequent erosion by pyroclastic flows, the $\lesssim$ 20-m thickness of the deposits at the rim is generally a minimum. In areas of low elevation at the time of the climactic eruption, particularly on the north caldera rim, nonwelded ash-flow deposits are interbedded with the air-fall pumice. These small-volume, local ash flows probably resulted from fallback of the margins of the eruptive column or from avalanching of near-vent air-fall material.

The widespread Mazama tephra has been used extensively as a stratigraphic marker (Powers and Wilcox, 1964; Kittleman, 1973). Much of the climactic pumice fall was carried to the northeast (Williams, 1942; Fisher, 1964; Lidstrom, 1971); only sparse, thin patches of redeposited pumice are present near the caldera rim around the southwest quadrant. The informally-named Mazama ash is > 50 cm thick at Newberry Volcano, 110 km from Crater Lake, and > 1 cm thick in southwestern Saskatchewan, 1200 km from its source. Part of the distant deposits may be fine ash from the ring-vent phase of the eruption (Sparks and Walker, 1977), and some minor layers (cf. Davis, 1978; Blinman et al., 1979; Mack et al., 1979) may represent the earlier eruption from the Llao Rock eruptive center. The total volume of magma erupted in the Plinian part of the single-vent phase was probably ~ 20—30 km^3 (see *The volume problem*).

Virtually all the juvenile material erupted during the single-vent phase was rhyodacite, containing between 70 and 71% SiO_2 (Fig. 7). A small proportion consists of phenocryst-rich rhyodacite pumice and andesitic scoria. Although the scoria is rare, bombs as large as 90 cm in diameter have been found as low as ~ 4 m above the base of the deposit within relatively coarse and incipiently welded air-fall beds; these bombs are apparently not merely vesiculated mafic inclusions of the variety found in the Llao Rock flow. Very rare banded pumice blocks and partially fused plutonic clasts are also present ~ 4 m above the base. Analyzed scoria bombs from the climactic pumice fall contain 59 to 62% SiO_2 (Fig. 7). Their presence indicates that mafic magma somehow was tapped early in the climactic eruption, perhaps in the manner suggested by Blake (1981). Vesiculating magma left the vent with sufficient velocity that pumice reached adequate height to cool enough to prevent significant welding upon deposition at what became the caldera rim. Nevertheless, pumice blocks record a paleomagnetic-pole position indistinguishable from that of the ash-flow deposits of the ring-vent phase (D.E. Champion, written commun., 1982).

The distribution of coarse pumice blocks and lithic fragments suggests that the vent for the single-vent phase was situated slightly northeast of the center of the caldera (Fig. 6C). Exposures of related nonwelded ash-flow deposits are generally restricted to the north wall of the caldera. The most compelling evidence, however, for the position of this vent, and the implication that there was only one vent for the single-vent phase, comes from the distribution of the Wineglass Welded Tuff.

Wineglass Welded Tuff

Although Williams (1942, pp. 60—63) recognized that the Wineglass Welded Tuff was an ash-flow tuff deposited soon after the underlying air-fall pumice, he failed to link these deposits to the climactic eruption or to recognize their extent. Pumice of the Wineglass Welded Tuff is virtually identical in composition to that of the underlying air-fall-pumice bed (Fig. 7). The lithologic homogeneity of exotic blocks in the Wineglass Welded Tuff argues for a single vent. Ash flows that deposited the Wineglass locally scoured out shallow channels, but field relations show that little time elapsed between cessation of air-fall and onset of ash-flow deposition.

The Wineglass Welded Tuff consists of several flow units that form a single cooling unit (Smith, 1960a). Red oxidized andesitic lithic fragments, as large as ~ 1 m in diameter, commonly are concentrated near the bases of flow units. The tuff is partly to densely welded throughout all but the lowermost ~ 30 cm. In most places, the top has been eroded down to rather densely welded tuff by subsequent ash flows; locally, the Wineglass grades upward into proximal-facies ash-flow deposits of the ring-vent phase containing a variety of coarse lithic fragments. In partly welded exposures on the caldera rim, the tuff appears to be relatively poor in fines. The Wineglass is thickest (max ~ 10 m on the caldera rim) and most densely welded in old valleys, and thin or absent on topographically high areas; it is present in topographic depressions from Pumice Point clockwise around the caldera rim to Skell Head.

Partly welded Wineglass is exposed beneath proximal-facies ash-flow deposits of the ring-vent phase in gullies northeast of the caldera as far as ~ 3 km from the rim and at the mouths of canyons as much as ~ 11 km east of the caldera. The Wineglass is several m thick in the canyon-mouth exposures and can be traced a short distance eastward where it disappears beneath pumiceous ring-vent-phase ash-flow deposits. It would seem likely that the Wineglass is a much more voluminous deposit than might be deduced from study of the caldera-rim exposures alone. The volume cannot be accurately estimated, but a value of several km^3 is probably appropriate; this volume is of necessity included in the total ash-flow volume given in Table III.

The distribution of the Wineglass Welded Tuff provides the main constraint on vent location and the topography of Mount Mazama during the single-vent phase. Although deposition of the climactic pumice fall was subject to the whim of wind direction, that of the ash-flow tuff was determined by the topography around the vent. Ash flows swept freely to the north from the amphitheater bounded by the Llao Rock and Redcloud flows, but were channeled between these lava flows and a topographic high to the south (Fig. 6C). The ash flows were apparently incapable of surmounting the barrier of eroded stratocones that extended from Hillman Peak to Sentinel Rock.

The compositional continuity between pumice of the Wineglass Welded

Tuff and the climactic pumice fall and the similarity of lithic fragments in the two deposits suggest that the ash flows that deposited the Wineglass formed by collapse of the Plinian column (Sparks and Wilson, 1976) that issued from a single vent. Because the pumice compositions are identical it is unlikely that a decrease in volatile content caused the column to collapse; rather, increase in vent radius, owing to erosion or slumping, may have been responsible for the change in mode of eruption (Sparks et al., 1978). Comparatively intense welding occurred because the material was not ejected to sufficient height for effective cooling before cascading down topographic lows; restricted mobility in comparison to ring-vent phase ash flows was also a consequence of relatively low height of collapse. Cessation of eruption of the Wineglass was evidently brought about by the sudden inception of caldera collapse (see *Timing of caldera collapse*).

Ring-vent phase

The voluminous ash-flow deposits of the climactic eruption, emplaced on top of the Wineglass Welded Tuff, flowed down all major drainages around Mount Mazama (Fig. 6D). Field relations show that these ash flows were erupted from a ring of vents as the caldera began to collapse. Deposition of phenocryst-rich mafic scoria at the close of the ring-vent phase suggests that the eruption was terminated when magma became too viscous to be disrupted by explosive fragmentation; i.e., the maximum viscosity barrier (Smith, 1979) was reached. The ash-flow deposits can be divided laterally into proximal, medial, and distal facies and vertically into zones of dominantly silicic, mixed, and mafic flows (Fig. 9). The total volume of magma represented by these ash-flow deposits and the Wineglass Welded Tuff, which cannot be distinguished from them where buried, was probably $\sim$ 13 km^3 (see *The volume problem* and Table III).

Proximal facies

Proximal-facies deposits of the ring-vent-phase ash flows occur all around the caldera rim, on the slopes of Mount Mazama, and in the heads of valleys south and east of the caldera. These deposits grade abruptly into medial facies, valley-filling ash-flow tuff, generally within $\sim$ 2—11 km of the caldera rim. The proximal deposits consist of varied lithic fragments (max $\sim$ 7 m) and dense juvenile clasts (max $\sim$ 2 m) within a sparse crystal-rich matrix that grades from pumiceous upward to scoria-bearing; in some areas, the deposits are at least 20 m thick. These deposits resemble breccias described from the rims of other calderas, e.g.: the Kamewarizaka breccia. Aira Caldera, Japan (Aramaki, 1969; Yokoyama, 1974); proximal breccias of the Aso 4 pyroclastic flow, Aso Caldera, Japan (K. Ono, oral commun., 1981); and products of the Cape Riva eruption, Santorini, Greece (Druitt and Sparks, 1982). Although they evidently are more indurated, in many other respects, including being preceded by a Plinian air-fall deposit and welded ash-flow tuff, the Cape Riva deposits resemble those at Crater Lake. The Mazama proximal deposits differ from the "lag-fall deposits" of Wright and Walker (1977) in that they generally show evidence of transport in pyroclastic flows. Channels cut into earlier deposits of the climactic eruption and filled with proximal ash-flow deposits are common (see *Erosion by pyroclastic flows*). Some dense material, however, may have originally accumulated by a fall mechanism before being picked up by the ash flows. Williams (1942) thought the lithic-rich lower part of the proximal-facies deposits to be glacial till, although in many places he qualified his interpretation with "probable" or a query.

A layer of fine crystal fragments and ash, typically < 1 m thick, overlies the scoria deposits. This layer is thin bedded, appears to be well sorted, is oxidized to pinkish or purplish gray, may contain some dune structures, and probably represents fine ash derived from clouds associated with the climactic ash flows (equivalent to layer 3 of Sparks et al., 1973).

Well-sorted crystal and pumice lapilli beds above the proximal-facies deposits on the caldera rim record the explosive disruption of intracaldera debris as the caldera settled to rest; these beds are not included here in the proximal facies. Some of this explosive activity probably ensued from disturbance of the hydrologic regime of Mount Mazama and may have continued for a longer period than the entire magmatic part of the climactic eruption.

On the slopes of Mount Mazama, the proximal-facies deposits commonly form dunes. Crests of the dunes may be parallel (longitudinal) to or lie at an acute angle (transverse) to the presumed flow direction. Transverse dunes are most prominent upstream and in the lee of topographic obstructions, e.g. on the north side of Llao Rock and upstream of a dome in lower Scott Creek; such dunes commonly have amplitudes of ~ 10 m and wavelengths of ~ 100 m. Longitudinal dunes are most conspicuous in areas where ash flows may have accelerated and attained unusually high velocities, e.g. between Hillman Peak and Llao Rock.

As many as five lithic-rich flow units have been recognized in proximal-facies deposits on the caldera rim. In the best exposure, the bottom two flow units contain rhyodacitic pumice, the next two flow units contain hornblende-phyric andesitic pumice, and the top flow unit contains crystal-rich mafic andesitic scoria. Elsewhere, large clasts in the highest preserved flow unit consist almost entirely of the scoria. Many large scoria bombs have an outer zone, as much as a few centimeters thick, in which silicic pumice has mingled with the mafic scoria. Similar, but smaller and less abundant, lithic fragments occur in the medial-facies deposits, especially at the bases of flow units. In the heads of some valleys, where the abrupt gradation to the medial facies occurs, the transitional deposits show irregular lithic concentrations in crystal-rich ash similar to those described by Druitt and Sparks (1982, fig. 11), which these authors attribute to separation of lithic fragments from the fluidized bodies of pyroclastic flows (Wilson, 1980). The outer rinds of lithic fragments in the proximal deposits preserve a component of thermoremanent magnetization within the range expected for climactic ash-flow deposits (D.E. Champion, written commun., 1981). Corners of many clasts have been abraded, or spalled off owing to thermal shock. Moreover, evidence of fumarolic alteration in some of the proximal-facies deposits, even where they lie on glaciated andesite, indicates that they were hot when deposited.

The lithologies of clasts within the proximal-facies deposits correlate with their position around the caldera rim. Counts of several hundred clasts at each of seven sites show a predominance of altered andesitic clasts everywhere, whereas fresh dacite and andesite occur only from just south of The Watchman to Cloudcap; clasts of hornblende-bearing andesite agglutinate occur only from Munson Valley to Hillman Peak, in accord with a source in the Hillman Peak eruptive center. Thus, the climactic ash flows were erupted from either a vent several km in diameter or, more likely, from *multiple vents*, which probably formed a ring inboard of the present caldera wall (Fig. 6D).

Volumetrically insignificant but uniquitous in the proximal-facies deposits are partially melted plutonic clasts (Taylor, 1968; Lidstrom, 1971; Ritchey, 1979), from a few cm to ~ 2 m in diameter. They range from fine-grained felsite or aplite to coarse-grained granodiorite containing mafic inclusions; the clasts seem to vary more in lithology on the west rim. Most blocks on the east rim appear to be granodiorite. Although both biotite and hornblende have been identified in the blocks, the hydrous mafic minerals typically are so reconstituted as to be unidentifiable in the field.

Few of the plutonic blocks are dense; vesicular glass is present in most examples. Near the caldera, a breadcrusted rind of dense glassy rock is present; rounded vesicular blocks at more distant localities and in the medial-facies deposits apparently lost their rinds during transport. Rare millimeter-scale veins of green-brown glass cut some blocks. The partially fused plutonic blocks are interpreted as roof or wallrocks of the magma chamber that had been heated above their solidi by conduction. The small amount of water dissolved in the interstitial melt exsolved upon unloading during eruption to form vesicles, whereas the outer surfaces of the blocks were chilled in the eruption column. Many blocks are compositionally similar to Pleistocene intermediate volcanic rocks of Mount Mazama and may represent their intrusive equivalents; others may be pre-Quaternary.

A second, less abundant, more uniform population of angular, commonly breadcrusted blocks (max ~ 4 m) contains blocky euhedral plagioclase in vesicular glass. These blocks may have originated in a largely crystallized margin of the climactic magma chamber.

Another important constituent of the upper proximal-facies deposits is phenocryst-rich scoria. The proportions and species of phenocrysts vary widely among the blocks. These blocks include the most mafic climactic ejecta yet analyzed (Fig. 7; Ritchey, 1979, 1980); they probably represent border facies of the magma chamber in which crystals were mechanically concentrated or from which melt was been removed.

Medial facies

The medial facies of the ash-flow deposits best illustrates the spectacular compositional zonation of Mazama ejecta (Christiansen, 1979, fig. 1; Ritchey, 1980, fig. 1). The medial facies comprises three intergradational zones, each consisting of more than one flow unit: a lower zone (1) in which silky rhyodacitic pumice is dominant, the matrix is buff where nonwelded or gray where partly welded, and scoria and banded pumice also occur rarely; a middle zone (2), commonly partly welded, in which all three clast types are abundant and the matrix is gray; and an upper zone (3) with a dark gray matrix, dark scoria as the predominant juvenile clast, and silicic pumice blocks only as rare constituents. The silicic pumice in zones 2 and 3 is typically purplish gray owing to high emplacement temperature. Several percent of lithic fragments is present in most of the medial-facies deposits; locally derived lithic blocks occur in some exposures of zone 1. Medial deposits apparently are restricted to within $\lesssim$ 20 km of the caldera. The total thickness of medial deposits exceeds ~ 110 m in the deepest valleys, where the base is not exposed. Zone 1 commonly exceeds ~ 50 m in thickness, zone 2 reaches at least ~ 50 m, and zone 3 is commonly ~ 15 m thick.

Most exposures of medial-facies tuff are nonwelded to incipiently welded (indurated). Zones 1 and 2 are commonly partly welded as far as 17 km from the caldera (average flattening ratios of pumice clasts typically between 2.5 and 4 to 1). Widely spaced columnar joints that cut across juvenile clasts are characteristic of partly welded medial-facies tuff. Breaks between individual flow units are subtle, and the entire deposit consists of a single compound cooling unit (Smith, 1960a). Evisceration of the magma chamber must have been rapid, and deposition of the medial-facies tuff nearly continuous.

The entire compositional range of the climactic ejecta, except for the most mafic crystal-rich scoria blocks, appears to be represented by the medial-facies deposits. Chemical analyses of non-banded pumice and scoria blocks from all facies of the ring-vent phase deposits indicate a range in SiO_2 content of ~ 54 to ~ 71%, with a gap between 62 and 70%; one sample contained ~ 68% and another ~ 69% (Fig. 7). Ritchey (1979, 1980) reported lower SiO_2 contents in olivine-bearing crystal-rich blocks from the caldera rim (Fig. 7); literature data for other than mafic scoria were not plotted on Fig. 7 because possible heterogeneity of individual clasts could not be evaluated. The rhyodacitic pumice commonly contains glomeroporphyritic clots embedded in blebs of relatively coarsely vesicular brown glass more mafic than the surrounding colorless glass — similar clots are present in some samples from the preclimactic rhyodacitic lava flows.

Postemplacement crystallization appears to be minor in the Mazama ash-flow sheet, except for fumarolic alteration. Fumarolic pipes or sheets are abundant in the medial-facies deposits. Related crystallization has created erosion-resistant curtains and pinnacles in many canyons. White bleaching of pumice, owing to crystallization, surrounded by red oxidation is evident both in vertical and plan views, especially where the ash-flow deposits were laid down on wet ground in depressions. In some areas, vigorous fumarolic activity and secondary explosions caused silicic pumice blocks, as large as ~ 30 cm in diameter, to be blown up through zones 2 and 3 and deposited on top, and thus locally to obscure the mafic material underneath. A white bleached zone of fumarolic alteration (Smith, 1960a) several tens of cm thick is common near the caldera where fine ash overlies the poorly sorted ash-flow tuff. This fine ash, which can be as much as several meters thick, represents fines that settled from clouds containing elutriated and other airborne material associated with ash-flow emplacement (Williams, 1942). The fine ash served as a relatively impermeable cap on fumaroles in the underlying tuff (cf. Smith 1960a, p. 157).

Oxidized fine-grained crystal-rich deposits, $\lesssim$ 2 m thick, are present at the margins of the ash-flow sheet low on the sides of deep valleys near the caldera. Similar deposits on slopes east of the caldera consist of dense, but comparatively fine material left by passage of ring-vent-phase ash flows.

The heterogeneity of juvenile clast types in all zones of the medial facies demonstrates complex physical mixing of ejecta during eruption and emplacement. Some of the heterogeneity within flow units may be ascribed to incorporation of previously deposited air-fall and ash-flow material (see *Erosion by pyroclastic flows*). In addition, some rhyodacitic pumice blocks are composites of many smaller pumice clasts, as if derived by redeposition of partially agglutinated blocks and lapilli that may have fallen during the single-vent phase. Banded pumice indicates deeper physical mixing or commingling, probably near the interface between mafic and silicic magma in the chamber concurrently with the eruption, and in conduits during the eruptive process itself. Scoria bombs with outer zones of mingled silicic pumice and mafic scoria are present in zone 1 and common in zone 2. Either the eruption tapped many levels in the chamber at the same time (Blake, 1981) or mafic magma was somehow erupted through silicic magma.

Distal facies

Ash-flow deposits of the distal facies contain silicic pumice as the predominant juvenile component. This facies is thus equivalent to zone 1 of the medial facies. The transition between these facies is, however, gradational, and areas characterized by a poorly developed zone 2 and no zone 3 cannot be conveniently grouped with either facies.

The distal-facies deposits extend as far as ~ 60 km from the caldera; the deposits normally are nonwelded, and contacts between flow units are commonly obscure. Carbonized wood locally is abundant where emplacement temperature was low. Secondary explosion craters, with attendant surge deposits and related fumarolic alteration, are conspicuous where ash flows argued with the Rogue River, ~ 20 km west of the caldera (Williams, 1942, pl. 17, fig. 3; pl. 19, fig. 1).

Pumice blocks locally accumulated at the tops of flow units in the distal-facies deposits and near valley walls. Some deposits near the distal ends of the ash-flow deposits contain relatively small pumice clasts and a large proportion of fine material. Fine-grained well-sorted ash-cloud deposits overlie and many extend beyond the ash-flow deposits.

Mobility of the ash flows

The ash flows of the ring-vent phase were far more mobile than those that deposited the Wineglass Welded Tuff. The ring-vent phase flows surmounted a 200-m-high ridge ~ 15 km south-southwest of the caldera and poured into the valley of Red Blanket Creek. The mobility and carrying power of the ash flows is illustrated by: (1) common blocks of altered andesite (max ~ 2 m), from at least as deep as the lower caldera walls, lying ~ 200 m above the surface of the ash-flow deposits on Red Cone 4 km from the caldera; (2) scoria bombs (max ~ 30 cm) in the crater of the Crater Peak cinder cone 6 km from the caldera and 400 m above the ash-flow deposits in the adjacent valley; and (3) a lag of andesitic lithic fragments (max ~ 20 cm) overlying the climactic pumic fall on the summit of Lookout Butte 9 km from the caldera and ~ 270 m above nearby valley-filling ash-flow deposits.

Eruption mechanism

The mode of eruption of the ring-vent phase can be inferred from the degree of welding of medial-facies tuff and the distribution of lithic fragments in proximal-facies deposits. According to Ritchey (1980) and Hildreth (1981), juvenile clasts and pre-eruption Fe-Ti oxide phenocryst equilibration temperatures of ~ 850 to ~ 970°C, yet welding was limited. Thus, the ejecta must have mixed with air sufficiently to cool substantially before coming to rest (Smith, 1960b), but not so much as to preclude partial welding of medial-facies tuff. The column-collapse models of Sparks et al. (1978) appear to reproduce the observations on welding and mobility of both the single-vent and ring-vent-phase ash-flow deposits, recognizing that these models would have to be modified for direct application to a multiple-vent eruption. The ring-vent phase of the eruption would have been characterized by very high eruption rate and relatively great column height. The high column would have maximized cooling and would have provided sufficient fall to give the ash flows the momentum needed to surmount topographic obstacles.

Erosion by the ash flows

Many examples of erosion by pyroclastic flows can be documented at Crater Lake. Deposits of the climactic pumice fall are > 10 m thick on the upper surface of Llao Rock but thin at lower elevations, symmetrically about the summit, as viewed from the lake (Fig. 5). Bedding is highlighted by accumulation of dark material from the overlying proximal-facies ash-flow deposits and is truncated by mafic debris where air-fall pumice has been eroded by passage of ash flows in the lower areas on either side of the summit of Llao Rock. Similar relations are less conspicuous on other topographic highs around the caldera or where hills sloped toward the caldera. On the west surface of the Cleetwood flow, deep channels, now filled with proximal-facies ash-flow material, have been cut into climactic-pumice-fall deposits (Fig. 8). Such channels are also evident where ash flows scoured the surface of thick Wineglass Welded Tuff in old valleys on the north caldera walls.

The rhyodacitic lava flow and dome of Grouse Hill also show the erosive capability of the climactic ash flows (Fig. 10). When the climactic eruption began, Grouse Hill was apparently $\lesssim$ 200 yr old and must have been virtually uneroded, and its outer shell of pumiceous blocky lava intact.

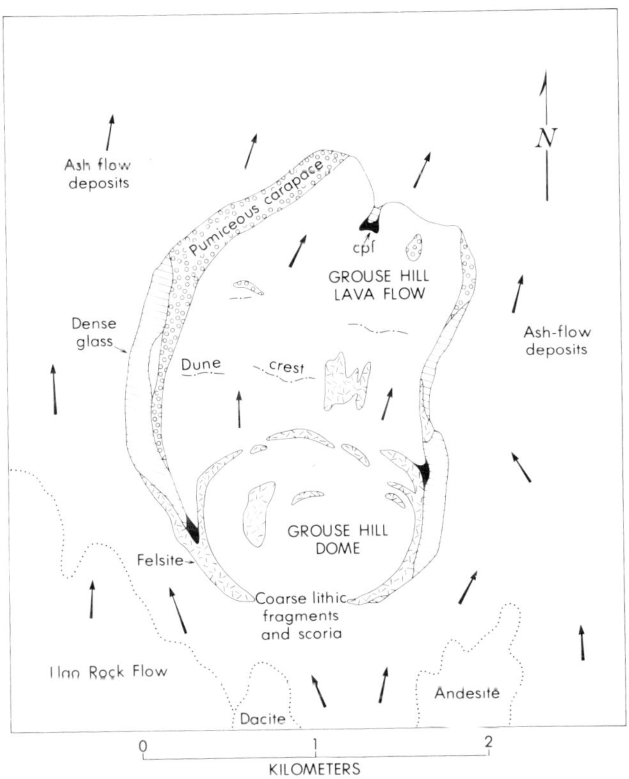

Fig. 10. Geologic map of Grouse Hill lava flow and dome. Arrows show inferred direction of transport of ash flows of ring-vent phase of climactic eruption; only most prominent dune crests are shown for clarity. Blocks of altered andesite reach 7 m in maximum dimension in area labeled "coarse lithic fragments and scoria". "Dacite" is dome of hill 7352′ (Fig. 6A); *cpf* = climatic pumice fall (area of outcrop exaggerated). Erosion by pyroclastic flows has exposed progressively deeper zones of crystallization in rhyodacite as caldera is approached.

During the single-vent phase, air-fall pumice blanketed Grouse Hill to a depth of several m. Shortly thereafter, ring-vent-phase ash flows stripped much of this air-fall deposit from the caldera-facing side of Grouse Hill and quarried all loose material from the southwest and southeast flanks of the lava flow. The north flanks of the flow retain their pumiceous carapace and are partly buried by air-fall and ash-flow deposits. Progressively deeper zones within the flow, exposed southward on either side, reveal first blocky dense porphyritic glass midway and then massive holocrystalline felsite nearest the caldera (Fig. 10). A major accumulation of immense lithic blocks (max ~ 7 m) and large scoria bombs lies banked against Grouse Hill on the caldera side. In the lee, a series of dunes of finer but lithologically similar proximal-facies material were constructed. Grouse Hill acted as a giant boulder in an ash-flow stream laden with ultracoarse abrasive. Similar features on the Cleetwood flow and Redcloud flow are less spectacular than at Grouse Hill. The erosion at low elevations is so impressive that it is no wonder that Williams (1942) thought the rhyodacite flows had been glaciated, although he was aware that pyroclastic flows are capable of eroding their substrate (e.g., Williams, 1942, pp. 85—86).

Timing of caldera collapse

Field relations near Cleetwood Cove document the collapse of Crater Lake caldera relative to the progress of the climactic eruption. The rhyodacitic Cleetwood flow (Fig. 6B) was still hot when the climactic pumice fall landed on its surface. Pumice high in the deposit is normal in appearance but becomes progressively oxidized, flattened, and welded downward, and is brick red and compressed on the flow surface and in cracks between lava spines (Fig. 8). These features are seen only on the Cleetwood flow and are not due to proximity to the vent for the climactic pumice fall. The entire climactic eruption and collapse of the caldera took place before complete cooling of the Cleetwood flow. Fumarolic alteration, brought about by degassing of the lava, affects not only the climactic pumice fall but also the overlying proximal-facies ash-flow deposits. Moreover, where the lava was thickest, its interior was sufficiently hot when the caldera formed that a great tongue oozed down the caldera wall at the apex of Cleetwood Cove; where thinner nearer the margins of the flow, the lava fractured to form the cliffs above either side of Cleetwood Cove (Fig. 11). This "backflow" moved as pasty lava because columnar joints formed approximately perpendicular to the sloping caldera wall, flow banding dips gently across the mass, and outcrops at the shoreline preserve the climactic paleomagnetic-pole position (D.E. Champion, written commun., 1981). Interestingly, J.S. Diller also interpreted this feature as a backflow (Diller and Patton, 1902); Williams (1942) argued that it was the feeder for the Cleetwood flow. The vent that fed the Cleetwood flow probably lies a few hundred meters north-northeast of the bend in the caldera rim at Cleetwood Cove

Fig. 11. Cleetwood Cove, showing symmetrical "wings" of Cleetwood flow and backflow in center. Arcuate rift valleys on surface of lava flow are visible beyond caldera rim. View northward.

(Fig. 6B), beneath low hills. One of these hills consists of oxidized agglutinated bombs of rhyodacite, suggesting a secondary vent which may have become active when the surface of the flow was disrupted.

The brittle surface of the Cleetwood flow broke up into a series of rift valleys as lava withdrew from the interior and flowed down the caldera wall. The rifts are as much as ~ 250 m wide and ~ 20 m deep, are concave toward the caldera, increase in width from the margin to the center of the lava flow, and are wider and more closely spaced near the caldera rim. Partly welded Wineglass ash-flow tuff, $\lesssim$ 2 m thick over part of the lava flow, had cooled to the point of brittle failure when the rifts formed because a horizontal bed of the tuff locally crops out on the walls of rifts of all widths. Collapse of the caldera thus began after some cooling of the thin partly welded Wineglass unit (although there may have been a slight lag in response of the lava flow crust north of the caldera to initiation of the backflow).

Proximal-facies climactic ash-flow deposits are present over much of the surface of the Cleetwood flow. These deposits are cut by rifts near the lateral margins of the lava flow but in the broadest rift valley are banked against rifted Wineglass Welded Tuff (Fig. 12). Thus, the caldera had begun to collapse, and the rifts had formed owing to backflow of Cleetwood lava into the caldera, before the ring-vent phase of the climactic eruption terminated.

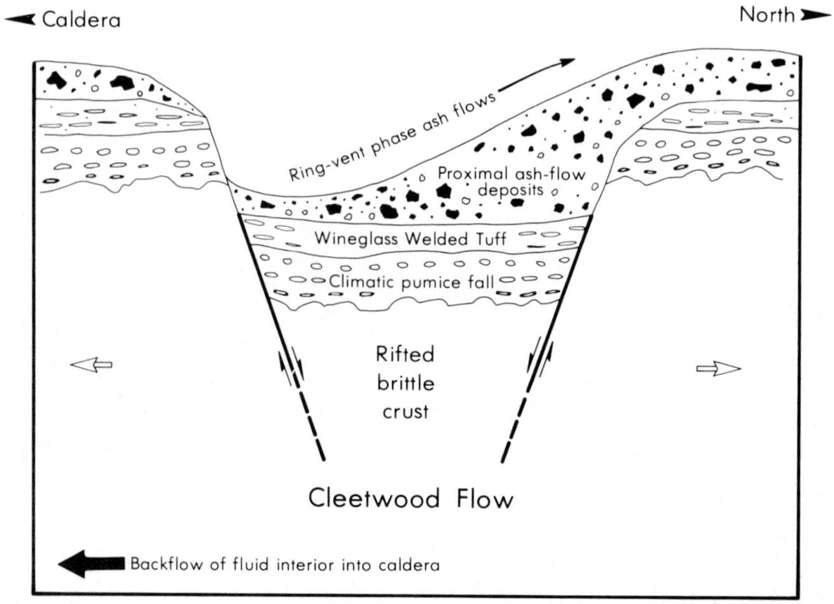

Fig. 12. Schematic north—south section through upper part of Cleetwood flow, showing relations between deposits in rift valleys. Overlap of faulted Wineglass Welded Tuff by ash-flow deposits of ring-vent phase indicates caldera began to collapse before last ash flows passed over Cleetwood flow.

Yet another clue to the timing of collapse is offered by outcrops of Wineglass Welded Tuff on the caldera rim near the Palisades. Gash fractures in the upper surface of the welded tuff parallel to the rim indicate that here, where the Wineglass was comparatively densely welded and thick, it too began to slump into the caldera before cooling to the point of complete rigidity. Moreover, at the Wineglass, the smooth caldera-facing joint surface of Wineglass Welded Tuff shows that the welded matrix had cooled enough to fracture under a high strain rate when the caldera collapsed; flattened pumice clasts, however, were still capable of flow and the dense supercooled melt oozed downward slightly over the joint surface (cf. Watanabe, 1978). The collapse of Crater Lake caldera clearly began virtually when emplacement of the Wineglass Welded Tuff ended and was well underway before the last ash flows of the ring-vent phase rushed over the Cleetwood flow.

Collapse versus enlargement of a funnel-shaped vent

Some controversy exists over the mode of formation of small- to medium-diameter calderas (e.g., Yokoyama, 1981, 1982; Self and Rampino, 1982). Some workers have described drill-hole and geophysical evidence for the funnel shape of caldera floors and have interpreted this evidence as indicating formation by enlargement of a vent (Yokoyama, 1969, 1981), as opposed to enlargement of a fault-bounded subsidence block by slumping of the walls (Lipman, 1976) or decreasing displacement along concentric faults toward caldera margins. Whatever the form of the floor below the postcaldera lavas and intracaldera tuff at Crater Lake, the caldera must have formed by collapse. Williams (1942) showed that the volume of lithic ejecta beyond the caldera was far less than even conservative estimates of that of the missing part of Mount Mazama; this conclusion is not affected by inclusion of the volume of lithic fragments in the proximal-facies ash-flow deposits. Moreover, the caldera walls are nearly vertical in many places for heights of as much as ~ 1000 m. The scalloped rim of the caldera apparently dates largely from the time of its formation, if the excellent preservation of the Cleetwood backflow (Fig. 11) can be taken as typical. Steeply dipping faults concentric to, and with displacement down toward, the caldera near Garfield Peak may also be related to failure of the walls during caldera formation. The distribution of lithic fragments in the proximal-facies ash-flow deposits and the locations of postcaldera vents indicate that major collapse of the roof of the climactic magma chamber probably occurred within a ring ~ 5 km in diameter (Fig. 6D), whether piecemeal or as a piston in a cylinder. Contemporaneous landsliding and caving of the walls enlarged the caldera to virtually its present size and created its scalloped outline. Calderas smaller than Crater Lake probably range from true collapse features, which erupted from many vents, to single vents enlarged by slumping of their walls (Novarupta?; Hildreth, 1983, this volume).

Radiocarbon ages

Radiocarbon ages of organic material associated with deposits of "Mazama ash" have been reported by many workers since the early days of ^{14}C dating. Most of these age determinations have been made either on (1) pieces of carbonized logs from ash-flow deposits or on (2) peat above or below ash layers. Results on both types of material are commonly open to question because of sampling problems and various assumptions. Charcoal fragments derived from twigs or small branches and intimately associated with pyroclastic deposits minimize opportunities for nonrandom error in the ^{14}C dating of volcanic events.

Table II lists six new age determinations relevant to Holocene Mount Mazama. Samples 1 through 5 constrain the date of the climactic eruption. Stratigraphic relations near the caldera indicate that the ring-vent phase of the eruption followed the single-vent phase within a very short time, certainly less than that resolvable by radiocarbon dating. Independent evidence from a study of pollen influx in Lost Trail Pass bog, Montana, places a maximum of 3 yr (Mehringer et al., 1977) on the period of ash fall associated with the climactic eruption; consideration of likely eruption rates (Wilson et al., 1978) suggests a duration of at most a few days if

TABLE II

Radiocarbon-age determinations

Sample	Laboratory No.	Location (lat. N., long W.)	Eruptive unit	Description	Age (yr ± 1σ)
1	W-4288	43° 15′, 122° 25′	CAF	Fragments in ash-flow deposit; flow unit overlies that of sample 2.	6780±100
2	W-4290	43° 15′, 122° 25′	CAF	Fragments in ash-flow deposit; flow unit overlain by that of sample 1.	6830±110
3	W-4255	43° 15′, 122° 25′	CAF	Small branch in ash-flow deposit; flow unit overlain by that of sample 2.	6880±70
4	W-4256	43° 17′, 122° 34′	CAF	Outermost 5 cm of log in ash-flow deposit.	7000±60
5	W-4295	43° 15′, 122° 21′	CPF	Twigs in air-fall tephra; tephra overlain by flow unit of sample 3.	6840±100
6	USGS-870	42° 58′, 122° 03′	LPF	Twigs in soil immediately below tephra	7015±45
7	USGS-872	42° 58′, 122° 04′	—	Soil immediately below tephra	25,480±580

CAF, climactic ash flows; CPF, climactic pumice fall; LPF, lower pumice fall. Samples 1—5 collected in 1978 by D.R. Crandell from roadcuts along Oregon Highway 138, North Umpqua River Valley, and analyzed in 1979 at the U.S. Geological Survey laboratory, Reston, Virginia, under the supervision of Meyer Rubin. Sample 6 collected in 1979 by C.R. Bacon from trench at the Wineglass; sample 7 collected in 1979 by C.R. Bacon from soil beneath mafic tephra below the Palisade flow, Crater Lake National Park. Samples 6 and 7 were analyzed in 1980 and 1982, respectively, at the U.S. Geological Survey laboratory, Menlo Park, California, under the supervision of S.W. Robinson; for sample 6, $\delta^{13}C$ was measured at —24.17 per mil, and the age was corrected for isotope fractionation. All ages based on Libby half-life (5568 ± 30 yr), referenced to A.D. 1950.

there were no lengthy pauses in the climactic eruption sequence. The weighted mean age of charcoal samples 1, 2, 3, and 5 is 6845 ± 50 yr; sample 4 was excluded from this calculation because its reported age differs significantly from those of the other four samples and because the log sampled could have died before the eruption (during the eruption of the lower pumice fall?).

The age of the lower pumice fall associated with the Llao Rock eruptive center is well constrained by the 7015 ± 45 yr date on sample 6 (Table II). The four radiocarbon ages reported by Mack et al. (1979) for peat samples bracketing two Mazama ash layers at Bonaparte Meadows, Washington, are consistent with the above ages for climactic and Llao Rock events. The finer grain size and lower percentage of crystals reported for the lower ash would follow from the much smaller magnitude of the earlier eruption. The nominal 170-yr interval between eruptions agrees within the quoted analytical uncertainties with the 142 yr deduced by Blinman et al. (1979), and confirmed at several other localities by P.H. Mehringer (oral commun., 1982), from palynological evidence.

The "volume problem"

Williams (1942, pp. 99—107) reported that the volume represented by the collapse of Mount Mazama to form Crater Lake caldera was more than twice the volume of magma ejected in the climactic eruption. He therefore suggested (p. 107) that "The summit of Mount Mazama was engulfed by withdrawal of magmatic support from below". Later upward revision of his estimate of the volume of ejecta (Williams and Goles, 1968) still led to a large discrepancy and resulted in retention of the original explanation. More recent data on the thickness and distribution of Mazama ash compiled by Lidstrom (1971), and my own work, prompt new estimates.

Table III lists the calculated volume of Mount Mazama lost by collapse and the estimated volumes of magma represented by products of the climactic magma chamber. The volume of the caldera was approximated by segments of frustums of inverted cones; the estimated ~ 38 km^3 is probably a minimum because the thicknesses of intracaldera tuff and collapse breccia and of postcaldera volcanic rocks are not known. The volume of Mount Mazama above the chosen bounds of the caldera was probably ~ 8—20 km^3; the accuracy of this estimate can be improved when current geologic investigations are completed. The volume of lithic fragments in the deposits of the climactic eruption is probably ~ 6 km^3 (Williams, 1942). A total of ~ 46—58 km^3 of rock must have been lost from Mount Mazama, of which ~ 40—52 km^3 were engulfed during caldera formation.

The volume of preclimactic rhyodacite is given as ~ 4—12 km^3 because the thicknesses of lava flows are poorly constrained, and the proportion of total Mazama tephra attributable to the lower pumice fall and to other explosive eruptions associated with the rhyodacitic lava flows is not yet

TABLE III

Estimates of volume

	Rock (km^3)		Magma (km^2)
Caldera	38	Preclimactic rhyodacites and lower pumice fall	4—12
Rest of Mount Mazama	8—20	Climactic pumice fall and distant ash	34
Lithic fragments in climactic deposits	6	Ash-flow deposits	13
Total lost by collapse	40—52	Total erupted from chamber	51—59

See *The volume problem* for explanation.

known. The volume of ash correlated with the lower pumice fall was estimated by assuming exponential decrease in thickness with distance and integrating over 180° of arc, using a range of estimates of original thicknesses of distant deposits; conversion to volume of magma, as discussed below for the climactic deposits, yielded ~ 2—10 km^3.

The volume of ejecta deposited as air-fall pumice and ash during the climactic eruption was calculated in two parts. The first part, based on Williams' (1942, fig. 16) isopach map, included the area within 100 km of the center of the caldera; isopachs at 1 and 2 inches were plotted by extrapolation, according to exponential variation in thickness with distance. Graphical integration yielded a volume of 20 km^3. This value was converted to a magmatic volume of ~ 7 km^3 by assuming densities of 2.4 and 0.8 g/cm^3 for rhyodacitic magma and unconsolidated pumiceous deposits, respectively; this value was reduced to ~ 5 km^3 because ~ 2 km^3 of magma may be attributed to the lower pumice fall within the area under consideration. In the second part, the volume of ejecta beyond 100 km was estimated to be ~ 102 km^3, from the compilation of Lidstrom (1971, fig. 5), extrapolated to 0.1-cm thickness and integrated graphically. Within this area, two ash layers are commonly identifiable, separated by other deposits, and the thickness of the lower one (lower pumice fall) is much less than that of the upper layer; consequently, all of the volume was attributed to ejecta of the climactic eruption. Assuming a density of 0.7 g/cm^3 for fine tephra yields a volume of magma of ~ 30 km^3. The volume of lithic ejecta included in these calculations is probably of the order of 1 km^3. The total volume is thus equivalent to ~ 34 km^3 of magma, disregarding any fine ash that might have fallen beyond the 0.1 cm isopach. Because some of this ~ 34 km^3 may have been derived from the ash-flow phases of the eruption

(Sparks and Walker, 1977), the volume erupted during the Plinian part of the single-vent phase was probably smaller. The volume of juvenile material in ash-flow deposits was taken from Williams (1942, p. 97) to be ~ 25 km^3; average density was arbitrarily taken as 1.2 g/cm^3, to give ~ 13 km^3 of magma. The total volume erupted from the climactic chamber was thus ~ 51—59 km^3 of magma. Uncertainty in this estimate could easily be ± 20% and it may be inappropriate to include the preclimactic rhyodacites in the erupted volume, but there is no "volume problem".

POSTCALDERA VOLCANISM

The only volcanic features in the Crater Lake area younger than the climactic eruption occur within the caldera. Williams (1961) interpreted the geology of the caldera floor on the basis of bathymetric data obtained by the U.S. Coast and Geodetic Survey (1959) and samples dredged by Nelson (1961). No subaerial tephra deposits correlative with postcaldera events have been recognized. Recent investigations (Nelson, 1967; Nelson et al., 1980; Williams and Von Herzen, 1983) lead to some important revisions.

Merriam Cone, a small (~ 0.1 km^3) olivine-bearing andesitic vent on the north margin of the caldera floor (Fig. 1), is the smallest of three distinguishable volcanic features. The caldera floor partly surrounding Merriam Cone owes its comparatively flat surface to sedimentary fill (Nelson et al., 1980), not to lava flows, as suggested by Williams (1961). Wizard Island forms the subaerial part of a voluminous pile of presumably similar andesitic material. The Wizard Island edifice accounts for ~ 1 to ~ 2 km^3 of magma, depending on where its base is drawn. Like Merriam Cone, the Wizard Island andesitic magmas were erupted near the inferred structural margin of the caldera.

The broad platform east of Wizard Island (Fig. 1) suggests silicic lava flows or coalescing domes. At the west end of this platform is a small dome composed of hornblende rhyodacite. The bathymetric map suggests that the platform may extend beneath the north side of the Wizard Island pile. Nelson (1961), however, did not report any Wizard Island tephra in dredge hauls from the platform or the rhyodacite dome, and so it may be that some silicic volcanism occurred after the last andesitic eruptions. The volume of silicic lava erupted was between ~ 0.5 and ~ 2 km^3, depending on its extent beneath and west of the dome. Erupted inboard of the two andesitic centers, the rhyodacite is about as differentiated as the most evolved magma of the climactic chamber (C.R. Bacon and R.L. Smith, unpubl. data, 1981). The postcaldera rhyodacite may represent either (1) uppermost climactic magma, which escaped eruption prior to formation of the caldera, or (2) the first leak from a new batch of silicic magma regenerated after the climactic eruption. A precedent exists for case 1 in the Novarupta dome that was emplaced in the vent area of the 1912 Valley of Ten Thousand Smokes eruption after the pyroclastic events (Hildreth, 1983). Case 2 seems unlikely but cannot be evaluated without more data.

Whatever the ages, relative or absolute, of the postcaldera eruptive units the Mazama magmatic system apparently remains the focus of volcanic activity for a large region. Areas of very high heat flow on the caldera floor provide evidence for ongoing hydrothermal circulation (Williams and Von Herzen, 1983), and the elevated chloride and sodium contents of lake water relative to other surface waters may indicate a component of geothermal fluid (J.M. Thompson, unpubl. data, 1981; Williams and Von Herzen, 1983).

A PROVISIONAL MODEL FOR THE MAZAMA MAGMATIC SYSTEM

The Mazama magmatic system apparently originated by focused injection of mafic magma into the crust. Complex feedback between heating of the lithosphere, in response to intrusion of asthenospheric magmas, and increased ease of "magmafracture" probably has been important (Shaw, 1980; Hildreth, 1981). A stress field, conducive to prolonged residence of magma in the crust once a focus developed, permitted growth and evolution of a long-lived open system. The Mazama focus may be the youngest in a series of volcanic foci extending eastward, subparallel to those described along the Brothers fault zone in central Oregon (MacLeod et al., 1976) and in the Lassen Peak area (L.J.P. Muffler, oral commun., 1981).

Evolution of the Pleistocene system

Prolonged heating of deep crustal rocks due to repeated injection of mafic magma, some of which lodges within the crust, leads to the growth of compositionally and thermally zoned magmatic systems (Smith and Shaw, 1975; Hildreth, 1981). Magmas of intermediate composition apparently form by concurrent differentiation of mafic magma, assimilation of partly or completely fused wall rocks, and magma mixing once a focal region reaches appropriate temperatures.

Vertical compositional zonation in high-level magma chambers has been documented for many silicic pyroclastic eruptions (see reviews by Smith, 1979, and Hildreth, 1981). Vertical zonation apparently existed within Pleistocene intermediate magma reservoirs at Mount Mazama, although only lava flows exist to record it. Evidence for such zonation consists of heterogeneities preserved by incomplete mixing of magmas of different compositions, whose inferred densities imply vertical zonation prior to the last mixing event (see *Mafic inclusions*). Mafic inclusions and disequilibrium features indicate that mixing, to varying degrees, in what was probably a system zoned from basalt below to andesite or dacite at its top was common during the evolution of Mount Mazama. Many eruptive sequences probably formed over a few hundreds or thousands of years as small reservoirs evolved, were partly emptied of low-density upper zones, and were repeatedly supplied with mafic magma from below. Shifting of comparatively short-lived eruptive centers created the complex of overlapping stratocones that built Mount Mazama. Preclimactic differentiation reached a maximum in dacites containing ~ 67% SiO_2, such as the "Pumice Castle" suite or the Watchman flow; these dacites show a complete gradation in major-element composition with less evolved intermediate magmas (C.R. Bacon, unpubl. data, 1981).

Compositional zonation of magma bodies is, above all, a consequence of thermal gradients imposed by protracted supply of heat from mantle-derived magma to a relatively cool environment, from which heat is lost by conduction, with or without the aid of fluid convection. Lateral gradients in composition follow, although they may be less extensive than vertical zonation in active crustal reservoirs. The shadow zone (Smith and Shaw, 1975), defined at any time by the inward limit of recently active mafic vents, shows the maximum extent of reservoirs with prolonged crustal residence of magma. Forgotten Crater provides a precise west limit for a time when olivine basalt interacted with the margin of a zoned intermediate system.

The climactic magma chamber

Crater Lake caldera provides the best evidence for a recent shallow magma chamber in the Cascade Range. Development of this chamber was a natural result of prolonged heating of the crustal environment of the Mazama magmatic system. Compositional data on the ejecta (Fig. 7; Williams, 1942; McBirney, 1968; Noble et al., 1969; Ritchey, 1979, 1980; Hildreth, 1981; C.R. Bacon, unpubl. data, 1982) document the zonation within the erupted part of the reservoir ~ 6900 yr B.P. The present tentative interpretation draws on these studies and, in particular, the conceptual models of Smith (1979), McBirney (1980), and Hildreth (1981).

The climactic magma chamber contained $\gtrsim$ 50 km^3 of moderately porphyritic rhyodacitic magma, which was zoned from ~ 72.5 to ~ 70% SiO_2; the bulk of the silicic magma contained ~ 70 to ~ 71% SiO_2.. Phenocryst-rich andesitic magma, zoned from ~ 62 to $\lesssim$ 56% SiO_2 (Fig. 7), underlay the rhyodacitic magma. The single-vent phase of the climactic eruption and the preclimactic lava flows and tephras tapped the rhyodacitic volume, with a small representation of the uppermost andesitic magma; the ring-vent phase brought up all types and, near its end, extremely crystal rich olivine-bearing blocks containing < 52% SiO_2 (Fig. 7). Despite the compositional gap, published Fe-Ti oxide equilibration temperatures appear to define a continuum (890–945°C, Ritchey, 1979; 875–975°C, Hildreth, 1981); however, most of the silicic pumice blocks studied by W. Hildreth (unpub. data, 1982) gave temperatures within ~ 10° of ~ 950°C. One low-silica scoria bomb yielded 850°C and a plutonic block 960°C (Ritchey, 1979). Recalculation of equilibration temperatures, using the Spencer and

Lindsley (1981) solution model, results in lower values but does not significantly change their relative order.

These temperature and compositional data imply that: (1) the interface between andesitic and rhyodacitic magma must have existed for a long time before the climactic eruption; and (2) the silicic volume was relatively well mixed, except near the boundaries of the chamber. The low-silica scoria blocks may represent either cumulates from a floored chamber or, more likely, mechanical accumulations of crystals from the margins of the system; the low oxide equilibration temperature and the high densities of the blocks suggest they were not derived from a crystal-enriched zone between andesitic and underlying basaltic magma. Scoria bombs in the climactic pumice fall, mafic inclusions in the rhyodacitic lava flows, and silicic andesitic scoria bombs in the ash-flow deposits all represent the uppermost mafic magma in the system (Fig. 7). Andesitic magma must have been injected into the overlying silicic magma, some of it $\gtrsim$ 140 yr before the climactic eruption, as indicated by mafic inclusions in the Llao Rock flow. The climactic eruption apparently drained virtually all the silicic magma from the system and stopped, not necessarily when all the magma was ejected, but when the maximum viscosity barrier was reached (Smith, 1979). Postcaldera eruption of moderately porphyritic andesite that lacks hornblende indicates, at the very least, resupply or regeneration of intermediate magma, if not the existence of some of this magma deeper in the system at the time of the climactic eruption.

The rhyodacitic magma probably developed as a boundary layer of low-density differentiated magma formed at the margins of the chamber and moved upward (Shaw, 1974) and, perhaps, as crustal rocks were partially fused deep in the system. Enrichment in H_2O of the entire erupted volume attended this process, as attested by the stability of amphibole in all the eruptive products (Ritchey, 1980). The most differentiated silicic magma accumulated at the top of the chamber; this magma was underlain by a much greater volume of convecting rhyodacitic magma represented by the bulk of the climactic ejecta.

Mafic magma apparently underlay the silicic magma. Mafic inclusions in the rhyodacitic lava, climactic scoria containing $> 58\%$ SiO_2, and climactic pumice containing $< 70\%$ SiO_2 may represent hybrid magmas, in part formed by mixing at the interface between mafic andesitic and rhyodacitic magmas, as suggested by resorption of olivine and growth of hornblende in mafic inclusions of the Llao Rock flow. Many of the petrographic features noted by Ritchey (1979, 1980) are consistent with instability of the crystals that grew in mafic andesitic or rhyodacitic magma, after mixing to form intermediate compositions: (1) phenocrysts in pumice and olivine and clinopyroxene in scoria are normally zoned, whereas reversely zoned hornblende (commonly mantling olivine or clinopyroxene) and orthopyroxene occur in scoria; (2) Fe-Ti oxides are common in scoria only in association with orthopyroxene and hornblende; and (3) plagioclase in pumice is relatively free of inclusions and is zoned from An_{55} to An_{30}, whereas plagioclase in scoria shows resorption, has glass inclusions, and ranges from An_{89} to An_{30}, rims being $> An_{70}$. The compositional gap between the silicic and underlying mafic magmas may thus be ~ 12% SiO_2. Presumably, basaltic magma, similar to that erupted at Forgotten Crater and other monogenetic basaltic cones, was present below the mafic andesitic magma and provided the source of heat necessary for growth and development of the rest of the system.

Approximately 15,000 to 40,000 yr elapsed during accumulation of the rhyodacitic magma. This period approximately agrees with that predicted by Smith (1979, fig. 12) on the basis of the estimated volume of magma expelled in the climactic eruption. Although the conditions that allowed the climactic magma chamber to develop are not known with certainty, the thermal-feedback focusing mechanism and continuity of magmatism in this region were evidently key factors. The compositional gap, not present to such an extent before ~ 50,000 yr B.P., suggests that a limiting minimum value of some key variable must be attained before efficient separation of silicic magma is achieved, and a large gravitationally stable silicic body can be established. The presence of hornblende in the climactic ejecta and its absence in most Pleistocene Mazama eruptive products suggests that the concentration of water may be the controlling parameter in promoting the style of convection (McBirney, 1980) envisioned for the climactic magma chamber.

ACKNOWLEDGMENTS

Howel Williams' classic study of the geology of Crater Lake National Park has provided a foundation for all who have studied Mount Mazama since 1942, and his encouragement during the early stages of my work is deeply appreciated. Collaboration with D.E. Champion and M.A. Lanphere has been exciting and fruitful. J.O. Davis, T.E.C. Keith, N.S. MacLeod, P.H. Mehringer, J.G. Smith, J.M. Thompson, R.E. Wilcox, and D.L. Williams have shared generously ideas and data. Work on the caldera walls was made possible through the help of W.A. Bartling. I thank Bartling, M.H. Moench, and S.W. Novak for able assistance in the field. Many of the observations and interpretations presented in this report have matured through discussion with numerous colleagues and friends, both on the outcrop and in the office; special thanks are owed to Wes Hildreth, R.L. Smith, and R.L. Christiansen. D.R. Crandell kindly provided 5 of the radiocarbon age determinations. Reviews by P.W. Lipman, W.E. Scott, and S.W. Novak helped to improve an early draft of the manuscript. The invaluable technical assistance of D.H. Johnson made timely preparation of the manuscript possible. Lastly, the cooperation of the staff of Crater Lake National Park, especially Mark Forbes and K.R. Cranson, has been instrumental in the field aspects of this study. I thank Superintendent J.S. Rouse for permission to work on the caldera walls.

REFERENCES

Adam, D.P., Sims, J.D. and Throckmorton, C.K., 1981. 130,000-yr continuous pollen record from Clear Lake, Lake County, California. Geology, 9: 373—377.

Aramaki, S., 1969. Geology and pyroclastic flow deposits of the Kokubu area, Kagoshima Prefecture. J. Geol. Soc. Japan, 75: 425—442 (in Japanese with English abstract).

Atwood, W.W., Jr., 1935. The glacial history of an extinct volcano, Crater Lake National Park. J. Geol., 43: 142—168.

Blinman, E., Mehringer, P.J., Jr. and Sheppard, J.C., 1979. Pollen influx and the deposition of Mazama and Glacier Peak tephra. In: P.D. Sheets and D.K. Grayson (Editors), Volcanic Activity and Human Ecology. Academic Press, New York, N.Y., pp. 393—425.

Blackwell, D.D., Bowen, R.G., Hull, D.A., Riccio, J. and Steele, J.L., 1982. Heat flow, arc volcanism, and subduction in Northern Oregon. J. Geophys. Res., 87: 8735—8754.

Blake, S., 1981. Eruptions from zoned magma chambers. J. Geol. Soc. London, 138: 281—287.

Campbell, N.P. and Bentley, R.D., 1981. Late Quaternary deformation of the Toppenish Ridge uplift in south-central Washington. Geology, 9: 519—524.

Christiansen, R.L., 1979. Cooling units and composite sheets in relation to caldera structure. In: C.E. Chapin and W.E. Elston (Editors), Ash-flow Tuffs. Geol. Soc. Am., Spec. Pap., 180: 29—42.

Colman, S.M. and Pierce, K.L., 1981. Weathering rinds on andesitic and basaltic stones as a Quaternary age indicator, western United States. U.S. Geol. Surv., Prof. Pap. 1210, 56 pp.

Couch, R., Gemperle, G., Connard, G. and Pitts, S., 1982. Structural and thermal implications of gravity and aeromagnetic measurements made in the Cascade volcanic arc. Geophysics, 47: 424.

Davis, J.O., 1978. Quaternary tephrochronology of the Lake Lahontan area, Nevada and California. Nevada Archeol. Surv., Res. Pap., 7, 137 pp.

Diller, J.S. and Patton, H.B., 1902. The geology and petrography of Crater Lake National Park. U.S. Geol. Surv., Prof. Pap. 3, 167 pp.

Druitt, T.H. and Sparks, R.S.J., 1982. A proximal ignimbrite breccia facies on Santorini, Greece. J. Volcanol. Geotherm. Res., 13: 147—171.

Eichelberger, J.C., 1980. Vesiculation of mafic magma during replenishment of silicic magma reservoirs. Nature, 288: 446—450.

Fiebelkorn, R.B., Walker, G.W., MacLeod, N.S., McKee, E.H. and Smith, J.G., 1982. Index to K-Ar age determinations for the state of Oregon. U.S. Geol. Surv., Open-File Rep. 82-596.

Fisher, R.V., 1964. Maximum size, median diameter, and sorting of tephra. J. Geophys. Res., 69: 341—355.

Heiken, G. and Eichelberger, J.C., 1980. Eruptions at Chaos Crags, Lassen Volcanic National Park. J. Volcanol. Geotherm. Res., 7: 443—481.

Heusser, L.E. and Shackleton, N.J., 1979. Direct marine-continental correlation: 150,000-year oxygen isotope-pollen record from the North Pacific. Science, 204: 837—839.

Hibbard, M.J., 1981. The magma mixing origin of mantled feldspars. Contrib. Mineral. Petrol., 76: 158—170.

Hildreth, W., 1981. Gradients in silicic magma chambers: Implications for lithospheric magmatism. J. Geophys. Res., 86: 10153—10192.

Hildreth, W., 1983. The compositionally zoned eruption of 1912 in the Valley of Ten Thousand Smokes, Katmai National Park, Alaska. In: S. Aramaki and S. Kushiro (Editors), Arc Volcanism. J. Volcanol. Geotherm. Res., 18: 1—56.

Hildreth, W. and Fierstein, J., 1983. Mount Adams volcano and its 30 "parasites". Geol. Soc. Am., Abstr. Programs, 15(5): 331.

Huppert, H.E., Sparks, R.S.J. and Turner, J.S., 1982. Effects of volatiles on mixing in calc-alkaline magma systems. Nature, 297: 554—557.

Kittleman, L.R., 1973. Mineralogy, correlation, and grain-size distributions of Mazama tephra and other postglacial pyroclastic layers, Pacific Northwest. Geol. Soc. Am. Bull., 84: 2957—2980.

Lawrence, R.D., 1976. Strike-slip faulting terminates the Basin and Range province in Oregon. Geol. Soc. Am. Bull., 87: 846—850.

Lidstrom, J.W., Jr., 1971. A New Model for the Formation of Crater Lake Caldera, Oregon. Ph.D. Thesis, Oregon State Univ., Corvallis, 85 pp.

Lipman, P.W., 1976. Caldera-collapse breccias in the western San Juan Mountains, Colorado. Geol. Soc. Am. Bull., 87: 1397—1410.

Mack, R.N., Okazaki, R. and Valastro, W., 1979. Bracketing dates for two ash falls from Mount Mazama. Nature, 279: 228—229.

MacLeod, N.S., Walker, G.W. and McKee, E.H., 1976. Geothermal significance of eastward increase in age of Upper Cenozoic rhyolitic domes in southeastern Oregon. In: Proceedings, Second United Nations Symposium on the Development and Use of Geothermal Resources, 1: 465—474.

McBirney, A.R., 1968. Compositional variations of the climactic eruption of Mount Mazama. In: H.M. Dole (Editor), Andesite Conference Guidebook. Oregon, Dep. Geol. Miner. Ind., Bull., 62: 53—56.

McBirney, A.R., 1978. Volcanic evolution of the Cascade Range. Annu. Rev. Earth Planet. Sci., 6: 437—456.

McBirney, A.R., 1980. Mixing and unmixing of magmas. J. Volcanol. Geotherm. Res., 7: 357—371.

Mehringer, P.J., Jr., Blinman, E. and Petersen, K.L., 1977. Pollen influx and volcanic ash. Science, 198: 257—261.

Nakamura, K., 1977. Volcanoes as possible indicators of tectonic stress orientation. J. Volcanol. Geotherm. Res., 2: 1—16.

Nelson, C.H., 1961. Geological Limnology of Crater Lake, Oregon. M.S. Thesis, Univ. Minnesota, Minneapolis, 175 pp.

Nelson, C.H., 1967. Sediments of Crater Lake, Oregon. Geol. Soc. Am. Bull., 78: 833—848.

Nelson, C.H., Larsen, M.C., Thor, D.R. and Wright, A.A., 1980. Sedimentary history of the floor of Crater Lake, Oregon: A restricted basin with non-channelized turbidites. Geol. Soc. Am., Abstr. Programs, 12(7): 491.

Noble, D.C., Drake, J.C. and Whallon, M.K., 1969. Some preliminary observations on compositional variations within the pumice- and scoria-flow deposits of Mount Mazama. In: A.R. McBirney (Editor), Proceedings of the Andesite Conference. Oregon, Dep. Geol. Miner. Ind., Bull., 65: 157—164.

Powers, H.A. and Wilcox, R.E., 1964. Volcanic ash from Mount Mazama (Crater Lake) and from Glacier Peak. Science, 144: 1334—1336.

Ritchey, J.L., 1979. Origin of Divergent Magmas at Crater Lake, Oregon. Ph.D. Thesis, Univ. Oregon, Eugene, 209 pp.

Ritchey, J.L., 1980. Divergent magmas at Crater Lake, Oregon: Products of fractional crystallization and vertical zoning in a shallow, water-undersaturated chamber. J. Volcanol. Geotherm. Res., 7: 373—386.

Rogers, G.C., 1979. Fault plane solutions. Juan de Fuca plate map 6, Dep. Energy, Mines, Resources. Sidney, British Columbia.

Sakuyama, M., 1981. Petrological study of the Myokyo and Kurohime volcanoes, Japan: Crystallization sequence and evidence for magma mixing. J. Petrol., 22: 553—583.

Scott, W.E., 1977. Quaternary glaciation and volcanism, Metolius River area, Oregon. Geol. Soc. Am. Bull., 88: 113—124.

Self, S. and Rampino, M.R., 1982. Comments on "A geophysical interpretation of the 1883 Krakatau eruption" by I. Yokoyama. J. Volcanol. Geotherm. Res., 13: 379—383.

Shackleton, N.J. and Opdyke, N.E., 1973. Oxygen isotope and paleomagnetic stratigraphy of equatorial Pacific core V28-238: Oxygen isotope temperatures and ice volumes on a 10^5 and 10^6 year scale. Quat. Res., 3: 39—55.

Shaw, H.R., 1974. Diffusion of H_2O in granitic liquids: Part I. Experimental data; Part II. Mass transfer in magma chambers. In: A.W. Hofmann, B.J. Giletti, H.S. Yoder, Jr. and R.A. Yund (Editors), Geochemical Transport and Kinetics. Carnegie Inst. Washington Publ., 634: 139—170.

Shaw, H.R., 1980. The fracture mechanisms of magma transport from the mantle to the surface. In: R.B. Hargraves (Editor), Physics of Magmatic Processes. Princeton Univ. Press, Princeton, N.J., pp. 201—264.

Smith, J.G., Page, N.J., Johnson, M.G., Moring, B.C. and Gray, F., 1982. Preliminary geologic map of the Medford $1^\circ \times 2^\circ$ quadrangle, Oregon and California, scale 1:250,000. U.S. Geol. Surv., Open-File Rep. 82-955.

Smith, R.L., 1960a. Zones and zonal variations in welded ash flows. U.S. Geol. Surv., Prof. Pap., 354-F: 149—159.

Smith, R.L., 1960b. Ash flows. Geol. Soc. Am. Bull., 71: 795—842.

Smith, R.L., 1979. Ash-flow magmatism. In: C.E. Chapin and W.E. Elston (Editors), Ash-flow Tuffs. Geol. Soc. Am., Spec. Pap., 180: 5—27.

Smith, R.L. and Shaw, H.R., 1975. Igneous-related geothermal systems. In: D.E. White and D.L. Williams (Editors), Assessment of Geothermal Resources of the United States — 1975. U.S. Geol. Surv., Circ., 726: 58—83.

Sparks, R.S.J. and Walker, G.P.L., 1977. The significance of vitric-enriched air-fall ashes associated with crystal-enriched ignimbrite. J. Volcanol. Geotherm. Res., 2: 329—341.

Sparks, R.S.J. and Wilson, L., 1976. A model for the formation of ignimbrite by gravitational column collapse. J. Geol. Soc. London, 132: 441—451.

Sparks, R.S.J., Self, S. and Walker, G.P.L., 1973. Products of ignimbrite eruptions. Geology, 1: 115—118.

Sparks, R.S.J., Wilson, L. and Hulme, G., 1978. Theoretical modeling of the generation, movement, and emplacement of pyroclastic flows by column collapse. J. Geophys. Res., 83: 1727—1739.

Spencer, K.J. and Lindsley, D.H., 1981. A solution model for coexisting iron-titanium oxides. Am. Mineral., 66: 1189—1201.

Taylor, E.M., 1968. Accidental plutonic ejecta at Crater Lake, Oregon. Geol. Soc. Am., Spec. Pap., 115: 221 (Abstr.).

Taylor, E.M., 1981. Central High Cascade roadside geology, Bend, Sisters, McKenzie Pass, and Santiam Pass, Oregon. In: D.A. Johnston and J.M. Donnelly-Nolan (Editors), Guides to Some Volcanic Terranes in Washington, Idaho, Oregon, and Northern California. U.S. Geol. Surv., Circ., 838: 55—58.

Thayer, T.P., 1937. Petrology of later Tertiary and Quaternary rocks of the north-central Cascade Mountains in Oregon. Geol. Soc. Am. Bull., 48: 1611—1652.

U.S. Coast and Geodetic Survey, 1959. Map of the floor of Crater Lake. In: Crater Lake National Park and vicinity, Oregon. U.S. Geol. Surv., Topogr. Map, scale 1:62,500.

Veen, C.A., Couch, R.W. and Pitts, G.S., 1981. Gravity anomalies and their structural implications for south-central Oregon. Trans., Am. Geophys. Union, 62(45): 1090.

Watanabe, K., 1978. Studies on the Aso pyroclastic flow deposits in the region to the west of Aso Caldera, Southwest Japan, I: Geology. Mem. Fac. Educ. Kumamoto Univ., Nat. Sci., 27: 97—120.

White, C.M. and McBirney, A.R., 1978. Some quantitative aspects of orogenic volcanism in the Oregon Cascades. In: R.B. Smith and G.P. Eaton (Editors), Cenozoic Tectonics and Regional Geophysics of the Western Cordillera. Geol. Soc. Am. Mem., 152: 369—388.

Williams, D.L. and Von Herzen, R.P., 1983. On the terrestrial heat flow and physical limnology of Crater Lake, Oregon. J. Geophys. Res., 88: 1094—1104.

Williams, H., 1941. Calderas and their Origin. Univ. California Publ., Dept. Geol. Sci., 25: 239—346.

Williams, H., 1942. The Geology of Crater Lake National Park, Oregon. Carnegie Inst. Washington Publ., 540, 162 pp.

Williams, H., 1961. The floor of Crater Lake, Oregon. Am. J. Sci., 259: 81—83.

Williams, H. and Goles, G., 1968. Volume of the Mazama ash-fall and the origin of Crater Lake caldera. In: H.M. Dole (Editor), Andesite Conference Guidebook. Oregon, Dep. Geol. Miner. Ind., Bull., 62: 37—41.

Wilson, C.J.N., 1980. The role of fluidization in the emplacement of pyroclastic flows: an experimental approach. J. Volcanol. Geotherm. Res., 8: 231—249.

Wilson, L., Sparks, R.S.J., Huang, T.C. and Watkins, N.D., 1978. The control of volcanic column heights by eruption energetics and dynamics. J. Geophys. Res., 83: 1829—1836.

Wright, J.V. and Walker, G.P.L., 1977. The ignimbrite source problem: Significance of a co-ignimbrite lag-fall deposit. Geology, 5: 729—732.

Yokoyama, I., 1969. Gravimetric studies and test drillings at three calderas in Japan. Assoc. Geofis. Ital., Atti Conv. Annu., 18: 659—671.

Yokoyama, I., 1981. A geophysical interpretation of the 1883 Krakatau eruption. J. Volcanol. Geotherm. Res., 9: 359—378.

Yokoyama, I., 1982. Author's reply to the comments by S. Self and M.R. Rampino. J. Volcanol. Geotherm. Res., 13: 384—385.

Yokoyama, S., 1974. Mode of emplacement of Ito pyroclastic flow from Aira Caldera, Japan. Sci. Rep. Tokyo Kyoiku Daigaku, Section C, 12: 17—62.

Journal of Volcanology and Geothermal Research, 18 (1983) 117—133
Elsevier Science Publishers B.V., Amsterdam — Printed in The Netherlands

LATERAL VARIATIONS IN THE TAUPO IGNIMBRITE

G.P.L. WALKER* and C.J.N. WILSON**

Department of Geology, University of Auckland, Auckland (New Zealand)
Department of Geology, Imperial College, London SW7 2BP (England)

(Received October 20, 1983)

ABSTRACT

Walker, G.P.L. and Wilson, C.J.N., 1983. Lateral variations in the Taupo ignimbrite. In: S. Aramaki and I. Kushiro (Editors), Arc Volcanism. J. Volcanol. Geotherm. Res., 18: 117—113.

The young non-welded Taupo ignimbrite shows remarkable lateral variations which are documented by granulometric and component analyses, and studies of maximum clast size and density. The grain size spans practically the entire known ignimbrite field, the coarser proximal ignimbrite having a median diameter 100 times greater than the finest distal ignimbrite. The content and maximum size of lithic fragments decrease also by a factor of 100 between proximal and distal parts. The content of free crystals first rises to reach a peak, but thereafter decreases to attain a very low value in far-distal exposures. The pumice maximum size decreases by a factor of about 10, and the most conspicuously coarse pumice rocks occur in a girdle nearly halfway out from vent to distal limit. The pumice in each grain size class decreases in density to half of its near-source value in distal ignimbrite. The overall outward trend is towards an ignimbrite which consists wholly of fine vitric ash; some distal exposures closely approach this condition.

These variations are accounted for by a combination of processes operating in the moving ash flow. One is a continuous fragmentation of pumice leading to a rounding of the clasts, a progressive decrease in maximum size, the generation of much vitric dust, and the liberation of crystals. Another is a continuous sedimentation of heavy constituents (lithics and crystals), and an antipathetic rise of lighter coarse pumice towards the top of the flow. These processes operated in a moving flow whose upper layers travelled progressively farther from source; it is the topmost layers, strongly depleted in heavy constituents and enriched in light pumice, which have travelled the farthest and constitute the far-distal parts of the ignimbrite.

A number of ignimbrite facies are characterized: the ignimbrite proper, with its proximal, distal, and pumice concentration zone facies; the deposits which form in the head and are then over-ridden by the body of the flow, including the fines-depleted ignimbrite variant and the heavies-enriched ground layer; and the ignimbrite veneer deposits which are left behind by the flow, which differ little from the ignimbrite except in their landscape-mantling form and the occurrence in them of lee-side coarse pumice lenses.

---

*Present address: Hawaii Institute of Geophysics, 2525 Correa Road, Honolulu, HI 96822 (U.S.A.).
**Present address: Department of Geology, University of Auckland, Auckland (New Zealand).

INTRODUCTION

The Taupo ignimbrite in New Zealand is a remarkably favourable one on which to make a study of lateral variations. The position of the source vent at Horomatangi Reef in Lake Taupo is known from a number of lines of evidence. Being young — the ^{14}C age is 1850 yr B.P. (Healy et al., 1964) and the true eruption date may be around 186 A.D. (Wilson et al., 1980) — the ignimbrite is largely preserved even where it is less than 1 m thick.

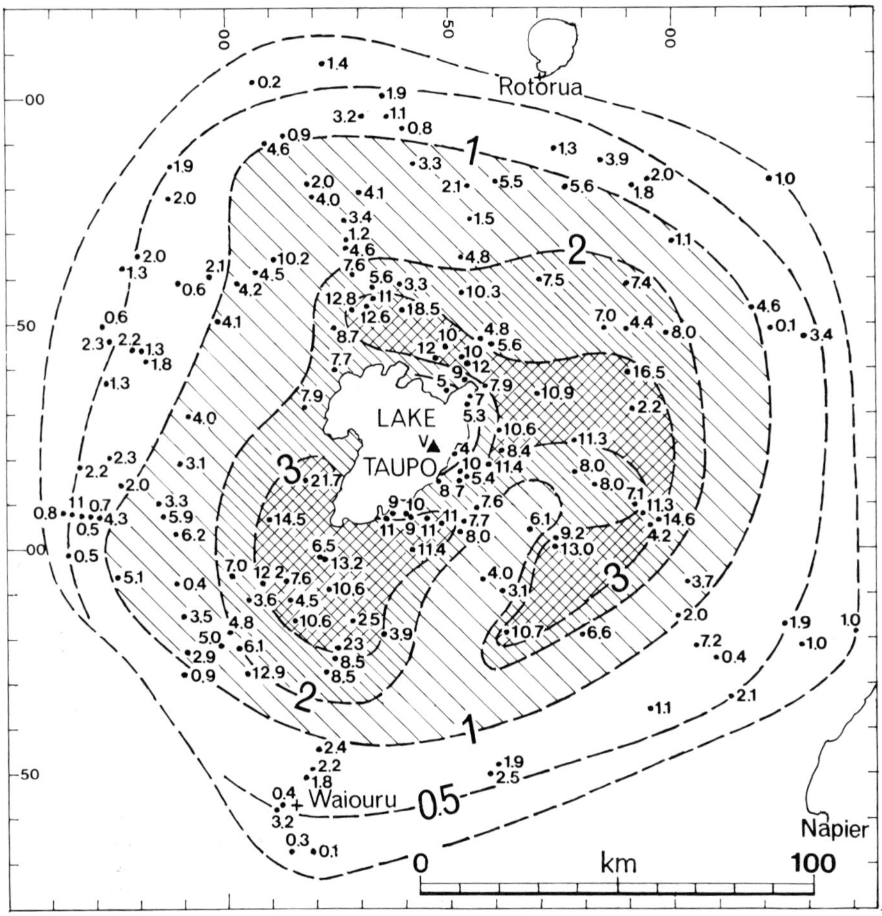

Fig. 1. Map showing the areal extent of the Taupo ignimbrite and locations of analyzed samples. Figures give the weight percentage of free crystals in the sub-2 mm fraction of the layer 2 samples (excluding pumice concentration zone), recalculated to 100%. When two or more samples were analysed from the same site, the average value is given. The generalized contours give the crystal content as a multiple of the 3.5% magmatic content; in the shaded areas the content exceeds magmatic. *V* = vent position. The grid is the 10,000 yard grid of the N.Z. Survey (N.Z.M.S. 1) topographic maps.

Having a relatively small volume of 30 km^3, full cross-sections extending down to the base are commonly seen. It is spread over a wide area — the envelope drawn around its extremities encloses about 20,000 km^2 (Fig. 1) — giving plenty of room for lateral variations. Being fresh and non-welded, the variations can be readily studied by means of mechanical analyses, which offer the best hope of being able to interpret lateral changes which are produced by mechanical processes.

Earlier studies of the Taupo ignimbrite, formally known as the Upper Taupo Pumice member of the Taupo Pumice Formation, determined its stratigraphy, delineated its approximate extent, and established its broad characteristics (Baumgart and Healy, 1956; Healy et al., 1964). The distinction between ignimbrite and redeposited pumice was not however always made, and the rocks which mantle the topography — our ground layer and ignimbrite veneer deposit — proved difficult to interpret and were tentatively regarded as airfall deposits. Our studies have resulted in a reinterpretation of specific aspects (Walker et al., 1980a, b, c, 1981a, b; Wilson and Walker, 1981). A revised stratigraphic scheme has been proposed by Froggatt (1981).

Few systematic studies have been made of lateral variations in grain size and constitution within the same ignimbrite. A significant decrease away from the vent of maximum lithic size and lithic content was established by Kuno et al. (1964), and has since been found elsewhere (Yokoyama, 1970, 1974; Sparks, 1975; Barberi et al., 1978; Taneda, 1979). It is attributed to a progressive loss of lithics from the moving flow. In some of these studies the maximum pumice size was also found to decrease, though less consistently. Systematic outward decreases in free-crystal contents attributed to gravitational sorting have been reported by Ono (1965), Fisher (1966) and Lipman (1967), and general outward decreases in grain size have also been noted.

Our main purpose for making this study was to seek to understand flow mechanisms operating in a pyroclastic flow, but a practical aim was to establish a pattern of lateral variations that might be applied to the problem of the location of source vents, and the determination of flow directions, for older or less favourably exposed ignimbrites.

METHODS OF STUDY

Ignimbrite samples were collected and sieved from some 150 well-scattered sites shown on Fig. 1, and field measurements were made at many other sites. The sieves had a one-phi spacing (that is, the aperture size of each sieve was half that of the sieve above) and for convenience of description, the size class is referred to by the aperture size of the sieve that retains it; thus the 1 mm class contains grains between 1 and 2 mm in size. Component analysis of each size class down to ¼ mm was then done by hand-picking and weighing of splits, and by grain-counting for the ¼ mm class. Pumice density was determined separately for each sieve class by placing each weighed fraction into a graduated container and reading the volume after gently compacting the sample by tapping the container; the apparent density was then corrected for inter-grain voids by applying an empirical factor. For a few samples the density was determined by cutting rectilinear blocks from each of ten pumice clasts in each selected size class.

IGNIMBRITE VARIANTS

The following facies occur in the Taupo ignimbrite (Wilson and Walker, 1982) and are significant in any study of lateral variations; grain size and component analysis histograms (Fig. 2) illustrate the characteristics of each.

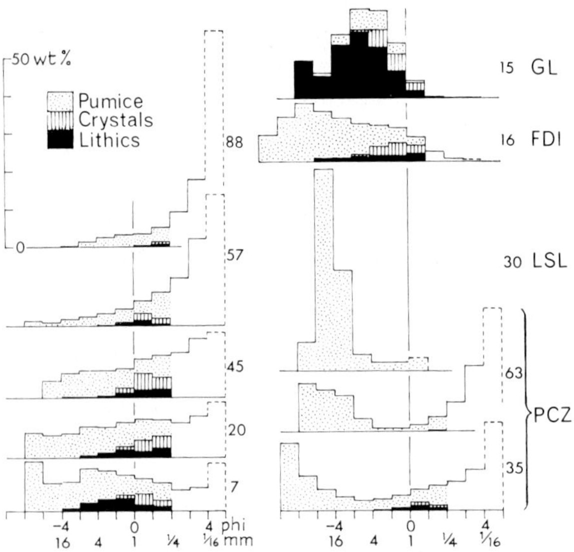

Fig. 2. Grain size histograms of typical samples of Taupo ignimbrite and its variants. The bars for ¼ mm and coarser classes are subdivided to show the proportions of components. Left: ignimbrite of layer 2, showing typical changes going from proximal ignimbrite (at the bottom) to distal ignimbrite (at the top). Numbers along right of histograms give distance from vent in kilometres. Right: PCZ (pumice concentration zone) at top of layer 2, showing bimodal size distribution and strong enrichment in pumice; LSL (lee-side pumice lenses) showing good sorting and strong enrichment in pumice; FDI (fines depleted ignimbrite) of layer 1 showing strong fines depletion; and GL (ground layer) of layer 1 showing strong enrichment in lithics and crystals.

The "standard flow unit"

The main part of the standard ignimbrite flow unit of Sparks et al. (1973) comprises layer 2, which often has a lower lithic concentration zone (layer 2bL) and an upper coarse pumice concentration zone (layer 2bP) attributed to the sinking and buoyant rising respectively of lithics and large low-density pumice clasts in the moving flow (Sparks, 1976). Layer 2bL is generally missing from the Taupo ignimbrite, but a pumice concentration zone (PCZ) is very widely developed. The PCZ has a strongly bimodal pumice population and is often clast-supported for clasts coarser than about 32 mm in size. The standard ignimbrite flow unit has a basal layer (layer 2a) which is distinctly finer than the main part of the unit (layer 2b); a well-defined basal layer is locally found but is generally absent from the Taupo ignimbrite.

Ignimbrite veneer deposits

An important distinction is made between Taupo ignimbrite (layer 2) which has been ponded in valleys to form a terrace having a flat and near-horizontal top surface, and the thin "ignimbrite veneer deposit" which mantles the topography between and is continuous with the material in the valley ponds (Walker et al., 1980a, b, 1981a). The veneer deposit shows a crude and discontinuous stratification nearer source and generally lacks coarser pumice, but is otherwise similar to the pond ignimbrite; it is interpreted as having been left behind on the ground by the tail of the pyroclastic flow; the effect is crudely analogous to the unrolling of a carpet.

The ignimbrite veneer deposit commonly contains lenticular bodies of well-sorted and well-rounded pumice; these lenses occur on the down-flow side of topographic obstacles and are called "lee-side lenses". As seen in cross-section they have an aspect ratio of the order of 1/10 or 1/20, and vary from about 1 cm to 1 m thick. A direct relationship exists between the lens size and the pumice size in it, and both tend to decrease away from source.

Lee-side lenses are interpreted to have developed in vortices beneath a fast-moving pyroclastic flow where it crossed obstacles. The ignimbrite veneer deposits within 20 km of vent locally contain similar lenses even where the underlying topography is flat, attributed to standing waves generated in the fast-moving flow. Note that we envisage such lenses as forming from a highly concentrated pyroclastic flow, having a high ratio of particulate material to gas, and not from a dilute pyroclastic surge (Fisher et al., 1980; Walker et al., 1980b).

Layer 1 deposits

Certain deposits occur beneath the main mass of the ignimbrite, from which they are separated by a sharp and often sheared contact, in circumstances which indicate that they are an integral part of the ignimbrite and were emplaced by the same eruptive act. These deposits are collectively referred to as layer 1 deposits and are characterised by their lack of lateral continuity, rapid variations in thickness, heterogeneity, and commonly low content of fine material. Two main types occur. One, the jetted deposit layer 1P, is very rich in coarse pumice and includes the distinctive fines-depleted ignimbrite variant (FDI: Walker et al., 1980c). The other, the ground layer (Walker et al., 1981b) or layer 1H, is very rich in heavy constituents (lithics and crystals). All gradations are found in layer 1 between FDI, ground layer, and material similar to normal ignimbrite.

Layer 1P is believed to have resulted from the forward jetting of portions of the flow head, and the other layer 1 deposits are interpreted to have been deposited in the pyroclastic flow head; they were then over-ridden by the main mass of the flow from which layer 2 was deposited. The common depletion in fine material indicates that the jets and the flow head were in a strongly fluidized condition. Towards the vent the ground layer apparently merges into deposits of co-ignimbrite lag fall type (Wright and Walker, 1977).

LATERAL GRAIN SIZE VARIATIONS

The main mass (about 80%) of the ignimbrite, that part constituting layer 2, shows a general progressive outward decrease in grain size, and Fig. 3 summarizes the gross changes. The content of material coarser than 16 mm

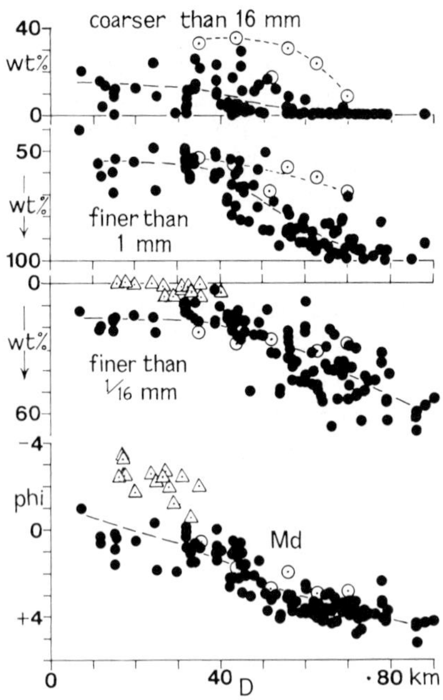

Fig. 3. Grain size variations in the Taupo ignimbrite plotted against distance D from vent. Md = median diameter; MP and ML = average maximum diameters of the three largest pumice and lithic clasts respectively • = ignimbrite of layer 2; ⊙ = pumice concentration zone at top of layer 2; ⊡ = lee-side pumice lenses; △ = fines-depleted ignimbrite; dashed = generalized trend line for ignimbrite; dotted = generalized trend line for pumice concentration zone.

varies from 20% near source to around 0% at 70 km or more. Much of the scatter can be attributed to the buoyant uprise of large pumice into the PCZ (PCZ samples have 25 to 30% more coarse pumice than average ignimbrite) and the corresponding depletion of the underlying ignimbrite.

The content of material finer than 1 mm shows a general increase from 40% near source to near 100% at 70 km and shows a considerable scatter between two extremes represented by the PCZ and basal layer.

The content of material finer than 1/16 mm has a roughly constant value of about 15% out to 45 km from source, suggesting either that a balance exists between fines loss and generation out to this distance, or that 15% is a minimum quantity required to sustain effective fluidization, that is to allow the material to flow and not be deposited. Beyond 45 km the ignimbrite shows a steady increase in fines content suggesting that thereafter the generation of fines always exceeded loss. The FDI variant, which occurs as layer 1 deposits out to 45 km, contains much less than 15% and evidently represents ignimbrite from which fines loss exceeded fines generation.

The median diameter, Md_ϕ is influenced by the contents of both coarse and fine material. It shows a steady and linear decrease outwards spanning more than 6 phi; in other words, the median diameter near source is nearly 100 times greater than it is at 85 km distance. PCZ samples have only a slightly greater median diameter than normal ignimbrite at the same distance, but FDI is systematically 2 to 3 phi (i.e. 4 to 8 times) coarser.

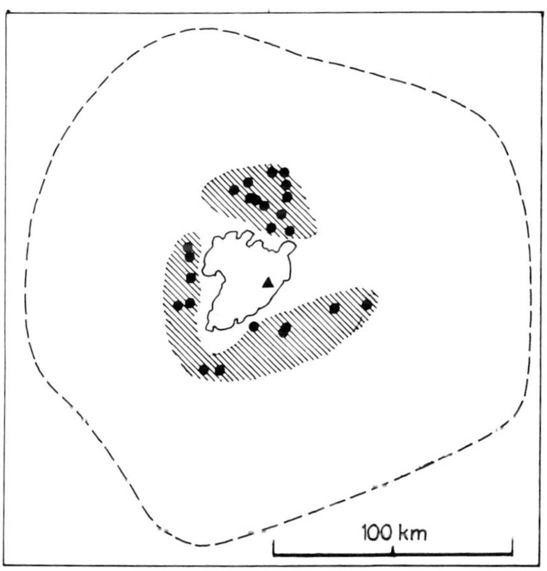

Fig. 4. Distribution map of the most conspicuously coarse pumice outcrops in the Taupo ignimbrite, either in the pumice concentration zone of layer 2 or the fines-depleted ignimbrite of layer 1. These outcrops lie in a crude girdle (generalised form shaded) concentric about the Horomatangi Reef vent; we have no evidence to support the idea, sometimes expressed, that ignimbrite source vents lie in this girdle.

The maximum lithic size in layer 2 varies from more than 100 mm near source to less than 1 mm at 80 km. The maximum pumice size ranges from nearly one metre to a few centimetres, and varies much less strikingly or consistently than the lithic size. The reason is believed to be that pyroclastic flows have an unlimited capacity to carry light-weight pumice but have a strictly limited and outwardly decreasing capacity to carry lithics, which tend to drop out; any outward decrease in pumice size is controlled by other factors such as fragmentation, and not by the ability of the flow to carry it.

An interesting relationship is that the most conspicuously coarse pumice concentrations are found in the PCZ of layer 2 and the FDI of layer 1 in a girdle lying 30 to 40 km out from the Horomatangi vent (Fig. 4). Various source vent positions have in recent years informally been proposed to lie within this girdle (e.g. Rijkse, 1977; Vucetich and Wells, 1978), but the explanation for the girdle appears to be that the pyroclastic flow needed to travel a distance of about 30 km to build up significant accumulations of coarse pumice; farther out than about 40 km the loss of coarse pumice by deposition and abrasion reduced the coarse pumice concentration.

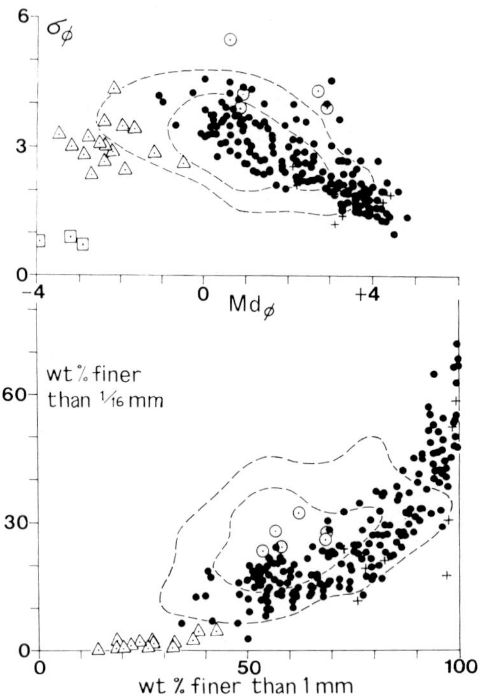

Fig. 5. Grain size characteristics of the Taupo ignimbrite and its variants. The upper diagram plots the graphic standard deviation (σ_ϕ) against median diameter (Md_ϕ). Symbols as Fig. 3. Dashed lines = 1% and 4% contours for the pyroclastic flow field. Note that on both diagrams the Taupo ignimbrite spans practically the entire known ignimbrite variation field. + = samples of Morrinsville ignimbrite from the locations shown on Fig. 8; the grain size characteristics of this ignimbrite parallel that of Taupo.

The form of the grain size histograms for layer 2 samples (Fig. 2) is commonly bimodal when crystals and lithics are removed, having one mode for coarse pumice and one for fine vitric ash. With increasing distance from vent the coarse pumice mode declines and eventually disappears while the fine ash mode increases. This pattern could be generated by abrasion within the pyroclastic flow, and is easily simulated by shaking coarse Taupo pumice in a container on a mechanical sieve shaker; the coarse pumice is rapidly rounded and reduced, and a great quantity of fine dust is generated from it.

Samples of Taupo ignimbrite cover practically the entire known ignimbrite grain size range (Fig. 5), and when variants such as fines-depleted ignimbrite, ground layer and lee-side lenses are included they extend farther the known range. Attention is drawn to the systematic decrease in graphic standard deviation, σ_ϕ, shown by normal ignimbrite samples, representing an improvement in sorting, as the median diameter decreases. Grain size studies of the sub-1/16 mm fractions by the pipetting technique reveal, however, that the content of finest material does not vary in any systematic way with median diameter; it can in part, however, be demonstrated to reflect variations in fluidization conditions within the pyroclastic flow. Material finer than 4 μm typically constitutes 0.5 to 2 wt.% of the sample.

INITIAL POPULATION

The nature of the initial population — the entire assemblage of fragments ejected from the vent — cannot be determined, and perhaps the closest approach to it is given by the average composition of the ten most proximal ignimbrite samples, Table I. This average is much coarser than the initial population of the plinian phase (P) of the same eruption (Walker, 1980). The average content of crystals (6.3%) in the proximal ignimbrite is however significantly greater than the magmatic content (assumed to be 3.5%, this being the crystal content of coarse pumice from the ignimbrite).

The concentration process which has operated to produce 6.3% of crystals is assumed to be the selective loss of fine vitric ash into the eruptive plume, by an amount equal to 45% of the total erupted material, and a correction has been made for this loss in column A of Table I. This corrected population is comparable with, but still slightly coarser than, the plinian one. A few exposures of fine ash are found resting on the plinian pumice at widely scattered sites east and north-east of the ignimbrite outcrop, and are interpreted to represent the resulting co-ignimbrite ash-fall (Sparks and Walker, 1977). Samples collected from 130 and 210 km E.N.E. of the vent contain 88% and 96% respectively of material finer than ½ mm.

LATERAL VARIATIONS IN CONSTITUTION

The ignimbrite shows a general progressive outward decrease in lithic content, and at first an increase followed by a progressive outward decrease

TABLE I

Estimated initial population of the Taupo ignimbrite

Sieve aperture mm	Taupo ignimbrite wt.%	P wt.%	A wt.%
64	0.6	0.7	0.3
32	5.4		0.3
16	5.6	1.5	3.1
8	6.4	2.3	3.6
4	8.1	3.1	4.5
2	8.6	3.5	4.8
1	9.2	3.8	5.1
½	9.5	4.4	5.3
¼	9.3	3.5	5.2
1/8	8.6	77.3	4.8
1/16	10.8		6.0
<1/16	17.9		9.9
Lost mass	—		44.4
wt.% crystals	6.3*	3.0**	3.5**
wt.% lithics			

The figure given is the average composition of the ten most proximal samples of Taupo ignimbrite, compared with the estimated initial population of the Taupo plinian pumice fall (P). Column A gives the Taupo ignimbrite average recomputed to reduce the crystal content to the 3.5% assumed magmatic content.

*Includes 5.1% free crystals plus 1.2% crystals estimated to be enclosed in 2 mm and coarser pumice.

**Assumed magmatic content = the content of crystals in coarse pumice clasts.

in free crystal content. Figure 6a shows the variation in lithic and free-crystal contents for the sum of the ¼ mm and coarser classes; microscopic examination reveals that these constituents are a minor component in classes finer than ¼ mm, so this figure reflects fairly accurately the true lithic and free-crystal variations.

The overall lithic content decreases more or less linearly, though there is a large amount of scatter due to localised fluidization effects, and is nearly 100 times greater near source than at 80 km. Near source, lithics commonly occur in the 16 mm and coarser classes; traced outwards the coarsest class in which lithics occur decreases until at 80 km there is only a trace in the 1 mm class. PCZ samples are systematically poorer in lithics, and have a smaller maximum lithic size, than normal ignimbrite at the same distance.

The content of free crystals shows a less simple variation since it at first rises slightly out to 30 or 40 km from vent before thereafter declining steadily in amount. The content of free crystals in the sum of the sub-2 mm classes (Figs. 1 and 5b) should be a better guide than the content in the entire

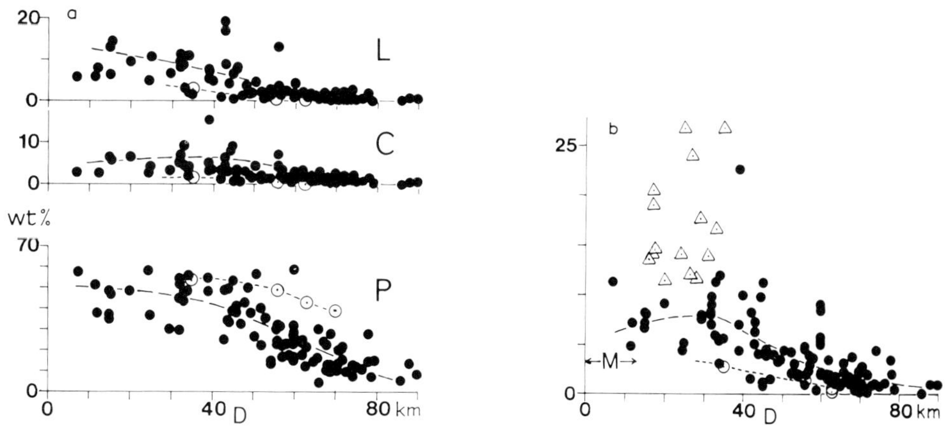

Fig. 6. a. Weight percentages of free crystals (C), lithics (L), and pumice (P) in the Taupo ignimbrite as they vary with distance D from vent for the fraction ¼ mm and coarser, expressed as a percentage of the whole sample.
b. Weight percentage of crystals in the sub-2 mm fraction, recalculated to 100% as in Fig. 1.
M = magmatic crystal content. Symbols as in Fig. 3. Note that pumice concentration zone samples are consistently richer in pumice and poorer in heavier components (crystals and lithics). Excessively high crystal contents characterise the depleted ignimbrite samples.

sample, since crystals have a maximum size of just over 2 mm in the Taupo ignimbrite and are presumably released when magma or pumice is reduced to fragments 2 mm or less in size. These plots show very clearly the slight but distinct increase out to a maximum at 30 or 40 km.

LATERAL VARIATIONS IN PUMICE DENSITY

In any given sample of Taupo ignimbrite, pumice shows an increase in density with decreasing grain size, as is typical for pyroclastic deposits in general (Walker, 1980), but in each size class the pumice also shows a general density decrease outwards to about half of its near-source value at a distance of 70 km, Fig. 7. This decrease is a significant factor when determining the origin of the other lateral variations.

The lee-side pumice lenses are anomalous in that the pumice density in them is either uniform for different co-existing grain size classes or shows a slight decrease as the grain size decreases. This is attributed to the very rapid abrasion of the large lower-density pumice clasts in the turbulent conditions existing in the vortex within which the lee-side lenses are thought to form. The good rounding of the pumice clasts points to the same conclusion.

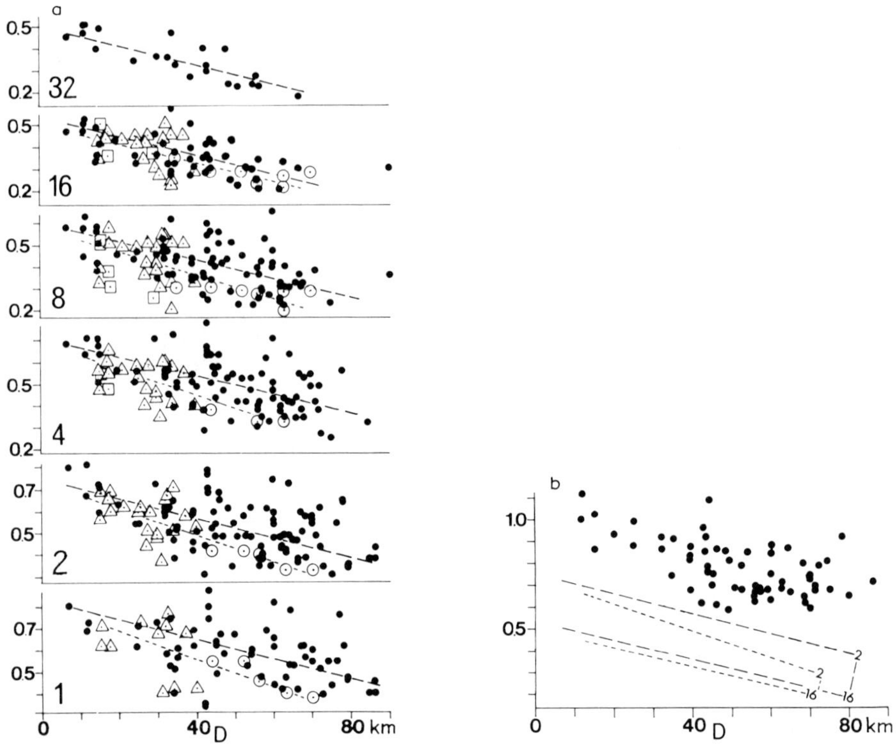

Fig. 7. a. Pumice density in g cm^{-3} in selected grain size classes showing the general trend to decrease with increasing distance from the vent. Note that the pumice in pumice concentration zone samples is consistently less dense than in normal ignimbrite at the same distance from vent.
b. Bulk density in g cm^{-3} of sub-2 mm fraction of Taupo ignimbrite samples. Dashed and dotted lines show the range of generalised trend lines for 2 to 16 mm size classes, from Fig. 7a, which have a consistently lower density.

POSSIBLE MECHANISMS FOR THE LATERAL VARIATIONS

The progressive outward decrease in content and maximum size of lithic fragments and eventually also of free crystals, results in large part from the sedimentation of these heavy constituents into the ground layer and is not considered further here.

The general outward decrease in content and maximum size of coarse pumice and the concomitant increase in content of "fines" seems at first sight to be readily explained by the fragmentation of coarser pumice within the moving pyroclastic flow, either by abrasion or by bursting of vesicles resulting from continued vesiculation. The pumice is mechanically weak, and it is moderately to well rounded showing that significant abrasion has undoubtedly occurred; the initial outward increase in the free-crystal content (Fig. 7) moreover can be partly explained by the release of crystals from fragmented pumice. Laboratory simulation by shaking coarse Taupo pumice in a container rapidly generates fine dust and liberates crystals. Because of the scarcity of bubble-wall shards except in the sub-¼ mm size fractions, and the relatively low temperature character of the ignimbrite shown by the lack of welding, we favour fragmentation by abrasion over that by vesiculation.

Post-eruption fragmentation of coarse pumice cannot however account for the outward pumice density decrease found in each size class; indeed if fragmentation were the sole process, it is the mechanically weakest pumice (that having the lowest density) which should be selectively reduced, and the strongest and heaviest pumice in each class should survive into the distal ignimbrite. One could argue that the outward density decrease results from post-eruption inflation of the pumice by continued vesiculation, but we have seen little evidence for such inflation in the form of breadcrusted or internally more vesicular pumice, this seems unlikely in a relatively low-temperature flow such as Taupo, and inflation would be expected to affect the coarser pumice most instead of affecting equally the coarser and finer pumice.

Consider now whether the various processes which abstract coarse pumice from the pyroclastic flow could account for the lateral variations. The quantitatively most significant process is forming the jetted deposits. The resulting layer 1P contains much coarse pumice, but this pumice has if anything a slightly lower average density than in the associated layer 2 deposit, and its abstraction cannot therefore explain the outward pumice density decrease found in layer 2. Similarly the local stranding of "rafts" of pumice concentration zone material effectively removes the lighter pumice from the flow. The formation of lee-side lenses also abstracts coarse pumice from the fast-moving flow, and the pumice has a slightly higher density than in the associated layer 2, but the amount of pumice so removed is not quantitatively significant. Some pumice is sedimented with lithics and crystals to form the ground layer and this pumice has a systematically higher density than layer 2: this mechanism operates in the right direction to remove the heavier pumice, but is not quantitatively important.

While all the above mechanisms undoubtedly operate, the progressive outward pumice density decrease is considered to result mainly from a combination of two additional mechanisms: a general upwards movement of low density material within the flow, resulting from fluidisation processes (Wilson, 1983) and the laminar flowage of the pyroclastic flow so that the lighter upper part is translated to the distal parts of the flow.

The former process can be seen to have operated to produce the pumice concentration zone, where the pumice density averages 20% lower than in the same size fraction found in associated normal ignimbrite.

The PCZ is also notably impoverished in lithics and free crystals (after allowing for the diluting effect of adding coarse pumice), and lithics have a smaller maximum size than in associated normal ignimbrite. The buoyant rise of pumice has thus clearly been accompanied by an antipathetic settling of heavy grains.

One possible additional mechanism should be considered, in which the first material to condense near the vent from the eruptive column and form a pyroclastic flow is envisaged to have a higher content of coarser and heavier material than the later-condensed material; laminar flowage then results in the later-condensed part enriched in finer and lighter material over-riding and travelling farther than the earlier-condensed part and so forming the distal ignimbrite. To some extent this has occurred; for example there is evidence that the first-erupted fraction was richer in very coarse lithics. The vertical compositional variation seen in the IVD is, however, much smaller than the lateral variation seen in the same layer, suggesting that the gross lateral variations result more from processes operating within the flow.

CONCLUSIONS

We see evidence for a considerable number of mechanisms combining to produce the observed lateral heterogeneities in the Taupo ignimbrite. We conclude that the progressive fragmentation of coarse pumice to generate finer material and at the same time to liberate crystals is one mechanism, and the continual sedimentation of heavy constituents to form the ground layer is another. The abstraction of coarse pumice from the flow to generate layer 1P deposits, standed pumice concentration zone "rafts", and lee-side pumice lenses, also contributed to the general outward fining.

The observed lateral pumice density distribution cannot be explained by these mechanisms, and a general upward segregation of light constituents in the pyroclastic flow combined with some form of "laminar" flowage seems best able to account for this feature. The upwards variation became translated by "laminar" flowage into a lateral variation, the highest material in the flow (duly modified by pumice breakage and other processes) coming to constitute the most far-distal ignimbrite fraction.

The Taupo ignimbrite shows greater lateral variations than are normally attributed to ignimbrite, and the question arises as to how typical Taupo is of ignimbrites in general. The fact that the Taupo ignimbrite has an unusually low aspect ratio (Walker et al., 1980) and shows an unusually consicuous development of layer 1 deposits (Wilson and Walker, 1982) suggests that emplacement conditions were exceptional. The high flow velocity (Wilson and Walker, 1981) may thus have resulted in a higher degree of fluidization than normal — the flow was one of type 3 (Wilson, 1980) — so favouring the development of extreme lateral variations.

Two other ignimbrites are presently known to show lateral variations similar in type and comparable in degree to those of Taupo. One is the 1500 yr B.P. Rabaul ignimbrite, of dacitic composition, in Papua New Guinea (Walker et al., 1981c); the documented variations are less extreme than at

Taupo, although the Rabaul ignimbrite has not yet been traced to its distal extremities. The other is a flow in New Zealand informally referred to by us as the Morrinsville ignimbrite which is found extending as far as Auckland, 200 km northwest of the Taupo Zone wherein the source vent is presumed to lie. Grain size and component analyses have been made on samples from a series of isolated outcrops spanning the 90 to 200 km interval from the Taupo Zone, and the lateral changes shown in this interval (Fig. 8) are very closely comparable in type and degree to those shown by the Taupo ignimbrite.

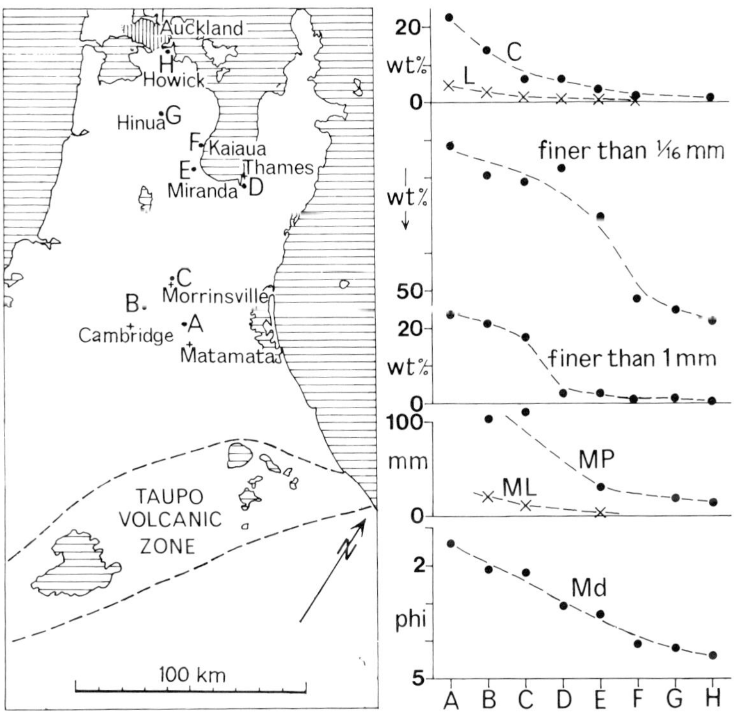

Fig. 8. Location of outcrops of the Morrisville ignimbrite showing grain size characters and composition of analyzed samples. C and L = contents of crystals and lithics respectively in the fraction ¼ mm and coarser expressed as weight percentages of the whole samples; MP and ML = average maximum diameters of the three largest pumice and lithic clasts respectively; Md = median diameter. Traced towards the distal limit in Auckland, the ignimbrite shows serial characters which are similar in nature and degree to those towards the distal end of the Taupo ignimbrite.

One final point deserves mention: towards their distal ends the Taupo, Rabaul and Morrinsville flows become so fine-grained and well sorted, that their deposits cease to have the characters normally regarded as typical of ignimbrites; the main criterion relied upon here for their ignimbrite origin is the gradual progressive change in characters that they show when traced outwards from outcrops which are clearly recognisable as ignimbrite. For older ignimbrites these distal parts, being generally thin and non-welded, have a rather low survival potential when exposed to erosion, and even if they do survive their characters may be such that they are not readily recognised as ignimbrite. The variation shown by the Taupo ignimbrite therefore may well prove not to be exceptional.

ACKNOWLEDGEMENTS

The research by G.P.L.W. was done at the University of Auckland as Captain James Cook research fellow of the Royal Society of New Zealand; C.J.N.W. acknowledges a British Natural Environment Research Council research studentship, and continuing support from a Royal Society Mr and Mrs. John Jaffé Donation Research Fellowship.

REFERENCES

Barberi, F., Innocenti, F., Lirer, L., Munno, R., Pescatore, T. and Santacroce, R., 1978. The Campanian ignimbrite: a major prehistoric eruption in the Neapolitan area (Italy). Bull. Volcanol., 41: 10—31.

Baumgart, I.L. and Healy, J., 1956. Recent volcanicity at Taupo, New Zealand. Proc. 8th Pacific Sci. Congr., 2: 113—125.

Fisher, R.V., 1966. Mechanism of deposition from pyroclastic flows. Am. J. Sci., 264: 350—363.

Fisher, R.V., Smith, A.L., Wright, J.V. and Roobol, M.J., 1980. Ignimbrite veneer deposits or pyroclastic surge deposits? Nature, 286: 912.

Froggatt, P.C., 1981. Stratigraphy and nature of Taupo Pumice Formation. N.Z. J. Geol. Geophys., 24: 231—248.

Hay, R.L., 1959. Formation of the crystal-rich glowing avalanche deposits of St. Vincent, B. W. I. J. Geol., 67: 540—562.

Healy, J., 1964. Volcanic mechanisms in the Taupo Volcanic Zone, New Zealand. N. Z. J. Geol. Geophys., 7: 6—23.

Healy, J., Vucetich, C.G. and Pullar, W.A., 1964. Stratigraphy and chronology of late Quaternary volcanic ash in Taupo, Rotorua and Gisborne districts. N. Z. Geol. Surv. Bull, New Ser., 73.

Kuno, H., Ishikawa, T., Katsui, Y., Yagi, K., Yamasaki, M. and Taneda, S., 1964. Sorting of pumice and lithic fragments as a key to eruptive and emplacement mechanism. Japan. J. Geol. Geograph., 35: 223—238.

Lipman, P.W., 1967. Mineral and chemical variations within an ash-flow sheet from Aso caldera, southwestern Japan. Contrib. Mineral. Petrol., 16: 300—327.

Ono, K., 1966. Glass lens in welded tuffs. Bull. Volcanol. Soc. Jpn., 11: 47—48.

Rijkse, W.C., 1977. Additional sources of Taupo Pumice. N.Z. Soil News, 25: 219—221.

Sparks, R.S.J., Self, S. and Walker, G.P.L., 1973. Products of ignimbrite eruptions. Geology, 1: 115—118.

Sparks, R.S.J., 1975. Stratigraphy and geology of the ignimbrites of Vulsini Volcano, Central Italy. Geol. Rundsch., 64: 497—523.

Sparks, R.S.J., 1976. Grain size variations in ignimbrites and implications for the transport of pyroclastic flows. Sedimentology, 23: 147—188.

Sparks, R.S.J. and Walker, G.P.L., 1977. The significance of vitric-enriched air-fall ashes associated with crystal-enriched ignimbrites. J. Volcanol. Geotherm. Res., 2: 329—341.

Taneda, S., 1979. Grain-size and -shape, median diameter, and proportion of pumice and lithic fragments as a key to movement and emplacement mechanism of pumice flow deposits, the "Shirasu" pumice flow deposits in Kyushu, Japan. Mem. Fac. Sci. Kyushu Univ. ser. D, 24: 109—154.

Vucetich, C.G. and Wells, N., 1978. Soils, agriculture, and forestry of Waiotapu region, Central North Island, New Zealand. N.Z. Soil Bureau Bull., 31, 100 pp.

Walker, G.P.L., 1980. The Taupo pumice: product of the most powerful known (ultraplinian) eruption? J. Volcanol. Geotherm. Res., 8: 69—94.

Walker, G.P.L., Heming, R.F. and Wilson, C.J.N., 1980a. Low-aspect ratio ignimbrites. Nature, 283: 286—287.

Walker, G.P.L., Heming, R.F. and Wilson, C.J.N., 1980b. Ignimbrite veneer deposits or pyroclastic surge deposits? Reply. Nature, 286: 912.

Walker, G.P.L., Wilson, C.J.N. and Froggatt, P.C., 1980c. Fines-depleted ignimbrite in New Zealand — the product of a turbulent pyroclastic flow. Geology, 8: 245—249.

Walker, G.P.L., Wilson, C.J.N. and Froggatt, P.C., 1981a. An ignimbrite veneer deposit: the trail-marker of a pyroclastic flow. J. Volcanol. Geotherm. Res., 9: 409—421.

Walker, G.P.L., Self, S. and Froggatt, P.C., 1981b. The ground layer of the Taupo ignimbrite: a striking example of sedimentation from a pyroclastic flow. J. Volcanol. Geotherm. Res., 10: 1—11.

Walker, G.P.L., Heming, R.F., Sprod, T.J. and Walker, H.R., 1981c. Latest major eruptions of Rabaul Volcano. In: R.W. Johnson (Editor), Cooke-Ravian Volume of Volcanological Papers. Geol. Surv., Papua New Guinea, Mem., 10: 181—193.

Wilson, C.J.N., 1980. The role of fluidization in the emplacement of pyroclastic flows: an experimental approach. J. Volcanol. Geotherm. Res., 8: 231—249.

Wilson, C.J.N., 1983. The role of fluidization in the emplacement of pyroclastic flows, 2: experimental results and their interpretation. J. Volcanol. Geotherm. Res., in press.

Wilson, C.J.N. and Walker, G.P.L., 1981. Violence in pyroclastic flow eruptions. In: S. Self and R.S.J. Sparks (Editors), Tephra Studies. Reidel, Dordrecht, pp. 441—448.

Wilson, C.J.N. and Walker, G.P.L., 1982. Ignimbrite depositional facies: the anatomy of a pyroclastic flow. J. Geol. Soc. London, 139: 581—592.

Wilson, C.J.N., Ambraseys, N.N., Bradley, J. and Walker, G.P.L., 1980. A new date for the Taupo eruption, New Zealand. Nature, 288: 252—253.

Wright, J.V. and Walker, G.P.L., 1977. The ignimbrite source problem: a co-ignimbrite lag fall deposit from Mexico. Geology, 5: 729—732.

Wright, J.V. and Walker, G.P.L., 1981. Eruption, transport and deposition of ignimbrite: a case study from Mexico. J. Volcanol. Geotherm. Res., 9: 111—131.

Yokoyama, S., 1970. Geomorphology of the Ito pyroclastic flow deposit to the north of the Aira caldera. Geogr. Rev. Jpn., 43: 464—482.

Yokoyama, S., 1974. Mode of movement and emplacement of the Ito pyroclastic flow from Aira caldera, Japan. Sci. Rep. Tokyo Kyoiku Daigaku, Sect. C, 12: 17—62.

Journal of Volcanology and Geothermal Research, 18 (1983) 135—150
Elsevier Science Publishers B.V., Amsterdam — Printed in The Netherlands

VOLCANIC DRY AVALANCHE DEPOSITS — IDENTIFICATION AND COMPARISON WITH NONVOLCANIC DEBRIS STREAM DEPOSITS

TADAHIDE UI

Department of Earth Sciences, Faculty of Science, Kobe University, Nada, Kobe 657 (Japan)

(Received October 20, 1982)

ABSTRACT

Ui, T., 1983. Volcanic dry avalanche deposits — identification and comparison with nonvolcanic debris stream deposits. In: S. Aramaki and I. Kushiro (Editors), Arc Volcanism. J. Volcanol. Geotherm. Res., 18: 135—150.

A "volcanic dry avalanche deposit" is defined as a volcaniclastic deposit formed as a result of a large-scale sector collapse of a volcanic cone associated with some form of volcanic activity. Avalanche transport occurred in response to the gravitational field, in a manner similar to the transport of nonvolcanic debris streams (e.g. Hsü, 1975). Such deposits are characterized by megablock structure — deformed and fractured large blocks up to several hundreds meters in diameter. A megablock preserves original layering, intrusive contacts or weathered surfaces of the source volcanic edifice. Surface topography of the deposit is characterised with hummocky relief. Ratios of fell height to travel distance for volcanic dry avalanche deposits are between 0.18 and 0.06. This range is similar but smaller than the value of 0.58 to 0.08 for nonvolcanic debris stream deposit. This similarity suggests similar transportation mechanisms. Excessive travel distances as defined by Hsü (1975), calculated for volcanic dry avalanche deposits, give values larger than for debris stream deposits of the same volume. The difference is explained by lower rigidity of the collapsing mass due to the existence of soft pyroclastic layers, alteration around the vent, development of fractures owing to new cryptodome intrusion, and boiling of supercritical fluid contained within the collapsed mass.

INTRODUCTION

A "volcanic dry avalanche deposit" is defined as a volcaniclastic deposit formed as a result of large-scale sector collapse of a volcanic cone associated with some form of volcanic activity. The term "volcanic dry avalanche" was first introduced in literature by Nakamura (1978). This type of volcaniclastic deposit has often been described in the past as a mud flow deposit, lahar or volcanic detritus. According to the classification of volcanic eruptions proposed by Macdonald (1972), an ultravulcanian eruption consists of the weak to violent ejection of solid fragments of old rocks. The 1888 eruption of Bandai-san — a historic example of

0377-0273/83/$03.00

a volcanic dry avalanche — was included within this category. According to Macdonald (1972), no juvenile magma is ejected during an eruption of this type. But in the March 1956 eruption of Bezymianny, Gorshkov (1959) concluded that the summit of the volcano was blown off by a large-scale directed blast, based on eyewitness reports of the eruption column inclined at about 30° to the horizon. According to Williams and McBirney (1979), some of the most powerful eruptions ever recorded were phreatic explosions resulting from the rise of magma into the water-soaked upper part of a volcano, such as the 1955—1956 eruptions of Bezymianny and the 1964 eruption of Shiveluch. They assumed that inclined blasts of immense energy blew out great volumes of lithic debris and were followed by purely magmatic eruptions.

In regard to nonvolcanic debris stream deposits, Heim (1882) explained the mode of transportation for the Elm deposit as follows: starting as a huge rock fall at a sudden steepening of slope the debris stream briefly left the ground, but the energy for motion was transmitted by block-to-block impact. Shreve (1968) later suggested that motion was accomplished at Elm by overriding and compressing a cushion of trapped air, which permitted the debris stream to traverse the gentle slope below with little friction. Howard (1973) insisted that an interstitial fluid, for example, a compressed gas or wet mud, was not necessary for transportation of debris streams in general because of the presence of deposits of a similar type on the lunar surface. Hsü (1975, 1978) reaffirmed Heim's conclusion, and suggested that the dispersion of fine debris and dust among the colliding blocks may have provided an uplifting stress during flowage, adopting the grain flow model proposed by Bagnold (1954). Thus, the energy for transportation of nonvolcanic debris streams is explained as gravitational.

A volcanic dry avalanche deposit was formed at the climactic May 18 eruption of Mount St. Helens in 1980. Observations of bulging prior to sliding and geological surveying of the deposit provides much information concerning the mode of eruption, transportation and deposition of the volcanic dry avalanche deposit (Christiansen, 1980; Lipman and Mullineaux, 1981). Vulcanian-type explosion column or a directed blast was not associated with the initiation of the dry avalanche. Instead, simple gravitational sliding was triggered by an earthquake, as suggested by eyewitness observation (Stoffel and Stoffel, 1980). Thus, the mode of transportation of volcanic dry avalanches is basically the same as that for debris streams.

For purpose of this paper, debris stream deposits are here termed as "nonvolcanic dry avalanche deposits". First, major criteria for the identification of volcanic dry avalanche deposits are given, and then the similarity of apparent coefficients of friction for volcanic and nonvolcanic dry avalanche deposits is discussed.

IDENTIFICATION OF VOLCANIC DRY AVALANCHE DEPOSIT

Several structural and topographical characteristics are commonly observed in volcanic dry avalanche deposits. Their source areas are characterised by collapsed topography, with a wide opening to one side of the former cones. The opening slope developes towards the flank of the volcano, where avalanche material was deposited. In places, large-scale radial grooves developed along this slope. The wide opening containing such collapsed topography, and its enclosing wall, is named as amphitheater (Voight et al., 1981). Usually, a young volcanic cone or dome arises from the floor of amphitheater. For example, a twin dome, 260 m in height, was grown until October 1956 in Bezymianny (Gorshkov, 1959). The amphitheater of the Chokai volcano, Japan, is now partially filled with younger lavas and pyroclastics, 2600 years after the formation of the Kisakata dry avalanche (Kato, 1978). At Yatsugatake volcano, Japan, the amphitheater is completely buried below younger cones (Kawachi, 1977).

A hummocky surface (the conical hills of Williams and McBirney, 1979) is a typical topographic feature of volcanic dry avalanche deposits (Fig. 1). In case of the 1980 deposit at Mount St. Helens, local channelling and

Fig. 1. Hummocky surface of the Kisakata dry avalanche deposit, Kamisaka, northern foot of the Chokai volcano. The area behind the car has been almost left intact by surface erosion. Source of the flow is to the right-hand side. Radiocarbon age of the deposit is ca. 2600 years before present (Kato, 1978).

slumping started just after deposition, and a flat surface made of secondary mud flow and water-laid deposits has developed, but thousands of hummocks protrude above the mean surface elevation of the deposit. Hummocks surrounded by a flat planar surface is a common topographic feature in various older deposits. The flat surface separating individual hummocks might be an effect of long-range reworking, infilling of water-laid sediments, and development of vegitation cover. Secondary water-influenced portions of a volcanic dry avalanche deposit consist of a mixture various kinds and sizes of rock fragments, with common sorting and grading effects. This part of a volcanic dry avalanche deposit is hard to distinguish from simple water-laid volcaniclastic deposits.

Each hummock consists of one or a few megablocks (Fig. 2). A megablock consists of a former portion of the volcanic edifice, transported to present site by dry avalanche (Mimura and Kawachi, 1981). The interior of a megablock is fractured into numerous clasts (Fig. 3). Adjacent clasts within part of megablock often show identical lithologic character, and outlines of nearby clasts occasionally fit each other, for example Pankechin dry avalanche, Shikaribetsu volcano (Yamagishi, 1977). It is sometimes possible to reconstruct the form of primary igneous body prior

Fig. 2. An example of megablock structure in the Nirasaki dry avalanche deposit. A section of hummocky hill at Anayamabashi, Yatsugatake volcano. Alteration of lavas and pyroclastics, dipping right-hand side is slightly deformed and displaced due to fracturing and minor normal faulting.

Fig. 3. Fracturing of lava flow within a megablock. Kamanashi-gawa-bashi, Nirasaki dry avalanche deposit, Yatsugatake volcano. Lower left corner is the base of a lava flow, dipping towards right. Irregular fractures are abundant within massive part of the lava flow. Cooling joint plane is observable at upper right.

to fracturing, if rotations and transformations can be deduced in three dimensions for each clast. Thus megablocks can be judged in relation to originally massive parts of lava flows or intrusives. Lithologic facies boundaries occur as contacts of air-fall tuffs and lavas, intrusion contacts, or weathered surfaces. Such boundaries are usually disturbed by fracturing. Shreve (1968) described similar fractured boundaries in the Blackhawk nonvolcanic dry avalanche deposit, and named it the jigsaw puzzle effect.

Mimura et al. (1971) showed by natural remanent magnetization measurements of the late-Quaternary Nirasaki dry avalanche deposit at Yatsugatake volcano that the inclination and declination of various clasts within a megablock show similar directions, but that declinations are variable among nearby megablocks. Inclination of all measured megablocks in the deposit shows a direction similar to that of the present magnetic field of the earth. Thus the megablocks were transported to their present sites at the foot of a volcano without appreciable rolling.

The average size of megablocks is variable among the various deposits, and variation in size is considerable even within a single deposit. The maximum size recorded is 500 m in diameter, in the Nirasaki dry avalanche deposit of Yatsugatake volcano (Mimura and Kawachi, 1981). An abrupt change of internal structure associated with the change of degree of fracturing or of lithologic character defines the boundary of adjacent megablocks. Sometimes, adjacent megablocks show evidence of accretion, separated by a flattened layer of soil or tuff, a few meters in thickness (equivalent to "soil schlieren" of Shreve, 1968). Thickness of the layer of separation is variable and depends on the shape of the megablocks.

The size and number of megablocks decreases towards the marginal or distal end of a deposit; the amount of matrix mixture correspondingly increases (Mimura and Kawachi, 1981). The matrix consists of various

Fig. 4. Matrix of volcanic dry avalanche deposit, Nagaoka, Kisakata dry avalanche deposit, Chokai volcano. Matrix mainly consists of a mixture of finer volcanic clasts; Gravel (upper right), chips of soil (center and left of scale), and fragment of twig (middle left). Small megablocks occur locally but are not visible within this figure. Banded stripe on the scale expresses 10 cm in length.

kinds of rock fragments, finer tuffaceous particles, chips of surface soil scraped off during flowage, alluvial gravels, and fragments of plants (Fig. 4). Megablocks of small size, several tens of centimeters to a few meters in diameter are usually included within the matrix mixture. Levees (lateral deposits) at lateral margins of the deposit are described in the case of the Mount St. Helens deposit (Voight et al., 1981). Similar topographic feature is developed at the northern margin of 1792 volcanic dry avalanche deposit of Unzen volcano. Interior of the levee shows similar structure to the matrix mixture.

APPARENT COEFFICIENT OF FRICTION FOR VOLCANIC DRY AVALANCHE DEPOSIT

Based on the criteria just described, I have searched for volcanic dry avalanche deposits in the literature descriptions of various volcanoes. Results of this search are listed in Table I. I have also made field observations at some of these deposits in order to verify the mode of emplacement as volcanic dry avalanches. Field-verified deposits are marked with asterisks in Table I. The inferred highest altitude of the source region and the lowest altitude of a deposit were located from topographic and published geologic maps, or by direct field work. The difference between high and low elevations define the maximum height difference during transportation (H). Maximum travel distance (L) was similarly determined. Volume (V) of the deposit was estimated from direct field data or from published geologic descriptions. L, H and V for nonvolcanic dry avalanche deposits (Hsü, 1975) and for nuée ardente deposits are listed in Tables II and III for comparison, and the relationships between H and L are plotted on Fig. 5.

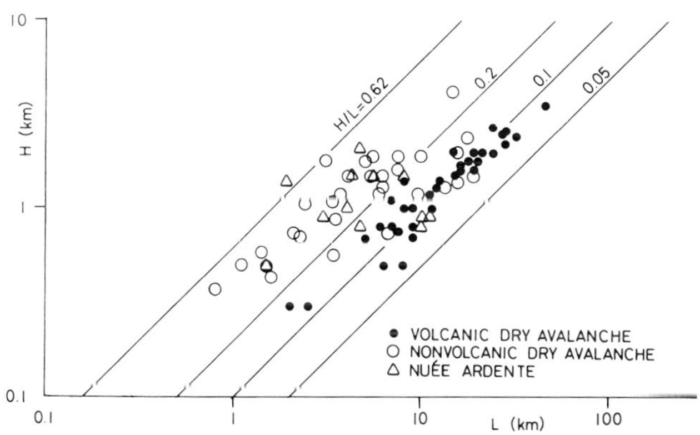

Fig. 5. Relationship between maximum fall height (H) and maximum travel distance (L). Data are listed in Tables I, II and III. The average H/L ratio for volcanic dry avalanche deposits are smaller than those for nonvolcanic dry avalanche deposits.

TABLE I

Volcanic dry avalanche deposits

Name of volcano	Name of deposit	H(km)	L(km)	H/L	L_e(km)	V(km^3)	Data source
16 Imuta		* 0.3	2	0.15	1.5		T. Kobayashi, pers. commun., 1982
18 Unzen	1792 Maeyama	* 0.7	6	0.12	3.9	0.48	Katayama, 1974
62 Fuji	Kofuji	2	24	0.083	21		Tsuya, 1940
64 Yatsugatake	Nirasaki	* 2.4	32	0.075	28	9.0	Mimura and Kawachi, 1981
64 Yatsugatake	Otsukigawa	1.4	12.5	0.112	10.2	0.27	Kawachi, 1981
77 Tomuro		* 0.3	2.5	0.12	2.0		Imai, 1959
85 Myoko	Sekikawa	2.0	19	0.105	16	0.8±0.2	Hayatsu, 1976
85 Myoko	Taguchi	1.4	8	0.175	5.7	0.23	Hayatsu, 1976
85 Myoko	Yashirogawa	1.8	18	0.1	15		Hayatsu, 1976
86 Kurohime	Nabewarigawa	0.8	6	0.133	4.7	0.12	Hayatsu, 1976
87 Iizuna		0.8	6	0.133	4.7		Hayatsu, 1976
91 Asama	Tsukahara	* 1.8	20	0.09	17	2	Aramaki, 1963
91 Asama	Okuwa	* 1.6	16	0.1	13		Aramaki, 1963
104 Akagi	Nashikizawa	1.6	19	0.084	16	4	Moriya, 1968
114 Nasu		* 1.6	19	0.084	16		Kato, 1964
120 Bandai-san	1888	* 1.2	11	0.109	9.1	1.5	Nakamura, 1978
120 Bandai-san	Okinajima	* 1.5	15	0.1	13		Mizuno, 1958
127 Shirataka		* 0.5	8	0.063	7.2		Ui and Shibahashi, in press
129 Gassan	Sasagawa	* 1.95	21	0.093	18		Ichimura, 1955; Ui, 1975
— Hayama	Hata	1.0	8	0.125	6.4		Ui and Shibahashi, in press

135 Chokai	Kisakata	* 0.2	25	0.088	21		Kato, 1978
142 Iwate	Gohyakumori	* 1.7	16	0.106	13		Tachibana, 1978
151 Tashiro	Itazawa	* 0.7	9	0.078	7.9		Sumi et al., 1962
153 Iwaki		1.5	15	0.1	13	1.3	Isshiki and Ozawa, 1967
187 Komagatake	Kurumizawa	1.0	11.5	0.087	9.9	0.25	Katsui et al., 1975
187 Komagatake	Bateikei-kako	1.0	9	0.111	7.4		Katsui et al., 1975
193 Usu	Zenkoji	* 0.5	6.5	0.077	5.7	0.3	Oba, 1966
194 Yotei		1.6	9.0	0.178	6.4		I. Moriya, pers. commun., 1982
— Shiribetsu		0.75	7.5	0.1	6.3		I. Moriya, pers. commun., 1982
216 Shikaribetsu	Eishin	* 0.8	7	0.114	5.7		Yamagishi, 1977
216 Shikaribetsu	Pankechin	* 0.8	9	0.089	7.7		Yamagishi, 1977
Kharimkotan		1.1	7	0.16	5.2		Gorshkov and Dubik, 1970
Bezymianny	1956	2.2	28	0.079	24	1.0	Gorshkov, 1959
Shiveluch	1964	2	15	0.13	12	1.5	Gorshkov and Dubik, 1970
St. Helens	1980	* 2.6	28	0.093	24	2.5±0.5	Voight et al., 1981
Shasta		3.5	45	0.078	39		
Mombacho		* 1.3	12	0.108	9.9	1	Ui, 1972a
Parinacota	Cotacotani I	* 2.0	19	0.105	16		Katsui and Gonzalez, 1968
White Island	1914	0.06	1	0.06	0.9		Cotton, 1944
Ruapehu		1.7	16	0.106	13		Grange, 1931
Egmont	Pungarehu	2.5	27	0.093	23	6	Grange, 1931
Galunggung	1822	1.8	20	0.09	17		Cotton, 1944
Etna		2.7	24	0.113	20		Tanguy and Kieffer, 1977

Note: Digits given preceding volcano name is reference number in "Volcanoes of Japan" (Ono et al., 1981). H = Collapse height; L = travel distance; L_e = excessive travel distance; V = deposit volume; * = field-verified deposit.

TABLE II

Nonvolcanic dry avalanche deposits (after Hsü, 1975)

Locality	H(km)	L(km)	H/L	L_e(km)	V(km^3)
Schächental	1.8	3.1	0.58	0.2	0.0005
Val Lagone	1.05	2.4	0.44	0.7	0.0005—0.0008
Mombiel	0.37	0.8	0.47	0.2	0.0008
Huascaran	4.2	14.5	0.30	11.8	0.002
Wengen 1	0.5	1.1	0.45	0.3	0.002—0.003
Wengen 2	0.59	1.4	0.42	0.45	0.005—0.006
Elm	0.71	2.3	0.31	1.15	0.01
Disentis	0.74	2.1	0.36	0.9	0.01—0.02
Corno di Dosde	1.2	3.7	0.32	1.8	0.02
Madison	0.43	1.6	0.27	0.9	0.029
Voralpsee	1.1	3.4	0.33	1.6	0.03
Frank	0.87	3.5	0.25	2.1	0.03
Sherman	1.3	6.2	0.21	4.1	0.03
Gordau	1.2	6.0	0.21	4.0	0.03—0.04
Gros Ventre	0.56	3.4	0.17	2.5	0.038
Diablerets	1.9	5.5	0.34	2.5	0.05
Scimada Saoseo	1.5	5.5	0.27	3.1	0.08
Obersee GL	1.8	5.0	0.36	2.1	0.12
Kandeltal	1.9	9.9	0.19	6.9	0.14
Poshivo	1.5	4.1	0.36	1.7	0.15
Vaiont	0.5	1.5	0.34	0.7	0.25
Blackhawk	1.2	9.6	0.13	7.6	0.28
Deyen, Glarus	0.74	6.6	0.11	5.4	0.6
Glarnish	1.9	7.5	0.25	4.5	0.8
Fernpass	1.4	15.6	0.09	13.3	1.0
Siders	2.4	17.4	0.14	13.5	1.0—2.0
Tamins	1.3	13.5	0.095	11.4	1.3
Pamir	1.5	6.2	0.24	3.8	2.0
Engelberg	1.6	7.4	0.22	4.8	2.5—3.0
Flims	2.0	15.6	0.13	12.3	12
Saidmarreh	1.5	18.9	0.08	16.5	20

The H/L ratio, equivalent to the apparent coefficient of friction for volcanic dry avalanche deposits, occurs within the range 0.18 to 0.06. The ratio slightly decreases with increasing H and L. The equivalent range for nonvolcanic dry avalanche deposits is between 0.58 and 0.08, decreasing with increasing H and L (Hsü, 1975). At the higher range of H and L, the ratio seems to be identical for both volcanic and nonvolcanic dry avalanches. Smaller H/L ratios for volcanic and nonvolcanic dry avalanches, compared with the case of sliding of rigid blocks (H/L = 0.62, (Hsü, 1975)), suggests lower rigidity for both kinds of dry avalanches during the process of transportation. The similarity of the H/L ratio for volcanic and nonvolcanic dry avalanches suggests that the mode of transportation of both are basically identical; both are related to gravitational sliding associated with

TABLE III

Nuée ardente deposits

Name of volcano	Name of deposit	H(km)	L(km)	H/L	L_e(km)	V(km^3)	Data source
60 Hakone	Kamiyama	* 0.8	4.7	0.17	3.4		Oki et al., 1978
76 Hakusan		0.5	1.5	0.33	0.7	0.002	Yamasaki et al., 1964
83 Shirouma-Oike	Kazefukidake	1	4	0.25	2.4	0.03	Sakuyama, 1980
91 Asama	Kanbara	* 1.45	8	0.18	5.7	$10^{-2\sim-3}$	Aramaki, 1963
135 Chokai	Odaino	* 0.9	10	0.09	8.5	1.0	Ui, 1972b
216 Shikaribetsu	Ojigahara	0.8	1.0	0.08	8.7		Yamagishi, 1977
216 Shikaribetsu	Shinkai	* 0.9	11	0.08	9.5		Yamagishi, 1977
Santiaguito	April 1973	1.5	4.2	0.36	1.8		Rose et al., 1977
Santiaguito	Sept. 1973	0.9	3.0	0.3	1.5	0.002	Rose et al., 1977
Ngauruhoe	1974	1.4	1.9	0.74	0		Sheridan, 1980
Mayon	1968	2.1	4.7	0.45	1.3	0.015	Moore and Melson, 1969
Hibok-Hibok	1951	1.5	5.6	0.27	3.2		Sheridan, 1980

Note: Digits given preceding volcano name are reference numbers in "Volcanoes of Japan" (Ono et al., 1981). H = collapse height; L = travel distance; L_e = excessive travel distance; V = deposit volume; * = field-verified deposit.

slope instability just as observed at Mount St. Helens (Christiansen, 1980; Voight et al., 1981) and in numerous examples of nonvolcanic dry avalanche (Hsü, 1975). Hsü demonstrated that the apparent coefficient of friction for nonvolcanic dry avalanches decreases with increasing volume. A similar trend is obtained for the data of volcanic dry avalanche deposits (Fig. 6).

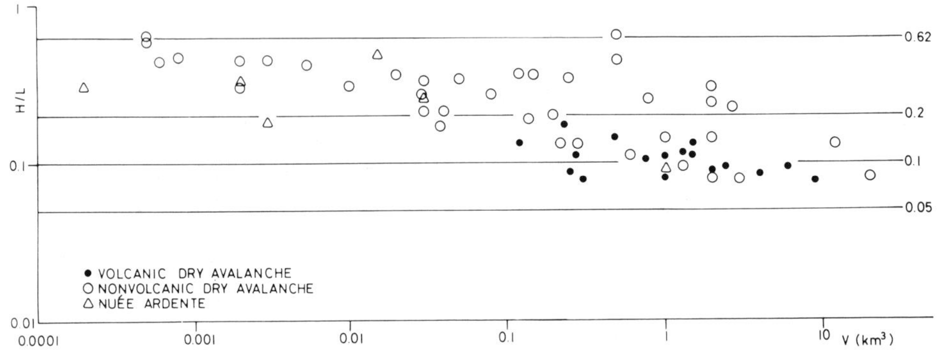

Fig. 6. Relationship between apparent coefficient of friction (H/L) and deposit volume (V). Symbols are identical with Fig. 5. H/L slightly decreases with increasing V.

A parameter called excessive travel distance (L_e) was defined by Hsü (1975) as the travel distance beyond the point obtainable by sliding of a body with an 0.62 friction coefficient. The relation is expressed by:

$$L_e = L - H/0.62$$

Volcanic dry avalanches have a slightly larger excessive travel distance (L_e) compared with nonvolcanic dry avalanches of the same volume range (Fig. 7).

Slightly smaller H/L ratios and slightly larger L_e values for volcanic dry avalanches are explained by the following factors:

(1) A volcanic edifice consists structurally of alternating soft pyroclastic layers, rigid lava flows and welded pyroclastic layers. Parts of the volcanic mass, especially around the vent, have suffered from fumarolic alteration.

(2) Fracturing of solid lava or welded pyroclastic layer (jigsaw puzzle effect) is common in parts of volcanic dry avalanche deposits. In case of Mount St. Helens, fractures are observed both in the dry avalanche deposit and in parts of the amphitheater wall (Voight et al., 1981). Bulging and fracturing of the surface due to an intrusion of a new cryptodome started at least 2 months before the climax eruption (Voight et al., 1981). This evidence suggests that fractures were already growing prior to the initiation of sliding.

(3) Heat discharge from the new cryptodome is helpful in converting

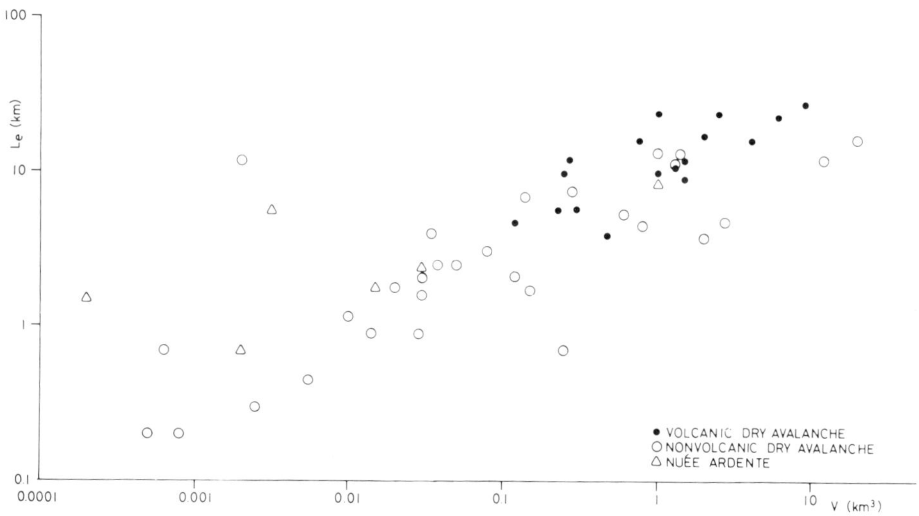

Fig. 7. Relationship between excess travel distance (L_e) and deposit volume (V). Symbols are identical with Fig. 5. L_e increases with increasing V. L_e for volcanic dry avalanche deposits are slightly larger than those for nonvolcanic dry avalanche deposits.

ground water to a supercritical fluid. Many projectiles with steam traillers were emitted from various portions of the sliding mass in the Mount St. Helens slide and consequent eruption. Sudden release of pressure due to initiation of sliding permitted the boiling of supercritical fluid within the volcanic edifice. This produced additional fracturing. A similar event was not documented for other similar historic eruptions, i.e. Bandai-san, Bezymianny and Shiveluch, but nevertheless possibly occurred.

These three factors explain the lower rigidity of the sliding block, the lower apparent coefficient of friction (H/L ratio) and the larger excessive travel distance (L_e) noted for the volcanic dry avalanche deposit.

CONCLUSIONS

Volcanic dry avalanche deposits are characterized by megablocks, jigsaw puzzle effects, hummocky surface, and amphitheater at the source. Similarity of the collapse height (H) to travel distance (L) ratio on volcanic and nonvolcanic dry avalanche (debris stream) deposits suggests the mode of transportation as gravitational sliding. Observations in the May 1980 eruption of Mount St. Helens support this model. The apparent coefficient of friction (H/L) is slightly smaller and the excessive travel distance (L_e) is slightly larger than those of nonvolcanic dry avalanches throughout the entire volume range. These differences are explained by smaller rigidity of the collapsing mass due to soft pyroclastic layers, alteration around the vent, fractures due to an intrusion of a cryptodome, and boiling of supercritical fluid contained within the collapsed mass.

ACKNOWLEDGEMENTS

I am much indebted to B. Voight for his critical review of the manuscript. Discussions with the following persons at the time of field observations of many volcanic dry avalanche deposits are helpful for this research: S. Aramaki, M. Brugman, D.R. Crandell, R.V. Fisher, H. Glicken, Y. Hayakawa, S. Kawachi, S. Kieffer, T. Kobayashi, P.W. Lipman, K. Mimura, H. Moreno, I. Moriya, Y. Nakamura, C. Newhall, M. Nolan, K. Ono, D.W. Peterson, J. Rosebaum, K. Sakaguchi, M. Sakuyama, K. Shibahashi, T. Sisson, T. Soya, K. Suzuki, K. Tadakuma, A. Taira, B. Voight, G.P.L. Walker and H. Yamagishi. Field work at Mount St. Helens in 1980 and 1981 were logistically supported by the David A. Johnston Cascade Volcano Observatory, U.S. Geological Survey. Expenses for field work at Mount Mombacho, Mount St. Helens and Mount Parinacota were supported by NASA Research Grant (NGR 038-033-012 to A.R. McBirney University of Oregon) and Scientific Research Grant, Ministry of Education, Japanese Government (Grant number 502401 to D. Shimozuru, University of Tokyo and grant number 56041015 to M. Kono, Tokyo Institute of Technology). Ms. H. Wada typed the manuscript.

REFERENCES

Aramaki, S., 1963. Geology of Asama volcano. J. Fac. Sci., Univ. Tokyo, Sec. II, 14: 229—443.

Bagnold, R.A., 1954. Experiments on a gravity-free dispersion of large spheres in a Newtonian fluid under shear. Proc. R. Soc. London, Ser. A, 225: 49—63.

Christiansen, R.L., 1980. Eruption of Mt. St. Helens: Volcanology. Nature, 285: 531—533.

Cotton, C.A., 1944. Volcanoes as Landscape Forms. Whitcombe and Tombs, Christchurch, 416 pp.

Gorshkov, G.S., 1959. Gigantic eruption of the volcano Bezymianny. Bull. Volcanol., 20: 77—109.

Gorshkov, G.S. and Dubik, Y.M., 1970. Gigantic directed blast at Shiveluch volcano (Kamchatka). Bull. Volcanol., 34: 261—288.

Grange, L.I., 1931. Conical hills on Egmont and Ruapehu volcanoes. N.Z. J. Sci. Technol., 13: 376—384.

Hayatsu, K., 1976. Geologic study of Myoko Volcanoes, central Japan, part I. Stratigraphy. Mem. Fac. Sci., Kyoto Univ., 42: 131—170.

Heim, A., 1882. Der Bergsturz von Elm. Z. Dtsch. Geol. Ges., 34: 74—115.

Howard, K., 1973. Avalanche mode of motion: Implications from lunar examples. Science, 180: 1052—1055.

Hsü, K.J., 1975. Catastrophic debris streams (Sturzstroms) generated by rock falls. Geol. Soc. Am. Bull., 86: 129—140.

Hsü, K.J., 1978. Albert Heim: Observations on landslides and relevance to modern interpretations. In: B. Voight (Editor), Rock Slides and Avalanches, 1. Natural Phenomena. Elsevier, Amsterdam, pp. 70—93.

Ichimura, T., 1955. Activities of the Gassan volcano. Bull. Earthq. Res. Inst., 33: 419—432.

Imai, I., 1959. Explanatory text of the geologic map of Japan, Kanazawa. Geol. Survey Japan, Kawasaki, 27 pp. (in Japanese with English abstract).
Isshiki, N. and Ozawa, A., 1967. Mudflow hills on the northeastern foot of Iwaki volcano (abstract). Bull. Volcanol. Soc. Jpn., 2nd ser., 12: 149 (in Japanese).
Katayama, N., 1974. Old records of natural phenomena concerning the "Shimabara Catastrophe". Rep. Shimabara Volcanol. Obs., Kyushu Univ., 9: 1—45 (in Japanese with English abstract).
Kato, M., 1978. Gigantic explosion of Chokai volcano, 2600 years B.P. and associated Kisakata mud flow (abstract). Abstr. Annu. Meet., Geol. Soc. Jpn., p. 271. (in Japanese).
Kato, Y., 1964. Petrology of Nasu volcano. J. Jpn. Assoc. Miner. Petrol. Econ. Geol., 51: 233—243 (in Japanese with English abstract).
Katsui, Y. and Gonzalez, O.F., 1968. Geologia del area neovolcanica de los Nevados de Payachata. Univ. Chile, Dept. Geologia, Publ. 29, 61 pp.
Katsui, Y., Yokoyama, I., Fujita, T. and Ehara, S., 1975. Komagatake, its volcanic geology, history of eruption, present state of activity and prevention of disaster. Committee for Prevention of Disaster of Hokkaido, Sapporo, 194 pp. (in Japanese).
Kawachi, S., 1977. Geology of Yatsugatake district. Quadrangle series, scale 1 : 50,000. Geol Surv. Jpn., Kawasaki, 92 pp. (in Japanese with English abstract).
Kawachi, S., 1981. Otsukigawa "mud flow" of Yatsugatake volcano (abstract). Abstr. Annu. Meet., Geol. Soc. Jpn., p. 295 (in Japanese).
Lipman, P.W. and Mullineaux, D.R. (Editors), 1981. The 1980 eruptions of Mount St. Helens, Washington. U.S. Geol. Survey Prof. Paper 1250, 844 pp.
Macdonald, G.A., 1972. Volcanoes. Prentice-Hall, Englewood cliffs, N.J., 510 pp.
Mimura, K. and Kawachi, S., 1981. Nirasaki debris avalanche, a catastrophic event at the Yatsugatake volcanic chain, central Japan (abstract). Abstr., 1981 IAVCEI Symposium, Tokyo and Hakone, p. 237.
Mimura, K., Fujimoto, U., Hyuga, T., Ichikawa, S., Kawachi, S. and Taneichi, M., 1971. N.R.M. of Nirasaki pyroclastic flow (abstract). Bull. Volcanol. Soc. Jpn, 2nd Ser., 15: 146—147 (in Japanese).
Mizuno, H., 1958. Geomorphology of Okinajima mud flow area, special reference to mud flow hills. Tohoku Geography, 11(1): 22—24 (in Japanese).
Moore, J.G. and Melson, W.G., 1969. Nuées ardentes of the 1968 eruption of Mayon volcano, Philippines. Bull. Volcanol., 33: 600—626.
Moriya, I., 1968. Geomorphology and geology of Akagi volcano. Maebashi Regional Forestry Office, Ministry of Agriculture and Forestry, 65 pp. (in Japanese).
Nakamura, Y., 1978. Geology and petrology of Bandai and Nekoma volcanoes. Sci. Rep., Tohoku Univ., Ser. 3, 14: 68—119.
Oba, Y., 1966. Geology and petrology of Usu volcano, Hokkaido, Japan. J. Fac. Sci., Hokkaido Univ., Ser. 4, 13: 185—236.
Oki, Y., Aramaki, S., Nakamura, K. and Hakamata, K., 1978. Volcanoes of Hakone, Izu and Oshima. Bull. Hot Spring Res. Inst., Kanagawa Pref., 9(5): 1—88.
Ono, K., Soya, T. and Mimura, K., 1981. Volcanoes of Japan, 2nd edition, 1 : 2,000,000 map series. Geol. Survey Japan, Yatabe.
Rose, W.I., Jr., Pearson, T. and Bonis, S., 1977. Nuée ardente eruption from the foot of a dacite lava flow, Santiaguito volcano, Guatemala. Bull. Volcanol., 40: 23—38.
Sakuyama, M., 1980. Geology of Shirouma-Oike volcano. J. Geol. Soc. Jpn., 86: 265—274 (in Japanese with English abstract).
Sheridan, M.F., 1980. Emplacement of pyroclastic flows: a review. Geol. Soc. Am., Spec. Pap., 180: 125—136.
Shreve, R.L., 1968. The Blackhawk landslide. Geol. Soc. Amer., Spec. Pap. 108, 47 pp.
Stoffel, D.B. and Stoffel, K.L., 1980. Mt. St. Helens seen close up on May 18. Geotimes, 25(10): 16—17.

Sumi, K., Ozawa, A. and Hirayama, J., 1962. Explanatory text of geologic map of Japan, Daira Kozan. Geol. Surv. Jpn., Kawasaki, 51 pp. (in Japanese with English abstract).
Tachibana, K., 1978. Volcanic activity and mud flow of Iwate volcano during Jomon period, concerning mud flow hills at Iwatemori and Gohyakumori. Annu. Rep., Fac. Education, Iwate Univ., 38: 63—90 (in Japanese).
Tanguy, J.C. and Kieffer, G., 1977. The 1974 eruption of Mount Etna. Bull. Volcanol., 40: 1—14.
Tsuya, H., 1940. Geological and petrological studies of volcano Huzi (Fuji). III. 3. Geology of southwestern foot of volcano Huzi. Bull. Earthq. Res. Inst., 18: 419—445 (in Japanese with English abstract).
Ui, T., 1972a. Recent volcanism in Masaya-Granada area, Nicaragua. Bull. Volcanol., 36: 174—190.
Ui, T., 1972b. Fault scarps on the slope of the Chokai volcano and genesis of pyroclastic rocks distributed at the flank. In: Chokai-san Tobishima, Yamagata Pref., 8—13 (in Japanese).
Ui, T., 1975. Origin of so-called "mud-flow deposit" in northwestern foot of Gassan volcano. In: Dewasanzan Hayama, Yamagata Pref., 15—20 (in Japanese).
Ui, T. and Shibahashi, K., in press. Quaternary volcanoes in Yamagata Prefecture, northern Japan — a review. In: Geology of Yamagata Prefecture (in Japanese with English abstract).
Voight, B., Glicken, H., Janda, R.J. and Douglass, P.M., 1981. Catastrophic rockslide avalanche of May 18. U.S. Geol. Surv., Prof. Pap., 1250: 347—377.
Williams, H. and McBirney, A.R., 1979. Volcanology. Freeman and Cooper, San Francisco, Calif., 397 pp.
Yamagishi, H., 1977. Pyroclastic flow deposits from the Shikaribetsu volcanoes, Hokkaido, Japan. Rep. Geol. Surv. Hokkaido, 49: 37—48 (in Japanese with English abstract).
Yamasaki, M., Nakanishi, N. and Keseno, Y., 1964. Nuée ardente deposit of Hakusan volcano. Sci. Rep., Kanazawa Univ., 9: 189—201.

Journal of Volcanology and Geothermal Research, 18 (1983) 151—164
Elsevier Science Publishers B.V., Amsterdam — Printed in The Netherlands

EXTENSIVE ASH FALLS IN AND AROUND THE SEA OF JAPAN FROM LARGE LATE QUATERNARY ERUPTIONS

HIROSHI MACHIDA and FUSAO ARAI

Department of Geography, Tokyo Metropolitan University, Tokyo (Japan)
Department of Geology, Gunma University, Maebashi (Japan)

(Received October 20, 1982)

ABSTRACT

Machida, H. and Arai, F., 1983. Extensive ash falls in and around the Sea of Japan from large, late Quaternary eruptions. In: S. Aramaki and I. Kushiro (Editors), Arc Volcanism. J. Volcanol. Geotherm. Res., 18: 151—164.

Five widespread ashes of late Quaternary age have been recognized on land and also in deep-sea sediments in and around the Sea of Japan. All are favorable for determining the magnitude and type of the great eruptions, and for establishing a Quaternary chronology around Japan.

Accurate determination of refractive indices of volcanic glass shards and phenocrysts, together with chemical data, have enabled successful characterization of the tephra layers and correlation between land and sea.

Of the five tephras two are alkalic and are the products of major eruptions of two Korean Volcanoes: the Baegdusan—Tomakomai ash (B-Tm) and the Ulreung-Oki ash (U-Oki), The B-Tm ash is recognized at the top of most cores in the northern part of the Sea of Japan and also on Hokkaido and the northern part of Honshu, where a reliable archeological age is given around the 11th century. The U-Oki ash occurs between the Kikai-Akahoya and Aira-Tn ashes in cores from adjacent to the Ulreung-do to central Honshu, where three radiocarbon ages of around 9300 yr B.P. were obtained.

The other three tephras are subalkalic and are correlated with large-scale eruptions producing voluminous pyroclastic-flows and calderas in Kyushu: the Kikai-Akahoya (K-Ah), Aira-Tn (AT) and Aso-4 ashes. The former two have been well described already but their fallout areas are partly revised in this manuscript. The AT ash is the most prominent marker, recognized not only on land of Japan but also in cores from the whole area of the Sea of Japan and part of the Pacific Ocean. The airfall Aso-4 ash, simultaneous with the Aso-4 pyroclastic-flow eruption is also described. It mantles very extensive areas and serves as an important time marker about 70,000 years ago.

INTRODUCTION

Machida and Arai (1976, 1978) reported two widespread tephras, the Akahoya and Aira-Tn ashes, which were produced by gigantic eruptions of the Kikai and Aira calderas respectively in southern Kyushu. These tephras are of great significance in volcanology as well as Quaternary research

0377-0273/83/$03.00

because they have a great volume and are distributed throughout western and central Japan.

These discoveries encouraged identification and correlation of other widespread tephras such as the Daisen-Kurayoshi pumice of late Pleistocene (Machida and Arai, 1979) and several middle Pleistocene tephras (Machida et al., 1980).

In Japan many tephra layers have been documented and a catalogue of Quaternary marker-tephras is used in studies of widespread tephra layers (Arai and Machida, 1980). Correlation of proximal tephras with distal ones is best achieved if, in addition to the stratigraphic position and the radiometric age, several petrographic parameters can be determined.

In this paper five major tephras are reported in and around the Sea of Japan. Two of them are alkalic and thickness and grain-size variation suggests that they have come from Korean volcanoes. Also, new data on the distribution of the subalkalic widespread tephras such as Kikai-Akahoya, Aira-Tn and Aso-4 ashes, originating from calderas in Kyushu, are given from the deep-sea studies.

CHARACTERIZATION AND IDENTIFICATION OF MARKER TEPHRA

Characterization and identification of tephras should be based on a combination of their features. The most important methods are stratigraphic and physical ones in proximal areas, but when we need to correlate deep-sea ashes and/or distal ashes with proximal ones, petrographic and chemical characterizations are indispensable. In practice we use all of the correlation data. Accurate determinations of refractive indices of glass and phenocrysts are used as the first and most efficient parameter for quick characterization of distal ashes, together with crystal assemblages and micrographic features of glass shards. Individual tephras may contain glass giving a range of indices that overlaps those of other tephras, and is then most usefully combined with such properties as indices of crystals. We have data for many Quaternary tephras in Japan (Arai, 1972; Machida et al., 1974; Yoshikawa, 1976; Machida and Arai, 1976; Arai and Machida, 1980). The chemical composition of glass and crystal is also used for discriminating tephras which are similar in other respects.

A list of the five widespread tephras reported in this paper is shown in Table I. Two, the Baegdusan-Tomakomai and Ulreung-Oki ashes are distinctive in being alkali feldspar bearing trachyte-rhyolite. Such ashes are rarely found in Japanese volcanoes. The other three tephras are subalkalic rhyodacite which commonly occurs in Japan.

TABLE I

A list of widespread tephras occurring in and around the Sea of Japan

Source and Tephra	Age (10^3 yr B.P.) Dating method	Mode of emplacement in proximal areas	Bulk volume* (km^3)	Volcanic glass		Phenocrystal minerals
				Type	Refractive index	
Baegdusan-Tomakomai (B-Tm)	0.8—0.9 Archeology	pfa, pfl	40—50	pm>bw	1.506—1.516	af: (ho, cpx) af n_1 = 1.522—1.524
Kikai-Akahoya (K-Ah)	6.3 C-14	pfa, pfl (Koya)	150	bw⩾pm	1.508—1.514	pl; opx, cpx opxγ = 1.709—1.712
Ulreung-Oki (U-Oki)	9.3 C-14	pfa, pfl	—10	pm	1.514—1.524	af; bi, ho, cpx af n_1 = 1.521—1.524 ho n_2 = 1.730—1.741
Aira-Tn (AT)	21—22 C-14	pfa (Osumi), pfl (Tsumaya, Ito)	300	bw>pm	1.498—1.501	pl; opx, cpx, (ho, qt) opxγ = 1.728—1.734
Aso-4	ca. 70 Stratigraphy	pfl (Aso-4)	200—	bw⩾pm	1.506—1.514	pl; ho, opx, cpx opxγ = 1.699—1.701 ho n_2 = 1.685—1.691

*Including pyroclastic-flow deposits.
pfa = pumice fall deposit; pfl = pyroclastic-flow deposit; pm = pumiceous shard; bw = bubble-walled shard.
af = alkali feldspar; pl = plagioclase; opx = orthopyroxene; cpx = clinopyroxene; ho = hornblende; bi = biotite; qt = quartz.

KOREAN WIDESPREAD TEPHRAS

Baegdusan-Tomakomai ash

Two vitric ashes comprising abundant micropumice and glass shards occur in strata bearing the human remains of the Heian age (AD794—1192) in northern Honshu (Fig. 1; *k—u*). Both have previously been thought to come from Towada caldera in northern Honshu, but have recently been shown to be completely different from each other in petrographic properties (Machida et al., 1981).

The lower tephra (Towada-a) is pyroxene bearing rhyodacite and is assigned to the phreatomagmatic eruption of Towada at the beginning of the 10th century. It is dispersed south of Towada caldera and is found almost everywhere in the Tohoku district.

The upper tephra (Baegdusan-Tomakomai ash) is characterized by fine-grained pumiceous and bubble-walled glass shards and small amounts of alkali feldspar. Refractive indices at several localities are shown in Fig. 2. It is very significant that the glass component in this ash is characterized by a high alkali content.

The upper ash occurs not only in the northern Tohoku district but also in the southern half of Hokkaido as a thin layer with maximum thickness of 5 cm and average maximum grain-size of 0.2 mm. It is named after the volcano, Baegdusan, and the type locality, Tomakomai, Hokkaido. Around Tomakomai it occurs above the Tarumai-c pumice (ca. 2000 yr B.P.) and below the Tarumai-b pumice (AD 1667) (Fig. 1; Section *t*).

This upper ash is also recognized in several piston cores from the northern Sea of Japan (Figs. 1 and 3). Thickness and grain-size gradually increase toward the west. The volcano Baegdusan (Korean name) or Changbaishan (Chinese name), on the boundary between North Korea and China (42° N,

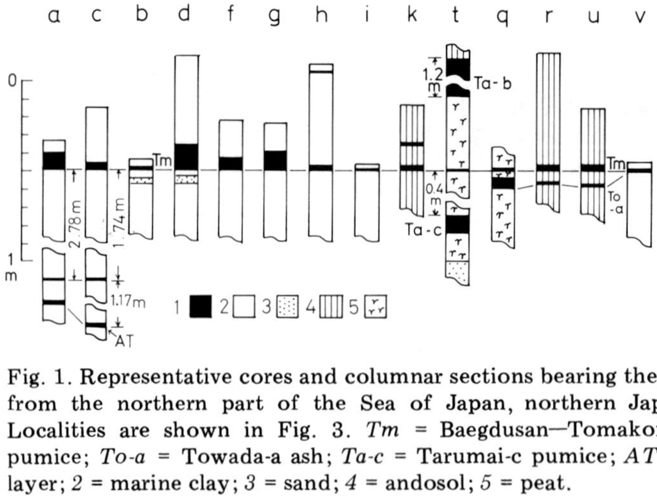

Fig. 1. Representative cores and columnar sections bearing the Baegdusan-Tomakomai ash from the northern part of the Sea of Japan, northern Japan and the Pacific Ocean. Localities are shown in Fig. 3. *Tm* = Baegdusan—Tomakomai ash; *Ta-b* = Tarumai-b pumice; *To-a* = Towada-a ash; *Ta-c* = Tarumai-c pumice; *AT* = Aira-Tn ash. *1* = tephra-layer; *2* = marine clay; *3* = sand; *4* = andosol; *5* = peat.

128°10′E), has a prominent caldera, with an approximate diameter of 4.5 km, at the top and is extensively covered by thick pumice falls and ignimbrite sheets (Asano, 1948). It seems probable that these ejecta represent the youngest major eruption of this volcano. Pumice blocks were found by Asano (1948) to be alkali rhyolite with alkali feldspar and micro-phenocryst of aegirineaugite. Correlation between the Baegdusan pumice and the distal B-Tomakomai ash is highly probable because of the similarity in petrographic and stratigraphic features.

As shown in Fig. 2, the variation of refractive index of glass in distal B-Tomakomai ash is 1.506—1.516, suggesting a heterogeneous composition. Relatively large variation among some trace elements is also recognized in glass separates, and will be reported elsewhere. This ash is younger than the Towada-a ash from the beginning of the 10th century, and is older than AD 1334, the documented age for construction of the old castle, Nejo in Hachinohe. The stratigraphic position suggests that it occurs in a time range between the end of the 10th century and the 11th century. Recent eruption of Baegdusan volcano is also suggested by the fact that vegetation at the summit has not yet recovered completely from a burial caused by pumice falls (Zhao, 1981): Buried forests are reported at several exposures high above the present timber line (Asano, 1948).

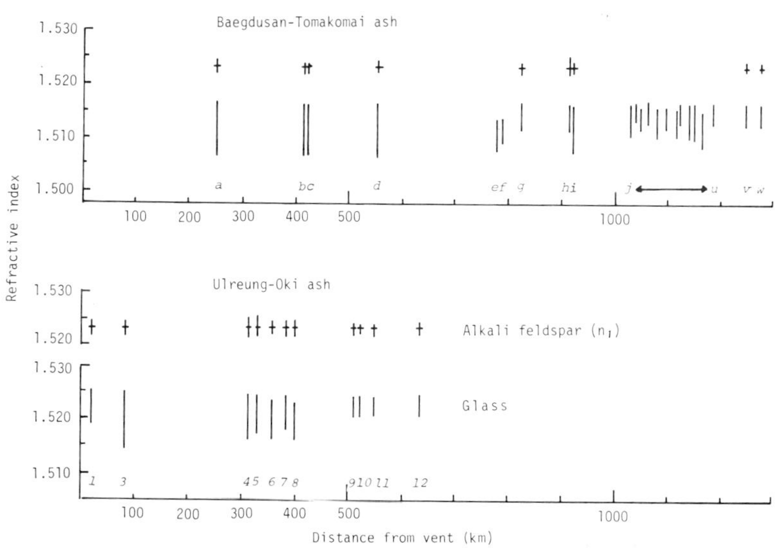

Fig. 2. Refractive indices of volcanic glass and alkali feldspar in the Baegdusan-Tomakomai ash and the Ulreung-Oki ash. Localities, *a—w* and *1—12*, are shown in Fig. 3.

On present data the approximate fallout limit is as in Fig. 3. Case histories from northern Japan, Northern Korea and Northeastern China will be presented elsewhere to demonstrate how useful this tephra is to archeologists as well as volcanologists.

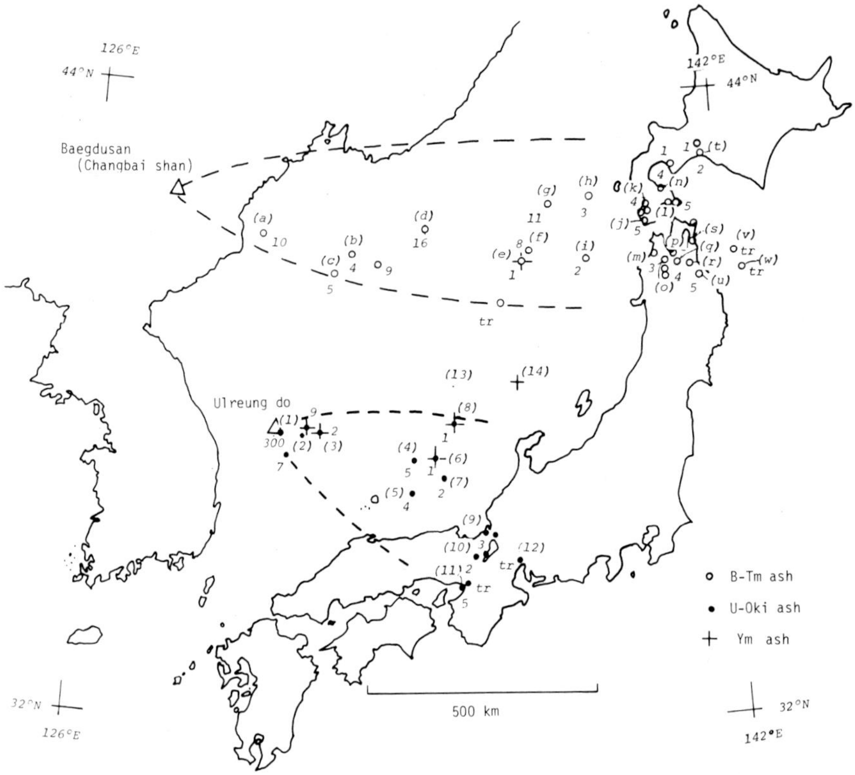

Fig. 3. Map showing localities where the marker-tephras from the Korean volcanoes have been identified. Thickness of the B-Tm ash and the U-Oki ash are shown in cm by numerals. Localities are shown by numerals in parenthesis: *a* = St. 6913; *b* = KH69-2-25; *c* = St. 6920; *d* = KH69-2-23; *e* = P129; *f* = KH77-3-M5; *g* KH79-3-C2; *h* = P130; *i* = V28-271; *j*—*u* = several localities in northern Japan; *v* = P82; *w* = P83. *1* = To-dong, Ulreung; *2* = P106; *3* = KH77-3-M3; *4* = KT79-6-19; *5* = KH77-3-M2; *6* = KH79-3-C3; *7* = P124; *8* = KH77-3-L4; *9* = Torihama site; *10* = Mizorogaike, Kyoto; *11* = South port of Osaka; *12* = Yatomi; *13* = P127; *14* = V28-268.

Ulreung-Oki ash

The other Korean tephra, the Ulreung-Oki ash, was first discovered at the Torihama archeological site, in northern Kinki (loc. *9* in Fig. 3), where it forms whitish thin ash layer in strata bearing human remains of the earliest Jomon culture. The ash is trachytic and is mainly composed of micropumice with a maximum diameter of 0.3 mm and alkali feldspar, biotite, amphibole (probably kaersutite) and clinopyroxene as phenocrysts. The refractive index of glass (n) ranges from 1.517 to 1.524, of alkali feldspar (n_1) is about 1.523 and of amphibole (n_2) is about 1.740 (Fig. 2; loc. *9*). Correlatives of this ash have been recognized at six localities in central Honshu (Fig. 3).

Occurrence of alkali feldspar and specific amphibole clearly shows that this ash is alkalic. Alkalic volcanoes occur in and around the Sea of Japan, but most of those situated in western Japan are too small and were too weakly explosive in the Quaternary to produce voluminous ashes. The ash occurs between the Kikai-Akahoya and Aira-Tn ashes in cores from adjacent to Ulreung island to Oki bank (Figs. 3 and 4), and the grain-size gradually increases toward Ulreung island.

Ulreung island, a large Quaternary volcano, is alkali-basaltic, with trachy-basaltic and trachytic differentiation products (Harumoto, 1970). Among upper Quaternary products, a thick plinian pumice, Ulreung pumice-1, is very prominent, covering almost every gentle slope in the island, and is about

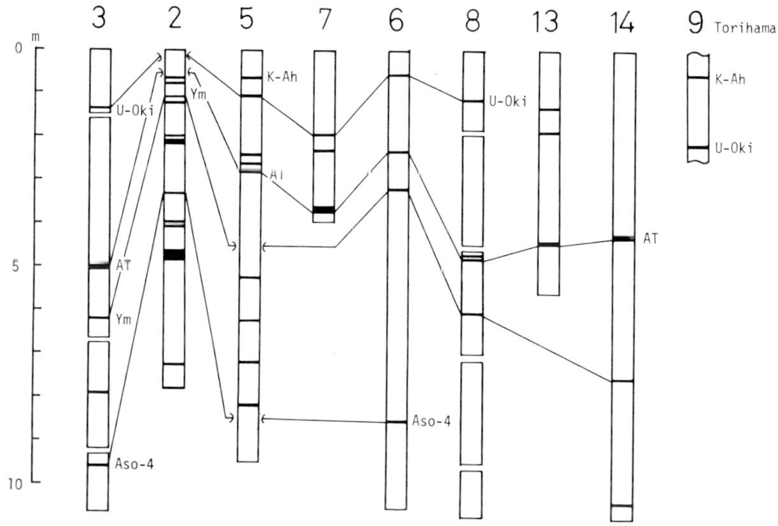

Fig. 4. Representative cores and columnar sections bearing several marker-tephras from the southern part of the Sea of Japan. *K-Ah* = Kikai-Akahoya ash; *U-Oki* = Ulreung-Oki ash; *AT* = Aira-Tn ash; *Ym* = Yamato ash; *Aso-4* = Aso-4 ash. Localities are shown in Fig. 3.

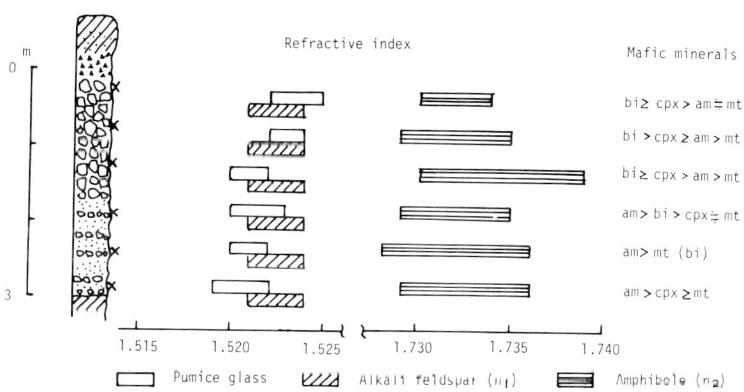

Fig. 5. Stratigraphy and variation of refractive indices of glass, alkali feldspar and amphibole and of assemblage of mafic minerals in the Ulreung-pumice-1. Locality, To-Dong, Ulreung-do.

3 m thick with a maximum grain-size about 20 cm near To-dong on the southeastern coast. Figure 5 shows a profile and compositional variation of this deposit. The refractive indices of glass and minerals clearly support the correlation with the distal ashes in the Sea of Japan and Honshu (Fig. 2). Hence the Ulreung-Oki ash (U-Oki) is named after the source volcano and type locality in the Sea of Japan.

Ulreung volcano has a caldera, Nari col, with a longer diameter of 3.5 km and a central cone, Arpong. It seems probable that the caldera has had a series of silicic tephra eruptions, forming a sequence of airfall deposits, and the Ulreung-Oki ash represents the largest recent eruption of this caldera. Three radiocarbon dates of ca. 9300 yr B.P. were obtained at the Torihama site and south port of Osaka for this ash (Arai et al., 1981). The youngest activity of Ulreung formed Arpong cinder cone and lava flows.

The U-Oki ash is used as an excellent time-marker in analyzing paleo-sea levels and paleo-oceanographical conditions in the Sea of Japan, together with the Kikai-Akahoya ash (6300 yr B.P.). Maeda (1976) reported that the ash correlated with the U-Oki ash was recognized in littoral or brackish clay about 30.5 m below present sea level at the south port of Osaka and that sea level rose slowly from −28 m to −20 m during 10,000 to 8000 yr B.P. Rapid rise is recorded in marine clay deposits above the U-Oki ash. Arai et al. (1981) discussed paleo-oceanographic conditions in the Sea of Japan during late Quaternary. Paleosalinity and paleotemperature estimates from the ^{18}O values of the bentic and planktonic foraminiferal tests in the cores from the Oki bank, localities Nos. *5* and *6* in Figs. 2 and 5, indicate that a remarkable inflow of the Tsushima warm current through the Straits of Korea into the Sea of Japan started between 9300 and 6300 yr B.P. indicated by the U-Oki and K-Ah ashes. Warm-water planktonic foraminifera appeared in the sediments immediately above the U-Oki ash horizon, and temperature of the surface water increased abruptly by 7—8° C. Influence of the inflow of the Tsushima current on local climate and vegetation along the coast of the Sea of Japan will be analyzed later by using these two excellent time-marker tephras.

Other Korean tephras

The cores Nos. *a* and *c* (Figs. 1 and 3) taken 70—250 km to the southeast of Chonjin, North Korea, contain an alkali rhyolitic tephra between the B-Tm and AT ashes. It is 2 cm to 10 cm thick composed of micropumice and glass shards with minor amounts of biotite, amphibole, clinopyroxene and alkali feldspar. Average maximum diameter of pumice is 1.0 mm. Refractive index of glass ranges from 1.507 to 1.518 (modal range: 1.510—1.515) and that of alkali feldspar 1.522—1.524. These petrographic features are similar to those of the B-Tm ash, suggesting that it also came from Baegdusan volcano. Stratigraphic position indicates that this ash is latest Pleistocene or earliest Holocene.

The other tephra is recognized between the AT and Aso-4 ashes in several cores from east of Ulreung island (Fig. 3). It was named the Yamato ash (Ym) after the Yamato bank by Arai et al. (1981) and is composed of whitish micropumice with small amounts of alkali feldspar, amphibole and biotite. Refractive index of glass (n) ranges between 1.521 and 1.524, modal index of alkali feldspar (n_1) is 1.522 and that of amphibole (n_2) 1.729. We suggest that Ulreung island is the source because the Yamato ash closely matches the U-Oki ash and is coarsest in the core No. 3 near Ulreung island. The age is around 30,000—50,000 yr B.P. from its stratigraphic position.

KYUSHUAN WIDESPREAD TEPHRAS

Five calderas in Kyushu have had major silicic tephragenic eruptions, forming a sequence of airfall and flow deposits, and some eruptions also produced remarkably extensive ashes. Of these at least three prominent ashes of late Pleistocene age have been identified in and around the Sea of Japan: the Kikai-Akahoya (K-Ah), Aira-Tn (AT) and Aso-4 ashes. All provide distinct datum-levels in Quaternary sediments and are significant time markers. The first two have already been reported (Machida and Arai, 1976, 1978) and short descriptions are given in this paper with special reference to deep-sea records.

Kikai-Akahoya ash

The Kikai-Akahoya ash (K-Ah, Machida and Arai, 1978) represents the largest recent eruption of Kikai caldera, 50 km south of Kyushu. The large-scale eruption approximately 6300 yr B.P. started with ejection of plinian airfall pumice, followed by the cataclysmic eruptions of two pyroclastic flows, and the widespread tephra, K-Ah, is assigned to an airfall part of the final and largest flows, the Koya flows.

The ash is rhyodacitic containing abundant bubble-walled glass shards and small amounts of orthopyroxene, clinopyroxene and opaques as mafic minerals. Refractive indices of glass and orthopyroxene are shown in Table I.

This ash mantles extensive areas from Kyushu to central Honshu and is also recognized and identified in several piston cores from parts of the Sea of Japan, the Pacific Ocean and the East China Sea (Fig. 6). Present data on areal distribution are incomplete, but the fallout area is now known to attain approximately 10^6 km^2 covering most of southwest to central Japan and adjacent seas. The bulk volume of the ejecta associated with the K-Ah ash is estimated to be more than 100 km^3.

More than thirty radiocarbon dates have been obtained, and the average is around 6300 yr B.P. This tephra has received major attention in archeological and geomorphological problems as well as in paleo-oceanography (Machida and Arai, 1978; Arai et al., 1981).

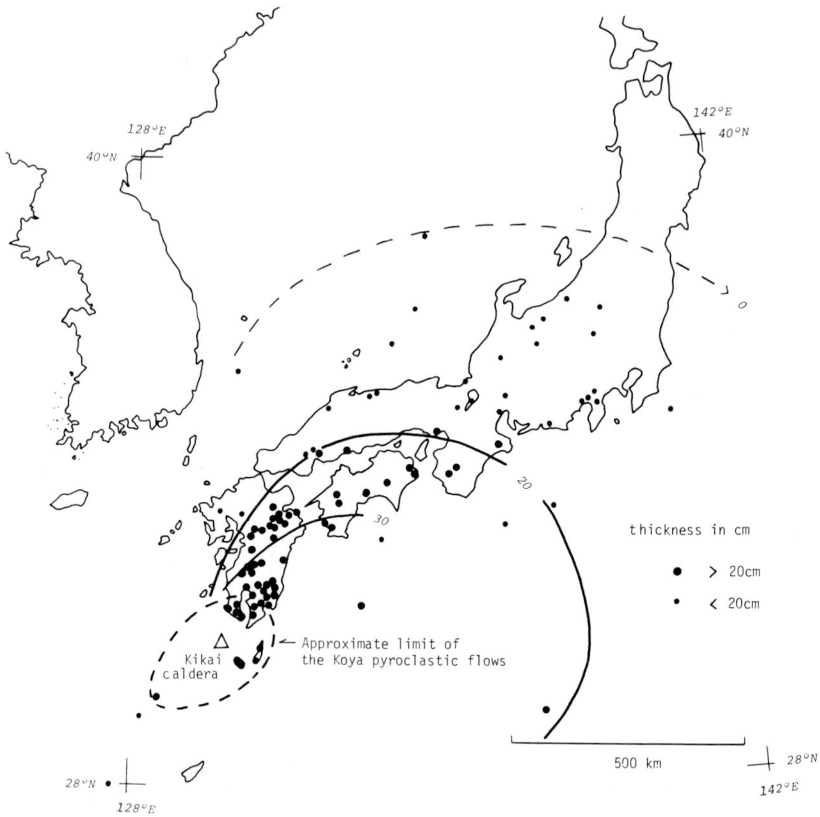

Fig. 6. Isopach map of the Kikai-Akahoya ash with thickness in cm.

Aira-Tn ash

The Aira-Tn ash (AT, Machida and Arai, 1976) is the most prominent marker-tephra of late Pleistocene, because it occurs most extensively on land and also in abyssal sediments around Japan (Fig. 7), where it is recognized in most piston cores and provides a distinct time-plane within deep-sea sediment in the whole Sea of Japan. More than twenty radiocarbon dates indicate that the cataclysmic eruption of Aira caldera occurred about 21,000—22,000 yr B.P.

The eruption of Aira caldera started with ejection of the great Osumi airfall pumice, followed by the Tsumaya pyroclastic flow and finally by the large-scale Ito pyroclastic flow. The AT ash is a coignimbrite airfall deposit of fine-grained materials from the upper part of an eruption column formed at the same time as the Ito flow.

This ash is composed of abundant rhyolitic bubble-walled glass shards with small amounts of pyroxenes, hornblende and plagioclase. Accurate determinations of the refractive indices of volcanic glass and orthopyroxene (Fig. 1), together with glass chemistry, have permitted identification of this tephra in abyssal sediment cores (Arai et al., 1981).

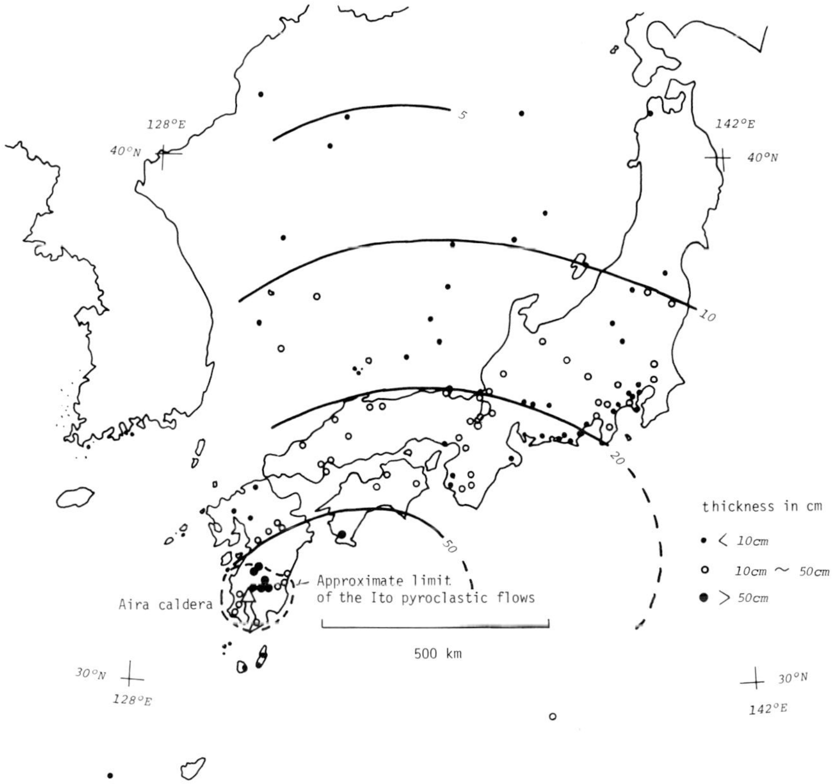

Fig. 7. Isopach map of the Aira-Tn ash with thickness in cm.

Aso-4 ash

The Aso-4 pyroclastic flow is one of the largest of late Pleistocene age in Japan covering extensive areas of Kyushu around the Aso caldera and western part of Honshu (Matumoto, 1943; Watanabe, 1978). Hitherto there were no data on airfall deposits simultaneous with the Aso-4 flows, but we have recently discovered a distal ash which has very similar petrographic and chemical properties to the flow deposits within four piston cores in the Sea of Japan (Arai et al., 1981), and is assumed to be an airfall part of the Aso-4 unit. It is named the Aso-4 ash. This discovery has further encouraged identification of this ash at many localities in central Honshu and in the Pacific Ocean as shown in Fig. 8. The piston core, KH80-3-12 bearing the Aso-4 ash in 143—145 cm level is located 1450 km east of the Aso caldera.

The Aso-4 ash is a crystal-vitric fine-grained ash, consisting of rhyodacitic bubble-walled glass shards and brown hornblende, orthopyroxene (bronzite) as mafic phenocrysts. Its identification is carried out by combined parameters; refractive index of glass ranges 1.506—1.511, low refractive index of

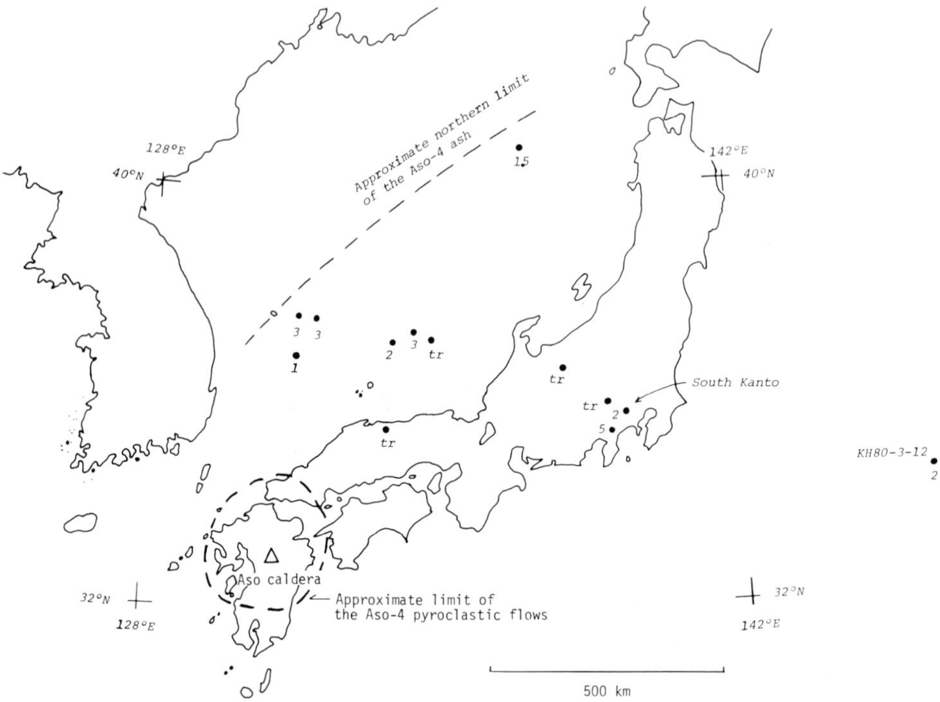

Fig. 8. Map showing localities where the Aso-4 ash has been identified. Numerals show thickness of the ash in cm.

orthopyroxene (γ = 1.699—1.701) and relatively high index of hornblende (n_2 = 1.687—1.688) (Table I). In addition glass in the ash is characteristic in color and bubble shape, and is alkali rich. These properties are similar to those of the Aso-4 pyroclastic flows (Watanabe, 1979). It is suggested that the ash fall is of coignimbrite type.

The bulk volume of the Aso-4 eruption should approximate to that of the Aira, over 200 km^3 including the Aso-4 ashes.

Available data for the radiometric age of the Aso-4 pyroclastic flows (Ono et al., 1977) suggest that it is older than 43,000 yr B.P. New data on the stratigraphic position in a standard tephra sequence of South Kanto show that it occurs below the Obaradai pumice of Hakone origin (66,000 yr B.P.) and above the Ontake Pm-I (ca. 80,000 yr B.P.) dated by the fission-track method on obsidian and zircon from the tephras (Machida and Suzuki, 1971). Thus the Aso-4 ash provides a very important marker for late Pleistocene events in and around Japan.

CONCLUSIONS

(1) Correlation of on-land and abyssal silicic tephra sequences based on accurate determination of refractive indices of glass and minerals, together

with other criteria, can be successful in establishing stratigraphy and chronology of widespread tephras in and around the Sea of Japan.

(2) The Ulreung-Oki ash was produced by plinian eruption. It is not certain, whether the Baegdusan-Tomakomai ash is associated with the pumice-fall, the ignimbrite, or both.

(3) Three widespread tephras which originated from Kyushu are all interpreted as coignimbrite ash falls. This type of tephra should be the dominant class of widespread tephra. Their specific features are: (a) very abundant fine-grained glass fragments, especially thin and platy bubble-walled glass shards caused by extraordinarily delayed vesiculation; and (b) very extensive fallout area and a huge volume, often equal to or more than the volume of pyroclastic flows ejected during the same cycle of eruption.

(4) Every tephra reported in this paper is favorable not only for establishing the magnitude and type of eruptions and their ages but also for studies in interdisciplinary subsciences including Quaternary research.

ACKNOWLEDGEMENTS

The authors wish to acknowledge the assistance given them by E. Honza, H. Moriwaki, T. Oba and H. Yamazaki in acquiring deep-sea tephra samples. S. Ishida and B. Lee are thanked for their instruction on geology of Ulreung island. T. Fukuoka provided valuable assistance with neutron activation analysis. The manuscript was substantially improved by the efforts of G.P.L. Walker and another referee. This work was performed under Grant in aid for scientific research of the Ministry of Education No. 56390016.

REFERENCES

Arai, F., 1972. Identification of particular tephras by means of refractive indices of orthopyroxenes or hornblendes — A fundamental study of tephrochronology. Quat. Res. Japan, 11: 254—270.

Arai, F. and Machida, H., 1980. Catalogue of marker-tephra layers of the late Quaternary age in Honshu and Kyushu, Japan. Karuishigaku Zasshi, 6: 65—76.

Arai, F., Oba, T., Kitazato, H., Horibe, Y. and Machida, H., 1981. Late Quaternary tephrochronology and paleo-oceanography of the sediments of the Japan Sea. Quat. Res. Jpn., 20: 209—230.

Asano, G., 1948. Several discoveries obtained from the Hakutosan (Baegdusan) Expedition in 1942—1943. Mineral. Geol., I: 27—31, 71—75, 128—130, 181—184, 228—231, 267—270.

Harumoto, A., 1970. Volcanic rocks and associated rocks of Ulreung island (Japan Sea). Dept. Geol. Mineral., Kyoto Univ., p. 39.

Machida, H. and Arai, F., 1976. The widespread tephra — The Aira-Tn ash. Kagaku, 46: 339—347.

Machida, H. and Arai, F., 1978. Akahoya ash — A widespread tephra erupted from the Kikai caldera, southern Kyushu, Japan. Quat. Res. Jpn., 17: 143—163.

Machida, H. and Arai, F., 1979. Daisen Kurayoshi pumice: Stratigraphy, chronology, distribution and implication to late Pleistocene events in central Japan. J. Geogr., 88: 313—330.

Machida, H. and Suzuki, M., 1971. A chronology of the late Quaternary tephras as established by fission-track dating. Kagaku, 41: 263—270.

Machida, H., Arai, F., Murata, A. and Hakamata, K., 1974. Correlation and chronology of the middle Pleistocene tephra layers in southern Kanto. J. Geogr., 83: 302—338.

Machida, H., Arai, F. and Sugihara, S., 1980. Tephrochronological study on the middle Pleistocene deposits in the Kanto and Kinki districts, Japan. Quat. Res. Jpn., 19: 233—261.

Machida, H., Arai, F. and Moriwaki, H., 1981. Two Korean tephras, Holocene markers in the Sea of Japan and the Japan islands. Kagaku, 51: 562—569.

Maeda, Y., 1976. The sea level changes of Osaka Bay from 12,000 B.P. to 6,000 B.P. J. Geosci., Osaka City Univ., 20: 43—58.

Matumoto, T., 1943. The four gigantic caldera volcanoes in Kyushu, Japan. J. Geol. Geogr., 19 (special issue): 1—57.

Ono, K., Matsumoto, Y., Miyahisa, M., Teraoka, Y., and Kanbe, N., 1977. Geology of the Taketa district. Quadrangle series, scale 1:50,000. Geol. Surv. Japan, 145 pp.

Watanabe, K., 1978. Studies on the Aso pyroclastic flow deposits in the region to the west of Aso caldera, southwest Japan, I: Geology. Memoirs, Fac. Educ. Kumamoto Univ., 27: 97—120.

Watanabe, K., 1979. Studies on the Aso pyroclastic flow deposits in the region to the west of Aso caldera, southwest Japan, II: Petrology of the Aso-4 pyroclastic flow deposits. Memoirs, Fac. Educ. Kumamoto Univ., 28: 75—112.

Yoshikawa, S., 1976. The volcanic ash layers of the Osaka Group. J. Geol. Soc. Jpn., 82: 497—515.

Zhao, D., 1981. Preliminary investigation on relationship between volcanic eruption of Changbai Mountain and the evolution of its vegetation. Res. Forest Ecosystem (Changbai Mountain Research Station of Forest Ecosystem, Academia Sinica), 2: 81—87.

Journal of Volcanology and Geothermal Research, 18 (1983) 165—190
Elsevier Science Publishers B.V., Amsterdam — Printed in The Netherlands

RELATIONSHIPS BETWEEN MINERALIZATION AND SILICIC VOLCANISM IN THE CENTRAL ANDES

P.W. FRANCIS[1], C. HALLS[2] and M.C.W. BAKER[3]

[1] *Lunar and Planetary Institute, 3303 NASA Road 1, Houston, TX 77058 (U.S.A.)*
[2] *Department of Mining Geology, Imperial College, Prince Consort Road, London SW7 (England)*
[3] *Hunting Geology and Geophysics, P.O. Box 363, Fyshwick, A.C.T. 2609 (Australia)*

(Received October 20, 1982)

ABSTRACT

Francis, P.W., Halls, C. and Baker, M.C.W., 1983. Relationships between mineralization and silicic volcanism in the Central Andes. In: S. Aramaki and I. Kushiro (Editors), Arc Volcanism. J. Volcanol. Geotherm. Res., 18: 165—190.

Studies of late Tertiary silicic volcanic centres in the Western and Eastern Cordilleras of the Central Andes show that three volcanic environments are appropriate sites for mineralization: (1) ring-fracture extrusions post-dating large calderas; (2) similar extrusions within ignimbrite shields; and (3) isolated, small silicic volcanoes. Subvolcanic tin mineralization in the Eastern Cordillera is located in silicic stocks and associated breccias of Miocene age. The Cerro Rico stock, Potosi, Bolivia, contains tin and silver mineralization and has an intrusion age apparently millions of years younger than that of the associated Kari Kari caldera. Similar age relationships between mineralization and caldera formation have been described from the San Juan province, Colorado. The vein deposits of Chocaya, southern Bolivia, were emplaced in the lower part of an ignimbrite shield, a type of volcanic edifice as yet unrecognized in comparable areas of silicic volcanism. The El Salvador porphyry copper deposit, Chile, is related to silicic stocks which may have been intruded along a caldera ring fracture. Cerro Bonete, Chile, provides a modern example of the volcanic superstructure which may have overlain isolated mineralized stocks and breccia pipes such as that of Salvadora at Llallagua, Bolivia.

Existing models for the genesis of porphyry copper deposits suggest that they formed in granodioritic stocks located in the infrastructure of andesitic stratovolcanoes. Sites of porphyry-type subvolcanic tin mineralization in the Eastern Cordillera of Bolivia are distinguished by the absence of such andesitic structures. The surface expression of a typical subvolcanic porphyry tin deposit was probably an extrusive dome of quartz latite porphyry, sometimes related to a larger caldera structure. Evidence from the El Salvador porphyry copper deposit in the Eocene magmatic belt in Chile suggests that it too may be more closely related to a silicic volcanic structure than to an andesitic stratovolcano.

The dome of La Soufriere, Guadeloupe is proposed as a modern analog for the surface expression of subvolcanic mineralization processes, the phreatic eruptions there suggesting the formation of hydrothermal breccia bodies in depth. Occurrence of mineralized porphyries, millions of years after caldera formation, does not necessarily indicate that intrusions and mineralization are not genetically related to the sub-caldera pluton, but may be a consequence of the long thermal histories (1—10 million years) of the lowermost parts of large plutons. Caldera formation can only inhibit mineralization by disper-

0377-0273/83/$03.00

sal of ore metals when these are of magmatic origin, and ignimbrites should not be taken as being unlikely to be associated with porphyry mineralization. Whether ore metals are of wall rock or magmatic origin, the key to understanding the relationships between silicic volcanism and mineralization lies in the fractionation of trace elements within large zoned magma chambers during their igneous history, and their subsequent hydrothermal migration. Small, highly mineralized intrusions formed late in a caldera cycle (such as the Cerro Rico) may be due to the introduction of fresh supplies of mafic magma into the lower parts of the main pluton.

INTRODUCTION

The Central Andes is one of the world's classic volcanic provinces, and the type locality of one of the main igneous rock types: andesite. The region is also well known for its mineral resources, which played a dominant role in the exploration and colonization of the continent. While major strides have been made over the last decade in understanding the petrogenesis of andesite and in detailed studies of individual ore deposits, much remains to be learned about the relationships between volcanism and mineralization. In this paper some important kinds of volcanic structures associated with ore deposition in the Central Andes are considered and a revised model for the volcanic relationships of certain porphyry-type deposits is presented. The relationships between caldera formation and ore deposition are also critically reviewed.

GEOLOGICAL BACKGROUND

Often considered as the standard example of a convergent plate margin, the Central Andes are more complex in detail than most models propose. Between latitudes 17° and 27°S, two volcanic belts are present, the Western and Eastern Cordilleras, separated by the 250-km wide Bolivian *altiplano*. The two cordilleras converge towards the southern end of the altiplano, the Eastern Cordillera becoming discontinuous and terminating in NW Argentina, while the Western Cordillera remains well defined as far as Lat. 28°S. During the late Tertiary, there is evidence that igneous activity migrated eastwards away from the trench (Farrar et al., 1970), but during the last million years or so this migration may have reversed (Francis et al., 1974; Kussmaul et al., 1977).

Radiometric dating surveys show that a major outbreak in volcanism was initiated some 20 m.y. ago in both Western and Eastern Cordilleras, perhaps in response to a change in spreading rate at the East Pacific Rise (Noble et al., 1974; Baker and Francis, 1978; Grant et al., 1979). In the Western Cordillera, construction of andesitic composite volcanoes and eruption of silicic pyroclastic flows have taken place concurrently from the initiation of volcanic activity up to the present day. Many fumarolically active andesitic volcanoes are present along the westernmost margin of the cordillera, while the youngest known major pyroclastic flow has been dated at 0.77 m.y. (Baker and Francis, 1978).

The situation in the Eastern Cordillera is complex and less well known, and it is probable that a different balance of geological processes was responsible for the distinctive petrological assemblage and lithophile mineralization found in this morphotectonically unique segment of the Andes (Angus et al., 1977; Halls, 1981). There is some evidence from both S. Bolivia and N. Argentina that construction of andesitic composite volcanoes preceeded the eruption of voluminous silicic pyroclastic flows (Grant et al., 1979; G.A. Kretschmar, unpubl. data) but it is clear that since about 20 m.y. ago silicic volcanism has been predominant over andesitic, especially in the northern part of the province. There are few old andesitic volcanoes, and no active ones, but silicic pyroclastic flows are so extensive as to form the Frailes ignimbrite plateau, covering some 13,000 km^2 and averaging about 1 km thick (Fig. 1). The age of the most recent activity is not well estab-

Fig. 1. Locality map of the central Andes, showing Western and Eastern Cordillera volcanic belts, Bolivian tin silver mineralization belt and key localities referred to in the text.

lished: the youngest dated ignimbrite from Bolivia has an age of 3.8 m.y. (Baker and Francis, 1978) but there are a number of extrusive domes which are presumed on morphological grounds to be less than 1 m.y. old. In the extreme south of the province in Argentina the youngest structure is Cerro Tuzgle, a small latitic composite volcano (Schwab and Lippolt, 1976).

The volcanoes of the Western Cordillera lie some 140 km above the well-defined eastward-dipping Benioff zone (Barazangi and Isacks, 1976), while those of the Eastern Cordillera lie some 300—350 km above it. Between the two cordilleras is situated the crustal "keel" of the Andes, where the continental crust reaches a thickness of some 70 km (James, 1971). The volcanic rocks of both cordilleras are overwhelmingly calc-alkalic and there is a general increase in K_2O content at standard SiO_2 eastwards away from the trench axis (Angus et al., 1977; Thorpe and Francis, 1979) but the pattern of silica-potash covariance is not straightforward, and the overall trend is complicated by the occurrence in SW Bolivia of more potassic rocks (shoshonites) (Kussmaul et al., 1977). The major, trace, REE and isotope chemistry of volcanic rocks from the Western Cordillera suggest that they were derived from asthenospheric mantle by partial melting followed by fractional crystallization and assimilation/contamination by ancient continental crust (Thorpe et al., 1981). The data from the more silicic and petrochemically specialized rocks of the tin belt of the Eastern Cordillera suggest that they were formed by somewhat different processes from those acting in the Western Cordillera. The rocks from the tin belt are strongly enriched in Li, Be, Rb and P_2O_5 relative to comparable andesites, dacites and rhyolites from the Western Cordillera (Angus et al., 1977) and the sum of petrological features considered together with the generally high initial strontium isotope ratios suggests that crustal contamination may have been a factor of greater importance in determining the character of the Eastern Cordillera magmas and their associated hydrothermal ore deposits than those of the Western Cordillera.

The zonation of mineral deposits in the Central Andes has been discussed by Sillitoe (1976, 1981) and by Halls (1981). Eastwards from the trench, belts of Fe; Cu- (Mo-Au); Cu-Pb-Zn-Ag; and Sn-W (Zn-Pb-Sb-Bi-Au) mineralization are successively encountered. The tin belt of Bolivia is distinctly polymetallic in addition to bearing the more crucial metallogenic association of tin and tungsten. The southern, subvolcanic, part of the tin province coincides with the area of Miocene and later volcanic activity in the Eastern Cordillera (Fig. 1), while the mineralization in northern sub-province is located within and at the margins of plutons of batholithic dimensions of predominantly Triassic age (McBride, 1977). The Sn-W vein deposits of this part of the belt are often associated with greisen alteration and show no evidence of volcanic associations in the strict sense. The northern part of the province thus displays a deeper level of erosion and a batholithic style of mineralization which contrasts with that of the southern subvolcanic province.

The Western Cordillera, made up in part of volcanic rocks of similar compositions erupted within the same period of time, has no conspicuously mineralized volcanic structures comparable with those of the Eastern Cordillera. There are some deeply eroded and fumarolically altered composite volcanoes such as Cerro Marquez (Sillitoe, 1973) which have undergone intense hydrothermal alteration. In the fumarolic deposits on Aucanquilcha volcano traces of covellite have been found in association with native sulphur (Clark, 1970), while at El Laco an unusual magnetite flow of volcanogenic origin is of some minor economic value (Park, 1961). All of the major economic porphyry copper deposits in the Central Andes pre-date the post-20 m.y. phase of volcanic activity, and are located in a narrow belt west of the Western Cordillera (Sillitoe, 1981).

TYPES OF VOLCANIC STRUCTURE ASSOCIATED WITH MINERALIZATION IN THE EASTERN CORDILLERA

The fullest accounts of the volcanic associated mineralization in the Eastern Cordillera have been presented by Sillitoe et al. (1975); and Grant et al. (1977, 1980). Grant et al. (1979) showed that in the tin-silver belt mineralization took place between 20 and 12 m.y. ago, while Sillitoe et al. (1975) argued that the tin silver deposits could be interpreted in terms of a genetic model similar to that proposed by Sillitoe (1973) for porphyry copper deposits; namely that the ore metals were deposited by hydrothermal fluids associated with the emplacement of porphyry stocks and breccia pipes into the lower part of a volcanic edifice. Accepting the general outline of this model, we examine in particular here the nature of the original volcanic structures associated with mineralization.

Three general characteristics of the igneous rocks associated with mineralization emerge from the studies of Grant et al. (1980):

(1) The mineralization is centered on small (typically 1—2 km^2), porphyry stocks of intermediate to silicic composition.

(2) The stocks are subvolcanic in character and were emplaced at shallow levels (<2 km) beneath or within the vent of a volcanic edifice.

(3) Hydrothermal breccia pipes are prominent: in some cases the stocks are essentially large breccia pipes. Typically, there were several phases of porphyry intrusion and brecciation. Boron-rich fluids clearly played an important part in the process of breccia formation, for tourmaline is abundant.

We have identified three volcanic environments which might correspond to the surface expression of these sub-volcanic features: late-stage ring-fracture extrusions adjacent to calderas; late-stage extrusions within ignimbrite shields; and small, isolated monogenetic silcic volcanic centers.

CALDERAS

The association of mineralization with large, ignimbrite-producing calderas is widely known, and has been particularly well documented in the San Juan province, Colorado, in southern New Mexico, Nevada and Oregon, and has been reviewed by Rytuba (1981). In many cases, mineralization was related to intrusion of small porphyry stocks into caldera ring fractures immediately after caldera formation, as in the case of Lake City caldera, but in others the mineralization followed caldera formation after a long interval; 6.5 m.y. after the formation of the Summitville and Platoro calderas, and more than 10 m.y. in other instances (Lipman et al., 1976).

Within the Central Andes, the Cerro Rico stock, Potosi, Bolivia provides a closely similar example of caldera-related mineralization. The stock is a dacite or quartz-latite porphyry which is pervasively altered throughout and which is located on the margins of the prominent Kari Kari caldera. The caldera has been described by Francis et al. (1981). It consists of a resurgent centre some 35 km in longest diameter consisting of a 1000-m-thick sequence of garnetiferous welded ignimbrites and surrounded on the north and northwest by outflow sheets of petrographically similar ignimbrites and a moat deposit of coarse breccia. Rocks from the resurgent center, outflow ignimbrites and moat deposit have yielded radiometric ages which are indistinguishable within error at 21.7 m.y. (Grant et al., 1979).

Historically, the Cerro Rico stock has been the site of one of the most important mineral deposits in South America, and is still an important source of tin. The upper part of the stock is silicified, while in the lower parts the alteration is sericitic and is characterized by the transformation of phenocrysts and groundmass to sericite and quartz with abundant pyrite. K-Ar studies suggest that the mineralization took place some 13 m.y. ago, about 7 m.y. after formation of the caldera (Grant et al., 1979). The age of the stock itself is not clear. On geological grounds it could be the same age as the caldera, but Grant et al. (1979) suggested that its emplacement preceeded mineralization by only 1—2 m.y. Since K-Ar dating of altered rocks may be unreliable, the age of the stock cannot be regarded as well determined.

Several studies of the mineralization at Cerro Rico have been carried out (see Grant et al., 1980, for review) but the stock itself has attracted less attention. Lindgren and Creveling (1928) first suggested that the stock was part of a volcano, whereas Turneaure (1960) suggested that its emplacement was not accompanied by volcanism. Sillitoe et al. (1975) concluded that "intrusion of the Cerro Rico stock could have represented one facet of activity at a volcanic center".

The stock is an oval intrusion measuring some 1700 × 1200 m at the surface and tapers downwards to a dyke-like body only 100 m wide at depth. The summit of the Cerro Rico is at an altitude of 4790 m and the youngest rocks which it intrudes are the grey shales of the Caracoles formation containing a fossil flora of probable Miocene age (Rivas and Carrasco, 1968). Francis et al. (1981) have interpreted the Caracoles formation as the product of sedimentation within a lake confined by the Kari Kari caldera rim. Consideration of the setting of the stock, its present geomorphological character as an isolated, conical hill and its intrusion into lake sediments all suggest that the stock was intruded at a very shallow level and is probably the root of a surface dacite dome of the sort that is characteristic of the ring-fracture zones of large resurgent calderas (Smith and Bailey, 1968). The uppermost part of the stock, above the level of intrusive contacts with sediments, may correspond with the originally subaerial part of the dome, but no trace of pyroclastic material that may have been erupted at the time of emplacement of the stock is preserved. The possibility that the stock was intruded beneath a lake may have been significant in the evolution of the hydrothermal system which gave rise to the associated mineralization.

Although many other mineral deposits within the Eastern cordillera are associated with silicic volcanic rocks and with ignimbrites, no other deposits associated with calderas have been identified at the present time. This probably reflects the lack of geological information of the region, rather than the uniqueness of the Kari Kari structure. Apart from "simple" calderas bounded by continuous ring fractures, future studies should also be aimed at the detection of less obvious structures, such as multiple calderas, and "trap-door" structures in which only one sector of the floor of a caldera exhibits subsidence. Such structures are well known from the San Juan province (Steven and Lipman, 1976), and from south western Bolivia (Baker, 1981).

SHIELDS

Ignimbrite shields are an important class of volcanic constructs in the Central Andes, and are amongst the most common sources of ash flows. Structures of this type identified to date range from the 1.5-m.y.-old, 25-km diameter Cerro Purico, north Chile (Baker, 1981; Hawkesworth et al., 1982) to the 9-m.y.-old, 70-km diameter Cerro Panizos, Bolivia (Baker, 1981). Ignimbrite "shields" have been recognized in other areas of recent volcanism, such as Tibesti (Vincent, 1963) and Italy (Sparks, 1975; Varekamp, 1980). These, however, are qualitatively different in that they possess calderas, possibly of explosive origin, and lack extensive post-caldera lavas and evidence for doming or elevation. The Andean shields are characterized by the absence of large subsidence calderas, and by the presence of conspicuous extrusive lavas and domes in their summit regions. Shield formation also appears to have been preceded by doming. Surprisingly, ignimbrite shields have not been recognized in other areas of silicic volcanism such as the San Juan province, Colorado. We consider that such structures are at least as common as calderas as sources of ash flows, but may be more difficult to identify in areas of older volcanism. In many respects they resemble Valles-type calderas in which the key stage of major caldera subsidence has not taken place (Fig. 2a, b).

Our recognition of shields as a structural class is based on small, young structures such as Cerro Purico, in which it is clear that neither large-scale

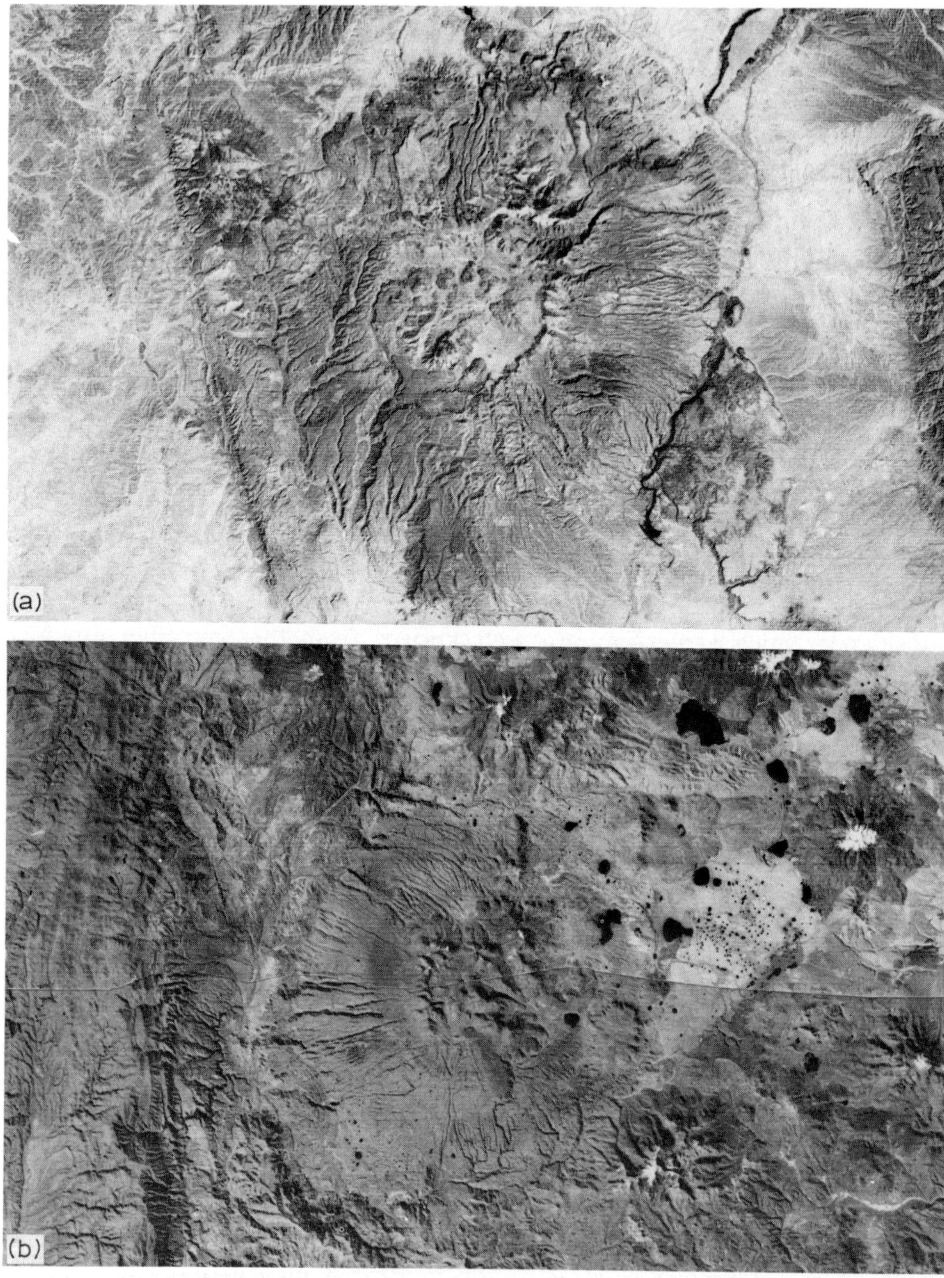

Fig. 2. a and b. LANDSAT images of Valles caldera (a),New Mexico, and Cerro Panizos ignimbrite shield (b), Bolivia/Argentina frontier on the same scale. Cerro Panizos exhibits little evidence of caldera subsidence, while in the Valles caldera the topographically defined structure is much broader (22 km) than the caldera defined by extrusive domes along ring fractures (15 km). Restored to its pre-erosion morphology, the Valles structure would resemble Cerro Panizos more closely. The summit area of Cerro Panizos appears to be largely occupied by dacitic extrusives. Tabular area to northwest is interpreted as erosional remnant of thick ignimbrite sheet. (Extracts from unenhanced scenes).

caldera subsidence nor subsequent resurgence have taken place. Larger, older shields (such as Cerro Panizos) may begin to resemble Valles-type calderas in which the effects of caldera subsidence have been masked by infilling of the caldera moat by younger extrusions. Many old "calderas" in fact may originally have had shield morphology. This possibility is enhanced by the

Fig. 2c. Explanation given on p. 174.

Fig. 2 c and d. LANDSAT images of the Cerro Purico shield (d) and Cerro Galan caldera (c). Cerro Purico is a small ignimbrite shield showing only minor evidence of subsidence along arcuate fractures in south of summit complex, and no evidence of resurgence. Scale bar 5 km. Cerro Galan is a major resurgent caldera 34 km in longest diameter. Aprons of ignimbrite are conspicuous around both in pale grey, deeply gullied terrain. (Extracts from unenhanced scenes).

fact that calderas defined structurally by their ring fractures are often much smaller than their visible, topographically defined counterparts (W.E. Elston, pers. commun., 1981; cf. Fig. 2a,b). While there is thus clearly a convergence between calderas and shields, we think that there is a sufficiently strong contrast between the end member types [e.g. between Cerro Purico and the Cerro Galan caldera (Fig. 2d,c)] that the distinction is worth making. The absence of large-scale subsidence in shield structures may imply the presence of a relatively thick roof above the pluton, which is borne out to some extent by the fact that shields are generally smaller than cal-

Fig. 3. LANDSAT image of Chocaya area, Bolivia, showing location of hydrothermally altered area at centre of eroded ignimbrite shield. Dissected ignimbrite apron is conspicuous in north-west. Hydrothermally altered areas have distinctive MSS signatures. (Extract from digitally processed scene).

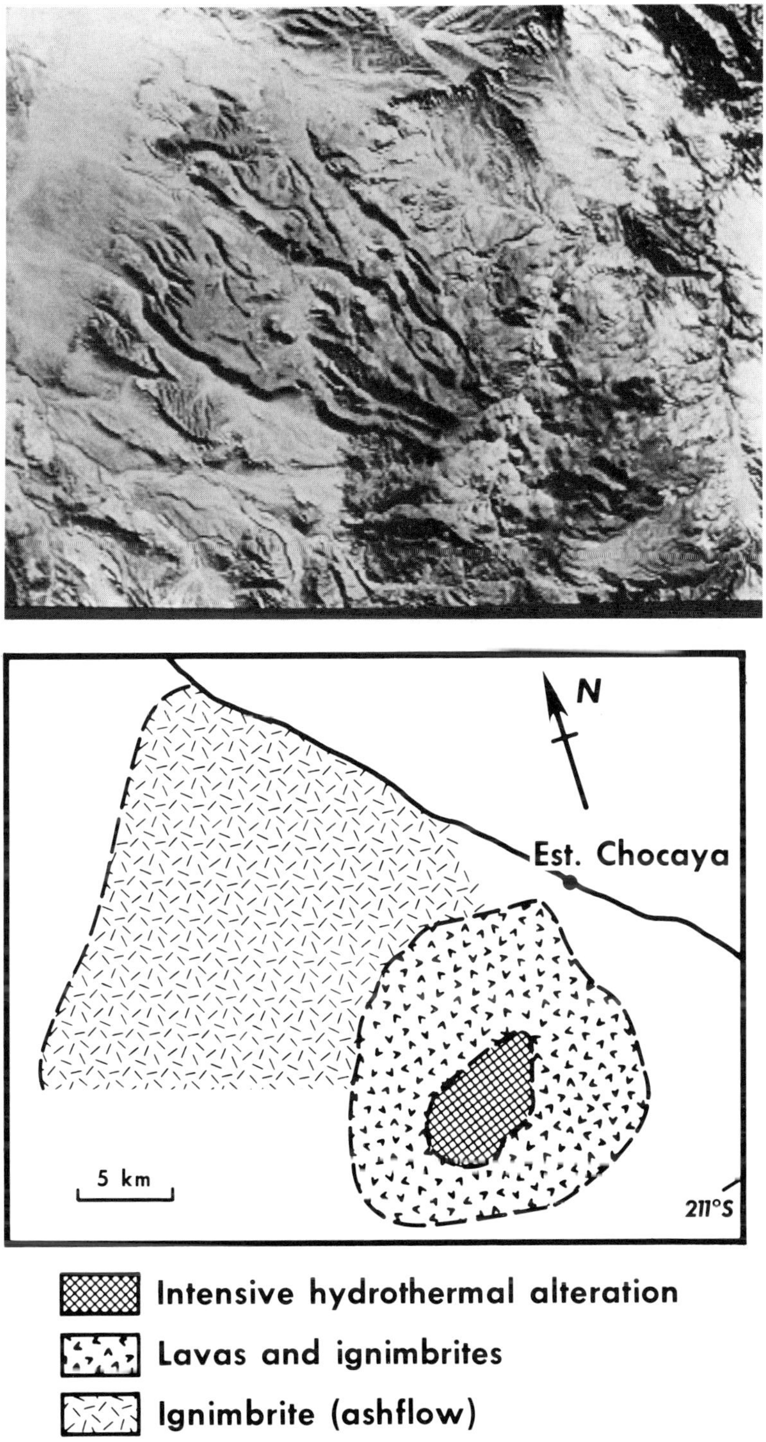
N
Est. Chocaya
5 km
21°S
Intensive hydrothermal alteration
Lavas and ignimbrites
Ignimbrite (ashflow)

deras. Alternatively, eruption of the ignimbrites may not have taken place in a single massive eruption, but rather in a sequence of smaller eruptions at a rate such that the underlying pluton was replenished from beneath rapidly enough to provide continuous roof support. The phenomenon of resurgence in Valles-type calderas has been attributed to re-supply of magma to the sub-caldera pluton (Smith and Bailey, 1968). In shield formation, the rate of supply of magma appears to have been sufficient to maintain roof support without bodily uplift of the central part of the structure.

Several of the Bolivian ore deposits exhibit an association of dacitic lavas and ignimbrites, which seem to have been erupted from small, shield-like structures rather than calderas. In particular, they lack evidence of central subsidence structures filled with massive ignimbrite units which are often taken to be diagnostic of calderas. In most instances, however, the volcanic geology has been only sketchily described, and our conclusions therefore are preliminary. Possible examples of deposits associated with such shields are Chocaya, Colquechaca, Oruro and Tatasi (Fig. 1) and of these Chocaya is much the most straightforward. This 13-m.y.-old structure (Grant et al., 1979) still retains relics of its original volcanic morphology and the 30 km diameter apron of pyroclastic flows is visible on LANDSAT images dipping radially away from the center of the structure (Fig. 3). An area some 6 × 3 km at the center of the complex has been pervasively altered and is cut by veins, some of which extend into the underlying Paleozoic sediments (Sillitoe et al., 1975). Apart from minor dykes, no intrusive rocks have been recognized, but this does not discount the possibility of their presence at shallow level beneath the surface. We suggest that the distribution of ignimbrites and the morphology of this structure indicate that it should be interpreted as an ignimbrite shield, comparable in size, morphology and structure to Cerro Purico.

SMALL, ISOLATED VOLCANIC STRUCTURES

A major difficulty in attempting to determine the volcanic superstructures overlying ore deposits is that the erosion level is often so deep that no vestige of the volcanic edifice remains. In many cases, the mineralization may have been associated with subsurface intrusions which had no significant surface volcanic expression, but these are outside the scope of the present contribution. The available descriptions of ore deposits from the Eastern Cordillera suggest that relatively small, isolated silicic volcanic structures may be the most common loci of subvolcanic mineralization.

A typical, and important example is the Salvadora stock, Llallagua, Bolivia (Fig. 1). This 9.4-m.y.-old stock was emplaced into Silurian—Devonian graywackes, sandstones and shales, and is some 5 km distant from the nearest outcrop of volcanic rocks, which are ignimbrites and lavas of the Frailes plateau (Sillitoe et al., 1975; Grant et al., 1980). The stock is 1.7 km × 1.05 km wide at the surface, and tapers downwards. It is so heavily altered that the nature of the original rock is difficult to determine, but textural evidence indicates that it was a quartz latite porphyry. Hydrothermal intrusion breccias are abundant, both at the contacts of the stock with the country rock, and within the stock itself. They occur as irregular pipelike bodies up to

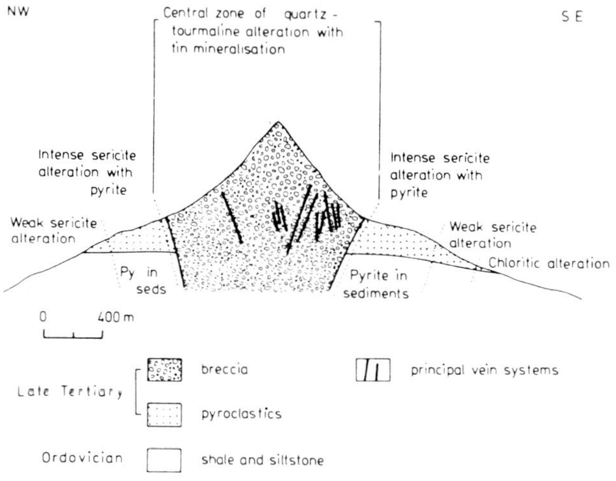

Fig. 4. Geological cross-section through Cerro Chorolque, Bolivia, interpreted as originally part of an isolated monogenetic acid volcano. (From Grant et al., 1977).

Fig. 5. Cerro Bonete (68°50′W, 28°50′). The 3—4-km diameter caldera containing lake and small extrusive dome is surrounded by dissected ignimbrites. The caldera is monogenetic and unrelated to other volcanic structures. (Extract from digitally processed LANDSAT scene).

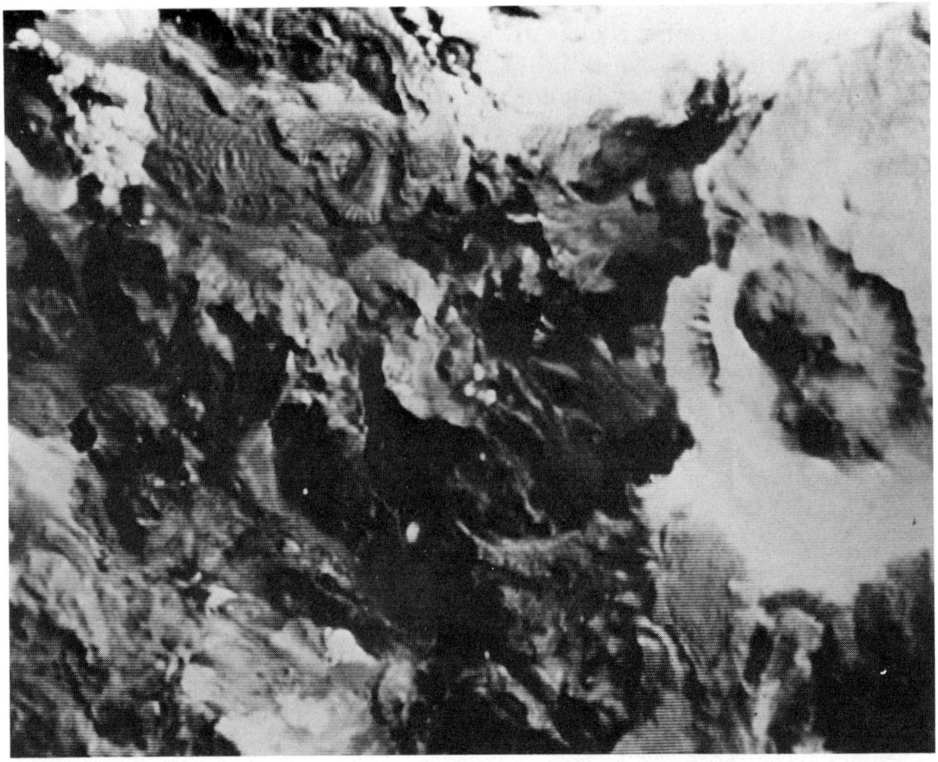

Fig. 6. Unnamed monogenetic acid volcano at 68°15′W 25°20′S. Pumice cone ca. 5 km in diameter contains large, younger dacite dome, postulated to be surface expression of stock. (Extract from digitally processed LANDSAT scene).

50 m across, and also as narrow dykes. The youthful age of the stock and the local geology suggest that it must have been intruded at shallow level, and that less than 1000 m of cover has been removed (Grant et al., 1980).

A slightly different example is provided by the Chorolque deposit (Grant et al., 1980), which consists of an isolated mass of porphyry and breccia, partly surrounded at the surface by silicic pyroclastic rocks and lavas (Fig.4). The vent is filled with hard, altered rocks, which form a steep triangular peak reaching 5600 m asl. The vent filling is composed of an intensively tourmalinized and silicified breccia in which the clasts are predominantly quartz porphyry with some altered sediments set in a matrix of comminuted rock and mineral fragments cemented by quartz and tourmaline. The extrusive rocks are interpreted as pyroclastic on account of their fragmental textures, uniformity of rock type and apparent absence of internal contacts (Grant et al., 1980). Petrographically, the rocks are quartz-latite or rhyodacite porphyries, and appear to be heavily altered ignimbrites.

We suggest that the original volcanic surface edifices surmounting these deposits probably resembled the modern examples in Figs. 5 and 6 which are relatively small, isolated silicic volcanoes, of which several examples are known in the Western Cordillera. Cerro Bonete, Chile (Figs. 1,5) is a 3-km diameter caldera excavated in pumiceous tephra and surrounded by the remains of pyroclastic flow deposits which can be traced for some 30 km from the caldera. A small dome is present at the center of the structure (Baker, 1981). A similar structure is shown in Fig. 6 which possesses a larger central dome, but lacks extensive pyroclastic flow deposits. Both appear to be composed exclusively of silicic ejecta and extrusives, are unrelated to other structures, are monogenetic in that they appear to have experienced only one *major* episode of activity, and have been the sites of vigorous explosive activity. Countless other small, isolated acid domes are present throughout the region, which might provide analogies to the smaller mineralized stocks and breccia pipes seen in the Eastern Cordillera, but it seems unlikely that the very small structures could have been the sites of extensive hydrothermal activity.

THE TOPS OF PORPHYRY ORE BODIES RECONSIDERED

In his seminal contribution, Sillitoe (1973) drew heavily on Andean examples when suggesting that andesitic stratovolcanoes represent the superstructure to porphyry copper ore bodies. His model for the "stratovolcano", however, was rather general, and allowed for an extremely wide range of possible volcanic edifices: "The stratovolcano above a porphyry copper deposit need not be a simple cone . . . but may be multiple in character and include the development of domes and collapse calderas, perhaps resurgent" (Sillitoe, 1973, p. 802). All of Sillitoe's discussions, however, seem to indicate *large* volcanic structures, and while this is applicable to some deposits, it is not applicable to all, or even most: there is no evidence, for example, that the 5-m.y.-old El Teniente orebody (Quirt et al., 1971) was ever overlain by a major volcanic edifice (Camus, 1975). The nature and dimensions of the Braden pipe suggest a structure akin to Cerro Bonete.

In a more recent study, Sillitoe showed that five widely different types of porphyry copper deposit exist, ranging from intrusive rocks of batholithic dimensions to mineralization in host rocks not cut by any intrusives, but still suggested that "many porphyry copper stocks are believed to be emplaced in a subvolcanic basement" (Sillitoe, 1981, p. B28).

While many major porphyry deposits (such as those associated with batholithic rocks) may have had *no* surface volcanic manifestation, and certainly none of an andesitic stratovolcano type, it is clear from the work of Sillitoe (1975, 1981) and Grant et al. (1980) that many porphyry-type deposits (both copper and tin) *are* associated with small, late subvolcanic stocks and breccia pipes. According to Sillitoe (1981), 51% of Andean porphyry deposits are centered on isolated stocks or groups of stocks. We have suggested here that such volcanic manifestations may occur in a wide variety of volcanic structures ranging from small to large. We emphasize, though, that the surface volcanic expression of the ore-related events themselves is likely to be small, silicic and associated with hydrothermal activity. It is useful to consider what kinds of modern volcanic activity provide the best analogs of these features. Wolfe (1981) discussed the formation of hydrothermal breccia pipes in a volcanic setting, and suggested that the phreatic eruptions of Taal volcano, Phillippines, may provide a good example of their mechanism of formation. He observed that some of the ejected material was totally replaced by silica and pyrite, and were indistinguishable from brecciated material found in pipes in some mining districts. While the phreatic eruptions of Taal provide a convincing analog of the formation of hydrothermal breccias, Taal is predominantly a basaltic andesite volcano, and one which is not sufficiently mature to have evolved rocks of more acid composition.

A better analog may be the dome of La Soufriere, Guadeloupe, which was emplaced onto a substratum of older volcanic rocks some 1—2 m.y. old (Briden et al., 1979). Semet and Vatin-Perignon (1981) state that the dome was emplaced into an older explosion crater excavated in massive andesites (Fig. 7) which was deepened and enlarged by the ejection of polylithologic ejecta. The ejecta accumulated to form a ring or "croissant" around the crater, similar to that shown in Fig. 6. Vertically directed explosions subsequently gave rise to pyroclastic flows which spread radially away along valleys, and the dome itself was finally emplaced into a crater which was then partially filled by pyroclastic ejecta. The last magmatic eruption took place shortly prior to the colonization of the island in 1635. Subsequent to its emplacement, the dome has been the site of a number of phreatic eruptions, most recently in 1976—1977 when a long series of violent explosions with ejection of blocks took place. These eruptions were described in detail by Heiken et al. (1980); one of their most interesting observations was that most of the fine-grained ejecta consisted of hydrothermally altered andesite or dacite fragments, the degree of alteration ranging from a trace of hematite on grain surfaces to a clay-pyrite mixture.

Although emplaced into an older andesite basement, the Soufriere dome is not part of an older andesitic stratovolcano, in the sense that the Santiaguito dome could be said to be part of the older Santa Maria volcano, or that the dome now growing in its crater is part of Mt. St. Helens. We consider that the Soufriere dome provides a close analogy to many of the late acid extrusions which are associated with mineralization, ranging from those extruded along caldera ring fractures to those apparently isolated from other volcanic centers. Its history of phreatic activity reflects the operation of hydrothermal processes (Heiken et al., 1980), and it may be associated with the emplacement of breccia pipes, as suggested by Wolfe for Taal.

A complicating factor in attempting to determine the nature of the volcanic surface expression of the Bolivian mineral deposits is that the hydrothermal breccias associated with the sub-volcanic quartz-latite stocks of Bolivia are characteristically tourmaline-rich and bear a strong similarity to tourmaline breccia bodies found in purely *plutonic* settings. The forma-

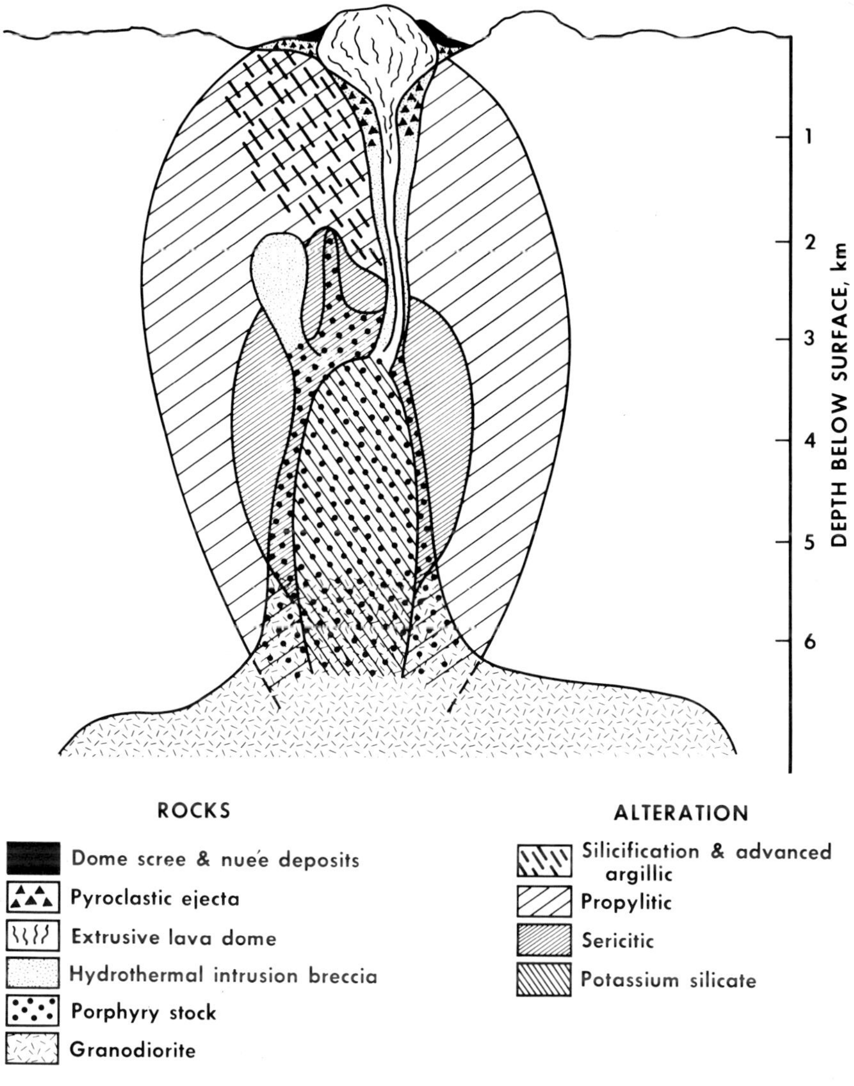

Fig. 7. Revised model for the volcanic superstructure of porphyry copper deposits. The essential surface expression is intermediate to acid dome exhibiting evidence of phreatic eruptions. The upper part of the structure is modelled on the dome of La Soufriere, Guadeloupe, from the work of Vincent et al. (1979) and the lower part from Sillitoe (1973).

tion of such bodies depends on the explosive escape of a boron- and silica-enriched fluid. Such fluids may evolve as a direct consequence of differentiation and progressive crystallization of magmas of granitic composition confined within the apical zones of stocks or batholithic apophyses. The relationship of the evolution and escape of such fluids to phreatic vents in the volcanic setting of the Bolivian tin porphyries is unlikely to be a simple one.

Despite the subtleties of hydrothermal breccia formation, it is clear that the porphyry-style mineralization of the Bolivian tin belt is emplaced in bodies of quartz-latite porphyry and related breccias whose surface expression was in extrusive domes within calderas, ignimbrite shields or isolated silicic structures. Similarly, it seems likely that in the porphyry copper belt the mineralization may often be related to dacite domes at the volcanic level. Dacite domes may form parts of the polygenetic andesitic "stratovolcanoes" considered by Sillitoe (1973), and such domes may well be mineralized. The significant factor, however, is that it is the domes themselves that provide the essential superstructures to porphyry ore bodies, and that these occur in a wide variety of volcanic structures. In Fig. 7 the conception of a porphyry copper system modelled by Sillitoe (1973) has been modified to take account of this proposed relationship.

CALDERAS AND THEIR RELATIONSHIPS TO MINERALIZATION

The Kari Kari caldera provides an unequivocal example of caldera-related mineralization in the Eastern cordillera. Within the Chilean porphyry copper province, the El Salvador deposit provides a further, more equivocal example, since in this case the caldera is ancient and not well preserved. Gustafson and Hunt (1975) have shown that the mineralization at El Salvador is associated with 41-m.y.-old porphyry intrusions which post-date by some 6 m.y. rhyolitic and quartz-porphyry extrusives which in turn post-date voluminous rhyolitic ignimbrites and andesitic volcanic rocks attributed to the Hornitos formation. Although not previously revealed by surface geological mapping, e.g. Garcia (1961), digitally enhanced LANDSAT images and our field studies suggest that the center of mineralization, Cerro Indio Muerto, lies on the edge of a circular structure which we interpret as the ring fracture surrounding a caldera (Fig. 8). This interpretation is strengthened by the fact that extensive ignimbrites predating the mineralization are exposed in the vicinity of the mine, and the mineralized porphyries themselves intrude some 400 m of ignimbrites. Formation of a caldera typically follows the eruption of voluminous ignimbrites: it seems inherently probable that the El Salvador circular structure is the caldera that was the source of the ignimbrites seen in the mine area. Furthermore, other rhyolitic ring fracture intrusions may be present along the circular structure where there is evidence of hydrothermal alteration, but no major mineralization (Fig. 8). The association of these features strongly suggests to us that the circular structure is indeed the relics of a caldera: it is unlikely that better evidence could be forthcoming from an old and deeply eroded structure. Accepting that the pre-mineralization ignimbrites were erupted from the caldera ring structure, there was apparently a large interval between caldera formation and mineralization, comparable in magnitude to that observed in Colorado, U.S.A. and at Cerro Rico, Bolivia. The ages of the Hornitos formation ignimbrites are unfortunately not as well determined as the other examples, but the interval appears to be of the order of 10 m.y.

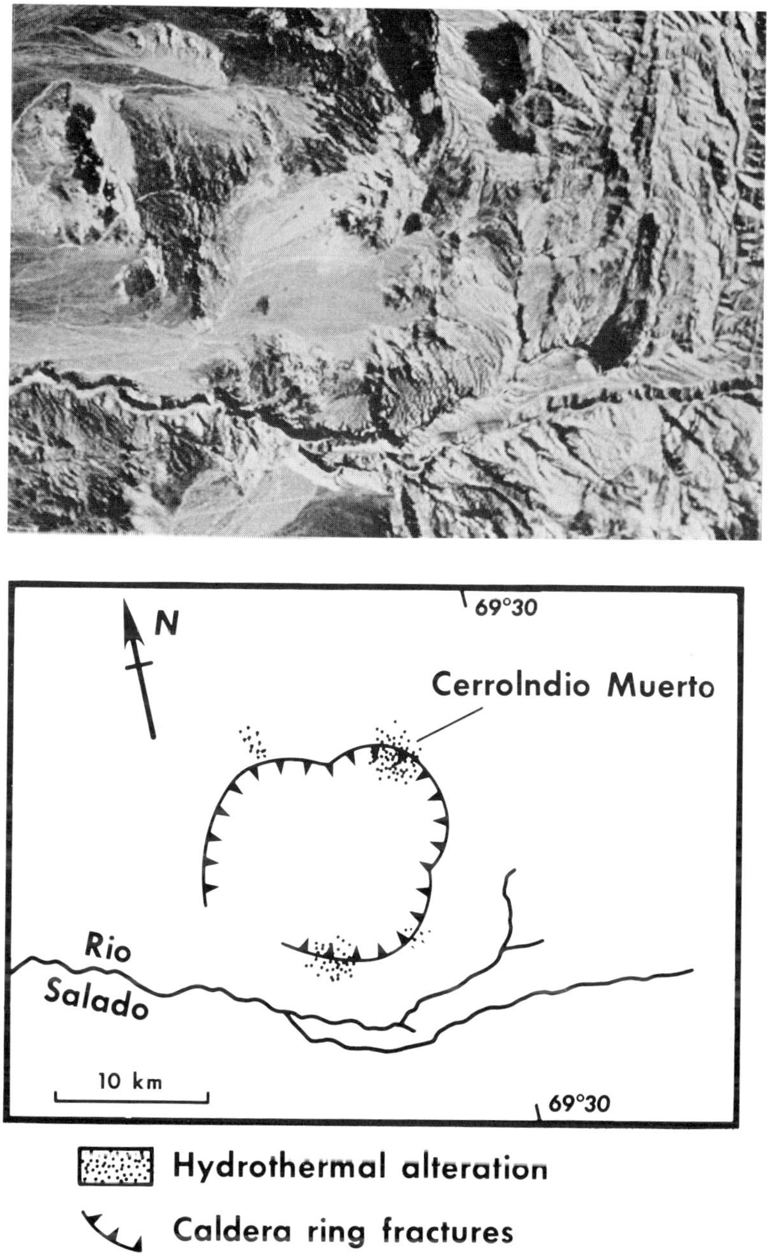

Fig. 8. El Salvador porphyry copper deposit, Chile. LANDSAT image showing location of mineralized area (Cerro Indio Muerte) on ring fractures of earlier caldera. Hydrothermal alteration is also pronounced in the southwest sector of caldera ring fractures, but no intrusives are exposed at surface. (Extract from digitally processed scene).

The widespread association between caldera formation, late-stage intrusion of acid or intermediate stocks, and mineralization, of which the Kari Kari and El Salvador calderas provide two South American examples, prompts us to comment on two other issues: the temporal relations between caldera formation and mineralization, and the possible inhibition of porphyry ore deposition by caldera formation.

The association between calderas and mineralization is a curious one. Some areas exhibit an apparently striking correlation between calderas and sites of mineralization, for example the San Juan Province, Colorado (Steven and Lipman, 1976) and southwestern New Mexico (Elston et al., 1976; Elston, 1978), whereas in others, such as Nevada there appears to be very little correlation (McKee, 1979). Where there is a spatial relationship between mineralization and caldera formation, the mineralization may be younger than the caldera formation by millions of years, as in the case of the Cerro Rico deposit (Francis et al., 1981). The large discrepancies in age between caldera formation and mineralization in a number of instances in Colorado have led Lipman et al. (1976) to suggest that the mineralization is associated with intrusions not genetically related to the calderas, which merely provided a structural control over the younger mineralization.

We start here from the standpoint first enunciated by Smith and Bailey that the "probable long duration of Stage VII (terminal solfataric and hot-spring activity) in large epicontinental cauldrons suggests long-lived hydrothermal systems and major ore-forming potential" (Smith and Bailey, 1968, p. 642). The fact that most calderas around the world are barren presumably reflects only the barrenness of their crustal or magmatic source regions. Considering only those that do exhibit associated mineralization, to what extent is it likely that the relationship is merely spatial and not genetic?

A specific aspect of this more general problem concerns mineralization associated with the intrusion of post-caldera stocks along caldera ring fractures. In some instances, as in the case of the Cerro Rico, the problem *may* be more apparent than real, in that the apparent discrepancy in age between caldera formation and mineralization may reflect only the difficulties in dating heavily altered materials by K-Ar techniques. In other instances, where K-Ar dates can be tested against fission track dates, it is clear that the discrepancy is real (Lipman et al., 1976). Unfortunately, such detailed geochronological data are not available for most mineralized provinces. In the San Juan province studied by Lipman et al, some of the mineralization is associated with stocks which significantly post-date the calderas they intrude. These age differences between caldera formation and stock intrusion may be due to the fact that the province is one with a long magmatic history, with multiple caldera formation, and superimposition of younger calderas on older ones. A number of interesting age relationships exists. For example, mineralization-related intrusions on the margins of the 27.5-m.y.-old Silverton caldera were intruded concurrently with the formation of the neighboring 22.5-m.y.-old Lake City caldera, which possesses its own mineralized ring fracture intrusives. Both calderas are developed within the 28-m.y.-old Uncompahgre caldera (Lipman et al., 1976). In this case, the mineralization has to be seen in relation to the evolution of the igneous complex as a whole, rather than in relation to an individual caldera. While there are important mineral deposits within the San Juan province which are much younger than *any* caldera, we are unable to agree with the rather general statement of Lipman et al. (1976) that "These time relations seem to preclude any significant genetic tie between the major mineralization and emplacement of the caldera-forming magmas, and the calderas seem to have served merely as structural controls for subsequent activity". This, we feel, is overstating the case. A significant proportion of the mineralization *is* related to one or more cycles of caldera formation.

The problem is slightly more difficult with calderas that do not seem to be parts of more complex, multiple structures such as the Kari Kari caldera, where the mineralization is associated with a single ring-fracture intrusion some 7 m.y. younger than the caldera, assuming that its radiometric age is meaningful. Although a very large pluton must underlie the caldera, the thermal models of Smith and Shaw (1975) and Norton (1977) show that cooling by conduction and convection (hydrothermal circulation) is likely to cool the upper parts of the pluton well below magmatic temperatures in only 1—2 m.y. The lower parts of a large pluton, however, are likely to remain at or near magmatic temperatures for much longer, perhaps of the order of ten million years, particularly if heating is maintained by the introduction of fresh batches of basaltic magma into the lower part of the crust, as Hildreth has convincingly suggested in his model for large zoned magma chambers (Hildreth, 1981). Thus, intrusion of a silicic body into caldera ring fractures long after caldera formation does not necessarily require that it be genetically unrelated to the caldera-related magmas — the opposite may in fact be the case. We return to this point in the next section.

Finally, having emphasized examples where the relationships between caldera formation and mineralization are not apparently straightforward, it is worth emphasizing that there are many instances where the relationships are indeed straightforward, such as Questa, New Mexico (Laughlin et al., 1969; Lipman, 1981) and Bursum, New Mexico (Elston, 1978).

CALDERA FORMATION AS AN INHIBITING FACTOR IN PORPHYRY MINERALIZATION

This problem is in essence a further facet of the previous one. Lipman et al. (1976) have argued that "the paucity of major mineral deposits that can be genetically related to the caldera cycle may indicate that explosive pyroclastic eruptions, which are the essence of the caldera-forming process, disperse rather than concentrate the minor elements of economic interest" (Lipman et al., 1976, p. 585).

This philosophy was embraced by Sillitoe, specifically in regard to porphyry copper deposits, arguing that porphyry copper deposits are inherently unlikely to be associated with calderas and that the presence of ignimbrites in an unexplored area could be taken as a "discouraging indication for the porphyry copper explorationist" (Sillitoe, 1980, p. 91). In view of the fact that the El Salvador deposit is intimately associated with thick ignimbrites, and may have been formed in a caldera, albeit after a long time interval, and that other "porphyry" deposits described by Sillitoe unquestionably *are* caldera related (e.g. the Cerro Rico tin deposit) we feel that this assertion may be unwise.

Two separate factors are involved: the source of the ore metals, and the geochemical relationships between ash flows, the sub-caldera pluton, and post-caldera ring-fracture intrusions. Both of these are complex general issues, but the Andean examples we have studied illustrate some facets of these larger problems.

Regarding the source of the ore metals, the important point is that whatever the relationship between caldera formation, ignimbrite eruption and mineralization, the mineralization invariably takes place *after* the eruption of the ignimbrites. If the source of the ore metals is in the wall rocks, rather than magmatic, then the "dispersal" of ore metals by large pyroclastic eruptions is irrelevant to the mineralization process if the ore metals are in fact leached out of the sub-volcanic rocks by circulating hydrothermal fluids long after ignimbrite eruption and caldera formation has taken place. In this context, the available oxygen isotope data from mineral deposits in the Eastern Cordillera of Bolivia show that both magmatic and meteoric waters were involved in the mineralization process, the magmatic component in the earliest, highest temperature stages (Grant et al., 1980). Both crustal and magmatic sources may have contributed to the ore metals. Since there is a striking geographic coincidence between the distribution of silver deposits in the Eastern cordillera and the presence of an extremely thick Paleozoic marine sedimentary sequence it is probable that these sediments provided the source for the silver, while the tin may be of sub-crustal origin. A crustal source has also been suggested for some of the San Juan metals (Lipman et al., 1978), and for leads in some southern Peruvian volcanic rocks (Tilton, 1979).

In the Western Cordillera, by contrast, where silver and tin deposits are absent, so too are thick Paleozoic sediments, the volcanic rocks resting on crystalline basement rocks or at best only thin sedimentary sequences. The major porphyry copper deposits in the belt immediately to the west are on isotopic grounds apparently of magmatic, sub-crustal origin (Sillitoe, 1981).

Where the ore metals in caldera-related mineral deposits are of magmatic origin, as for example the tin in Cerro Rico or the copper at El Salvador, the problem becomes one of understanding the fractionation of metals through different stages of the caldera cycle. The geochemical relationships between ignimbrites and their parental magma chamber have been addressed by Hildreth (1979). In the case of the Bishop Tuff, he showed that the early erupted ash flows were markedly enriched in some elements (e.g. Nb, Cl, Sb, Cs) and depleted in others (e.g. Sr, Ba and Eu) relative to the later erupted flows, and interpreted these results in terms of the tapping of successively deeper zones of a layered magma chamber.

Specifically discussing the problems of the relations of ore bodies to large silicic calderas, Smith (1979) provided some perceptive insights into the issues. He argued that for those elements that concentrate downward in magma chambers, eruptions of the most fractionated part of the system can only enhance their availability to the high-level hydrothermal systems that may be responsible for their concentration to ore grade. Furthermore, he pointed out that the apparently paradoxical occurrence of ore deposits of elements such as Au and Cu associated with granitic to granodioritic intrusive bodies is due to the fact that continental mafic magmas simply do not normally form large-volume magma bodies at high enough levels in the crust to support significant hydrothermal systems. The ore bodies in question may form during the late magmatic or hydrothermal stages of silicic systems from underlying or late injections of mafic magma.

It seems clear from the accumulating body of new work on large, zoned magma chambers that the key to the understanding of the origin of small, highly mineralized late intrusions associated with large calderas such as Kari Kari (or even El Salvador) lies in the processes outlined by Smith. Some aspects of the complex fractionation processes that can take place in a periodically replenished, continuously fractionated magma chamber have been discussed by O'Hara and Mathews (1981), who showed that extremely wide ranges of trace-element enrichments and depletions could be achieved. In the light of such studies, the chronological discrepancies between caldera formation and mineralisation become insignificant, if mineral rich intrusions such as the Cerro Rico stock are to be interpreted as resulting from late-stage injection of mafic magma into the base of a mature, complexly fractionated pluton. Unfortunately, however, as O'Hara and Mathews point out, it may not be possible to deduce the sequence of petrological events from a knowledge of the chemistry of the final products alone. Thus, the origins of many of the world's major ore bodies may remain perpetually rather mysterious.

ACKNOWLEDGEMENTS

Valuable comments on an early draft of this manuscript were received from W.E. Elston, P. Lipman, G. Wadge, and C.A. Wood, and their help is gratefully acknowledged. Fieldwork in the Andes is supported under a grant from the Natural Environment Research Council, and assisted in the field by the National Geological Services of Bolivia and Argentina. The manuscript was written while P.W. Francis was a visiting Scientist at the Lunar and Planetary Institute, Houston, and constitutes L.P.I. Contribution No. 480.

REFERENCES

Angus, J., Halls, C., Sillitoe, R.M., Meave, J. and Suarez, L., 1977. The petrochemical characterization of the eruptives in the subvolcanic province of the Bolivian tin belt: comparative aspects and implications for exploration. Proc. 2nd. Int. Tin Symposium, La Paz, Bolivia, 1977.

Baker, M.C.W., 1981. The nature and distribution of Upper Cenozoic ignimbrite centres in the Central Andes. J. Volcanol. Geotherm. Res., 11: 293—315.

Baker, M.C.W. and Francis, P.W., 1978. Upper Cenozoic volcanism in the Central Andes — ages and volumes. Earth Planet. Sci. Lett., 41: 175—187.

Barazangi, M. and Isacks, B.L., 1976. Spatial distribution of earthquakes and subduction of the Nazca plate beneath South America. Geology, 4: 686—692.

Briden, J.C., Rex, D.C., Faller, A.M. and Tomblin, J.F., 1979. Geochronology and paleomagnetism of volcanic rocks in the Lesser Antilles volcanic arc. Philos. Trans. R. Soc. London, Ser. A, 291: 485—528.

Camus, F., 1975. Geology of the El Teniente orebody with emphasis on wall rock alteration. Econ. Geol., 70: 1341—1372.

Clark, A.H., 1970. An occurrence of the assemblage native sulphur-covellite "Cu 5.5 × Fe × S 6.5× ", Aucanquilcha, Chile. Am. Mineral., 55: 913—918.

Elston, W.E., 1978. Mid-Tertiary cauldrons and their relationship to mineral resources, south western New Mexico: a brief review. N.M. Geol. Soc., Spec. Publ., 7: 107—113.

Elston, W.E., Rhodes, R.C. and Erb, E.E., 1976. Control of mineralization by mid-Tertiary volcanic centres, southwestern New Mexico: N.M. Geol. Soc., Spec. Publ., 5: 125—130.

Farrar, E., Clark, A.H., Haynes, S.J., Quirt, G.S., Conn, H. and Zentilli, M., 1970. K-Ar evidence for the post-Paleozoic migration of granitic intrusion foci in the Andes of northern Chile. Earth Planet. Sci. Lett., 10: 60—66.

Francis, P.W., Roobol, M.J., Walker, G.P.L., Cobbold, P.R. and Coward, M.P., 1974. The San Pedro and San Pablo volcanoes of North Chile and their hot avalanche deposits. Geol. Rundsch., 63: 357—388.

Francis, P.W., Baker, M.C.W. and Halls, C., 1981. The Kari Kari caldera, Bolivia, and the Cerro Rico Stock. J, Volcanol. Geotherm. Res., 10: 113—124.

Garcia, F., 1961. Geological map of the area west of the Salar de Pedernales, Chile. Empresa Nacional de Petroleo (unpubl.).

Grant, J.N., Halls, C., Avila, W. and Avila, G., 1977. Igneous geology and the evolution of hydrothermal systems in the sub-volcanic tin deposits of Bolivia. Geol. Soc. London Spec. Publ., 7: 117—127.

Grant, J.N., Halls, C., Avila, W. and Snelling, N.J., 1979. K-Ar ages of igneous rocks and mineralization in part of the Bolivian tin belt. Econ. Geol., 74: 838—851.

Grant, J.N., Halls, C., Sheppard, S.M.F. and Avila, W., 1980. Evolution of the porphyry tin deposits of Bolivia. Min. Geol., Spec. Iss., 8: 151—173.

Gustafson, L.B. and Hunt, J.P., 1975. The porphyry copper deposit at El Salvador, Chile. Econ. Geol., 70: 857—912.

Halls, C., 1981. Chronology and mechanism in the formation of the Bolivian tin province. Access (University College Cardiff: Mineral Exploitation Society, pp. 107—122.

Hawkesworth, C.J., Hammill, M., Gledhill, A.R., van Calsteren, P. and Rogers, G., 1982. Isotope and trace element evidence for late stage intra-crustal melting in the high Andes. Earth Planet. Sci. Lett., 58: 240—254.

Heiken, G., Crowe, B., McGetchin, T., West, F., Eichelberger, J., Bartram, D., Peterson. R. and Wohletz, K., 1980. Phreatic eruption clouds: the activity of La Soufriere de Guadeloupe, F.W.I., August-October, 1976. Bull. Volcanol., 43(2): 383—396.

Hildreth, W., 1979. The Bishop Tuff: evidence for compositional zonation in silicic magma chambers. In: C.C. Chapin and W.E. Elston (Editors), Ash flow Tuffs. Geol. Soc. Am., Spec. Pap., 180: 43—72.

Hildreth. W., 1981. Gradients in silicic magma chambers: Implications for lithospheric magmatism. J. Geophys. Res., 86 (B 11): 10153—10192.

James, D.E., 1971. Plate tectonic model for the evolution of the Central Andes. Geol. Soc. Am. Bull., 82: 3325—3346.

Kussmaul, S., Hormann, P.K., Ploskonka, E. and Subieta, T., 1977. Volcanism and structure of southwestern Bolivia. J. Volcanol. Geotherm. Res., 2: 73—111.

Laughlin, A.W., Rehrig, W.A. and Mauger, R.L., 1969. K-Ar Chronology and sulfur and strontium isotope ratios at the Questa Mine, New Mexico. Econ. Geol., 64: 903—909.

Lindgren, W. and Creveling, J.G., 1928. The ores of Potosi, Bolivia. Econ. Geol., 23: 233—262.

Lipman, P.W., 1981. Volcano-tectonic setting of Tertiary ore deposits, southern Rocky Mountains. In: W.L. Dickinson and W.D. Payne (Editors), Relations of Tectonics to Ore Deposits in the Southern Cordillera. Ariz. Geol. Soc. Dig., 14: 199—213.

Lipman, P.W., Fisher, F.S., Mehnert, H.H., Naeser. C.W., Luedke, R.G. and Steven, T.A., 1976. Multiple ages of mid-Tertiary mineralization and alteration in the western San Juan mountains, Colorado. Econ. Geol., 71: 571—588.

Lipman, P.W., Doe, Br.R., Hedge, C.E. and Steven, T.A., 1978. Petrologic evolution of the San Juan volcanic field, southwestern Colorado: Pb and Sr isotope evidence. Geol. Soc. Am. Bull., 89: 59—82.

McBride, S.L., 1977. A K-Ar Study of the Cordillera Real, Bolivia and its Regional Setting. Thesis, Queens University, Kingston, Ontario, 230 pp.

McKee, E.H., 1979. Ash flow sheets and calderas: their genetic relationships to ore deposits in Nevada. In: C.C. Chapin and W.E. Elston (Editors), Ash Flow Tuffs. Geol. Soc. Am., Spec. Publ., 180: 205—210.

Noble, D.C., Mckee, E.H., Farrar, E. and Petersen, U., 1974. Episodic Cenozoic volcanism and tectonism in the Andes of Peru. Earth Planet. Sci. Lett., 21: 213—220.

Norton, D., 1977. Fluid circulation in the Earth's crust. Am. Geophys. Union, Geophys. Monogr., 20: 693—703.

O'Hara, M.J. and Mattheus, R.E., 1981. Geochemical evolution in an advancing, periodically replenished, periodically tapped, continuously fractionated magma chamber. J. Geol. Soc. London, 138: 237—288.

Park, C.F., 1961. A magnetite flow in northern Chile. Econ. Geol., 56: 431—436.

Quirt, S., Clark, A.H., Farrar, E. and Sillitoe, R.H., 1971. Potassium-argon ages of porphyry copper deposits in northern and central Chile. Geol. Soc. Am., Abstr. with Programs, 3 (7): 676—677.

Rivas, S. and Carrasco, R., 1968. Geologia y yaciemientos minerales de la region de Potosi. Tomo 1. Bol. Serv. Geol. Bolivia. 11, 95 pp.

Rytuba, J.J., 1981. Relation of calderas to ore deposits in the Western United States. Ariz. Geol. Soc. Dig., XIV: 227—236.

Schwab, K. and Lippolt, H., 1976. K-Ar mineral ages and late Cenozoic history of the Salar de Cauchari area (Argentine Puna). In: O. Gonzalez Ferran (Editor), Proceedings Symposium "Andean and Antarctic Volcanology Problems", Santiago, Chile, 1974. IAVCEI, Naples, pp. 698—714.

Semet, M.P. and Vatin-Perignon, N., 1981. A triggering mechanism for explosive calc-alkaline eruptions: The "magma mixing" hypothesis. Part II: The 16th Century eruption of La Soufriere, Guadeloupe (F.W.I.). IAVCEI, Int. Symposium "Arc Volcanism" Tokyo, Japan, 1981, Abstr.

Sillitoe, R.H., 1973. The tops and bottoms of porphyry copper deposits. Econ. Geol., 68: 799—815.

Sillitoe, R.H., 1976. Andean mineralization — a model for the mineralization of convergent plate margins. Geol. Assoc. Can., Spec. Pap., 14: 59—100.

Sillitoe, R.H., 1980. Cauldron subsidence as a possible inhibitor of porphyry copper formation. Min. Geol., Spec. Iss., 8: 89—93.

Sillitoe, R.H., 1981. Regional aspects of the Andean porphyry copper belt in Chile and Argentina. Trans. Inst. Min. Metall., 90: B15—B36.

Sillitoe, R.H., Halls, C. and Grant, J.N., 1975. Porphyry tin deposits in Bolivia. Econ. Geol., 70: 913—927.
Smith, R.L., 1979. Ash flow magmatism. In: C.E. Chapin and W.E. Elston (Editors), Ash Flow Tuffs. Geol. Soc. Am. Spec. Pap., 180: 5—28.
Smith, R.L. and Bailey, R.A., 1968. Resurgent cauldrons. Mem. Geol. Soc. Am., 116: 613—662.
Smith, R.L. and Shaw, H.R., 1975. Igneous related geothermal systems. In: Assessment of Geothermal Resources of the United States. U.S. Geol. Surv., Circ., 726: 58—84.
Sparks, R.S.J., 1975. Stratigraphy and geology of the Vulsini volcano, Central Italy. Geol. Rundsch., 64: 497—523.
Steven, T.A. and Lipman, P.W,, 1976. Calderas of the San Juan volcanic field, southwestern Colorado. U.S. Geol. Surv., Prof. Pap. 958, 35 pp.
Thorpe, R.S. and Francis, P.W., 1979. Variations in Andean andesite compositions and their petrogenetic significance. Tectonophysics, 57: 53—70.
Thorpe, R.S., Francis, P.W. and Harmon, R.S., 1981. Andean andesites and crustal growth. Philos. Trans. R. Soc. London, Ser. A., 301: 305—320.
Tilton, G.R., 1979. Isotopic studies of Cenozoic Andean calc-alkalic rocks. Carnegie Inst. Washington, Yearb., 78: 298—304.
Turneaure, F.S., 1960. A comparative study of major ore deposits of Central Bolivia. Econ. Geol., 55: 217—254 (Part I) and 574—606 (Part II).
Varekamp, J.C., 1980. The geology of the Vulsinian area, Lazio, Italy. Bull. Volcanol., 43: 487—505.
Vincent, P.M., 1963. Le volcanisme ignimbritique du Tibesti (Sahara Tchadien), Essai d'interpretation dynamique. Bull. Volcanol., 26: 259—272.
Vincent, P.M., Vatin-Perignon, N., Semet, M. and Cheminee, J.L., 1979. Le dome de la Soufriere (Guadeloupe): son age et son mode de mise en place. C.R. Acad. Sci. Paris, Ser. D, 288: 51—54.
Wolfe, J.A., 1980. Fluidization versus phreatomagmatic explosions in breccia pipes. Econ. Geol., 75: 1105—1111.

Journal of Volcanology and Geothermal Research, 18 (1983) 191—214
Elsevier Science Publishers B.V., Amsterdam — Printed in The Netherlands

ORIGIN OF DACITES OF TAUPO VOLCANIC ZONE, NEW ZEALAND

F.W. REID and J.W. COLE

Geology Department, Victoria University of Wellington, Private Bag, Wellington (New Zealand)

(Received October 20, 1982)

ABSTRACT

Reid, F.W. and Cole, F.W., 1983. Origin of dacites of Taupo Volcanic Zone, New Zealand. In: S. Aramaki and I. Kushiro (Editors), Arc Volcanism. J. Volcanol. Geotherm. Res., 18: 191—214.

Dacites form a relatively small proportion of lavas in the Taupo Volcanic Zone, New Zealand (~5km^3), and occur mainly on the eastern side. In this paper their origin is considered in terms of three processes: (a) partial melting of crustal rocks; (b) fractional crystallisation of basalt and andesite; and (c) sub-surface mixing of basic and acid magma. Modelling techniques are used to calculate the most acceptable degree of fractional crystallisation and magma mixing to fit major-element data, and these values are used to compare calculated and observed trace-element values. The success or failure of the model is determined by the closeness of the two sets of values. For partial-melt models, trace-element values alone are calculated by the batch-melting equation.

Results indicate that White Island dacite can best be modelled by fractional crystallisation; Manawahe by fractional crystallisation plus limited crustal contamination; Maungaongaonga by partial melting of Western Basement greywacke, and Tauhara by partial melting of this greywacke together with minor mixing with a more basic magma. Results from Parekauau and Horohoro indicate that these lavas are unlikely to have formed by any of the processes examined.

INTRODUCTION

The Taupo Volcanic Zone (T.V.Z.) extends approximately 300 km NNE across central North Island, New Zealand, from Ohakune to White Island (Fig. 1). It represents the volcanic part of the Taupo-Hikurangi arc-trench or subduction system (Cole and Lewis, 1981) and can be divided into an andesite-dacite volcanic arc on the eastern side and a bimodal high-alumina basalt-rhyolite assemblage within an ensialic marginal basin on the western side.

Dacite is largely restricted to the volcanic arc on the eastern side of the zone and, in terms of volume (total volume of dacite in T.V.Z. ~5km^3; Cole, 1979) most occurs in the northern part of the zone within White Island, Whale Island, Manawahe and Edgecumbe volcanoes (Fig. 1). In this paper

0377-0273/83/$03.00

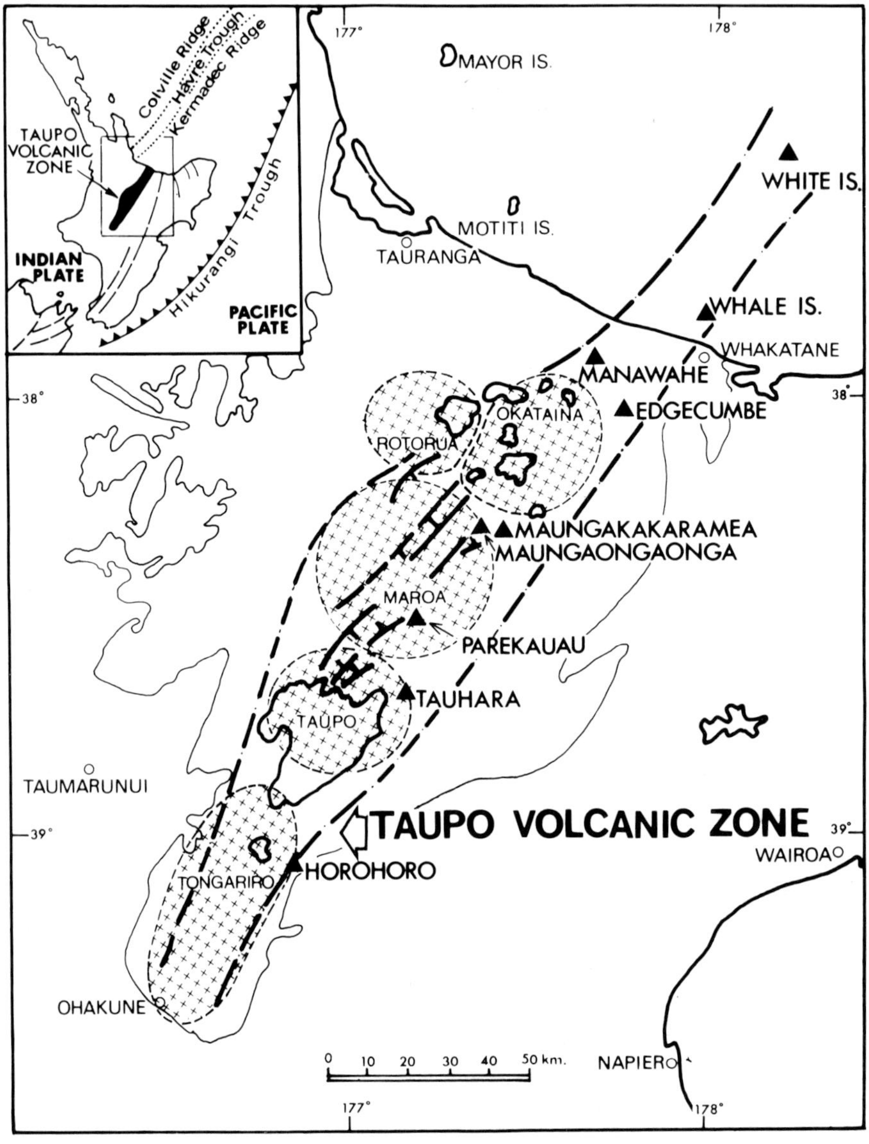

Fig. 1. Distribution of dacites (shown by closed triangle) in T.V.Z. Five main Volcanic Centres — Rotorua, Okataina, Maroa, Taupo and Tongariro — are also shown. The solid line indicates maximum extent of ignimbrites.

these cones are referred to as the Bay of Plenty dacites. Details of their form are described in Duncan (1970). Smaller dacite domes occur between Rotorua and Taupo at Maungaongaonga and Maungakakaramea (Rainbow Mountain) on the northeastern side of the Maroa Volcanic Centre, and Tauhara on the northeastern side of the Taupo Volcanic Centre (Fig. 1). A small dike occurs at Tutukau Road, Parekauau in the Maroa Volcanic Centre and a small dome at Horohoro, east of Rangipo. These lavas are referred to in the paper as Rotorua-Taupo dacites.

Small amounts of dacite also occur within the andesitic Tongariro Volcanic Centre. Some dacitic flows occur on Ruapehu (W.R. Hackett, pers. commun., 1981); dacite lapilli occur in the Te Rato Lapilli, probably erupted from North Crater, Tongariro (Topping, 1973) and a small dome forms an island in the centre of Lake Rotoaira.

This paper describes the petrography and chemistry of some of the dacites from the Bay of Plenty and Rotorua-Taupo areas, T.V.Z., and using modelling techniques examines possible origins of the lavas. Lavas within the Tongariro Volcanic Centre are directly related to the associated andesites and are not discussed further in the paper.

PETROGRAPHY

All the dacites of T.V.Z. are porphyritic with augite, orthopyroxene (hypersthene-bronzite), zoned plagioclase and titanomagnetite in a hyalopilitic or pilotaxitic ground-mass containing small laths of plagioclase, orthopyroxene, augite and titanomagnetite with occasional accessory apatite, zircon or rutile. Cristobalite is common at Edgecumbe and Tauhara and rare at Manawahe. Specific mineralogy of each occurrence is given in Table I and composition of pyroxenes and plagioclases in Figs. 2 and 3, respectively*.

Lavas at Tauhara contain embayed quartz crystals, which are normally surrounded by a rim of pyroxene, and commonly coexist with small magnesian olivine phenocrysts (Fo_{81-86}).

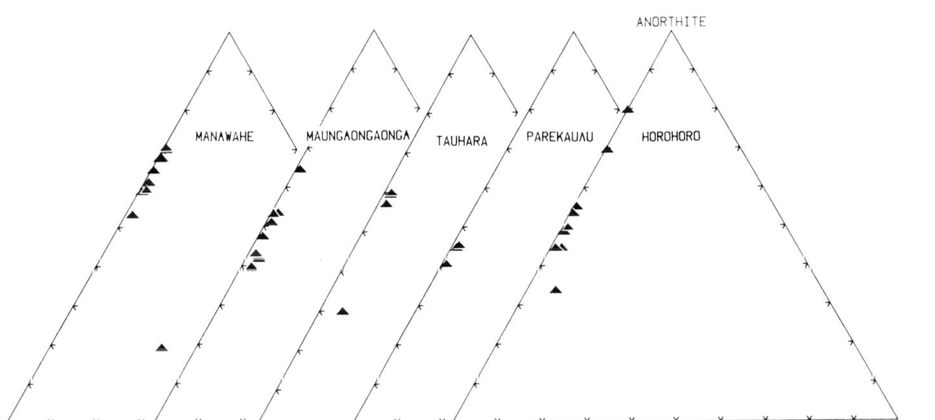

Fig. 2. Composition of feldspar (by EPMA) of dacites within the Taupo Volcanic Zone.

*A full list of analyses obtained by electron probe microanalysis may be obtained from the authors.

TABLE I

Summary of Features of dacites of T.V.Z.

Location	Types of occurrence	Approximate age (million years)	Mineralogy	Range in SiO_2 content	Main References
White Island	Two overlapping stratocones with breached crater. Active: Last eruption of lava 1977	>0.016 (1)	Plag, aug, hypbronz, titanomag. Most lavas slightly altered	63—66% 55—58% (1977 eruption)	Hamilton and Baumgart, 1959; Duncan, 1970; Clark and Cole, 1976; Clark et al., 1979
Whale Island	Eroded cone with domes and breccia fans	>0.008 (2)	Plag, aug, hypbronz, titanomag. ± biot, hble always	63.9—67.3%	Duncan, 1970
Manawahe	Eroded dome?	0.750 ? (2)	oxidised and altered.	63.2—65.2%	Duncan, 1970
Edgecumbe	Central cone and surrounding domes	<0.008 (2)	Xenoliths of microdiorite common.	59.8—66.1%	Duncan, 1970

Maungaongaonga	Dome	0.154 (3)	Plag, opx, titano-mag. occasionally aug.	67.2—67.4%	Grange, 1937
Parekauau	Dike intruded into rhyolite	Pleistocene (4)	Plag, hyp, aug, titanomag. Shows slight alteration	62.5—64%	Lloyd, 1972
Tauhara	Series of overlapping domes	0.031 (3)	Plag, hble, bronz, aug, titanomag. Embayed qtz with reaction rim. Occasional magnesian oliv. (Fo_{81-86}) Xenoliths common	64.6—67.2%	Lewis, 1968a,b
Horohoro	Single dome	>0.014 (5)	Plag (2 types — large anhedral and small euhedral) aug, hyp.	65.8—75%	

References to age: (1) Kohn, Glasby (1978); (2) Duncan (1970); (3) Stipp (1968); (4) Lloyd (1972); (5) Topping (1973).

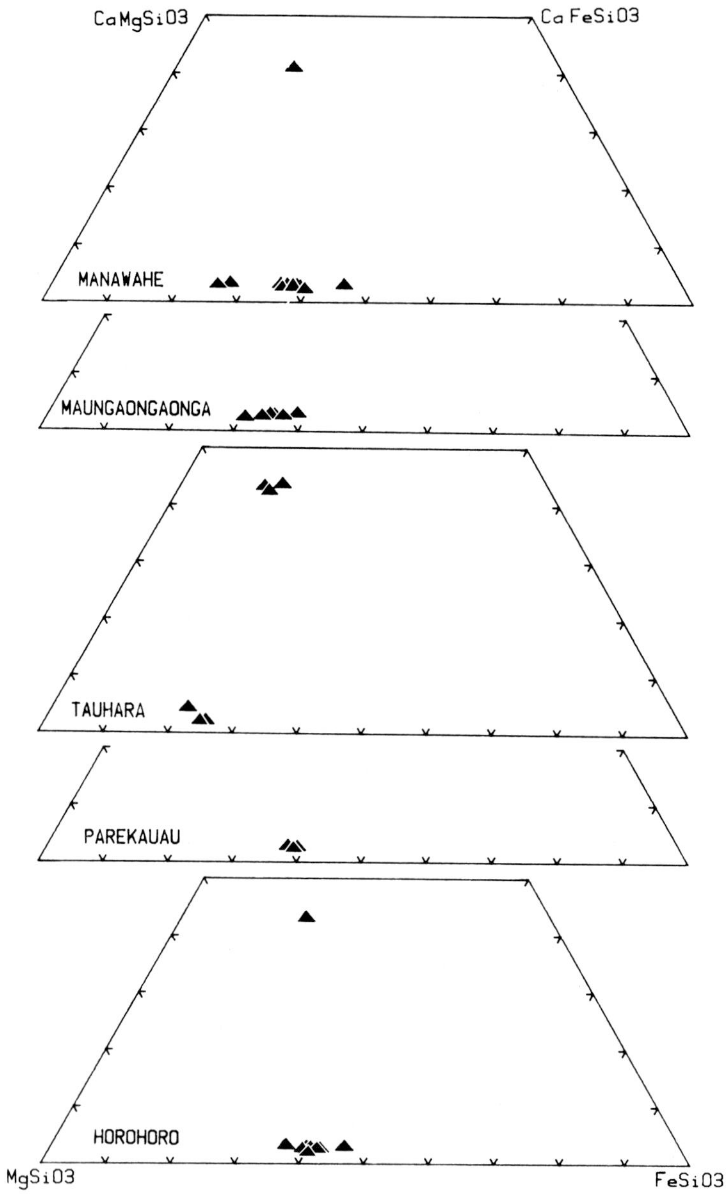

Fig. 3. Composition of pyroxene (by EPMA) of dacites within the Taupo Volcanic Zone.

At Horohoro, two types of lava occur, dacite (Table III, analyses 7—9) and a rhyolite. Contacts between the two types are not seen, but both are common on the surface of the dome and there is a clear chemical difference between them. Two types of plagioclase phenocryst occur in the dacite:

one is large, anhedral and full of glass inclusions, and the other is small, euhedral and normally zoned.

CHEMISTRY

Analytical techniques

Major- and trace-element chemistry for unpublished analyses were determined by X-ray fluorescence spectrometry. The method of Norrish and Hutton (1969) was used for major elements and details of the analytical procedures used, together with the method for trace-element analysis are given in Kennedy et al. (1981). Volatile contents are reported as loss on ignition at 1000°C and total iron is reported as Fe_2O_3. R.E.E. were determined by spark source mass spectrometry using procedures developed by Taylor (Nance and Taylor, 1976). The average precision of the line ratio measurements was ± 12% and concentration accuracy was determined by the reliability of the reference standard (BCR-1) data which was considered to be ± 10%.

Major and trace elements

All the dacites of T.V.Z. can be regarded as calcalkaline (Fig. 4). Average compositions of Bay of Plenty dacites are listed in Table II together with an average analysis of dacites from Fonualei, Tonga. Compositions of Rotorua-Taupo dacites, including 7 new analyses are listed in Table III. The Bay of Plenty lavas show a complete range from andesite to dacite (58—68% SiO_2: Duncan, 1970; Cole, 1979) while those from Rotorua-Taupo are generally more silicic (range 64—68% SiO_2). There is some compositional variation within the Bay of Plenty dacites; for example, K_2O content is highest at White Island and Manawahe and lowest at Whale Island (Table II and Fig. 5). All are quite different from the dacites of Fonualei, Tongan arc, shown for comparison in Figs. 4 and 5.

Rare earth elements (REE)

Dacites of T.V.Z. are all more strongly enriched in light REE (LREE) than heavy REE (HREE) and the Rotorua-Taupo dacites are more enriched in REE than those of White Island (Fig. 6; Table IV). The shape of the chondrite-normalised curves for LREE is similar to those of Chilean and New Zealand andesites (Thorpe et al., 1976; Lopez-Escobar et al., 1977; Cole, 1979) although HREE are generally somewhat higher. The dacite from Parekauau has a pronounced negative Ce anomaly. The reason for this is not certain, but the location is near the Orakeikorako thermal area and the sample is slightly altered (3.76 wt.% water). Ludden and Thompson (1979) note that LREEs, particularly Ce, show considerable mobility during low-temperature weathering processes and this may be the reason.

TABLE II

Major- and trace-element analyses of dacites from Bay of Plenty, New Zealand and Fonualei, Tonga (normalised to 100% water free)

	1	2	3	4	5	6	7
SiO_2	64.82	64.49	64.16	66.02	65.25	65.11	65.34
TiO_2	0.66	0.54	0.74	0.61	0.59	0.58	0.58
Al_2O_3	16.20	15.83	14.86	15.87	16.68	16.36	14.17
Fe_2O_3	5.94*	5.70*	6.30*	5.72*	5.78*	5.69*	1.88
FeO							6.15
MnO	0.09	0.09	0.09	0.10	0.13	0.09	0.18
MgO	2.16	2.66	3.48	1.47	1.58	1.48	1.53
CaO	5.47	5.91	5.57	4.48	4.24	4.73	5.89
Na_2O	3.23	2.84	3.18	3.65	3.26	3.54	2.98
K_2O	1.99	1.85	2.23	2.41	2.33	2.28	1.11
P_2O_5	0.12	0.09	0.11	0.14	0.11	0.13	0.19
ΣFeO	5.34	5.13	5.67	5.15	5.20	5.12	7.84
Rb	63.2	63.7	74	80	81.6	85	16
Sr	246	225	174	218	215	223	300
Ba	523	712	900	637	783	649	280
Zr	106	100	150	157	154	149	47
Cu	17.3	17.0	72	14	17	13	27
Ni	7.6	7.3	49	n.d.	2	4	<2
Sc	19.4	17.4	19	18	—	—	26
V	142	130	180	100	104.9	94.7	93
Cr	28.2	30.0	115	2	9.9	8.8	4
Ga	16.9	14.9	16	17	—	—	—
ΣREE	—	—	55.04	—	—	—	41.3
K/Rb	236	241	255	241	236	223	576
Rb/Sr	0.26	0.28	0.43	0.37	0.38	0.38	0.05
Cr/V	0.199	0.231	0.639	0.020	0.094	0.093	0.043
Sr^{87}/Sr^{86}	0.7061	—	0.7061	—	—	—	0.7043

*FeO + Fe_2O_3 expressed as ΣFe_2O_3.

1. Average of 9 analyses from Edgecumbe (Duncan, 1970; Ewart and Stipp, 1968).
2. Average of 16 analyses from Whale Island (Duncan, 1970).
3. Average of 6 analyses from White Island (Duncan, 1970; Ewart and Stipp, 1968).
4. Average of 3 analyses from Manawahe (Duncan, 1970).
5. 16059 Manawahe Quarry (unpubl.).
6. 16060 Manawahe Quarry (unpubl.).
7. Average of 11 analyses from Fonualei Island, Tonga (Ewart and Bryan, 1973; Ewart et al., 1973, 1977).

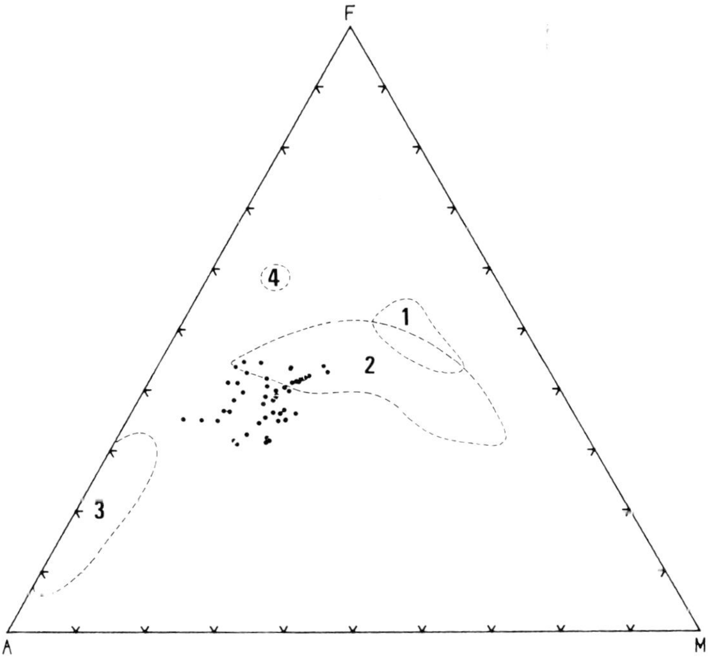

Fig. 4. Dacites of T.V.Z. plotted in terms of $Na_2O + K_2O$ (A): $FeO + Fe_2O_3$ (F) and MgO (M). Dashed lines enclose fields of: *1* = high-alumina basalt; *2* = andesites; *3* = rhyolites and ignimbrites; *4* = dacites from Fonualei, Tonga.

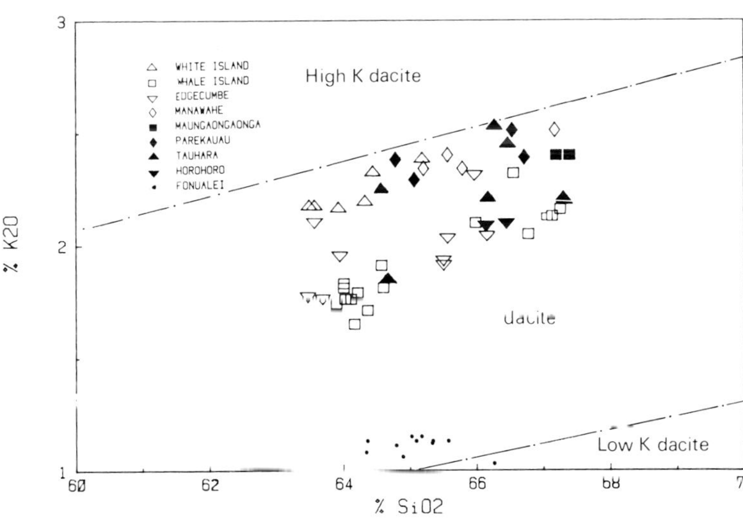

Fig. 5. Dacites of T.V.Z. plotted in terms of K_2O v. SiO_2. Dashed lines indicate boundaries given by Peccerillo and Taylor (1975).

TABLE III

Major- and trace-element analyses of dacites of Rotorua-Taupo area (normalised to 100% water free)

	1	2	3	4	5	6	7	8	9
SiO_2	67.76	66.54	66.72	64.79	66.55	67.25	66.19	66.50	66.62
TiO_2	0.63	0.83	0.85	0.95	0.40	0.39	0.74	0.75	0.75
Al_2O_3	15.70	15.61	15.34	15.65	15.37	14.84	15.71	15.63	15.67
Fe_2O_3	3.12	2.64	5.66*	6.28*	3.13	4.29*	4.81*	4.85*	4.78*
FeO	1.31	2.79			1.21				
MnO	0.10	0.09	0.08	0.11	0.07	0.10	0.13	0.10	0.11
MgO	1.11	1.63	1.43	1.78	2.53	2.70	1.63	1.70	1.56
CaO	3.52	3.91	3.89	4.49	4.85	4.58	4.18	4.40	4.43
Na_2O	4.13	3.24	3.44	3.38	3.52	3.57	4.32	3.81	3.81
K_2O	2.42	2.51	2.39	2.38	2.27	2.19	2.09	2.10	2.09
P_2O_5	0.20	0.21	0.20	0.19	0.10	0.09	0.20	0.16	0.18
ΣH_2O	1.25	3.74	3.76	3.13	1.71	0.50	0.42	0.41	0.50
ΣFeO	4.12	5.16	5.09	5.65	4.03	3.86	4.32	4.36	4.30
Rb	73.3	82	90.3	86.1	63	67.6	74.7	71.9	98.5
Sr	242	229	226	230	315	305	240	237	181
Ba	680	520	565	569	450	—	493	503	608
Zr	215	198	201	180	143	110	191	187	173

	1	2	3	4	5	6	7	8	9
Cu	3	8.1	13	20	23	22	7	11	6
Ni	2.8	6.3	6	6	43	33	3	2	5.7
Sc	11	—	—	—	13	—	—	—	—
V	26	—	96.6	150.8	80	33	97.5	101.9	38.4
Cr	0.9	—	13.3	10.1	56	12	10.1	5.2	14.7
Y	31	—	34.0	43.4	17	20.5	31.0	28.1	30.7
ΣREE	—	—	161.95	—	—	123.75	139.45	—	—
K/Rb	274	254	220	229	299	269	232	242	176
Rb/Sr	0.30	0.36	0.40	0.37	0.20	0.22	0.31	0.30	0.54
Cr/V	0.035	—	0.138	0.067	0.695	0.382	0.104	0.051	0.383
$^{87}Sr/^{86}Sr$	0.7054	—	—	—	0.7049	—	—	—	—

1. Average of 2 (1)[+] analyses from Maungaongaonga. (Ewart et al., 1968; Steiner, 1958; Ewart and Stipp, 1968).
2. 24112 Dike, Tutukau Road, Parekauau (Cole, 1979).
3. 16057 Dike, Tutukau Road, Parekauau (unpubl.).
4. 16058 Dike, Tutukau Road, Parekauau (unpubl.).
5. Average of 6 (1) analyses from Tauhara (Lewis, 1968b; Duncan, 1970; Ewart and Stipp, 1968).
6. 30007 Burrows Quarry, N.E. end Rotokawa Road, Tauhara. (unpubl.).
7. 16051 Horohcro Dome, E. of Rangipo (unpubl.).
8. 16052 Horohcro Dome, E. of Rangipo (unpubl.).
9. 16054 Horohcro Dome, E. of Rangipo (unpubl.).

* FeO + Fe_2O_3 expressed as ΣFe_2O_3.

[+] Numbers in brackets refer to numbers of trace-element analyses.

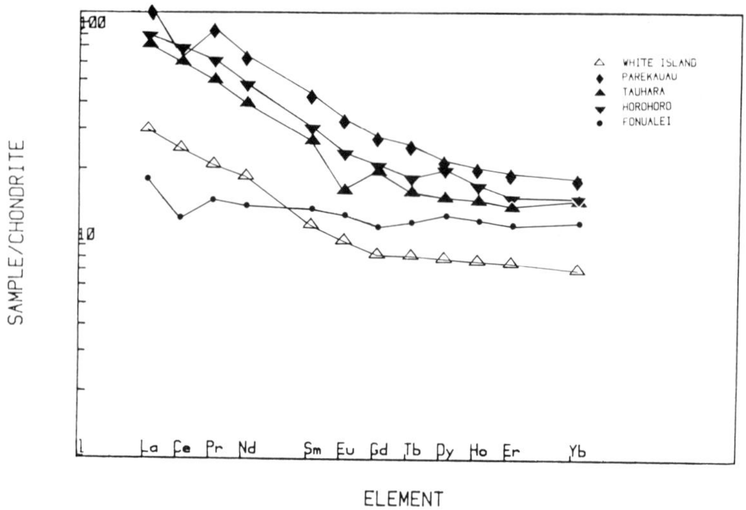

Fig. 6. REE contents of dacites of T.V.Z. with dacite from Fonualei, Tonga for comparison.

GENESIS

Three petrogenetic processes:
(a) partial melting of crustal rocks;
(b) fractional crystallisation of basalt or andesite; and
(c) sub-surface mixing of basic and acid magma, or a combination of these processes could account for the formation of dacitic magma in T.V.Z. These processes have been evaluated by petrogenetic modelling techniques and the results are detailed below.

Partial melting models

Rocks exposed on either side of T.V.Z. are Triassic—Jurassic greywackes and these probably form the basement under much of T.V.Z. Reid (1983) has distinguished between a "Western Basement" (~62% SiO_2) and an "Eastern Basement" (~75% SiO_2) and considers that the voluminous rhyolitic volcanics of T.V.Z. are produced by approximately 35% partial melting of crustal rocks, the bulk composition of which is similar to the Western Basement. The dacites could be formed by slightly higher degrees of partial melting (40—50%) and this process is tested by the batch-melting model. Because of the high degree of partial melting, differences between batch and fractional melting are minimal, so the effects of the latter process are not considered.

The proposed source mineralogy is based on the modal mineralogy of greywackes exposed at the surface. The proportions of quartz and feldspar

TABLE IV

Rare-earth element analyses of dacites of Taupo Volcanic Zone

	1	2	3	4	5
La	9.1	21.5	31	23	5.3
Ce	20	48	49	54	9.5
Pr	2.5	5.9	9.6	7.0	1.7
Nd	11	23	36	27	7.85
Sm	2.3	5.4	8.4	6.0	2.6
Eu	0.72	1.25	2.4	1.7	0.9
Gd	2.5	5.8	8.0	6.0	3.2
Tb	0.42	0.80	1.25	0.9	0.57
Dy	2.6	4.8	6.8	6.2	3.9
Ho	0.6	1.1	1.5	1.25	0.88
Er	1.7	3.0	4.1	3.2	2.4
Yb	1.6	3.2	3.9	3.2	2.5
La/Sm	3.96	3.98	3.69	3.83	2.03
La/Yb	5.69	6.72	7.95	7.19	2.12
Eu/Eu*	0.93	0.69	0.89	0.87	0.96
Ba/La	99.12	—	18.22	21.43	54.7
ΣREE	55.04	123.75	161.95	139.45	41.3
SiO_2%	63.45	67.25	66.72	66.19	65.34
K_2O%	2.17	2.19	2.39	2.09	1.13
CaO%	5.91	4.58	3.89	4.18	5.87
Na_2O%	3.24	3.57	3.44	4.32	3.06
Rb	74	67.6	90.3	74.7	17
Sr	183	305	226	240	300
Ba	920	—	565	493	290

1. 11242 White Island
2. 30007 Burrows Quarry, Tauhara.
3. 16057 Dike, Tutukau Road, Parekauau.
4. 16051 Horohoro Dome, E. of Rangipo.
5. 11548-39 Young lava N.W. corner Fonualei, Tonga (from Ewart et al., 1973).

(Analyst for 1—4: Dr. P. Rankin, Soil Bureau, DSIR).

should not change markedly during prograde metamorphism and the ferromagnesian phases are those which might be expected to occur with high-grade metamorphism of a chlorite-rich rock. The flat HREE patterns of the dacites (Fig. 6) are not consistent with the presence of garnet or amphibole in the source rock (Hanson, 1980), therefore these phases are not considered in the model.

The original modal mineralogy of the source used in the model is given in Table V (column 1), together with the proportions in which the minerals would enter the melt (column 2). Modal composition of the restite with 30%, 40% and 50% melting is given in columns 3, 4, and 5. From this information it is evident that with increasing degrees of partial melting the restite

TABLE V

Proportions of minerals produced in melt and restite phases by partial melting of greywacke

Mineral	Original proportion	Proportion in melt	Proportion in restite		
			30% melt	40% melt	50% melt
Quartz	0.28	0.33	0.272	0.266	0.258
Plagioclase	0.37	0.45	0.413	0.409	0.404
K. feldspar	0.11	0.00	0.000	0.000	0.000
Biotite	0.05	0.07	0.049	0.047	0.044
Clinopyroxene	0.06	0.05	0.084	0.088	0.093
Orthopyroxene	0.08	0.05	0.103	0.109	0.116
Magnetite	0.02	0.02	0.035	0.037	0.039
Cordierite	0.02	0.03	0.025	0.024	0.024
Apatite	0.004	0.005	0.005	0.005	0.005

becomes progressively enriched in ferromagnesian phases relative to felsic phases which go into the melt.

The effects of 50% partial melting on the concentration of Rb, Sr, Ba, Cr and V are given in Table VI and of the values obtained the most acceptable correlation is that for Maungaongaonga. Values for Cr and V are lower at Maungaongaonga, but this could easily be explained by subsequent fractionation of magnetite and/or pyroxene. The value for Ba obtained in the model is high, but this could be due to an underestimation of K_{Ba}^{biot} (a value of 7.0 was used in the model). The Rb content of the Tauhara lavas is consistent with partial melting, but values for Sr and Cr are too high. When considered in conjunction with the petrography, this could, however, be due to addition of small amounts of magma to a partial melt of Western Basement greywacke.

The effects of 40—50% melting on REE is shown in Fig. 7, and the general slope of the chondrite normalised values is similar to that of the dacites. Ab-

TABLE VI

Trace-element values of T.V.Z. dacites compared to predicted value for 50% partial melt of greywacke

	50% partial melt.predict.	White Island	Manawahe	Maunga-ongaonga	Tauhara	Parekauau	Horohoro
Rb	74.6	74	82	73	67	88	82
Sr	254	174	218	242	305	228	219
Ba	741	900	633	680	450	567	535
Cr	10.5	115	5	0.9	12	12	10
V	51	130	100	2.6	33	124	79

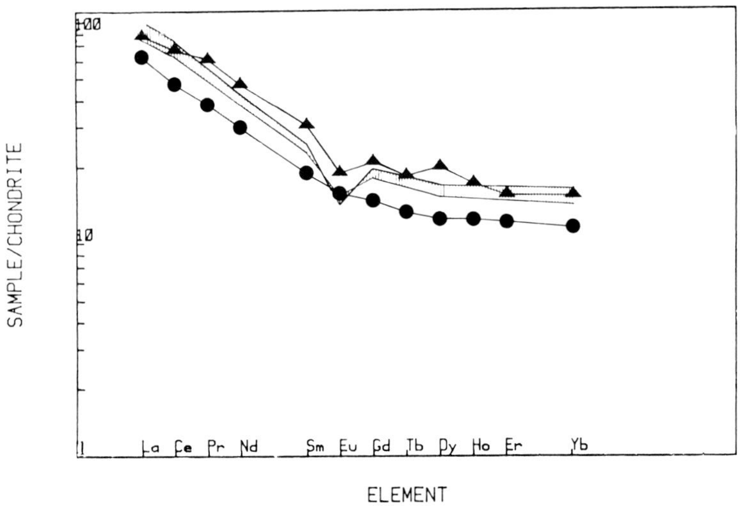

Fig. 7. Effects of partial melting on REE values. Circle = greywacke, triangle = Tauhara dacite; shaded area = variation in values found for 40—50% partial melt.

solute values are very dependent on apatite because of the extreme $K_{REE}^{apatite}$ value (Nagasawa, 1970) and small changes in content of this mineral will, therefore, change the enrichment substantially. It is significant that the model produces a marked negative Eu anomaly similar but more pronounced to that of Tauhara. This tends to confirm evidence from other trace elements that White Island, Parekauau and Horohoro have not formed by partial melting of Western Basement greywacke. No REE data is available for Maungaongaonga.

Fractional crystallisation

The degree of fractional crystallisation of mineral proportions from basalt or andesite of T.V.Z. that is required to produce the best matches with observed compositions of the dacites has been estimated by using the least squares mixing procedure of Wright and Doherty (1970). The basic lava which was used as a starting point for each model is given in Tables VII and VIII and the composition of phenocryst phases fractionated were taken from this "potential parental lava". Where strongly zoned plagioclase was present, end member compositions were used, and recalculated to check whether the plagioclase compositions required by the models were realistic. Otherwise, a single average plagioclase composition was used. Results show that for most elements major-element compositions can be approximated by fractionation of a phenocryst assemblage whose proportions are consistent with the assemblage in the parental rocks (Tables VII and VIII).

Using the F' values obtained from the least-squares models and published K_d values (Gill, 1978; Hanson, 1980) trace-element concentrations were

TABLE VII

Fractional crystallisation models for dacites of T.V.Z. from an andesite parent

	White island		Manawahe		Tauhara		Parekauau		Horohoro	
	Parent	Residual	Parent	Residual	Parent	Residual	Parent	Residual	Parent	Residual
SiO_2	55.49	0.0075	59.22	0.0000	55.49	0.0031	59.22	0.0001	59.22	0.0000
TiO_2	0.62	0.1101	0.72	0.0000	0.62	0.0707	0.72	0.0007	0.72	0.0004
Al_2O_3	13.26	0.0057	16.83	0.0000	13.26	−0.0126	16.83	0.0000	16.83	0.0000
ΣFeO	7.38	−0.0232	6.45	0.0000	7.38	−0.0103	6.45	0.0000	6.45	0.0000
MnO	0.12	0.0026	0.13	0.0006	0.12	−0.0216	0.13	−0.0239	0.13	−0.0130
MgO	9.62	−0.0020	3.78	0.0001	9.62	−0.0006	3.78	0.0003	3.78	0.0002
CaO	8.72	−0.0086	6.77	0.0000	8.72	−0.0039	6.77	−0.0001	6.77	−0.0001
Na_2O	2.33	−0.1289	3.27	0.0001	2.33	0.0774	3.27	−0.0004	3.27	−0.0002
K_2O	1.16	−0.0209	1.50	0.0000	1.16	−0.1188	1.50	−0.0002	1.50	0.0000
P_2O_5	0.02	−0.0149	—	—	—	—	—	—	—	—
Sum Res^2		0.0301		0.0000		0.0259		0.0006		0.0002
Liquid	50		61		46		57		61	
Plag: or			5				1		1	
Plag: ab	18		11		20		11		7	
Plag: an			12				15		15	
Opx	5		5		—		7		8	
Cpx	17		10		20		7		6	
Tm: mag	2		2				2		2	
Tm: ilm	—		4		4		<0.1		0.1	
Oliv	8		—		10		—		—	

TABLE VIII

Fractional crystallisation models for Rotorua-Taupo dacites from a high-alumina basalt parent

		Parent composition	Tauhara residuals	Parekauau residuals	Horohoro residuals
SiO_2		50.43	0.0000	0.0001	0.0001
TiO_2		1.02	0.0011	0.0009	0.0008
Al_2O_3		17.13	0.0001	0.0001	0.0001
ΣFeO		9.31	−0.0001	0.0000	0.0000
MnO		0.16	−0.0372	−0.0304	−0.0268
MgO		6.30	0.0008	0.0006	0.0005
CaO		10.99	−0.0002	−0.0005	−0.0003
Na_2O		2.67	−0.0001	−0.0005	−0.0006
K_2O		0.51	−0.0001	−0.0003	−0.0004
Sum Res^2			0.0014	0.0009	0.0007
Liquid			19	17	19
Plag	or		<0.2	<0.2	<0.4
	ab		16	17	16
	an		29	28	28
Cpx			24	27	26
Tm	mag		5	5	5
	ilm		1	1	1
Oliv			6	5	5

calculated using the Rayleigh fractionation equation:

$$\frac{C_1}{C_0} = F^{(D-1)}$$

which assumes the liquid is a uniform reservoir and crystals are removed as soon as they are formed. The results are given in Tables IX and X. White Island dacite can be successfully modelled by fractional crystallisation of a primitive olivine-bearing basaltic andesite of the same composition as that erupted in 1977 (Clark et al., 1979). Trace-element values for Manawahe are consistent with fractionation of a similar andesite, but the predicted levels of Rb and Ba are too low and Cr and V too high. This could, however, be explained by subsequent crustal contamination which could equally explain the wide range of Mg values in orthopyroxenes (Fig. 2). Trace-element data for Horohoro dacite are also consistent with fractionation of a two-pyroxene andesite from the Tongariro Volcanic Centre with similar contamination, but Maungaongaonga, Tauhara and Parekauau cannot be modelled by fractional crystallisation of the specific andesite chosen (Table IX). Basalt is not a suitable parent for any of the dacites (only Tauhara, Parekauau and Horohoro are shown in Table X) as the required clinopyroxene fractionation reduces the predicted Cr value well below that observed in the dacites.

TABLE IX

Comparison of predicted and observed trace-element values for fractional crystallisation models from an andesite parent

	White island		Manawahe		Maungaongaonga	
	Observed	Predicted	Observed	Predicted	Observed	Predicted
Rb	74	83	82	59	73	58
Sr	174	215	218	267	242	210
Ba	900	875	633	436	680	533
Cr	115	105	5	17	0.9	20
V	130	165	100	177	26	188

	Tauhara		Parekauau		Horohoro	
	Observed	Predicted	Observed	Predicted	Observed	Predicted
Rb	67	92	88	63	82	58
Sr	305	216	228	197	219	266
Ba	450	767	567	569	535	421
Cr	12	16.8	12	15	10	19
V	33	166	124	198	79	167

TABLE X

Comparison of predicted and observed trace-element values for fractional crystallisation models from high-alumina basalt parent

	Tauhara		Parekauau		Horohoro	
	Observed	Predicted	Observed	Predicted	Observed	Predicted
Rb	67	68	88	74	82	66
Sr	305	154	228	149	219	159
Ba	450	541	567	586	535	579
Cr	12	0	12	0	10	0
V	33	40.6	124	65	79	41

Examples of REE concentrations predicted by the models are shown in Fig. 8. Basalt is used as a parent in Fig. 8a and it is clear that the degree of enrichment required by the model is insufficient to match the REE patterns of the dacites. This is equally true of all other dacites examined. White Island (Fig. 8b) produces a reasonably good match, except that HREE for the model are higher than those of the dacite, but this is a feature that is common to such calculations (Gill, 1981). Two-pyroxene andesites from Tongariro typically have a negative Eu anomaly (Cole et al., 1981) and this

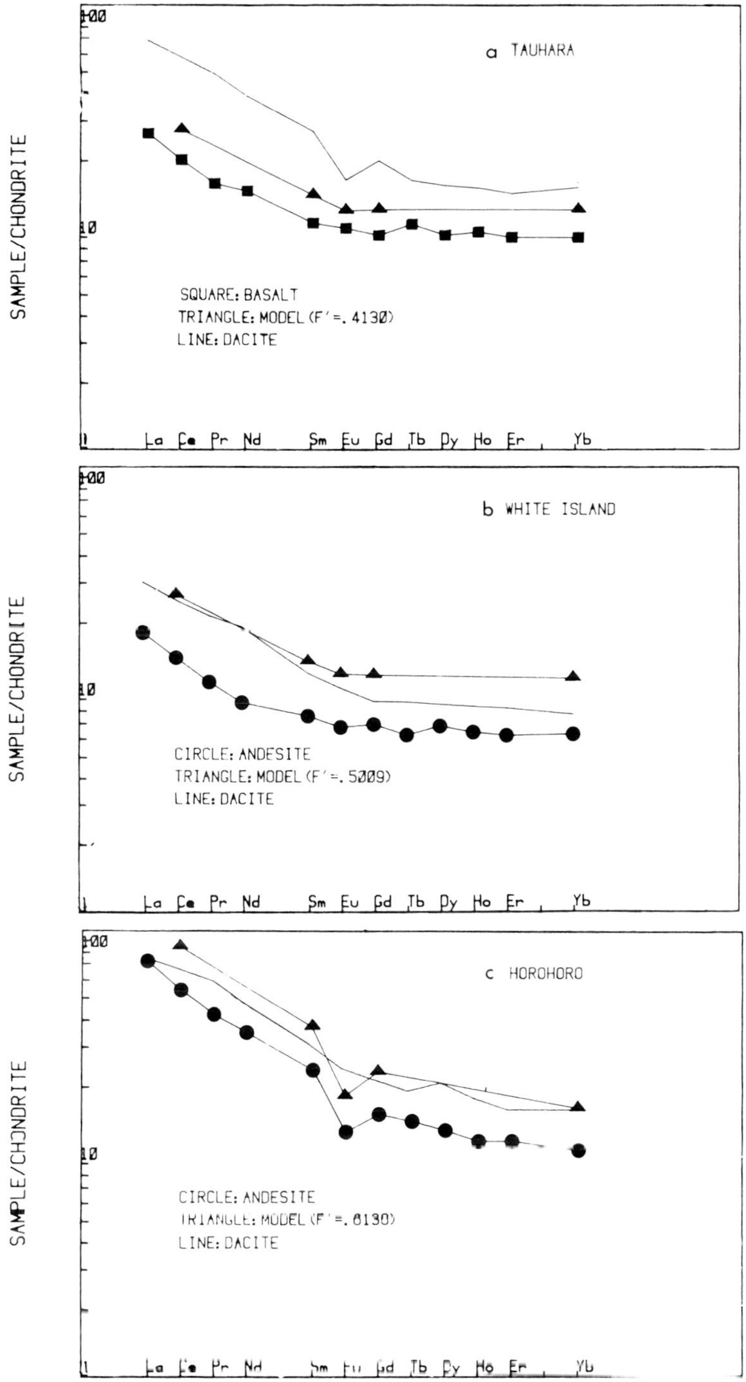

Fig. 8. Effects of modelling on REE values of three dacites of T.V.Z. a. Tauhara. b. White Island. c. Horohoro.

TABLE XI

Major-element mixing models for Rotorua-Taupo dacites

	Tauhara			Parekauau			Horohoro		
	K. trig. basalt	Av. rhyolite	Residuals	Orakeikorako h.a. basalt	Av. rhyolite	Residuals	Tongariro andesite	Horohoro rhyolite	Residuals
SiO_2	51.03	73.73	0.0022	50.43	73.73	0.0976	59.22	74.97	0.0523
TiO_2	1.17	0.29	0.1963	1.02	0.29	−0.3013	0.72	0.23	−0.2358
Al_2O_3	16.71	13.30	−0.2057	17.13	13.30	−0.6227	16.83	13.04	−0.3740
ΣFeO	10.10	2.00	0.2192	10.31	2.00	−0.8757	6.45	2.10	0.1243
MnO	0.00	0.06	−0.0592	0.16	0.06	0.0093	0.13	0.07	0.0035
MgO	6.39	0.36	−0.3193	6.30	0.36	0.8379	3.78	0.15	0.4950
CaO	10.81	1.67	0.1613	10.99	1.67	0.8118	6.77	1.41	0.0422
Na_2O	3.07	4.28	0.3688	2.67	4.28	0.3404	3.27	4.70	0.1078
K_2O	0.41	3.17	0.1097	0.51	3.17	−0.0607	1.50	2.92	0.0373
Sum Res^2			0.4084			2.7355			0.4734
Component %	33%	67%		32%	68%		56%	44%	

would be retained during fractionation (Fig. 8c). Horohoro dacite, the nearest to Tongariro does not have such an anomaly and would, therefore, seem unlikely to be derived from this andesite.

Magma mixing

The third possibility for the generation of T.V.Z. dacites is by large-scale sub-surface mixing of rhyolitic and basic magmas. The results of major element calculations for mixing models of Rotorua-Taupo dacites are given in Table XI. Models for Tauhara and Parekauau use spatially related high-alumina basalt and average rhyolite of T.V.Z. as components while the model for Horohoro uses the average composition of Tongariro two-pyroxene andesite and Horohoro rhyolite. The residuals calculated by least-squares mixing programme are substantially larger than for fractional crystallisation models, but this is to be expected as only two components are involved.

The trace-element calculations are based on mineral proportions derived from a major-element least-squares mixing programme and assume that for two-magma mixing in an essentially liquid state the following relationship is approximately correct:

$$xC^{\text{basic}}_{\text{trace}} + yC^{\text{silicic}}_{\text{trace}} = C^{\text{dacite}}_{\text{trace}}$$

where x and y are the proportions of basic and silicic lavas from the least squares models.

The results are given in Table XII and it is obvious that correlation is very poor. Even allowing for the variation in composition of rhyolitic magmas of T.V.Z. and of possible basic magmas, it seems unlikely that widespread magma mixing is viable and this conclusion is supported by the REE patterns which are all strongly depleted, compared to actual values of the dacites.

TABLE XII

Trace-element values for mixing models of Rotorua-Taupo dacites, T.V.Z.

	Tauhara		Parekauau		Horohoro	
	Observed	Predicted	Observed	Predicted	Observed	Predicted
Rb	67	82	88	7	82	69
Sr	305	187	228	136	219	185
Ba	450	571	567	628	535	503
Cr	12	103	12	23	10	38
V	33	92	124	58	79	92
Parents:	High-al. basalt Av. rhyolite		High-al. basalt Av. rhyolite		Tongariro andesite Horohoro rhyolite	

CONCLUSIONS

We therefore suggest that some of the dacites can be modelled by partial melting and fractional crystallisation, but none by widespread mixing of basic and rhyolitic magmas. However, we are aware of the problems in evaluating magma mixing without a complete knowledge of temporal relations between potential parents.

Data for Maungaongaonga are consistent with partial melting of Western Basement greywacke. Cr and V values are low, but this could be explained by limited late-stage magnetite or clinopyroxene + magnetite fractionation from the melt. Partial melting also seems to be the most likely origin at Tauhara but the lower Rb and Ba values and the higher Cr value all argue for addition of a small amount of more basic magma. This may result from intrusion of basalt or andesite which may have induced crustal melting and/ or eruption of the resultant silicic magma as suggested for other areas by Eichelberger (1980) and Hildreth (1981). Cole (1981) has suggested that the alignment of dacites along the eastern side of T.V.Z. and the eruption of the Taupo and Rotoiti ignimbrites may reflect intrusion of low-Si andesite magma within the last 10,000 years related to the Taupo-Hikurangi subduction system.

Several authors have recently suggested that large bodies of silicic magma originated by fractional crystallisation of more basic magma (e.g. Cameron et al., 1980; Berman, 1981). In T.V.Z. only the dacites at White Island can be satisfactorily modelled by this process and even with this example some other process has to be operative to account for the HREE values. Manawahe dacite can also be modelled by fractional crystallisation but only if there is crustal contamination. When all data are taken into account no other occurrence examined is likely to have formed principally by fractional crystallisation.

Horohoro and Parekauau do not fit any of the possible origins. The association of dacite and rhyolite at Horohoro suggests that their origin may be linked and it is possible that partial melting of sediments other than "Western Basement" greywacke may occur. Parekauau is sited close to the Orakeikorako thermal area. Slight alteration has probably affected the Ce content and may well have affected other elements.

ACKNOWLEDGMENTS

We thank Dr. Peter Rankin (Soil Bureau, D.S.I.R., Lower Hutt) for previously unpublished analyses of rare-earth elements, and Dr. John Gamble (Geology Department, Victoria University of Wellington) for discussion and reviewing of this paper. Financial support for FWR came from D.S.I.R. Research Contract No. UV/6/25.

REFERENCES

Berman, R.G., 1981. Differentiation of calc-alkaline magmas: evidence from the Coquihalla Volcanic Complex, British Columbia. J. Volcanol. Geotherm. Res., 9: 151—179.

Cameron, M., Bagby, W.C. and Cameron, K.L., 1980. Petrogenesis of voluminous mid-Tertiary ignimbrites of the Sierra Madre Occidentale, Chihuahua, Mexico. Contrib. Mineral. Petrol., 74: 271—284.

Clark, R.H. and Cole, J.W., 1976. Prediction studies on White Island Volcano, Bay of Plenty, New Zealand. In: R.W. Johnson (Editor), Volcanism in Australasia. Elsevier, Amsterdam, pp. 375—384.

Clark, R.H., Cole, J.W., Nairn, I.A. and Wood, C.P., 1979. Magmatic eruption of White Island Volcano, New Zealand, December 1976—1977. N.Z. J. Geol. Geophys., 22(2): 175—190.

Cole, J.W., 1979. Structure, petrology and genesis of Cenozoic volcanism, Taupo Volcanic Zone, New Zealand — a review. N.Z. J. Geol. Geophys., 22(6): 631—657.

Cole, J.W., 1981. Genesis of lavas of the Taupo Volcanic Zone, North Island, New Zealand. J. Volcanol. Geotherm. Res., 10: 317—337.

Cole, J.W. and Lewis, K.B., 1981. Evolution of the Taupo-Hikurangi subduction system. Tectonophysics, 72: 1—21.

Cole, J.W., Cashman, K.V. and Rankin, P.C., 1981. Rare Earth Element Geochemistry and the Origin of Andesites and Basalts of Taupo Volcanic Zone. Geol. Soc. N.Z., Hamilton Conference Programme, Abstr., p. 26.

Duncan, A.R., 1970. Eastern Bay of Plenty Volcanoes. Ph.D Thesis, lodged in the Library, Victoria University of Wellington (unpubl.).

Eichelberger, J.C., 1980. Vesiculation of mafic magma during replenishment of silicic magma reservoirs. Nature, 268: 446—450.

Ewart, A. and Stipp, J.J., 1968. Petrogenesis of the volcanic rocks of the central North Island, New Zealand, as indicated by a study of $^{87}Sr/^{86}Sr$ ratios and Sr, Rb, K, U and Th abundances. Geochim. Cosmochim. Acta, 32(7): 699—735.

Ewart, A., Taylor, S.R. and Capp, A.C., 1968. Trace and minor element geochemistry of rhyolitic volcanic rocks, Central North Island, New Zealand — total rock and residual liquid data. Contrib. Mineral. Petrol., 18: 76—104.

Ewart, A., Bryan, W.B. and Gill, J.B., 1973. Mineralogy and geochemistry of the younger volcanic islands of Tonga, S.W. Pacific. J. Petrol., 14(3): 429—465.

Ewart, A., Brothers, R.N. and Mateen, A., 1977. An outline of the geology and geochemistry and the possible petrogenetic evolution of the volcanic rocks of the Tonga-Kermadec-New Zealand arc. J. Volcanol. Geotherm. Res., 2: 205—250.

Gill, J.B., 1978. Role of trace element partition coefficients in models of andesite genesis. Geochim. Cosmochim. Acta, 42: 709—724.

Gill, J.B., 1981. Orogenic andesites and Plate Tectonics. Springer-Verlag, Berlin, Heidelberg, New York, 390 pp.

Grange, L.I., 1937. The geology of the Rotorua-Taupo Subdivision. N.Z., Geol. Surv., Bull. N.S. 37, 138 pp.

Hamilton, W.M. and Baumgart, I.L., 1959. White Island. Dep. Scientific and Industrial Research Bulletin, Wellington, 127, 84 pp.

Hanson, G.N., 1978. The application of trace elements to the petrogenesis of igneous rocks of granitic composition. Earth Planet. Sci. Lett., 38: 26—43.

Hanson, G.N., 1980. Rare earth elements in petrogenetic studies of igneous systems. Annu. Rev. Earth Planet. Sci., 8: 371—406.

Hildreth, W., 1981. Gradients in silicic magma chambers: implications for lithosphere magmatism. J. Geophys. Res., 86(B 11): 10153—10192.

Kennedy, P.C., Palmer, K. and Roser, B.P., 1981. Chemical analysis of the eight U.S.G.S. rock standards: BHVO-1, MAG-1, QLO-1, RGM-1, SCo-1, SDC-1, SGR-1, STM-1. Publ. Geol. Dep., Victoria Univ. of Wellington, 21: 1—16.

Kohn, B.P. and Glasby, G.P., 1978. Tephra distribution and sedimentation rates in the Bay of Plenty, New Zealand. N.Z. J. Geol. Geophys., 21(1): 49—70.

Lewis, J.F., 1968a. Tauhara Volcano, Taupo Zone. Part 1 — Geology and Structure. N.Z. J. Geol. Geophys., 11(1): 212—224.

Lewis, J.F., 1968b. Tauhara Volcano, Taupo Zone. Part 2 — Mineralogy and Petrology. N.Z. J. Geol. Geophys., 11 (3): 651—684.

Lloyd, E.F., 1972. Geology and Hot Springs of Orakeikorako. N.Z., Geol. Surv., Bull. n.s. 85, 164 pp.

Lopez-Escobar, L., Frey, F.A. and Vergara, M., 1977. Andesites and high-alumina basalts from the Central—South Chile High Andes: Geochemical evidence bearing on their petrogenesis. Contrib. Mineral. Petrol., 63 (3): 199—228.

Ludden, J.N. and Thompson, G., 1979. An evaluation of the behaviour of the rare earth elements during the weathering of seafloor basalt. Earth Planet. Sci. Lett., 43: 85—92.

Nagasawa, H., 1970. Rare earth concentrations in zircons and apatites and their host dacites and granites. Earth Planet. Sci. Lett., 9: 359—364.

Nance, W.B. and Taylor, S.R., 1976. Rare earth element patterns and crustal evolution. I. Australian post-Archean sedimentary rocks. Geochim. Cosmochim. Acta, 40: 1539—1551.

Norrish, K. and Hutton, J.T., 1969. An accurate X-ray spectrographic method for the analysis of a wide range of geological samples. Geochim. Cosmochim. Acta, 33: 431—453.

Peccerillo, A. and Taylor, S.R., 1975. Geochemistry of Upper Cretaceous volcanic rocks from the Pontic Chain, Northern Turkey. Bull. Volcanol., 39(4): 1—13.

Reid, F.W., 1983. The origin of the rhyolitic rocks of the Taupo Volcanic Zone, New Zealand. J. Volcanol. Geotherm. Res., 15: 315—338.

Steiner, A., 1958. Petrogenetic implication of the 1954 Ngauruhoe lava and its xenoliths. N.Z. J. Geol. Geophys., 1(2): 325—363.

Stipp, J.J., 1968. The Geochronology and Petrogenesis of the Cenozoic Volcanics of the North Island, New Zealand. Ph.D. Thesis. Australian National University, Canberra, A.C.T. (unpubl.).

Thorpe, R.S., Potts, P.J. and Francis, P.W., 1976. Rare earth data and petrogenesis of andesite from the North Chilean Andes. Contrib. Mineral. Petrol., 54: 65—78.

Topping, W.W., 1973. Tephrostratigraphy and chronology of Late Quaternary eruptions from the Tongariro Volcanic Centre, New Zealand. N.Z. J. Geol. Geophys., 16(3): 397—423.

Wright, T.L. and Doherty, P.C., 1970. A linear program and least-squares computer method for solving petrologic mixing problems. Geol. Soc. Am. Bull., 81: 1995—2008.

Journal of Volcanology and Geothermal Research, 18 (1983) 215—247
Elsevier Science Publishers B.V., Amsterdam — Printed in The Netherlands

OPHIOLITE-CONTAMINATED ANDESITES, TRACHYBASALTS, AND COGNATE INCLUSIONS OF MOUNT LAMINGTON, PAPUA NEW GUINEA: ANHYDRITE-AMPHIBOLE-BEARING LAVAS AND THE 1951 CUMULODOME

R.J. ARCULUS[1], R.W. JOHNSON[2], B.W. CHAPPELL[3], C.O. McKEE[4] and H. SAKAI[5]

[1] *Research School of Earth Sciences, Australian National University, P.O. Box 4, Canberra, A.C.T. 2600 (Australia)*
[2] *Bureau of Mineral Resources, P.O. Box 378, Canberra City, A.C.T. 2601 (Australia)*
[3] *Department of Geology, Australian National University, P.O. Box 4, Canberra, A.C.T. 2600 (Australia)*
[4] *Volcanological Observatory, P.O. Box 386, Rabaul (Papua New Guinea)*
[5] *Institute for Thermal Spring Research, Okayama University, Misasa, Tottori-ken, 682-02 (Japan)*

(Received October 20, 1982)

ABSTRACT

Arculus, R.J., Johnson, R.W., Chappell, B.W., McKee, C.O. and Sakai, H., 1983. Ophiolite-contaminated andesites, trachybasalts, and cognate inclusions of Mount Lamington, Papua New Guinea: anhydrite-amphibole-bearing lavas and the 1951 cumulodome. In: S. Aramaki and I. Kushiro (Editors), Arc Volcanism. J. Volcanol. Geotherm. Res., 18: 215—247.

The catastrophic explosive eruptions in January 1951 from Mount Lamington were followed by a year-long episode of activity which was thoroughly described by Taylor (1958). However, only preliminary major-element analyses and petrographic observations have previously been reported from this volcano. The present study was prompted by the description of anhydrite and numerous ultramafic inclusions in the lavas of Mount Lamington, together with reports of similar inclusions and anomalously high Ni and Cr contents in other volcanics of southeastern Papua.

Mount Lamington, together with three other major andesitic centers — Mount Victory, Mount Trafalgar and Hydrographers Range — rest on a northeast-dipping ophiolite sheet, the Papuan Ultramafic Belt (PUB); there is no evidence for a Benioff-Wadati zone beneath this part of Papua. The lavas of Mount Lamington are porphyritic trachybasalt and high-Si, medium-to-high-K andesite. Phenocrysts are plagioclase, amphibole, and biotite, and less common minerals are magnetite, pyroxene, apatite, cristobalite, zircon and sulphide. Olivine and orthopyroxene grains and clusters are found as foreign fragments, and anhydrite is a minor constituent, particularly of pristine samples. Numerous inclusions of xenolithic harzburgite and amphibole-biotite-plagioclase-dominated, coarse-grained igneous inclusions are also present. The lava suite is characterized by enrichments in several incompatible elements relative to intermediate lavas from Bougainville Island and New Britain. Products of the 1951 eruption have Ni and Cr concentrations in the ranges 23—45 ppm and 54—111 ppm, respectively.

There are dramatic compositional zonations in the constituent phases of both lavas and inclusions. For example, amphibole zones range from compositions compatible with crystallization from pristine intermediate magmas, to Mg and Si-rich and alkali-poor components. Spinels range from refractory, Al-poor chromite to magnetite in composition. Contamination of rising andesitic magmas by the PUB is demonstrated conclusively for Mount Lamington lavas, and overall interaction with crustal rocks in other andesitic provinces may be more generally prevalent than has previously been recognized. The high Ni and Cr contents of the lavas have resulted from partial digestion of PUB material, and are not the result of direct partial melting of an upper-mantle source; neither do the high values exclude derivation of the intermediate magmas by fractional crystallization of basalt below the ophiolite.

Isotopic analysis indicates the source of the sulphur in anhydrite was of primary magmatic rather than sedimentary origin. Two possible sources for the sulphur are either remobilized sulphides in the PUB, or, more speculatively, low-pressure breakdown of sulphate meionite (scapolite), possibly present in the lower crust of southeastern Papua.

INTRODUCTION

A cumulodome of hornblende andesite was protruded from the crater of Mount Lamington, southeastern Papua (Fig. 1), throughout most of 1951, following catastrophic Peléean eruptions on Sunday 21st January. About 180 km^2 of countryside was devastated by the paroxysmal explosions, and almost 3000 people killed. Most of the 1951 activity was closely observed by the late G.A.M. Taylor, and his subsequent Bulletin* on the eruption (Taylor, 1958) has served as a classic in the field of observational and predictive volcanology, particularly with regard to the growth of volcanic domes and the emplacement of pyroclastic-flow deposits. Some preliminary petrological results were published in the Mount Lamington Bulletin (petrography and 19 major-element whole-rock analyses), but no other rock studies were undertaken. Our study began because of (1) the reported abundance of cognate-plutonic and ultramafic inclusions in the new volcanic dome, and (2) the intriguing observation by W.B. Dallwitz of the presence of anhydrite in the 1951 lavas.

This paper is a documentation of the petrology and geochemistry of the 1951 cumulodome and its inclusions, as well as of older lavas. Standard X-ray fluorescence, spark-source mass spectrography, electron microprobe and sulphur isotopic analysis techniques (see Geochemistry and Mineralogy sections, below) were employed. The results reported bear on three controversial aspects of arc-trench type magmatism:

(1) The magmas erupted from Mount Lamington and the neighbouring volcanoes of Mounts Victory and Trafalgar, have unusually high Ni and Cr values, but we will show these features are caused by contamination by an ultramafic-mafic source in the upper crust of this part of Papua New Guinea.

(2) There is no Benioff-Wadati zone beneath this part of Papua New Guinea, despite the presence of active andesitic volcanism. A relationship

*Bulletin 38 has recently been reissued by the Bureau of Mineral Resources.

between andesite generation and subduction of oceanic lithosphere is therefore unclear.

(3) The presence of anhydrite in the Mount Lamington rocks is a mineralogical curiosity, but in theory at least, it could have significance in relation to the origin of sulphide mineralization in arc-trench systems — in particular, whether the sulphides of porphyry-copper deposits derive from sulphur originally deposited on the seafloor and carried into subduction zones (see, for example, Sillitoe, 1972).

REGIONAL GEOLOGY

Mount Lamington (1585 m) is one of four, large Quaternary stratovolcanoes on the north coast of southeastern Papua, at the "tail" of New Guinea island; Hydrographers Range, Mount Trafalgar, and Mount Victory, which is believed to have been active early this century, are the other three centres (Fig. 1b). Mount Lamington was not even recognized as a volcano by Australian government geologists before the 1951 eruption, and the local Orakaiva people had no lore of any eruptive activity. However, Van Bemmelen (1939) not only identified the mountain as a volcano, but also reported it to be active.

All four stratovolcanoes are on the Solomon Sea side of the PUB, which was emplaced onto thick crust during an episode of plate convergence in the early Tertiary (Davies, 1971, 1981). Cretaceous sediments originally deposited on the continental margin of Australia were metamorphosed, and now form the elevated Owen Stanley Range to the southwest of the PUB. The immediate basement of Mount Lamington consists of Quaternary alluvium, derived mainly from the Owen Stanley Range, and underlying Tertiary sediments that fill part of the Cape Vogel Basin.

The uppermost, northeasterly dipping, ultramafic layers of the PUB are about 8 km beneath Mount Lamington and about 21 km below sea level at the coastline, according to the results of seismic-refraction results (Finlayson et al., 1976, 1977). The ultramafic zone of the PUB is 4—8 km thick (Davies, 1981) which means that the lowermost harzburgite and dunite parts of the PUB may be between 12 and 16 km beneath the volcano. However, these depths must be regarded as rough estimates because of complications caused by the presence of major faults in the Mount Lamington area. Indeed, the site of the volcano itself may have been determined by the intersection of a southeast-trending fault, known as the Gira Fault, and an unnamed northeast-trending fault that appears to displace the surface outcrop of the PUB to the southwest of Mount Lamington. Motion on both faults has been predominantly left-lateral strike slip, but components of vertical displacement may also be present (H.L. Davies, pers. commun., 1982).

Rare intermediate-depth earthquakes are known in southeastern Papua (e.g., Ripper, 1982). These do not define a Benioff-Wadati zone beneath

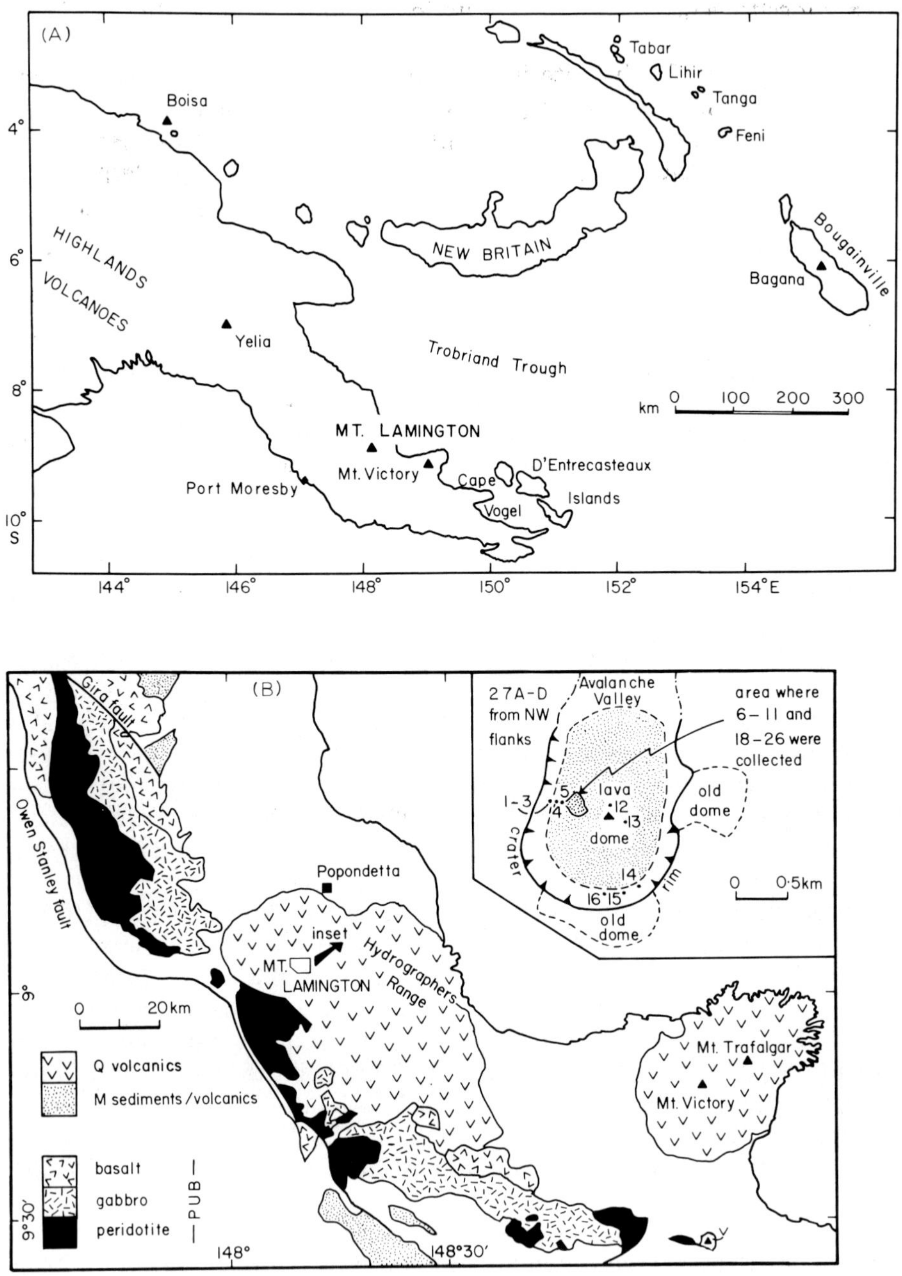

Fig. 1. a. Papua New Guinea localities. b. Simplified geology of the region around Mount Lamington (after Davies, 1971), and (inset) sample-locality map of the summit region of the volcano. Q = Quaternary; M = Miocene. Those parts of the map without ornamentation refer to Quaternary alluvium. Sample L17 is not included in the sample locality map as it consists of assorted rock chips from different parts of the dome.

Mount Lamington, however, although there is such a zone dipping westwards beneath Mount Yelia, the most easterly of the Highlands volcanoes, 300 km to the northwest (Fig. 1a). Shallow earthquakes in southeastern Papua define the boundary between the major Indo-Australian plate to the southwest and the much smaller Solomon Sea plate to the northeast. This plate boundary probably has a component of convergence, but it is not obviously a subduction zone. Hamilton (1979) believed that the Trobriand Trough, 250 km northwest of Mount Lamington, was the convergent plate-boundary related to the andesitic volcanism, and that it was characterized by slow or quiescent subduction. However, the Trobriand Trough is well within the margins of the Solomon Sea plate and a tectonic relationship between the trough and the southeast Papuan volcanic activity is at best obscure (e.g. Johnson, 1979). The present andesitic activity of Mount Lamington and Victory, therefore, represents an example of active orogenic volcanism without contemporaneous subduction.

1951 ERUPTION AND ROCK COLLECTIONS

Mount Lamington is a stratovolcano made up of lava flows, pyroclastic-flow and fall deposits and, in the upper part of the volcano, a cluster of disrupted cumulodomes. A small dome in an overgrown crater is shown on pre-1951 aerial photography (Taylor, 1958), and a breach in the northern wall of the crater is seen leading out to a valley forming the headwaters of the Ambogo River. Many pyroclastic flows erupted in 1951 were channelled through this breach and down the course of the Ambogo, giving rise to the name "Avalanche Valley". The extrusion of the cumulodome began soon after the major eruption and continued with intermittent partial destruction and pyroclastic-flow emission, for about one year. The volume of the dome is about 0.1 km^3, and it reaches a height of about 240 m above the foot of the crater wall (Fig. 2).

Taylor and his co-workers recognized three varieties of dome-forming andesite, which they correlated with different temperatures of extrusion: "Type I: anhydrite-bearing red-lamprobolite andesite; Type II: anhydrite-bearing brown-lamprobolite andesite; Type III: anhydrite-hornblende andesite" (Taylor, 1958, p. 75; the term lamprobolite has since been abandoned in favour of oxyhornblende — Leake, 1978). A glassy, green-hornblende andesite similar to the Type III dome-forming andesite was the main component of the pyroclastic-flow deposits, and seems to have been the principal magma type of the eruption. Ultramafic inclusions of partially serpentinised dunite were recognized and, together with an anhydrite-bearing hornblendite inclusion, were interpreted as pieces of country rock.

The rock samples described by Taylor and co-workers were mislaid following his premature death in 1972. Collections were undertaken again in July 1979 by R.W.J. and C.O.M., and were made primarily to obtain a new suite of inclusions from the 1951 cumulodome. However, samples of the

Fig. 2. 1972 photograph of the 1951 cumulodome inside the summit crater of Mount Lamington.

dome lava were also taken, as well as some from older lavas in the western and southern crater walls and on the northwestern flank of the volcano (Fig. 1b).

The effect of nearly 30 years of tropical weathering of the 1951-dome samples collected in 1979, is obvious from the apparent total absence of anhydrite, although its presence was confirmed (by microprobe analysis) in samples from the southern wall of the crater. Some of the original lava samples of Taylor, including some from the 1951 dome, were fortuitously relocated after completion of our study of the 1979 collection, and the presence of anhydrite was again confirmed.

LAVA SAMPLES

Rock types

All samples studied were collected in the 1951 crater area or northern slopes of the volcano, and the full range of rock compositions present is not known. The lavas are porphyritic intermediate rocks containing phenocrysts of (in order of decreasing abundance) plagioclase, amphibole, and biotite, and lesser amounts of magnetite, clinopyroxene, orthopyroxene, and apatite. Absolute and relative abundances of phenocrysts differ somewhat from one sample to another but are of the order of 30% plagioclase, 10% amphibole and 5% biotite. Accessory minerals are cristobalite, zircon, chalcopyrite and, in pristine samples, rare anhydrite; olivine is present as foreign fragments (see below). Groundmass makes up about 40—50% of each rock. The abundance of the hydrous phenocrysts amphibole and biotite in these intermediate rocks and in others from mainland New Guinea, is in strong contrast to the pyroxene-dominated (anhydrous) andesites of the Bismarck volcanic arc and Bougainville Islands in northern Papua New Guinea (Johnson, 1981).

Mount Lamington lavas are high-Si andesites and trachybasalts (terminology of Johnson et al., 1978) (Table I; Taylor, 1958, table II). 1951 rocks, including those from the cumulodome, straddle the medium-to-high-K-andesite boundary defined by Gill (1978), whereas three pre-1951 trachybasalts (columns 1, 13 and 14 in Table I) have more potassium than do the high-K andesites of Gill (1978). Relatively high abundances of several incompatible elements is a characteristic of the Mount Lamington rocks, and is also a feature of many of the intermediate rocks of the Fly-Highlands volcanic province to the west (Fig. 1a; Hamilton et al., 1983, this volume). However, interpretation of the whole-rock chemical analyses is to some extent made difficult by the known presence of PUB fragments in the andesites. Abundances for the transition elements Ni and Cr are the most obviously affected, but Mg/Fe values are probably modified too. Contamination by the ophiolite may in part explain why so many volcanic rocks from southeastern Papua fall below — that is, on the Mg-rich side of — the field of "hypersthenic-series" rocks (Kuno, 1968) in an AFM diagram (Johnson et al., 1978, fig. 4).

Petrography

Plagioclase phenocrysts range up to 8 mm in diameter and are strongly oscillatory zoned. Some phenocrysts have zones in which numerous inclusions of amphibole, biotite, apatite, and gaseous-fluid mixtures are prominent. Many clusters of plagioclase phenocrysts have synneusis textures (Vance, 1969), and glomerocrysts of plagioclase, amphibole, and magnetite are common. Most plagioclase phenocrysts are euhedral. There is little evidence for fragmentation of the crystals, although rare broken fragments of phenocrysts can be identified.

Amphibole phenocrysts are strikingly euhedral and fresh; few rock samples have amphiboles showing any degree of resorption or breakdown. Maximum phenocryst dimensions are of the order 5—6 mm. Many phenocrysts contain inclusions (less than about 0.5 μm) of plagioclase, biotite, magnetite, and apatite. Pleochroism is in shades of deep red-brown in some lavas to light-brown/grey-green in other samples. Many amphiboles have a distinct zoning from dark cores to lighter coloured rims.

Biotite is also generally found as remarkably fresh, euhedral phenocrysts, up to 5 mm in length. Inclusions of magnetite, apatite, and plagioclase are present. Small biotite flakes in some lavas have distorted crystal shapes, suggesting a degree of movement of the lava during crystallization.

Magnetite crystals have a considerable range in size and modal abundance, but in general they appear to be slightly more abundant (though still less than about 3 modal percent) in lavas characterized by reddish-brown amphibole phenocrysts. Maximum crystal sizes are about 0.4 mm. Modal abundances of magnetite are notably much greater in aggregates of phenocrysts, and also in the coarse-grained plutonic inclusions. A few magnetite phenocrysts contain apatite inclusions.

Pyroxene is comparatively rare in the Mount Lamington lavas, and crystals are relatively small (less than about 1 mm across). Many clinopyroxene microphenocrysts have amphibole reaction rims, and there appears to be little evidence for the production of clinopyroxene or orthopyroxene from amphibole-breakdown reactions (cf. Taylor, 1958).

The *groundmass* of most lavas is finely microcrystalline and predominantly feldspathic, containing minute crystals of amphibole, biotite, magnetite, and possibly apatite. Nearly hexagonal microphenocrysts of partially inverted cristobalite are found in most samples. The interiors of the hexagons are made up of numerous, irregular, interlocking fragments, some of which have low birefringence. Anhydrite, where present, is found as somewhat ragged microphenocrysts (less than 0.05 mm across) surrounded by magnetite and feldspar or by biotite and feldspar. The anhydrite does not give the appearance of being a breakdown or replacement product of another mineral.

Geochemistry

Major and trace-element whole-rock chemical analyses of 13 samples collected in 1979 are presented in Table I, together with three representative major-element analyses published by Taylor (1958), and their normative compositions are plotted in Fig. 3. The thirteen 1979 rocks were analysed by X-ray-fluorescence spectrometry using the methods outlined by Norrish and Chappell (1977). Values for additional trace elements, including the rare-earth elements (REE), in a high-Si andesite and a trachybasalt (Table II) were determined by spark source mass spectrography using the techniques described by Taylor and Gorton (1977). These two samples have fractioned, light-REE-enriched, chondrite-normalized abundance patterns (Fig. 4); La/Yb values range from 13 (sample L13) to 19 (sample L1). The trace-element composition of the two trachybasalts contrasts strongly with those of the other analyzed samples (including rocks of the 1951 eruption). For example, they have considerably higher Ba, Rb, Zr, and total REE values, and higher La/Yb values.

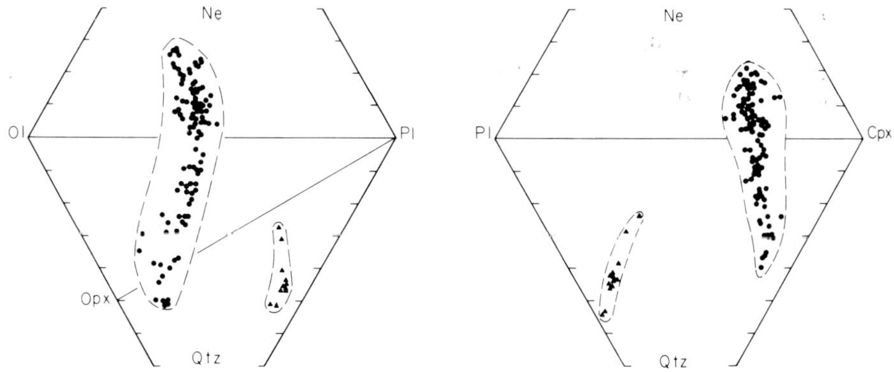

Fig. 3. Whole-rock (triangles) and amphibole (circles) compositions. Molecular components are: nepheline (Ne), olivine (Ol), plagioclase (Pl), clinopyroxene (Cpx), and quartz (Qtz). Opx represents orthopyroxene. Amphibole compositions include those of phenocrysts, inclusion grains, groundmass crystallites, and reaction rims on PUB xenoliths and crystal fragments.

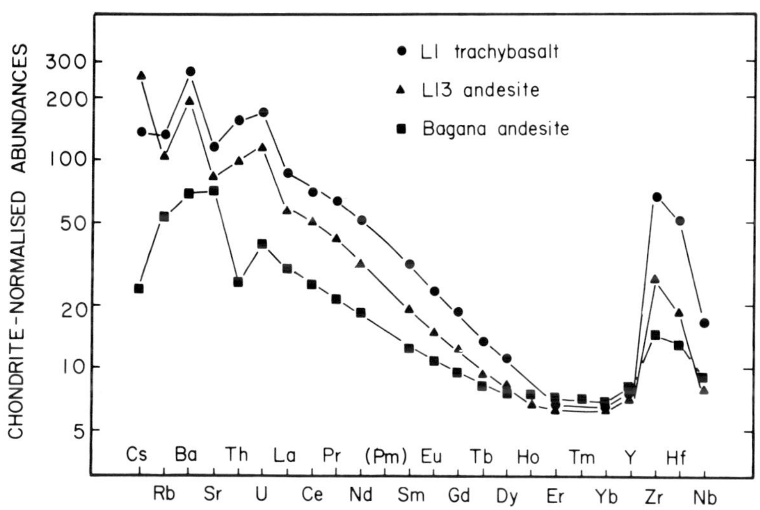

Fig. 4. REE and other trace elements for trachybasalt L1 and high-Si andesite L13 from Mount Lamington, compared with the pattern for the Bagana reference andesite composition. Chondrite normalizing values are those of Taylor and McLennan (1981).

Some useful geochemical comparisons can be made between the Mount Lamington rocks and a reference average composition for andesites from Mount Bagana on Bougainville Island (Fig. 1). Bultitude et al. (1978) showed that the major-element values of this average were close to those for almost 900 volcanic rocks from different late Cainozoic volcanic provinces in Papua New Guinea (Johnson et al., 1978). Abundances of elements such as Cs, Rb, Ba, Th, U, Zr, and Hf are all considerably greater in the Mount Lamington samples compared with those from Bagana. Further, the Mount Lamington samples are enriched overall in REE and have higher La/Yb values compared to those of intermediate rocks elsewhere in southeastern

TABLE I

Chemical analyses of rocks from Mount Lamington[a]

Sample	L1[b]	L3	L6A	L11	L13	L14	L15	L16
SiO_2	55.06	60.65	60.02	60.49	60.66	58.70	60.05	59.48
TiO_2	1.11	0.61	0.61	0.60	0.61	0.66	0.58	0.57
Al_2O_3	15.99	17.31	17.22	17.35	17.47	16.98	17.02	17.16
Fe_2O_3	2.74	3.01	2.96	2.99	2.92	2.48	2.86	2.85
FeO	3.57	1.96	2.08	1.86	2.06	2.93	1.89	1.75
MnO	0.12	0.11	0.11	0.11	0.11	0.12	0.09	0.10
MgO	5.89	3.24	3.38	3.31	3.16	4.07	3.42	2.97
CaO	7.41	5.76	5.90	5.66	5.81	6.37	5.00	5.41
Na_2O	3.44	4.43	4.46	4.49	4.42	4.24	4.16	4.08
K_2O	3.17	2.14	2.10	2.14	2.13	2.12	1.86	1.85
P_2O_5	0.64	0.24	0.23	0.25	0.25	0.24	0.22	0.23
S	<0.02	<0.02	0.05	<0.02	<0.02	0.05	1.19	1.28
H_2O^+	0.40	0.41	0.63	0.43	0.38	0.72	0.96	0.96
H_2O^-	0.21	0.08	0.10	0.09	0.11	0.12	0.48	0.65
CO_2	0.01	0.07	0.18	0.24	0.22	0.17	0.65	0.78
rest	0.42	0.27	0.28	0.27	0.27	0.29	0.28	0.28
			100.31			100.26	100.71	100.40
O = S			0.02			0.02	0.59	0.64
Total	100.18	100.29	100.29	100.28	100.58	100.24	100.12	99.76
Ba	970	710	705	705	695	660	675	685
Rb	64	50	49.5	50	49.0	47.5	41.0	39.5
Sr	1285	946	950	948	949	913	981	998
Pb	33	28	28	28	29	40	30	34
Zr	384	154	154	154	153	151	151	149
Nb	7.5	3.5	3.5	3.5	3.5	3.5	3.5	3.5
Y	16	16	16	15	15	16	15	15
La	31	20	20	19	19	18	20	20
Ce	75	44	45	46	44	44	48	47
Nd	29	18	18	19	19	19	19	17
V	166	102	103	99	104	124	107	106
Cr	169	57	68	54	65	111	67	63
Ni	68	23	27	24	26	45	28	24
Cu	40	13	15	8	6	19	6	18
Zn	77	63	66	64	64	103	69	65
Ga	19.0	19.0	19.5	19.5	19.5	19.0	19.5	19.0

[a]In this table major elements are reported in wt.% and trace elements as p.p.m.
[b]Samples prefixed L are the 1979 collection. Those prefixed T are analyses given by Taylor (1958), and tr stands for trace. T4 is a trachybasalt, T9 the "Type I" andesite, and T17 the anhydrite-bearing hornblendite.

Papua. Differences in certain incompatible element ratios between rocks from Mount Lamington and Bagana are evidence that source compositions, and possibly fractionation histories, must differ between the two regions. For example, the values Th/U, Cs/Rb and Zr/Hf are 3.1—3.3, 0.02—0.04

Sample	L26	L27A	L27B	L27C	L27D	T4	T9	T17
SiO_2	60.01	59.17	59.33	60.14	54.95	55.19	59.82	47.48
TiO_2	0.61	0.67	0.66	0.61	1.12	1.84	0.79	2.44
Al_2O_3	17.06	17.37	17.21	17.24	16.43	15.69	17.27	10.42
Fe_2O_3	2.78	3.15	2.78	2.91	3.13	2.72	4.64	3.52
FeO	2.29	2.21	2.41	1.96	2.95	3.02	0.50	4.97
MnO	0.11	0.11	0.11	0.11	0.12	0.14	0.11	0.18
MgO	3.42	3.51	3.47	3.05	5.24	5.52	3.37	13.71
CaO	5.90	5.95	5.97	5.82	6.83	7.36	5.84	10.86
Na_2O	4.40	4.34	4.38	4.46	3.47	3.36	4.22	1.97
K_2O	2.10	2.17	2.15	2.14	3.33	3.35	2.63	1.30
P_2O_5	0.24	0.27	0.25	0.23	0.63	1.44	0.79	0.60
S	< 0.02	< 0.02	< 0.02	< 0.02	< 0.02	0.05	0.04	0.93
H_2O^+	0.50	0.91	0.62	0.57	0.72	tr	tr	tr
H_2O^-	0.09	0.22	0.09	0.47	0.49	0.82	0.04	0.03
CO_2	0.53	0.01	0.37	0.26	0.42	nil	nil	nil
rest	0.28	0.29	0.28	0.27	0.41			
						100.50	100.06	98.41
O=S						0.02	0.02	0.46
Total	100.32	100.35	100.08	100.24	100.24	100.48	100.04	97.95
Ba	700	700	680	690	940			
Rb	50	48.5	49.0	49.0	72			
Sr	936	980	961	931	1240			
Pb	28	28	28	28	32			
Zr	154	165	156	149	405			
Nb	4.0	4.0	4.0	4.0	8.0			
Y	16	17	17	16	17			
La	19	20	20	20	32			
Ce	46	51	48	44	78			
Nd	19	20	19	19	31			
V	107	120	121	106	163			
Cr	78	58	74	63	152			
Ni	30	27	25	26	63			
Cu	11	16	9	10	42			
Zn	67	74	70	68	80			
Ga	19.0	20.0	20.0	19.0	19.5			

TABLE II

Additional trace-element data (in ppm) for two Mount Lamington rocks obtained by spark-source mass spectrography

Sample	L1	L13
Cs	1.5	2.8
Th	7.9	5.0
U	2.4	1.6
Hf	8.6	3.2
La	32	21
Ce	67	48
Pr	9.1	5.7
Nd	37	23
Sm	7.3	4.4
Eu	2.0	1.3
Gd	5.7	3.9
Tb	0.77	0.53
Dy	4.3	3.2
Ho	0.62	0.56
Er	1.7	1.5
Yb	1.7	1.6

and 45—127, respectively, for Mount Lamington and 2.1—2.3, about 0.01 and 34—41, respectively, for Bagana.

The relatively high abundances of Ni and Cr in the analyzed rocks of Table I are particularly noteworthy. For example, the products of the 1951 eruption have concentrations in the range 23—45 ppm Ni and 54—111 ppm Cr. These levels are up to an order of magnitude larger than, for example, most andesites containing about 60% SiO_2 from the Bismarck volcanic arc (e.g. Johnson, 1981), but are similar to those in lavas from Mounts Victory and Trafalgar (Fig. 1; Jakeš and Smith, 1970). If PUB harzburgites have been melted into Mount Lamington magmas, then about 1—5 w. % of added harzburgite would account for the Ni and Cr abundances. These are obviously rough estimates that are dependent on assuming the contents of Ni and Cr in the pristine magma. Furthermore, larger proportions of lherzolite and gabbros, relative to magma, would be required to satisfy the abundance levels in the magma because of the lower abundances of Ni and Cr in these rocks. These calculations are probably unrealistic in that partial reaction and incorporation of solid fragments of the PUB is also visualized for the Mount Lamington rocks rather than just complete incorporation of PUB by melting (see below). Nevertheless, the direct correlation between the presence of PUB fragments and high Ni and Cr contents in the lavas of Lamington, Victory and Trafalgar is convincing proof of the effects of contamination on the andesitic lavas of these volcanoes.

Two anhydrite-pyrite-bearing lava samples from the southern crater wall of Mount Lamington have been analyzed for $\delta^{34}S$ values and for their concentrations of sulphide and sulphate sulphur (Table III). Three significant features are shown by these data. First, there is a predominance of sulphide over sulphate in these samples which accords with the relative abundances of pyrite and anhydrite in the rocks. Second, the isotopic fractionation factor between sulphate and sulphide is 9.8‰ in sample L15 and 11.1‰ in sample L16. These values correspond (assuming equilibrium between sulphate and sulphide) to temperatures of about 800 and 700°C, respectively, and are compatible with slight cooling from near-solidus temperatures. Thirdly, although the $\delta^{34}S_{\Sigma S}$ values are not the same for both samples, the values are close to zero ‰, and this excludes the possibility that marine sulphate ($\delta^{34}S_{\Sigma S}$ value of +20.3‰ on average) is involved in the genesis of the anhydrite.

TABLE III

Sulphur isotopic and abundance data for two Mount Lamington samples

	S (ppm)			δ (‰)			CO_2 (ppm)
	S^{2-}	SO_4	ΣS	S^{2-}	SO_4^{2-}	δΣS	
L15	9380	1990	11370	−3.3	+6.5	−1.6	12
L16	6980	4420	11400	−1.6	+9.5	+2.7	1133

INCLUSIONS

Coarse-grained inclusions of amphibole, plagioclase, biotite, magnetite, and groundmass glass are common in the 1951 dome and in the older lavas of Mount Lamington; olivine, surrounded by reaction rims of amphibole, is also present in many of them. Accessory minerals are mostly small crystals (less than 0.1 mm across) of apatite, zircon, rutile, cristobalite, and sulphide.

The inclusions are mineralogically similar in many respects to the host lavas, but there are some marked differences in modal abundances and textures. For example, amphibole, biotite, and magnetite in most inclusions form a greater percentage of the mode than in the lavas — up to about 40% (amphibole), 10% (biotite) and 5% (magnetite). Further, although some pronounced compositional zoning of amphibole and plagioclase is present in the inclusions, the marked oscillatory zoning of plagioclase phenocrysts in the lavas is generally absent. The plagioclase in one sample in particular is found as large, poikilitic, but unzoned grains within an inclusion, but they project into the host lava where strong oscillatory zoning is developed on the projecting parts of the crystals.

The pleochroism of the amphiboles in the inclusion and host lava of any one pair is generally the same, but in some rare samples, there is a zonation of amphibole colours within the inclusion towards the contact with, and towards the colour of the amphibole in, the host lava. Accessory minerals are also present in the inclusions — zircon, apatite, sulphide, and cristobalite. Additional accessories are rutile, found within amphibole, and chromian magnetite which is also generally included in amphibole.

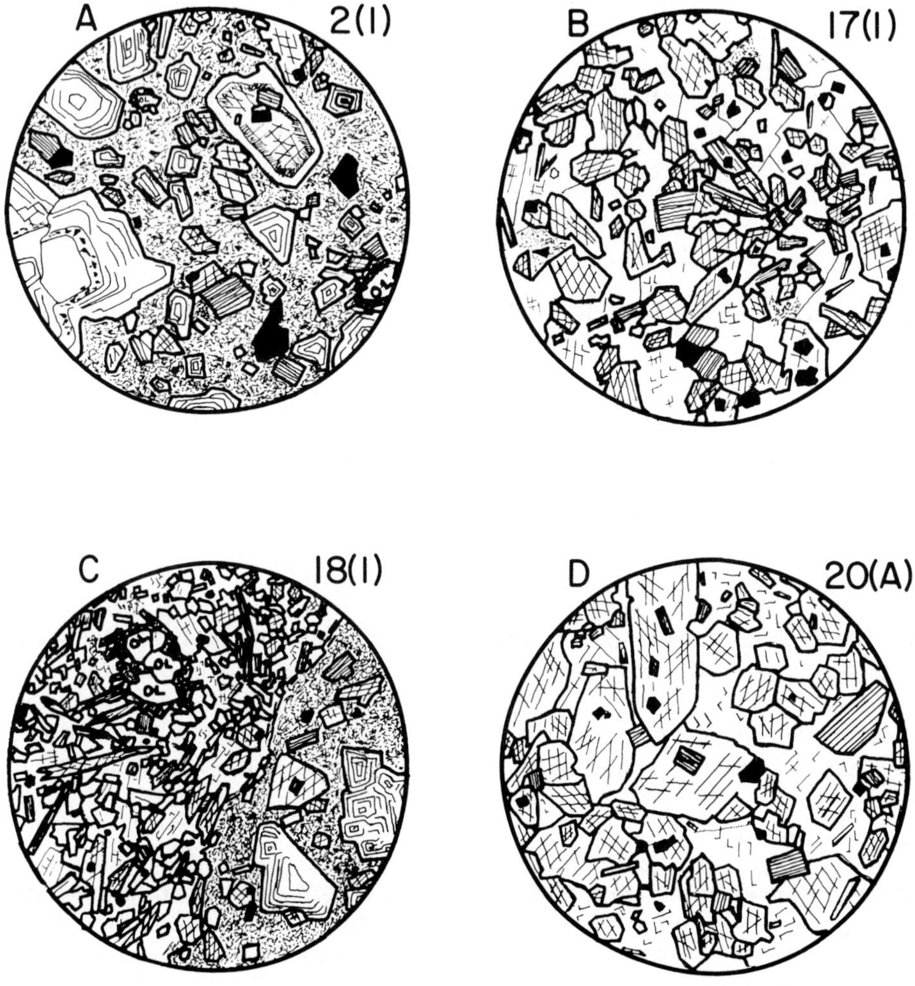

Fig. 5. Drawings of thin-sections of four different inclusions, projected with the aid of a Nikon Profiler. All are drawn at the same scale. Diameter of field of view is about 10 mm. Olivine grains surrounded by amphibole and biotite are labelled *OL*. Microcrystalline lava groundmass in A and the fine-grained (host lava) part of C is represented by fine shading, and phenocryst plagioclase by oscillatory-zoning patterns. Poikilitic and adcumulate-textured plagioclase in B and D forms the matrices of the inclusions. Biotite is represented by parallel cleavage and amphibole by 60° cleavage. Small needles of apatite are present in B, C, and D. Sample numbers shown in top right-hand corner of each projection.

Textures in the inclusions range from coarsely adcumulate (using the terminology, but not the genetic connotations, of Wager et al., 1960), to poikilitic, to a fairly even-sized mass of euhedral crystals that do not interlock and which are confined by a glassy matrix. Examples of these textural types are given in Fig. 5. Many of the inclusions are characterized by a truly glassy matrix, unlike the microcrystalline feldspathic groundmass present in the porphyritic host lavas. However, a degree of crystallinity is developed in some inclusions which approaches that of the host lavas, and has a similarly microcrystalline assemblage of feldspar, amphibole, biotite, magnetite, and apatite. Some inclusions give the appearance of being formed from two generations of crystals: earlier and larger (less than 1 mm across) crystals consisting of plagioclase, amphibole, olivine, biotite, and clinopyroxene, are set in a matrix of finely interlocking plagioclase, amphibole, biotite, magnetite, apatite, and a felsic glass. Olivine has amphibole reaction rims, and some of the plagioclase has a dusty appearance caused by many small inclusions within the crystals. A few inclusions have a lineation caused by the alignment of lath-like (2—5 mm long) amphibole and biotite. In contrast, another rare inclusion type has a microcrystalline hornfels texture.

Enclosed within the andesite of the 1951 dome are small (less than 50 mm across) altered harzburgite fragments, typically containing veins and reaction rims dominated by mica and feldspar. The constituent grains of highly magnesian olivine and orthopyroxene are large (about 10—15 mm), and many have strain lamellae and dislocation boundaries. Anhedral chromite is also present (less than about 4 modal percent), and numerous CO_2-H_2O fluid inclusions are found in both the olivine and orthopyroxene. The overall fabric of these harzburgite inclusions in general matches those shown in published photomicrographs of rocks from the harzburgite-zone of the PUB (Davies, 1971, plates 3 and 5).

There are small fragments of coarse-grained plutonic inclusions in some lavas, as well as single olivine grains and polycrystalline clusters of olivine and orthopyroxene that seem to have been disrupted from larger masses. Both the olivine and orthopyroxene clusters have reaction rims of amphibole and biotite.

MINERALOGY

About 1500 microprobe analyses were obtained in the course of the study, of which only a representative selection were used in the following Figures and Tables. All of the electron-microprobe analyses were made with the TPD instrument, equipped with an Ortec EDS, housed at the Research School of Earth Sciences, Australian National University. Analytical conditions and techniques were those described by Reed and Ware (1973, 1975) and Ware (1981).

Plagioclase

Plagioclase-phenocryst compositions range from labradorite to oligoclase, but the majority of core compositions are in the range An_{60-30} (Fig. 6). Most of the plagioclase has a low orthoclase component (less than Or_2 at An_{30} to less than Or_1 at An_{60}), but the two trachybasalts (samples L1 and L13 in Table I) have plagioclase containing 3—5 mole percent orthoclase in phenocrysts of An_{60-30} composition.

Many plagioclase phenocrysts have reverse zoning and quite marked Ca enrichment near crystal rims. The plagioclase in some inclusion/host-lava pairs has an extensive compositional overlap between the inclusion plagioclase and the host plagioclase, whereas in other pairs a gap in the lava-plagioclase compositions appears to be filled by the inclusion-plagioclase compositions. There is little development of potassium-rich rims on the

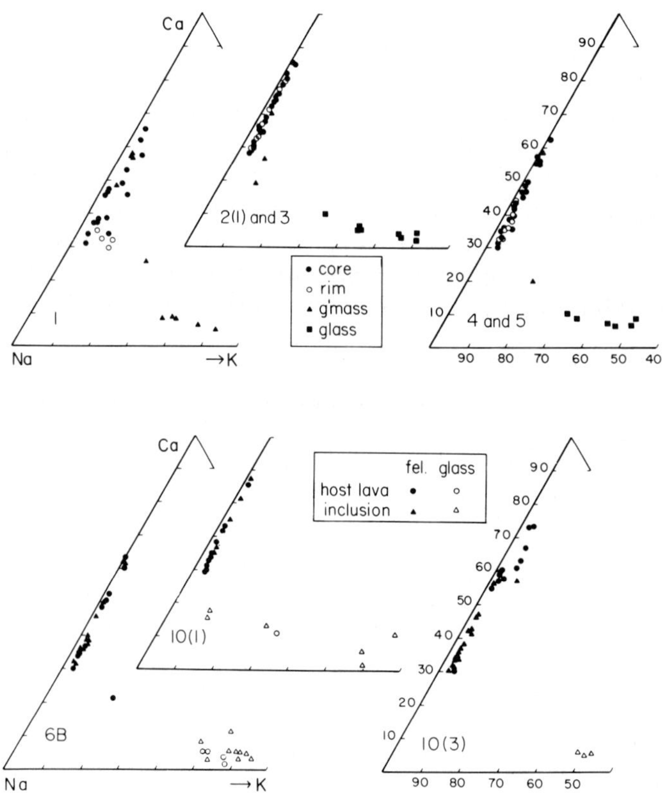

Fig. 6. Plagioclase and groundmass-glass compositions in lavas and inclusions. The upper part of the figure shows compositional ranges in lavas 1, 2(1), 3, 4 and 5. In the lower part of the figure considerable compositional overlap is shown between plagioclase in host lava and inclusion.

plagioclase phenocrysts; most potassium feldspar appears to be in the groundmass.

Amphibole

Differences in the compositions of amphibole in the lavas and, in particular, the inclusions, are spectacular, and amphibole compositions cover a considerable part of the known range of all calcic-amphibole compositions (see, for example, Cawthorn, 1976). In order that appropriate names can be assigned, and for the purposes of graphical projections of normative components, an oxidation state for microprobe analyzed amphibole has been assumed. Based on the oxidation state of the wet-chemically analyzed hornblendite sample described by Taylor (1958), and also by analogy with those of cumulus amphiboles from the Lesser Antilles (Arculus and Wills, 1980), an oxidation state of $Fe_2O_3 = 0.39 \times$ all-Fe-as-FeO has been adopted. This oxidation state may be too high for some Mount Lamington amphiboles, but the main outline of the discussion below is little affected by any errors in projection (or nomenclature) produced by it.

There is generally strong zoning within individual phenocrysts in both lavas and in inclusions, as well as some overall differences from grain to grain. The core compositions range predominantly from magnesiohastingsite to magnesiohastingsitic hornblende (nomenclature of Leake, 1978). This is similar in many respects to the amphibole-phenocryst compositions found within lavas and cumulates from the Lesser Antilles (Arculus and Wills, 1980), and, to some extent, to the potassic suite of the Tabar, Lihir, Tanga and Feni Islands (Fig. 1) of Papua New Guinea (Arculus and Johnson, 1981a). However, the zoning trends of the amphibole from Mount Lamington are unusual in that the overall compositional changes include a strong *increase* in $Mg/\Sigma Fe$, Si, and Mg, a marked *decrease* in total alkalies — in particular, K — and a decrease in Al and Ti (Table IV, Figs. 3 and 7). Expressed in terms of the nomenclature of Leake (1978), the changes are from magnesiohastingsite cores, through edenitic and pargasitic hornblende, to intermediate and high-Si magnesiohornblende, and finally to actinolitic hornblende. Within any particular inclusion/host-lava pair, the amphibole of the inclusions is generally compositionally zoned, and there is a more pronounced trend towards magnesiohornblende and actinolitic hornblende rims.

An important point to note is that these types of compositional trends are not typical of igneous amphiboles, which normally have decreasing $Mg/\Sigma Fe$ values and typically steady, or even increasing, K contents during crystallization of magma. The explanation for the Mount Lamington trends is that the andesite has reacted with and assimilated PUB ultramafic and mafic materials during crystallization. The arrested process of reaction and assimilation can be seen in some samples as amphibole rims around magnesian olivine or orthopyroxene grains, but in other samples there is no

TABLE IV

Amphibole compositions from the Mount Lamington rock suite chosen to cover the range present in both lavas and inclusions

wt.%	1H3	1H12	1H20	2(1)H1	2(1)H4	5H6	5H1	5H4
SiO_2	41.33	40.11	41.61	44.49	46.03	43.17	46.33	49.78
TiO_2	3.26	3.42	1.70	1.43	1.39	1.65	1.22	0.85
Al_2O_3	13.03	12.17	13.06	9.95	10.00	11.36	8.67	6.61
V_2O_3	0.12	0.19	0.00	0.00	0.11	0.00	0.00	0.00
Cr_2O_3	0.00	0.00	0.26	0.17	0.00	0.00	0.00	0.00
FeO*	10.61	12.15	11.39	14.80	12.15	15.89	13.56	12.48
MnO	0.00	0.16	0.00	0.27	0.32	0.37	0.21	0.34
MgO	15.15	15.09	18.04	13.57	14.30	11.93	14.04	15.27
CaO	11.90	11.75	8.67	11.13	10.96	11.59	11.31	11.47
Na_2O	2.38	2.33	1.97	2.08	1.85	1.84	1.30	1.13
K_2O	1.45	1.21	0.76	0.71	0.60	0.90	0.51	0.43
Total	99.29	98.59	97.45	98.69	97.80	98.78	97.17	98.45

	6BI8	6BI4	6BI40	6BH21	6BH23	10(1)I4	10(1)I2	10(1)H14
SiO_2	42.22	46.12	51.77	43.93	46.94	43.39	44.08	46.23
TiO_2	1.53	1.16	0.47	1.71	1.23	1.54	1.06	1.03
Al_2O_3	12.36	8.97	4.84	10.66	8.37	11.26	13.95	8.57
V_2O_3	0.00	0.00	0.00	0.00	0.00	0.12	0.00	0.00
Cr_2O_3	0.00	0.00	0.00	0.00	0.00	0.00	0.00	0.00
FeO*	14.84	13.70	11.32	14.76	12.96	14.03	11.68	13.63
MnO	0.26	0.13	0.17	0.23	0.18	0.31	0.13	0.30
MgO	12.31	14.55	17.24	13.25	15.01	13.40	14.05	14.36
CaO	11.22	11.39	11.61	11.52	11.44	11.19	10.71	11.12
Na_2O	2.01	1.45	0.56	1.72	1.37	2.03	2.04	1.61
K_2O	0.95	0.56	0.14	0.67	0.39	0.69	0.46	0.57
Total	97.77	98.03	98.12	98.55	97.89	97.97	98.15	97.43

	10(3)H9	10(3)H30	10(3)H11	10(3)H19	18(2)H44	18(2)H33	18(2)H36	18(2)H47
SiO_2	43.01	40.64	45.25	44.49	49.30	48.33	43.82	43.15
TiO_2	1.52	2.48	1.37	1.38	0.39	0.40	0.92	1.35
Al_2O_3	12.57	14.21	9.16	10.59	7.94	8.33	13.20	12.53
V_2O_3	0.00	0.00	0.00	0.00	0.00	0.00	0.00	0.00
Cr_2O_3	0.00	0.00	0.00	0.00	0.00	0.00	0.00	0.00
FeO*	9.31	14.79	14.48	11.43	7.07	9.23	8.57	12.66
MnO	0.00	0.28	0.25	0.00	0.00	0.00	0.00	0.00
MgO	16.49	11.61	14.29	15.21	19.35	17.67	16.01	14.00
CaO	11.79	11.01	11.35	11.03	11.41	11.63	11.34	11.92
Na_2O	2.65	2.40	1.98	2.16	1.26	1.50	1.69	2.02
K_2O	0.65	0.94	0.63	0.49	0.16	0.21	0.60	0.72
Total	97.99	98.44	98.75	96.79	96.88	97.30	96.26	98.35

TABLE IV (continued)

	10(3)I16	18(3)H9	18(3)I17	20(A)H12	20(A)H8	26H5	26H7	26H17
SiO_2	42.96	43.40	46.37	43.65	47.99	41.31	45.33	52.96
TiO_2	1.33	1.68	1.10	1.57	0.69	1.90	1.00	0.00
Al_2O_3	14.00	13.39	10.67	11.40	8.54	12.41	9.91	3.60
V_2O_3	0.00	0.00	0.00	0.00	0.00	0.00	0.00	0.00
Cr_2O_3	0.00	0.17	0.00	0.00	0.00	0.00	0.00	0.00
FeO*	8.81	11.29	10.96	13.82	10.89	15.37	13.04	9.98
MnO	0.00	0.00	0.00	0.26	0.22	0.29	0.18	0.47
MgO	16.75	14.21	15.62	13.32	16.27	12.18	14.72	19.20
CaO	12.06	11.75	11.54	11.63	11.44	11.05	10.90	9.85
Na_2O	2.32	1.76	1.58	1.91	1.36	2.42	1.97	0.97
K_2O	0.67	0.72	0.28	0.67	0.30	0.89	0.54	0.09
Total	98.89	98.36	98.12	98.31	97.72	97.82	97.60	97.13

*Total Fe as FeO.
H or I in the column headings refers to host lava or inclusion, respectively. The sets of numbers and letters preceding an H or I are rock-sample numbers, and the numbers following an H or I refer to specific mineral-analysis numbers. Sample 18(2) consists wholly of an amphibole-orthopyroxene-rich inclusion, and sample 26 includes clusters of orthopyroxene grains.

sign of pre-existing anhydrous ferromagnesian minerals within the amphibole, which appears rather to have grown from a melt of steadily changing composition.

Another indication of an unusual paragenesis for the Mount Lamington inclusions, in particular, is the presence of small chromian magnetite grains in the amphibole (see below). This mineral is not characteristic of high-Si andesites, and can be regarded as part of the same general process of partial reaction and assimilation of portions of the PUB that accounts for the major compositional changes of the host amphibole.

Compositions more extreme than even actinolitic hornblende are found in some samples. These are Si-Mg-rich ("refractory") and K-poor "pyribole" zones that have neither regular pyroxene nor amphibole stoichiometry. These compositions are not included in Figs. 3 and 7, but would project — for example in Fig. 7a — at points even higher in Mg/ΣFe and lower in K than the actinolitic hornblende compositions.

The two trachybasalts of Mount Lamington contain amphibole of magnesiohastingsite composition which is considerably more K-rich than the amphibole cores found in the 1951 rocks. These high-K amphiboles are also zoned towards non-stoichiometric, low-K, Mg-rich pyribole-rim compositions

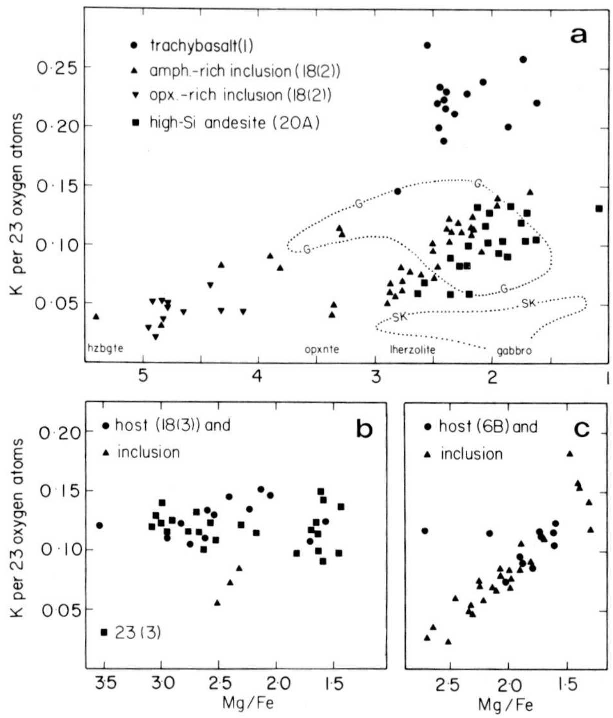

Fig. 7. Potassium versus Mg/Fe plot for amphiboles in lavas and inclusions. In a, dotted-line fields G and SK refer to amphibole compositions in plutonic blocks from the Lesser Antilles — field G is for the relatively alkalic blocks of Grenada, and field SK is for the tholeiitic blocks of St. Kitts (Arculus and Wills, 1980). Rock names along the abcissa represent compositional ranges of different PUB rock types (Jaques and Chappell, 1980) — *hzbgte* refers to harzburgite, and *opxnte* to orthopyroxenite. In b and c, note that the absolute value of the Mg/Fe ratio is displaced relative to a, but the scales of the axes are the same.

Spinel and other oxides

A low-Ti magnetite is the most common oxide phenocryst within both the lavas and inclusions, but is absent from the harzburgite fragments. However, rare grains of spinel within amphibole, as well as separate crystals, have a range of compositions that is in many ways as dramatic as the overall differences in amphibole composition. Thus, spinel compositions range from the highly refractory, Al-poor chromite in harzburgite inclusions, through relatively Al-rich ferrian chromite and chromian magnetite, to low-Ti magnetite (Table V, Fig. 8). The published chromite analyses from the harzburgite zone of the PUB (Jaques and Chappell, 1980) are generally more Cr-rich and Al-poor than those present in the harzburgite inclusions of Mount Lamington. Rutile has been detected as small (less than 30 μm across) inclusions in amphibole hosts in two inclusions.

TABLE V

Selected spinel compositions

wt.%	1H5	1H2	4H1	4H2	10(1)I20	10(1)I17	10(1)I10
SiO_2	0.28	0.00	0.35	0.75	0.72	0.40	0.42
TiO_2	2.02	4.84	1.28	5.75	1.79	3.44	3.61
Al_2O_3	10.43	1.70	4.40	2.19	1.56	4.18	3.54
V_2O_3	0.34	0.54	0.48	0.46	0.24	0.54	0.41
Cr_2O_3	35.58	3.26	14.26	0.00	23.61	14.67	5.37
Fe_2O_3*	19.46	54.99	47.46	54.02	36.62	42.58	51.52
FeO	25.78	32.01	29.36	35.74	31.05	32.10	32.79
MnO	0.77	0.46	0.89	0.25	0.68	0.68	0.58
MgO	6.03	2.23	2.15	1.15	2.12	1.79	1.26
CaO	0.07	0.00	0.38	0.14	0.32	0.35	0.22
Total	100.77	100.03	101.01	100.45	101.71	100.70	99.73
	12II5	18(3)H2	20(A)H6	20(A)H9	20(A)H14	22H1	25H6
SiO_2	0.29	0.26	0.21	0.45	0.25	0.75	0.43
TiO_2	0.00	0.61	3.39	1.48	1.05	0.00	0.64
Al_2O_3	10.02	14.71	3.88	3.60	3.82	3.62	3.91
V_2O_3	0.23	0.00	0.34	0.29	0.19	0.18	0.17
Cr_2O_3	59.16	47.67	22.14	30.82	15.29	57.70	27.32
Fe_2O_3*	2.74	8.47	36.27	31.10	48.03	13.84	36.70
FeO	16.46	14.56	31.81	30.50	28.51	7.36	28.91
MnO	0.58	0.45	0.78	0.48	0.94	0.36	0.88
MgO	11.25	12.88	1.74	1.92	2.19	16.91	2.40
CaO	0.00	0.15	0.47	0.36	0.38	0.07	0.30
Total	100.74	100.04	101.03	100.95	100.96+	100.79	101.66

*Fe_2O_3 calculated from AB_2O_4 spinel stoichiometry.
+Includes 0.30 weight percent NiO.
Column headings and abbreviations as in Table IV. Samples 12 and 22 are harzburgite host rocks.

Co-existing, unexsolved magnetite and ilmenite in one andesite define an equilibration temperature of about 650°C and an oxygen-fugacity (f_{O_2}) value of about −16.0 $\log_{10}$ units. These values clearly correspond to sub-solidus conditions, but are of interest in that they plot close to the Ni-NiO buffer curve in a f_{O_2}-versus-temperature diagram, coinciding with values determined for a large number of andesitic suites (Haggerty, 1976). There is no indication of a major decrease in the oxidation state of the magmas through incorporation of parts of the PUB. Much of the groundmass and phenocryst magnetite has exsolution lamellae of ilmenite.

Mica

The abundant biotite of both the inclusions and host lavas generally has a limited compositional range close to Mg/(Mg + ΣFe) values of 0.63—

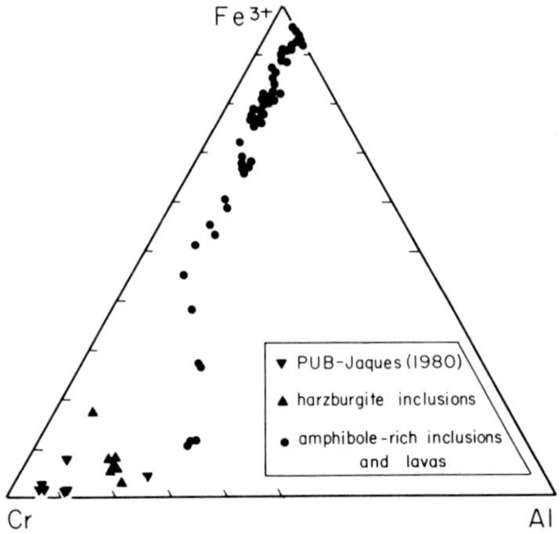

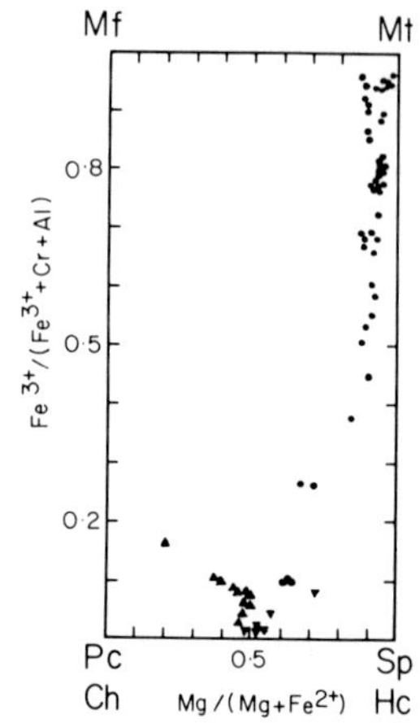

Fig. 8. Selected spinel compositions for lavas and amphibole-rich inclusions, and for fragments of the PUB (unpublished Ph.D. thesis by A.L. Jaques). Molecular percentages of Fe^{3+} (calculated from total Fe according to spinel stoichiometry), Al, and Cr are shown in the triangular plot. End-member spinel components in the rectangular plot are as follows: Mf = magnesioferrite; Mt = magnetite; Pc = picrochromite; Sp = spinel; Ch = chromite; Hc = hercynite.

0.68. This complete compositional range in a few samples is present within a single phenocryst, and there is reverse zoning from a relatively Mg-poor core to Mg-rich rim. Chlorine contents reach a maximum of about 0.15 wt.% (Table VI).

TABLE VI

Selected biotite compositions chosen to represent the range present in the lavas and inclusions

wt.%	3I2	6BH1	6BI2	10(3)H5	10(3)I4	12H1	18(2)H4
SiO_2	36.89	37.36	38.01	36.64	37.26	42.13	39.55
TiO_2	2.96	4.35	4.01	3.95	3.94	0.00	0.83
Al_2O_3	15.09	14.93	14.23	14.48	14.39	12.86	15.69
Cr_2O_3	0.26	0.00	0.00	0.00	0.00	0.00	0.24
FeO*	15.02	14.93	15.33	14.71	15.00	4.35	7.52
MgO	15.61	15.25	15.76	15.62	15.95	24.76	22.15
CaO	0.15	0.00	0.00	0.23	0.08	0.00	0.00
Na_2O	1.22	0.57	0.69	1.02	0.88	0.00	0.76
K_2O	8.71	8.75	9.04	8.47	9.08	9.72	8.41
Cl	0.15	0.10	0.15	0.13	0.10	0.00	0.08
Total	96.23	96.23	97.23	95.24	96.80	93.82	95.23

*Total Fe as FeO.
Column headings and abbreviations as in Table IV. Sample 12 is a harzburgite host rock.

Phlogopite is developed around and within veins of the harzburgite inclusions, but there is a major compositional break between this phlogopite and the mica present in the amphibole-rich inclusions and lavas.

Olivine

Olivine in the harzburgite inclusions range from $Fo_{94 \cdot 5}$ to Fo_{91}, and is compositionally similar to olivine in the PUB harzburgite zones (Jaques and Chappell, 1980). Olivine in the amphibole-rich inclusions and lavas, on the other hand, is never as forsteritic and ranges from a maximum of Fo_{90} down to Fo_{80}. These magnesian compositions are not equilibrium crystallization products of high-Si andesite magmas, and juxtaposition of groundmass cristobalite and amphibole-rimmed olivine in some lavas is a striking example of the disequilibrium mineralogy.

A curious feature of some of the olivine crystals, both in the lavas and inclusions, is the preservation of a fairly euhedral and regular outline, despite the presence of amphibole reaction rims. This might be unexpected if the gabbroic or ultramafic portions of the PUB were the source of the olivine, because of the mainly cumulate and metamorphic textures present in these rocks. However, judging from published photomicrographs (Davies, 1971, plate 10) as well as from the PUB thin sections that we have examined, the PUB gabbros at least contain olivine which, allowing for some amphibole rimming, closely match the features of the olivine crystals present in the Mount Lamington rocks.

Clinopyroxene and orthopyroxene

Pyroxenes are a relatively minor component of the 1951 andesites, but are more abundant in the older trachybasalts. Clinopyroxene in the 1951 rocks is in the range diopside-augite-salite, and there is a limited overall trend of Fe-enrichment both in core-to-rim traverses and, in a more general sense, from grain to grain (Table VII). Co-existing pyroxenes appear to be in equilibrium (see tie lines in Fig. 9) and define temperatures in the approximate range 900—975°C based on the solidus of Ross and Huebner (1975) and the algorithm of Wells (1977). The report of pigeonite (J.K. Lovering; in Taylor, 1958) appears to be an error.

Relatively magnesian orthopyroxene (En_{83-66}) is found as polycrystalline aggregates surrounded by amphibole and biotite in some lavas (sample L26, Table VII, Fig. 9). The more magnesian end of this range does not appear to be an equilibrium crystallization composition of the host andesite magmas, and the presence of chromian magnetite in the clusters is probably evidence that most of each orthopyroxene cluster has been derived from the gabbroic parts of the PUB. The orthopyroxene in the harzburgite inclusions has a refractory Al-free composition of En_{95-84}, and corresponds with the orthopyroxene compositions of the PUB harzburgite zones (samples L12 and L22, Table VII).

TABLE VII

Representative pyroxene analyses

wt.%	1H1	1H4	1H7	1H13	4H1	12H15	12H13	18(2)H1
SiO_2	51.08	51.01	50.06	53.66	50.47	58.73	57.68	56.67
TiO_2	0.77	0.69	0.24	0.41	0.60	0.00	0.00	0.00
Al_2O_3	3.69	3.20	4.83	1.47	4.39	0.00	0.48	1.14
Cr_2O_3	0.31	0.00	0.00	0.00	0.32	0.11	0.00	0.00
FeO*	5.48	7.85	17.49	14.11	6.01	3.62	9.63	11.05
MnO	0.00	0.18	0.44	0.45	0.00	0.00	0.27	0.21
MgO	15.96	15.72	23.29	27.56	16.44	37.53	33.00	31.63
CaO	22.41	20.32	2.40	1.74	20.76	0.00	0.33	0.41
Na_2O	0.28	0.38	0.33	0.15	0.28	0.00	0.00	0.00
Total	99.97	99.35	99.08	99.56	99.27	99.98	101.39	101.10
	18(2)H15	18(3)H1	18(3)I5	22H6	26H3	26H4	26H5	26H10
SiO_2	56.08	50.59	50.75	59.00	56.45	55.09	49.81	53.30
TiO_2	0.00	0.44	0.40	0.00	0.00	0.00	0.75	0.00
Al_2O_3	0.25	4.20	4.31	0.00	0.29	0.37	4.82	0.58
Cr_2O_3	0.00	0.00	0.00	0.00	0.00	0.00	0.39	0.00
FeO*	11.82	8.31	6.20	4.67	11.27	16.20	7.41	21.30
MnO	0.33	0.20	0.00	0.00	0.45	0.74	0.12	1.32
MgO	31.32	14.17	15.04	36.96	32.40	28.35	14.27	23.90
CaO	0.28	22.30	22.16	0.00	0.00	0.22	21.25	0.49
Na_2O	0.00	0.26	0.00	0.00	0.00	0.00	0.83	0.15
Total	100.07	100.47	98.85	100.62	100.85	100.96	99.65	101.04

*Total Fe as FeO.
Column headings and abbreviations as in Table IV. Samples 12 and 22 are harzburgite host rocks.

Other phases

The groundmass glass present in amphibole-rich inclusions has the high-K, high-Si composition of a potassic peraluminous rhyolite (Table VIII, Fig. 6). Apatite contains 1.0—1.5 wt.% Cl.

SUMMARY OF EVIDENCE FOR CONTAMINATION BY PUB

Several distinctive petrographic, mineralogical, and geochemical features of the Mount Lamington rock suite are strong lines of evidence that magmas have been contaminated by parts of the PUB. Taylor (1958) recognized the presence of serpentinized dunite in the lavas, but, in the era before electron-microprobe analysis permitted examination of mineralogical differences, he was unaware of the degree of assimilation of ultramafic and mafic materials.

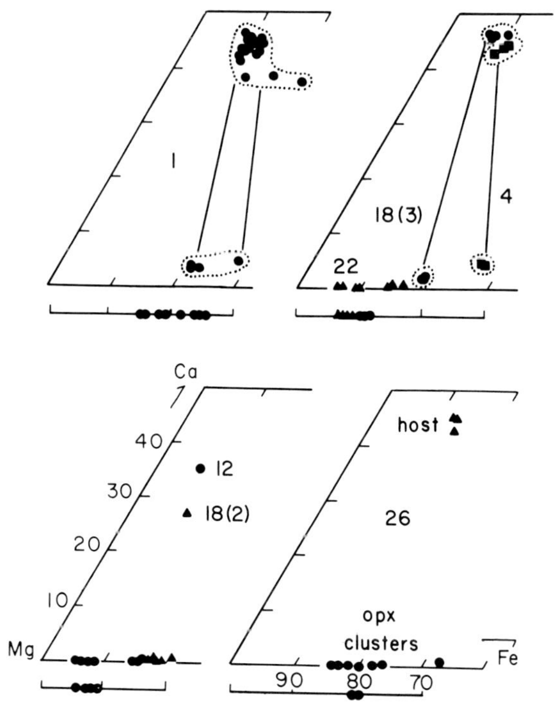

Fig. 9. Pyroxene and olivine compositions in lavas (samples 1, 4, 18), harzburgites (samples 12 and 22), and host lava and orthopyroxenite inclusions (26). Numbers in each of the four parts of the figure refer to different samples.

TABLE VIII

Representative analyses of glass matrices of inclusions

wt.%	6B.2	18(2).2	20(A).3	23(3).2	26.3
SiO_2	76.41	73.07	76.16	76.27	77.35
TiO_2	0.00	0.00	0.00	0.00	0.15
Al_2O_3	12.76	14.94	13.51	12.63	11.83
FeO*	0.86	0.95	0.56	0.76	0.63
MnO	0.00	0.00	0.00	0.00	0.00
MgO	0.00	0.38	0.23	0.00	0.31
CaO	0.42	1.68	0.35	0.39	0.48
Na_2O	2.17	2.57	2.01	2.36	2.67
K_2O	6.40	3.37	5.74	6.23	5.10
Total	99.09	96.96	98.66	98.64	98.53

*Total Fe as FeO.

Numbers preceding the point in column headings are rock-sample numbers, and those following the point refer to specific mineral-analysis numbers.

All stages of assimilation of the PUB can be found in the Mount Lamington lavas — from partially digested harzburgite fragments to polycrystalline clusters and grains of magnesian olivine and orthopyroxene that are heavily rimmed by amphibole and biotite. Further, there is evidence that a distinctive process has taken place during the evolution of these magmas because of the extraordinary zoning trends of amphibole and spinel, in particular. Amphibole compositions range from the expected equilibrium products of intermediate-melt crystallization (magnesiohastingsite) to highly magnesian and alkali-poor compositions that are thoroughly atypical of such melt compositions. Similarly, spinel zonation from chromite, through ferrian aluminous chromite to chromian magnetite, is not characteristic of the mineralogy of intermediate melts and can be best accounted for by incorporation and subsequent reaction of refractory chromite from the PUB with the host magma. An important point to note is that the core-to-rim trends for amphiboles reflect changes in host-*liquid* composition. The spinel trends, on the other hand, may well have been caused by reactions of PUB seed spinels with magma.

The anomalously high Ni and Cr contents in the Mount Lamington lavas can be recognized, given the above petrographic and mineralogical evidence, as the result of interactions between magma and the underlying ophiolite. There is no reason to invoke any hypothesis of direct ascent of a Ni-Cr-rich primary andesite melt that was in equilibrium with an upper-mantle-peridotite source. Nor is there any reason on the basis of high Ni and Cr abundances to rule out the evolution of andesite by fractional crystallization from more primitive melt types below the PUB.

We have examined samples of lavas and their inclusions from Mounts Victory and Trafalgar which are also underlain by the PUB (Fig. 1). Interactions between zones of the PUB and rising intermediate magmas have evidently taken place, and therefore the high Ni and Cr contents in these lavas (Jakeš and Smith, 1970) may be accounted for by processes similar to those proposed for the Mount Lamington rocks.

Some of the mineralogical features of the Mount Lamington rocks are similar to those described for high-magnesia andesites or boninites (e.g. Kuroda et al., 1978) — particularly the magnesian spinel, pyroxene, and olivine. However, the near-identity of the compositions and petrography of the inclusions within the lavas with the published analyses and descriptions of the PUB, together with the presence of the ophiolite (but no known boninitic rocks) beneath the volcano, are indications that involvement of boninitic material is unlikely. Furthermore, clinoenstatite has not been found in any of the Mount Lamington samples that we have studied. High-magnesia andesites have, however, been reported from Cape Vogel (Dallwitz et al., 1966) about 200 km southeast of Mount Lamington (Fig. 1). Given the evidence presented above for partial fusion of parts of the PUB, future studies of the high-magnesia andesites should consider the possibility that they represent a greater degree of fusion of the ophiolite than that resulting from the rise of the Mount Lamington magmas through the PUB. The refractory nature of the sources for high-magnesia andesites or boninites (e.g. Jenner, 1981) corresponds well with that of the lower parts of an ophiolite such as the PUB.

CONSEQUENCES OF CONTAMINATION

The profound influence of the studies made by N.L. Bowen (e.g. 1928) has persuaded most petrologists to abandon hypotheses involving partial or total assimilation of solid crustal material by silicate liquids, as an explanation for major chemical differences within igneous suites. Bowen eloquently demonstrated that the latent heat of fusion (ΔH_f) required to melt country rock results from crystallization of the invading magma, so inhibiting additional partial fusion. He believed that such interactions in single melt-solid equilibration events would be inadequate to generate a range of intermediate magmas.

More recent investigators have suggested that melting of conduit and chamber walls in spreading-axis rifts and below oceanic-islands can be achieved by fluxes of mafic magmas (e.g. O'Hara, 1980; Taylor, 1980; O'Hara and Mathews, 1981). There is also a burgeoning literature on geochemical modelling (e.g. Allègre and Minster, 1978; DePaolo, 1981) and on the consequences of magma-crust interactions in continental flood-basalt (e.g. Carlson et al., 1981) and arc-trench-type environments (e.g. Arculus and Johnson, 1981b). However, the thermal budget of melting of the solid crustal component remains the key problem, and is acute in the case of the PUB because of the refractory nature of the harzburgite layers. Recent estimates of the ΔH_f of forsterite and enstatite are of the order of 120 to 160 J mole^{-1}, respectively (Bottinga and Richet, 1978). These are considerably higher than ΔH_f values for intermediate plagioclase, the most likely liquidus mineral of the invading magma.

The degree of preservation of PUB inclusions in Mount Lamington lavas corresponds fairly closely to the extent to which the inclusions are refractory. Thus, the Mg-rich, Al-poor harzburgites are relatively well preserved, whereas more Fe-rich rocks — and especially those containing partially serpentinized olivine — appear to have been thoroughly digested. Major factors in determining the degree of melting achieved by passage of a single pulse of intermediate magma through the PUB are: (1) the temperature of the wall-rocks before introduction of magma; (2) the mineralogy of the low-melting-point component (which may include, for example, serpentine, carbonate, uralite, and altered feldspar), and hence volatile content, of the wall-rocks; and (3) the volatile content and extent of crystallization of the magma during contact with PUB material.

These factors are not well constrained, and only general assessments can be made of their relative effects. For example, the volatile content of the magma can be estimated if the phenocryst assemblage — including hornblende and biotite, together with groundmass quartz — is assumed to accurately reflect the indigenous volatile concentrations, independent of any gain from, or loss to, the PUB. Sakuyama (1979, 1981) suggested that the presence of hornblende and biotite preceding crystallization of quartz in island-arc andesites corresponds to water contents of about 3—4 wt.%, at least. If this is so, liberation of some of this water would drastically lower the fusion temperature of horizons within the PUB.

The amphiboles in the coarse-grained igneous inclusions retain evidence for a much greater degree of interaction between magma and PUB material than do the amphibole phenocrysts in the lavas themselves, because they are much more strongly zoned towards mafic compositions (Fig. 7). Further, the widely different textures of the coarse-grained inclusions — some have groundmass glass and others have numerous crystals that are small compared with the large phenocrysts in the host lavas — are evidence for differences in the rates of quenching in different parts of the magma-PUB interaction zone. All of these features are consistent with the view that partial fusion of the PUB and crystallization of the magma proceeded in a decidedly non-equilibrium fashion. The coarse-grained inclusions appear to represent those parts of the magma that had more opportunity to react with PUB material. However, they and pieces of the PUB were caught up, and to some extent became disintegrated, in the more pristine parts of the magma which, as a result of incorporation of small, disintegrated crystal fragments and clusters in particular, also acquired the contamination and non-equilibrium signature.

The popular concept of the depth of magma reservoirs beneath volcanoes is strongly influenced by comprehensive data on the configuration of reservoirs beneath Hawaiian volcanoes, which have depths of about 3—5 km. However, other lines of evidence, such as the depths of equilibration of cumulates in Lesser Antilles lavas (about 4—10 kbar, Arculus and Wills, 1980), and the fact that the Mount Lamington intermediate magmas had formed *before* they rose into the PUB (pressures greater than 4 kbar) are indications that reservoirs in which magmatic crystallization and fractionation can take place, may be deeper beneath at least some volcanoes of arc-trench systems than beneath ocean-island shield volcanoes.

ORIGIN OF THE ANHYDRITE

Anhydrite is not commonly found in volcanic rocks. However, it may be more common in freshly erupted lavas and inclusions than is generally realized. Anhydrite may be overlooked in thin section because of its low abundance (although its bright, high birefringence is quite distinctive), it may not survive more than a few years after an eruption (because of rain-water leaching), and if whole-rock samples are not analyzed for sulphur, then its presence may not even be suspected. Nevertheless, several Japanese anhydrite localities are known, apart from the Mount Lamington occurrence, and anhydrite-bearing-inclusion localities have been reported for Santorini volcano, Greece (Nicholls, 1971).

Yoshiki (1933) described xenoliths in the products of the 1929 Komagatake eruption (Hokkaido) which were anhydrite-bearing assemblages of cordierite, augite, anorthite, and pyrrhotite (see also Kozu, 1934). Anhydrite phenocrysts in a rhyolite from Himeshima have been interpreted as an early formed magmatic product (Kawano, 1948), but Taneda (1949) proposed they were of deuteric origin. Anhydrite has also been observed in the groundmass of an alkali olivine basalt groundmass from Rishiri volcano (Hokkaido) by Katsui (1958). and anhydrite-bearing, tridymite-dominated xenoliths were contained in volcanic bombs ejected during the 1970 eruption of Akita-Komagatake in northern Honshu (Yagi et al., 1972). These xenoliths were regarded by Yagi and his co-workers to be the result of metasomatic alteration of pre-existing sublimation products or silica-rich rocks. A sulphur-isotope analysis of the Akita-Komagatake anhydrite by N. Nakai (reported in Yagi et al., 1972) gave a $\delta^{34}S$ value of +2.3‰ compared with the Canyon Diablo standard.

There are known to be local concentrations of magmatic sulphides in parts of the PUB ophiolite (H.L. Davies, pers. commun., 1982), and this, together with non-marine isotopic character of the Mount Lamington sulphur (Table III) raises the possibility that the high concentrations of sulphur in some analyzed rocks from Mount Lamington are related to the overall crustal contamination process outlined above, rather than derivation from a subduction zone. The abundances of sulphur in some of the lavas (Table I) are considerably higher than known saturation levels (Haughton et al., 1974; Ueda and Katiya, 1981), and may be in part due to fumarolic additions. However, abundance levels in other rocks are more appropriate to those of original magmatic concentrations, and these lavas, where sampled shortly after the 1951 eruption, were also found to contain anhydrite. Thus, the presence of anhydrite does not seem to be closely related to late-stage fumarolic additions or replacement.

The relatively high Fe_2O_3/FeO values of the bulk rocks (Table I), together with the single set of T-f_{O_2} values given by the coexisting ilmenite-magnetite assemblage, are evidence that the host lavas were comparatively oxidized compared with, for example, sulphide-bearing ocean-floor basalts (e.g. Mathez, 1976; Haggerty, 1976). A considerable proportion of the sulphur in such magmas will probably exist as sulphate at oxidation levels of one $\log_{10}$ unit or more above the Ni-NiO buffer at magmatic (about 1000°C) temperatures (Haughton et al., 1974). Oxidation of sulphides in the PUB is a possible explanation for the presence of anhydrite in the Mount Lamington lavas, but details of the reactions are obscure. Alternatively, a more speculative possibility is that anhydrite in volcanic rocks results from breakdown of sulphur rich scapolite ($3CaAl_2Si_2O_8$-$CaSO_4$-sulphate meionite), a phase stable in granulite assemblages characteristic of the lower crust (Lovering and White, 1964; Goldsmith, 1976). Obviously further study of the Lamington and other occurrences is needed for definitive determination of the origin of anhydrite in andesitic lavas.

CONCLUDING REMARKS

The principal conclusion of this study is that upper crustal materials have demonstrably contaminated the andesitic and trachybasaltic magmas of an arc-trench-type volcano, Mount Lamington. The contaminant is clearly identifiable — in the field, in thin section, and on mineral-composition variation diagrams — on account of its highly refractory nature relative to the host lava, but it would be much less obvious had the magma risen through, and been contaminated by for example, a thick pile of lavas that had a similar composition to the host magma. Melting of the PUB by intermediate magmas is demonstrated, in particular, by the amphibole-composition trends and the textural evidence that the amphiboles have grown progressively from a melt of changing composition. However, much of the whole-rock contamination effect may be accounted for by incorporation of solid fragments in different degrees of disintegration, which were so refractory that they could not be melted. Incorporation of solid fragments into Si-rich melts was also proposed recently to account for the compositions of andesites on Boisa Island, Papua New Guinea (Fig. 1; Gust and Johnson, 1981).

Contamination by the upper crust cannot, however, be used to explain the incompatible-element-enriched character of the Mount Lamington rocks, as the contaminant is depleted in incompatible elements relative to the pristine magma. The processes of incompatible-element enrichment must have taken place below the ophiolite, and our data cannot be used to choose from a range of possible hypotheses. This same impasse is reached in the isotopic study of rocks from the Fly-Highlands province (Hamilton et al., 1983, this volume). These isotopic data are consistent with some contamination of mafic and intermediate magmas by upper continental crust, but those rocks that seem to have been the least affected isotopically by the upper crust, are nevertheless notably rich in incompatible elements. Whether or not the process of enrichment beneath eastern Papua or the Fly-Highlands province is subduction-related remains an open question, because the supporting geochemical evidence for this hypothesis in arc-trench-type volcanic areas is at best ambiguous (Arculus and Johnson, 1981b).

ACKNOWLEDGEMENTS

Considerable help in pursuit of this study was received from A.C.M. Mackinlay and F. Dira, both formerly of the Geological Survey of Papua New Guinea. We are also indebted to W.B. Dallwitz, H.L. Davies, and A.L. Jaques for acting as valuable sources of information and advice. S.R. Taylor and P. Oswald-Sealy are thanked for their assistance in acquisition of spark-source mass spectrographic data. Professor A.E. Ringwood displayed his usual degree of enthusiasm for projects of this nature, and provided R.J.A. with the opportunity to pursue problems of island-arc petrogenesis. R.J.A. and R.W.J. gratefully acknowledge the financial assistance provided by the organisers of the 1981 Arc Volcanism Symposium and by UNESCO, respectively, which made possible their attendance at the symposium. R.W.J. publishes with the permission of the Director of the Bureau of Mineral Resources, and C.O.M. with that of the Chief Government Geologist of the Geological Survey of Papua New Guinea.

REFERENCES

Allègre, C.J. and Minster, J.F., 1978. Quantitative models of trace element behaviour in magmatic processes. Earth Planet. Sci. Lett., 38: 1—25.

Arculus, R.J. and Johnson, R.W., 1981a. Mineralogy of strongly silica undersaturated volcanic rocks from the Tabar, Lihir, Tanga, and Feni islands, Papua New Guinea. 1981 I.A.V.C.E.I. Symposium — Arc Volcanism, Volc. Soc. Japan and I.A.V.C.E.I., Tokyo, pp. 18—19.

Arculus, R.J. and Johnson, R.W., 1981b. Island-arc magma sources: a geochemical assessment of the roles of slab-derived components and crustal contamination. Geochem. J., 15: 109—133.

Arculus, R.J. and Wills, K.J.A., 1980. The petrology of plutonic blocks and inclusions from the Lesser Antilles island arc. J. Petrol., 21: 743—799.

Bottinga, Y. and Richet, P., 1978. Thermodynamics of liquid silicates, a preliminary report. Earth Planet. Sci. Lett., 40: 382—400.

Bowen, N.L., 1928. The Evolution of the Igneous Rocks. Princeton University Press, Princeton, N.J., 322 pp.

Bultitude, R.J., Johnson, R.W. and Chappell, B.W., 1978. Andesites of Bagana Volcano, Papua New Guinea: chemical stratigraphy, and a reference andesite composition. B.M.R. J. Aust. Geol. Geophys., 3: 281—295.

Carlson, R.W., Lugmair, G.W. and Macdougall, J.D., 1981. Columbia River volcanism: the question of mantle heterogeneity or crustal contamination. Geochim. Cosmochim. Acta, 45: 2483—2499.

Cawthorn, R.G., 1976. Some chemical controls on igneous amphibole compositions. Geochim. Cosmochim. Acta, 40: 1319—1328.

Dallwitz, W.B., Green, D.H. and Thompson, J.E., 1966. Clinoenstatite in a volcanic rock from the Cape Vogel area, Papua. J. Petrol., 7: 375—403.

Davies, H.L., 1971. Peridotite-gabbro-basalt complex in eastern Papua: an overthrust plate of oceanic mantle and crust. Bur. Miner. Resour. Aust. Bull., 128.

Davies, H.L., 1981. Crustal structure and emplacement of ophiolite in southeastern Papua New Guinea. Colloques Int. C.N.R.S., 272: 17—33.

DePaolo, D.J., 1981. Trace element and isotopic effects of combined wallrock assimilation and fractional crystallization. Earth Planet. Sci. Lett., 53: 189—202.

Finlayson, D.M., Drummond, B.J., Collins, C.D.N. and Connelly, J.B., 1976. Crustal structure under the Mount Lamington region of Papua New Guinea. In: R.W. Johnson (Editor), Volcanism in Australasia. Elsevier, Amsterdam, pp. 259—274.

Finlayson, D.M., Drummond, B.J., Collins, C.D.N. and Connelly, J.B., 1977. Crustal structures in the region of the Papua Ultramafic Belt. Phys. Earth Planet. Inter., 14: 13—29.

Gill, J.B., 1978. Role of trace element partition coefficients in models of andesite genesis. Geochim. Cosmochim. Acta, 42: 709—724.

Goldsmith, J.R., 1976. Scapolites, granulites, and volatiles in the lower crust. Geol. Soc. Am. Bull., 87: 161—168.

Gust, D.A. and Johnson, R.W., 1981. Amphibole-bearing inclusions from Boisa island, Papua New Guinea: evaluation of the role of fractional crystallization in an andesite volcano. J. Geol., 89: 219—232.

Haggerty, S.E., 1976. Opaque mineral oxides in terrestrial igneous rocks. In: D. Rumble III (Editor), Oxide Minerals — Short Course Notes, Vol. 3. Min. Soc. Am., Washington, D.C., pp. 101—301.

Hamilton, P.J., Johnson, R.W., Mackenzie, D.E. and O'Nions, R.K., 1983. Pleistocene volcanic rocks from the Fly-Highlands province of western Papua New Guinea: a note on new Sr and Nd isotopic data and their petrogenetic implications. In: S. Aramaki and I. Kushiro (Editors), Arc Volcanism. J. Volcanol. Geotherm. Res., 18: 000—000.

Hamilton, W., 1979. Tectonics of the Indonesian Region. U.S. Geol. Surv., Prof. Pap., 1078: 345 pp.

Haughton, D.R., Roeder, P.L. and Skinner, B.J., 1974. Solubility of sulfur in mafic magmas. Econ. Geol., 69: 451—467.

Jakeš, P. and Smith, I.E.M., 1970. High potassium calc-alkaline rocks from Cape Nelson, eastern Papua. Contrib. Mineral. Petrol., 28: 259—271.

Jaques, A.L. and Chappell, B.W., 1980. Petrology and trace element geochemistry of the Papuan ultramafic belt. Contrib. Mineral. Petrol., 75: 55—70.

Jenner, G.A., 1981. Geochemistry of high-Mg andesites from Cape Vogel, Papua New Guinea. Chem. Geol., 33: 307—332.

Johnson, R.W., 1979. Geotectonics and volcanism in Papua New Guinea: a review of the late Cainozoic. B.M.R. J. Aust. Geol. Geophys., 4: 181—207.

Johnson, R.W., 1981. Papua New Guinea. In: R.S. Thorpe (Editor), Andesites. Wiley, London, pp. 225—244.

Johnson, R.W., Mackenzie, D.E. and Smith, I.E.M., 1978. Volcanic rock associations at convergent plate boundaries; reappraisal of the concept using case histories from Papua New Guinea. Geol. Soc. Am. Bull., 89: 96—106.

Katsui, Y., 1958. Groundmass anhydrite in the olivine basalt from the Rishiri volcano, Hokkaido. J. Japan. Assoc. Min. Petr. Econ. Geol., 42: 188—191.

Kawano, Y., 1948. A new occurrence of phenocrystic anhydrite crystals in the glassy rocks from Himeshime, Oita Prefecture, Japan. Rep. Geol. Surv. Japan, No. 126, 7 pp.

Kozu, S., 1934. The great activity of Komagatake in 1929. Miner. Petrogr. Mitt., 45: 2/3.

Kuno, H., 1968. Origin of andesite and its bearing on the island arc structure. Bull. Volcanol., 32: 141—176.

Kuroda, N., Shiraki, K. and Urano, H., 1978. Boninite as a possible calc-alkalic primary magma. Bull. Volcanol., 41(4): 563—575.

Leake, B.E., 1978. Nomenclature of amphiboles. Can. Mineral., 16: 501—520.

Lovering, J.F. and White, A.J.R., 1964. The significance of primary scapolite in granulite inclusions from deep-seated pipes. J. Petrol., 5: 195—218.

Mathez, E.A., 1976. Sulfur solubility and magmatic sulfides in submarine basalt glass. J. Geophys. Res., 81: 4269—4276.

Nicholls, I.A., 1971. Calcareous inclusions in lavas and agglomerates of Santorini volcano. Contrib. Mineral. Petrol., 30: 261—276.

Norrish, K. and Chappell, B.W., 1977. X-ray fluorescence spectrometry. In: J. Zussman (Editor), Physical Methods in Determinative Mineralogy, 2nd ed. Academic Press, London, pp. 201—272.

O'Hara, M.J., 1980. Non-linear nature of unavoidable long-lived isotopic, trace and major element contamination of a developing magma chamber. Philos. Trans. R. Soc. London, Ser. A, 297: 215—227.

O'Hara, M.J. and Matthews, R.E., 1981. Geochemical Evolution in an advancing, periodically replenished, periodically tapped, continuously fractionated magma chamber. J. Geol. Soc. London, 138: 237—277.

Reed, S.J.B. and Ware, N.G., 1973. Quantitative electron microprobe analysis using a lithium-drifted silicon detector. X-ray spectrometry, 2: 69—74.

Reed, S.J.B. and Ware, N.G., 1975. Quantitative electron microprobe analysis of silicates using energy-dispersive X-Ray spectrometry. J. Petrol., 16: 499—519.

Ripper, I.D., 1982. Seismicity of the Indo-Australian/Solomon Sea plate boundary in the southeast Papua region. Tectonophysics, 87: 355—369.

Ross, M. and Huebner, J.S., 1975. A pyroxene geothermometer based on composition-temperature relationships of naturally occurring orthopyroxene, pigeonite and augite. In: Extended Abstracts of the International Conference of Geothermometry and Geobarometry, Pennsylvania State University.

Sakuyama, M., 1979. Lateral variations of H_2O contents in Quaternary magmas of northeastern Japan. Earth Planet. Sci. Lett., 43: 103—111.

Sakuyama, M., 1981. Phenocryst assemblages and H_2O contents in circum-Pacific arc magmas. 1981 I.A.V.C.E.I. symposium — Arc Volcanism, Volc. Soc. Japan and I.A.V.C.E.I., Tokyo, pp. 316—317.

Sillitoe, R.W., 1972. A plate tectonic model for the origin of porphyry copper deposits. Econ. Geol., 67: 184—197.

Taneda, S., 1949. On the anhydrite from Himeshima lava. J. Jpn. Assoc. Min. Petr., Econ. Geol., 33: 69—73.

Taylor, H.P., Jr., 1980. The effects of assimilation of country rocks by magmas on $^{18}O/^{16}O$ and $^{87}Sr/^{86}Sr$ systematics in igneous rocks. Earth Planet. Sci. Lett., 47: 243—254.

Taylor, G.A., 1958. The 1951 eruption of Mount Lamington, Papua. Bur. Miner. Resour., Geol. Geophys. Bull., 38, 117 pp.

Taylor, S.R. and Gorton, M.P., 1977. Geochemical application of spark source mass spectrometry — III. Element sensitivity, precision and accuracy. Geochim. Cosmochim. Acta, 41: 1375—1380.

Taylor, S.R. and McLennan, S.M., 1981. The composition and evolution of the continental crust: rare earth element evidence from sedimentary rocks. Philos. Trans. R. Soc. London, Ser. A, 301: 381—399.

Ueda, A. and Itaya, T., 1981. Microphenocrystic pyrrhotite from dacite rocks of Satsuma-Iwojima, Southwest Kyushu, Japan and the solubility of sulfur in dacite magma. Contrib. Mineral. Petrol., 78: 21—26.

Van Bemmelen, R.W., 1939. The geotectonic structure of New Guinea. De Ingenieur in Nederlandsch-Indië, IV. Mijnbouw Geol., 6: 17—27.

Vance, J.A., 1969. On synneusis. Contrib. Mineral. Petrol., 24: 7—29.

Wager, L.R., Brown, G.M. and Wadsworth, W.J., 1960. Types of igneous cumulates. J. Petrol., 1: 73—85.

Ware, N.G., 1981. Computer programs and calibration with the PIBS technique for quantitative electron probe analysis using a lithium drifted silicon detector. Computers Geosci., 7: 167—184.

Wells, P.R.A., 1977. Pyroxene thermometry in simple and complex systems. Contrib. Mineral. Petrol., 62: 129—139.

Yagi, K., Takeshita, H. and Oba, Y., 1972. Petrological study on the 1970 eruption of Akita-Komagatake volcano, Japan. J. Fac. Sci. Hokkaido University, Ser. IV, 15: 109—138.

Yoshiki, B., 1933. Activity of Akita-Komagatake volcano. J. Jpn. Assoc. Min. Petr., Econ. Geol., 9: 153—160.

Journal of Volcanology and Geothermal Research, 18 (1983) 249—278
Elsevier Science Publishers B.V., Amsterdam — Printed in The Netherlands

THE GABBROIC ROCKS OF THE PENINSULAR RANGES BATHOLITH, SOUTHERN CALIFORNIA: CUMULATE ROCKS ASSOCIATED WITH CALC-ALKALIC BASALTS AND ANDESITES

T.E. SMITH[1], C.H. HUANG[1], M.J. WALAWENDER[2], P. CHEUNG[1] and C. WHEELER[1]

[1] *Department of Geology, University of Windsor, Windsor, Ont. N9B 3P4 (Canada)*
[2] *Department of Geological Sciences, San Diego State University, San Diego, CA 92182 (U.S.A.)*

(Received October 20, 1982)

ABSTRACT

Smith, T.E., Huang, C.H., Walawender, M.J., Cheung, P. and Wheeler, C., 1983. The gabbroic rocks of the Peninsular Ranges batholith, southern California: cumulate rocks associated with calc-alkalic basalts and andesites. In: S. Aramaki and I. Kushiro (Editors), Arc Volcanism. J. Volcanol. Geotherm. Res., 18: 249—278.

Gabbroic rocks occur only in the west, and are the oldest intrusions in the Peninsular Ranges Cordilleran batholith. They comprise an olivine-pyroxene gabbronorite series and an amphibole gabbro series both of which contain abundant plagioclase and amphibole. They formed by crystal accumulation and in situ differentiation, in multiple intrusive complexes, and are not considered to be related by fractionation to the granitoid rocks of the batholith.

Pure mineral separates of plagioclase, olivine, clinopyroxene, orthopyroxene, and amphibole were obtained by magnetic and heavy-liquid methods from a representative suite of gabbroic rocks. Their major- and trace-element contents were determined by X-ray fluorescence, and the data used to test hypotheses on the genesis and fractionation of the gabbros.

The plagioclases range from An_{98} to An_{65} in composition, olivines, Fo_{79} to Fo_{70}, occur in rocks where $An>_{86}$. All clinopyroxenes are augite with Mg #'s varying from 81.1 to 64.7. Orthopyroxene occurs where $An<_{92}$, and is generally inverted pigeonite or bronzite, and has Mg #'s ranging from 77.9 to 52.1. The amphiboles include tschermakite, tschermakitic hornblende, pargasite, pargasitic hornblende, ferroan pargasite, magnesio-hornblende, and magnesio-taramite, Mg #'s range from 80.4 to 62.5. Systematic chemical and mineralogical changes confirm that differentiation, controlled by mineral assemblages of plagioclase, olivine, spinel, and clinopyroxene initially, and orthopyroxene, amphibole, and magnetite later, took place between intrusive episodes and in situ.

The highly clacic plagioclase coexisting with olivine and amphibole suggests that the gabbros were formed from hydrous mafic magmas. The modal mineralogy of the gabbros, and the chemistry of the minerals is very similar to that of the cumulate blocks of the Lesser Antillean volcanoes. These features confirm that the gabbros were derived from a hydrous mafic magma, with high Al_2O_3 and low TiO_2 contents, typical of orogenic environments.

Cumulate minerals from the gabbros show little or no zoning and are considered to have formed in equilibrium with the evolving melts. Selected trace-element contents and distribution coefficients are used to calculate the compositions of the melts. The calculations show that the melts in equilibrium with the olivine-pyroxene gabbronorite series

*Complete chemical analyses of all mineral separates are given in the Appendix on p. 633.

0377-0273/83/$03.00

contain approximately 100—200 ppm Ba, 200—400 ppm Sr, 30—10 ppm Ni, 20—10 ppm Co, and 300—100 ppm V. K/Rb ratios of the melts, derived from post-cumulus and prismatic amphiboles, are generally in the range 550—250. These values are typical of calc-alkalic basalts and andesites, and it is suggested that they may have erupted at the surface to form a coeval calc-alkalic volcanic sequence.

INTRODUCTION

Controversy persists concerning the origin and evolution of the various plutonic intrusive rocks making up Cordilleran batholiths. These immense intrusions consist of a wide range of rock types varying from ultramafics and anorthosites, through gabbros and diorites, to tonalites, granodiorites, and granites. Several hypotheses have been suggested to explain the origin and relationships of the rock types in Cordilleran batholiths (Larsen, 1948; Albarede, 1976; Nishimori, 1976; Erikson, 1977; Mullan and Bussell, 1977; Price and Sinton, 1978; Walawender et al., 1979; Walawender and Smith, 1980). This debate has an important bearing on the broader issue as to whether granitoid rocks are derived from crustal or mantle sources and on the origin and evolution of the continental crust (Brown, 1979; Pitcher, 1979; Tarney and Saunders, 1979; Leake et al., 1980).

In this study, we describe the petrology and mineralogy of a series of gabbroic plutons occurring within the Peninsular Ranges batholith in Southern California. The majority of the rocks in these plutons have the characteristics of cumulates (see p. 253), probably none represent the composition of the parental melt, and few if any, lie on a liquid line of descent. Without knowledge of the composition of the parental melt or the liquid line of descent, it is difficult to test models of various processes designed to explain the evolution of such batholiths.

Pure mineral separates, of all of the principal minerals occurring in the major rock types of the gabbroic plutons, have been analyzed for their major- and trace-element contents. The major-element chemistry is used to identify the minerals precisely and to determine the cryptic variation in their composition. This data is compared with similar data, available from other well described mafic intrusions and volcanic associations, in an attempt to place some constraints on the possible compositions of the parental melt. The trace-element compositions of the cumulate minerals are used, with distribution coefficients, to calculate the trace-element contents of the melts with which the minerals were in equilibrium during crystallization. Where the results of these calculations are consistent, the composition of the most primitive melt is considered to be similar to that of the parental magma, and less primitive melts are considered to lie on the liquid line of descent. A comparison of this data with available tabulations of typical compositions of orogenic rock associations is used to confirm the nature of the parental melt and its differentiation products.

GEOLOGICAL SETTING

The Peninsular Ranges batholith has an average width of more than 100 km and extends for more than 1000 km in length in the Peninsular Ranges geological province (Fig. 1). It is a typical Cordilleran- or Andino-type (Pitcher, 1979) batholith made up of a large number of gregarious plutons, ranging in composition from peridotite to granite, intruded over a considerable time span. Its temporal and spatial petrographic and geochemical patterns suggest that it was formed by magmatic activity occurring at a convergent plate margin (Baird et al., 1974; Gastil, 1975; Krummenacher et al., 1975; Silver et al., 1975). The batholith intrudes the Julian Schist, a sequence of Triassic or Jurassic metasediments, and the Santiago Peak Volcanics, a basalt-andesite-dacite sequence of Jurassic age.

Larsen (1948) and Nishimori (1976) have suggested that all of the pluton-

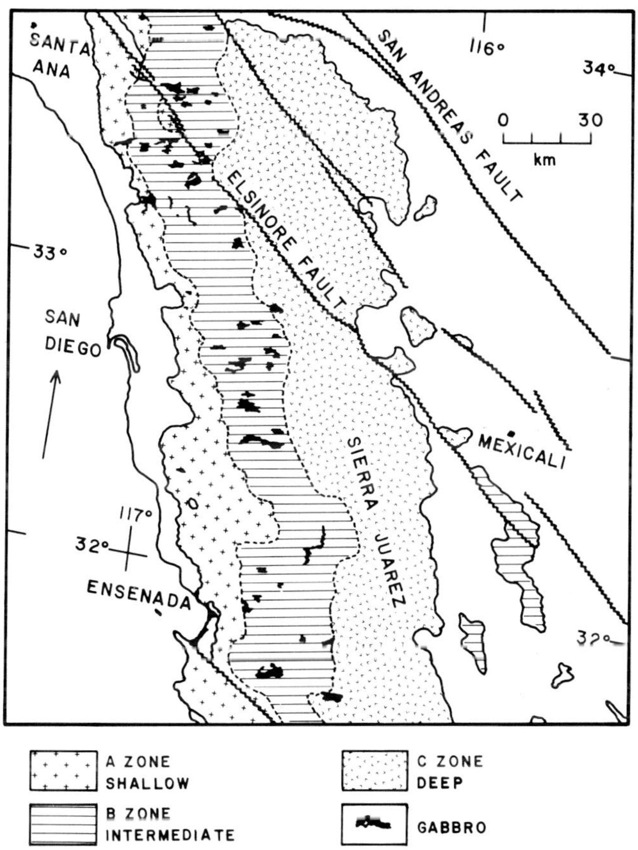

Fig. 1. Geological map of the Peninsular Ranges in southern California, U.S.A., and northern Baja California, Mexico, adapted from Gastil, 1975. A zone shallow metamorphic and plutonic phenomena; B zone intermediate-depth phenomena; C zone deep phenomena.

ic rocks of the batholiths differentiated from a gabbroic magma. An alternative model (Albarede, 1976) is that the gabbroic rocks are mafic cumulates of a tonalitic magma and that the granodiorites formed from the residual melt. It has also been suggested that the gabbros are not cogenetic with the granitoids, but formed separately from a parental high-alumina basaltic melt (Walawender et al., 1979; Walawender and Smith, 1980).

The third model is preferred by the present authors for several reasons. The percentage of gabbroic rocks exposed in the batholith is much smaller than would be anticipated if all of the rocks differentiated from a mafic melt, and geophysical evidence does not indicate that a much larger mafic fraction is present at depth (for more complete discussion see Cheung, 1981). In addition, the gabbroic rocks generally occur in discrete plutons containing only mafic rocks and there are big gaps in mineralogical and geochemical characteristics between these rocks and the granitoids. In the few mafic plutons (e.g., Target Range, Walawender et al., 1979; and Corte Madera, Lillis et al., 1979) which do contain rocks intermediate between the gabbros and the granitoids, in mineralogy and major-element chemistry, the trace-element distributions and trends suggest that the intermediate rocks are the result of marginal contamination of the mafic magma.

The gabbroic plutons formed at pressures of less than 5 m.p.a. (Nishimori, 1976; Bergreen and Walawender, 1977), and experimental studies (Piwinskii and Wyllie, 1968) have shown that at such pressures the liquidus minerals present in a tonalitic melt are plagioclase, hornblende, and biotite. The gabbroic rocks comprise mixtures of plagioclase (An_{98-60}), olivine, clinopyroxene, orthopyroxene, and amphibole and could not have formed as cumulates from a tonalitic melt as suggested by Albarede (1976).

In a summary paper Walawender and Smith (1980) show that each gabbroic pluton in the Peninsular Ranges is the result of multiple intrusion and may consist of a heterogeneously distributed mixture of peridotites, anorthosites, troctolites, gabbronorites, norites, hornblende gabbros, and in a few instances diorites and quartz diorites (nomenclature after Streckeisen, 1973). Field evidence, in most gabbroic plutons, indicates that they are made up of rocks differentiated at relatively shallow levels from similar batches of gabbroic melt. The compositional and petrographic continuity suggests that the plutons are cogenetic and can be related by similar fractionation processes. A few plutons (i.e., the Target Range pluton and the Corte Madera pluton) seem to have less complex intrusive histories and some gradational contacts.

Mineralogy and petrography can be used to define an olivine-pyroxene gabbronorite series, divided into an olivine-clinopyroxene and an amphibole-orthopyroxene subseries, and a much less abundant amphibole gabbro series. Mineral paragenesis (Fig. 2) shows that plagioclase always crystallizes early, in the olivine-clinopyroxene subseries with olivine, spinel and clinopyroxene, and in the amphibole-orthopyroxene subseries with clinopyroxene and orthopyroxene, together with interstitial and poikilitic amphibole. The amphibole-gabbro series is bimineralic and plagioclase crystallization initial-

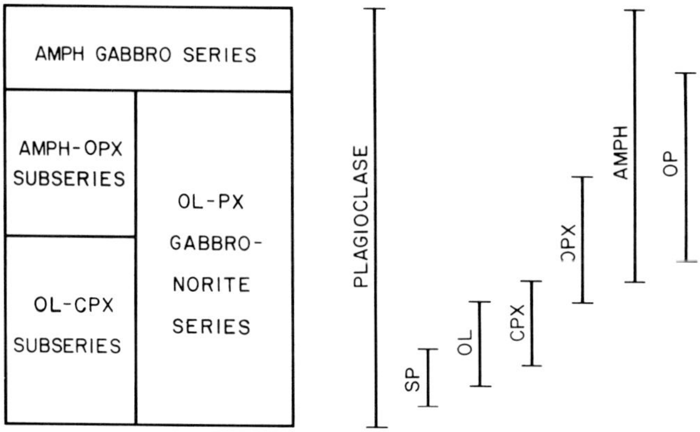

Fig. 2. The rock series and mineral paragenesis of the Peninsular Ranges gabbros.

ly preceded that of amphibole, but later both minerals crystallized together.

Many of the gabbroic rocks are layered and banded containing euhedral to subhedral, unzoned plagioclase, olivine, clinopyroxene, and amphibole, and are regarded as accumulates. The origin of such rocks is currently under debate (see McBirney and Noyes, 1979), however, the terminology of Wager et al. (1960) is retained in its descriptive connotations. The unzoned, euhedral and subhedral cumulus minerals are considered to have formed in equilibrium with the parental melt (Arculus and Wills, 1980) regardless of the precise way in which they formed.

Walawender (1976) suggested that some of the rocks in the Los Pinos pluton formed in a volcanic pipe, and also noted the presence of comb layered structures in the complex.

This evidence together with the numerous mafic dykes penetrating the gabbroic plutons suggests that the magmas may have vented to the surface to form a coeval volcanic sequence.

METHODS

A total of 90 pure mineral separates, obtained from 40 representative samples collected from 3 plutons, were analyzed. These consisted of 33 plagioclases, 12 olivines, 11 clinopyroxenes, 13 orthopyroxenes, and 21 amphiboles. The samples selected in this study were all unweathered and the separation procedures outlined by Hutchison (1974) were followed. Full details of all the analyses are given in Tables I—V. These data are supplemented by 42 major-element mineral analyses (11 plagioclase, 7 olivine, 7 clinopyroxene, 9 orthopyroxene, 8 amphibole) from 3 additional plutons (Nishimori, 1976). The locations of the plutons studied here and by Nishimori (1976) are shown in Fig. 3.

The samples were crushed, ground, and sieved then the 100—200 mesh and 200—400 mesh fractions were washed in an ultrasonic cleaner. These

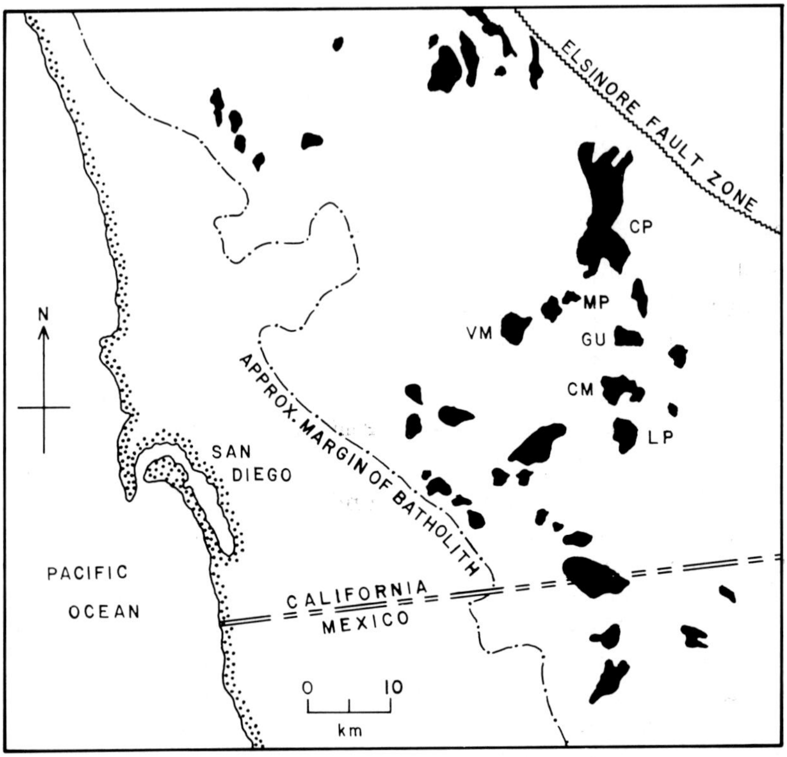

Fig. 3. Distributuon of the gabbroic rocks of the Peninsular Ranges batholith: MP = Mount Poser; CM = Corte Madera; LP = Los Pinos; GU = Guatay; VM = Viejas Mountain; CP = Cuyumaca Peak.

fractions were separated into their constituent minerals using a combination of magnetic (Frantz Isodynamic Separator) and heavy liquid (methylene iodide, and s-tetrabromoethane) methods. Every effort was made to ensure the purity and suitability of the minerals separated, including thin section study to determine the presence or absence of zoning, binocular examination, as well as final hand picking. The major- and trace-element analyses were examined for any features attributable to alteration or contamination. Where such features were recognized, the analysis was not used in the study. Full details of the procedures are given in Cheung (1981).

The major- and trace-element analyses were carried out using a Philips PW1410 Universal Vacuum X-ray Spectrometer. The methods and their precision have been previously described (Norrish and Hutton, 1969; Abbott and Smith, 1978).

MINERALOGY

In the following descriptions, the chemical characteristics of all of the mineral separates are generalized. The actual analyses of selected typical samples are given in Tables I and II.

Plagioclase

Plagioclase is commonly the most abundant phase in the gabbroic rocks of the Peninsular Ranges. The crystal form is generally subhedral and there is little or no visible zoning. Major-element (wt.%), trace-element (ppm) and structural formulae of selected representative compositions of the plagioclases are given in Table I. They range in composition from anorthite ($An_{97.5}$) to labradorite ($An_{51.7}$). (All ranges quoted include Nishimori's analyses). A characteristic feature of all of the plagioclases is their extremely low Or content which is generally less than 1%, except in the 3 most sodic feldspars where it reaches approximately 3%. Plagioclase in the olivineclino-pyroxene subseries ranges from anorthite ($An_{97.5}$) to bytownite ($An_{86.9}$). In addition, labradorite was separated from two samples allocated to this subseries, but these samples are considered to be hornfelsed inclusions of an earlier mafic igneous phase. Plagioclase ranges in composition from bytownite ($An_{88.9}$) to labradorite ($An_{59.0}$) in the amphibole-orthopyroxene subseries. The amphibole gabbro series contains bytownite ($An_{89.4}$) to labradorite ($An_{54.6}$).

Plagioclases from the olivine-clinopyroxene series of the Mt. Poser and Corte Madera plutons typically contain around 500 to 600 ppm Sr and 15 ppm Ba while those from the Los Pinos pluton generally contain in excess of 1000 ppm Sr and 25 ppm Ba. Rb in all of these minerals ranges between 1 and 2 ppm. No systematic variation in the concentration of any of these elements, with the An content of the plagioclase, is noted.

The plagioclases from the amphibole-orthopyroxene series of the Corte Madera and Mt. Poser plutons generally contain about 600 ppm Sr, 25 ppm Ba, and 1 to 2 ppm Rb.

In the amphibole gabbro series, the plagioclases from the Los Pinos pluton often contain more than 1200 ppm Sr, 30 ppm Ba, and 2 ppm Rb, and those of the Corte Madera and Mt. Poser plutons contain about 700 ppm Sr, 20 ppm Ba, and 2 ppm Rb.

Olivine

Olivine only occurs in the rocks of the olivine-clinopyroxene subseries and is generally much less abundant than plagioclase. Chemical analyses and structural formulae of selected samples are given in Table I. It is present as subhedral to anhedral unzoned crystals and ranges in composition from Fo_{79} to Fo_{70}. The MnO content of the olivines increases systematically with the FeO content and ranges from about 0.24 to 0.37%. There is a rapid decrease in the Ni content of the olivines from about 500 ppm in the most forsteritic to about 100 ppm in the least forsteritic olivine. The Co content does not show a similar decrease, remaining steady at about 90 ppm. The Cr and V contents vary erratically and may be controlled by the presence or absence of very small inclusions of spinel and opaques.

TABLE I

Representative plagioclase and olivine analyses and structural formulae selected to represent typical and extreme compositions present

OLIVINE-CLINOPYROXENE SUB SERIES

Plagioclase

	MP-22		C-16		C-36		C-34		C-9-A	
SiO_2	45.03		45.16		45.95		44.12		46.41	
Al_2O_3	33.89		34.59		35.16		33.65		34.73	
TiO_2	—		0.01		—		—		—	
Fe_2O_3	0.46		0.75		0.58		0.60		0.80	
MnO	0.02		0.02		0.01		0.01		0.02	
MgO	—		0.72		—		—		—	
CaO	18.35		18.12		18.06		18.02		18.04	
Na_2O	0.14		0.62		0.72		0.81		0.76	
K_2O	0.18		0.19		0.17		0.18		0.17	
P_2O_5	—		0.03		—		—		—	
Numbers of ions on the basis of 32 (0)										
Si	8.455		8.324		8.408		8.373		8.471	
Al	7.501		7.514		7.582		7.525		7.472	
Ti	—	16.02	0.001	15.94	—	16.07	—	15.98	—	16.05
Fe^{3+}	0.065		0.105		0.080		0.086		0.110	
Mn	0.002		0.003		0.001		0.002		0.003	
Mg	—	3.79	0.197	4.05	—	3.84	—	4.01	—	3.84
Ca	3.692		3.578		3.541		3.665		3.528	
Na	0.052		0.222		0.256		0.297		0.269	
K	0.042		0.044		0.040		0.043		0.041	
P	—		0.005		—		—		—	
Ab	1.4		5.5		6.7		7.5		7.0	
An	97.5		93.4		92.3		91.4		92.0	
Or	1.1		1.1		1.0		1.1		1.0	
Name	Anorthite		Anorthite		Anorthite		Anorthite		Anorthite	
Trace elements (ppm)										
Rb	2		2		1		1		1	
Sr	487		669		487		474		546	
Ba	15		25		12		15		11	

Olivines

	MP-22	C-16	C-36	C-34	C-9-A
SiO_2	37.13	37.76	38.61	38.82	37.80
Al_2O_3	1.61	2.92	1.74	0.51	0.92
TiO_2	0.03	0.03	0.04	0.07	0.03
FeO	19.53	19.50	20.48	20.83	23.38
MnO	0.27	0.26	0.30	0.33	0.37
MgO	39.28	37.61	36.51	33.61	34.62
CaO	0.32	0.42	0.52	3.01	1.20

TABLE I (*continued*)

Olivines — continued

	MP-22		C-16		C-36		C-34		C-9-A	
Numbers of ions on the basis of 4(0)										
Si	0.973		0.982		1.011		1.037		1.008	
Al	0.050	2.03	0.090	1.99	0.054	1.95	0.016	1.92	0.029	1.97
Ti	0.001		0.001		0.001		0.001		0.001	
Fe^{2+}	0.428		0.424		0.449		0.466		0.521	
Mn	0.006		0.006		0.007		0.007		0.008	
Mg	1.535		1.458		1.425		1.339		1.376	
Ca	0.009		0.012		0.015		0.086		0.034	
Mg#	78.2		77.5		76.0		74.2		72.5	
Name	Chrysolite		Chrysolite		Chrysolite		Chrysolite		Chrysolite	
Trace elements (ppm)										
Nb	4		4		6		5		3	
Ni	369		504		229		92		104	
Co	94		95		91		197		211	
Cr	73		93		42		110		34	
V	8		17		15		41		14	

AMPHIBOLE-ORTHOPYROXENE SUB SERIES

Plagioclase

	CM-68		CM-119		C-53		CM-165		CM-152	
SiO_2	44.34		50.30		50.36		54.66		63.15	
Al_2O_3	32.01		31.44		30.86		28.66		22.09	
TiO_2	0.07		0.09		0.01		0.10		0.03	
Fe_2O_3	0.79		0.78		0.61		0.43		0.61	
MnO	—		—		0.01		—		0.01	
MgO	0.85		0.64		—		0.60		0.15	
CaO	18.76		14.92		13.24		11.54		8.45	
Na_2O	1.29		2.72		3.89		4.57		5.11	
K_2O	0.13		0.19		0.22		0.24		0.38	
P_2O_5	—		—		—		—		—	
Numbers of ions on the basis of 32 (0)										
Si	8.397		9.099		9.257		9.801		11.192	
Al	7.145	15.67	6.703	15.92	6.685	16.04	6.057	15.92	4.615	15.89
Ti	0.010		0.012		0.002		0.014		0.004	
Fe^{3+}	0.113		0.106		0.084		0.058		0.082	
Mn	—		—		0.001		—		0.001	
Mg	0.240		0.173		—		0.160		0.038	
Ca	3.807	4.55	2.892	4.06	2.608	4.05	2.217	4.02	1.604	3.49
Na	0.474		0.954		1.386		1.589		1.756	
K	0.031		0.044		0.051		0.055		0.086	
P	—		—		—		—		—	
Ab	10.4		23.5		34.2		39.5		50.3	
An	88.9		75.4		64.4		59.1		47.0	
Or	0.7		1.1		1.3		1.4		2.5	

(*continued*)

TABLE I (*continued*)

Plagioclase — continued

	CM-68	CM-119	C-53	CM-165	CM-152
Name	Bytownite	Bytownite	Labradorite	Labradorite	Andesine
Trace elements (ppm)					
Rb	1		—		
Sr	595		734		
Ba	47		50		

AMPHIBOLE-GABBRO SERIES

Plagioclase

	LP-200		LP-184		CM-28		LP-112	
SiO_2	45.28		49.59		49.68		54.11	
Al_2O_3	33.11		33.32		30.44		28.54	
TiO_2	0.05		0.02		0.03		0.03	
Fe_2O_3	0.68		0.79		1.27		0.90	
MnO	—		0.01		0.01		0.01	
MgO	0.12		0.20		0.16		0.31	
CaO	17.72		15.87		15.24		10.59	
Na_2O	1.11		1.93		2.73		4.79	
K_2O	0.11		0.08		0.13		0.12	
P_2O_5	0.01		0.09		0.01		—	
Numbers of ions on the basis of 32 (0)								
Si	8.509	16.00	8.893	16.05	9.137	15.98	9.823	16.06
Al	7.335		7.044		6.600		6.106	
Ti	0.062		0.003		0.004		0.004	
Fe^{3+}	0.096		0.106		0.143		0.123	
Mn	—	4.03	0.002	3.81	0.002	4.06	0.002	3.86
Mg	0.034		0.053		0.044		0.084	
Ca	3.568		3.049		3.003		2.060	
Na	0.404		0.671		0.974		1.686	
K	0.026		0.018		0.031		0.028	
P	0.002		0.013		0.002		—	
Ab	10.0		17.8		24.0		44.7	
An	89.4		81.7		75.1		54.6	
Or	0.6		0.5		0.9		0.7	
Name	Bytownite		Bytownite		Bytownite		Labradorite	
Trace elements (ppm)								
Rb	1		11		2		7	
Sr	1012		579		795		1199	
Ba	26		104		90		59	

Clinopyroxene

Clinopyroxenes occur in relatively small proportions as prismatic, unzoned, subhedral crystals in the rocks of the olivine-pyroxene gabbronorite series. All of the calcium-rich pyroxenes analyzed in this study are classified as augite (Fig. 4) with the exception of one sub-calcic augite from the Corte Madera pluton. Chemical analyses and structural formulae of selected samples are given in Table II. The clinopyroxenes show a systematic decrease in Mg# (100 × Mg/Mg + ΣFe) from 81.1 to 75.5 in the olivine-clinopyroxene subseries, and from 74.7 to 64.7 in the amphibole-orthopyroxene subseries. The Al_2O_3 and TiO_2 contents of all of the clinopyroxenes generally fall within the range 3.0% to 5.5%, and 0.3% to 0.7%, respectively, and no systematic variation is evident. Nishimori (1976) has reported diopside and salite from these gabbros (Fig. 4) with Al_2O_3 and TiO_2 contents varying from 1.5% to 4.5%, and 0.2% to 0.8%, respectively. These pyroxenes are generally similar to those analyzed in this study but are distinctly more calcic.

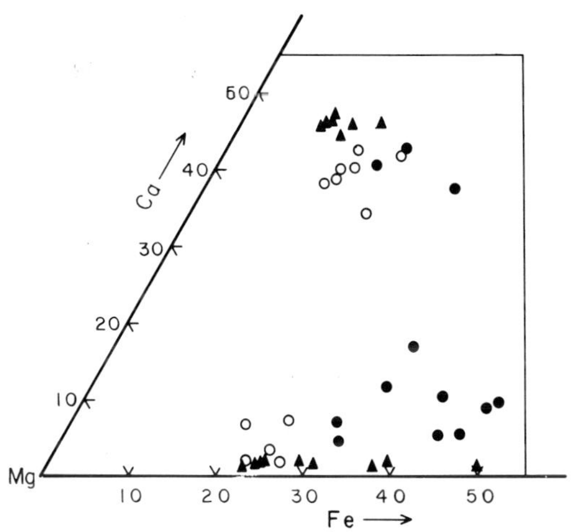

Fig. 4. Compositions of the high-calcium and low-calcium pyroxenes of the Peninsular Ranges gabbros. Open circles = olivine-clinopyroxene subseries; filled circles = amphibole-orthopyroxene subseries; filled triangles = analyses by Nishimori (1976).

There is no overall systematic trend of variation of Al^{iv} with Mg#, but within the Mt. Poser pluton, there is a clear decrease of Al^{iv} with Mg#.

There is insufficient trace element data to identify any specific trends in their distribution; Rb ranges from 3 to 6 ppm, Sr from 70 to 100 ppm, Ba from 11 to 65 ppm, and V is constant at about 200 ppm.

Orthopyroxene

Orthopyroxenes occur in small amounts in the rocks of the olivine-pyroxene gabbronorite series as prismatic bronzite and inverted pigeonite (Fig. 4).

TABLE II

Representative clinopyroxene, orthopyroxene, and amphibole analyses and structural formulae selected to represent typical and extreme compositions present

CLYNOPYROXENE

	Olivine-clinopyroxene subseries								Amphibole-orthopyroxene subseries							
	C-44		MP31		C-34		CM63A		CM68		C-53		CM165		CM88	
SiO_2	51.39		51.35		53.12		51.26		52.35		50.87		52.78		49.72	
Al_2O_3	4.17		4.34		4.24		5.60		4.51		2.98		5.75		4.97	
TiO_2	0.59		0.62		0.40		0.63		0.44		0.52		0.44		0.66	
FeO	7.70		7.24		8.42		7.80		7.90		10.62		10.84		16.51	
MnO	0.31		0.29		0.18		—		—		0.32		—		—	
MgO	18.55		17.21		16.78		13.54		13.03		14.31		11.94		16.98	
CaO	20.47		20.53		21.08		19.68		20.13		19.42		16.35		7.50	
Na_2O	—		0.28		0.18		0.58		0.51		0.53		1.03		0.29	
K_2O	—		0.02		—		0.11		0.12		—		0.16		0.12	
Numbers of ions on the basis of 6 (0)																
Si	1.843	2.00	1.861	2.00	1.883	2.00	1.901	2.00	1.945	2.00	1.914	2.00	1.956	2.00	1.903	2.00
Al^{iv}	0.157		0.139		0.117		0.099		0.055		0.086		0.044		0.097	
Al^{vi}	0.019		0.047		0.060		0.146		0.142		0.046		0.208		0.128	
Ti	0.016		0.017		0.011		0.018		0.012		0.015		0.012		0.019	
Fe^{2+}	0.231		0.219		0.250		0.242		0.245		0 301		0.336		0.529	
Mn	0.009	2.05	0.009	2.04	0.005	2.02	—	1.98	—	1.97	0.010	2.02	—	1.95	—	1.98
Mg	0.991		0.930		0.886		0.748		0.722		0.831		0.660		0.969	
Ca	0.786		0.797		0.800		0.782		0.801		0.783		0.649		0.308	
Na	—		0.020		0.012		0.042		0.037		0.039		0.074		0.022	
K	—		0.001		—		0.005		0.006		—		0.008		0.006	
Mg	48.3		45.6		43.9		37.8		36.6		41.1		33.8		48.9	
Fe	13.4		14.3		15.9		20.3		20.6		18.2		28.7		34.1	
Ca	38.3		40.1		40.2		41.9		42.8		40.7		37.5		17.0	

Name	Augite	Augite	Augite	Augite	Augite	Augite	Augite	Sub Ca-Augite
Mg#	81.1	80.9	78.0	75.5	74.7	73.4	66.3	64.7
Trace elements (ppm)								
Rb	3	5	5	3	—	5	12	3
Sr	71	102	29	52	51	73	93	50
Ba	49	65	39	37	19	69	45	27
Y	72	31	12	18	15	11	41	17
Zr	29	—	11	—	27	—	15	—
Nb	2	6	4	9	4	—	4	1
Ni	32	14	72	6	2	23	15	39
Co	22	31	30	27	32	54	35	59
Cr	41	16	478	—	—	29	—	—
V	203	208	279	—	—	202	—	—

ORTHOPYROXENE

	Olivine-clinopyroxene subseries				Amphibole-orthopyroxene subseries			
	MP-31	LP107	CM92	C-44	C-48	C-53	CM60	CM109
SiO_2	51.78	51.60	49.91	50.43	50.46	51.34	49.30	50.82
Al_2O_3	3.13	3.94	3.22	3.23	2.56	1.55	4.93	2.26
TiO_2	0.37	0.35	0.16	0.32	1.31	0.25	0.61	0.93
FeO	14.75	14.01	15.88	14.79	18.15	18.83	23.37	24.61
MnO	0.52	—	0.35	0.55	0.57	0.62	—	—
MgO	29.15	26.62	28.77	25.02	22.66	22.59	14.41	15.02
CaO	3.82	1.49	0.97	3.58	3.67	2.28	3.84	3.56
Na_2O	0.05	0.14	—	0.10	—	—	0.33	0.28
K_2O	0.02	0.07	0.12	—	—	—	0.11	0.14

(continued)

TABLE II *(continued)*

	Olivine-clinopyroxene subseries								Amphibole-orthopyroxene subseries							
	MP-31		LP107		CM92		C-44		C-48		C-53		CM60		CM109	
Numbers of ions on the basis of 6 (0)																
Si	1.827		1.889	2.00	1.833		1.878	2.00	1.996	2.00	1.947	2.00	1.924	2.00	1.978	2.00
Al^{iv}	0.130		0.111		0.139		0.122		0.004		0.053		0.076		0.022	
Al^{vi}	—	2.00	0.059		—	2.00	0.020		0.109		0.017		0.151		0.081	
Ti	0.010		0.010		0.004		0.009		0.037		0.007		0.018		0.027	
Fe^{iii}	0.033		—		0.024		—		—		—		—		—	
Fe^{ii}	0.402		0.429		0.464		0.461		0.566		0.597		0.763		0.801	
Mn	0.016		—	2.02	0.011		0.017	2.05	0.018	2.02	0.020	2.01	—		—	
Mg	1.533	2.10	1.453		1.575	2.09	1.389		1.260		1.277		0.838		0.871	
Ca	0.144		0.058		0.038		0.143		0.147		0.093		0.161	1.96	0.148	1.96
Na	0.003		0.010		—		0.007		—		—		0.025		0.021	
K	0.001		0.003		0.006		—		—		—		0.005		0.007	
Mg	73.0		71.9		75.4		67.8		62.4		63.5		42.8		44.4	
Fe	20.0		24.6		22.5		24.9		30.3		31.9		47.5		46.6	
Ca	7.0		3.5		2.1		7.3		7.3		4.6		9.7		9.0	
Name	Inverted pigeonite		Bronzite		Bronzite		Inverted pigeonite		Inverted pigeonite		Inverted pigeonite		Inverted pigeonite		Inverted pigeonite	
Mg#	77.9		77.2		76.3		75.1		69.0		68.1		52.3		52.1	
Trace elements (ppm)																
Rb	1		—		7		2		3		4		18		17	
Sr	14		20		13		16		4		9		112		19	
Ba	33		11		6		24		30		26		35		55	
Y	—		5		1		—		1		1		27		54	
Zr	—		—		—		—		3		—		3		47	
Nb	3		11		—		—		1		4		23		15	
Ni	19		29		1181		24		1		16		48		47	
Co	76		34		92		96		45		46		55		67	
Cr	17		—		264		17		44		51		—		—	
V	102		—		—		89		130		105		—		—	

	Olivine-clinopyroxene subseries													
	C-16		MP-22		MP-18		C-36		MP-87		LP-179		CM-47	
SiO_2	43.25		43.87		41.37		41.13		42.62		41.96		46.85	
Al_2O_3	13.78		14.51		19.78		17.88		15.66		13.87		8.46	
TiO_2	0.45		0.13		0.09		0.28		1.20		2.00		1.18	
FeO	8.04		8.21		9.41		9.26		10.28		11.52		14.07	
MnO	0.15		0.15		0.14		0.14		0.16		—		—	
MgO	18.47		17.16		15.59		14.97		14.41		14.66		13.18	
CaO	10.51		10.94		10.46		10.93		11.07		11.32		12.29	
Na_2O	1.74		1.08		1.78		0.98		3.30		1.80		0.56	
K_2O	0.12		0.11		0.06		0.12		0.35		0.24		0.30	
Numbers of ions on the basis of 23 (0)														
Si	6.259	8.00	6.345	8.00	5.872	8.00	6.018	8.00	6.118	8.00	6.149	8.00	6.919	8.00
Al^{iv}	1.741		1.655		2.128		1.982		1.382		1.851		1.081	
Al^{vi}	0.609		0.818		1.181		1.101		0.768		0.544		0.391	
Ti	0.049		0.016		0.010		0.031		0.130		0.220		0.131	
Fe^{2+}	0.973	5.63	0.993	5.55	1.117	5.62	1.133	5.55	1.234	5.23	1.412	5.38	1.738	5.16
Mn	0.018		0.018		0.017		0.017		0.019		—		—	
Mg	3.984		3.699		3.298		3.265		3.083		3.202		2.901	
Ca	1.629		1.695		1.591		1.713		1.703		1.777		1.945	
Na	0.488	2.14	0.303	2.02	0.490	2.09	0.278	2.01	0.918	2.69	0.511	2.33	0.160	2.16
K	0.022		0.020		0.011		0.022		0.064		0.045		0.057	
Name	Pargasitic hornblende		Tschermakitic hornblende		Pargasite		Tschermakite		Magnesio-taramite		Ferroan pargasite		Magnesio-hornblende	
Mg#	80.4		78.8		74.7		74.2		71.4		69.4		62.5	
Trace elements (ppm)														
Rb	4				1				5				4	

(continued)

TABLE II *(continued)*

	Olivine-clinopyroxene subseries						
	C-16	MP-22	MP-18	C-36	MP-87	LP-179	CM-47
Sr	86		114		72		51
Ba	53		12		141		59
Y	21		—		67		17
Zr	31		—		66		36
Nb	10		—		6		9
Ni	170		34		101		29
Co	29		74		37		—
Cr	144		44		106		490
V	121		31		291		—

	Amphibole-orthopyroxene subseries				Amphibole-gabbro series		
	C-48	CM-68	CM-88	CM-119	CM-28	LP-184	LP-117
SiO_2	48.36	46.77	44.70	49.05	50.04	42.48	42.68
Al_2O_3	10.94	9.19	10.98	7.15	8.17	13.29	13.29
TiO_2	0.89	1.86	2.15	0.97	0.60	2.38	2.12
FeO	9.48	12.18	12.26	13.51	12.56	12.40	12.40
MnO	0.23	—	—	—	—	—	—
MgO	13.95	13.39	12.58	13.27	15.78	14.11	13.40
CaO	13.94	12.46	12.68	12.34	11.38	11.14	11.25
Na_2O	1.01	1.62	1.98	0.72	0.55	1.83	1.57
K_2O	0.05	0.38	0.31	0.49	0.26	0.21	0.29

Numbers of ions on the basis of 23 (0)

	1		2		3		4		5		6		7	
Si	6.855	8.00	6.813	8.00	6.559	8.00	7.162	8.00	7.089	8.00	6.212	8.00	6.287	8.00
Al^{iv}	1.145		1.187		1.441		0.838		0.911		1.788		1.713	
Al^{vi}	0.683		0.391		0.458		0.393		0.454		0.503		0.594	
Ti	0.095		0.204		0.237		0.107		0.064		0.262		0.235	
Fe^{2+}	1.124	4.88	1.484	4.99	1.504	4.95	1.650	5.04	1.488	5.34	1.516	5.36	1.527	5.30
Mn	0.028		—		—		—		—		—		—	
Mg	2.947		2.907		2.751		2.888		3.332		3.076		2.942	
Ca	2.117		1.945		1.993		1.931		1.727		1.745		1.775	
Na	0.278	2.40	0.458	2.47	0.563	2.62	0.204	2.23	0.151	1.93	0.519	2.30	0.448	2.28
K	0.009		0.071		0.058		0.091		0.047		0.039		0.054	

Name	Magnesio-hornblende	Edenite	Edenitic hornblende	Magnesio-hornblende	Magnesio-hornblende	Ferroan pargasite	Ferroan paragasitic hornblende
Mg#	72.4	66.2	64.6	63.6	69.1	67.0	65.8
Trace elements (ppm)							
Rb	5	15	—	19	4	3	3
Sr	107	99	120	41	33	181	130
Ba	181	44	52	61	59	79	101
Y	25	39	57	51	8	27	25
Zr	—	71	45	70	36	32	46
Nb	5	1	4	19	6	8	8
Ni	12	62	41	20	18	35	13
Co	60	48	45	47	—	—	—
Cr	25	—	—	—	466	338	507
V	482	—	—	—	—	—	—

Optical examination of these minerals revealed no evidence of zoning. Chemical analyses and structural formulae of selected samples are given in Table II. The bronzite occurs in the most differentiated members of the olivine-clinopyroxene subseries. Inverted pigeonite occurs in dykes of the olivine-clinopyroxene subseries and in the least differentiated members of the amphibole-orthopyroxene subseries. The bronzites and inverted pigeonites of the olivine-bearing group all have high Mg#'s which decrease from 77.9 to 72.8 (Table II). In the amphibole-orthopyroxene subseries, inverted pigeonites have Mg#'s ranging from 69.0 to 52.1. The orthopyroxenes analyzed by Nishimori all contain less than 5.0% Ca and are classified as bronzite and hypersthene. They show a wide range of Mg# from 77.6 to 50.3.

No systematic trends of variation in trace-element contents can be identified from the limited data. However, the orthopyroxenes of the olivine-clinopyroxene subseries may contain lower Rb (1—2 ppm) and Cr (~17 ppm) and higher Sr (13—16 ppm), Ni (19—24 ppm), and Co (76—96 ppm), than those of the amphibole-orthopyroxene subseries (Rb 3—4 ppm, Cr 44—51 ppm, Sr 26—30 ppm, Co 45—46 ppm, Ni ~16 ppm). The V (89—130 ppm) and Ba (24—33 ppm) contents of the orthopyroxenes from each series are similar.

Amphibole

The amphiboles in the olivine-pyroxene gabbronorite series occur as anhedral, interstitial and poikilitic crystals. Minor zoning was detected in a few samples of these post-cumulus minerals. Prismatic, unzoned, amphiboles occur in the amphibole-gabbro series and they are considered to be cumulus minerals. All of the amphiboles with one exception are classified as calcic (nomenclature after Leake, 1978) and they generally belong to the tschermakitic and pargasitic groups according to their alkali content. The one exception is a sodic-calcic amphibole, magnesio-taramite, obtained from the olivine-pyroxene gabbronorite series of the Mt. Poser pluton. Chemical analyses and structural formulae of selected samples are given in Table II. The magnesium #'s of these amphiboles show a considerable range which overlaps between the units; in the olivine-clinopyroxene subseries it is 80.4 to 62.5, in the amphibole-orthopyroxene subseries it is 73.4 to 63.6, and in the amphibole gabbro series it is 69.1 to 64.6. A characteristic feature of all of the amphiboles, both cumulus and post-cumulus is their low K_2O content ranging from 0.06% to 0.35%.

Tschermakite, tschermakitic hornblende, pargasite, pargasitic hornblende, ferroan pargasite, and magnesio-hornblende occur in the olivine-clinopyroxene subseries. Most of these mineral types occur in the Mt. Poser pluton, except magnesio-hornblende from Corte Madera, and ferroan pargasite from Los Pinos. These amphiboles indicate that the post-cumulus material is increasing in alkali content from tschermakitic varieties ($(Na + K) < 0.50$) to pargasitic varieties ($(Na + K) > 0.50$) and to some degree in silica content

(Si < 6.25 in tschermakite and pargasite and 6.25—6.50 in pargasite and tschermakitic hornblende) in the olivine-bearing rocks of Mt. Poser. The most alkali enrichment is shown by the magnesio-taramite from Mt. Poser where (Na + K) $\cong$ 0.98. Al^{iv} ranges from 1.66 to 2.13 in these amphiboles.

In the amphibole-orthopyroxene subseries magnesio-hornblende, edenitic hornblende, and edenite occur. The majority of these amphiboles are from the Corte Madera pluton, and in addition to the decrease in Mg#, they show that the principle variation taking place in the intercumulus material of the Corte Madera pluton is an increase in the total alkalis (Na+K~0.30—0.62). Al^{iv} varies from 0.80 to 1.44 in these amphiboles.

In the amphibole-gabbro series magnesio-hornblende, ferroan pargasite, and ferroan pargasitic hornblende occur. The magnesio-hornblende is from Corte Madera, and the ferroan pargasite and ferroan pargasitic hornblende from Los Pinos. One sample, a ferroan pargasitic hornblende was analyzed from Mt. Poser. The (Na + K) values of the minerals from Los Pinos only vary from 0.50 to 0.58 and Al^{iv} from 1.71 to 1.93.

Nishimori (1976) reported a very similar series of amphibole analyses including tschermakitic hornblende, pargasitic hornblende, magnesio-hornblende, ferroan pargasite, edenitic hornblende, and secondary actinolitic hornblende. Their Mg#'s varied from 78 to 64, Al^{iv} from 0.88 to 1.78, and (Na +K) from 0.34 t 0.68.

Crystallization of the interstitial and poikilitic amphiboles may have taken place in both open and closed systems and therefore their trace-element contents may be expected to show considerable variation. The Rb contents are generally low (1—8 ppm), Sr ranges from 50—130 ppm, Ba from 12—277. The Ni, Co, Cr, and V show a great deal of fluctuation and may reflect the presence of minor amounts of accessory minerals not removed during the separation.

The trace-element contents of the prismatic amphiboles are more constant but there is insufficient data to identify any systematic trends of variation. The ranges of various trace-element contents are Rb 3—6 ppm, Sr 130—167 ppm, Ba 79—167 ppm, Y 27—64 ppm, Nb 25—32 ppm, Ni 9—35 ppm, Cr 338—507 ppm.

DIFFERENTIATION

The variations in the chemistry of the minerals reported above support the interpretation (Walawender and Smith, 1980) that the Peninsular Ranges gabbros were formed by multiple intrusion of a sequence of similar but not identical magmas, and that the magmas differentiated in situ.

There are systematic decreases in anorthite content of the plagioclase, and in the Mg#'s of the various mafic phases present in each of the three rock series described above (Tables I and II). In the Mt. Poser pluton, there is a systematic decrease in Al^{iv} with decreasing Mg# in the clinopyroxene (Table II), a trend which has been interpreted as resulting from cooling (Arculus and Wills, 1980). This is accompanied by the successive appearance and disappearance of the various mineral phases in the rocks (Fig. 2). The continuous changes in cumulus mineral composition, in each series, are consistent with down temperature trends of the parental magma and indicate fractional crystallization in a generally closed system.

There are overlapping ranges of anorthite content in the plagioclase of the three rock series (Table I). The increase in the anorthite content between the most differentiated plagioclase of the olivine-clinopyroxene subseries and the least differentiated plagioclase of the amphibole-orthopyroxene subseries and the complete overlap of the range of anorthite content of the plagioclases of the latter by those of the amphibole-gabbro series cannot be explained by fractional crystallization. These reversals of trends and overlapping compositional ranges together with the different mineralogy and paragenetic sequence (Figs. 2 and 3) of the three suites indicates that each was derived from separate batches of mafic melt, and that there were significant differences in composition between the melts.

There is also evidence that the rocks of each series in each pluton have some distinct characteristics of their own. For instance, there are differences in the compositions of the calcium-rich and calcium-poor pyroxenes occurring in different plutons. Diopside and salite, together with bronzite, occur in the plutons described by Nishimori (1976); augite, and inverted pigeonite occur in rocks of the same series in the plutons described in this study (Fig. 4). There are also much higher Sr contents in the plagioclases of the olivine-clinopyroxene series of the Los Pinos pluton as compared to those of the same series in the Mt. Poser and Corte Madera plutons (p. 255). This evidence suggests that each pluton was formed by multiple intrusion of a series of melts. Each of these melts show subtle compositional differences within the various plutons, in addition to the larger differences recognized between the overall characteristics of each rock series.

PETROGENESIS

The Peninsular Ranges gabbros contain the unusual mineral assemblage of amphibole and olivine coexisting with very anorthitic plagioclase, and are similar to rock sequences described at several localities (Kuno, 1950; Helsley, 1960; Longshore, 1966; Yamazaki et al., 1966; Baker, 1968; Nicholls, 1971; Lewis, 1973; Nishimori, 1976; Arculus and Wills, 1980). A comparison between these suites and other well documented rock associations places some constraints on the magma types involved.

Table III lists the mineralogical characteristics of several of the rock suites containing amphibole, olivine, and anorthitic plagioclase and compares them with those of high Mg (Thessaloniki), tholeiitic (Skaergaard and Kiglapait) and calc-alkalic (Guadalupe) suites. These three suites generally show much greater ranges in the composition of all of their minerals than does the Peninsular Ranges and similar sequences. Figure 5 shows that at any given

TABLE III

A comparison of mineral compositions and ranges between the Peninsular Ranges gabbros and selected well described suites

	Peninsular ranges				
	ol-cpx s.s.	amph-opx s.s.	amph-gabbros	Longwoods[1]	Virgins[2]
Plagioclase	An 97.5—86.9	88.9—59.0	89.4—54.6	90—45	98—75
Olivine	Fo 79—70			80—75	80—76
Clinopyroxene	Mg# 83.7—75.5	75.7—64.7		83—68.5	Augite Pigeonite
Orthopyroxene	Mg# 77.9—72.8	69.0—50.3		76.7—61.5	75—73
Amphibole	Mg# 80.4—62.5	73.4—63.6	69.1—64.6	76.9—53.1	Hornblende
	Lesser Antilles[3]	Skaergaard[4]	Kiglapait[5]	Guadalupe[6]	Thessalonki[7]
Plagioclase	100—36	66—30	62—28	82—9	95—91
Olivine	90—59	40—02	72—4	83—70	81.5—73.5
Clinopyroxene	83.0—70.0	66.1—1.6	73-15 (Mg%)	78.6—53.8	86.8—77.8
Orthopyroxene	77.0—43.0	61.5—49.5	—	70.9—26.5	81.4—66.3
Amphibole	86.0—38.0	—	—	66.7—19.6	—

[1] Price and Sinton (1978); [2] Helsley (1960); Longshore (1966); [3] Arculus and Wills (1980); [4] Wager and Brown (1978); [5] Morse (1979a, b); [6] Best and Mercy (1967); [7] Sapountzis (1979).

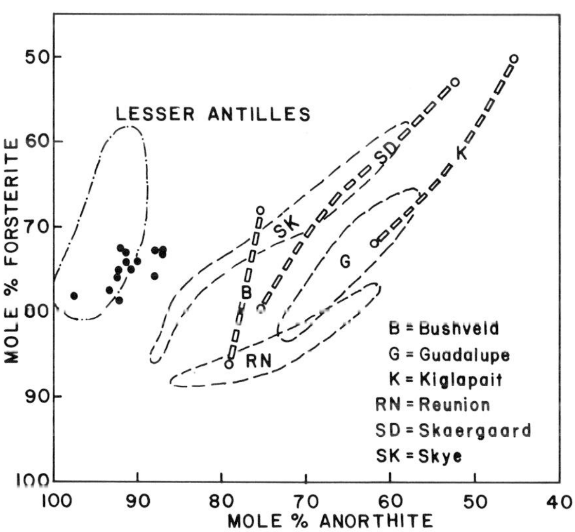

Fig. 5. Compositions of coexisting plagioclase and olivine from the Peninsular Ranges gabbros and selected suites. *B* (Wager and Brown, 1969), *G* (Best and Mercy, 1967), *K* (Morse, 1979a, b), *RN* (Upton and Wadsworth, 1972), *SD* and *SK* (Wager and Brown, 1968), Lesser Antilles (Arculus and Wills, 1980).

forsterite content, the plagioclases of the Peninsular Ranges gabbros and the Lesser Antilles plutonic blocks are much more anorthitic than those of the Skaergaard, Kiglapait, and Guadalupe complexes. The coexisting plagioclase-olivine compositions in the Thessaloniki suite are similar to those of the Peninsular Ranges, but amphibole is lacking in the rocks. Amphibole is abundant in the Peninsular Ranges and similar sequences, and absent from the Skaergaard complex. The presence of amphibole in mafic rocks indicates crystallization from magmas having relatively high H_2O contents ($\geqslant$ 2—3 wt.% H_2O in the melt, Holloway and Sykes, 1979). Arculus and Wills (1980) have suggested that the presence of H_2O in mafic melts also causes the crystallization of unusually anorthitic plagioclases by disturbing the relationship between the chemical potentials of the An and Ab components in the melt. The Peninsular Ranges gabbros, and similar sequences are considered to have crystallized from mafic magmas with a relatively high water content.

The relatively high Al_2O_3 content, and low proportion of Al^{IV} and TiO_2 in the clinopyroxenes suggests that they were derived from tholeiitic, high-alumina or calc-alkaline parental magmas and not alkaline or peralkaline magmas (LeBas, 1962). While, as noted above, the overall range of composition of the clinopyroxenes (Table II) is less than that observed in tholeiitic and alkaline intrusions, it is similar to the limited range observed in andesitic suites (Arculus and Wills, 1980).

The plutonic blocks associated with the Lesser Antillean volcanics (Arculus and Wills, 1980) are the best documented of the similar sequences and show very similar ranges of mineral composition to those of the Peninsular Ranges (Table III). This similarity extends to the compositional ranges of the coexisting minerals. In the Lesser Antillean blocks, olivine lies predominantly in the range Fo_{82-64}, and coexists with orthopyroxene where Fo $<$ 71, and with plagioclase where An $>$ 89. In the Peninsular Ranges gabbros, these figures are Fo_{79-70}, Fo $<$ 77, and An $>$ 86. Orthopyroxene and plagioclase coexist where An $<$ 83 in the Lesser Antilles and where An $<$ 92 in the Peninsular Ranges. Both of these sequences contain similar types of tschermakitic and pargasitic calcic amphiboles. The amphiboles and the plagioclases in both sequences contain very little potassium. The prismatic amphiboles of the amphibole-gabbro series plot in the field of amphiboles derived from island-arc andesites as defined by Jakes and White (1972a) (Fig. 6) and in the field defined by amphiboles from St. Kitts (Arculus and Wills, 1980, fig. 8).

These comparisons suggest that the Peninsular Ranges gabbros and Lesser Antilles plutonic blocks were derived from similar magmas. However, Brown et al. (1977) have identified island-arc tholeiitic, calc-alkalic, and alkalic rock sequences, Jackson (1980) recognized rocks of tholeiitic and calc-alkali affinities. Others (Rea and Baker, 1980; Smith et al., 1980) suggest that the majority of the rocks are low K_2O calc-alkaline in nature, and alkalic rocks are limited to Grenada. These differences of interpretation are the result of trying to classify gradationally varying sequences into neat pigeonholes (Johnson et al., 1978; Jackson et al., 1981).

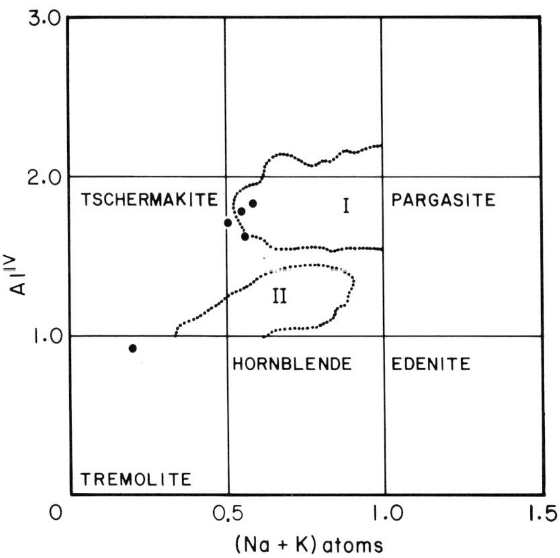

Fig. 6. Compositions of prismatic amphiboles from the amphibole-gabbro series of the Peninuslar Ranges gabbros. Fields: I = island-arc amphiboles; II = continental orogenic amphiboles after Jakes and White (1972b).

The comparison of mineral types and their chemical compositions suggests only that the Peninsular Ranges gabbros were derived from an orogenic magma (island-arc tholeiitic, calc-alkalic, or alkalic), characterized by high Al_2O_3 and H_2O, and low TiO_2 and K_2O contents. A more precise identification of the magma may be derived from a study of the trace element compositions of the minerals.

TRACE-ELEMENT CONTENTS OF THE MELTS

Distribution coefficients may be used to reconstruct the trace element concentration of a magma from the analyses of minerals equilibrated with it (Hart and Brooks, 1977; Hamilton, 1977; Cox et al., 1979). The cumulate minerals selected for this study are optically unzoned and show no evidence of alteration or recrystallization. They vary in composition systematically within each series, and there is an absence of crossing tie lines in compositional plots, suggesting that they crystallized in equilibrium with differentiating melts, and are suitable for this type of treatment. Arculus and Wills (1980) have emphasized that such minerals are better indicators of fractionation processes than are the phenocrysts of lava flows and pyroclastics.

Distribution coeffcients for various trace elements vary with temperature, pressure, and composition of the melt (Irving, 1978). However, the consistency of much of the distribution coefficient data suggests that many of these problems are not as severe as might be supposed (Philpotts and Schnetzler, 1972, p. 53). Correct interpretation of the results of calculations involving mineral trace element analyses and distribution coefficients depends on the use of representative data, and on the full understanding of the relationships between the rocks from which the mineral separates were obtained. Such detailed data and knowledge is available for Mt. Poser and further treatment is generally limited to its samples. Samples from other plutons generally confirm the compositions indicated by the Mt. Poser samples. The calculations do not yield precise compositions, largely because of the range of values of distribution coefficients, but they are sufficiently consistent to show the general range of trace element contents and elemental ratios characteristic of the melts, and the ranges are small enough to indicate the most probable magma type. The greatest reliance is placed on the results derived from trace elements which are present in fairly high concentration which have been determined with precision, having distribution coefficients between 10 and 0.1 and a limited, systematic range of concentration. Subject to these constraints, good values for Sr and Ba (plagioclase), Ni and Co (olivine), and Co and V (clinopyroxene and orthopyroxene), of the melts have been obtained (Table IV).

The Sr. and Ba contents of the melts in equilibrium with the cumulate plagioclases of the olivine-clinopyroxene gabbronorite series are in the range 200—400 ppm and 100—200 ppm, respectively. Neither element shows a systematic trend of variation with An content of the plagioclase in the olivine-clinopyroxene subseries, but Sr may decrease and Ba increase with decrease in An content in the amphibole-orthopyroxene subseries. In the amphibole-gabbro series, the Sr and Ba contents of the equilibrium melts calculated from the plagioclases range from 556—200 ppm and 104—22 ppm, and the Sr content does tend to decrease with An. The Ni ($K_D = 18$) and Co ($K_D=4.5$) contents of the equilibrium melts calculated from the olivines range from 28—11 ppm and 21—18 ppm, respectively, and decrease with the forsterite content. Co and V contents calculated from orthopyroxenes and clinopyroxenes in the olivine-clinopyroxene gabbronorite series are in the range 16 to 11 and 254 to 81 respectively (lower values for V are obtained from the orthopyroxenes).

Comparison of the general ranges of trace-element contents, with the exception of Sr, with those of typical values from orogenic volcanic associations (Jakes and White, 1972a) shows them to be similar to the chemistry of island-arc tholeiite and calc-alkalic basalts and andesites (Table V), and to those of the plutonic blocks of Dominica, Lesser Antilles. The Sr values which are believed to be the most reliable of the whole data set, suggest that the equilibrium melts had compositions similar to calc-alkalic basalts and andesites. The differentiation trends preserved in the post-cumulus amphiboles, limited Fe enrichment with alkali and silica enrichment, support this suggestion.

K/Rb ratios of calc-alkalic rocks are typically between 560 and 220 and those of island arc tholeiitic rocks lie between 1300 and 500. The post-cumulus amphiboles, which are believed to have crystallized from interstitial

TABLE IV

Calculated trace-element composition of equilibrium melts

	Range of concentration in crystals		Range of distribution coefficient		Range of concentration in melts	
Plagioclase:						
	Sr (ppm)	Ba (ppm)	Sr[1]	Ba[1]	Sr (ppm)	Ba (ppm)
Ol-Cpx s-s	669—474	85—8			446—306 352	167—50 105
Amph-Opx s-s	734—660	50—23	1.5—3.3 varies with An content	0.16—0.31	440—193	192—127
Amph-Ga series	695	22			409	122
Olivine:						
	Ni (ppm)	Co (ppm)	Ni[2]	Co[2]	Ni	Co
Ol-Cpx s-s	504—195	95—82	18—9	4.5—3.0	56—21 32 28—11 16	31—27 30 21—18 20
Clinopyroxene:						
	Co (ppm)	V (ppm)	Co[3]	V[3]	Co (ppm)	V (ppm)
Ol-Cpx s-s	31—22	279—203	2—8	0.9—18	16—11	254—185
Amph-Opx s-s	54	202	2[a]	1.1[a]	27	184
Orthopyroxene:						
	Co (ppm)	V (ppm)	Co[3]	V[3]	Co (ppm)	V (ppm)
Ol-Cpx s-s	96—76	102—89	3—12	0.9—18	16—13	93—81
Amph Opx s-s	45	130 105	6[a]	1.1[a]	8	118 95

Source of distribution coefficients: [1] Korringa and Noble (1971); [2] Dale and Henderson (1972); [3] Gill (1978).
[a] Recommended values (Gill, 1978).

melt, have K/Rb ratios between 622 and 218. The ratios do not vary systematically with Mg# which suggests open system crystallization, in part. The prismatic amphiboles, from the amphibole gabbro series have K/Rb ratios between 803 and 160. Because Rb is preferentially excluded from amphi-

TABLE V

Typical trace-element contents and ratios of orogenic rock associations (Jakes and White, 1972a) and Lesser Antilles volcanic arc (Brown et al., 1977)

Orogenic rock associations

	Island-arc tholeiite			Calc-alkaline			Shoshonitic		
	Basalt	Andesite	Dacite	Basalt	Andesite	Dacite	Basalt	Andesite	Dacite
SiO_2 (wt.%)	52	58	63	52	58	63	—	—	—
Rb (ppm)	5	6	15	10	30	45	75	100	120
Ba (ppm)	75	100	175	115	270	520	1000	850	900
Sr (ppm)	200	220	90	330	385	460	700	850	850
Co[1] (ppm)		17—32			16—38				
Ni (ppm)	30	20	1	25	18	5	20	—	—
V (ppm)	270	175	19	255	175	68	200	—	—
K/Rb	1000	890	870	340	430	380	200	200	200

Lesser Antilles arc

	St. Kitts (Tholeiitic)			Dominica (calc-alkalic)			Grenada (alkalic)		
	Basalt	Andesite	Dacite	Basalt	Andesite	Dacite	Basalt	Andesite	Dacite
SiO_2 (wt.%)	50—52	56—58	62—64	50—52	56—58	62—64	52—54	56—58	62—64
Rb (ppm)	5	11	14	16	36	51	25	36	61
Ba (ppm)	126	165	218	105	210	248	378	466	640
Sr (ppm)	290	309	292	310	292	276	652	726	780
Ni (ppm)	11	2	3	36	35	42	142	104	53
V (ppm)	235	123	97	—	—	—	195	152	92
K/Rb	598	423	451	254	129	124	266	281	223

[1] Co values from Dupuy et al. (1977).

bole as compared to K, the melts from which the minerals formed would have even lower K/Rb than the crystals, such (low) ratios are characteristic of calc-alkalic sequences.

It is concluded from the trace-element data that the Peninsular Ranges gabbros formed as cumulates in equilibrium with melts having compositions similar to calc-alkalic basalts and andesites. This is in agreement with and extends the conclusions drawn from the study of major-element chemistry, including the limited Fe enrichment and increase in total alkalis and silica shown by the amphiboles.

The gabbroic plutons studied contain a limited range of mafic rocks, and there is field evidence to suggest that the residual melts were vented at the surface to form a coeval volcanic association. The limited range of iron enrichment shown by the mafic minerals of these plutons is probably the result of the venting of the residual melts at the surface at regular intervals in the unstable orogenic environment. The Santiago Peak volcanics are a sequence of basalts, andesites, and dacites of the same general age as the gabbros, Jurassic (Fife et al., 1967), and may represent a coeval volcanic sequence.

CONCLUSIONS

The data presented above on the petrography, mineralogy, and mineral chemistry of the Peninsular Ranges gabbros lead to the following conclusions. The gabbros are cumulate rocks formed from mafic melts, and the major-element chemistry of their minerals shows that each pluton is formed by multiple intrusion and in situ differentiation of a series of similar melts. The presence of amphibole and highly anorthitic plagioclase in these rocks indicates that the melts contained $\geqslant$ 2—3 wt.% H_2O. Pyroxene compositions show that the melts were high in Al_2O_3 and low in TiO_2, and plagioclase and amphibole compositions suggest that the melts were low in K_2O. Hydrous mafic magmas with high Al_2O_3, and low TiO_2 and K_2O contents are common in orogenic environments but their exact nature is not defined by these features.

Calculations using the trace element contents of the cumulus minerals and distribution coeffcients show that the melts typically contained 200—400 ppm Sr, 100—200 ppm Ba, 28—11 ppm Ni, ~20 ppm Co, 254—81 ppm V, and had K/Rb ratios generally in the 500—200 range. The values are typical of those found in the basalts and andesites of the calc-alkalic rocks association. It is suggested that the melts having compositions in this range were erupted at the surface to form a coeval calc-alkalic rock sequence.

REFERENCES

Abbott, P.L. and Smith, T.E., 1978. Trace-element comparison of clasts in Eocene conglomerates, southwestern California and northwestern Mexico. J. Geol., 86: 753—762.

Albarede, F., 1976. Some trace element relationships among liquid and solid phases in the course of the fractional crystallization of magmas. Geochim. Cosmochim. Acta, 40: 667—673.

Arculus, R.J. and Wills, K.J.A., 1980. The petrology of the plutonic blocks and inclusions from the Lesser Antilles Island Arc. J. Petrol., 21: 743—799.

Baird, A.K., Baird, K.W. and Welday, E.E., 1974. Chemical trends across Cretaceous batholithic rocks of Southern California. Geology, 2: 493—495.

Baker, P.E., 1968. The petrology of Mt. Misery Volcano, St. Kitts, West Indies. Lithos, 1: 124—150.

Bergreen, R.G. and Walawender, M.J., 1977. Petrology and metamorphism of the Morena Reservoir roof pendant, Southern California. Calif., Div. Mines Geol., Spec. Rep., 129: 61—66.

Best, M.G. and Mercy, E.L.P., 1967. Composition and crystallization of mafic minerals in the Guadalupe igneous complex, California. Am. Mineral., 52: 436—474.

Brown, G.C., 1979. The changing pattern of batholith emplacement during earth history. In: M.P. Atherton and J. Tarney (Editors), Origin of Granite Batholiths: Geochemical Evidence. Shiva Publishing Ltd., Orpington, pp. 106—115.

Brown, G.M., Holland, J.G., Sigurdsson, H., Tomblin, J.F. and Arculus, R.J., 1977. Geochemistry of the Lesser Antilles volcanic island arc. Geochim. Cosmochim. Acta, 41: 785—801.

Cheung, P.C.T., 1981. The Mineralogy, Geochemistry, and Petrogenesis of the Mount Poser Gabbroic Pluton, Southern California. M.Sc. Thesis, Univ. of Windsor, 136 pp. (unpubl.).

Cox, K.G., Bell, J.D. and Pankhurst, R.J., 1979. The Interpretation of Igneous Rocks. George Allen and Unwin, London, 450 pp.

Dale, I.M. and Henderson, P., 1972. The partition of transition elements in phenocryst-bearing basalts and the implications about melt structure. 24th Int. Geol. Congr. Sect. 10, Geochemistry. Convenors R.W. Boyle and D.M. Shaw, pp. 105—111.

Dupuy, C., Dostal, J. and Vernières, J., 1977. Genesis of volcanic rocks related to subduction zones, geochemical point of view. Bull. Soc. Géol., Fr., 19: 1233—1243.

Erikson, E.M., 1977. Petrology and petrogenesis of the Mount Stuart batholith — plutonic equivalent of the high alumina basalt association. Contrib. Mineral. Petrol., 60: 183—207.

Fife, D.L., Minch, J.A. and Crampton, P.J., 1967. Late Jurassic age of the Santiago Peak Volcanics, California. Geol. Soc. Am. Bull., 28: 299—304.

Gastil, R.G., 1975. Plutonic zones in the Peninsular Ranges of Southern California and Northern Baja California. Geology, 3: 361—363.

Gill, J.B., 1978. Role of trace element partition coefficients in models of andesite genesis. Geochim. Cosmochim. Acta, 42: 709—724.

Hamilton, P.J., 1977. Isotope and trace element studies of the Great Dyke and Bushveld mafic phase and their relation to Early Proterozoic magmatism. J. Petrol., 18: 24—52.

Hart, S.R. and Brooks, C.J., 1977. The geochemistry and evolution of the early Pre-Cambrian mantle. Contrib. Mineral. Petrol., 61: 109—128.

Helsley, C.E., 1960. Geology of the British Virgin Islands. Ph.D. Thesis, Princeton University, 219 pp. (unpubl.).

Holloway, J.R. and Sykes, M.L., 1979. Melting of biotite and hornblende in the crust. Dep. Nat. Resour., Bur. Miner. Resour., Geol. Geophys., Record 1979/2: 54—55.

Hutchison, C.S., 1974. Laboratory Handbook of Petrographic Techniques. Wiley and Sons, New York, N.Y., 527 pp.

Irving, A.J., 1978. A review of experimental studies of crystal/liquid trace element partitioning. Geochim. Cosmochim. Acta, 42: 743—770.

Jackson, T.E., Smith, T.E. and Huang, C.H., 1981. The petrology and petrogenesis of the basaltic rocks of Carriacou, Grenadine Islands, West Indies. IAVCEI Symposium-Arc volcanism. Abstracts, pp. 153—154.

Jakes, P. and White, A.J.R., 1972a. Major and trace element abundances in volcanic rocks of orogenic zones. Geol. Soc. Am. Bull., 83: 29—40.

Jakes, P. and White, A.J.R., 1972b. Hornblendes from calc-alkaline volcanic rocks of island arcs and continental margins. Am. Miner., 57: 887—902.

Johnson, R.W., Mackenzie, D.E. and Smith, I.E.M., 1978. Volcanic rock associations at convergent plate boundaries: Reappraisal of the content using case histories from Papua New Guinea. Geol. Soc. Am. Bull., 89: 96—106.

Korringa, M.K. and Noble, D.C., 1971. Distribution of Sr and Ba between natural feldspar and igneous melt. Earth Planet. Sci. Lett., 11: 147—151.

Krummenacher, D., Gastil, R.G., Bushee, J. and Doupont, J., 1975. K/Ar apparent ages, Peninsular Ranges batholith, Southern California. Geol. Soc. Am. Bull., 86: 760—768.

Kuno, H., 1950. Petrology of Hakone volcano and the adjacent areas, Japan. Geol. Soc. Am. Bull., 61: 957—1020.

Larsen, E.S., 1948. Batholith and associated rocks of the Corona, Elsinore, and San Luis Rey quadrangles. Geol. Soc. Am. Mem., 29: 1—182.

Leake, B.E., 1978. Nomenclature of amphiboles. Mineral. Mag., 42: 533—563.

Leake, B.E., Brown, G.C. and Halliday, A.N., 1980. The origin of granite magmas: A discussion. J. Geol. Soc. London, 137: 93—97.

LeBas, M.J., 1962. The role of aluminum in igneous clinopyroxenes with relation to their parentage. Am. J. Sci., 260: 267—288.

Lewis, J.F., 1973. Petrology of the ejected plutonic blocks of the Soufriere Volcano, St. Vincent, West Indies. J. Petrol., 14: 81—112.

Lillis, P.G., Walawender, M.J., Smith, T.E. and Wilson, J., 1979. Petrology and emplacement of the Corte Madera gabbro pluton, Southern California. In: P.L. Abbott and V.R. Todd (Editors), Mesozoic Crystalline Rocks. Guideb., 92nd Annul. Meet. Geol. Soc. Am., pp. 143—150.

Longshore, J.D., 1966. Chemical and Mineralogical Variations in the Virgin Island Batholith and its Associated Wall Rocks. Ph.D. Thesis, Princeton Univ., Princeton, N.J., 94 pp. (unpubl.).

McBirney, A.R. and Noyes, R.M., 1979. Crystallization and layering of the Skaergaard intrusion. J. Petrol., 20: 487—554.

Morse, S.A., 1979a. Kiglapait geochemistry I: Systematics, sampling, and density. J. Petrol., 20: 555—590.

Morse, S.A., 1979b. Kiglapait geochemistry II: Petrography. J. Petrol., 20: 591—624.

Mullan, H.S. and Bussell, M.A., 1977. The basic rock series in batholithic associations. Geol. Mag., 114: 265—280.

Nicholls, I.A., 1971. Petrology of Santorini Volcano, Cyclades, Greece. J. Petrol., 12: 67—119.

Nishimori, R.K., 1976. The petrology and geochemistry of gabbros from the Peninsular Ranges batholith, California, and a model for their origin. Ph.D. Thesis, Univ. California, San Diego, Calif., 234 pp. (unpubl.).

Norrish, K. and Hutton, J.T., 1969. An accurate X-ray spectrometric method for the analysis of a wide range of geological samples. Geochim. Cosmochim. Acta, 33: 431—453.

Philpotts, J.A. and Schneztler, C.C., 1972. Large trace cation partitioning in igneous processes. 24th Int. Geol. Congr., Section 10, Geochemistry, Convenors R.W. Boyle and D.M. Shaw, 51—59.

Pitcher, W.S., 1979. Comments on the geological environments of granites. In: M.P. Atherton and J. Tarney (Editors), Origin of Granite Batholiths: Geochemical Evidence. Shiva Publishing Ltd., Orpington, pp. 1—8.

Piwinskii, A.J. and Wyllie, P.J., 1968. Experimental studies of igneous rock series: A zoned pluton in the Wallowa batholith, Oregon. J. Geol., 76: 205—234.

Price, R.C. and Sinton, J.M., 1978. Geochemistry of granitoids and gabbros from Southland, New Zealand. Contrib. Mineral. Petrol., 67: 267—278.

Rea, W.J. and Baker, P.E., 1980. The geochemical characteristics and conditions of petrogenesis of the volcanic rocks of the northern Lesser Antilles — A review. Bull. Volcanol., 42: 325—336.

Sapountzis, E.S., 1979. The Thessalonki gabbros. J. Petrol., 20: 37—70.

Silver, L.T., Early, T.E. and Anderson, T.M., 1975. Petrological, geochemical, and geochronological assymetries of the Peninsular Ranges batholith. Geol. Soc. Am., Abstr. with Programs, 7(3): 375—376.

Smith, A.L., Roobol, M.J. and Gunn, B.M., 1980. The Lesser Antilles — A discussion of the island arc magmatism. Bull. Volcanol., 43: 287—302.

Streckeisen, A.L., 1973. Plutonic rocks —classification and nomenclature. Geotime, 18: 26—30.

Tarney, J. and Saunders, A.D., 1979. Trace element constraints on the origin of cordilleran batholiths. In: M.P. Atherton and J. Tarney (Editors), Origin of Granite Batholiths: Geochemical Evidence, pp. 90—105.

Upton, B.G.J. and Wadsworth, W.J., 1972. Peridotitic and gabbroic rocks associated with the shield forming lavas of Reunion. Contrib. Mineral. Petrol., 35: 139—158.

Wager, L.R. and Brown, G.M., 1967. Layered Igneous Rocks. Freeman, San Francisco, Calif., 588 pp.

Wager, L.R., Brown, G.M. and Wadsworth, W.J., 1960. Types of igneous cumulates. J. Petrol., 1: 73—75.

Walawender, M.J., 1976. Petrology and emplacement of the Los Pinos pluton, southern California. Can. J. Earth Sci., 13: 1288—1300.

Walawender, M.J. and Smith, T.E., 1980. Geochemical and petrologic evolution of the basic plutons of the Peninsular Ranges batholith, Southern California. J. Geol., 88: 232—242.

Walawender, M.J., Smith, T.E., Hoppler, H. and Riddle, C., 1979. Trace element evidence for contamination in a gabbronorite-quartz diorite sequence in the Peninsular Ranges batholith. J. Geol., 87: 87—97.

Yamazaki, T., Onuki, M. and Tiba, T., 1966. Significance of hornblende gabbroic inclusions in calc-alkaline rocks. J. Miner. Petrol. Econ. Geol., 55: 87—103.

Journal of Volcanology and Geothermal Research, 18 (1983) 279—295
Elsevier Science Publishers B.V., Amsterdam — Printed in The Netherlands

MAGMA MIXING IN THE ALEUTIAN ARC: EVIDENCE FROM COGNATE INCLUSIONS AND COMPOSITE XENOLITHS

WALTER K. CONRAD*, SUZANNE M. KAY and ROBERT W. KAY

Department of Geological Sciences, Cornell University, Ithaca, NY 14853 (U.S.A.)
(Received October 20, 1982)

ABSTRACT

Conrad, W.K., Kay, S.M. and Kay, R.W., 1983. Magma mixing in the Aleutian arc: evidence from cognate inclusions and composite xenoliths. In: S. Aramaki and I. Kushiro (Editors), Arc Volcanism. J. Volcanol. Geotherm. Res., 18: 279—295.

The complexity of igneous processes in the Aleutian calc-alkaline magma series can be inferred from study of xenolithic fragments. Composite xenoliths and cognate inclusions provide direct evidence for magma—magma and wall-rock—magma mixing processes. Using distributions of Cr in clinopyroxene, compositional endmembers involved in mixing are identified within the xenoliths. The basaltic mixing endmember is more mafic than calc-alkaline lavas in the arc. Magma mixing and wall-rock assimilation within calc-alkaline basaltic to andesitic magmas is identified in phenocrystic assemblages as well as in xenoliths, and appears to be a widespread phenomenon in Aleutian calc-alkaline magmas.

INTRODUCTION

Petrographers have often noted that calc-alkaline andesites from oceanic as well as continental convergent plate-margins contain a variety of texturally distinct inclusions. Many of these inclusions are xenolithic wall-rock fragments of upper mantle, lower and upper crustal rock types (e.g., Kuno and Aoki, 1970; Aoki, 1971; Conrad and Kay, 1981, 1983; Pope et al., 1981, 1983). Several workers have proposed that other types of inclusions record magmatic processes important in the genesis of the host lavas (e.g., Stewart, 1975; Eichelberger, 1978). Implicit in such studies is the recognition that equilibrium assemblages in lavas erupted at the surface contain little evidence for earlier stages of magmatic history. Instead, they reflect low-pressure crystallization in shallow magma chambers prior to eruption. Inclusions, on the other hand, often contain relict minerals clearly out of equilibrium with the host lava, some of which crystallized at greater depths. Study of xenolith mineralogy allows identification of magmas that are in many cases much more primitive than observed surficial lavas. Inclusions preserve textural relationships reflecting their mode of origin, which may be used with composi-

*Present address: School of Geology and Geophysics, University of Oklahoma, Norman, OK73019, U.S.A.

0377-0273/83/$03.00

tional data to infer specific magmatic histories, including processes such as assimilation of wall-rock and magma mixing.

The diversity of xenolith types found in the aleutian calc-alkaline series magmas is shown in Table I. Xenoliths from Aleutian tholeiitic series magmas (see Kay et al., 1982) have not been studied. This paper presents a summary of previous investigations of Aleutian cumulate xenoliths and of the phenocryst mineralogy of calc-alkaline magmas. This provides a background for discussion of the compositional features of texturally and mineralogically heterogeneous "composite xenoliths".

SAMPLE LOCATIONS AND ANALYTICAL METHODS

All of the xenoliths and calc-alkaline lavas discussed in this report are from islands in the oceanic part of the Aleutian arc (Central Aleutians, 175° W). The xenoliths are found in composite pyroclastic flow units at two Plio-Pleistocene volcanic centers on Adak Moffett and Adagdak, (Coats, 1956), and a dome (sill?) of probable Miocene age in the interior of Kanaga Island (Fraser and Barnett, 1959). Calc-alkaline lavas and flows are described from the Plio-Pleistocene volcanic centers Moffett, Adagdak and Great Sitkin. Mineral analyses and whole-rock analyses of fused glasses were done at Cornell University with a JEOL 733 electron microprobe, using Tracor Northern software with a Bence-Albee matrix correction program. Silicate standards were obtained from the Smithsonian Institute (Jarosewich et al., 1979). Cr analyses in cognate inclusions and composite xenoliths were analyzed at 0.2 microamps current with 100-sec counting times, standardized on a synthetic Corning glass standard. The statistically calculated minimum detection limit (Reed, 1975) for Cr is less than 100 ppm (95% CI). Precision, based on counting statistics (twice 2σ at 500 ppm) is 30—35 relative percent.

SUMMARY OF MINERAL CHEMISTRY IN THE CUMULATE XENOLITHS

Conrad and Kay (1981, 1983) have described a suite of cumulate mafic and ultramafic xenoliths in calc-alkaline lavas from Mt. Moffett, a Plio-Pleistocene volcano on Adak Island. The xenoliths are interpreted to have crystallized from magmas at various stages along a liquid line-of-descent, from olivine basalt to olivine-free basaltic andesite. Olivine, clinopyroxene and pargasitic hornblende in the cumulate xenoliths show progressive variation in FeO/MgO and depletion of compatible elements (Ni and Cr) from ultramafic olivine-clinopyroxenites to hornblende-bearing and hornblende-free gabbros with minor or no olivine.

Ultramafic cumulate xenoliths are olivine clinopyroxenites of two types, one with abundant (>50%) cumulus amphibole and the other with only minor, post-cumulus amphibole (Table I, IA). The cumulus pargasitic hornblende is Cr-rich (0.05—0.55 wt% Cr_2O_3), indicating near-liquidus amphibole crystallization from a basaltic melt before substantial clinopyroxene

TABLE I

Summary of xenoliths and inclusions from Moffett and Adagdak Volcanoes, Adak Island and Miocene Volcanics, Kanaga Island (Modal mineralogy in order of abundance)

Rock type	Modal mineralogy	Location	Comments (Reference)
I. *Cumulate xenoliths*			
A. Olivine clinopyroxenites	(1) AMP, CPX, OL, CRM	Moffett	Crystallized from primitive magmas (1).
	(2) CPX, OL, AMP, MT, CRM	Moffett	(1)
B. Hornblende gabbros	(1) PLAG, AMP, CPX, MT, OL	Moffett	(1)
	(2) AMP, PLAG, CPX, MT	Moffett	Also found on Bogoslof Island (1, 2)
	(3) AMP, PLAG, MT, ± CPX	Moffett	(1)
C. Orthopyroxene gabbros	(1) PLAG, CPX, OPX, MT, OL	Moffett	Crystallized from evolved magmas (1).
	(2) PLAG, OPX, CPX, QTZ, MT	Moffett	(1)
II. *Metamorphic xenoliths*			
A. Spinel Lherzolites	(1) OL, CPX, OPX, SP, AMP	Kanaga	Upper mantle origin, highly strained (3).
	(2) CPX, OL, OPX, SP	Kanaga	(3)
B. Wehrlites	(1) CPX, OL, SP	Kanaga, Adagdak	(3, 4)
C. Orthopyroxene gabbros	(1) PLAG, CPX, OPX, MT	Kanaga	Lower crustal origin, low strain (3, 4).
Primary assemblage	Secondary assemblage		
III. *Composite xenoliths*			
A. OL, CPX?, CRM	CPX, AMP, PLAG, OPX, MT (+LIQ.)	Moffett	(1)
B. OL, CPX, CRM	AMP, CPX, PLAG, MT, OPX, (+LIQ.)	Moffett	(1)
C. CPX, OL, MT, CRM (+LIQ.)	AMP, CPX, PLAG, MT (+LIQ.)	Moffett	(1)
IV. *Cognate inclusions*			
A. OL, PLAG	CPX, PLAG, OPX, MT (+LIQ.)	Moffett	(1)

(1) Conrad and Kay (1981, 1983); (2) Arculus et al. (1977); (3) Pope et al. (1981, 1983); (4) Delong et al. (1975).

fractionation could deplete the melt in Cr. Olivine in the ultramafic xenoliths is more forsteritic (Fo_{83-91}) and Ni-rich (0.15—0.49 wt.% NiO) than any previously reported in xenolithic assemblages from oceanic island arcs. Clinopyroxene in the ultramafic cumulates is primarily diopside, with low Na_2O, variable Al_2O_3 and TiO_2, and high Cr_2O_3 (to 0.65 wt.%) contents.

Hydrous melting and crystallization experiments under varying f_{O_2} at pressures up to 10 kbar have duplicated the mineralogies and crystallization sequences observed in the ultramafic xenoliths (Yoder and Tilley, 1962; Holloway and Burnham, 1972; Helz, 1973). The ultramafic xenoliths are inferred to have crystallized from hydrous, but vapor-undersaturated, basaltic magmas at pressures greater than 3 to 5 kbar, with liquidus temperatures less than 1150°C.

The gabbroic xenoliths from Mt. Moffett are typical of those found in many magmatic arcs (Aoki, 1971; Powell, 1978; Arculus and Wills, 1980; Gust and Johnson, 1981). Two clinopyroxene Fe-enrichment trends are identified in the suite of cumulate xenoliths from Mt. Moffett: (a) a salitic trend in the hornblende gabbros (Table I, IB), marked by slight enrichment in CaO, greatly increasing Al_2O_3 and TiO_2, and high recalculated Fe_2O_3/FeO; and (b) an augitic trend in hornblende-free, orthopyroxene-bearing gabbros (Table I, IC), marked by CaO depletion, lower recalculated Fe_2O_3 and slightly increasing Al_2O_3 and TiO_2 contents. Both clinopyroxene trends are consistent with crystallization of the gabbros from progressively fractionated melts with varying water content and f_{O_2}, as discussed by Conrad and Kay (1983).

The hornblende gabbro cumulates are inferred to have crystallized from hydrous basaltic magmas, at minimum pressures of about 2kbar, with oxygen fugacities above the nickel-nickel-oxide buffer by up to two log units (Conrad and Kay, 1983). These estimates are based on comparison with experimental results using hydrous basaltic (Yoder and Tilley, 1962; Holloway and Burnham, 1972; Helz, 1973) and andesitic (Eggler, 1972; Eggler and Burnham, 1973) melts.

SUMMARY OF PHENOCRYST MINERAL CHEMISTRY IN CALC-ALKALINE LAVAS

Phenocrysts with free-growth forms are interpreted to have crystallized directly from magmas, and represent the composition of magmas in the last stages of their development. As described in Kay and Kay (1982), the mafic mineralogy in basalts and andesites from the calc-alkaline volcanoes Buldir, Adagdak, Moffett, Great Sitkin and Kasatochi correlate with whole-rock FeO/MgO, and indicate two general types of lavas. Lavas in one group (49—63 wt.% SiO_2, FeO/MgO ratio near one) contain some forsteritic, Ni-rich olivine and Cr-rich clinopyroxene as phenocrysts, xenocrysts, or both. The Cr and Mg-rich clinopyroxenes rimming orthopyroxenes and clinopyroxenes in the groundmass of some of these lavas suggest that magma mixing occurred shortly before eruption. Lavas in the second group (49—66 wt.%

SiO_2, FeO/MgO ratio approximately 2) generally do not contain Ni or Cr-rich phenocrysts and show no obvious evidence of mixing with primitive magmas. Basalts in this second group are characterized by salitic/calcic augitic clinopyroxene and less-forsteritic, Ni-poor olivine (Fig. 3). Orthopyroxene replaces olivine in lavas with > 57 wt.% SiO_2 (cf. Marsh, 1976), and amphibole is variably present in both groups (Kay and Kay, 1982; and unpublished data).

TEXTURAL EVIDENCE FOR MIXING: COGNATE INCLUSIONS AND COMPOSITE XENOLITHS

In addition to holocrystalline xenoliths, several classes of texturally distinct inclusions and xenoliths are found in lavas and pyroclastic flows from Mt. Moffett (Conrad and Kay, 1981, 1983). "Cognate inclusions" and "composite xenoliths" contain mineral assemblages recognized in the cumulate xenoliths and phenocrysts discussed above. However, the assemblages show textural evidence for mixing.

Cognate inclusions

An essential feature of all cognate inclusions is that the majority of the inclusion crystallized within the host lava. Three types of cognate inclusions showing three relatively distinct petrogenetic processes are recognized:

(a) Intergrown phenocrystic assemblages that are minerals occurring as equilibrium phenocrysts and groundmass phases in the host lava are best described as crystal-clots. The origins of crystal-clots have been debated by Stewart (1975) and Garcia and Jacobson (1979).

(b) Relict phases (e.g., resorbed pargasitic hornblende) and relict mineral assemblages with disequilibrium textures are also common. These presumably crystallized in a different (higher *P-T*) environment from the equilibrium phenocryst assemblage in the host lava.

(c) Cognate inclusions can also result from reaction when small amounts of xenolithic material are introduced into the host lava. Crystallization within the host produces very distinctive textural assemblages which indicate assimilation of contaminant material. These are described below.

A photomicrograph and petrographic description of typical cognate inclusions in a basaltic-andesite host lava from Mt. Moffett is shown in Fig. 1. Megacrysts of clinopyroxene, with large magnetite inclusions, and plagioclase are texturally distinct from the phenocrysts and groundmass of the host. Several textural and compositional relationships suggest the megacrysts grew during an assimilation reaction:

(a) Resorbed olivine and calcic plagioclase are preserved only at the cores of clinopyroxene, surrounded with concentrations of magnetite.

(b) Clinopyroxene megacrysts are compositionally homogeneous, and plagioclase megacrysts are relatively unzoned in comparison to phenocryst-sized plagioclase.

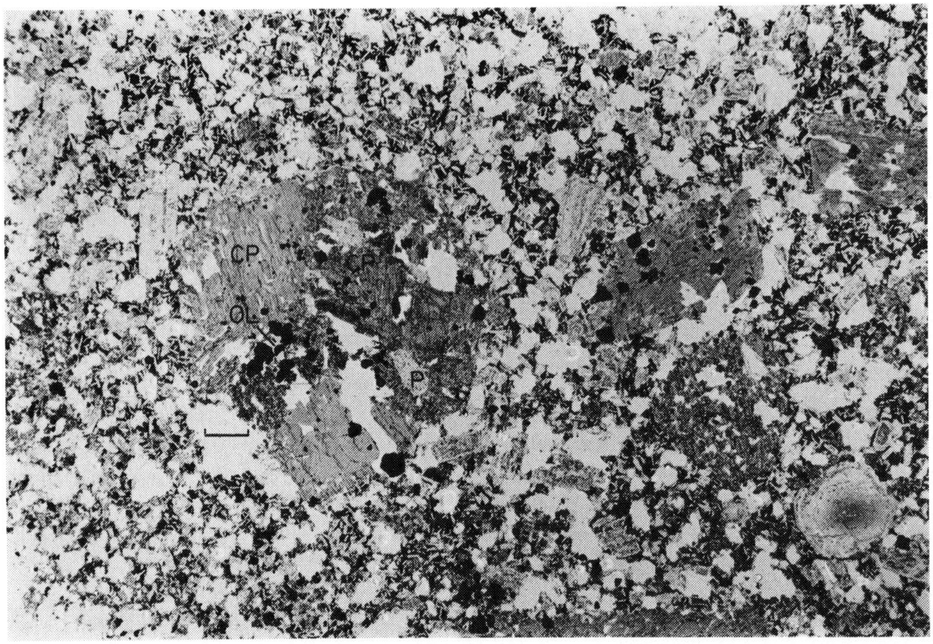

Fig. 1. Photomicrograph of typical cognate inclusions in the groundmass of a basaltic andesite from Moffett volcano (MM-79-A). Scale bar is 1 mm in length. The inclusions consist primarily of unzoned, sparsely twinned clinopyroxene megacrysts (*CP* on figure; Ca_{42-45}, Fe_{13-14}, Mg_{42-45}; Cr_2O_3 < 0.01 wt.%) intergrown with sparsely twinned, relatively unzoned (An_{84-91}) plagioclase megacrysts (*P* on figure). Plagioclase on the exterior margins in contact with the groundmass is near An_{50}. Partially resorbed olivine (*OL* on figure; Fo_{72-75}; NiO < 0.01 wt.%) intergrown with resorbed An_{93} plagioclase occurs in the cores of some clinopyroxenes. Magnetite occurs interspersed in the clinopyroxene and in clusters surrounding resobred olivines. Megacrysts of clinopyroxene with magnetite inclusions and plagioclase also occur individually in the groundmass. The host lava contains phenocrysts of subhedral clinopyroxene and zoned plagioclase (An_{53-81}). Small orthopyroxenes, small anhedral clinopyroxenes, acicular zoned plagioclase laths with compositions as low as An_{35}, and magnetite appear to be quenchphases in the groundmass. The groundmass is a red-brown, vesicular hyalopilitic matrix. The megacrysts are interpreted to have crystallized upon assimilation of olivine and calcic plagioclase by the host lava saturated with magnetite, clinopyroxene and plagioclase. See text for discussion.

(c) Plagioclase compositions in the megacrysts range from An_{84} to An_{91}, intermediate to the assimilated plagioclase (An_{93}), and the maximum anorthite content of the zoned phenocrysts (An_{81-42}).

(d) Equilibrium with the host lava was maintained following the assimilation reaction, evidenced by marginal zoning of plagioclase in the megacrysts to compositions similar to phenocryst rims (An_{52}).

As discussed by McBirney (1970) and Bowen (1928), if crystalline material that is early in a reaction series is added to a more evolved magma, the magma will react with the more refractory material (not melt it). During reaction, the magma crystallizes more of the liquidus phases while dissolving the added refractory phases. In the cognate inclusions from Mt. Moffett, the large grain sizes and compositional homogeneity of magnetite, clinopyroxene and plagioclase suggest an extended period of crystal growth. This could have occurred if magnetite, clinopyroxene and plagioclase were liquidus phases in the host lava when it assimilated olivine and more calcic plagioclase (probably from wall-rock). Shortly after the assimilation-reaction, the host was quenched upon eruption. The quench assemblage contains fine-grained orthopyroxene in addition to fine-grained magnetite, clinopyroxene and plagioclase. The occurrence of orthopyroxene only in the groundmass suggests it was not a liquidus phase at the moment olivine and calcic plagioclase were assimilated.

Composite xenoliths

Composite xenoliths are inclusions of quenched magma that are foreign to, and texturally distinct from, the host. The xenoliths from Mt. Moffett are nodules containing coarsely porphyritic assemblages within a finely crystalline to glassy matrix, representing former liquid. The nodules are distinct from the host lava and vary from 2 to 6 cm in diameter.

Textural features in composite xenoliths give evidence for assimilation of wall-rock or magma ("primary assemblage") by another magma ("secondary assemblage"), all of which occurs prior to incorporation of the xenolith in the host lava. As in the reaction described above, phases in the primary assemblage are consumed by increased crystallization of liquidus phases in the assimilating magma. Minerals found in primary assemblages are typically Cr-spinel, Fo_{81-93}) olivine (NiO to 0.85 wt.%), diopsidic clinopyroxene CrO_3 to 0.98 wt.%), and magnetite. All composite xenoliths show textural evidence that olivine is in reaction to form amphibole. Several parageneses have been identified for assemblages within the composite xenoliths (Table I).

Primary assemblages with relatively unzoned phases are interpreted to have been solid wall-rock, analogous to the cumulate xenoliths (Table I, III A and B). In the xenolith described in Fig. 2 (Table I, III B), olivine enclosed within amphibole is highly variable in size, but not in composition. Olivine ranges from small inclusions within single acicular amphibole laths to large grains surrounded by several amphibole crystals. Even the largest olivine grains are unzoned from core to rim, and all grains analyzed are within a narrow compositional range (Fo_{87-89}; NiO 0.16 wt.%). These features indicate that the olivine was added to the secondary assemblage magma, and did not crystallize as phenocrysts that later reacted to amphibole. A whole-rock analysis of the same xenolith (Table II) is Mg-rich and Si-poor, correlating with the large modal abundances of olivine and amphibole within the inclusion. The whole-rock composition is clearly not that of a melt, and is interpreted to reflect accumulation of both olivine and amphibole.

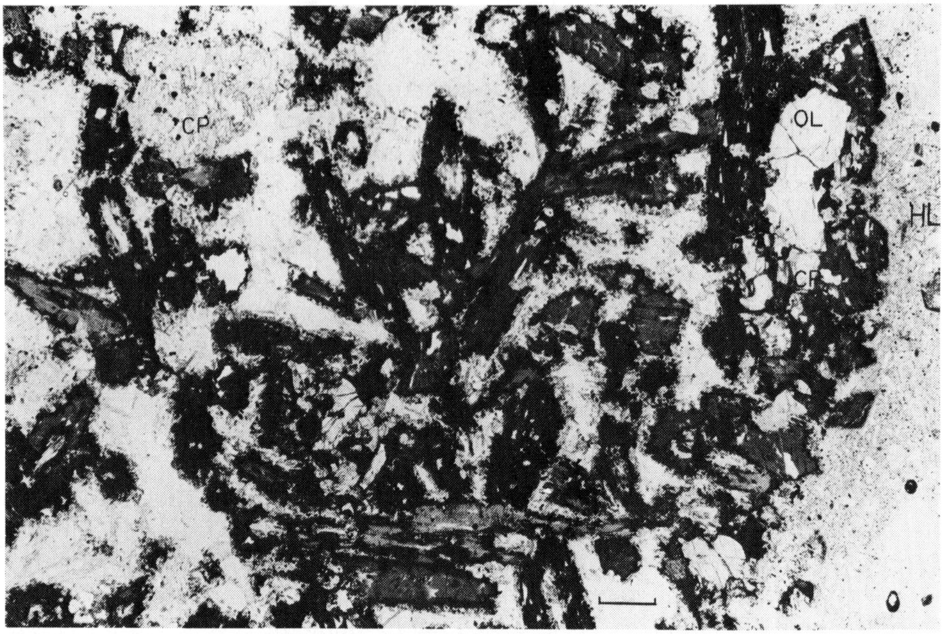

Fig. 2. Photomicrograph of composite xenolith in andesite from Moffett volcano (MM-78E). Scale bar is 1 mm in length. The xenolith is comprised primarily of amphibole and clinopyroxene surrounded by a fine-grained crystalline matrix that is distinct from the host lava (*HL* on figure). The amphibole is euhedral to acicular pargasite that is rimmed and partially replaced by fine grained opaques. Included in the amphibole is abundant partially resorbed olivine (*OL* on figure; Fo_{87-89}; NiO 0.16 wt%), which contains small inclusions of chromite (43—46 wt.% Cr_2O_3). Lesser amounts of partially resorbed clinopyroxene (*CP* below olivine in figure) are intergrown with resorbed olivine. Resorbed chromian magnetite (14—18 wt.% Cr_2O_3) is found in the amphibole near resorbed olivine. A second generation of clinopyroxene, (*CP* on figure to upper left) containing subhedral magnetite, occurs in aggregates. Although not shown on figure, resorbed olivine sometimes occurs in these clinopyroxenes. The clinopyroxenes are zoned from core (Ca_{46}, Fe_5, Mg_{49}; Cr_2O_3 to 0.77 wt.%) to rim (Ca_{45}, Fe_{12}, Mg_{43}; Cr_2O_3 <0.03 wt.%). The fine-grained crystalline matrix is dominated by quench-textured, often acicular, plagioclase (not distinguishable on figure) with lesser clinopyroxene, magnetite and orthopyroxene. Assimilation of chromite-bearing olivine and associated clinopyroxene (primary assemblage) by a magma saturated with magnetite, clinopyroxene and amphibole (secondary assemblage) produced the reaction textures. The hybrid magma was then quenched and brought to the surface in the host lava. See text for discussion.

In another type of composite xenolith (Table I, III C), zoned olivine and clinopyroxene in the primary assemblage are interpreted to have coexisted with a melt. In such inclusions, olivine and clinopyroxene are concentrically zones, and often have compositionally discontinuous overgrowths filled with melt inclusions and opaque microphenocrysts. Clinopyroxene is sometimes both oscillatory and sector-zoned. This is evidence for free growth of these primary assemblage phases in a melt.

TABLE II

Whole-rock compositions

	1	2	3	4
SiO_2	46.65	48.97	49.41	49.48
TiO_2	1.30	0.72	0.82	0.76
Al_2O_3	13.44	16.27	19.88	19.07
FeO*	10.31	8.79	9.15	8.97
MnO	0.14	0.18	0.17	0.17
MgO	13.18	9.62	5.59	6.61
CaO	11.01	12.86	11.98	11.95
Na_2O	2.22	2.00	2.48	2.29
K_2O	0.86	0.54	0.68	0.57
P_2O_5	0.10	0.11	—	—
Total	99.21	100.06	100.16	99.87

1: Composite inclusion MM 78-E.
2: Tholeiitic basalt, analysis OK4, Byers (1961)
3: Calculated steady-state residual basalt, after fractionating CPX:OL = 70:30, from analysis 2. Conrad and Kay (1983).
4: Calculated steady-state residual basalt, after fractionating CPX:OL:AMP = 55:15:30, from analysis 2. Conrad and Kay (1983).

Secondary assemblages in all composite xenoliths contain acicular amphibole, magnetite and clinopyroxene. These phases plus plagioclase and glass make up a fine-grained, often vesicular groundmass. Some composite xenoliths have groundmass orthopyroxene, indicating the secondary assemblage to be an evolved basaltic-andesitic magma. Acicular amphibole and hollow plagioclase crystals give morphological evidence for quenching, as do patches of glass in the fine-grained groundmass.

Acicular hornblende-plagioclase xenoliths with variable amounts of clinopyroxene and interstitial glass/scoria are found in andesites in island arc and continental settings. These are probably similar in origin to the secondary assemblages in the Mt. Moffett composite xenoliths, although evidence for assimilation reactions having occurred within the inclusions has not been described. Photographs and detailed descriptions are given for inclusions from Ecuador and California (Eichelberger and Gooley, 1977) and the Lesser Antilles (Lewis, 1973; Eichelberger, 1978). Several possible mechanisms have been proposed to account for the xenolithic occurrence and phenocrystic texture of these inclusions, including chilling of basaltic magma in a cooler host (e.g., Sparks et al., 1977; Eichelberger, 1978) and quenching of cognate cumulus minerals plus intercumulus liquid (Lewis, 1973; cf. Hermes and Cornell, 1981). Microprobe analyses of the glassy groundmass areas in the Mt. Moffett xenoliths indicate variable compositions, often highly silicic, which suggests that the xenoliths were quenched and that the glassy areas are residual after rapid crystal growth, as observed in the groundmass of Hawaiian basalts (Evans and Moore, 1968).

COMPOSITIONAL EVIDENCE FOR MAGMA MIXING

The textural evidence for mixing must now be evaluated in terms of mixing criteria used by other workers. Mixing in andesitic magmas has been inferred from observation of mixed, non-equilibrium mineral assemblages and reversely zoned minerals (e.g., Eichelberger and Gooley, 1977; Sakuyama, 1978). A more specific definition of magma mixing is given by Anderson (1976), which requires the identification of compositionally distinct endmembers and proof that each component contained some melt. The textural evidence described in one type of Mt. Moffett composite xenolith (i.e., assimilation of zoned olivine, clinopyroxene and melt, Table I, III C) establishes the possibility of true magma mixing at depth, prior to incorporation of the xenolith in the host lava. However, compositions of the component magmas have yet to be identified. The mineralogy of the composite xenoliths suggests mixing of a primitive basaltic endmember, crystallizing chromite, olivine and clinopyroxene, with a more evolved basaltic-andesitic endmember, crystallizing magnetite, clinopyroxene, amphibole, plagioclase and sometimes orthopyroxene.

Mixing between compositionally distinct magmas can be confirmed by the identification of a distinct mafic mineralogy corresponding to each magma in the mixed assemblage. Quadrilateral compositions of clinopyroxenes from recent lavas, ultramafic cumulate xenoliths, and composite xenoliths are plotted in Fig. 3. Clinopyroxene in the ultramafic xenoliths is primarily diopside, and defines a field for clinopyroxene crystallizing from primitive, mantle-derived melts. The phenocryst and groundmass clinopyroxenes are from several evolved calc-alkaline Plio-Pleistocene lavas ($FeO/MgO > 1.8$, 49—66 wt.% SiO_2) which do not show any obvious evidence for magma mixing. Several composite xenoliths show a range of clinopyroxene compositions that span both fields, supporting the textural evidence that the composite xenoliths are mixed assemblages.

If mixing occurs between magmas closely related by progressive crystallization, only highly compatible elements are likely to show sufficiently large variations in concentration to distinguish compositional endmembers. Magmatic concentrations of the compatible elements Ni and Cr can be inferred from analyses of these elements in olivine (e.g., Sato, 1977) and clinopyroxene. In the Mt. Moffett xenoliths, clinopyroxene is the most useful mineral for demonstrating magma mixing because it is found in cumulate assemblages throughout a liquid line-of-descent.

Figure 4 shows the same three clinopyroxene groups from Fig. 3 in terms of a highly compatible element, Cr, versus Fe-enrichment (FeO/MgO ratio) in clinopyroxene. Clinopyroxene from the recent lavas with whole-rock $FeO/MgO > 1.8$ (both basalts and andesites) is uniformly depleted in Cr, and

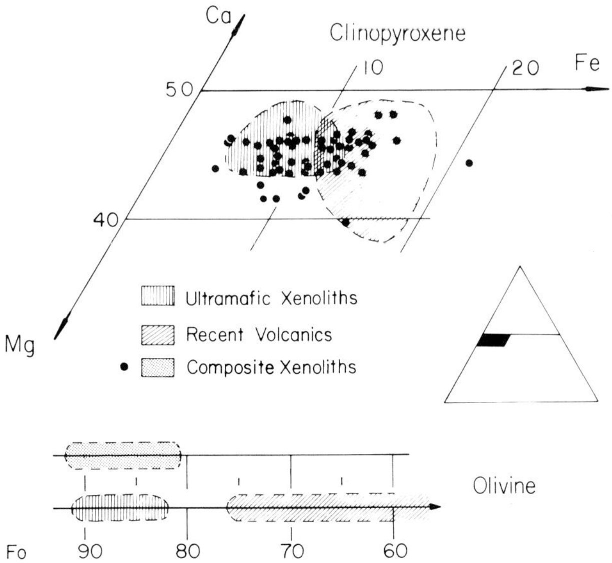

Fig. 3. Clinopyroxene and olivine compositions for ultramafic xenoliths, recent volcanics, and composite xenoliths. Clinopyroxene endmembers are plotted as Mg, Ca, and total Fe as Fe^{+2}. Olivine is plotted as Fe^{+2} and Mg endmembers. The ultramafic fields are for clinopyroxene and olivine from cumulate ultramafic xenoliths from Mt. Moffett. The recent volcanics field includes clinopyroxene and olivine from calc-alkaline Plio-Pleistocene basalts and andesites from Moffett, Adagdak, and Great Sitkin, with whole-rock FeO/MgO >1.8 (Kay and Kay, 1982, and unpubl. data). Points plotted are representative compositions of clinopyroxene in 4 composite xenoliths. The fields for diopside from the cumulate xenoliths and augites/salites from the recent volcanic rocks overlap only slightly. The clinopyroxene compositions from the composite xenoliths span both fields and suggest fractionation and/or mixing between mantle-derived primitive melts and evolved basaltic-andesitic melts. Clinopyroxene in individual composite xenoliths spans the indicated compositional range.

compositionally similar to that in the cumulate gabbroic xenoliths (Conrad and Kay, 1983). Clinopyroxene in the composite xenoliths shows a wide range of compositions, spanning the fields for primitive melts and the recent lavas. Many composite xenoliths also have oscillatory-zoned clinopyroxene fluctuating between Cr-rich, Fe-poor and Cr-poor, Fe-rich compositions, with each zone typically 0.05—0.25 mm in thickness (Conrad and Kay, 1983). These coarsely-spaced oscillatory-zones are reflected by a bimodal clustering of data points on Fig. 4.

The data suggest two important points: (a) evolved (high FeO/MgO) lavas erupted at the surface are uniformly depleted in compatible elements, indicating they are fractionates of more primitive melts; and (b) clinopyroxene in composite xenoliths records the fractionation of primitive melts to more evolved melts and/or mixing between these two kinds of melts. It is also possible that the oscillatory-zoned clinopyroxene records cyclic fractionation

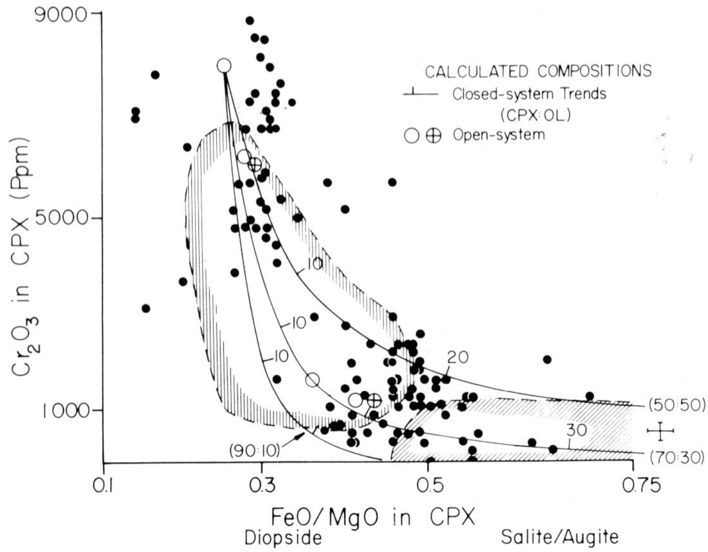

Fig. 4. Cr_2O_3 (in ppm) versus FeO/MgO ratio in clinopyroxenes from ultramafic xenoliths, recent volcanics, and composite xenoliths. Fields and points are the same as in Fig. 3. Calculated compositions are from an equilibrium fractionation model (see Stern, 1979 and Hanson and Langmuir, 1978) described in Conrad and Kay (1983). The initial composition used in both closed- and open-system models is given in Table II. The Cr_2O_3 concentration in the initial melt was calculated using a mineral/melt distribution coefficient of 15 and a value of 8000 ppm Cr_2O_3 in the first clinopyroxene. Closed-system trends assuming clinopyroxene to olivine ratios of 90:10, 70:30, and 50:50 are shown as solid lines. Compositions at 10% crystallization intervals are shown with hatch marks. Open-system compositions were calculated assuming a clinopyroxene to olivine ratio of 70:30 and 15% crystallization, 55% residual melt removal, and 70% primitive melt replenishment in each cycle. Intermediate cycles are indicated by open circles and steady-state compositions, which are reached in three cycles, are indicated by circles with crosses. The calculated whole-rock composition of the steady-state residual melt is a high-Al basalt (Table II, analysis 3). Clinopyroxene compositions in the composite xenoliths reflect both fractionation of the primitive melt and mixing of these primitive clinopyroxenes with more evolved melt. Clinopyroxene dominated fractionation of the primitive melt can yield the evolved melt compositions found in the recent calc-alkaline andesites in the Aleutian arc.

and melt replenishment, although there is considerable uncertainty about the mechanisms causing minor element zonation in clinopyroxene (Downes, 1974; Deer et al., 1978; Shimizu and le Roex, 1982).

Modelled compositions for clinopyroxene in open and closed magma systems crystallizing both clinopyroxene and olivine are consistent with the above conclusions (Conrad and Kay, 1982). Figure 4 also shows a series of closed-system fractionation models and one cyclic open-system model. The primitive melt composition (Table II, analysis 2) for the model calculations is a tholeiitic basalt from Okmok volcano (Byers, 1961) that is used by Kay et al. (1982) to represent the most primitive known basalt in the Aleutian arc. For the open-system model, if relatively large percentages of crystallization and primitive melt replenishment occur (cf. O'Hara, 1977; Stern, 1979), the system will rapidly approach a steady-state (O'Hara and Mathews, 1981). Interruption of the open-system process at any point, and progressive fractionation in a closed-system, would yield clinopyroxene compositions along the indicated paths.

The relative importance of open- or closed-system fractionation processes to the clinopyroxene Cr-depletion trend shown in Fig. 4 is difficult to constrain. The model of cyclic fractionation and replenishment of primitive melts must therefore be regarded as speculative, but such models also yield reasonable bulk melt compositions. Conrad and Kay (1983) have proposed that high-Al basalts may be the steady-state residual products of open-system olivine and clinopyroxene fractionation from magnesian basalts, consistent with the conclusions of Perfit et al. (1980a) and Kay et al. (1982). The calculated whole-rock composition of the steady-state residual melt using the model system shown in Fig. 4 (Table II, analysis 3) is similar to Aleutian high-Al basalts which are considered by some workers to be primary (e.g., Baker and Eggler, 1981). Also given in Table II (analysis 4) is a second calculated residual melt for an open-system, steady-state magma chamber crystallizing clinopyroxene, olivine and amphibole (Conrad and Kay, 1983). Ultramafic residuum from these magma chambers would be similar in modal mineralogy and mineral chemistry to both the Mt. Moffett ultramafic xenoliths, and the large accumulations of olivine and clinopyroxene found in Alaskan-type ultramafic complexes (Findlay, 1969; Murray, 1972; Irvine, 1974).

SUMMARY AND CONCLUSIONS

The compositional data and theoretical considerations discussed above support textural evidence that composite xenoliths result from multiple stages of magma mixing. Mixing between primitive and slightly fractionated (10 to 15% crystallization, Fig. 4) basaltic melts, probably in lower crustal magma chambers, is a possible mechanism for producing both ultramafic olivine-clinopyroxenite cumulates and discontinuously zoned olivine and clinopyroxene found in composite xenoliths. Further mixing of these melts with more evolved, cooler melts (crystallizing amphibole) produces the distinctive reaction assemblages in the composite xenoliths. Several mechanisms have a possible role in quenching of the incompletely hybridized melts, including: rapid crystallization of the liquid during reaction, depressurization and vesiculation of the hydrous magma during ascent, and chilling of the hybrid magma in the host magma (Eichelberger and Gooley, 1977; Hermes and Cornell, 1981).

The mineral chemistry of composite xenoliths records fractional crystallization and magma mixing of the same primitive magmas which crystallize deep-seated cumulate ultramafic and gabbroic xenoliths. In some cases, these mafic magmas appear to reach shallow crustal levels. When the locations of magma mixing and assimilation are near the surface, these processes are recorded in cognate inclusions. Cognate inclusions related to mixing and assimilation are found in a limited number of lavas and flows at every calc-alkaline volcanic center we have studied. Even inclusion-free calc-alkaline lavas are likely to be hybridized melts, products of a complex magmatic history involving magma mixing and assimilation of wall-rock.

The lack of primary calc-alkaline basaltic melts among lava flows in the arc may be explained by the xenolithic evidence for magma chambers at both lower crustal as well as shallow depths. Magmas from depth intrude these chambers, which act as density filters. A crustal column inferred from the xenolith suite indicates crystallization of igneous rocks at all levels of the crust; no xenolith with a sedimentary protolith has been found. The xenolith population from Mt. Moffett and Kanaga Island indicates that gabbroic rocks, many of them hornblende gabbros, dominate the lower crust (cf. Anderson, 1980). As discussed in Conrad and Kay (1983) lower crustal gabbros are intruded in approximately equal volumes to more silicic dioritic to granodioritic plutons added at levels of 10 to 15 km or less (Perfit and Lawrence, 1979; Citron, 1980; Perfit et al., 1980b). Similar crustal profiles have been proposed for the Japanese island arc (e.g., Aoki, 1971; Zashu et al., 1980).

ACKNOWLEDGEMENTS

This research has been supported by the National Science Foundation (grants EAR 8007751, EAR7919719) and COPSTOC (Cornell Program for the Study of Continents). We also acknowledge the U.S. Geological Survey, Branch of Western Mineral Resources, for partial student support for Conrad. We thank personnel from the U.S. Naval Base, Adak; the U.S. Coast Guard; NOAA; and the U.S. Fish and Wildlife service, Aleutian Wildlife Refuge. H. Breuckner, B. Hausbach, G. Citron, J. Rubenstone, W. Snedden, K. Dodd, J. Reynolds, and D. Coehlo assisted in the field. R. Pope is thanked for valuable discussion. We also thank the editors of this volume and a reviewer for their patience and valuable assistance.

REFERENCES

Anderson, A.T., 1976. Magma mixing-petrological process and volcanological tool. J. Volcanol. Geotherm. Res., 1: 3—33.

Anderson, A.T., 1980. Significance of hornblende in calc-alkaline andesites and basalts. Am. Mineral., 65: 837—851.

Aoki, K., 1971. Petrology of mafic inclusions from Itinome-gata, Japan. Contrib. Mineral. Petrol., 30: 314—331.

Arculus, R.J. and Wills, K.J.A., 1980. The petrology of plutonic blocks and inclusions from the Lesser Antilles arc. J. Petrol., 21: 743—799.

Arculus, R.J., DeLong, S.E., Kay, R.W., Brooks, C. and Sun, S.S., 1977. The alkalic rock suite of Bogoslof island, eastern Aleutian arc, Alaska. J. Geol., 85: 177—186.

Baker, D.R. and Eggler, D.H., 1981. Water contents and low pressure differentiation paths in basaltic magma from the Aleutian and Mariana arcs: experimental constraints (abstract). IAVCEI 1981 Symposium: Arc Volcanism: 24.

Bowen, N.L., 1928. The evolution of igneous rocks. Princeton University Press, Princeton, N.J., 332 pp.

Byers, F.M., 1961. Petrology of three volcanic suites, Umnak and Bogoslof Islands, Aleutian Islands, Alaska. Geol. Soc. Am. Bull., 72: 93—128.

Citron, G.P., 1980. The Hidden Bay Pluton, Alaska: Geochemistry, Origin and Tectonic Significance of Oligocene Magmatic Activity in the Aleutian Island Arc. Ph.D. Thesis, Cornell University, Ithaca, N.Y., 240 pp.

Coats, R.R., 1956. Geology of Northern Adak Island, Alaska. U.S. Geol. Surv., Bull., 1028-C.

Conrad, W.K. and Kay, R.W., 1981. Xenoliths from Adak island: evidence for hydrous basaltic primary magmas in the Aleutian arc (abstract). EOS, Trans. Am. Geophys. Union, 62: 1092.

Conrad, W.K. and Kay, R.W., 1983. Ultramafic and mafic inclusions from Adak island: crystallization history, and implications for the nature of primary magmas and crustal evolution in the Aleutian arc. J. Petrol. (in press).

Deer, W.A., Howie, R.A. and Zussman, J., 1978. Rock-Forming Minerals, Volume 2A: Single Chain Silicates. Wiley, New York, N.Y., 668 pp.

DeLong, S.E., Hodges, F.N. and Arculus, R.J., 1975. Ultramafic and mafic inclusions, Kanga Island, Alaska, and the occurrence of alkaline rocks in island arcs. J. Geol., 83: 721—736.

Downes, M.J., 1974. Sector and oscillatory zoning in calcic augites from M. Etna, Sicily. Contrib. Mineral. Petrol., 47: 187—190.

Eggler, D.H., 1972. Water-saturated and undersaturated melting relations in a Paracutin andesite and an estimate of water content in the natural magma. Contrib. Mineral. Petrol., 34: 261—271.

Eggler, D.H. and Burnham, C.W., 1973. Crystallization and fractionation trends in the system andesite H_2O-CO_2-O_2 at pressures to 10 kb. Geol. Soc. Am. Bull., 84: 2517—2532.

Eichelberger, J.C., 1978. Andesite in island arcs and continental margins: relationship to crustal evolution. Bull Volcanol., 41: 480—500.

Eichelberger, J.C. and Gooley, R., 1977. Evolution of silicic magma chambers and their relationship to basaltic volcanism. Am. Geophys. Union, Geophys. Monogr., 20: 57—77.

Evans, B.W., and Moore, J.G., 1968. Mineralogy as a function of depth in the prehistoric Makaopuhi tholeiitic lava lake, Hawaii. Contrib. Mineral. Petrol., 17: 85—115.

Findlay, D.C., 1969. Origin of the Tulameen ultramafic-gabbro complex, southern British Columbia. Can. J. Earth Sci., 6: 399—425.

Fraser, G.D. and Barnett, H.F., 1959. Geology of the Delarof and westernmost Andreanof Islands, Aleutian Islands, Alaska. U.S. Geol. Surv., Bull., 1028—I.

Garcia, M.O. and Jacobson, S.S., 1979. Crystal clots, amphibole fractionation and the evolution of calc-alkaline magmas. Contrib. Mineral. Petrol., 69: 319—327.

Gust, D.A. and Johnson, R.W., 1981. Amphibole-bearing inclusions from Boisa Island, Papua New Guinea: Evaluation of the role of fractional crystallization in an andesitic volcano. J. Geol., 89: 219—232.

Hanson, G.N. and Langmuir, C.H., 1978. Modelling of major elements in mantle-melt systems using trace element approaches. Geochim. Cosmoch. Acta, 42: 725—741.

Helz, R.T., 1973. Phase relations of basalts in the melting range at Ph20=5kb as a function of oxygen fugacity. J. Petrol., 14: 249—302.

Hermes, D.O. and Cornell, W.C., 1981. Quenched crystal mush and associated magma compositions as indicated by intercumulus glasses from Mt. Vesuvius, Italy. J. Volcanol. Geotherm. Res., 9: 133—149.

Holloway, J.R. and Burnham, C.W., 1972. Melting of basalt with equilibrium water pressure less than total pressure. J. Petrol., 13: 1—29.

Irvine, T.N., 1974. Petrology of the Duke Island ultramafic complex, southeastern Alaska. Geol. Soc. Am., Mem., 138.

Jarosewich, E., Nelson, J.A. and Norburg, J.A., 1979. Electron microprobe reference samples for mineral analysis. Smithsonian Contrib. Earth Sci., 22: 68—72.

Kay, S.M. and Kay, R.W., 1982. Mafic mineralogy in calc-alkaline Aleutian volcanic rocks: evidence for primitive magmas and magma mixing (abstract). Geol. Soc. Am., Programs with Abstr., 14: 526.

Kay, S.M., Kay, R.W. and Citron, G.P., 1982. Tectonic controls on tholeiitic and calc-alkaline magmatism in the Aleutian arc. J. Geophys. Res., 87: 4051—4072.

Kuno, H. and Aoki, K., 1970. Chemistry of ultramafic nodules and their bearing on the origin of basaltic magmas. Phys. Earth Planet. Inter., 3: 273—301.

Lewis, J.F., 1973. Petrology of the ejected plutonic blocks of the Soufriere volcano. St. Vincent, West Indies. J. Petrol., 14: 81—112.

Marsh, B.D., 1976. Some Aleutian andesites: their nature and source. J. Geol., 84: 45.

McBirney, A.R., 1979. Effects of assimilation, In: H.S. Yoder (Editor), The Evolution of Igneous Rocks: Fiftieth Anniversary Perspective. Princeton University Press, Princeton, N.J., 588 pp.

Murray, C.G., 1972. Zones ultramafic complexes of the Alaskan type: Feeder pipes of andesitic volcanoes. In: R.E. Shargam et al. (editors), Studies in Earth and Space Sciences—Hess Volume, Geol. Soc. Am., Mem., 132: 313—335.

O'Hara, M.J., 1977. Geochemical evolution during fractional crystallization of a periodically refilled magma chamber. Nature, 266: 503—507.

O'Hara, M.J. and Mathews, R.E., 1981. Geochemical evolution in an advancing, periodically replenished, periodically tapped, continuously fractionating magma chamber. J. Geol. Soc. London, 138: 237—278.

Perfit, M.R. and Lawrence, J.R., 1979. Oxygen isotope evidence for meteoric water interaction with the Captains Bay pluton, Aleutian Islands. Earth Planet. Sci. Lett., 65: 16—22.

Perfit, M.R., Gust, D.A., Bence, A.E., Arculus, R.J. and Taylor, S.R., 1980a. Chemical characteristics of island-arc basalts: implications for mantle sources. Chem. Geol., 30: 227—256.

Perfit, M.R., Breuckner, H.K., Lawrence, J.R. and Kay, R.W., 1980b. Trace element and isotopic variations in a zoned pluton and associated volcanic rocks. Unalaska Island: a model for fractionation in the Aleutian calc-alkaline suite. Contrib. Mineral. Petrol., 73: 69—87.

Pope, R.R., Kay, S.M. and Kay, R.W., 1981. Ultramafic xenoliths from Kanaga Island: insight into mantle composition and dynamics below the island arc (abstract). EOS, Trans. Am. Geophys. Union, 62: 1092.

Pope, R.R., Kay, R.W. and Kay, S.M., 1983. Thermal structure and lithosphere—asthenosphere boundary beneath an oceanic island arc. Tectonics, in prep.

Powell, M., 1978. Crystallization conditions of low-pressure cumulate nodules from the Lesser Antilles island arc. Earth Planet. Sci. Lett., 39: 163—172.

Reed, S.J.B., 1975. Electron Microprobe Analysis. Cambridge Univ. Press, Cambridge, England, 400 pp.

Reid, M. and Nye, C., 1981. Geochemistry of least fractionated basalts from Okmok volcano, central Aleutians (abstract). EOS, Trans. Am. Geophys. Union, 62: 1092.

Sakuyama, M., 1978. Petrographic evidence of magma mixing in Shirouma-Oike volcano, Japan. Bull. Volcanol., 41: 501—512.

Sato, H., 1977. Nickel content of basaltic magmas: Identification of primary magmas and a measure of the degree of olivine fractionation. Lithos, 10: 113—120.

Shimizu, N. and le Roex, A., 1982. Trace element zoning of augite phenocrysts in alkalic basalt from Gough island (abstract). EOS, Trans. Am. Geophys. Union: 63: 453.

Sparks, S.R.J., Sigurdsson, H. and Wilson, L., 1977. Magma mixing: a mechanism for triggering acid explosive eruptions. Nature, 267: 315—318.

Stern, C.R., 1979. Open and closed system igneous fractionation within two Chilean ophiolites and the tectonic implications. Contrib. Mineral. Petrol., 68: 243—258.

Stewart, D.C., 1975. Crystal clots in calc-alkaline andesite as breakdown products of high-Al amphiboles. Contrib. Miner. Petrol., 53: 195—204.

Yoder, H.S. and Tilley, C.E., 1962. Origin of basaltic magmas: an experimental study of natural and synthetic rock systems. J. Petrol., 3: 342—532.

Zashu, S., Kaneoka, I. and Aoki, K., 1980. Sr isotope study of mafic and ultramafic inclusions from Itinome-gata, Japan. Geochemical J., 14: 123—128.

Journal of Volcanology and Geothermal Research, 18 (1983) 297—320
Elsevier Science Publishers B.V., Amsterdam — Printed in The Netherlands

PETROLOGY OF ARC VOLCANIC ROCKS AND THEIR ORIGIN BY MANTLE DIAPIRS

M. SAKUYAMA

Geological Institute, Faculty of Science, University of Tokyo, Hongo, Bunkyo-ku, Tokyo 113 (Japan)

(Received October 20, 1982)

ABSTRACT

Sakuyama, M., Petrology of arc volcanic rocks and their origin by mantle diapirs. In: S. Aramaki and I. Kushiro (Editors), Arc Volcanism. J. Volcanol. Geotherm. Res., 18: 297—320.

Calc-alkalic chemical trends characteristic of arc volcanic rocks mainly result from three mechanisms which act additively: (1) fractional crystallization involving separation of titanomagnetite; (2) selective concentration of plagioclase phenocrysts and selective depletion of titanomagnetite phenocryst compared with the actually fractionated proportion; and (3) mixing of magmas on continuous fractionation trends. The association of calc-alkalic and tholeiitic trends in a single composite volcano may not represent different fractional crystallization processes or different chemistries of primary magmas, but the calc-alkalic chemical trend can be considered as a mixing trend resulting from mixing of various magmas on associated tholeiitic chemical trends. Chemical variations of most arc volcanic rocks, including calc-alkalic ones, can accordingly be essentially accounted for by the low-pressure fractional crystallization of phenocrystic phases from primary basaltic magmas.

Crystallization sequences of arc magmas which are strongly dependent on water content in magmas are deduced from the phenocryst assemblages. The crystallization sequence changes laterally across-arc, suggesting increasing water contents in magmas toward the back-arc side, as is also seen for other incompatible elements such as K and Rb. Systematic differences in the characteristic crystallization sequence are also observed among arcs, roughly correlating with the crustal thickness. Water content in magma, like other incompatible elements, tends to increase with increasing crustal thickness. The variation of incompatible elements including water roughly represents that of the degree of partial melting of the upper mantle, which is broadly controlled by the crustal thickness.

The variation of water content indicates that arc magmas are not saturated with water during differentiation to late differentiates such as dacite or rhyolite. This strongly constrains the maximum water contents in primary basaltic magma, at most 2.5 wt.%. This value suggests that magma generation beneath arcs is dependent on dry solidus of peridotite. Diapiric uprise of the hot deeper mantle and associated adiabatic decompression would be necessary for mantle peridotite to attain the temperature as high as dry solidus. Diapirs that begin to rise from the subduction zone may stop at or near the crust—mantle boundary because of the surrounding density change, and their degree of partial melting is roughly controlled by their stopped depth assuming their similar temperature. Across-arc variation is also explained by the stopped depth of diapirs, but is not controlled by crustal thickness.

0377-0273/83/$03.00

INTRODUCTION

The source materials of "arc magma" (a general term designating magmas at convergent plate boundaries in this paper) have been discussed by several authors from the view point of trace-element abundances, their ratios and isotopic ratios (e.g., Perfit et al., 1980; Arculus and Johnson, 1981; Gill, 1981). The main components affecting the trace-element and isotopic features of arc magmas are subducted oceanic crust composed of oceanic sediments and altered or unaltered MORB, wedge upper mantle and crust whose contribution widely changes in different arcs (Kay, 1980; Arculus and Johnson, 1981; Gill, 1981; Langmuir and Goldstein, 1981)..

In spite of the wide variety of source components, the major chemical features, especially that of basalts, seem to be rather homogeneous with some systematic differences (Perfit et al., 1980; Arculus and Johnson, 1981), suggesting that similar processes act on the generation of the magmas in various arcs since major chemical compositions are largely controlled by the phase relations which are strongly dependent on the physical conditions during melting and differentiation. Nevertheless, major chemical compositions and petrologic features of differentiated varieties are systematically different in different arcs (Kuno, 1966; Miyashiro, 1974; Coulon and Thorpe, 1981; Gill, 1981; Sakuyama, in press). The variations of arc volcanic rocks must be dependent on processes during differentiation of the arc magmas which might reflect the slight difference in chemical and physical conditions of the magma-generation process.

Petrologic features and variations of arc magmas are synthesized and their implications for magma-generation and differentiation processes are discussed from the broader view point in this paper. Based on these discussions, the importance of mantle diapirs on the generation of arc magmas will be emphasized.

ORIGIN OF CALC-ALKALIC TREND

Effect of magma mixing

The "calc-alkalic chemical trend" is one of the most characteristic chemical features of arc volcanic rocks. In spite of the confusion on the definition of this trend, the essential characteristic is a suppressed iron-enrichment with respect to the increasing silica content which is expressed in the FeO*/MgO-SiO diagram (Miyashiro, 1974). Many hypotheses were proposed for the interpretation of the trend, for example, magnetite fractionation (Osborn, 1959), hornblende fractionation (Cawthorn and O'Hara, 1976; Boettcher, 1977), eclogite fractionation (Green and Ringwood, 1968) and magma mixing (Eichelberger, 1975). Among these, the fractionation of magnetite is likely to be the most essential mechanism for generating the calc-alkalic trend since this mechanism seems to give the most consistent explanation for the chemical and petrographic observations (e.g., Gill, 1981).

The understanding about the calc-alkalic trend, however, is still not enough, since the problem of the association of calc-alkalic and tholeiitic trends in a single composite volcano remains. As an explanation for this association, two principal mechanisms were proposed; two different fractionation series from the common basaltic parent depending on the different f_{O_2}, water content or degree of interaction with crustal materials (e.g., Kuno, 1950, 1968a), and different basaltic parents for each fractionation series (e.g., Kushiro and Sato, 1978; Masuda and Aoki, 1979).

Recently, Sakuyama (1981) studied volcanic rocks of Myoko and Kurohime volcanoes, central Japan, within which calc-alkalic and tholeiitic magmas are erupted from a common vent. Volcanic rocks of the Myoko and Kurohime volcanoes show a variety ranging from high-alumina basalts according to Kuno (1960)'s definition, through two-pyroxene basic andesites to hornblende andesites. The SiO_2 content ranges from 49.5 to 63.9 wt.%, belonging to medium-K variety, 0.87 to 2.13 wt.% K_2O, according to Gill (1981).

From a petrographic view point, the volcanic rocks were divided into two types; N-type and R-type. All the mafic phenocrysts show the normal zoning on their rim and groundmass pyroxenes are enriched in Fe compared with the cores of the phenocrysts in the N-type rocks. The R-type rocks contain mafic phenocrysts with reverse zoning on their rim. Groundmass pyroxenes are often more enriched in Mg than the core of the phenocrysts. In some samples, two kinds of phenocrysts coexist, one with normal zoning and the other with reverse zoning. The Fe-Mg apparent partition coefficients $[(1 - X_{Mg})/X_{Mg}]^{\text{olivine}}/[(1 - X_{Mg})/X_{Mg}]^{\text{pyroxene}}$ between phenocrystic olivine and pyroxenes are plotted in Fig. 1, in which each point represents a rock sample. In N-type rocks phenocrystic olivine and pyroxenes are in equilibrium relationship with respect to Mg-Fe partitioning. In R-type rocks, they show a disequilibrium relationship. Although most of the pyroxene phenocrysts of the R-type plotted in Fig. 1 show reverse zoning, there are some normally zoned magnesian augite phenocrysts which can coexist with olivine in equilibrium relationship.

Chemical compositions of plagioclase phenocrysts also show the equilibrium and disequilibrium relationships, respectively in N-type and R-type rocks. The wide compositional variation (An_{90} – An_{50}) in R-type rocks is notable in contrast with the narrow one in N-type rocks. Bimodal distribution, An_{80-90} and An_{50-60}, is also common in R-type rocks. Groundmass plagioclases of R-type rocks, in general, have the composition within the variation of phenocrysts.

Based on these petrographic observations, Sakuyama (1981) concluded that N-type rocks are made from simple cooling of the stagnant magma in the reservoir where all the phenocrysts are nearly in equilibrium. On the other hand, disequilibrium petrographic features of R-type rocks are concluded to be formed by mixing of high-T basic magma and low-T acidic

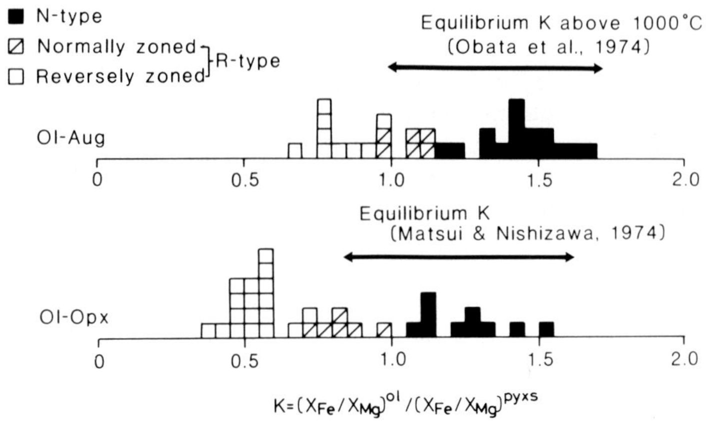

Fig. 1. Apparent partition coefficient of Mg and Fe between olivine and pyroxene phenocrysts in Myoko and Kurohime volcanoes. Normally zoned olivine and pyroxenes in N-type and R-type rocks are in equilibrium relation. Olivine and reversely zoned pyroxenes in R-type rocks are not in equilibrium. Data are taken from Sakuyama (1981).

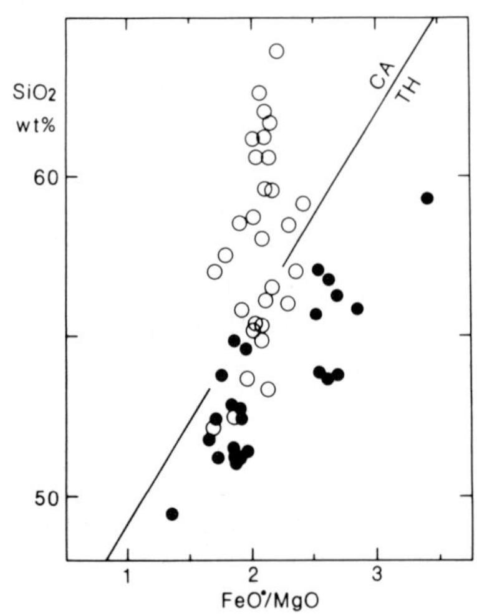

Fig. 2. FeO*/MgO-SiO_2 diagram. Closed circles are N-type rocks; open circles are R-type rocks. Boundary between CA and TH series is taken from Miyashiro (1974). Apparent two differentiation trends can be seen, but that of the R-type is just a mixing trend.

magma before eruption. Therefore, the whole-rock chemical trend shown by R-type rocks is not a differentiation trend but possibly a mixing trend. This is partly proven by the strong correlation between whole-rock major chemistry and the compositional variation of plagioclase phenocryst.

Whole-rock major chemical compositions are plotted on a FeO*/MgO vs. SiO_2 diagram (Fig. 2). On this diagram, each type makes a different chemical trend; N-type rocks make the iron-enrichment trend compared to the suppressed iron-enrichment trend of R-type rocks. Without any petrographic data, one might conclude that both tholeiitic and calc-alkalic differentiation trends coexist in Myoko volcano. As shown above, however, "calc-alkalic trend" is most likely a mixing trend.

Moreover, from the examination of the genetic relationships among magmas before mixing, Sakuyama (1981) concluded that mixing has occurred between magmas of various stages on the continuous crystallization differentiation course corresponding to the gradual decrease of magmatic temperature. Therefore, the calc-alkalic trend can be made by the mixing of cogenetic magmas whose differentiation trend is initially tholeiitic. After the development of the tholeiitic trend (Fig. 2), FeO*/MgO ratio must be constant or decrease slightly for producing the silicic end products in mixed magmas. This is possible because of the effective fractionation of FeO* by titanomagnetite.

In northeastern Japan as is the case of the Myoko volcano group, calc-alkalic rock series are commonly associated with tholeiitic (including high-alumina basaltic) or alkali olivine basaltic rock series (Kuno, 1968a; Masuda and Aoki, 1979). Calc-alkalic rocks of northeastern Japan have some distinct contrasts to rocks of the associated series. The well-known suppressed iron-enrichment trend on AFM or FeO*/MgO vs. SiO_2 diagrams, and the higher contents of highly compatible elements, such as Ni, Cr and Mg at the comparable contents of incompatible elements such as Rb, Ba, K and Th (Katsui et al., 1978; Masuda and Aoki, 1979) of calc-alkalic ones, are distinct chemical contrasts indicating the impossibility of the generation of calc-alkalic magma through crystallization differentiation from magma of associated series. From petrographic view point, groundmass Ca-poor pyroxenes are hypersthene in calc-alkalic rocks in contrast to pigeonite or to Ca-poor pyroxene in associated series. The equilibration temperatures of phenocrystic phases are generally higher in tholeiitic than in calc-alkalic rock series at the comparable Fe/Mg ratio of minerals (Ishii, 1975). Various disequilibrium petrographic features such as plagioclase phenocrysts with dusty zones or honey-comb structures, coexistence of Mg-rich olivine and quartz phenocrysts and reverse zonings on the rim of phenocrysts of plagioclase and orthopyroxene are common in calc-alkalic rock series (Kuno, 1936, 1950; Ishikawa, 1958; Yoder, 1969; Wada, 1981).

In spite of these chemical and petrographic contrasts, the presence of the chemical coherence between calc-alkalic and associated series must be emphasized. K_2O contents in volcanic rocks at comparable SiO_2 content increase toward the back-arc side in northeastern Japan not only in calc-alkalic andesites or dacites but also in basalts of associated series. The chemical differences shown in different volcanoes are generally larger than those between different rock series in a single composite volcano. This suggests that all the volcanic rocks of a single volcano are derived from common parental basaltic magma through crystallization differentiation. This is consistent with the constant ratios of Sr and O isotopes in a single cone (Katsui et al., 1978; Matsuhisa, 1979). The above-mentioned contradictory petrological features of calc-alkalic and associated series can be totally interpreted by the internal mixing processes as revealed in the Myoko volcano. Apparently, two distinct chemical trends do not necessarily indicate two discrete differentiation trends, but calc-alkalic is a mixing trend and all the rocks of a single composite volcano are cognate.

Concentration of phenocrysts

The effect of magma mixing was evaluated in the previous section with special reference to the problem of the association of the calc-alkalic and tholeiitic series. However, in the considerations mentioned above still remains the unevaluated assumption that the whole-rock bulk chemical compositions correctly represent the liquid line of descent. According to Ewart (1976), 70% of the arc lavas compiled by him have a total phenocryst content of more than 20 vol. % and such a porphyritic nature seems to be one of the distinctive characteristics of arc volcanic rocks.

The results of the modal analysis on the N-type rocks of the Myoko and Kurohime volcanoes and those of the Adatara volcano (Fujinawa, 1980), are shown in Fig. 3. The total volume of mafic phenocrysts such as olivine, pyroxenes and titanomagnetite, is rather constant in spite of the wide variation in volume ratio of plagioclase phenocrysts and groundmass. The volume of titanomagnetite phenocryst in all the phenocryst is also nearly constant and less than 5% against the wide variation in the ratio of plagioclase and mafic silicate phenocrysts. These tendencies are crudely consistent with the accumulation of plagioclase phenocryst and the depletion of titanomagnetite relative to mafic silicate phases as will be discussed later.

Based on these modal data and whole-rock chemical compositions, the chemical compositions of the groundmass have been calculated. Known chemical compositions of phenocrystic phases are used in the case of Myoko volcano (Sakuyama, 1981). For the Adatara rocks, chemical compositions and densities of phenocrysts are fixed as follows; An_{70} and ρ=2.72 for plagioclase, Fo_{70} and ρ=3.57 for olivine, n_{60} and ρ=3.44 for orthopyroxene, Wo_{40}, En_{40}, Fs_{20} and ρ=3.37 for augite, Usp_{60} and ρ=4.95 for titanomagnetite and ρ=2.42 for magmatic melt. Calculated groundmass compositions are plotted on an AFM diagram (Fig. 4). Although these data points show a large scatter due to the low accuracy of the modal analysis, it is notable that they all fall in the Fe-rich region compared to the whole-rock data. The deviation from the whole-rock trend is conspicuous in the rocks with a total phenocryst content larger than 30 vol. %. This systematic difference indicates that the FeO*/MgO ratio of the presently observed total phenocrysts is smaller than the ratio of bulk solid actually subtracted from the magmatic melt for generating the liquid line of descent (=the chemical trend of groundmass). If the proportion of the observed phenocrysts keeps that of crystals which are already crystallized, whole-rock bulk chemical compositions must make a continuous chemical trend with the groundmass compositions. It is most

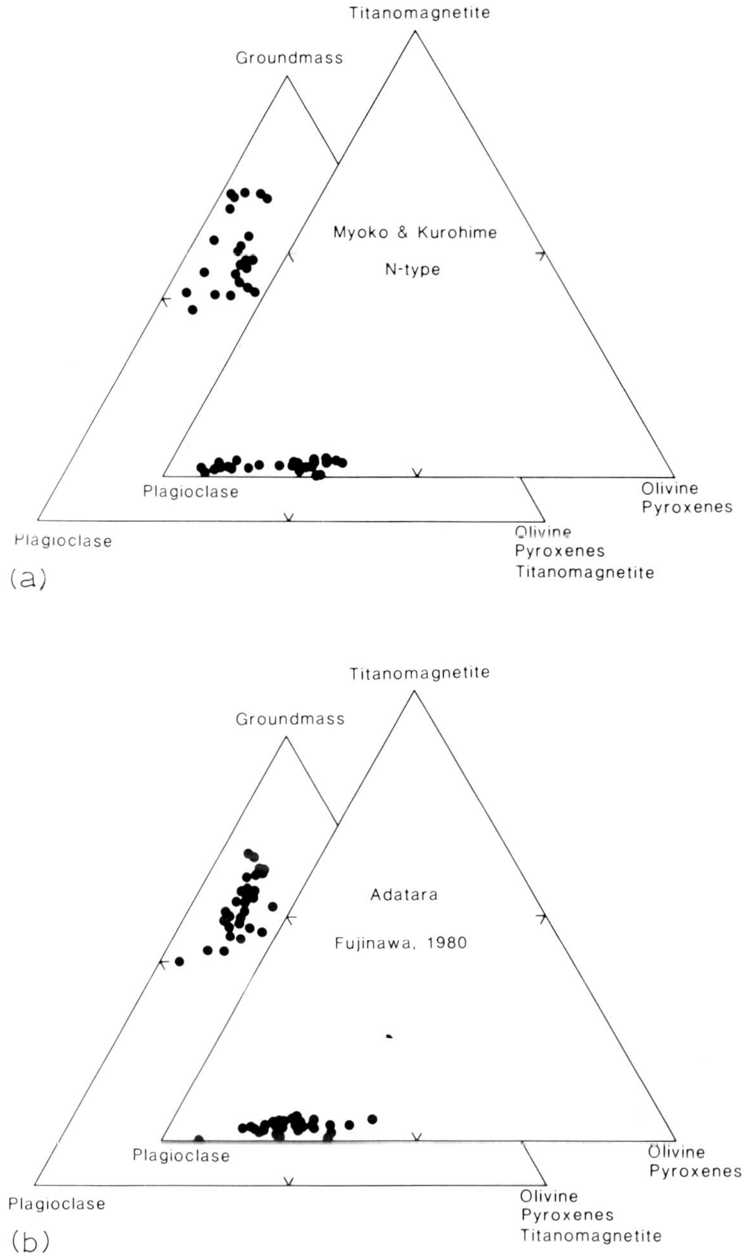

Fig. 3. Volume proportion of phases in volcanic rocks plotted on groundmass-plagioclase-mafic phases and titanomagnetite-plagioclase-mafic silicates triangular diagrams. a. N-type rocks of the Myoko and Kurohime volcanoes. b. Rocks of Adatara volcano, northeastern Japan (Fujinawa, 1980).

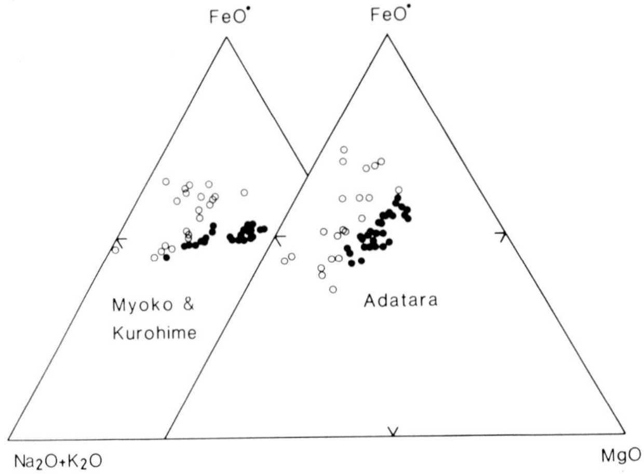

Fig. 4. Whole-rock and calculated groundmass compositions of volcanic rocks of the Myoko, Kurohime and Adatara volcanoes on AFM diagram. Detailed procedure for groundmass calculation is seen in the text. Solid circles are whole-rock compositions; open circles are calculated groundmass compositions.

reasonable to consider that titanomagnetite is selectively depleted in the phenocrysts compared with the volume proportion with which the magma has fractionated.

The concentration of plagioclase phenocryst, which is the most dominant phenocrystic phase in arc volcanic rocks (Ewart, 1976), also shifts the whole-rock chemical compositions toward calc-alkalic affinity. Since plagioclase is free from Fe and Mg, the concentration of plagioclase phenocrysts, especially sodic ones, enlarge the abundance ratio of alkalies relative to Fe and Mg in the whole-rock composition without changing their ratio.

Ewart (1976) pointed out the negative correlation between the total phenocryst content and FeO*/MgO ratio of whole-rock bulk composition of circum-Pacific arc volcanic rocks. This strongly suggests that the selective depletion of phenocrystic titanomagnetite and concentration of phenocrystic plagioclase commonly take place in arc magmas.

In summary, three processes would be operating additively to make the so-called calc-alkalic chemical trend. Firstly, crystallization of titanomagnetite suppresses the iron-enrichment in magmatic melt. Effective fractionation of titanomagnetite would be favored by higher contents of water and alkalies and resulting in higher f_{O_2} in arc magmas compared with other provinces. The crystallization of titanomagnetite would possibly change within arc magmas according to the variation of water and alkalies contents. Secondly, the modal proportion among phenocrystic phases does not truly represent the proportion of crystallized minerals: depleted titanomagnetite and concentrated plagioclase. Accordingly, the FeO*/MgO ratio of porphyritic bulk composition becomes lower, and thus more calc-alkalic. Thirdly, mixing of magmas on a continuous curved chemical trend on a AFM diagram makes a mixing trend with a much lower FeO*/MgO ratio.

CRYSTALLIZATION SEQUENCE AND VARIATION OF WATER

Variation of crystallization sequence

As most arc magmas are essentially differentiated from basaltic parents through low-pressure crystallization differentiation of phenocrystic phases as previously discussed, the crystallization differentiation process can be discussed from the standpoint of crystallization sequence of minerals which can be deduced from phenocryst assemblages. Only a few general considerations on crystallization sequence of minerals in arc magmas have been stated (e.g., Best, 1963; Kuno, 1968b; Cawthorn, 1976a; Marsh, 1981), but the comparative works on its differences and their petrologic implications have not been attempted except for the late differentiates (Sakuyama, 1978a, in press). In this paper, all crystallization sequences from basalt through andesite and dacite to rhyolite are considered and four typical sequences which are believed to cover almost all variations of crystallization sequences of arc magmas are proposed mainly based on the results of Japanese volcanic rocks.

The disequilibrium phenocryst assemblages must be carefully divided into equilibrium pairs (Sakuyama, 1981). Since olivine phenocrysts in arc volcanic rocks are generally magnesian above Fo_{50} (Ewart, 1976), the assemblage of olivin and quartz undoubtedly represents disequilibrium. Olivine-hornblende assemblage is also considered to be in disequilibrium since all the petrographically well-analyzed hornblende andesites or dacites containing olivine phenocyrsts are, without exception, products of magma mixing of andesitic-to-dacitic magma and basaltic magma (Eichelberger, 1978; Sakuyama, 1979b, 1981; Luhr and Carmichael, 1980; Heiken and Eichelberger, 1980). Equilibrium coexistence of olivine and hornblende is possible in water-saturated basaltic magma (Yoder and Tilley, 1962; Cawthorn et al., 1973; Helz, 1973; Allen et al., 1975; Cawthorn, 1976b) at pressures of about 5 to 20 kbar. In all these experiments, augite is associated with them in contrast to their common coexistence with orthopyroxene in natural andesites and dacites.

The typical crystallization sequences are constructed by arranging the deduced equilibrium phenocryst assemblages as each mineral crystallizes continuously (Kuno, 1968b) (Fig. 5). The type-I sequence is deduced mainly from phenocryst assemblages of Zao volcano on the volcanic front. In this volcano, olivine-orthopyroxene-plagioclase basalt, olivine-augite-orthopyroxene-plagioclase (-titanomagnetite) andesite, augite-orthopyroxene-plagioclase (-titanomagnetite) andesite, and augite-orthopyroxene-plagioclase-quartz-titanomagnetite dacite are observed (Ichimura, 1951, 1953, 1955, 1957, 1960). They correspond to the phenocryst assemblages with decreasing temperature in above order. While the assemblage which represents a temperature lower than the augite-orthopyroxene-plagioclase-quartz titano magnetite assemblage can not be seen in this volcano, it would be repre-

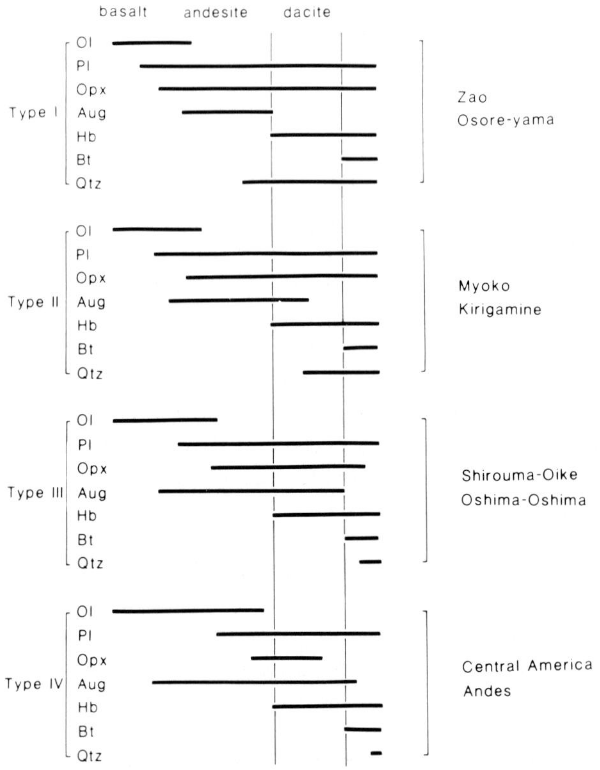

Fig. 5. Typical crystallization sequences of arc magmas. Those of olivine (Ol), plagioclase (Pl), orthopyroxene (Opx), augite (Aug), hornblende (Hb), biotite (Bt) and quartz (Qtz) are indicated. Water content of magmas would increase from type I sequence to type IV sequence. Beginnings of crystallization of hornblende and biotite are set in the same position since they are inferred to be constant on temperature, about 950 and 850°C respectively.

sented by the orthopyroxene-plagioclase-quartz-hornblende-titanomagnetite assemblage associated with augite-orthopyroxene-plagioclase-quartz-titanomagnetite assemblage in other volcanoes, for example Osore-yama volcano (Togashi, 1977). Biotite containing the assemblage shown in Fig. 5 which is deduced from the results of hydrous experiments on andesite or dacite (e.g., Maaløe and Wyllie, 1975; Wyllie, 1977) can not actually be observed.

The type-II sequence is taken from the results of the Myoko and Kurohime volcanoes (Sakuyama, 1981). The biotite-bearing assemblage which is not actually seen in the Myoko volcano is deduced from the general coexistence of biotite with orthopyroxene, hornblende and quartz in rocks associated with Myoko-type quartz-bearing assemblage in other volcanoes, for example, Kirigamine volcano (Yamazaki, 1965a,b). Two different points from the type-I sequence are obvious. First, the beginning of crystallization of orthopyroxene precedes that of augite in the type-I sequence, but it is reversed in the type-II sequence. Second, quartz begins to crystallize later than hornblende in the type-II sequence in contrast to the type-I sequence.

The type-III sequence is taken from that of Shirouma-Oike volcano, central Japan (Sakuyama, 1979b). There are two distinct points from the type-II sequence. One is that augite begins to crystallize earlier than plagioclase in the type-III sequence. This is deduced from the olivine-augite (± plagioclase) assemblage of basalts which contaminate biotite-quartz-bearing dacites. The other is that quartz crystallizes later than biotite in this sequence.

The type-IV sequence is essentially the same as the type-III sequence, but orthopyroxene ceases to crystallize earlier than augite. This is deduced from the augite-hornblende-plagioclase-titanomagnetite and augite-hornblende-biotite-plagioclase-titanomagnetite assemblages which do not coexist with orthopyroxene. These assemblages are rarely observed in andesite and dacite of high alkalies contents; Central America (Mooser et al., 1958; Williams, 1960), Andes (Hantke and Parodi, 1966; Francis et al., 1974), Papua New Guinea (Fisher, 1957; Taylor, 1958), Bougainville Islands (Blake and Miezitis, 1968; Blake, 1968), Indonesia (Neuman van Padang, 1951), Kamchatka (Vlodavetz and Piip, 1959), Aleutian (Coats, 1952; Byers, 1959). Since all these assemblages do not coexist with quartz, the liquidus of quartz is strongly depressed below that of biotite.

The implications of the variation of crystallization sequence, especially that of quartz, hornblende and biotite, are discussed in Sakuyama (1979a, in press). According to the results of hydrous experimental works on andesite or dacite (e.g., Piwinskii and Wyllie, 1968; Eggler, 1972; Eggler and Burnham, 1973; Maaløe and Wyllie, 1975; Wyllie, 1977), liquidus temperatures (= upper temperature stability limit) of hornblende and biotite are approximately independent of the total pressure within 2 to 10 kbar, bulk chemical composition within normal calc-alkalic andesite to dacite, and water content, being about 950°C and 850°C, respectively. Based on the concept that the temperatures of beginning of crystallization of hornblende and biotite are essentially constant, the left ends of the bars of hornblende and biotite in Fig. 5 are fixed in all the sequences.

On the other hand, liquidus temperature of quartz is drastically depressed with increasing water content (Whitney, 1975; Maaløe and Wyllie, 1975; Wyllie, 1977). The position of beginning of crystallization of quartz in comparison with those of hornblende and biotite in Fig. 5, therefore, strongly depends on the water content in magma. Detailed discussion on the effect of other elements, especially K, is seen in Sakuyama (in press).

All variations of crystallization sequence can be consistently interpreted by the variation of water content in magma. The lower crystallization temperature of plagioclase than that of pyroxenes in type-III and -IV sequences would be a result of enriched water, and thus depressed plagioclase liquidus (Yoder, 1969). Orthopyroxene begins to crystallize at a higher temperature than augite in the type-I sequence. Since there is no systematic difference in major-element chemistry between basalts with the type-I sequence and those with the type-II sequence except for a slight difference in K_2O contents (Kawano et al., 1961), this change should not be the effect of bulk major chemistry. According to Kushiro (1969), the primary phase volume of augite and olivine expands against that of orthopyroxene by the introduction of water into the diopside-forsterite-silica system. The reversal of the crystallization order of orthopyroxene and augite is consistent with this experimental result; higher water content in magma of the type-II sequence enhances the stability of augite and olivine, resulting in the shrinkage of orthopyroxene phase volume.

The variation of water content also seems to affect the lower temperature stability limits of pyroxenes. The stability of augite expands downward with increasing water. Augite ceases to crystallize at or above the liquidus temperature of hornblende in the type-I sequence. In contrast to this, augite is still stable at or below the liquidus temperature of biotite in the type-III and -IV sequences. On the other hand, the lower temperature stability limit of orthopyroxene becomes higher from type-III to type-IV sequence and is reversed with that of augite. Augite seems to have a reaction relation with quartz when they coexist with hornblende. Augite may react with a liquid to precipitate hornblende and quartz. If the stability field of quartz might shrink with increasing water content, the lower stability limit of augite may also enlarge. The disappearance of orthopyroxene without any reaction relations with other phases is problematical (Cawthorn, 1976a).

Regional distribution

If volcanic rocks composing a single composite cone are assumed to be comagmatic as previously discussed, the type of the crystallization sequence of magma of the single composite cone can be deduced from the distinct phenocryst assemblages as to the coexisting relations among hornblende, biotite and quartz (Fig. 5). The type-I sequence is characterized by assemblage of quartz without hornblende and biotite. The type-II sequence is characterized by two assemblages; hornblende without quartz and biotite, and hornblende-quartz without biotite. The type-III sequence including the type-IV sequence is characterized by the assemblage of hornblende-biotite without quartz. Only the assemblage of quartz-hornblende without biotite characterizes both the type-I and -II sequences, but distinguishes it from the type-III sequence. Only the assemblage of hornblende without quartz and biotite characterizes both the type-II and -III sequences, but distinguishes it from the type-I sequence. They are named the type I—II and type II—III, respectively.

Volcanoes are classified into 5 types based on the published petrographic data mentioned above. It is needless to mention that the classification is given up when there is no rock containing hornblende, biotite or quartz in a volcano. Figure 6 shows the result of the compilation on the northeastern Japan arc. Volcanoes of the types I, I—II and II are distributed along the vol-

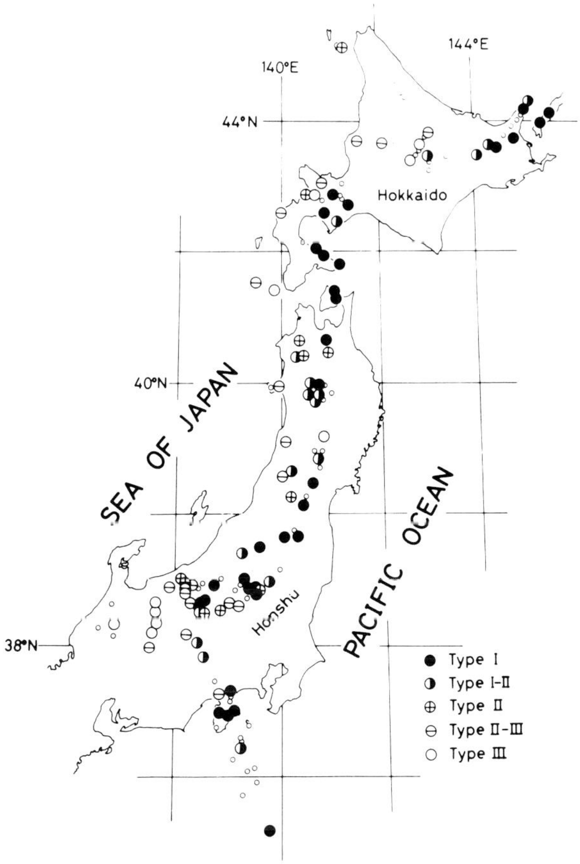

Fig. 6. Distribution of crystallization sequences in Quaternary volcanoes of northeastern Japan. Types I, I—II and II sequences are distributed along the volcanic front; type II—III and III sequences are on the back arc side (modified from Sakuyama, 1979a).

canic front and those of the types II—III and III are in the back-arc side. From those observations, Sakuyama (1979a) concluded that water contents in magmas of northeastern Japan increase away from the volcanic front likewise so-called incompatible elements such as K, Rb, Sr, Ba, Th, U. Figure 7 shows the positive correlation between the type of volcano and the K 60 value (K_2O contents at 60% SiO_2) determined by Ui and Aramaki (1978). These lateral variations can be most reasonably interpreted by the gradual decrease of degree of partial melting of the upper mantle toward back-arc side (Sakuyama, 1979a). The lateral change of the incompatible-element abundances across of partial melting tends to decrease toward the back-arc side also in general arcs.

The result of compilation of the circum-Pacific area is shown in Fig. 8. The number of data points is too small to clarify the inter-arc lateral varia-

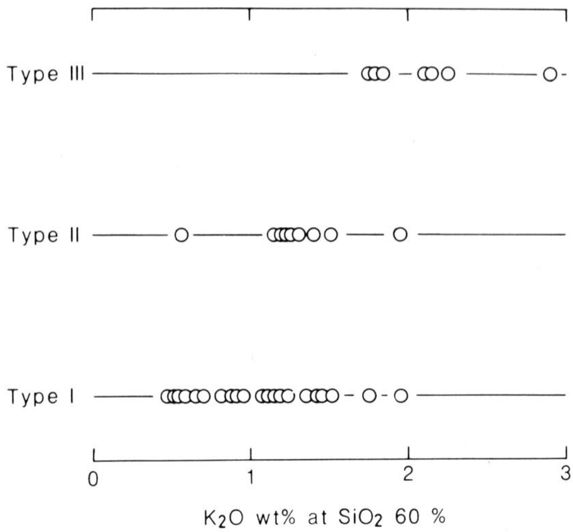

Fig. 7. Correlation between the type of crystallization sequence and K_2O wt.% at 60 wt.% SiO_2.

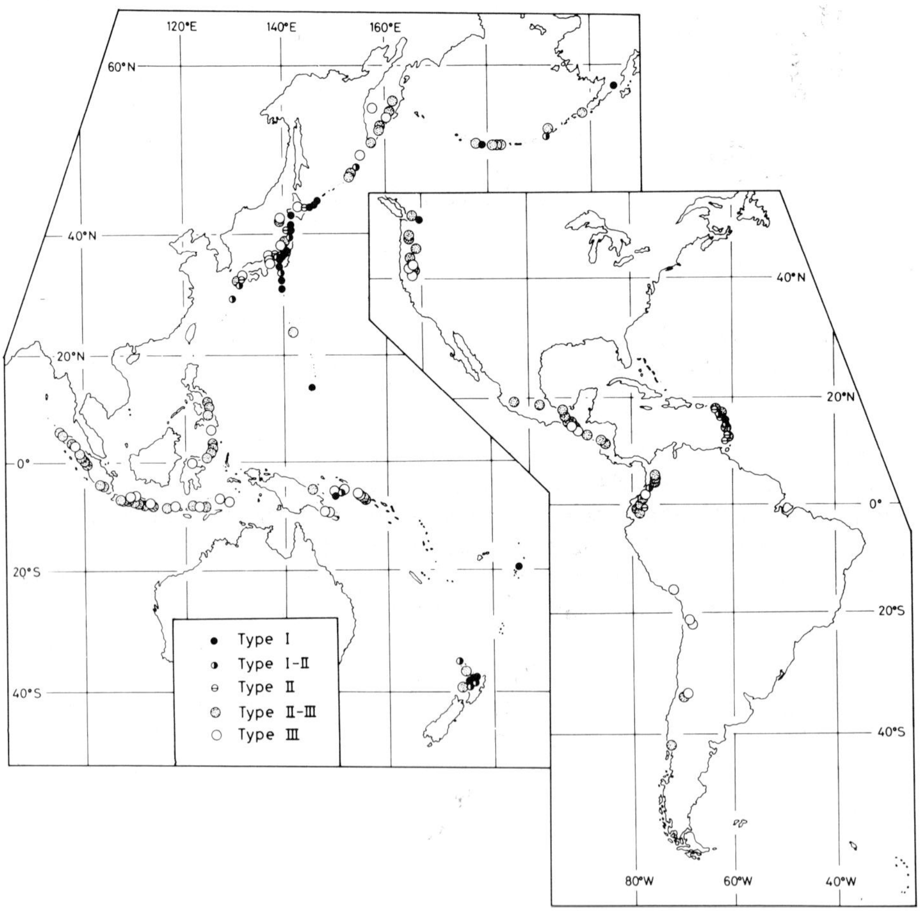

Fig. 8. Distribution of crystallization sequences in Quaternary volcanoes of circum-Pacific arcs (after Sakuyama, in press).

tion as shown in Fig. 6, but the intra-arc variation can be clearly seen. There are few or no type I or I—II volcanoes in the Andes and Indonesia arcs. In contrast, Northeastern Japan and Izu-Mariana arc, are dominated by type I or I—II. Sakuyama (in press) classified arcs into 3 types based on the frequency distributions of volcano type; types A, B and C dominated by volcanoes of types I, II and III, respectively. Due to the previous considerations, the water content in characteristic magma to each arc increases from type-A arc to type-C arc. This tendency is consistent with the variation of the average K_2O content at fixed SiO_2 contents. The close positive correlation between arc type and average K_2O contents at SiO_2=57.5% (Gill, 1981) is shown in Fig. 9, indicating that water in magma behaves just like K and other incompatible elements also in intra-arc variation.

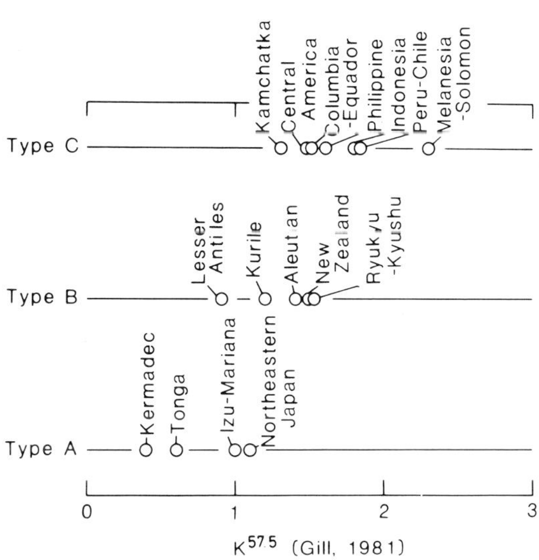

Fig. 9. Correlation between the type of arc and average K_2O content at 57.5 wt.% SiO_2 taken from Gill (1981).

Crustal thickness control

As previously shown, most of the arc magmas are derived from parental basaltic magmas through crystallization differentiation, especially fractionation of olivine, pyroxenes and magnetite (Fig. 5) (Gill, 1981). Therefore, relative abundances of K_2O, water and other incompatible elements in andesitic or dacitic magmas could be a reflection of those in primary basaltic magmas. Since most of the incompatible elements including water would be partitioned into melt phase during melting in the mantle at least in a conceivable degree of partial melting (>10%), their abundances in magmas are dependent on the degree of partial melting of the upper mantle if the upper mantle is roughly homogeneous with respect to these elements.

Abundance of the incompatible elements is correlated to thickness of the crust through which magma was extruded (Miyashiro, 1974; Ui and Aramaki, 1978; Coulon and Thorpe, 1981; Gill, 1981). Arcs with thick crusts are characterized by magmas with enriched incompatible elements. Correlation between a distribution of the type of crystallization sequence and the crustal thickness is inevitably resulted from a positive correlation between water and K_2O contents in magma as previously shown (Figs. 7 and 9) (Sakuyama, in press). These correlations strongly suggest that the degree of partial melting of the upper mantle is roughly controlled by the crustal thickness. The across-arc change of the degree of partial melting, however, can not be accounted for by the crustal thickness control since no systematic increase of the crustal thickness of the arc can be observed.

There remains a possibility that the above correlation might be caused by the interaction of the mantle-derived magma with crustal material (Coulon and Thorpe, 1981; Gill, 1981). In general, arc volcanic rocks on continental margins have higher $^{87}Sr/^{86}Sr$ and $^{18}O/^{16}O$ (e.g., Gill, 1981), suggesting the possible interaction of magmas with continental crust with respect to other elements in addition to Sr and O. Intra- and inter-arc variation of incompatible elements in many arcs, especially across-arc variation, is coherently observed on whole compositional variation from basalt through andesite to dacite, that is, basalt associated with high-K andesite or dacite is also high-K. If the variation of incompatible elements should be attributed to the interaction with crustal material, the interaction should have essentially taken place at least in the basaltic stage of the magma. Since crustal material enriched in incompatible elements is granodioritic to granitic, the interaction must be selective so as not to cause the appreciable change of major chemistry. Briqueu and Lancelot (1979) and Francis et al. (1980) showed that Sr isotope variation can not be explained by the bulk contamination of crustal material but Sr is selectively contaminated. However, it is impossible to enrich all the incompatible elements including water as much as five times by the selective contamination. The variation of incompatible elements thus should be considered to reflect the difference of degree of partial melting of the upper mantle.

ARC MAGMAS BY MANTLE DIAPIR

Water contents in primary arc magmas

Arc magmas, especially calc-alkalic ones, used to be considered hydrous due to their explosive eruptive fashion and common occurrence of hydrous phenocrysts in them (e.g., Yoder, 1969). Along this context, the possibility that primary arc magma itself originates by the partial melting of the upper mantle under relatively wet conditions has been discussed (e.g., Kushiro, 1974). Some of the high-Mg andesites such as boninite in Mariana (Kuroda et al., 1978), sanukite in Southwestern Japan (Tatsumi and Ishizaka, 1981) and

Papuan clinoenstatite andesite (Dallwitz et al., 1966) might be actually originated from wet partial melting of the upper mantle. However, the experimental reproduction of the natural crystallization sequences (e.g., Eggler, 1972; Eggler and Burnham, 1973; Maaløe and Wyllie, 1975; Sekine et al., 1979) and thermodynamic considerations (e.g., Ewart et al., 1975; Luhr and Carmichael, 1980) clarified that most of the arc magmas are not saturated with water. The variation of water content in magmas which is reflected in the difference in crystallization sequence also supports the water-undersaturation of arc magmas. Moreover, arc magmas are not saturated with water even in the later differentiates such as dacites. Considering that most of arc magmas are, as previously shown, differentiated from primary basaltic magmas through fractionation mainly of olivine, pyroxenes, plagioclase and titanomagnetite, the role of water during generation of arc magmas is strongly constrained (Sakuyama, in press).

Mg-rich olivine and plagioclase, generally An-rich one, coexist in arc basaltic magmas (Fig. 5). Since other phenocrysts would crystallize at the similar depth to them, the depth of the differentiation of arc magmas is at most 10 kbar. The solubility limit of water in andesitic to dacitic melt at 10 kbar is about 10 wt.% (Hamilton et al., 1964; Sakuyama and Kushiro, 1979) which is nearly independent of temperature and bulk chemistry. It is thus possible to obtain the maximum water content in primary basaltic magma for the condition of water-undersaturation still in dacite after fractionation; 2.5 wt.% estimated by Sakuyama (in press). Actual water content in primary basaltic magmas would be much smaller than this. Based on the careful assessment of experimental and petrographic data, Gill (1981) deduced the maximum water content in andesitic magmas to be 5 wt.% which corresponds to the value less than 3% in basaltic magmas, being consistent with the above estimation from the solubility limit.

For the generation of the basaltic magma containing so little water, the upper mantle must melt essentially under dry condition. If high-K olivine basalt with 2.5 wt.% of water which would generate the type IV sequence of crystallization is made by 20% partial melting of the upper mantle, initial water content in the pre-melting upper mantle can be estimated to be at most 0.5 wt.%. Actual water content of the pre-melting upper mantle would be much smaller than this value since deduced degree of partial melting for generating alkali olivine basalt is less than 20% (Gast, 1968; Frey et al., 1978). The comparison between the melting fashion of dry and wet (containing 1.9 wt.% water) peridotites by Mysen and Kushiro (1977) suggests that the melting of peridotite with water content less than 0.5% should be essentially dry.

Mantle diapir

Arc magmas are, without doubt, generated by a process taking place along the Benioff zone in relation to a descending movement of an oceanic plate beneath other plate. The depth of the Benioff zone beneath a volcanic front is rather constant in different arcs, 124 ± 38 km according to Gill (1981), suggesting that the process is controlled by a pressure-dependent mechanism. The width of volcanic zone negatively correlates to the dip angle of the Benioff zone; the more gently the Benioff zone dips, the wider the volcanic zone is (Marsh, 1979; Shimozuru, 1981). This may suggest that the source region of arc magmas is a layer with some thickness along the Benioff zone. In addition to these deeply determined aspects, arc magmas have shallowly determined aspects. First, as previously pointed out, the degree of partial melting of the upper mantle is roughly correlated to the crustal thickness. This suggests that magma segregation takes place in the uppermost part of the mantle, and requires a magma-generation process in which the crustal thickness can control the degree of partial melting of the upper mantle. Second, results of experimental works show that basaltic magmas ranging from tholeiitic basalt to alkali olivine basalt can equilibrate with peridotitic upper mantle at most below 20 kbar under dry conditions (e.g., Green and Ringwood, 1967; Takahashi et al., 1981). The equilibration pressure can not exceed 25 kbar even if only 2.5 wt.% water is added (Tatsumi, in press). Third, chemical compositions of volcanic rocks, especially those with incompatible element concentration, are different in different volcanoes (Kawano et al., 1961; Katsui et al., 1978; Onuma et al., 1981). The systematic difference can be observed even if the distance apart of volcanoes is only 10 km (Sakuyama, 1981), suggesting that each volcano is developed from magma separately segregated in the relatively shallower region of the upper mantle. Fourth, activity of composite volcanoes has commonly the life span of several tens to hundreds of thousand years, suggesting intermittent magma generation and its limited total heat energy. The magma-generation model beneath arcs must explain these dual aspects on the depth of generation.

The previous considerations on water contents in arc magmas show the prime importance of dry solidus of peridotite for magma generation. Experimentally determined dry solidus of peridotite is about 1200°C at 10 kbar and about 1500°C at 30 kbar (Kushiro et al., 1968; Jaques and Green, 1980; Takahashi et al., 1981). A rather high temperature of 1200 to 1300°C is, thus, necessary for magma generation in the uppermost mantle. But, this temperature is twohundred degrees higher than that estimated beneath arcs (Hasebe et al., 1970; Oxburgh and Turcotte, 1970; 1971; Toksöz et al., 1971). Accordingly, it can be concluded that primary arc magmas are not continuously generated on the steady geotherm. Diapiric uprise and associated adiabatic decompression of the hot deeper mantle would likely be the most reasonable mechanism for increasing the temperature of the uppermost mantle above 1200°C (Green and Ringwood, 1967; Wyllie, 1971; Ramberg, 1972).

The dual aspects of arc magmas about their depth of generation can be explained by the mantle diapir. Gravitational instability would be developed along the subduction zone by influx of water-rich fluid or magmatic melt from descending slab into wedge mantle, or by temperature increase of frictional heating, from where the mantle begins to uprise diapirically. The source region for the diapir may be a thin (1—10 km), partially dipping melted layer along the Benioff zone as deduced from volcano spacing by Marsh (1979). The volcanic front would be an upper limit of the layer which is determined by a pressure-dependent process such as the upper pressure stability limit of hornblende in peridotite.

The temperature of the diapir would exceed the dry solidus of peridotite at some depth. Above this level, the degree of partial melting drastically increases because of the eutectic-like melting fashion of peridotite (Mysen and Kushiro, 1977). According to the estimated *P-T* path of the mantle diapir in consideration of the latent heat of fusion by Cawthorn (1975), the initial temperature of the mantle diapir whose temperature is 1300°C at 10 kbar must be more than 1400°C at a depth of about 100 km. This might indicate that the temperature of the upper mantle wedge near the subduction zone is at least 1400°C beneath the volcanic zone.

Diapiric uprise would be stopped by the decrease of the surrounding density below that of the diapir. The crust—mantle boundary is the most important boundary through which the diapir can not move upward only by buoyant force. Assuming the similar initial temperature of the diapir, the degree of partial melting of the diapir strongly depends on the stopped depth of the diapir; the shallower the diapir goes up beneath thinner crust, the higher its degree of partial melting is. Magma segregated from a stopped diapir, which has distinct chemical features mainly determined by the degree of partial melting, goes up through the crust, fractionates and extrudes making a composite cone. The life span of activity of a composite volcano might be limited by the total heat content of the diapir.

The decrease of degree of partial melting of the upper mantle across many arcs, however, can not be explained by the variation of crustal thickness. There is no systematic increase of the crustal thickness toward the back-arc side. On the contrary, the thickness of the back-arc associated with a back-arc basin (e.g., northeastern Japan, Izu-Bonin, Aleutian) is rather thin compared with that of the frontal side. This suggests that the diapir possibly ceases to uprise within the upper mantle due to other factors rather than the surrounding density change. The deeper stopped depth of the diapir in the back-arc region is supported by the experimental petrology on various types of basaltic magma (e.g., Green and Ringwood, 1967) and the common occurrence of the mantle nodule in alkalic lavas of the back-arc region (e.g., Takahashi, 1978). Green et al. (1968) considered that the diapir of the back-arc region, with a higher temperature than that of the frontal side, would reach to some melting degree for segregation at deeper level. But, this model conflicts with the smaller melting degree in the back-arc region. Physical properties, such as temperature and viscosity, of the surrounding upper mantle might be laterally different.

Recent trace-element and isotopic data on arc volcanic rocks suggest that the geochemical feature of arc magma is determined by the multiple contribution from various source components such as descending slab, heterogeneous wedge mantle and lower and upper crust (Perfit et al., 1980; Kay, 1980; Arculus and Johnson, 1981; Gill, 1981). It is thus evident that the compositional homogeneity of the source mantle tacitly assumed in this paper is strictly incorrect. While the proper evaluation of the contribution from each component should be necessary, the essential petrological feature of arc volcanic rocks can be explained by the generation of primary basaltic magma by the adiabatic decompression associated with diapiric uprise of the mantle and crystallization differentiation from that primary magma.

ACKNOWLEDGEMENTS

The first draft of the manuscript was critically read by Professor I. Kushiro, University of Tokyo. The discussions with Dr. H. Fukuyama and Dr. Y. Tatsumi were very fruitful. The author is indebted deeply to them for their constructive criticism and improvement of the manuscript.

REFERENCES

Allen, J.C., Boettcher, A.L. and Marland, G., 1975. Amphiboles in andesite and basalt: I. Stability as a function of P-T-f_{O_2}. Am. Mineral., 60: 1069—1085.

Arculus, R.J. and Johnson, R.W., 1981. Island-arc magma sources: a geochemical assessment of the roles of slab derived components and crustal contamination. Geochem. J., 15: 109—133.

Best, M.G., 1963. Petrology of the Guadalupe igneous complex, south-western Sierra Nevada Foothills, California. J. Geol., 4: 223—259.

Blake, D.H., 1968. Post-Miocene volcanoes in Bougainville Island, Territory of Papua and New Guinea. Bull. Volcanol., 32: 121—138.

Blake, D.H. and Miezitis, Y., 1967. Geology of Bougainville and Buka Islands, New Guinea. Bur. Miner. Resour., Geol. Geophys. (Aust.) Bull., 93, 56 pp.

Boettcher, A.L., 1977. The role of amphiboles and water in circum-Pacific volcanism. In: M.H. Maghnami and S. Akimoto, (Editors), High-pressure Research, Applications in Geophysics. Academic Press, London, pp. 107—125.

Briqueu, L. and Lancelot, J.R., 1979. Rb-Sr systematics and crustal contamination models for calc-alkaline igneous rocks. Earth Planet. Sci. Lett., 43: 385—396.

Byers, F.M., 1959. Geology of Umnak and Bogoslof islands, Aleutian Islands, Alaska. U.S. Geol. Surv., Bull., 1028—L: 257—369.

Cawthorn, R.G., 1975. Degrees of melting in mantle diapirs and the origin of ultrabasic liquids. Earth Planet. Sci. Lett., 27: 113—120.

Cawthorn, R.G., 1976a. Calcium-poor pyroxene reaction relations in calc-alkaline magmas. Am. Mineral., 61: 907—912.

Cawthorn, R.G., 1976b. Melting relations in part of the system CaO-MgO-Al_2O_3-SiO_2-Na_2O-H_2O under 5 kb pressure. J. Petrol., 17: 44—72.

Cawthorn, R.G., Curran, E.B. and Arculus, R.J., 1973. A petrogenetic model for the origin of the calc-alkaline suite of Grenada, Lesser Antilles. J. Petrol., 14: 327—337.

Cawthorn, R.G. and O'Hara, M.J., 1976. Amphibole fractionation in calc-alkaline magma genesis. Am. J. Sci., 276: 309—329.

Coats, R.R., 1952. Magmatic differentiation in Tertiary and Quaternary volcanic rocks from Adak and Kanaga islands, Aleutian Islands, Alaska. Geol. Soc. Am. Bull., 63: 485—514.

Coulon, C. and Thorpe, R.S., 1981. Role of continental crust in petrogenesis of orogenic volcanic associations. Tectonophysics, 77: 79—93.

Dallwitz, W.B., Green, D.H. and Thompson, J.E., 1966. Clinoenstatite in a volcanic rock from the Cape Vogel area, Papua. J. Petrol., 7: 375—403.

Eggler, D.H., 1972. Water-saturated and undersaturated melting relations in a Paricutin andesite and an estimate of water content in the natural magma. Contrib. Mineral. Petrol., 34: 261—271.

Eggler, D.H. and Burnham, C.W., 1973. Crystallization and fractionation trends in the system andesite-H_2O-CO_2-O_2 at pressure to 10 kb. Geol. Soc. Am. Bull., 84: 2512—2532.

Eichelberger, J.C., 1975. Origin of andesite and dacite; Evidence of mixing at Glass Mountain in California and at other circum-Pacific volcanoes. Geol. Soc. Am. Bull., 86: 1381—1391.

Eichelberger, J.C., 1978. Andesites in island arcs and continental margins; Relationship to crustal evolution. Bull. Volcanol., 41: 480—500.

Ewart, A., 1976. Mineralogy and chemistry of modern orogenic lavas — some statistics and implications. Earth Planet. Sci. Lett., 31: 417—432.

Ewart, A., Hildreth, W. and Carmichael, I.S.E., 1975. Quaternary acid magma in New Zealand. Contrib. Mineral. Petrol., 51: 1—27.

Fisher, N.H., 1957. Catalogue of the active volcanoes of the world including solfatara fields, Part V, Melanesia. Int. Volcanol. Assoc., Napels, 105 pp.

Francis, P.W., Roobol, M.H., Walker, G.P.L., Cobbold, P.R. and Coward, M., 1974. The San Pedro and San Pabro volcanoes of northern Chile and their hot avalanche deposits. Geol. Rundsch., 63: 357—388.

Francis, P.W., Thorpe, R.S., Moorbath, S., Kretzschmar, G.A. and Hammill, M., 1980. Strontium isotope evidence for crustal contamination of calc-alkaline volcanic rocks from Cerro Galan, northwest Argentina. Earth Planet. Sci. Lett., 48: 257—267.

Frey, F.A., Green, D.H. and Roy, S.D., 1978. Integrated models of basalt petrogenesis; a study of quartz tholeiite to olivine melilitites from southeastern Australia utilizing geochemical and experimental petrological data. J. Petrol., 19: 463—513.

Fujinawa, A., 1980. Geology and petrology of Adatara volcano (in Japanese with English abstract). J. Jpn. Assoc. Miner. Petrol., Econ. Geol., 75: 385—395.

Gast, P.W., 1968. Trace element fractionation and the origin of tholeiitic and alkaline magma types. Geochim. Cosmochim. Acta, 32: 1057—1086.

Gill, J.B., 1981. Orogenic Andesites and Plate Tectonics. Springer-Verlag, Berlin, 390 pp.

Green, D.H. and Ringwood, A.E., 1967. The genesis of basaltic magmas. Contrib. Mineral. Petrol., 15: 103—190.

Green, T.H. and Ringwood, A.E., 1968. Genesis of the calc-alkalic igneous rock suite. Contrib. Mineral. Petrol., 18: 105—162.

Green, T.H., Green, D.H. and Ringwood, A.E., 1967. The origin of high-alumina basalts and their relationships to quartz tholeiites and alkali basalts. Earth Planet. Sci. Lett., 2: 41—51.

Hamilton, D.L., Burnham, C.W. and Osborn, E.F., 1964. The solubility of water and effect of oxygen fugacity and water content on crystallization in mafic magmas. J. Petrol., 5: 21—39.

Hantke, G. and Parodi, I.A., 1966. Catalogue of the active volcanoes of the world including solfatara fields, Part XIX, Colombia, Ecuador and Peru. Int. Volcanol. Assoc., Napels, 73 pp.

Hasebe, K., Fujii, N. and Uyeda, S., 1970. Thermal processes under island arcs. Tectonophysics, 10: 335—355.

Heiken, G. and Eichelberger, J.C., 1980. Eruptions at Chaos Crags, Lassen Volcanic National Park, California. J. Volcanol. Geotherm. Res., 7: 443—481.

Helz, R.T., 1973. Phase relations of basalts in their melting range at P_{H_2O}=5 kb as a function of oxygen fugacity. Part I. Mafic phases. J. Petrol., 14: 249—302.

Ichimura, T., 1951. Geological investigations on the Zao volcanoes. I. Goshiki-dake, a central cone of the Zao Proper. Bull. Earthq. Res. Inst., Tokyo Univ., 29: 327—339.
Ichimura, T., 1953. Geological investigations on the Zao volcanoes. II. Aoso volcano. Bull. Earthq. Res. Inst., Tokyo Univ., 31: 129—150.
Ichimura, T., 1955. Geological investigations on the Zao volcanoes. III. Byobu volcano. Bull. Earthq. Res. Inst., Tokyo Univ., 33: 593—629.
Ichimura, T., 1957. Geological investigations on the Zao volcanoes. IV. Ryu-zan volcano. Bull. Earthq. Res. Inst., Tokyo Univ., 35: 573—593.
Ichimura, T., 1960. Geological investigations on the Zao volcanoes. V. Kumano volcano. Bull. Earthq. Res. Inst., Tokyo Univ., 38: 255—290.
Ishii, T., 1975. The relations between temperature and composition of pigeonite in some lavas and their application to geothermometry. Mineral. J., 8: 48—57.
Ishikawa, T., 1958. Petrological characteristics of the Chokai and Norikura volcanic zones (in Japanese). Jubilee Publ. Commem. Prof. Jun Suzuki, Sixtieth Birthday, pp. 263—277.
Jaques, A.L. and Green, D.H., 1980. Anhydrous melting of peridotite at 0—15 kb pressure and the genesis of tholeiite basalts. Contrib. Mineral. Petrol., 73: 287—310.
Katsui, Y., Oba, Y., Ando, S., Nishimura, S., Masuda, Y., Kurasawa, H. and Fujimaki, H., 1978. Petrochemistry of the Quaternary volanic rocks of Hokkaido, north Japan. J. Fac. Sci., Hokkaido Univ., Ser. IV, 18: 449—484.
Kawano, Y. and Aoki, K., 1960. Petrology of Hachimantai and surrounding volcanoes, northeastern Japan. Sci. Rep. Tohoku Univ., Ser. III, 6: 409—429.
Kawano, Y., Yagi, K. and Aoki, K., 1961. Petrography and petrochemistry of the volcanic rocks of Quaternary volcanoes of northeastern Japan. Sci. Rep. Tohoku Univ., Ser. III, 7: 1—46.
Kay, R.W., 1980. Volcanic arc magmas; implications of a melting-mixing model for element recycling in the crust—upper mantle system. J. Geol., 88: 497—522.
Kuno, H., 1936. Petrological notes on some pyroxene andesites from Hakone volcano, with special reference to some types with pigeonite phenocrysts. Japan. J. Geol. Geogr., 13: 107—140.
Kuno, H., 1950. Petrology of Hakone volcano and the adjacent areas, Japan. Geol. Soc. Am. Bull., 61: 957—1020.
Kuno, H., 1966. Lateral variation of basalt magma type across continental margins and island arcs. Bull. Volcanol., 29: 195—222.
Kuno, H., 1968a. Origin of andesite and its bearing on the island arc structure. Bull. Volcanol., 32: 141—176.
Kuno, H., 1968b. Differentiation of basaltic magmas. In: H.H. Hess and A. Poldervaart (Editros), Basalts: The Poldervaart Treatise on Rocks of Basaltic Composition, vol. 2. Wiley, New York, N.Y., pp. 623—688.
Kuroda, N., Shiraki, K. and Urano, H., 1978. Boninite as a possible calc-alkaline primary magma. Bull. Volcanol., 41: 563—575.
Kushiro, I., 1969. The system forsterite-diopside-silica with and without water at high pressure. Am. J. Sci., 267A: 269—294.
Kushiro, I., 1974. Melting of hydrous upper mantle and possible generation of andesitic magma: An approach from synthetic systems. Earth Planet. Sci. Lett., 22: 294—299.
Kushiro, I. and Sato, H., 1978. Origin of some calc-alkalic andesites in the Japanese Islands. Bull. Volcanol., 41: 576—585.
Kushiro, I., Syono, Y. and Akimoto, S-I., 1968. Melting of a peridotite nodule at high pressures and high water pressures. J. Geophys. Res., 73: 6023—6029.
Langmuir, C. and Goldstein, S., 1981. Constraints on arc petrogenesis from the Aleutians and Central America. Abstracts 1981 IAVCEI Symposium Arc Volcanism, pp. 198—199.
Luhr, J.F. and Carmichael, I.S.E., 1980. The Colima volcanic complex, Mexico, I. Post-caldera andesites from Volcan Colima. Contrib. Mineral. Petrol., 71: 343—372.

Maaløe, S. and Wyllie, P.J., 1975. Water content of a granite magma deduced from the sequence of crystallization determined experimentally with water-undersaturated conditions. Contrib. Mineral. Petrol., 52: 175—191.
Marsh, B.D., 1979. Island arc development: some observations, experiments and speculations. J. Geol., 87: 687—713.
Marsh, B.D., 1981. On the crystallinity, probability of occurrence and rheology of lava and magma. Contrib. Mineral. Petrol., 78: 85—98.
Masuda, Y. and Aoki, K., 1979. Trace element variations in the volcanic rocks from the Nasu zone, northeast Japan. Earth Planet. Sci. Lett., 44: 139—149.
Matsuhisa, Y., 1979. Oxygen isotopic compositions of volcanic rocks from the East Japan island arc and their bearing on petrogenesis. J. Volcanol. Geotherm. Res., 5: 271—296.
Matsui, Y. and Nishizawa, O., 1974. Iron (II)—magnesium exchange equilibrium between olivine and calcium-free pyroxene over a temperature range 800°C to 1300°C. Bull. Soc. Fr. Minéral. Crystallogr., 97: 122—130.
Miyashiro, A., 1974. Volcanic rock series in island arcs and active continental margins. Am. J. Sci., 274: 321—355.
Mooser, F., Meyer-Abich, H. and McBirney, A.R., 1958. Catalogue of active volcanoes of the world including solfatara fields, Part VI, Central America. Int. Volcanol. Assoc., Napels, 146 pp.
Mysen, B.O. and Kushiro, I., 1977. Compositional variations of coexisting phases with degree of melting of peridotite in the upper mantle. Am. Mineral., 62: 843—865.
Neumann van Padang, M., 1951. Catalogue of the active volcanoes of the world including solfatara fields, Part I, Indonesia. Int. Volcanol. Assoc., Napels, 271 pp.
Obata, M., Banno, S. and Mori, T., 1974. The iron-magnesium partitioning between naturally occurring coexisting olivine and Ca-rich clinopyroxene: an application of the simple mixture model to olivine solid solution. Bull. Soc. Fr. Minéral. Cristallogr., 97: 101—107.
Onuma, N., Hirano, M. and Isshiki, N., 1981. Sr/Ca—Ba/Ca systematics in Quaternary volcanoes of Izu Islands, Japan. Abstracts 1981 IAVCEI Symposium, Arc Volcanism, pp. 281—282.
Osborn, E.F., 1959. Role of oxygen pressure in the crystallization and differentiation of basaltic magmas. Am. J. Sci., 257: 609—647.
Oxburgh, E.R. and Turcotte, D.L., 1970. Thermal structure of island arcs. Geol. Soc. Am. Bull., 81: 1665—1688.
Oxburgh, E.R. and Turcotte, D.L., 1971. Origin of paired metamorphic belts and crustal dilation in island arc regions. J. Geophys. Res., 76: 1315—1327.
Perfit, M.R., Gust, D.A., Bence, A.E., Arculus, R.J. and Taylor, S.R., 1980. Chemical characteristics of island arc basalts: implications for mantle sources. Chem. Geol., 30: 227—256.
Piwinskii, A.J. and Wyllie, P.J., 1968. Experimental studies of igneous rock series: a zoned pluton in the Wallowa batholith, Oregon. J. Geol., 76: 205—234.
Ramberg, H., 1972. Mantle diapirism and its tectonic and magmagenetic consequences. Phys. Earth Planet. Inter., 5: 45—60.
Sakuyama, M., 1979a. Lateral variations of H_2O contents in Quaternary magmas of northeastern Japan. Earth Planet. Sci. Lett., 43: 103—111.
Sakuyama, M., 1979b. Evidence of magma mixing: petrological study of Shirouma-Oike calc-alkaline andesite volcano, Japan. J. Volcanol. Geotherm. Res., 5: 179—208.
Sakuyama, M., 1981. Petrological study of the Myoko and Kurohime volcanoes, Japan: crystallization sequence and evidence for magma mixing. J. Petrol., 22: 553—583.
Sakuyama, M., in press. Phenocryst assemblages and H_2O contents in circum-Pacific arc magmas. Geodynamics Series A.G.U. and G.S.A.
Sakuyama, M. and Kushiro, I., 1979. Vesiculation of hydrous andesitic melt and transport of alkalies by separated vapor phase. Contrib. Mineral. Petrol., 71: 61—66.

Sekine, T., Katsura, T. and Aramaki, S., 1979. Water saturated phase relations of some andesites with application to the estimation of the initial temperature and water pressure at the time of eruption.Geochim. Cosmochim. Acta, 43: 1367—1376.

Shimozuru, D., 1981. Volcano spacing and subduction. Abstracts 1981 IAVCEI Symposium, Arc Volcanism, pp. 336—337.

Takahashi, E., 1978. Petrologic model of the crust and upper mantle of the Japanese island arcs. Bull. Volcanol., 41: 529—547.

Takahashi, E., Kushiro, I. and Tatsumi, Y., 1981. Melting of a peridotite at high pressures and its bearing on island arc magmas. Abstracts 1981 IAVCEI Symposium, Arc Volcanism, pp. 365—366.

Tatsumi, Y., in press. Origin of high magnesian andesites in the Setouch volcanic belt, southwest Japan. Part II. Melting phase relations at high pressures.

Tatsumi, Y. and Ishizaka, K., 1981. Existence of andesitic primary magma: an example from southwest Japan. Earth Planet. Sci. Lett., 53: 124—130.

Taylor, G.A., 1958. The 1851 eruption of Mount Lamington, Papua. Aust., Bur. Miner. Resour., Geol. Geophys. Bull., 38, 117 pp.

Togashi, S., 1977. Petrology of Osore-yama volcano, Japan (in Japanese with English abstract). J. Jpn. Assoc. Miner. Petrol., Econ. Geol., 72: 45—60.

Toksöz, M.N., Minear, J.W. and Julian, B.R., 1971. Temperature field and geophysical effects of a downgoing slab. J. Geophys. Res., 76: 1113—1138.

Ui, T. and Aramaki, S., 1978. Relationship between chemical composition of Japanese island-arc volcanic rocks and gravimetric data. Tectonophysics, 45: 249—259.

Vlodavetz, V.I. and Piip, B.I., 1959. Catalogue of the active volcanoes of the world including solfatara fields, Part VIII, Kamchatka and continental areas of Asia. Int. Volcanol. Assoc., Napels, 110 pp.

Wada, K., 1981. Contrasted petrological relations between tholeiitic and calc-alkalic series from Funagata volcano, northeastern Japan. J. Jpn. Assoc. Miner. Petrol., Econ. Geol., 76: 215—232.

Whitney, J.A., 1975. The effects of pressure, temperature and X_{H_2O} on phase assemblage in four synthetic rock compositions. J. Geol., 83: 1—31.

Williams, H., 1960. Volcanic history of the Guatemalan Highlands. Calif. Univ. Publ. Geol. Sci., 38: 1—86.

Wyllie, P.J., 1971. The Dynamic Earth: Textbook in Geosciences. Wiley, New York, N.Y., 416 pp.

Wyllie, P.J., 1977. Crustal anatexis: An experimental review. Tectonophysics, 43: 41—71.

Yamazaki, T., 1965a. Petrography and petrochemistry of Kirigamine volcano, central Japan (I). J. Jpn. Assoc. Miner. Petrol., Econ. Geol., 54: 39—54.

Yamazaki, T., 1965b. Petrography and petrochemistry of Kirigamine volcano, central Japan (II). J. Jpn. Assoc. Miner. Petrol., Econ. Geol., 54: 92—103.

Yoder, H.S., 1969. Calc-alkalic andesites: experimental data bearing on the origin of their assumed characteristics. In: A.R. McBirney (Editor), Proceedings of the Andesite Conference. Oreg., Dep. Geol. Mineral. Ind., Bull., 65: 77—89.

Yoder, H.S. and Tilley, C.E., 1962. Origin of basalt magmas: an experimental study of natural and synthetic rock systems. J. Petrol., 3: 342—532.

Journal of Volcanology and Geothermal Research, 18 (1983) 321—335
Elsevier Science Publishers B.V., Amsterdam — Printed in The Netherlands

POTASSIC LAVAS IN THE OROGENIC VOLCANISM OF THE MEDITERRANEAN AREA

JÖRG KELLER

Mineralogisches Institut der Universität, D-7800 Freiburg (Federal Republic of Germany)

(Received October 20, 1982)

ABSTRACT

Keller, J., 1983. Potassic lavas in the orogenic volcanism of the Mediterranean area. In: S. Aramaki and I. Kushiro (Editors), Arc Volcanism. J. Volcanol. Geotherm. Res.,: 18: 321—335.

Volcanism in the Mediterranean fold belts is dominantly related to plate-convergence processes. The occurrence of highly potassic and often leucite-bearing lavas is a typical feature of many Neogene to Quaternary volcanic provinces in the Mediterranean area.

The paper describes possible relationships between the Aeolian island arc and the Roman comagmatic region and presents petrological evidence for an ultrapotassic province within the Cypriot-Taurus arc of Turkey. It is thus a contribution to the controversy about tectonic setting and geodynamic significance of the highly potassic volcanism of the Mediterranean suite.

INTRODUCTION

A long-known and special feature of Quaternary Mediterranean volcanism is the occurrence of ultrapotassic volcanoes in central Italy which includes Sabatini, Vico and Vulsini volcanoes north of Rome, Alban Hills, Ernici and Roccamonfina between Rome and Naples and the Campanian province with the Phlegrean Fields and Somma-Vesuvius. This is the "Roman Comagmatic Region" of Washington (1906) and the "Mediterranean Clan" of Niggli (1923).

The Mediterranean region is dominated by geological features resulting from the highly complex convergence of Africa and Europe. Folding, collision and subduction resulted from this plate convergence throughout the alpidic history of the Mediterranean border areas. Volcanism connected with plate consumption is typically represented by the andesite series of Mediterranean arc systems of Tertiary and Quaternary age. Two active island arcs are recognized, the Aeolian Islands and the Hellenic arc.

An extremely large body of petrological and geochemical data — including isotope studies for strontium, lead, neodymium and oxygen (Cox et al., 1976; Hawkesworth and Vollmer, 1979; Taylor et al., 1979; Vollmer and

0377-0273/83/$03.00

Hawkesworth, 1980; Civetta et al., 1981) and experimental results (Thompson, 1977) — are available for the central Italian potassic volcanoes. Nevertheless, as to tectonic setting and geodynamic significance a very controversial discussion is going on (e.g. Cundari, 1980; Edgar, 1980; Vollmer et al., 1981).

The problem, whether the petrogenetic processes involved in the origin of Mediterranean leucite-bearing volcanic rocks are similar to continental rift-valley situations as e.g. Birunga in the East African rift, or are rather related to particular subduction or postsubduction phenomena, is approached in this paper by showing that highly potassic series are widespread and represent a typical feature in several areas of generally plate-convergence tectonics and orogenic volcanic activity in the Mediterranean.

ROMAN COMAGMATIC REGION

Magmatologic explanations for the Quaternary potassic volcanoes around Rome and Naples vary between the model of a classical island arc with increasing K_2O as Benioff depth increases, as published by Ninkovich and Hays (1972) in a generalized picture, and the view that the Italian leucitic volcanoes can best be compared with intraplate rift-valley situations (Vollmer et al., 1981). Cundari (1980) stresses that a genetic relationship between leucite-bearing rocks and "subduction-related processes is remote". It became obvious from the controversy that the large data collection does not provide unambiguous criteria for distinguishing whether a high-K volcano belongs to a subduction/arc situation or to continental rifting.

In this situation, several authors used conventional simple discriminants to suggest an arc affiliation. Di Girolamo (1978) shows with the aid of discriminant function F1 of Pearce (1976) that Vesuvius and Roccamonfina, as well as the Aeolian Islands and Batu Tara, Indonesia, exhibit "converging plate-margin characteristics" whereas the Birunga volcanoes in the East African Rift, using the same criteria, show "within-plate" values. The same conclusion is reached by Rittmann (1981) using his τ-value (τ = Al_2O_3-Na_2O/TiO_2) to discriminate between orogenic and non-orogenic volcanic series. The most efficient single-element discriminant is TiO_2. Thompson (1977) has shown that the central Italian potassic rocks (and this holds true for all similar Mediterranean potassic series) are significantly lower in TiO_2 than continental rift-valley type potassic rocks, as East African Rift or Rhinegraben. Low titanium is a prominent arc signature and although it seems difficult to accept such a simplistic distinction in a difficult magma-genesis problem, the clear separation of cratonic from orogenic potassic lavas on the basis of TiO_2 has yet to be explained by supporters of a common origin.

Several genetic hypotheses have been proposed which try to explain the variable but high LIL elements, high $^{87}Sr/^{86}Sr$, low $^{143}Nd/^{144}Nd$ and high $^{16}O/^{18}O$ in the Roman comagmatic region. Crustal contamination (Turi and Taylor, 1976; Taylor et al., 1979) and/or large-scale mantle heterogeneities produced by metasomatism with LIL-enriched fluids (Cox et al., 1976; Vollmer and Hawkesworth, 1980) play the major role in explaining the geochemistry of the potassic series. In the mixing model of Hawkesworth and Vollmer (1979) and Vollmer and Hawkesworth (1980) at least two components are necessary where one component is a metasomatizing fluid variably enriched in LIL elements and radiogenic isotopes and the other is a mantle melt. Mixing occurred in the mantle source area although metasomatism introduces definitely some crustal characteristics. There is a decreasing influence of the added fluid from north to south which leads to the marked regional trend noted in Italy. An important addition to this hypothesis is the possibility that the fluids causing the mantle heterogeneities are derived from the slab as proposed by Cox et al. (1976) and Civetta et al. (1981).

THE AEOLIAN ISLAND ARC

The Aeolian Islands in the southern Tyrrhenian Sea form an island arc with high-Al basalt-andesite-dacite series of a typical calc-alkaline arc association (Barberi et al., 1974; Keller, 1974). However, the occurrence of high-K calc-alkaline series and of K-rich shoshonitic and leucite-tephritic lavas on Vulcano (Keller, 1980) and Stromboli (Rosi, 1980) might be a key for the understanding of the still more pronounced potassic rocks northwards along the same Tyrrhenian Sea border.

Structural features of the Aeolian arc are a steep but discontinuous Benioff plane dipping NW, and active deep-focus seismicity with earthquake centers c. 250 km underneath the volcanic islands (Keller, 1982). The seismically active arc structure extends to the edge of the Campanian potassic province (Fig. 1).

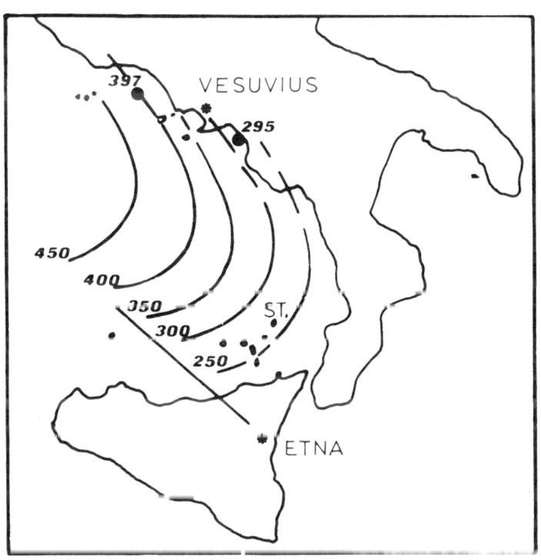

Fig. 1. Relationship of Vesuvius to the active Aeolian island arc. Benioff-depth contours in km after Gasparini et al. (1982). The deep-focus earthquakes closest to the Campanian province are marked with their focal depth. *ST.* = Stromboli volcano.

In the evolution of the Aeolian arc an increase in K_2O with time from calc-alkaline (CA) to high-K calc-alkaline (HKCA), shoshonitic and leucite-tephritic has been established. Shoshonitic and leucite-rephritic lavas appear in a late or senile stage of the arc evolution (Barberi et al., 1974). The recent volcanic activity belongs to this K-rich trend (Fig. 2).

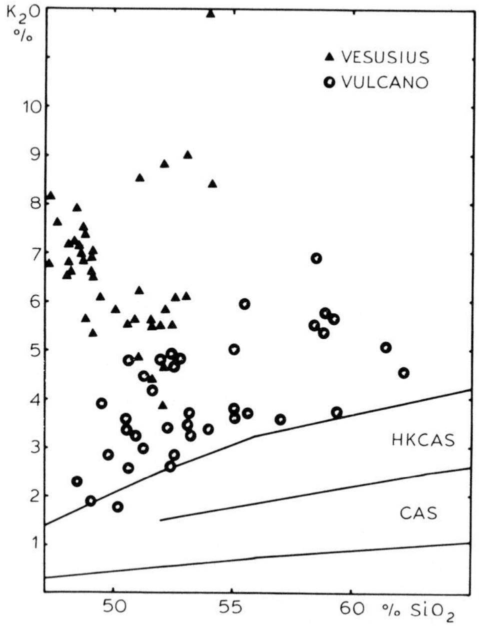

Fig. 2. K_2O vs. SiO_2 variation for K-rich lavas from the Aeolian island arc (circles = shoshonitic and leucite-tephritic series of Vulcano/Vulcanello, Keller, 1980) and from the Somma-Vesuvius evolution (triangles, Savelli, 1967).

On a K_2O-free basis, basaltic members of the CA, HKCA, shoshonitic and leucite-tephritic series are very similar in geochemical and isotopic composition. There is general agreement (e.g., Barberi et al., 1974; Rosi, 1980; Vollmer et al., 1981) that all these magmas belong to the arc evolution of the Aeolian Islands. Different degrees in K_2O and LIL-element concentrations are best explained by mantle-source heterogeneities induced by slab-derived LIL-rich fluids. A rather constant $^{87}Sr/^{86}Sr$ ratio of 0.7040—0.7045 for members of all series (Barberi et al., 1974) indicates low radiogenic strontium in the metasomatizing fluids. In the case of low $^{87}Sr/^{86}Sr$ in the K_2O and Rb-rich shoshonitic lavas, mantle metasomatisim must be rather contemporaneous with the arc-magma genesis. It can be assumed that the subducted oceanic crust of the Ionian sea-floor did not introduce large amounts of radiogenic crustal material into the source area of the arc magmas. This crustal influence increases in a northward direction along the Italian peninsula. Even within the limited extension of the Aeolian Islands this trend in the isotopic heterogeneity is present, as Stromboli lavas are significantly more radiogenic than the islands to the south.

Petrogenetic and geochemical features exhibit a gradual transition from the more K-rich Aeolian arc lavas to the central Italian volcanic rocks. This is shown with the potassium-silica diagram of Fig. 2 and by comparing some representative analyses of Vulcano/Vulcanello with Somma-Vesuvius (Table I). The potassic nature and LIL-element concentration are still higher in the extremely leucitic lavas of e.g. Vesuvius. However, the compositional features show a distinct overlap and gradual differences with geochemical correlations as given in Fig. 3. Finally, shoshonitic and leucite-tephritic lavas of the Aeolian island arc find their counterpart in the potassium series (as opposed to "high potassium series") of Roccamonfina, Ernici and Phlegrean Fields (Appleton, 1972; Ghiara et al., 1979; Civetta et al., 1981).

TABLE I

Comparison between leucite-bearing lavas from the Aeolian island arc and from Somma-Vesuvius

	Vulcano, Aeolian Arc			Somma-Vesuvius		
	1	2	3	4	5	6
SiO_2	55.4	52.5	50.7	50.5	48.6	51.0
TiO_2	0.7	0.7	0.85	0.83	0.6	0.9
Al_2O_3	18.3	15.5	16.3	18.5	18.9	18.4
Fe_2O_3	3.85	4.1	4.35	5.2	3.5	3.0
FeO	2.75	4.9	6.1	3.2	5.0	4.6
MnO	0.13	0.16	0.18	0.11	0.13	—
MgO	2.5	4.6	4.4	3.4	3.2	4.7
CaO	4.9	8.3	7.7	8.0	8.6	9.2
Na_2O	4.2	3.6	3.6	3.4	2.8	2.6
K_2O	5.8	4.6	4.7	4.8	7.5	5.6
P_2O_5	0.52	0.47	0.54	0.52	0.5	—
H_2O	0.9	0.5	0.6	1.2	0.9	—
Total	99.95	99.93	100.02	99.66	100.23	100
$Mg/(Mg + Fe^{2+})$	0.47	0.52	0.47	0.47	0.47	0.57
Rb	230	170		250	300	
Sr	1400	1330		800	1000	
Ba	1318	1180		1850	2200	
$^{87}Sr/^{86}Sr$ (range)	0.7041—0.7046			0.7069—0.7080*		

Data sources: 1—3: Keller, 1980; 4 + 5: Savelli, 1967; 6: average of 29 phonolitic tephrites, Somma (Cundari and Le Maitre, 1970).
*Cortini and Hermes, 1981.
$Mg(Mg + Fe^{2+})$ (mol.) with $Fe_2O_3/FeO = 0.15$.
Values for oxides in %, those for elements in ppm.

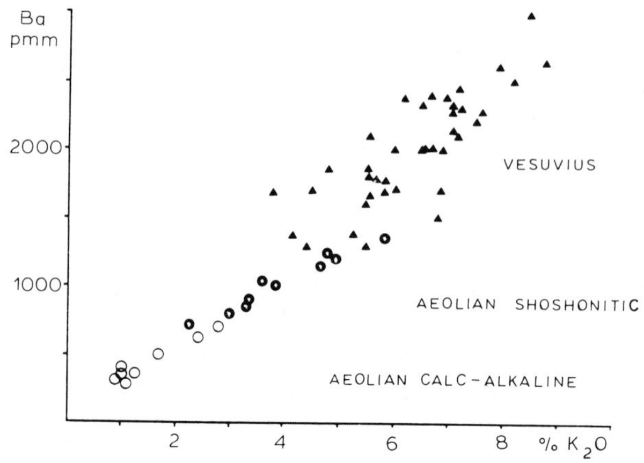

Fig. 3. Ba vs. K_2O correlation of calcalkaline and shoshonitic series from the Aeolian arc and leucite-bearing lavas of Somma-Vesuvius. Data sources as in Fig. 2.

HELLENIC ARC AREA

The evolution of the island arc from Corinth to Nisyros across the Aegean Sea has been summarized by Keller (1982). In its eastern part the modern calc-alkaline arc lavas contrast with Miocene K-rich rocks of Kos, Patmos and Bodrum (Robert, 1976; Innocenti et al., 1982; Keller, 1982), which belong to a Late Tertiary subduction configuration. No leucite-bearing lavas are known but trachytic rocks reach $K_2O > 10$ wt.%.

HIGH POTASSIC SERIES OF AFYON/ANATOLIA

Geological setting

An important, yet poorly known potassic province extends north and south of the town of Afyon in western Anatolia (Fig. 4). Keller and Villari (1972) reported the association of K-rich latites and trachytes with leucite-bearing basaltic to phonolitic rocks and leucitites.

The volcanic area of Afyon is part of the Neogene volcanic belt which follows the Tauride fold chains related to Late Tertiary collision and subduction processes of the Cypriot-Taurus arc. Present seismicity shows scattered centers and does not define an active Benioff plane. However, some deep-focus centers up to 200 km deep have been observed (Harsch et al., 1980). It is generally accepted that Neogene subduction was from south to north (Brunn et al., 1971; Harsch et al., 1981; Sengör and Yilmaz, 1981).

Extensive andesitic volcanism developed in Anatolia from ca. 30 Ma B.P. to the Quaternary (Fig. 4) with its maximum between 20—10 Ma B.P. (Innocenti et al., 1975, 1982; Besang et al., 1977). The dominant associa-

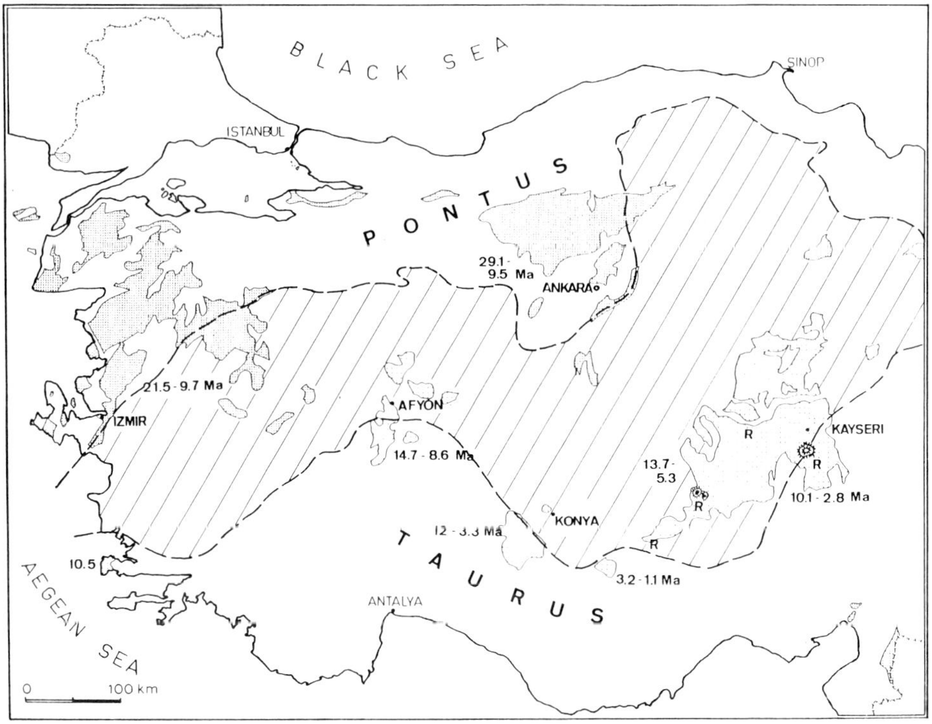

Fig. 4. Volcanic areas of western and central Anatolia showing the location of Afyon volcanism. Age values in Ma from Besang et al. (1977), Innocenti et al. (1975, 1982). *R* = Recent to sub-Recent age of volcanic activity.

tions of Neogene Anatolian volcanism are high-K calc-alkaline andesites and dacites connected with silicic ignimbrites (Innocenti et al., 1975; Keller et al., 1977). The general characterization is that of a continental or "Andean-type" calc-alkaline province. The volcanic area of Konya is a typical example (Figs. 4,5) where K_2O is high in terms of a calc-alkaline suite and LIL elements are distinctly enriched in comparison with calc-alkaline series from other regions. It has been concluded (Keller et al., 1977) that the source material for these andesites has been anomalously enriched by slab-derived emanations which may become further enriched in LIL elements by their scavenging uprise through the mantle wedge.

Age

Potassium-argon age determinations were carried out on 10 samples from the Afyon area (Besang et al., 1977). The age range is 8.6—14.75 Ma B.P. Over this time span there seems to have been continuous volcanic activity. The age of the Afyon volcanism belongs to the general period of andesitic activity in western and central Anatolia and coincides with a period of maxi-

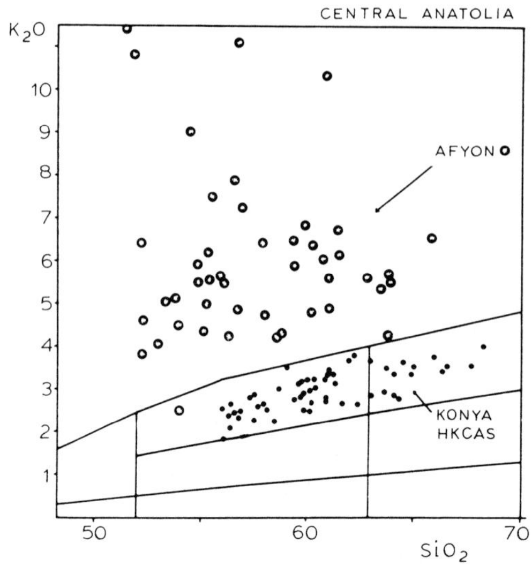

Fig. 5. Potassium variation diagram showing the ultrapotassic nature of analyzed samples from Afyon. For comparison the high-K calc-alkaline series from the nearby and mostly coeval Konya volcanic area (Keller et al. 1977).

mum compressive movements in the Taurus fold belt. In detail, the tectonic evolution in this generally convergent zone is complex and characterized by the superposition of compressional and extensional phases, as shown for the Aegean area and western Anatolia (Angelier et al., 1981; Sengör and Yilmaz, 1981). It is conceivable that this complex tectonic regime provides the conditions for the ascent of the most unusual melts generated in the plate-convergence magmatic systems.

Geochemical and petrological characterization

The K_2O-SiO_2 diagram describes the degree of potassium concentration and variation in Afyon volcanic rocks (Fig. 5). Examples for different representative lava compositions are reported in Table II. High K_2O is connected with very variable but always high LIL-element concentrations including Rb, Sr, Ba, Zr. Additional data to Table II are the range for La 55—200 ppm, Ce 88—231 ppm and Nb 21—41 ppm. K/Rb ratios are dominantly between 200—350. TiO_2 is consistently low and this low TiO_2 content is an important magmatological arc feature in addition to the regional setting of the Afyon region in a zone of compressional tectonic features (Fig. 7).

The most mafic compositions have high molecular $Mg/(Mg + Fe^{2+})$ (> 0.70 with normalized iron oxidation, see Fig. 6) and Ni and Cr contents up to 300 ppm and 500 ppm, respectively. According to these criteria the most magnesian lavas are likely to be primary mantle melts. Olivine phenocrysts in these lavas have extreme Fo contents of Fo_{90-94} (Table III). These primi-

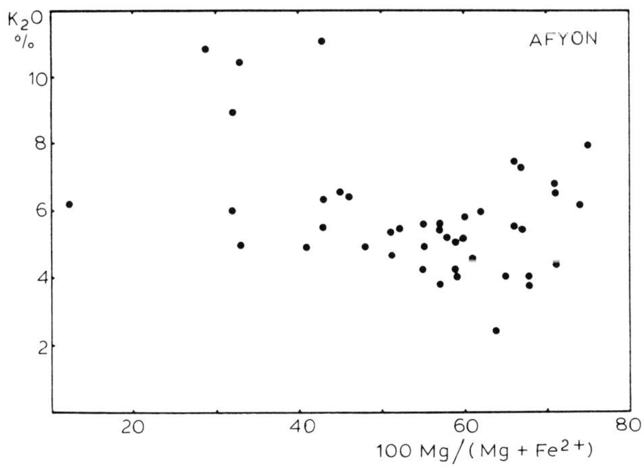

Fig. 6. Magnesium-potassium relationship for potassic rocks from Afyon. Samples with high Mg(Mg + Fe^{2+}) (up to 0.75) are already high in K_2O. Highest potassium values in fractionated rock types (leucitites, phonolites). Mg-values with Fe_2O_3/FeO = 0.15.

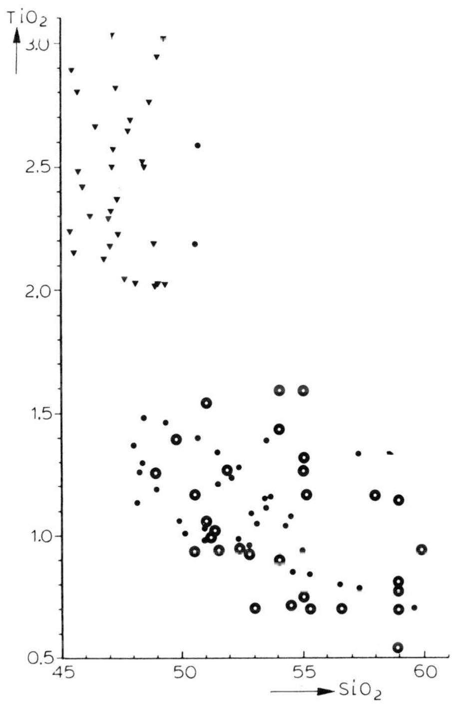

Fig. 7. TiO_2 in Neogene-Quaternary volcanic rocks from Anatolia, used for discrimination between low-TiO_2 arc-related magmas and high-TiO_2 within-plate basalts. Afyon potassic series have the low TiO_2 of the arc series. Triangles: within-plate basalts of eastern Anatolia. Dots: calc-alkaline basalts and related series in central Anatolia. Heavy circles: Afyon potassic series.

TABLE II

Chemical composition of variably potassic rocks from the Afyon province (western Anatolia) (XRF-data)

	Af10	Af110	Af173	Af173	Af119	Af45	Af112
SiO_2	53.0	52.25	50.6	53.15	51.44	51.24	51.94
TiO_2	1.44	1.61	1.0	0.9	0.94	1.25	0.96
Al_2O_3	11.94	10.81	15.38	17.14	16.8	14.5	16.76
Fe_2O_3	1.58	1.75	3.73	3.88	4.03	3.2	3.55
FeO	4.63	3.71	2.45	2.2	1.90	4.5	2.48
MnO	0.1	0.01	0.12	0.12	0.12	0.13	0.12
MgO	8.67	7.97	6.0	3.4	3.98	6.72	4.15
CaO	5.54	4.8	10.19	7.34	8.29	8.67	7.94
Na_2O	1.92	1.26	3.51	2.9	3.9	2.13	3.52
K_2O	5.9	7.3	3.7	4.8	4.86	3.91	4.93
P_2O_5	1.03	0.68	0.63	0.59	0.60	0.68	0.62
CO_2	2.94	5.35	0.15	0.16	0.42	0.87	0.07
H_2O	0.6	1.67	1.12	1.69	1.04	2.25	1.40
Total	99.29	99.25	98.58	98.27	98.32	100.05	98.44
$Mg/(Mg + Fe^{2+})$*	0.74	0.75	0.68	0.55	0.59	0.65	0.60
Ni	270	315	48	12	28	55	26
Cr	511	475	176	23	74	266	89
Rb	189	294	229	116	130	142	120
Sr	737	618	2233	2593	2018	568	2104
Ba	1688	601	2244	3094	2426	864	2533
Zr	618	827	303	414	396	282	398

*Mg-values with $Fe_2O_3/FeO = 0.15$.
Af 10 — olivine-melatrachyte, Pasaköykorusu; Af110 — olivine-biotite lamprophyric dike, Senirköy; Af173 — phonolitic leucite-tephrite, Gökceli; Af178 — latitic lava, Agin; Af119 — leucite-bearing phonolitic tephrite, Ilyasli/Suhut; Af45 — shoshonitic basalt

tive compositions are already K_2O-rich with K_2O/Na_2O = 2—3 and have the high LIL-element concentration of the whole province. This reduces the possibility of significant degrees of crustal contamination.

Strontium isotopic ratios are reported in a preliminary set of 9 determinations in Table IV. Initial ratios range from 0.70388 to 0.70719. Sample Af 10 is among the most primitive compositions and has a $^{87}Sr/^{86}Sr$ of 0.7094 which is one of the most radiogenic ratios. The variable $^{87}Sr/^{86}Sr$ values are taken as evidence of large geochemical heterogeneities in the mantle source. The geochemical variation within the Afyon province cannot be explained with different degrees of partial melting and subsequent fractionation from a broadly homogeneous, nor from a "normal" mantle. The source mantle for Afyon magmas (and on a broader scale for the Anatolian volcanism) was anomalously and heterogeneously enriched in LIL elements and radiogenic

Af40	Af64	Af114	Af116	Af181	Af106	Af32	Is155	Is156
57.65	58.25	49.8	58.14	62.7	63.12	62.01	53.1	62.7
1.17	0.78	1.55	0.49	0.76	0.68	0.57	0.82	0.39
13.73	16.62	16.78	17.94	14.51	15.92	15.78	16.45	16.45
3.51	4.1	5.3	2.86	2.67	1.60	3.47	3.88	2.04
1.6	1.5	1.86	0.86	1.37	2.2	0.67	2.11	0.80
0.10	0.11	0.22	0.07	0.07	0.11	0.08	0.12	0.05
5.67	2.36	1.51	0.83	2.47	2.18	1.97	3.02	1.69
5.01	5.31	5.46	2.75	4.03	4.46	3.85	6.86	2.89
1.94	3.19	2.68	1.39	3.5	3.67	3.42	3.52	4.91
6.27	4.79	11.12	9.88	5.55	4.19	5.18	5.32	5.39
0.56	0.61	0.58	0.18	0.55	0.26	0.43	0.73	0.23
0.11	0.09	0.34	0.17	0.08	0.55	0.10	0.24	0.07
1.2	1.36	2.47	3.18	0.60	1.0	1.1	2.25	0.9
99.04	99.07	99.67	98.74	98.86	99.94	98.63	98.42	98.51
0.71	0.48	0.31	0.33	0.57	0.55	0.51	0.52	0.57
139	0	33	6	35	20	10	11	33
268	30	7	8	91	58	38	8	32
239	160	917	352	183	129	166	110	150
783	1284	2201	3229	1240	787	1289	6021	2973
1207	1634		2639	1718	183	1871	3299	1984
552	349	767	773	427	1239	320	582	356

dike, Bayat; Af112 — phonolitic tephrite, Balcikhisar; Af40 — trachyte lava, Egerli Dag; Af64 — latitic lava, Kislecik; Af114 — phonolitic leucitite, Cicektepe/Cölovasi; Af116 — trachyte, Kinik/Cölovasi; Af181 — latite, dome near Bektas; Af106 — latitic lava, Servan; Af32 — latite, dome of Afyon; Is155 — shoshonite, maar crater Gölcük/Isparta; Is156 — latite, Gölcük/Isparta.

strontium. Processes of mantle metasomatism can be related to the migration of metasomatic fluids from subduction of a lithospheric slab with involvement of crustal material in the subduction process, as proposed for Italy by Cox et al. (1976), Thompson (1977), Civetta et al. (1981). Addition of crustal material to the partial melting in the mantle through subduction was suggested by Keller et al. (1977) in order to explain the HKCA series of Konya.

CONCLUSIONS

High-potassic volcanic rocks are common in Neogene to Quaternary volcanism in the Mediterranean fold chains. Besides the well-known Roman comagmatic region, important examples are the Aeolian island arc, the

TABLE III

Mineral data from Afyon potassic rocks

	1	2	3	4
SiO_2	42.07	40.83	41.8	54.0
TiO_2			3.23	
Al_2O_3			13.73	23.75
FeO	8.37	5.96	3.2	0.80
MgO	49.56	53.2	25.96	
Na_2O			0.27	0.25
K_2O			9.82	20.5
Total	100	100	98	99.5
Si	1.02	0.984	5.724	3.924
Ti			0.332	
Al			2.216	2.034
Fe	0.17	0.12	0.365	0.048
Mg	1.79	1.911	5.3	
Na			0.071	0.037
K			1.715	1.892
Total	2.98	3.015	15.73	7.935
Oxygens	4	4	22	12

1 — olivine in olivine-trachyte lava Af 10.
2 — olivine phenocryst from lamprophyric dyke Af 110.
3 — biotite phenocryst form lamprophyric dyke Af 110.
4 — leucite, large phenocryst from phonolitic leucitite Af 114.

TABLE IV

Strontium isotopic composition of high-K lavas from the Afyon area in western Anatolia. See Table II for geochemical data

No.	$^{87}Sr/^{86}Sr$	$^{87}Sr/^{86}Sr_I$	σ	$^{87}Rb/^{86}Sr$	σ	Rb (ppm)	Sr (ppm)
AF 10	0.70706	0.70694	5	0.700*	—	189⁺	777
AF 32	0.70581	0.70575	5	0.347*	—	166⁺	1376
AF 40	0.70734	0.70719	4	0.88046	86	252	830
AF 106	0.70543	0.70535	10	0.48183	87	137	826
AF 114	0.70597	0.70577	5	1.160*	—	917⁺	2276
AF 119	0.70485	0.70482	4	0.172*	—	130⁺	2173
AF 173	0.70393	0.70388	4	0.275*	—	229⁺	2396
AF 178	0.70487	0.70485	6	0.118*	—	116⁺	2829
AF 181	0.70512	0.70505	4	0.408*	—	183⁺	1291

All samples normalized to NBS 987 = 0.71014. Sr_I calculated with an average age of 12 m.y.
⁺ Rb with XRF, *calculated with XRF Rb.
Analyst: Dr. H. Schleicher.

Hellenic arc and Afyon area in Anatolia. In addition, several similar examples of potassic rocks are quoted in the literature, e.g. from Turkey (Lacroix, 1881; Ota and Dincel, 1975) and from Iran (Riou et al., 1981). An affiliation with plate-convergence processes and the association with typical arc magmas on a regional scale contrasts with the setting of cratonic leucite-bearing rocks in continental-rift situations (Gupta and Yagi, 1980). The problem of distinction between rift-valley type ultrapotassic rocks and the orogenic leucitites on geochemical arguments is far from being solved. Although fundamental differences exist, as e.g. in TiO_2, the main emphasis for distinction is tectonic setting and magmatic evolution in continuity with typical arc volcanism. This is best shown in the Aeolian Islands. While most potassic lavas in the Mediterranean are distinctly fractionated with Mg-values <60, primary magma compositions from Afyon with Mg-values >70 have high K_2O and related elements and $^{87}Sr/^{86}Sr \sim 0.707$, providing important constraints for mantle composition.

Geochemical characteristics of Mediterranean potassic series, with high but variable concentrations of K_2O and LIL elements, including variably radiogenic strontium point to marked geochemical anomalies in the upper mantle. In accordance with the models proposed by Cox et al. (1976), Thompson (1977), Civetta et al. (1981), the source mantle is believed to be strongly metasomatized and enriched in LIL elements and radiogenic isotopes by metasomatic fluids whose ultimate source is the subducted slab. Crustal components in varying amounts can be added by this mechanism to the partial melting process deep in the mantle.

ACKNOWLEDGEMENTS

Samples from Afyon stem from a common project with Letterio Villari (Catania). I am also indebted to H. Schleicher (Freiburg) for the possibility to refer to our unpublished strontium isotope data.

REFERENCES

Angelier, J., Dumont, J.F., Karamanderesi, H., Poisson, A., Şimşek, Ş. and Uysal, Ş., 1981. Analysis of fault mechanism and expansion of southwestern Anatolia since the late Miocene. Tectonophysics, 75: T1—T9.

Appleton, J.D., 1972. Petrogenesis of potassium rich lavas from the Roccamonfina volcano, Roman Region, Italy. J. Petrol., 13: 425—456.

Barberi, F., Innocenti, F., Ferrara, G., Keller, J. and Villari, L., 1974. Evolution of Eolian arc volcanism. Earth Planet. Sci. Lett., 21: 269—276.

Besang, C., Eckhardt, F.-J., Harre, W., Kreuzer, H. and Müller, P., 1977. Radiometrische Altersbestimmungen an neogenen Eruptivgesteinen der Türkei. Geol. Jahrb., Reihe B, 25: 3—36.

Borsi, S., Ferrara, G., Innocenti, F. and Mazzuoli, R., 1972. Geochronology and petrology of recent volcanics in the Eastern Aegean Sea (West Anatolia and Lesvos Island). Bull. Volcanol., 36: 473—496.

Brunn, J.H., Dumont, J.F., Graciansky, P.Ch. de, Gutnic, M., Juteau, Th., Marcoux, J., Monod, O. and Poisson, A., 1971. Outline of the geology of the Western Taurids. In: A.S. Campbell (Editor), Geology and History of Turkey. Petrol. Explor. Society of Libya, Tripoli, pp. 225—256.

Civetta, L., Innocenti, F., Manetti, P., Peccerillo, A. and Poli, G., 1981. Geochemical characteristics of potassic volcanics from Mts. Ernici (Southern Latium, Italy). Contrib. Mineral. Petrol., 78: 37—47.

Cortini, M. and Hermes, O.D., 1981. Sr isotopic evidence for a multi-source origin of the potassic magmas in the Neapolitan area (South Italy). Contrib. Mineral. Petrol., 77: 47—55.

Cox, K.G., Hawkesworth, C.J., O'Nions, R.K. and Appleton, J.D., 1976. Isotopic evidence for the derivation of some Roman Region volcanics from anomalously enriched mantle. Contrib. Mineral. Petrol., 56: 173—180.

Cundari, A., 1980. Role of subduction in the genesis of leucite-bearing rocks: facts on fashion? — Reply to A.D. Edgar's discussion paper. Contrib. Mineral. Petrol., 73: 432—434.

Cundari, A. and Le Maitre, R.W., 1970. On the petrogenesis of the leucite-bearing rocks of the Roman and Birunga volcanic regions. J. Petrol., 11: 33—47.

Dewey, J.E., Pittman III, W.C., Ryan, W.B.F. and Bonnin, J., 1973. Plate tectonics and evolution of the Alpine System. Geol. Soc. Am., Bull., 84: 3137—3180.

Di Girolamo, P., 1978. Geotectonic settings of Miocene-Quaternary volcanism in and around the eastern Tyrrhenian Sea border (Italy) as deduced from major element geochemistry. Bull. Volcanol., 41: 229—250.

Edgar, A.D., 1980. Role of subduction in the genesis of leucite-bearing rocks: discussion. Contrib. Mineral. Petrol., 73: 429—431.

Gasparini, C., Iannacone, G., Scandone, P. and Scarpa, R., 1982. Seismotectonics of the Calabrian Arc. Tectonophysics, 84: 267—286.

Ghiara, M.R., Lirer, L. and Munno, R., 1979. Mineralogy and geochemistry of the "low-potassium series" of the Campania volcanics (South Italy). Chem. Geol., 26: 29—49.

Gupta, A.K. and Yagi, K., 1980. Petrology and Genesis of Leucite-bearing Rocks. Springer, Berlin-Heidelberg-New York, 252 pp.

Harsch, W., Küpfer, T., Rast, B. and Sägesser, R., 1981. Seismotectonic considerations on the nature of the Turkish-African plate boundary. Geol. Rundsch., 70: 368—384.

Hawkesworth, C.J. and Vollmer, R., 1979. Crustal contamination versus enriched mantle: $^{143}Nd/^{144}Nd$ and $^{87}Sr/^{86}Sr$ evidence from the Italian volcanics. Contrib. Mineral. Petrol., 69: 151—165.

Innocenti, F., Manetti, P., Mazzuoli, R., Pasquaré, G. and Villari, L., 1982. Anatolia and NW Iran. In: R.S. Thorpe (Editor), Andesites. John Wiley, London, pp. 327—349. Wiley, London, pp. 327—349.

Innocenti, F., Mazzuoli, R., Pasquaré, G., Radicati di Brozolo, F. and Villari, L., 1975. The Neogene calcalkaline volcanism of Central Anatolia: geochronological data on Kayseri-Nigde area. Geol. Mag., 112: 349—360.

Keller, J., 1974. Petrology of some volcanic rock series of the Aeolian arc, Southern Tyrrhenian Sea: Calcalkaline and shoshonitic associations. Contrib. Mineral. Petrol., 46: 29—47.

Keller, J., 1980. The island of Vulcano. Rend. Soc. Ital. Mineral. Petrol., 36: 396—414.

Keller, J., 1982. Mediterranean island arcs. In: R.S. Thorpe (Editor), Andesites. John Wiley, London, pp. 307—325.

Keller, J., Jung, D., Burgath, K. and Wolff, F., 1977. Geologie und Petrologie des neogenen Kalkalkali-Vulkanismus von Konya (Erenler Dag-Alaca Dag-Massiv, Zentral Anatolien). Geol. Jahrb., Reihe B, 25: 37—117.

Keller, J. and Villari, L., 1972. Rhyolithic Ignimbrites in the region of Afyon (Central Anatolia). Bull. Volcanol., 36: 342—358.

Lacroix, A., 1881. Les roches à leucite de Trabzonde. Bull. Soc. Géol. Fr.

Niggli, P., 1923. Gesteins- und Mineralprovinzen. Bornträger, Berlin, pp. 602.

Ninkovich, D. and Hays, J.D., 1972. Mediterranean island arcs and origin of high potash volcanoes. Earth Planet. Sci. Lett., 16: 331—345.

Ota, R. and Dincel, A., 1975. Volcanic rocks of Turkey. Bull. Geol. Surv. Jpn., 26: 393—419.

Pearce, J.A., 1976. Statistical analysis of major element patterns in basalts. J. Petrol., 17: 15—43.

Riou, R., Dupuy, C. and Dostal, J., 1981. Geochemistry of coexisting alkaline and calc-alkaline volcanic rocks from Northern Azerbaijan (N.W. Iran). J. Volcanol. Geotherm. Res., 11: 253—275.

Rittmann, A., 1981. Vulkane und ihre Tätigkeit. 3rd Ed., Enke, Stuttgart, 399 pp.

Robert, U., 1976. Données nouvelles sur le volcanisme du Sud-Est de la Mer Egée: existence d'une épisode à caractère alcalin. In: "Proc. Int. Congr. on Thermal Waters, Geothermal Energy and Vulcanism of the Mediterranean Area", Athens, Vol. 3, Vulcanism, pp. 211—224.

Rosi, M., 1980. The island of Stromboli. Rend. Soc. Ital. Mineral. Petrol., 36: 345—368.

Savelli, C., 1967. The problem of rock assimilation by Somma-Vesuvius magma. I. Composition of Somma and Vesuvius lavas. Contrib. Mineral. Petrol., 16: 328—353.

Sengör, A.M.C. and Yilmaz, Y., 1981. Tethyan evolution of Turkey: a plate tectonic approach. Tectonophysics, 75: 181—241.

Taylor Jr., H.P., Giannetti, B. and Turi, B., 1979. Oxygen isotopic geochemistry of the potassic igneous rocks from the Roccamonfina volcano, Roman comagmatic region, Italy. Earth Planet. Sci. Lett., 46: 81—106.

Turi, B. and Taylor Jr., H.P., 1976. Oxygen isotope studies of potassic volcanic rocks of the Roman Province, Central Italy. Contrib. Mineral. Petrol., 55: 1—31.

Thompson, R.N., 1977. Primary basalts and magma genesis. III. Alban Hills, Roman Comagmatic Province, Central Italy. Contrib. Mineral. Petrol., 60: 91—108.

Vollmer, R. and Hawkesworth, C.J., 1980. Lead isotopic composition of the potassic rocks from Roccamonfina (South Italy). Earth Planet. Sci. Lett., 47: 91—101.

Vollmer, R., Johnston, K., Ghiara, M.R., Lirer, L. and Munno, R., 1981. Sr isotope geochemistry of megacrysts from continental rift and converging plate margin alkaline volcanism in South Italy. J. Volcanol. Geotherm. Res., 11: 317—327.

Washington, H.S., 1906. The Roman comagmatic region. Carnegie Inst. Washington, Publ., 57: 199 pp.

Journal of Volcanology and Geothermal Research, 18 (1983) 337—359
Elsevier Science Publishers B.V., Amsterdam — Printed in The Netherlands

POTASSIUM-RICH VOLCANIC ROCKS OF THE MURIAH COMPLEX, JAVA, INDONESIA: PRODUCTS OF MULTIPLE MAGMA SOURCES?

I.A. NICHOLLS and D.J. WHITFORD

Department of Earth Sciences, Monash University, Clayton, Victoria 3168 (Australia)
CSIRO, Divison of Mineralogy, North Ryde, N.S.W. 2113 (Australia)

(Received October 29, 1982)

ABSTRACT

Nicholls, I.A. and Whitford, D.J., 1983. Potassium-rich volcanic rocks of the Muriah complex, Java, Indonesia: products of multiple magma sources? In: S. Aramaki and I. Kushiro (Editors), Arc Volcanism. J. Volcanol. Geotherm. Res., 18: 337—359.

The extinct Pleistocene volcano Muriah, situated behind the main Pleistocene—Recent Sunda magmatic arc in north-central Java, has erupted at least two contrasted groups of lavas. One group forms a well-defined compositional series (Anhydrous Series) from leucite basanite to tephritic phonolite, with olivine and tschermakitic clinopyroxene the main phenocrysts. The other group, the "Hydrous Series", includes compositionally variable tephrites and high-K andesites with common plagioclase, biotite and amphibole. Lavas of the Anhydrous Series are much richer in LIL trace elements than the most potassic lavas of neighbouring active volcanoes, but relative HFS element enrichment is less pronounced. REE patterns have almost constant slopes from La (250—600 times chondrites) to Yb (5—10 times chondrites), while those of lavas of active centres are less light-enriched, and show flattening in the heavy REE. Anhydrous Series initial $^{87}Sr/^{86}Sr$ ratios (0.7043—0.7046) are lower than those of active centres (0.7047—0.7053). Hydrous Series lavas are intermediate in all these geochemical characteristics.

The most mafic A-series leucite basanite, with $Mg/(Mg + Fe^{2+}) \simeq 0.69$, 140 ppm Ni and 620 ppm Cr was probably derived from the primary magma for the series by fractionation of only ~ 5 wt.% olivine. Its REE pattern suggests derivation from a garnet-bearing source. Experiments on this basanite, with up to 10% olivine and 20% orthopyroxene added, and in the presence of H_2O and H_2O/CO_2 mixtures, have shown that for all but very high magma water contents, the olivine and garnet liquidus fields are widely separated by fields of phlogopite and clinopyroxene. There is no liquidus field of orthopyroxene. Hence, if magma production involved an equilibrium melting process alone, the most probable sources are of garnet-bearing phlogopite clinopyroxenite type. Alternatively, this magma may represent the end-product of interaction between a low-K basanite magma from a garnet lherzolite source in the asthenosphere and a phlogopite-bearing lherzolite zone in the lower lithosphere. Its production was probably related to crustal doming and extension superimposed on the dominant subduction regime. Hydrous Series magmas may have resulted from mixing between Anhydrous Series magmas and high-K calc-alkaline basaltic to andesitic magmas more directly related to subduction processes.

0377-0273/83/$03.00

INTRODUCTION

Discussion by Edgar (1980) and Cundari (1980) has shown that wide divergence of opinion exists on the origin and tectonic significance of potassium-rich (especially leucite-bearing) volcanic rocks in island arc and continental margin settings. A major question is whether K-rich magma production in these settings is directly related to subduction of oceanic crust, to crustal uplift and rifting, or both.

In the Sunda volcanic arc of Indonesia, K-rich lavas have been erupted from several Pleistocene-Recent centres above the active northward-dipping Benioff zone. The best known of these are the extinct Muriah and Ringgit Beser in Java (van Bemmelen, 1947, 1949) and the extinct Sangenges and Soromundi and active Sangeang Api and Tambora in Sumbawa (Foden and Varne, 1980). Both Edgar and Cundari have selectively used geochemical data for Muriah lavas to support their views of the significance of K-rich arc volcanism. Edgar, citing the suggestion of Whitford and Nicholls (1976) that tephritic to andesitic lavas of Muriah with initial $^{87}Sr/^{86}Sr$ ratios similar to those of calc-alkaline basaltic to dacitic lavas of nearby active volcanoes could be genetically related to the latter group, states that both groups are related to subduction of oceanic crust. Cundari, noting that the lower $^{87}Sr/^{86}Sr$ ratios of leucite-bearing lavas of Muriah suggest that they are genetically distinct, favours an origin related to rifting. Neither author discusses the possible significance of these two groups of lavas within the same volcanic complex, or attempts to reconcile their opposed viewpoints.

The Muriah centre has yet to be mapped and sampled in detail. In this paper, some petrological and geochemical features of a reconnaissance collection of lavas and intrusive rocks, made as part of a study of geochemical variation in lavas across the Sunda arc in Java and Bali (Whitford et al., 1979; Nicholls et al., 1980) are presented. A full-scale geochemical and petrological study must await further data on the stratigraphy, structure and eruptive history of the Muriah volcano.

TECTONIC AND GEOLOGICAL SETTING OF MURIAH

In Java and Bali (Fig. 1) the crust of the Sunda Arc is 20—25 km thick, with an essentially continental seismic velocity structure. The submerged Sunda Shelf to the north has similar seismic velocity characteristics (Ben Avraham and Emery, 1973). The Sumatra to Java sector of the arc has apparently lain close to the margin of the Asian continent since at least late Mesozoic times. Katili (1973) and Hamilton (1979) have suggested that from Cretaceous to Miocene times, it extended from central Java to eastern Kalimantan (Borneo). The formation of the modern arc eastward from Java, largely flanked on both sides by oceanic crust, required a post-Miocene rotation, accompanied by the opening of the Flores and Banda Seas.

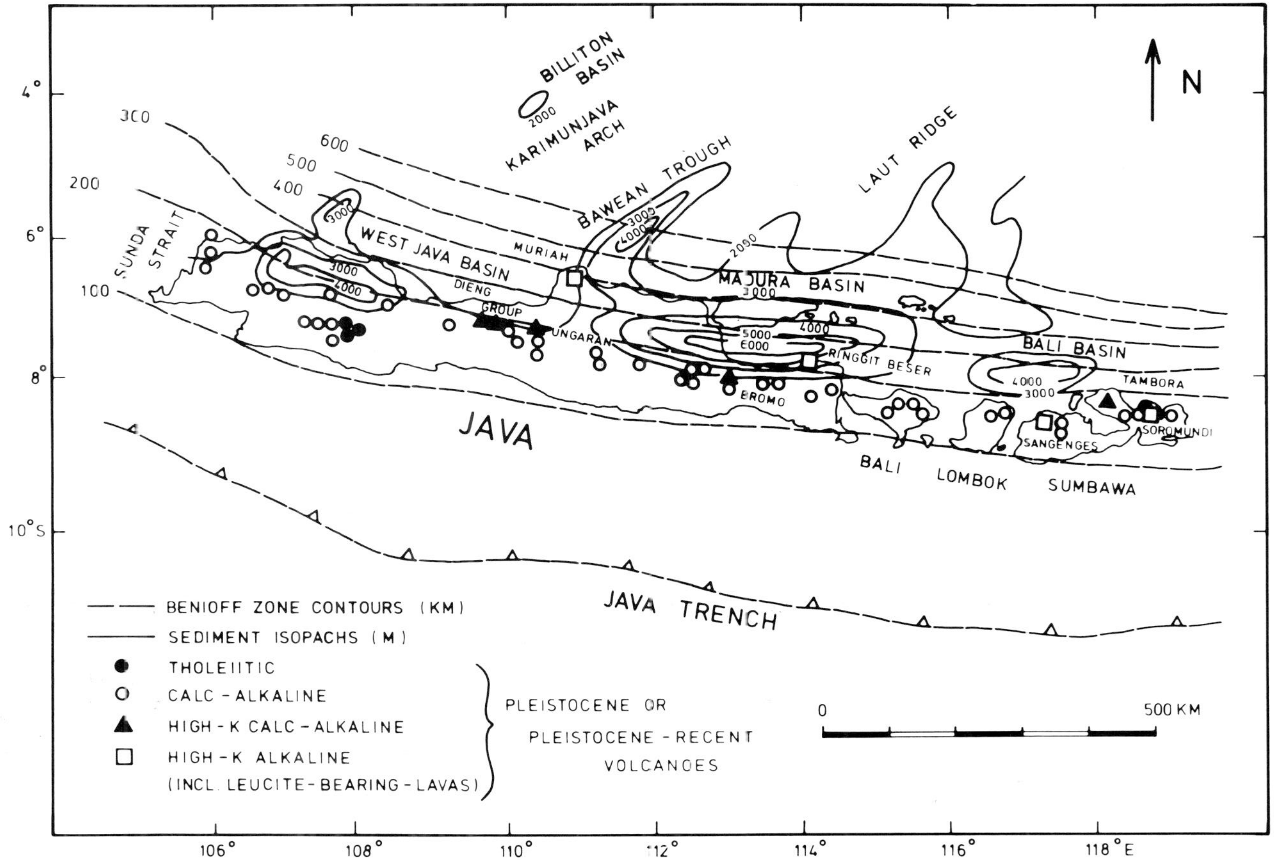

Fig. 1. Tectonic setting of Pleistocene and younger volcanoes of the Java-Sumbawa sector of the Sunda arc. Benioff zone contours and sediment isopachs after Hamilton (1979).

Structures associated with the pre-Miocene trend of the arc appear to be reflected by a series of northeasterly trending basins and rises in the Java Sea (Fig. 1), including, in the Central to East Java sector, the Karimunjava Arch and Laut Ridge, and intervening Bawean Trough. However, during the Oligocene, two east—west-trending foreland basins, the West Java and Madura Basins, (Fig. 1) developed at the southern margin of the Sunda Shelf, and large thicknesses of marine sediments accumulated before progressive emergence began in the late Miocene (Hamilton, 1979). Folded shales, sandstones and limestones of the Madura Basin now comprise much of the broad anticlinorial zone of north-central Java, and the maximum thickness of sediments exceeds 6000 m. The Muriah complex lies at the western end of the Madura Basin, near the intersection of the basin axis with that of the Bawean Trough. This location strongly suggests that major extensional structures within the crust or lithosphere controlled magma uprise.

The present-day Benioff zone is very clearly defined in the Java-Bali arc sector, extending to depths of 600 km. It lies approximately 360 km beneath Muriah (Fig. 1).

NATURE AND AGE OF MURIAH DEPOSITS

According to van Bemmelen (1947, 1949) the earliest products of Muriah volcanism are calc-alkaline ("Pacific Suite") basaltic to andesitic lavas and fragmental deposits of the Genuk cone, at the northern foot of the volcano. The age of these deposits is unknown. No samples were collected for the present study. The Genuk deposits are partly covered by leucite-bearing breccias of the main cone. The latter contains several "sector graben", and it has been deeply incised by the approximately north—south Gelis River. A Middle Pleistocene vertebrate fauna has been found beneath the southern foot of the cone, and gives a maximum age for the beginning of eruption of leucite-bearing rocks.

There are also two younger parasitic cones (Bako and Ragas cones) consisting of trachytes or phonolites (van Bemmelen, 1947). The volcano is now apparently totally extinct.

Most of the samples studied, like those of van Bemmelen (1947) were taken from 0.5—1 m boulders in the floor of the Gelis River valley. They include basanites, tephrites and rarer trachytes and phonolites, some clearly intrusive. Corresponding lavas and fragmental deposits exposed in the valley walls are heavily weathered, and rarely sufficiently fresh for geochemical study. However, a small collection of very fresh tephrites was taken from a lava-flow sequence on the southern flank of the main cone.

Initial subdivision of the available samples into two groups — the Anhydrous (A) and Hydrous (H) Series (Whitford, 1975a; Nicholls and Whitford, 1976a; Whitford and Nicholls, 1976a, b) was based on the abundance of phenocrysts of hornblende and biotite relative to clinopyroxene. This subdivision was later refined using features of bulk rock and mineral chemistry (see below). Samples belonging to the Anhydrous Series show much less evidence of low-temperature alteration and weathering, suggesting that they are the younger group. No stratigraphic control is yet available to confirm this observation.

MINERALOGY

Features of the normative and modal mineralogy of the two series are summarized in Table I.

TABLE I

Summary of normative and modal mineralogy of Muriah Anhydrous and Hydrous Series rocks

	A-series	H-series
Compositional range	Ol-rich basanite—tephritic phonolite	Ol-poor basanite—high-K andesite
Normative mineralogy (wt.%):		
Ol	15.0— 2.2	9.5— 0
Di	37.1— 6.8	25.0— 1.8
Hy	—	0 — 7.0
Ab	0 —15.2	0 —41.0
Ne	23.8— 7.9	17.8— 0
Lc	17.1— 0	2.2— 0
Qz	—	0 — 2.0
Major modal mineral compositions:		
Olivine (Mg No.)[a]	90—39	(80)[c]
Ca-Cpx (Mg No.)	92—36	80—52
[cat.%(Al+Fe^{3+}+Ti)]	3—20	3—14
Amph (Mg No.)	30—26[b]	72—51
Bi (Mg No.)	81—72	71—46
Plag Fsp (mol.% An)	74—29	86—39
Alk Fsp (mol.% Or)	37—89	52—73
Ti-Mt (mol.% Usp)	59— 4	40—17

[a]Mg No.: 100 Mg/(Mg+Fe^{2+}).
[b]Tephritic phonolite 72 - 940 only.
[c]Optical estimate, basanite 72 - 937.

Anhydrous Series

Lavas of the A-series are mainly basanites and tephrites, containing up to 20 vol.% phenocrysts, 1—10 mm in diameter, set in fresh, dark, finely crystalline groundmasses. Green clinopyroxenes, typically with strong normal, reverse or oscillatory zoning from diopsidic compositions to tschermakitic compositions with up to 20 cation % (Al^{3+} + Fe^{3+} + Ti^{4+}) are the dominant phenocrysts, while up to 10% olivine phenocrysts are present in the basanites and more mafic tephrites, with rare fresh leucite and Ti-magnetite.

Phenocryst assemblages of the more felsic members of the series are more diverse, usually including biotite, plagioclase, alkali feldspar, leucite and occasionally pargasitic amphibole and Ti-magnetite. The most felsic analysed sample (tephritic phonolite 72-940) contains 1—5 mm phenocrysts of red-brown Ti-andradite.

Groundmass assemblages are dominated by deep green aluminous clinopyroxene, plagioclase and Ti-magnetite, with olivine in more mafic types. The pseudoinvariant assemblage alkali feldspar, leucite, nepheline and sodalite is present in most samples (cf. Baldridge et al., 1981).

Hydrous Series

The most mafic sample (72-937) of the Hydrous Series is an intrusive basanite, rich in augite and plagioclase and with minor olivine and biotite. Most other samples are tephrites (18—6% Ne, $\leqslant$ 2% Lc) or high-potassium andesites (3% Ne to 2% Qz) with up to 30% phenocrysts. The dominant clinopyroxene phenocrysts are weakly zoned pale green augites. Plagioclase is more abundant (usually > 5%) than in Anhydrous Series equivalents and alkali feldspar, biotite and pargasitic amphibole are ubiquitous minor phenocrysts.

All the H-series lavas have partly altered groundmasses, with subhedral to anhedral augite, plagioclase, alkali feldspar and Ti-magnetite predominant, and original nepheline and rare leucite extensively replaced by analcite and calcite.

GEOCHEMISTRY

Major element, trace element and isotopic data for representative A- and H-series rocks are given in Table II. All available data for selected elements are given in Figs. 2—4. The most obvious differences between the series are the much greater degree of silica-under-saturation and more mafic character of the A-series, its higher K contents (Fig. 2) and K/Na ratios, its higher P contents and its relative enrichment in LIL (large ion lithophile ion) and HFS (high field strength ion) trace elements. Variation in particularly the LIL trace elements Rb, Ba and Sr (Fig. 3a) within and between the A- and H-series is much less regular than that of K (Fig. 2). In the case of the H-series, this may in part reflect alteration. Variation in the HFS trace elements, particularly Zr (Fig. 3b), is relatively well defined.

All the Muriah rocks are enriched in K, P and LIL and HFS trace elements with respect to lavas of the nearest active volcanoes, Ungaran and the Diëng group (Whitford et al., 1979). All Muriah, Ungaran and Diëng rocks have the high Al and low Ti contents usually associated with arc lavas.

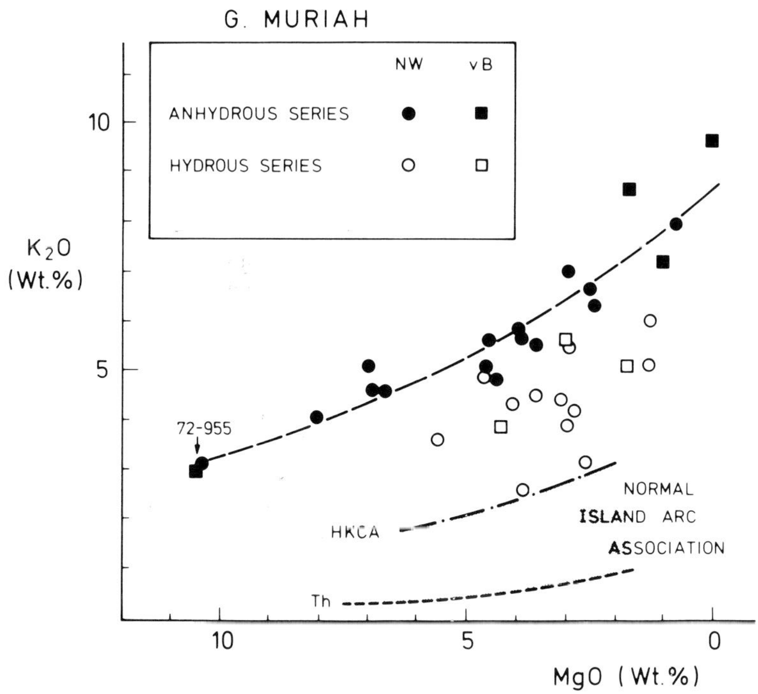

Fig. 2. K_2O—MgO (wt.%) relationships between Muriah A- and H-series lavas, and lavas of the active volcanoes of Java and Bali (*Th* = tholeiitic series; *HKCA* = high-K-calc-alkaline series). Data sources: *NW* = this study. *vB* = van Bemmelen (1947).

TABLE II

Analytical data for representative samples from Muriah and nearby active volcanoes

	Muriah			Ungaran High-K	Diëng calc-alkaline
	A-series		H-series	71-982[b]	71-1010[b]
	72-955[a]	72-940[a]	72-939[a]		
SiO_2	45.2	54.5	55.5	48.3	58.5
TiO_2	1.26	0.29	0.70	1.16	1.08
Al_2O_3	13.63	22.8	18.27	17.38	17.41
ΣFeO^c	9.4	3.38	7.05	10.14	7.73
MnO	0.17	<0,12[d]	<0,12[d]	0.20	0,19
MgO	10.4	0.79	2.86	6.24	2.19
CaO	13.28	4.6	7.09	11.72	5.64
Na_2O	2.6	5.48	3.86	2.58	4.00

(continued)

TABLE II (continued)

	Muriah			Ungaran High-K	Diĕng calc-alkaline
	A-series		H-series	71-982[b]	71-1010[b]
	72-955[a]	72-940[a]	72-939[a]		
K_2O	3.1	8.0	4.23	1.62	2.74
P_2O_5[b]	1.02	0.12	0.46	0.33	0.48
Loss[e]	0.62	2.44	5.71	1.36	1.01
Fe_2O_3/FeO[f]	0.62	1.10	h	0.63	0.82
$Mg/(Mg+\Sigma Fe)$	0.663	0.294	0.420	0.523	0.335
K_2O/Na_2O	1.2	1.5	1.10	0.63	0.69
Cs[g]	3.7	8.3	3.1	4.1	4.7
Rb	136	215	125	55.1	78.1
Ba	950	920	1200	480	700
Sr	809	1690	1000	456	405
Pb	4.5	21	14	5.7	26
La	74.4	189	81.9	23.1	40.4
Yb	0.89	1.05	1.67	1.52	5.24
Y	27	34	28	25	51
Th	11	49	20	6.5	12
U	2.1	9.8	4.3	1.2	2.5
Zr	224	367	228	93	188
Hf	5.2	4.4	4.6	2.0	5.9
Sn	2.4	1.2	1.5	1.2	2.9
Nb	53	120	36	11	12
Cr	620	3		71	25
V	290	130		290	120
Sc	38	3		31	20
Ni	140	6.3		24	<2[d]
Co	44	6.5		36	12
Cu	83	24		125	20
$^{87}Sr/^{86}Sr$[i]	0.70446	0.70431		0.70479	0.70524

[a]Major-element analyses by energy-dispersive microprobe on glass prepared by Ir-strip fusion.

[b]Major-element analyses by X-ray fluorescence spectrometry. All analyses normalized to 100%, volatile-free.

[c]All Fe calculated as FeO.

[d]Below limit of detection.

[e]Loss on ignition includes H_2O, CO_2 halogens, etc.

[f]Fe_2O_3 and FeO determined colourimetrically.

[g]Trace-element analyses (ppm) by: X-ray fluorescence spectrometry (Rb, Sr, Zr, Y); emission spectrography (Cr, V, Sc, Ni, Co, Cu, Ba); spark-source mass spectrography (Th, U, Pb, Hf, REE, Ba, Cs, Sn, Nb — see Whitford et al., 1979).

[h]Space = not determined.

[i]For $^{87}Sr/^{86}Sr$ data — see Whitford (1975a). For 72-940, $^{143}Nd/^{144}Nd$ = 0.51270 — see Whitford et al. (1981).

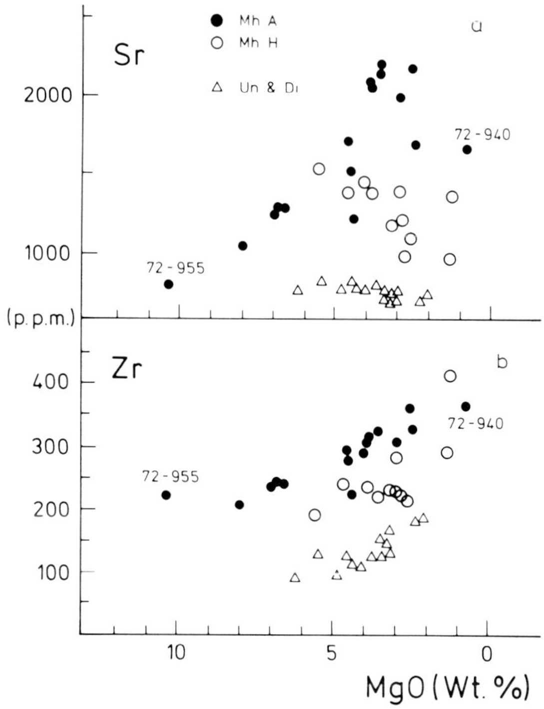

Fig. 3. Sr—MgO and Zr—MgO (ppm.-wt.%) relationships between Muriah A- and H-series lavas, and lavas of the nearby active high-K calc-alkaline volcanoes of Ungaran and the Diëng group.

Representative rare earth element patterns for Muriah, Ungaran and Diëng lavas are given in Fig. 4. Muriah A-series patterns show strong light enrichment, with La ranging from 250 times chondrites for the most mafic basanite (72-955) to 600 times chondrites for tephritic phonolite 72-940, and almost constant slopes through the REE to low Yb levels (4—12 times chondrites). The single H-series pattern, for tephrite 72-939, has lower LREE levels than those of similar A-series tephrites, and it shows a slight flattening in the HREE. Patterns for Ungaran and Diëng lavas are still less enriched in LREE, and flattening in the HREE is more pronounced. There are no pronounced Eu anomalies in Muriah patterns, and only small positive or negative anomalies in some Ungaran patterns.

As noted by Whitford (1975a), the Muriah A-series lavas have lower initial $^{87}Sr/^{86}Sr$ ratios (0.7043—0.7046) than almost all other Pleistocene—Recent volcanic rocks of Java and Bali. The three available H-series determinations (0.7046—0.7048) are intermediate between those of the A-series and those of Ungaran and Diëng lavas (0.7047—0.7053). A single $^{143}Nd/^{144}Nd$ determination for the A-series tephritic phonolite 72-940 (0.51270) places this rock within the Sr-Nd "mantle array" (Whitford et al., 1981).

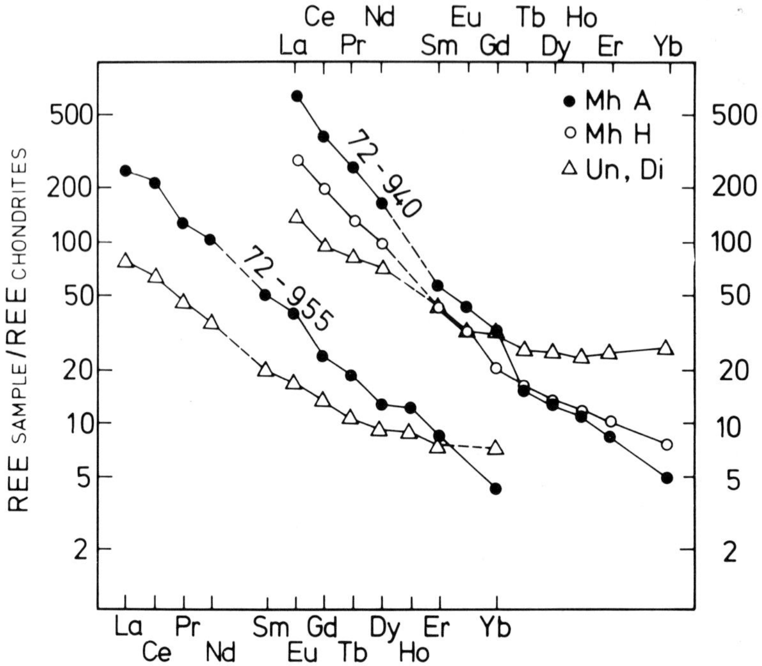

Fig. 4. Rare earth element patterns for the representative samples of the Muriah A- and H-series and high-K calc-alkaline series of Ungaran and Diëng given in Table II. A-series patterns are labelled. The remaining patterns are: H-series—tephrite 72-939; high-K calc-alkaline series—basanite 71-982, Ungaran (left) and andesite 71-1010, Diëng (right).

Figures 2–4 show clearly that in both major- and trace-element abundances, the Muriah H-series rocks fall between the Muriah A-series and the lavas of Ungaran and Diëng. This is also the case for Sr-isotope ratios. The possible significance of these features is discussed below.

K-*h* RELATIONSHIPS

A K_2O-MgO diagram (Fig. 2) has advantages over the more conventional K_2O-SiO_2 diagram as the basis on which to examine potassium content-depth to Benioff zone (K-*h*) relationships in the Java-Bali arc sector (Fig. 5). While the use of a constant SiO_2 value (e.g., 55% SiO_2 for K_{55} — Dickinson and Hatherton, 1967) for comparison of series of contrasting silica-saturation effectively compares the more mafic members of oversaturated series with more felsic members of undersaturated series, use of a constant MgO content (e.g., 6% MgO) allows comparison of lavas of similar maficity and degree of fractionation from the parent magma. Use of the Mg-number, 100 $Mg/(Mg + Fe^{2+})$, a true differentiation index (Green and Ringwood, 1967) should in fact be preferable, but this requires assumptions concerning Fe^{3+}/Fe^{2+} ratios to be made.

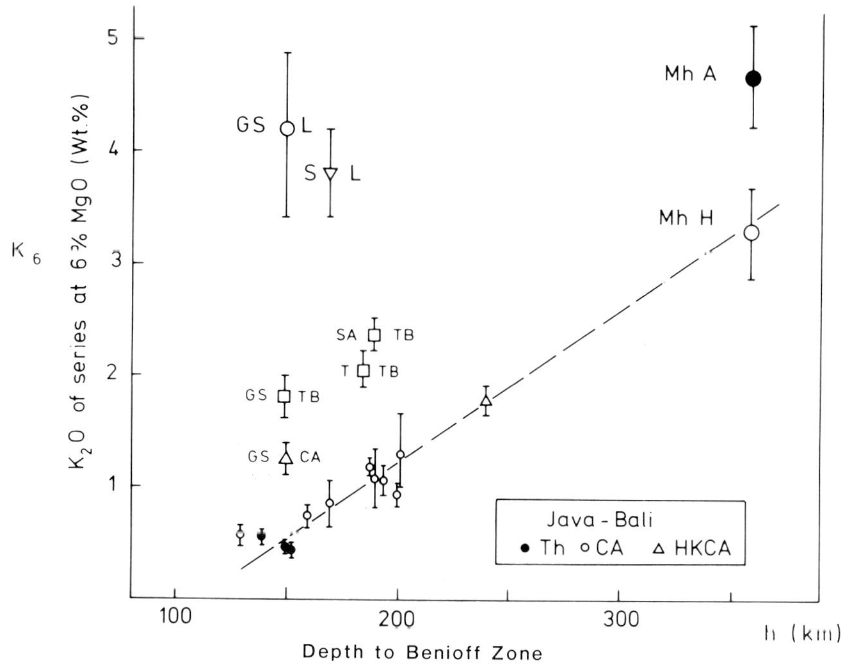

Fig. 5. K_{Mg} vs. *h* diagram for lava series of the normal island arc association in Java and Bali, Muriah A- and H-series lavas, and lavas of extinct Pleistocene and active high-K centres of Sumbawa (Foden and Varne, 1980). *GS* = Sangenges; *S* = Soromundi; *SA* = Sangeang Api; *T* = Tambora; *CA* = calc-alkaline; *TB* = trachy-basalts; *L* = leucitites.

Values of K_6 (wt.% K_2O at 6 wt.% MgO) for tholeiitic, calc-alkaline and high-K calc-alkaline lava series of Java and Bali (Whitford, 1975a; Whitford et al., 1979) form a well-defined trend (Fig. 5) when plotted against corresponding depths to Benioff zone of 130—250 km (see also Hutchison, 1975; Whitford and Nicholls, 1976). The very imprecise K_6 value for Muriah H-series lavas plots almost on this trend, while the A-series has a K_6 value far above an extrapolation of this trend to h = 360 km. Values of K_6 for the Pleistocene—Recent K-rich volcanic rocks of Sumbawa, with corresponding depths to Benioff zone of only 150—190 km, are also given. As noted by Foden and Varne (1980), these rocks, like the Muriah A-series, violate the K-*h* trend for the active volcanoes of the Java Bali arc sector.

Because they appear to conform to the K-*h* relationship for lavas of the active volcanoes of Java and Bali, the Muriah H-series rocks might simplistically be interpreted as "related to subduction" in the same sense as the former lavas (Edgar, 1980). However, the presence in at least two of the K-rich centres of Java and Sumbawa (Muriah, Sangenges) of two or more groups of lavas with differing degrees of silica-saturation, K-contents and K/Na ratios, and probably, Sr-isotope ratios, indicates either multiple magma sources, or change in conditions of magma genesis, or both. These contrasted groups probably reflect changes in tectonic environment during arc evolution, the possible nature of which is discussed below.

EVOLUTION OF A-SERIES LAVAS

As illustrated in Figs. 2 and 3, the Muriah A-series lavas show well-defined serial relationships within the compositional range leucite basanite to tephritic phonolite. Assuming that it had a low pre-eruptive $Fe^{3+}/(Fe^{2+} + Fe^{3+})$ ratio (~ 0.10), compatible with a reduced mantle source (Arculus and Delano, 1981), the most mafic leucite basanite, 72-955 (Table II), has $Mg/(Mg + Fe^{2+}) \simeq 0.69$. Using a minimum value of $K^{Ol\text{-}Liq}_{Fe^{2+}-Mg} = 0.30$ (for discussion, see Grover et al., 1980), its liquidus olivine should have a maximum $Mg/(Mg + Fe^{2+})$ of 0.88. However, cores of olivine phenocrysts reach $Mg/(Mg + Fe^{2+}) = 0.90$. The presence of olivine slightly more magnesian than that expected from the bulk composition, plus the relatively low Ni content (140 ppm) of 72-955, suggest that it has fractionated olivine. Cores of clinopyroxene phenocrysts reach $Mg/(Mg + Fe^{2+}) \simeq 0.91$ (using Fe^{2+} contents calculated from charge balance), approximately the composition expected to be in equilibrium with the most magnesian olivine cores (Nicholls, 1974). However, the high Cr content (620 ppm) of 72-955 indicates that it has not fractionated appreciable clinopyroxene.

The composition of a parental magma for 72-955, in equilibrium with olivine, $Mg/(Mg + Fe^{2+}) = 0.90$, may be calculated by incrementally restoring fractionated olivine to the 72-955 composition (for procedure, see Nicholls and Whitford, 1976b). The required proportion of olivine is 5.2 wt.%, and the parental magma composition has 11.9 wt.% MgO. This result can be checked using Ni systematics. A well-constrained value of $D^{Ol\text{-}Liq}_{Ni}$ of 10.25, determined by Nabelek (1980) for an olivine tholeiite at 1267°C, probably represents a minimum value for more potassic liquids at similar temperatures (see Takahashi, 1978). Using this value, a parent magma which could give rise to 72-955 by Rayleigh fractionation of 5.2 wt.% olivine must have contained a minimum of 230 ppm Ni. This is within the concentration range (200—350 ppm Ni) for primary magmas produced by $<$ 20% melting of a lherzolitic mantle with 1500—2000 ppm Ni (Hanson and Langmuir, 1978; Nicholls, 1978). These calculations indicate that the composition produced by adding 5 wt.% olivine (~ $Fo_{88.5}$) to 72-955 is a close approximation to a primary mantle derived magma composition.

In Fig. 6, the A-series lavas are plotted in logarithmic cation-percentage variation diagrams using Mg as abscissa. Plotted for comparison are the A-series primary magma composition (PM), the 72-955 composition (related to PM by olivine fractionation alone) and calculated liquid composition trends for single-mineral fractionation of olivine and simplified equivalents of the phenocryst clinopyroxenes of the A-series from a parent of 72-955 composition.

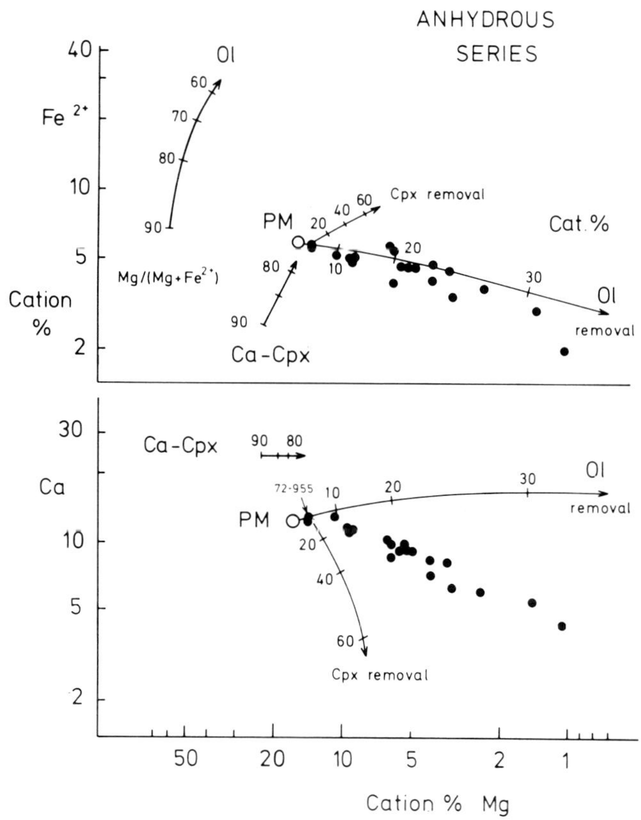

Fig. 6. Cation percentage variation diagrams for Muriah A-series lavas, with comparative calculated fractionation trends for olivine and calcic clinopyroxene, using leucite basanite 72-955 as parental composition. Also shown is the estimated composition of the A-series primary magma (PM).

The Fe^{2+}-Mg trends were calculated using published $K^{\text{Cryst-Liq}}_{Fe^{2+}\text{-}Mg}$ relationships for olivine and clinopyroxene (Nicholls, 1974). The Ca-Mg trend for clinopyroxene was calculated assuming a constant atomic proportion of Ca in the clinopyroxene (a close approximation to the behaviour of the natural phenocrysts). Also given are the corresponding compositional trends for the fractionated olivine and clinopyroxene.

In the Ca-Mg diagram, the rapid depletion of Ca through the series shows that clinopyroxene is the dominant fractionating phase (plagioclase fractionation is precluded by the gradual rise in Al through the series, and the lack of negative Eu anomalies in REE patterns; Fig. 4). In the Fe^{2+}-Mg diagram, the Fe^{2+} depletion through the series is more rapid than that produced by fractionation of either olivine or clinopyroxene, suggesting fractionation of an additional Fe-rich phase such as Ti-magnetite. Least-squares mixing calculations using natural phenocryst compositions yield mixtures of olivine, clinopyroxene and Ti-magnetite (weight ratio approximately 5:15:1) as fractionates for the steps leucite basanite to mafic tephrite, and mafic tephrite to felsic tephrite (Whitford, 1975b).

EXPERIMENTAL PHASE RELATIONS OF LEUCITE BASANITE 72-955

Liquidus phase relations of the 72-955 composition in the presence of water and carbon dioxide have been studied experimentally, using glass starting materials prepared from analysed rock powder. Aliquots of 10—20 mg of glass were sealed into Ag-Pd alloy or Pt capsules with accurately measured amounts of water and/or silver oxalate, and crystallized at pressures of 5—35 kbar (0.5—3.5 GPa) in a 0.5″-diameter piston-cylinder apparatus. For experiments with only water added, simple talc furnace assemblies which produce fH_2 between the NNO and WM buffers (Mysen et al., 1974) were used. The results of these experiments are described below. For experiments with carbon dioxide, internal talc parts were replaced by Alsimag crushable ceramic, to yield $fH_2 \simeq$ NNO. Experiments with CO_2 and H_2O/CO_2 mixtures are still in progress, and will be reported elsewhere.

The near-liquidus phase relations of 72-955 with added water are summarized in Fig. 7. Major features are the relatively low maximum pressure of olivine crystallization (< 15 kbar) and the high minimum

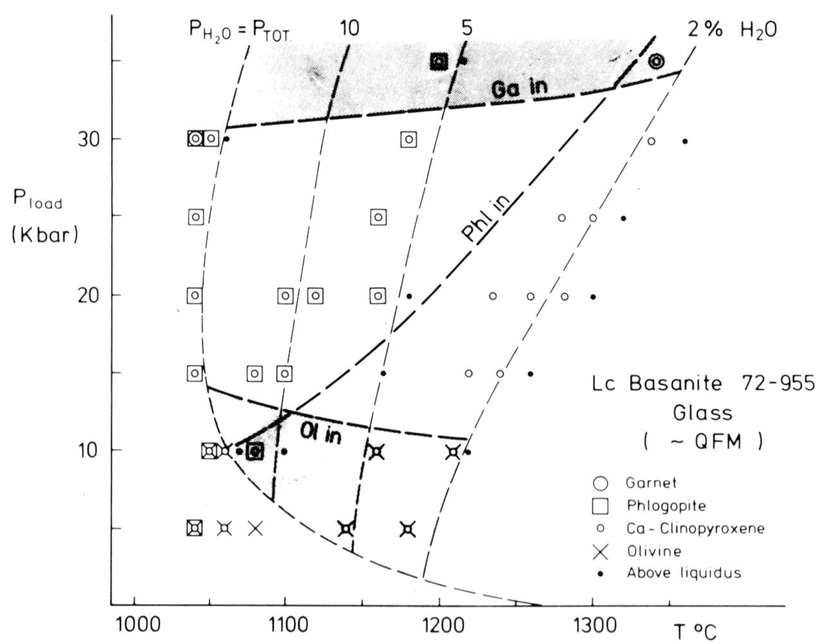

Fig. 7. P-T-$X_{H_2O}^{Liq}$ liquidus surface diagram for the 72-955 composition.

pressure of garnet crystallization (> 25 kbar). The appearance of neither mineral is very sensitive to variation in melt H_2O content. Diopsidic clinopyroxene is a ubiquitous near-liquidus phase, becoming enriched in jadeite with increasing pressure and decreasing temperature. The other major phase is phlogopite, which is on or close to the liquidus for melt water contents > 5 wt.%. Amphibole is not a liquidus phase, but it occurs at temperatures near 950°C, about 80°C below the H_2O-saturated liquidus, at 15—20 kbar (Nicholls and Whitford, 1978). Most notably, orthopyroxene has failed to crystallize under any P-T-xH_2O conditions used.

Further experiments have been carried out to assess the effects of addition of synthetic crystalline olivine and orthopyroxene on liquidus phase relations. In each case, added material more Fe-rich than the equilibrium composition has been used to offset Fe loss to the sample capsule. Addition of 5 wt.% of ~ 5 μm grain size olivine Fo_{80}, in experiments in which the run temperature was approached from below, increased the maximum pressure of olivine growth by about 5—10 kbar. When these experiments were "reversed" by first holding olivine-seeded charges above the liquidus before approaching run temperature, the maximum pressure of olivine growth was only 2—5 kbar above that for unseeded charges with the same water content. The stability fields of the other phases remained almost unaffected. Addition of 10 wt.% of 10—20 μm grain size orthopyroxene En_{70}, in charges taken to near-liquidus temperatures from below, caused slight increases in the maximum pressure of olivine crystallization, but all orthopyroxene disappeared. Similar orthopyroxene-seeded experiments in which H_2O was replaced by CO_2 on a mole-for mole basis produced garnet at pressures (20—25 kbar) below those at which it appeared in the presence of H_2O alone (an effect also observed in an olivine melilitite by Brey and Green, 1975, 1977), but orthopyroxene again completely reacted out.

These phase relations indicate that if a simple equilibrium partial melting process was involved in the production of the A-series primary magma, then:

(a) The residue from this process is unlikely to have contained orthopyroxene. Hence the source was not a lherzolite, harzburgite or websterite.

(b) If the source contained garnet (as indicated by REE evidence) it is unlikely to have also contained olivine.

(c) By elimination, the most probable source was of garnet-bearing phlogopite clinopyroxenite type.

Alternatively, a more complex magma genesis process was involved (see below).

DISCUSSION

Tectonic regime during Muriah magmatism

The close association between intracontinental K-rich alkaline volcanism and regions of crustal uplift and extension is well documented, (Lloyd and

Bailey, 1975). Similar settings for K-rich volcanism occur in active continental margins, a relatively small-scale example being the Colima Graben, within the western Mexican Volcanic Belt (Luhr and Carmichael, 1980, 1981).

As discussed by Luhr and Carmichael (1981), the Mexican Volcanic Belt is influenced by two tectonic regimes, crustal uplift and approximately east—west extension being superimposed upon the dominant north-northeastward subduction of the Cocos plate into the Middle America Trench. Within the 80 km long, north—south-trending Colima Graben, eleven cinder and lava cones composed of strongly silica-undersaturated, K-rich basanites and minettes are grouped about the Pleistocene—Recent stratovolcano of Volcán Colima. Material similar to the basanites and minettes occurs as a component of young pyroclastics of Volcán Colima. This is thought to indicate injection of basanitic magma into its dominantly andesitic magma system. Major- and trace-element data for rocks of Volcán Colima and the cinder cones indicate that there is no direct genetic relationship between the two groups. There are strong geochemical similarities between the Colima Graben basanites and the mafic members of the Muriah A-series , particularly in the "arc-related" characteristics of high Al and low Ti and HFS-element abundances. As in the case of the Colima Graben volcanics, geochemical evidence (particularly Sr-isotopic ratios) indicates that the A-series magmas came from sources distinct from those associated with less alkaline types both at Muriah and at the adjacent active calc-alkaline centres. There is a strong possibility that the contrasted products of Muriah also reflect the overprinting of the dominant subduction regime by one of crustal doming and extension. The apparently brief period of Muriah activity suggests that the process was short-lived.

Origin of the A-series primary magma

The compositions and liquidus phase relations of basanite 72-955, and hence the A-series primary magma, differ considerably from those of K-rich mafic to ultramafic magmas in continental rift environments (Edgar et al., 1976, 1980; Edgar and Condliffe, 1978; Ryabchikov and Green, 1978). In East African occurrences, "mafurite" and "ugandite" compositions with very high K and Ti, relatively low Al and Ca, and extremely low Na (by comparison with 72-955) crystallize orthopyroxene in addition to at least two of olivine, clinopyroxene and garnet, at pressures near 30 kbar, in the presence of H_2O and CO_2. In each case, the corresponding magmas have been interpreted as equilibrium partial melts of lherzolitic sources. Wendtlandt and Eggler (1980a,b) have discussed the nature of phase assemblages and melting reactions applicable to the production of these magmas.

"Madupites" of another well-known continental occurrence, the Leucite Hills of Wyoming, show no liquidus stability field for orthopyroxene in the presence of water, and Barton and Hamilton (1979) conclude that a lherzolitic source is unlikely, even in the presence of CO_2. They propose a phlogopite pyroxenite source, but note the possibility that the madupite represents a K-enriched modification (by volatile transfer in a magma chamber) of a precursor magma derived from a lherzolitic source, and that its phase relations may not correctly indicate the nature of the source residue.

There is a strong possibility that the A-series primary magma also represented a modification of a less potassic magma. This is indicated by apparently contradictory evidence for garnet in the source from REE patterns (indicating a garnet-bearing source, Fig. 4) and liquidus phase relations of 72-955 (indicating no garnet in the source, at least if this is olivine-rich Fig. 7). A possible explanation is that the A-series primary magma represented the end product of progressive reaction between a relatively K-poor basanitic magma, produced by melting of garnet lherzolite at 25–30 kbar (Green, 1973a; Windom and Boettcher, 1981) i.e. within the asthenosphere beneath Muriah, and K-enriched, phlogopite-bearing lherzolite in the lower lithosphere (see Cox et al., 1976). The processes envisaged are similar to the "flow cumulate" model outlined by Irving (1980) to explain the geochemistry and mineralogy of pyroxene-rich dyke or vein phases within spinel lherzolite nodules. In this model, a slowly ascending basanite magma loses water to the wall-rocks of conduits through which it is flowing, causing preferential partial melting of minerals of lower stability. The "contact melt" formed is extracted by the flowing magma, thus changing the latter's composition. This interaction continues until the modified magma has precipitated a lining of its liquidus phases on the conduit walls.

If the wall-rocks contain phlogopite as the major aluminous phase, largely at the expense of amphibole, the "contact melt" will be derived mainly from phlogopite and clinopyroxene, the phases of lowest temperature stability (Green, 1973b; Wendtlandt and Eggler, 1980b). Minor apatite may also contribute (Frey et al., 1980). At 20–25 kbar the liquidus phases of the modified basanite magma will be mainly olivine and clinopyroxene (Irving, 1980). Clinopyroxene will probably be less aluminous than that in the wall rocks due to co-precipitation of Al-spinel. A simplified representation of the overall magma modification reaction envisaged is:

$$2KMg_3[AlSi_3O_{10}](OH)_2 + \underline{CaAl_2SiO_6 + Mg_2Si_2O_6 + 2CaMgSi_2O_6} =$$

phlogopite (in lherzolite wall-rocks) — sub-calcic aluminous clinopyroxene

$$= 3Mg_2SiO_4 + 3CaMgSi_2O_6 + MgAl_2O_4 + 2KAlSi_2O_6 + 2H_2O$$

Olivine — Diopsidic cpx — Al spinel (in cumulates from magma) — Lc (in magma)

Eventually, increased normative Lc in the modified magma will cause phlogopite to crystallize, and K-enrichment of the magma will cease. Change in the $Mg/(Mg + Fe^{2+})$ ratio of the modified magma will probably be slight since, with the exception of Al-spinel ($Mg/(Mg + Fe^{2+}) \simeq 0.80$ (Fujii, 1977) the mafic phases in the modification reaction will all have $Mg/(Mg + Fe^{2+})$ values close to 0.90.

The equilibrium liquidus phase relations of a K-enriched, Lc-normative basanite produced by these processes will show major differences with respect to the parental basanite, especially expansion of the stability fields of phlogopite and diopsidic pyroxene at the expense of olivine and garnet, with the olivine field contracting to lower pressures and the garnet field to higher pressures. The orthopyroxene liquidus field, typically very limited in P-T-xH_2O-xCO_2 space for K-poor basanitic compositions (Green, 1973a) will be lost. These are the features shown by the liquidus diagram of 72-955 (Fig. 7). However, simple calculations show that while undergoing appreciable changes in major-element chemistry and liquidus phase relations, the modified magma may preserve original features of its trace element geochemistry, particularly the REE pattern, which indicate the nature of its source. This is mainly because clinopyroxene, the only crystalline phase in the magma modification reaction which has relatively high REE coefficients, dominates the REE patterns of both the "contact melt" component entering the modified magma and the cumulates precipitated from it.

Calculated REE patterns for the Muriah A-series primary magma (MhAP = 72-955 + 5 wt.% olivine), a possible basanite precursor (X), and the crystalline assemblages (CPA, COS) involved in the proposed magma modification reaction are given in Fig. 8. The calculations assume:

(a) The phlogopite lherzolite wall rocks were enriched in light REE, as well as K (see Nicholls et al., 1980; for evidence of LREE enrichment in the source of basaltic magmas of the Ungaran volcano). A linear pattern with La at 10 times and Yb at 2 times chondritic abundances was chosen as a model.

(b) Olivine, orthopyroxene, clinopyroxene, phlogopite, spinel and apatite were present in the wall rocks, and they had equilibrium REE distributions, as calculated from published partition coefficients (Hanson, 1980; Irving and Price, 1981).

(c) Production of the wall rock "contact melt" (CPA, Fig. 8) involved batch melting of clinopyroxene phlogopite and apatite. This is clearly only an approximation of the (presumably) disequilibrium melting process which actually occurred. The precipitation of near-liquidus clino-pyroxene, olivine and Al-spinel (COS) from the modified magma was also a "batch fractionation" (bulk equilibrium) process.

(d) As given by the simplified magma modification reaction, CPA contained 50 wt.% clinopyroxene and 49% phlogopite (+ 1% apatite), while COS was 53% clinopyroxene, 35% olivine and 12% Al-spinel. The ratio CPA/COS was 1.4.

Figure 8 shows that due to the dominance of clinopyroxene in CPA and

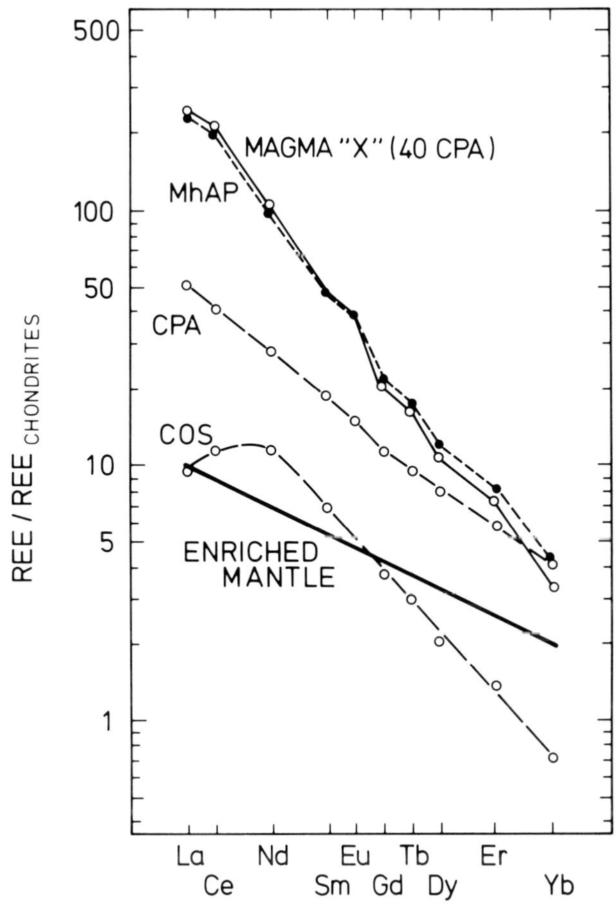

Fig. 8. Calculated REE patterns for the mineral assemblage (CPA) entering "contact melt" produced by the passage of a basanite magma through phlogopite-bearing lherzolite wall rocks, for the near-liquidus precipitate from this magma after modification (COS) and for the initial composition (X) of this magma required to produce the A-series primary magma (MhAP) by 40 wt.% addition of CPA component. For further explanation, see text.

COS, the REE patterns are almost parallel. Provided that the lherzolite wall-rocks are enriched in LREE, both patterns have similar slopes to that of the MhAP pattern. Hence, even when MhAP is produced from a basanitic precursor X by up to 40 wt.% CPA addition, the precursor pattern is almost identical to the MhAP pattern. In spite of large changes in the major element composition of the precursor magma (for 40% CPA addition, magma K_2O content changes from 1.4 to 3.0%), the REE pattern may still correctly indicate the nature of the magma source, especially if this contains garnet.

The only other trace elements which may be appreciably affected by the proposed magma modification process are those associated with phlogopite, particularly Rb and Ba (see biotite-liquid partition coefficients of Irving and Price, 1981). Phlogopite may also be relatively rich in Ni, but increase in Ni content of the modified magma due to phlogopite addition will be offset by precipitation of near-liquidus olivine.

Origin of the Muriah H-series lavas

In contrast to the A-series, the H-"series" lacks well-defined serial relationships for both major and trace elements, especially K (Fig. 2) and Rb. No simple fractionation model can be used, as for the A-series, to explain compositional variation. In the absence of information on the stratigraphic and temporal relationships of H-series samples it is possible that they are not a cogenetic group. However, their consistent petrographic features argue against this, as do the limited Sr-isotopic data.

Assuming that the geochemical features of the H-series rocks have not been grossly modified during groundmass recrystallization (see above), the simplest explanation for their varied compositions is that they represent products of mixing of primary or differentiated magmas of the A-series type with less mafic and K-rich magmas more like those of the active volcano Ungaran. The evidence in support of this suggestion is entirely geochemical — notably that the H-series samples are intermediate between A-series and high-K calc-alkaline series samples (from Ungaran and Diëng) in K (Fig. 2) most trace element abundances (Fig. 3) and Sr-isotopic ratios. The more compelling petrographic indicators of magma mixing in K-rich lavas, such as compositionally discordant cores in clinopyroxene phenocrysts (Barton et al., 1982) have not been recognized, and discordant cores in plagioclase are no more common than in the tholeiitic, calc-alkaline and high-K calc-alkaline lavas of Java. Clearly, detailed mapping and sampling of the Muriah complex, and thorough petrographic and geochemical studies of a wider range of well-characterized samples are urgently needed to solve the problems posed by the H-series rocks.

ACKNOWLEDGEMENTS

Field work at Muriah was sponsored by the Indonesian Institute of Sciences (LIPI), and carried out with the assistance of geologists of the Geological Survey of Indonesia, Bandung, the Bandung Institute of Technology (M.T. Zen) and Gadjah Mada University, Jogjakarta (Iman Soemarinda).

Analytical studies and early crystallization experiments were carried out at the Research School of Earth Sciences, Australian National University. We thank Dr S.R. Taylor, and Professors D.H. Green and A.E. Ringwood for access to equipment. Experimental studies have continued at Monash University with the support of the Australian Research Grants Committee (ARGC). The manuscript was prepared while I.A.N. held an Alexander von

Humboldt Research Fellowship at the Max-Planck-Institut für Chemie, Mainz. Facilities provided by Professor Dr H. Wänke are gratefully acknowledged. Suggestions by Drs. M. van Bergen and S.S. Sun and reviews by Drs. M. Barton and G. Brey, helped improve the manuscript.

REFERENCES

Arculus, R.J. and Delano, J.W., 1981. Intrinsic oxygen fugacity measurements: techniques and results for spinels from upper mantle periodities and megacryst assemblages. Geochim. Cosmochim. Acta, 45: 899–913.

Baldridge, W.S., Carmichael, I.S.E. and Albee, A.L., 1981. Crystallization paths of leucite-bearing lavas: examples from Italy. Contrib. Mineral. Petrol., 76: 321–335.

Barton, M. and Hamilton, D.L., 1979. The melting relations of a madupite from the Leucite Hills, Wyoming, to 30 kb. Contrib. Mineral. Petrol., 69: 133–142.

Barton, M., Varekamp. M. and van Bergen, M., 1982. Complex zoning of clinopyroxenes in the lavas of Vulsini, Italy: evidence for magma mixing? J. Volcanol. Geotherm. Res., 14: 361–388.

Bemmelen, R.W., van, 1947. The Muriah volcano (Central Java) and the origin of its leucite-bearing rocks. Proc. Kl. Ned. Akad. Wet., 50: 653–658.

Bemmelen, R.W., van, 1949. The Geology of Indonesia, Vol. 1. Martinus Nijhoff, Den Haag, 732 pp.

Ben Avraham, Z. and Emery, K.O., 1973. Structural framework of the Sunda Shelf. Bull., Am. Assoc. Pet. Geol., 57: 2328–2366.

Brey, G. and Green, D.H., 1975. The role of CO_2 in the genesis of olivine melilitite. Contrib. Mineral. Petrol., 49: 93–103.

Brey, G. and Green D.H., 1977. Systematic study of liquidus phase relations in olivine melilitite + H_2O + CO_2 at high pressures, and petrogenesis of an olivine melilitite magma. Contrib. Mineral. Petrol., 61: 141–162.

Cox, K.G., Hawkesworth, C.J., O'Nions, R.K. and Appleton, J.D., 1976. Isotopic evidence for the derivation of some Roman Region Volcanics from anomalously enriched mantle. Contrib. Mineral. Petrol., 56: 173–180.

Cundari, A., 1980. Role of subduction in the genesis of leucite-bearing rocks: Facts or fashion? (Reply to A.D. Edgar's discussion paper). Contrib. Mineral. Petrol., 73: 432–434.

Dickinson, W.R. and Hatherton, T., 1967. Andesitic volcanism and seismicity around the Pacific. Science, 157: 801–803.

Edgar, A.D., 1980. Role of subduction in the genesis of leucite-bearing rocks: Discussion. Contrib. Mineral. Petrol., 73: 429–431.

Edgar, A.D., Green, D.H. and Hibberson, W.O., 1976. Experimental petrology of a highly potassic magma. J. Petrol., 17: 339–356.

Edgar, A.D. and Condliffe, B., 1978. Derivation of K-rich ultramafic magmas from a peridotitic mantle source. Nature, 275: 639–640.

Edgar, A.D., Condliffe, B., Barnett, R.L. and Shirran, R.J., 1980. An experimental study of an olivine ugandite magma and mechanisms for the formation of its K-enriched derivatives. J. Petrol., 21: 475–497.

Foden, J.D. and Varne, R., 1980. The petrology and tectonic setting of Quaternary–Recent volcanic centres of Lombok and Sumbawa, Sunda arc. Chem. Geol., 30. 201–226.

Frey, F.A., Roden, M.F. and Zindler, A., 1980. Constraints on mantle source compositions imposed by phosphorus and the rare earth elements. Contrib. Mineral. Petrol., 75: 165–173.

Fujii, T., 1977. Fe-Mg partitioning between olivine and spinel. Carnegie Inst. Washington, Yearb., 76: 563—569.

Green, D.H., 1973a. Conditions of melting of basanite magma from garnet peridotite. Earth Planet. Sci. Letts., 17: 456—465.

Green, D.H., 1973b. Experimental melting studies on model upper mantle compositions at high pressures under both water-saturated and water-undersaturated conditions. Earth Planet. Sci. Letts., 19: 37—53.

Green, D.H. and Ringwood, A.E., 1967. The genesis of basaltic magmas. Contrib. Mineral. Petrol., 15: 103—190.

Grover, J.E., Lindsley, D.H. and Bence, A.E., 1980. Experimental phase relations of olivine vitrophyres from breccia 14321: the temperature- and pressure-dependence of Fe-Mg partitioning for olivine and liquid in a highlands melt rock. Proc. Lunar Planet. Sci. Conf., 11th, pp. 179—196.

Hamilton, W., 1979. Tectonics of the Indonesian region. U.S. Geol. Surv., Prof. Pap. 1078, 345 pp.

Hanson, G.N., 1980. Rare earth elements in petrogenetic studies of igneous systems. Annu. Rev. Earth. Planet. Sci., 8: 371—406.

Hanson, G.N. and Langmuir, C.H., 1978. Modelling of major elements in mantle-melt systems using trace element approaches. Geochim. Cosmochim. Acta, 42: 725—741.

Hutchison, C.S., 1975. Correlation of Indonesian active volcanic geochemistry with Benioff zone depth. Geol. Mijnbouw, 54: 157—168.

Irving, A.J., 1980. Petrology and geochemistry of composite ultramafic xenoliths in alkalic basalts and implications for magmatic processes within the mantle. Am. J. Sci., 280-A: 389—426.

Irving, A.J. and Price, R.C., 1981. Geochemistry and evolution of lherzolite-bearing phonolitic lavas from Nigeria, Australia, East Germany and New Zealand. Geochim. Cosmochim. Acta, 45: 1309—1320.

Katili, J.A., 1973. Geochronology of West Indonesia and its implication on plate tectonics. Tectonophysics, 19: 195—212.

Lloyd, F.E. and Bailey, D.K., 1975. Light element metasomatism of the continental mantle: the evidence and consequences. Phys. Chem. Earth, 9: 389—416.

Luhr, J.F. and Carmichael, I.S.E., 1980. The Colima volcanic complex, Mexico. Part I. Post-caldera andesites from Volcán Colima. Contrib. Mineral. Petrol., 71: 343—372.

Luhr, J.F. and Carmichael, I.S.E., 1981. The Colima volcanic complex, Mexico. Part II. Late Quaternary cinder cones. Contrib. Mineral. Petrol., 76: 127—147.

Mysen, B.O., Kushiro, I., Nicholls, I.A. and Ringwood, A.E., 1974. A possible mantle origin for andesitic magmas: discussion of a paper by Nicholls and Ringwood. Earth Planet. Sci. Lett., 21: 221—229.

Nabelek, P.I., 1980. Nickel partitioning between olivine and liquid in natural basalts: Henry's Law behaviour. Earth Planet. Sci. Lett., 48: 293—302.

Nicholls, I.A., 1974. Liquids in equilibrium with peridotitic mineral assemblages at high water pressures. Contrib. Mineral. Petrol., 45: 289—316.

Nicholls, I.A., 1978. Primary basaltic magmas for the pre-caldera volcanic rocks of Santorini. In: C. Doumas (Editor), Thira and the Aegean World, I. London, pp. 109—120.

Nicholls, I.A. and Whitford, D.J., 1976a. Potassium-rich volcanic rocks of the Sunda island arc, Java, Indonesia. Proc. Int. Geol. Congr., 1: 58 (abstract).

Nicholls, I.A. and Whitford, D.J., 1976b. Primary magmas associated with Quaternary volcanism in the western Sunda arc. In: R.W. Johnson (Editor), Volcanism in Australasia. Elsevier, Amsterdam, pp. 77—90.

Nicholls, I.A. and Whitford, D.J., 1978. Geochemical zonation in the Sunda volcanic arc, and the origin of K-rich lavas. Bull. Aust. Soc. Explor. Geophys., 9: 93—98.

Nicholls, I.A., Whitford, D.J., Harris, K.L. and Taylor, S.R., 1980. Variation in the geochemistry of mantle sources for tholeiitic and calc-alkaline mafic magmas, western Sunda volcanic arc, Indonesia. Chem. Geol., 30: 177—199.

Ryabchikov, I.D. and Green, D.H., 1978. The role of carbon dioxide in the petrogenesis of highly potassic magmas. In: Problems of Petrology of the Earth's Crust and Upper Mantle. Issue 403, Nauka, Novosibirsk: Tr. Inst. Geol. Geofiz. Akad.

Takahashi, E., 1978. Partitioning of Ni^{2+}, Co^{2+}, Fe^{2+}, Mn^{2+} and Mg^{2+} between olivine and silicate melts: compositional dependence of partition coefficient. Geochim. Cosmochim. Acta, 42: 1829—1844.

Wendtlandt, R. and Eggler, D.H., 1980a. The origins of potassic magmas: 1. Melting relations in the systems $KAlSiO_4$—Mg_2SiO_4—SiO_2 and $KAlSiO_4$—MgO—SiO_2—CO_2 to 30 kilobars. Am. J. Sci., 280: 385—420.

Wendtlandt, R. and Eggler, D.H., 1980b. The origins of potassic magmas: 2. Stability of phlogopite in natural spinel lherzolite and in the system $KAlSiO_4$—MgO—SiO_2—H_2O—CO_2 at high pressures and high temperatures. Am. J. Sci., 280: 421—458.

Whitford, D.J., 1975a. Strontium isotopic studies of the volcanic rocks of the Sunda arc, Indonesia, and their petrogenetic implications. Geochim. Cosmochim. Acta, 39: 1287—1302.

Whitford, D.J., 1975b. Geochemistry and Petrology of Volcanic Rocks from the Sunda Arc, Indonesia. Ph. D. Thesis, Australian National University, Canberra, 449 pp.

Whitford, D.J. and Nicholls, I.A., 1976. Potassium variation in lavas across the Sunda arc in Java and Bali. In: R.W. Johnson (Editor), Volcanism in Australasia. Elsevier, Amsterdam, pp. 63—76.

Whitford, D.J., Nicholls, I.A. and Taylor, S.R., 1979. Spatial variations in geochemistry of Quaternary lavas across the Sunda arc in Java and Bali. Contrib. Mineral. Petrol., 70: 341—356.

Whitford, D.J., White, W.M. and Jezek, P.A., 1981. Neodymium isotopic composition of Quaternary island arc lavas from Indonesia. Geochim. Cosmochim. Acta, 45: 989—995.

Windom, K. and Boettcher, A.L., 1981. Phase relations for the joins jadeite-enstatite and jadeite-forsterite at 28 kb and their bearing on basalt genesis. Am. J. Sci., 281: 335—351.

Wyllie, P.J., 1979. Magmas and volatile components. Am. Mineral., 64: 469—500.

Journal of Volcanology and Geothermal Research, 18 (1983) 361—385
Elsevier Science Publishers B.V., Amsterdam — Printed in The Netherlands

THE SOLUBILITY MECHANISMS OF VOLATILES IN SILICATE MELTS AND THEIR RELATIONS TO CRYSTAL-ANDESITE LIQUID EQUILIBRIA

BJØRN O. MYSEN

Geophysical Laboratory, Carnegie Institution of Washington, Washington, DC 20008 (U.S.A.)

(Received October 20, 1982)

ABSTRACT

Mysen, B.O., 1983. The solubility mechanisms of volatiles in silicate melts and their relations to crystal-andesite liquid equilibria. In: S. Aramaki and I. Kushiro (Editors), Arc Volcanism. J. Volcanol. Geotherm. Res., 18: 361—385.

The anionic structure of magmatic liquids has been estimated at 1 atm and at pressures corresponding to those of the upper mantle. These estimates are based predominantly on spectroscopic data on binary metal oxide-silica and ternary metal oxide-silica-alumina melts. Structural information on melt compositions in aluminate-silica joins has been used to provide detailed information on the role of Al^{3+} in natural magma at atmospheric and high pressure.

Regardless of pressure, andesitic melts may be described as combinations of chain, sheet, and three-dimensional network units. Nearly all Al^{3+} in the magmatic liquid resides in the three-dimensional network units. This Al^{3+} is locally charge-balanced with Na^+, K^+, Ca^{2+}, and Mg^{2+}. In the latter two cases, Al^{3+} and Si^{4+} are ordered, whereas for Na^+ and K^+, Si^{4+} and Al^{3+} are randomly mixed. Solution of water in natural magma results in the formation of new nonbridging oxygens in addition to OH groups attached to Si^{4+} and metal cations.

On the basis of determined solution mechanisms of CO_2 and H_2O in silicate melts, thermodynamic properties of H_2O+CO_2 fluids and hydrous silicate melts and melting phase relations in peridotite-H_2O-CO_2 systems, it is found that natural andesitic magma in equilibrium with spinel lherzolite in the upper mantle (10—20 kbar) must contain at least 5—7 wt.% H_2O. Andesitic magma with 5—7 wt.% H_2O in solution may be described as a mixture of Al-free three dimensional units, sheets, and chains with a small proportion (less than 10%) of monomers.

INTRODUCTION

Experimental phase-equilibrium data relevant to the petrogenesis of magmatic rocks in and beneath island arcs (e.g., O'Hara, 1965; Kushiro et al., 1968; Yoder, 1969; Eggler, 1972, 1974; Kushiro, 1972, 1974; Eggler and Burnham, 1973; Mysen and Boettcher, 1975) indicate that H_2O and CO_2 significantly affect the activity of silicate components in magmatic liquids. In

0377-0273/83/$03.00

brief, solution of H_2O in melts at high pressures and temperatures results in an expansion of liquidus fields of minerals with lower activity of silica than in the absence of water (see Carmichael et al., 1974, for discussion). Such reasoning helps explain why quartz-normative silicate melts may be in equilibrium with four-phase peridotite at pressures above 10 kbar in the presence of water, whereas the partial melts from peridotite in the upper mantle are olivine normative in the absence of H_2O. Dissolved carbon dioxide in solution in silicate melts, on the other hand, results in an enhancement of the activity of silica. Consequently, partial melting in the presence of CO_2 yields melt compositions that are more silica-deficient than those formed in the absence of CO_2 (Eggler and Rosenhauer, 1978).

On the basis of the simplified observations summarized above, it has been suggested that H_2O/CO_2 in the source region of magma in the upper mantle may control the bulk composition of various types of intrusive and extrusive rocks found on or near the surface of the earth (Mysen and Boettcher, 1975). For example, island arc volcanism resulting in andesite and more acidic rocks tends to be favored by high activity of water. High CO_2 activity in the source region may lead to the formation of highly alkalic, silica-deficient magmas such as those often found in continental regions. Intermediate mixing ratios of CO_2 and H_2O or no volatiles at all are the conditions in the upper mantle under which typical oceanic tholeiite magma is formed (Mysen and Boettcher, 1975; Mysen and Kushiro, 1977).

The liquidus mineralogy of most basaltic and more acidic rocks under pressure conditions corresponding to those of the upper mantle may be described in terms of simple, yet importantly different, combinations of olivine, orthopyroxene solid solution, clinopyroxene solid solution, and an aluminous phase (plagioclase, spinel, or garnet). In addition, there may be small amounts of hydrated minerals (amphibole or phlogopite) and carbonated minerals (calcite, dolomite, or magnesite). Consequently, for a given mineral assemblage (that defines the activity of relevant major-element components), the structure of possible magmatic liquids will exert an important control on the composition of the partial melt. In order to understand the relationships between magma composition and the pressures and temperatures of their formation and subsequent evolution, it is first necessary to determine their structures. In the present paper relevant data on simple and complex silicate systems are reported to illuminate some important features of the structure of volatile-free and volatile-bearing magma at 1 atm and at high pressure. A few examples of relationships between melt structure and element partitioning between coexisting melt and minerals are also given.

SIMPLE SILICATE SYSTEMS

Silicate melts may be treated in terms of network modifiers and network formers. The latter cations are building blocks in the tetrahedral network and may include Si^{4+}, Al^{3+}, Fe^{3+}, Ti^{4+}, and P^{5+} among the major and minor elements found in natural magma. The major-element network modifiers (K^+, Na^+, Ca^{2+}, Mg^{2+}, and Fe^{2+}) serve as the interconnecting tissue to obtain local charge-balance in the melts.

Present research has been concentrated on the network formers so as to form models of the anionic structure of the silicate melts. In order to determine possible structural variability as a function of melt bulk composition, data on simple metal oxide-silica binaries have been treated at some length. Brawer and White (1975, 1977) and Furukawa and White (1980), on the basis of Raman spectroscopic data on quenched melts on M_2O-SiO_2 (M = Li, K, and Na) and CaO-SiO_2 joins, concluded that such melts may be treated in terms of a few simple anionic complexes similar to three-dimensional network, chain, and sheet units in the melts. These concepts were expanded by Virgo et al. (1980) and Mysen et al. (1980a), who summarized the available spectroscopic data in systems such as Li_2O-SiO_2, Na_2O-SiO_2, K_2O-SiO_2, PbO-SiO_2, CaO-SiO_2, and MgO-SiO_2 (see their papers for detailed references). They concluded that binary silicate melts in the bulk compositonal range between ortho- and tectosilicate may be described in terms of coexisting simple anionic units such as SiO_4^{4-}, $Si_2O_7^{2-}$, $Si_2O_6^{4-}$, $Si_2O_5^{2-}$, and SiO_2. The abundance, rather than the type of anionic units, changed as a function of bulk composition of the systems.

The identification of anionic units in the Raman spectra is based on Raman scattered light from chemical bonds involving a tetrahedrally coordinated cation (T) and oxygen (bridging, O°, or nonbridging, O^-). The frequency of characteristic rocking, bending, and stretch vibrations, their polarization characteristics, comparison with spectra of known mineral structure, and theoretical calculations have been used for this purpose (Etchepare, 1972, Lazarev, 1972; Brawer, 1975; Brawer and White, 1975, 1977, Verweij and Konijnendijk, 1976; Furukawa et al., 1978, 1981; Verweij, 1979a, b). On the basis of studies of several hundred compositions formed under a range of conditions, it is now concluded that the frequency of the symmetric stretch bands is most sensitive to the degree of polymerization of a given anionic structural unit (degree of polymerization is defined by the ratio of nonbridging oxygen per silicon, NBO/Si).

In the summary of some of the most important Raman frequencies shown in Fig. 1, the bands near 860, 900, 960, and 1100 cm^{-1} are symmetric stretch bands of T-O^- bonds from structural units with NBO/Si = 4 (monomer), 3 (dimer), 2 (chain), and 1 (sheet), respectively. The band near 1050 cm^{-1} is due to bridging oxygen occurring in the melts. The 1150 and 1200 cm^{-1} bands are antisymmetric stretch bands that occur either as the 1150 cm^{-1} band alone or together with the 1200 cm^{-1} band whenever three-dimensional network units exist in the melt. Additional bands near 800, 650, and 500 cm^{-1}, due to bending and rocking motions of the Si-O bonds, may be correlated further with the existence of the various sorts of units. Interested readers are referred to Brawer (1975), Furukawa et al.

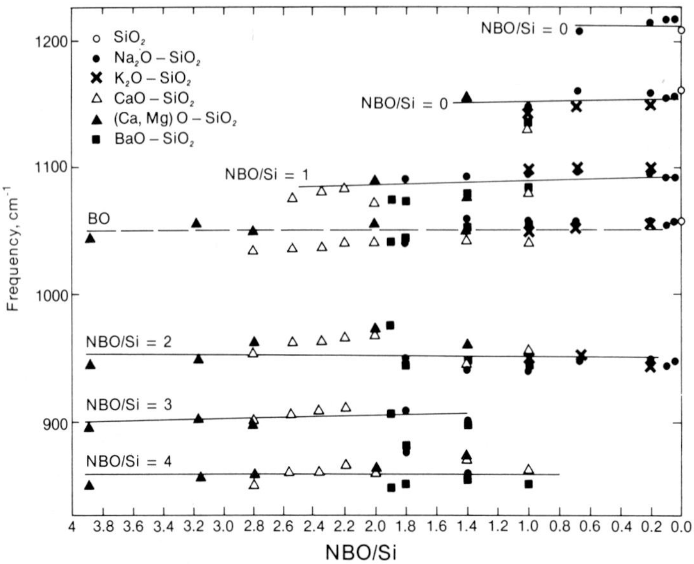

Fig. 1. Summary of frequencies of stretch vibrations as observed in Raman spectra of binary silicate melts on the joins Na_2O-SiO_2 (closed circles), K_2O-SiO_2 (crosses), CaO-SiO_2 (open triangles), (Ca,Mg)O-SiO_2 with Ca/Mg = 1 (closed triangles), and BaO-SiO_2 (closed squares). Labels refer to anionic units with NBO/Si as identified. NBO/Si (x axis) refers to calculated NBO/Si of the bulk compositions.

(1981), and Mysen et al. (1980a, 1982) for detailed description and discussion of the spectroscopic features and their relationships to the structure of the melt.

Although spectroscopic features such as half-width of the individual bands differ for different metal cations (cf. Na^+ and Ca^{2+}), the positions of individual bands do not depend on the type of metal cations or on the overall degree of polymerization of the melt. This observation led to the conclusion (Virgo et al., 1980; Furukawa and White, 1980; Mysen et al., 1980a, 1982; Furukawa et al., 1981) that the types of coexisting anionic silicate units in binary silicate melts do not depend on either the type of network modifier or the degree of polymerization within bulk compositional ranges of a certain number of coexisting anionic units. Only their relative proportions differ. Without consideration of metal cations present, the anionic equilibria that describe the coexisting structural units in melts with bulk NBO/Si of the melt between 4 and 0 may then be expressed with the following equations:

$$2\ SiO_4^{4-} \rightleftharpoons Si_2O_7^{6-} + O^{2-} \tag{1}$$

$$2Si_2O_7^{6-} \rightleftharpoons Si_2O_6^{4-} + 2SiO_4^{4-} \tag{2}$$

$$3Si_2O_6^{4-} \rightleftharpoons 2Si_2O_5^{2-} + 2SiO_4^{4-} \tag{3}$$

$$Si_2O_6^{4-} \rightleftharpoons SiO_2 + SiO_4^{4-} \quad (4)$$

$$2Si_2O_5^{2-} \rightleftharpoons Si_2O_6^{4-} + 2SiO_2 \quad (5)$$

The bulk compositional ranges within which each of these equilibria is operative and the relative proportions of the individual units as a function of bulk NBO/Si of the melt depend on the type of metal cation present. Metal cations with large field strength tend to favor coexistence of highly depolymerized units such as SiO_4^{4-} and $Si_2O_7^{6-}$ over $Si_2O_6^{4-}$ and $Si_2O_5^{2-}$ together with fully polymerized units such as SiO_2 (see also Liebau, 1981). In the case of Mg^{2+}, for example, sheet units do not occur at all. Instead of eq. 3, eq. 4 describes the anionic equilibrium in the compositional range less depolymerized than that of approximately metasilicate (Mysen et al., 1982). For example, melt of $MgSiO_3$ composition consists of coexisting SiO_4^{4-}, $Si_2O_6^{4-}$, and SiO_2 units. Melt of $CaSiO_3$ composition, on the other hand, is made up of SiO_4^{4-}, $Si_2O_6^{4-}$, and $Si_2O_5^{2-}$ units (see also Fig. 1).

The relative abundance of anionic units as a function of types of cations may also be noted from the data in Fig. 2. In the case of the disilicate bulk compositions (NBO/Si = 1), for example, the relative abundance of the $Si_2O_5^{2-}$ units decreases from about 40% through 30 and 20% as the metal cation is changed from Ba^{2+} to Na^+ and Ca^{2+}, respectively.

ALUMINUM IN SILICATE MELTS

Except for SiO_2, Al_2O_3 is the most important major element oxide in most natural magma. Mysen et al. (1981a, 1982), on the basis of a survey of approximately 16,000 published analyses of Cenozoic extrusive rocks (F. Chayes, 1975; pers. commun., 1981), concluded that 95% or more of the tetrahedrally coordinated cations in igneous rocks are Si^{4+} and Al^{3+}. The remainder are Fe^{3+}, Ti^{4+}, and P^{5+}.

In order to determine the structural role of Al^{3+} in magmatic liquids, it is necessary to determine the compositional ranges (in terms of degree of polymerization) of magma. With the possible exception of uncommon rocks such as olivine nephelinite, basanite, and peridotitic komatiite, natural magmas have NBO/T (T = tetrahedrally coordinated cation) in the range between 1 and about 0. In this calculation it is assumed that Al^{3+} is tetrahedrally coordinated whenever there are sufficient charge-balancing cations (K^+, Na^+, Mg^{2+}, and Ca^{2+}) for local charge-balance of Al^{3+} in tetrahedral coordination (see Wood and Hess, 1980, and Seifert et al., 1981, for further discussion of charge-balance of Al^{3+} in silicate melts). Average Cenozoic andesite (Chayes, 1969) has NBO/T = 0.4, for example. It is likely, therefore, that most naturally occurring silicate melts have coexisting chain, sheet, and three-dimensional units. If random substitution of Al^{3+} and Si^{4+} is assumed, at least 50% of these melt structures consist of three-dimensional units.

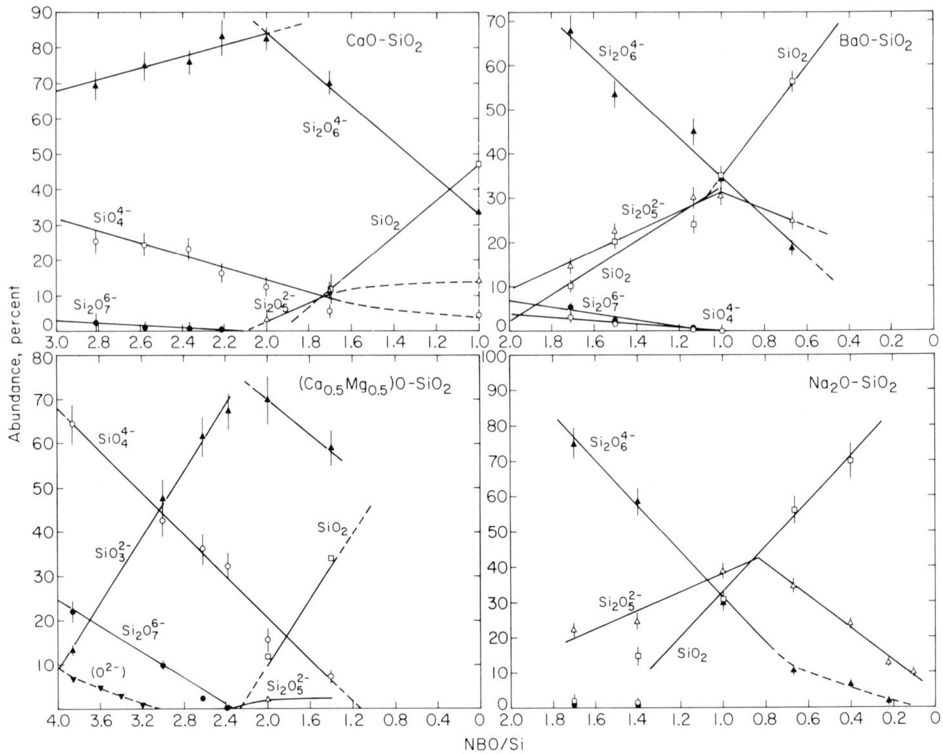

Fig. 2. Relative abundance (mole %) of anionic units as a function of bulk degree of polymerization of binary metal oxide-silica melts. Closed triangles, O^{2-} (free oxygen in the sense of Toop and Samis, 1962) as calculated by mass balance (see Mysen et al., 1982); open circles, SiO_4^{4-} (monomer) units; closed circles, $Si_2O_7^{6-}$ (dimer) units; closed triangles, $Si_2O_6^{4-}$ (chain) units; open triangles, $Si_2O_5^{2-}$ (sheet) units; open squares, SiO_2 (three-dimensional) units.

Geometric and charge-balance considerations lead to the suggestion that Al^{3+} may show a preference for individual types of structural units. The extent to which this feature may be observed can be assessed with Raman spectroscopy. The frequencies of Raman bands decrease as systematic functions of Al/(Al + Si) of the particular structural units through coupling of the vibrations provided that the Al^{3+} and Si^{4+} are in close proximity in a given anionic unit in the sample (probably less than 50 A apart). Typical rates of change of stretch band frequencies as a function of Al/(Al + Si) in the systems $Na_2Si_2O_5$-$NaAlO_2$ and $CaSi_2O_5$-$CaAl_2O_4$ are shown in Fig. 3. Aluminum partition coefficients derived from such data (Mysen et al., 1981a) are shown in Table I. Evidently, Al^{3+} always shows preference for the most polymerized structural unit. The extent of preference appears more pronounced with a large charge-balancing cation such as Na^+ as compared with Ca^{2+}.

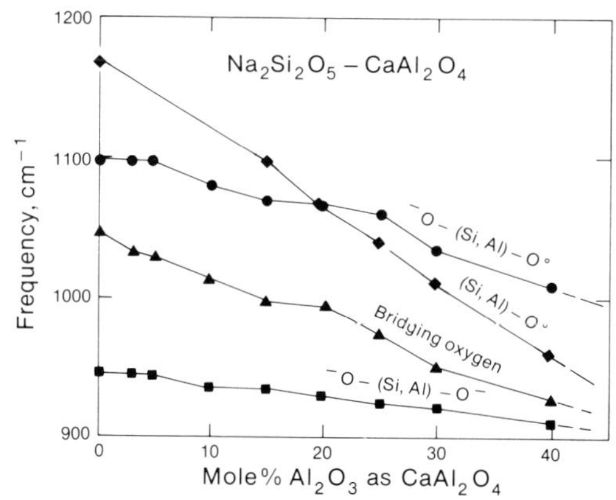

Fig. 3. Rate of change of frequencies of stretch vibrations as observed in Raman spectra of melts on the join $Na_2Si_2O_5$-$CaAl_2O_4$ as a function of Al_2O_3 content of the melt. -O-(Si, Al)-O°, symmetric stretch in units with NBO/T = 1; (Si,Al)-O°, antisymmetric stretch vibration of unit with NBO/T = 0; $^-$O-(Si,Al)-O$^-$, symmetric stretch vibration in unit with NBO/T = 2; "bridging oxygen", antisymmetric stretch vibration from (Si,Al)-O-(Si,Al) bridges in the melt. Data from Mysen et al. (1981a).

TABLE 1

Partition coefficients of Al^{3+} between anionic units in the system Na_2O-CaO-Al_2O_3-SiO_2

	3D/sheet	3D/chain
$(Ca_{0.5}Al)^{4+}$*	2.4	5.3
$(NaAl)^{4+}$**	3.1	9.4

*Aluminum is charge-balanced exclusively with Ca^{2+} in tetrahedral coordination.
**Aluminum is charge-balanced exclusively with Na^+ in tetrahedral coordination.
Source of data: Mysen et al. (1981a).

Most igneous melts have $M^+/(M^+ + M^{2+})$ less than 0.5. Typical andesite shows a value of about 0.3. The average partition coefficient for Al^{3+} between the anionic units must be adjusted for this ratio. The above information leads to the conclusion that a melt of the average Cenozoic andesite composition determined by Chayes (1969) will have about 75% three-dimensional network unit in which 99% of all Al^{3+} in the melt is located. The Al/(Al + Si) is about 0.25 for the three-dimensional units in this melt at 1 atm. It is clear, therefore, that the structure of three-dimensional aluminate-silica melts must be understood in order to derive a complete model of the anionic structure of natural magma.

Taylor and Brown (1979), with the aid of X-ray radial distribution calculations (XRDF), concluded that melts of feldspar compositions have a three-dimensional network structure that generally does not differ dramatically from the structure of feldspars themselves. A similar suggestion has been made by Burnham (1979, 1981) on the basis of thermodynamic properties of aluminosilicate melt derived from melt-mineral equilibria.

Vibrational spectroscopic methods are, as already indicated, often sensitive to local (several Angstroms to several tens of Angstroms) variations in electron density (see Brawer, 1975; Long, 1977). As a result, Al-Si distribution between possible three-dimensional units and within units may be discerned. On the basis of Raman spectroscopic data, Seifert et al. (1981) found that in melts on the join $NaAlO_2$-SiO_2, Si^{4+} and Al^{3+} substitute randomly for each other. They also found, however, that the force constants of *T*-O-*T* bonds (bridging oxygen bonds) decreased as linear functions of bulk Al/(Al + Si) in two distinctly different three-dimensional aluminosilicate melt units (Fig. 4). This difference in force constants may be related, with the aid of the complete Raman spectra, to a difference in the *T*-O-*T* bond angle and, therefore, the *T*-O bond distance in the network. The angle differs by 5—10°. In summary, Seifert et al. (1981) concluded that in aluminosilicate melts with Al^{3+} locally charge-balanced by Na^+, and probably K^+, there are two different three-dimensional structural units. One unit has a

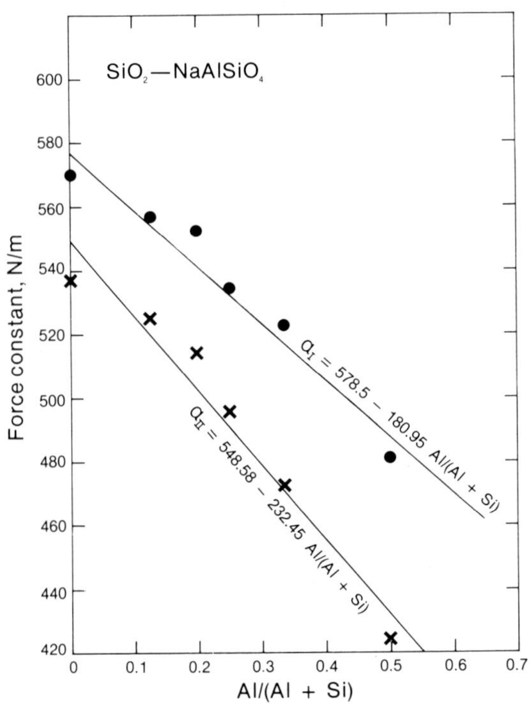

Fig. 4. Rate of change of force constants, N/m, for vibrations over oxygen bridges in the system $NaAlO_2$-SiO_2 as a function of Al/(Al + Si) of the melt. Unit I is the unit with the largest *T*-O-*T* angle. Force constants calculated from the central force model (Sen and Thorpe, 1977; Galeener, 1979; Martin and Galeener, 1981). Data from Seifert et al. (1981).

slightly smaller intertetrahedral angle and a slightly longer T-O bond than the other unit. The former unit probably also shows a slight preference for Al^{3+} over the other unit. Silicon and aluminum are, however, randomly mixed within each three-dimensional unit.

In the systems $CaAl_2O_4$-SiO_2 and $MgAl_2O_4$-SiO_2 there is no evidence of varying degrees of (Al, Si) coupling of the vibrational bands as a function of bulk Al/(Al + Si) of the system. Rather, it has been suggested (Seifert et al., 1981) that melts on these joins, at least at 1 atm pressure, may be described as mixtures of Al-free SiO_2 units, $Al_2Si_2O_8{}^{2-}$ units, and Si-free AlO_2^- units. The proportions of these units vary as a function of Al/(Al + Si) of the melt (Fig. 5).

With the aid of the structural data given above, the three-dimensional portion of magmatic liquids may be subdivided into two different types. With Na^+ or K^+ charge-balanced or both, randomly mixed three-dimensionally interconnected (Al, Si) units occur. Aluminosilicate melts in which aluminum is charge-balanced with Ca^{2+} or Mg^{2+} or both can be described with coexisting Al-free, interconnected SiO_2 units, three-dimensional interconnected $Al_2Si_2O_8{}^{2-}$ units, and possibly AlO_2 units. The extent of this subdivision will depend on the $(Na^+ + K^+)/(Na^+ + K^+ + Ca^{2+} + Mg^{2+})$ of the magma and on the temperature and pressure. In addition, dissolved volatiles such as H_2O and CO_2 may affect the distribution of anionic units and the Al/Si in each unit.

THE EFFECT OF PRESSURE

A number of physical and chemical properties of highly polymerized aluminosilicate melts indicate that at pressures as low as 10 kbar, significant changes in their values occur in the melt (Fig. 6). The viscosity of $NaAlSi_3O_8$ melt decreases by several orders of magnitude between 5 and 15 kbar, for example (Kushiro, 1978a). Certain trace-element partition coefficients (Ni, K, Rb, Sr) involving upper mantle minerals and silicate melt decrease rapidly in this pressure range (Shimizu, 1974; Mysen and Kushiro, 1979). The redox ratio of iron in melts on the join $NaAlSi_2O_6$-$NaFeSi_2O_6$ decreases to a minimum in the range between 10 and 20 kbar and then increases as the pressure is increased further (Mysen and Virgo, 1978). Observations such as these have been used to suggest that some or all Al^{3+} in the aluminosilicate network in natural magma at 1 atm pressure may undergo a coordination transformation with increasing pressure (Waff, 1975). If this is the case, the 1-atm data have limited applicability to the structure of silicate melts at high pressure. Inasmuch as many important petrogenetic processes occur at pressures greater than 1 atm, it is necessary to determine the effect of pressure on the structure of silicate melts.

Spectroscopic methods suitable for the study of silicate melts quenched at high pressure are insensitive to direct observation of non-tetrahedral Al-O bonds. An aluminum coordination transformation from network-former to

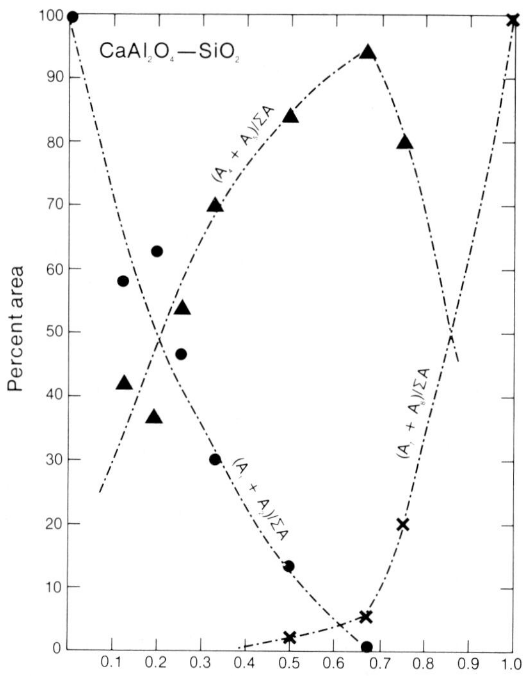
CaAl$_2$O$_4$—SiO$_2$
Percent area
$(A_4 + A_5)/\Sigma A$
$(A_1 + A_2)/\Sigma A$
$(A_7 + A_8)/\Sigma A$
Al/(Al + Si)

(a)

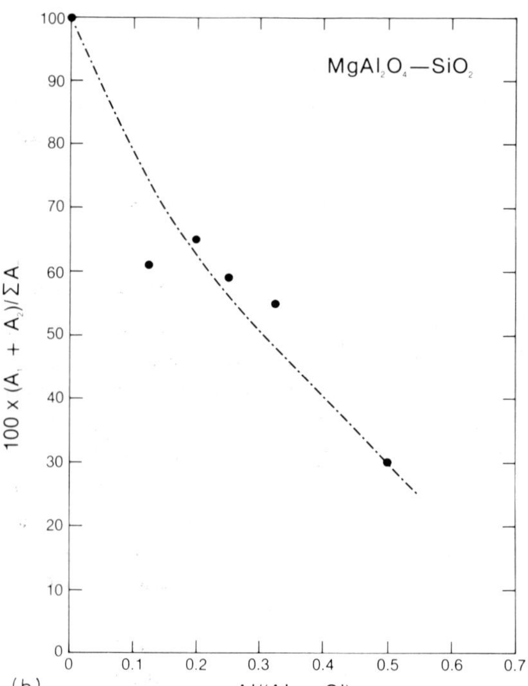
MgAl$_2$O$_4$—SiO$_2$
$100 \times (A_1 + A_2)/\Sigma A$
Al/(Al + Si)

(b)

network modifier results, however, in the formation of non-bridging oxygen (O^-). These T-O^- bonds are easily detected with Raman spectroscopy. Raman spectroscopic data on melts on binary metal oxide-silica joins, combined with data on defect-bearing vitreous SiO_2 and water-bearing vitreous SiO_2 (Stolen and Walrafen, 1976; Lucovsky, 1979), may be used to assess the sensitivity of the method to the abundance of nonbridging oxygen in highly polymerized silicate melts. The Raman spectroscopic technique applied to the problem has a sensitivity of about 1% nonbridging oxygens. In other words, if Al^{3+} were to leave the network owing to increased pressure, for example, after 1% nonbridging oxygen was formed as result of this process, the attendant nonbridging Si-O bonds would be detected.

Because the spectroscopic data were obtained on glasses formed by quenching the samples at high pressure, it is necessary to determine whether possible quenching problems may preclude the observation of structural changes in the liquids as a function of pressure. Increasing pressure results in a decrease in the specific volume of the melt. As a result, at given pressure, decreasing temperature will tend to enhance the pressure effect. Furthermore, quenching at high pressure is essentially isobaric. It is unlikely, therefore, that quenching by turning off the power to the furnace will result in transformation of the resulting glass to the higher volume state. Put differently, if transformation of Al^{3+} coordination occurred in the melt, the new Al^{3+} coordination probably would be quenched into the glass used for the spectroscopic analyses.

In order to determine the effects of pressure on the structural role of Al^{3+} in silicate melts, Mysen et al. (1980a) conducted Raman spectroscopic studies of melts of $NaAlSi_2O_6$ and $NaAlSi_3O_8$ composition quenched from high temperature at pressures up to 38 kbar. No evidence was found for Al^{3+} leaving the tetrahedral network in this pressure range. Sharma et al. (1979), also on the basis of Raman spectroscopic studies, suggested that increased pressure resulted in a slight decrease of the local symmetry around the (Al, Si) tetrahedra.

Results from molecular dynamics calculations of highly polymerized compositions such as SiO_2 (Woodcock et al., 1976; Mitra, 1982) suggest that the melt responds to pressure simply by a decrease in cavity volume (about 15% between 1 atm and 5.5 kbar) and a slight decrease in Si-O-Si angle (3–4°). These conclusions are consistent with those from Raman spectroscopy. The only observed changes are subtle decreases in the proportion of the three-dimensional units with the shorter T-O bond (unit I) compared with those with the longer T-O bond (Fig. 7). In addition, the 580 cm^{-1}

Fig. 5. Relative intensities ("relative areas") of antisymmetric T-O° stretch bands in the systems $CaAl_2O_4$ SiO_2 (a) and $MgAl_2O_4$ SiO_2 (b). Area $A_1 + A_2$, three-dimensional SiO_2 unit without aluminum; $A_4 + A_5$, three-dimensional unit with Al/Si = 1 ("anorthite ring"); $A_7 + A_8$, three-dimensional aluminate unit without silicon. Data from Seifert et al. (1981).

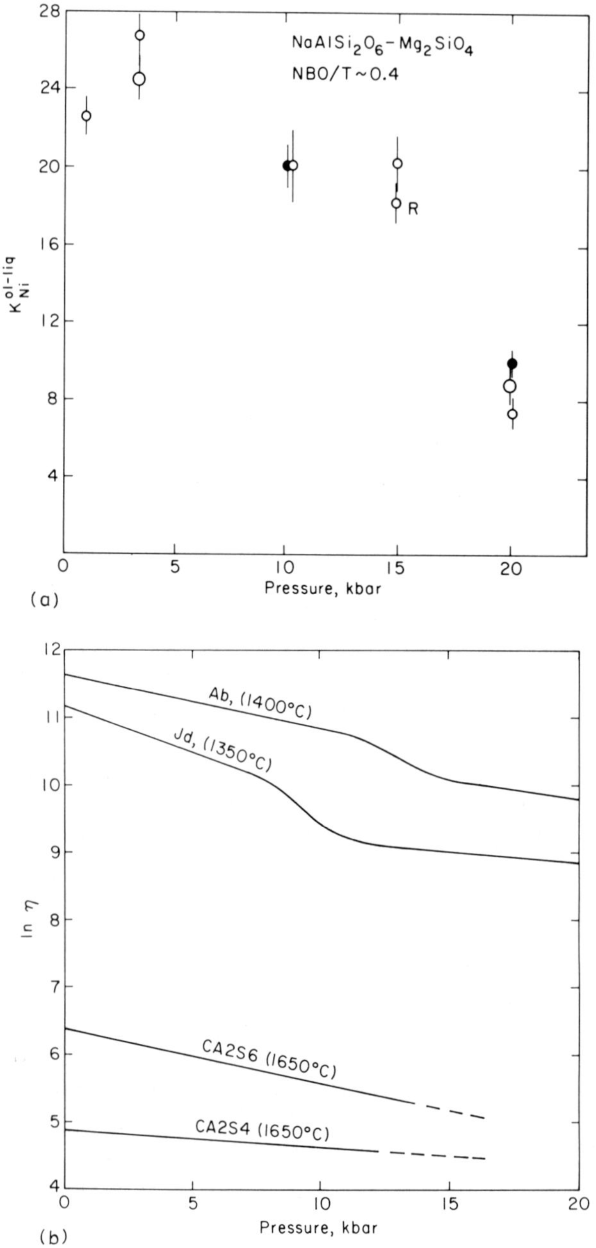

Fig. 6. Properties of aluminosilicate melt and aluminosilicate melt-mineral systems that depend on pressure. a. Partition coefficient (C^{fo}/C^{liq}) of nickel as a function of pressure at 1300°C (after Mysen and Kushiro, 1979). b. Viscosity of melts with three-dimensional network structure as a function of pressure (after Kushiro, 1976, 1978a, 1981). Ab, $NaAlSi_3O_8$; Jd, $NaAlSi_2O_6$; CA2S6, $CaAl_2Si_6O_{16}$; CA2S4, $CaAl_2Si_4O_{12}$.

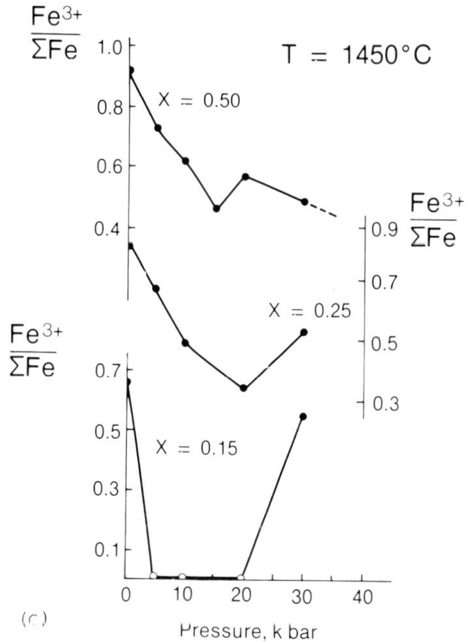

Fig. 6. c. $Fe^{3+}/\Sigma Fe$ of melts on the composition join $NaAlSi_2O_6$-$NaFeSi_2O_6$ as a function of pressure and mole fraction (x) of $NaFeSi_2O_6$ component in the starting material (after Mysen and Virgo, 1978).

band, indicative of less than 0.1% broken oxygen bridges in the 1-atm glass, has diminished further in intensity.

For Ca^{2+} and Mg^{2+} charge-balanced aluminate-silica glasses, increasing pressure from 1 atm to 15 kbar results in a distinct decrease in the proportion of Al-free, interconnected SiO_2 units with the largest Si-O-Si angle (see also Gaskell, 1975, Soules, 1979, Gaskell and Tarrant, 1980; Seifert et al., 1981; for detailed discussion of the structure of vitreous SiO_2) relative to the SiO_2 unit with the smallest Si-O-Si angle (Fig. 7). The response of the SiO_2 portion of the network in these melts resembles that of the randomly mixed (Al,Si) interconnected network in melts on the join $NaAlO_2$-SiO_2. The melts on the Ca- and Mg-aluminate—silica joins differ from those on the Na join, however, in that Al^{3+}-bearing polyhedra do not seem to respond to pressure at all.

It is concluded, therefore, that the effect of pressure on the structure of volatile-free aluminate-silica melts in the pressure range where physical and chemical properties show distinct changes is not to change the structure greatly. Thermochemical data on $NaAlSi_3O_8$ melt (Fukuyama and Navrotsky, 1980) quenched in the comparable pressure range lead to similar conclusions. In fact, for the purpose of describing the structure of magmatic liquids under upper-mantle pressure conditions, the error introduced by using 1-atm data is not significant. On the basis of this conclusion, the anionic structure of the (Al + Si) network of magma under both near-surface and upper mantle conditions may be calculated. An example of such cal-

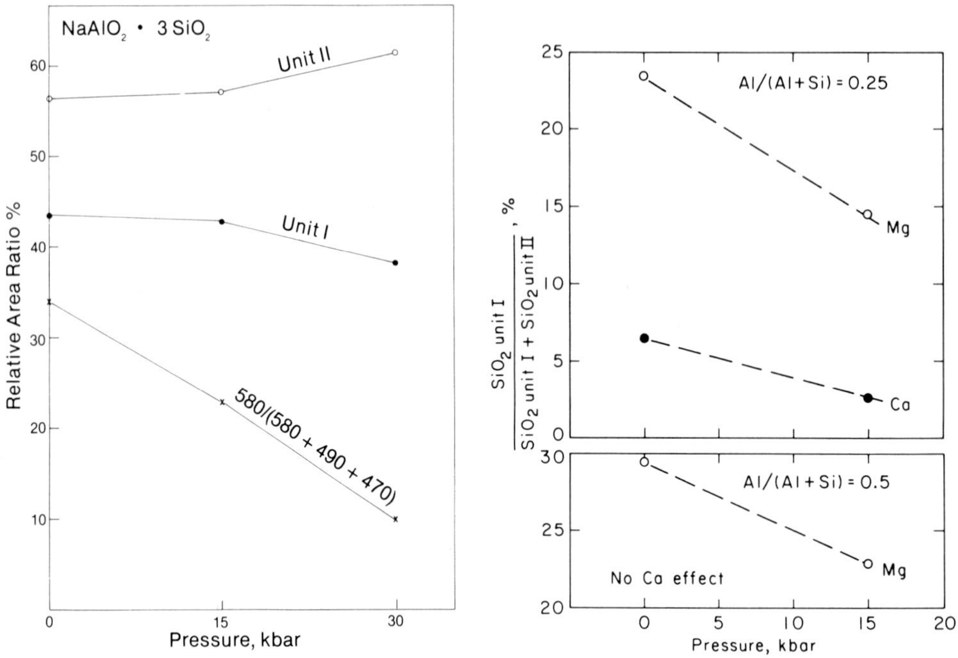

Fig. 7. Summary of pressure effects on structure of aluminate-silica melts (three-dimensional). Units I (1170 and 1020 cm^{-1} bands) and II (1100 and 970 cm^{-1} bands); 580/(580 + 490 + 470), relative contribution of broken oxygen defects in structure.

culations with the average andesite of Chayes (1969) as the bulk composition is shown in Table II. Details of the method of calculation may be found in Mysen et al. (1981a, 1982). The anionic structure of other natural melt compositions may be calculated similarly. A few examples of the anionic structure of basalt, andesite, and granite melt are given by Mysen et al. (1981a, 1982).

TABLE II

Anionic constitution of anhydrous andesite*

Chain units with Al/(Al + Si) < 0.005	12%
Sheet units with Al/(Al + Si) < 0.01	13%
Three-dimensional units with Al/(Al + Si)~0.25	75%
(1) SiO_2 units with no Al^{3+}	20%
(2) Three-dimensional units with Al/(Al + Si)~0.25	25%
(3) Three-dimensional $Al_2Si_2O_8^{2-}$	25%

*Recalculated to 100% aluminate-silica. Approximately 5% of the structure is Fe^{3+}, Ti^{4+}, and P^{5+} where Fe^{3+} and Ti^{4+} probably occur as separate, three-dimensional clusters and P^{5+} most likely is found in linear clusters (Seifert et al., 1979; Mysen et al., 1980b, c, 1981b; Fox et al., 1982).
Bulk composition: average Cenozoic andesite of Chayes (1969).

OTHER TETRAHEDRALLY COORDINATED CATIONS

Phase-equilibrium data (e.g., Kushiro, 1975; Visser and Koster van Groos, 1979; Ryerson and Hess, 1980; Wood and Hess, 1980) indicate that solution of a few percent P_2O_5, TiO_2, or Fe_2O_3 in silicate melts may affect the liquidus phase equilibria significantly. It has been shown that both Ti^{4+} and P^{5+} are network formers in silicate melts (Furukawa and White, 1979; Mysen et al., 1980b, 1981b). Titanium shows a small tendency to mix with Si^{4+}, but formation of separate titanate clusters is more common. Most likely, these clusters are in the form of interconnected TiO_2 tetrahedra. Phosphorus does not mix with Si^{4+} but forms phosphate complexes instead (Mysen et al., 1981b). Ferric iron occurs in the network, probably as a separate, highly polymerized ferrite complex, whenever alkali metals can be found for local charge-balance (Seifert et al., 1979; Mysen et al., 1980c; Fox et al., 1982). Because of the low abundance of these three cations in melt compositions relevant to most island arc volcanics, their structural role will not be considered further here.

VOLATILES (H_2O AND CO_2) IN SILICATE MELTS AT HIGH PRESSURE

Island arc andesites contain several weight percent H_2O. Liquidus phase relations of so-called high-magnesian andesite (Tatsumi, 1981) indicate that as much as 7 wt.% must be dissolved in such melt in order for this liquid composition to be in equilibrium with spinel lherzolite in the upper mantle. Partial melting experiments with natural and synthetic lherzolite compositions indicate that under water-saturated conditions the partial melt resembles andesite to more than 20 kbar pressure (Kushiro et al., 1968, 1972; Kushiro, 1972, 1974). Melting experiments in the peridotite-H_2O-CO_2 system indicate that even with $CO_2/(CO_2 + H_2O)$ in a vapor coexisting with melt and minerals as high as about 0.4—0.5 (molar ratio), the partial melt will be andesitic at least to 15 or 20 kbar pressure (Mysen and Oettcher, 1975).

Experimental and natural evidence suggest, therefore, that high water activities although not necessarily water-saturated conditions, are needed to produce andesitic melts by partial melting in the upper mantle. Furthermore, physical and chemical properties of magma depend on their water content (e.g., Shaw, 1963; Burnham and Davis, 1971, 1974; Kushiro et al., 1976; Kushiro, 1978b). Consequently, experimental determination of such properties under different conditions may not be applicable to magma genesis unless these properties can be related to water and other intensive and extensive variables. It is necessary, therefore, to determine the solubility mechanisms of water in relevant melt compositions.

Burnham (1975) concluded, on the basis of *P-V-T* studies in the system $NaAlSi_3O_8$-H_2O, that albite may be used to model acidic magmas. The data summarized in Table II indicate why this is so as the data in that table show that about 75% of the andesite melt structure may be described in terms of feldspar compositions. On the basis of such observations, one may also use the $NaAlSi_3O_8$ composition as a model for direct determination of the solution mechanisms of H_2O in acidic magma with the aid of spectroscopic methods.

It is known that H_2O dissolves in silicate melts principally to form OH complexes (Uys and King, 1963; Burnham and Davis, 1971; Velde and Kushiro, 1976). Burnham (1975) concluded that for aluminosilicate melts such as that of $NaAlSi_3O_8$ composition, the OH groups were formed by exchanging equimolal amounts of alkali metal (Na^+) with H^+ to form one OH group. Additional H_2O dissolved in the form of a more loosely held $Na(OH)^\circ$ complex. This model has been very successful in describing a number of physical and chemical features of aluminosilicate melts.

Raman spectroscopic studies of water-bearing aluminosilicate melts have been carried out for compositions on the joins $NaAlSi_3O_8$-$CaAl_2Si_2O_8$-H_2O and $CaNaAlSi_2O_7$-H_2O (Mysen et al., 1980d; Mysen and Virgo, 1980a). In these studies, it was possible to determine: (1) the form of water in solution; (2) the type of bonds between water and the silicate network; (3) changes in the aluminosilicate network resulting from the dissolved water. Mysen et al. (1980d) concluded that water dissolves in the form of OH complexes. There is no conclusive spectroscopic evidence for molecular H_2O such as suggested by Hodges (1973) for the system Mg_2SiO_4-H_2O or for H_2 as suggested by Freund (1981) on the basis of studies of silicate and oxide minerals. The Si-OH symmetric stretch vibration with a band near 970 cm^{-1} (Stolen and Walrafen, 1976) is evident, for example. Its intensity relative to the antisymmetric stretch bands that stem from the three-dimensional network structure increases with increasing water content of the melt. It was also found that the antisymmetric, (Si,Al)-coupled stretch band near 1100 cm^{-1} in $NaAlSi_3O_8$ shifted to higher frequency (Fig. 8) with increasing water content. This observation indicates that the Al(Al + Si) of the three-dimensional network decreases with increasing water content. In addition, a new band occurs between 850 and 900 cm^{-1}, identified as a symmetric, (Al,Si)-coupled stretch band from a depolymerized structural unit *without OH.* Its intensity increases with increasing water content (Fig. 8).

On the basis of these and related observations, it is concluded that water dissolves in aluminosilicate melts in the form of OH groups that are bonded exclusively to Si^{4+} in the network and to M^+ (or M^{2+}) outside the network. An amount of Al^{3+} in the network of the anhydrous melt equivalent to the amount of metal cation used to form M(OH) complexes is no longer in the network owing to the lack of a charge-balancing cation. This expulsion of Al^{3+} leads to the formation of new non-bridging oxygens in the hydrous aluminosilicate melt. The band near 850--900 cm^{-1} is one of those that are detected for T-O^- vibrations from this new unit.

The complete solution mechanism for H_2O in an aluminosilicate melt such as that of $NaAlSi_3O_8$ composition is, therefore,

$$14AlSi_3O_8^- + 13H_2O \rightleftharpoons 9Si_4O_7(OH)_2 + 6AlSiO_6^{5-} + 8OH^- + 8Al^{3+} \quad (6)$$

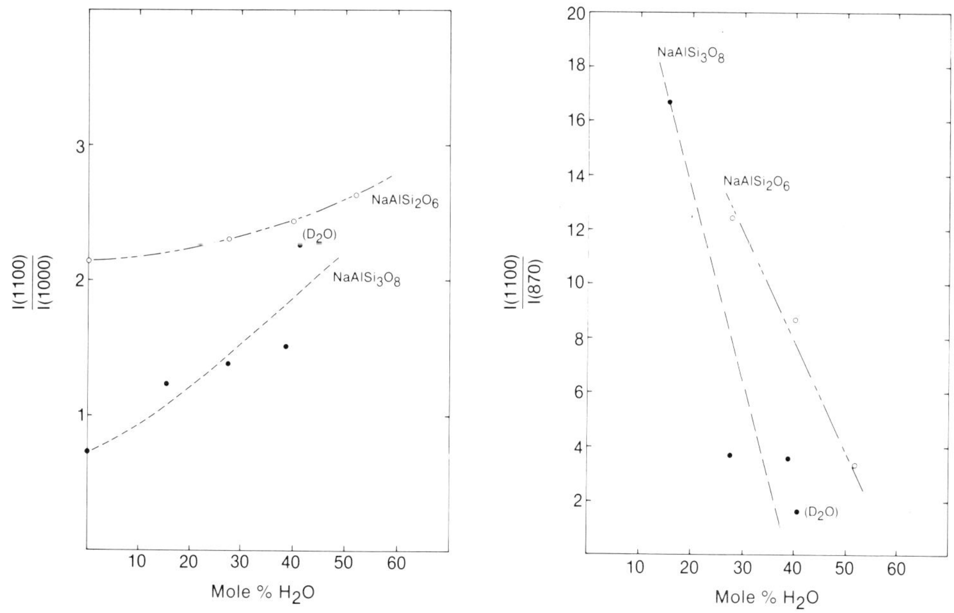

Fig. 8. Relative intensities of relevant (Si,Al)-O stretch bands in $NaAlSi_3O_8$-H_2O and $NaAlSi_2O_6$-H_2O melts as a function of water content (O = 8 in all molar calculations). D_2O, 95% D_2O + 5% H_2O in volatile component. The ratio, I(1100)/I(1000) represents relative abundance of the two, (Al,Si)-bearing, three-dimensionally interconnected units, where I(1100) represents the abundance of the unit with the lowest Al/(Al + Si). The increase in the value of this intensity ratio with increasing water content of the melt results from decreasing Al/(Al + Si) of the three-dimensional network units. The ratio I(1100)/I(870) represents the abundance of Al-bearing, anhydrous structural units with nonbridging oxygen, relative to fully polymerized (NBO/T = 0) aluminosilicate units. The decrease in the value of this intensity ratio is a reflection of increasing NBO/T of the melts with increasing water content of the melts. Data from Mysen et al. (1980d).

Similar equations may be written for M^{2+}-balanced Al^{3+} and for melts with nonbridging oxygens in their anhydrous form. The rate change of NBO/T of aluminosilicate melts with anhydrous NBO/T between 0.8 and 0 is 0.8—1% per mole percent dissolved H_2O (Mysen and Virgo, 1980b).

Although CO_2 probably plays a subordinate role in the formation of quartz-normative andesitic liquids beneath islands arcs, the solubility mechanisms of this volatile component in silicate melts are relevant, partly because of the use of CO_2 to lower the activity of H_2O in peridotite-H_2O-CO_2 melting experiments. In order to relate these experimental data to water-undersaturated melting in the absence of CO_2 beneath island arcs, it is necessary to determine the solubility mechanisms of both H_2O and CO_2 in silicate melts.

Experimental carbon dioxide solubility mechanisms in silicate melts at high pressures and temperatures indicate that this component is dissolved as

CO_3^{2-}. The formation of CO_3^{2-} from CO_2 is accompanied by a reduction of the activity of non-bridging oxygens in the silicate melt:

$$CO_2 + O^{2-} \rightleftharpoons CO_3^{2-} \quad (7)$$

Spectroscopic studies of CO_2-saturated silicate melts show that the principal mechanism indicated by eq. 7 is correct. Mysen and Virgo (1980c) found that this mechanism may be integrated with an equation such as eq. 5 to illustrate the solubility mechanism of CO_2 in magmatic liquids:

$$5Si_2O_6^{4-} + 2CO_2 \rightleftharpoons 2SiO_4^{4-} + 4Si_2O_5^{2-} + 2CO_3^{2-} \quad (8)$$

This mechanism indicates the NBO/T will decrease at the rate of about 0.8% per mole percent dissolved CO_2.

As a result of the above considerations (see, for example, eqs. 6 and 8), it follows that if both CO_2 and H_2O dissolve in an aluminosilicate melt, the NBO/T of this melt will remain constant if the molar CO_2/H_2O is about 1.2.

IMPLICATIONS FOR THE STRUCTURE OF VOLATILE-FREE AND VOLATILE-BEARING NATURAL MAGMA

A description of the structure of hydrous andesitic magma beneath island arcs requires knowledge of its bulk composition as well as its water content. The information discussed above can then be used to calculate the minimum water content of such hydrous magma in equilibrium with peridotite in the upper mantle. The structure of such magma may then be estimated.

Phase-equilibrium studies in natural and synthetic peridotite-H_2O-CO_2 systems (Boettcher, 1973; Eggler, 1974; Mysen and Boettcher, 1975) indicate that quartz-normative liquids that resemble andesitic magma can coexist with spinel peridotite in the upper mantle with molar $CO_2/(CO_2 + H_2O)$ in the coexisting vapor between 0.4 and 0.5. The activity of H_2O in this vapor and, therefore, the coexisting melt may be estimated from the derived thermodynamic data on CO_2 + H_2O vapors at comparable pressure and temperatures (Eggler and Kadik, 1979). The activity coefficient of H_2O in the vapor is about 1.5. Consequently, the activity of H_2O, a_{H_2O}, is about 0.75–0.8. On this basis the minimum water content of CO_2-saturated andesitic melt in equilibrium with spinel peridotite in the upper mantle may be estimated provided that the activity coefficient of H_2O in the silicate melt is known. By using the method of Burnham (1975) to calculate molecular weight of the melt, $a^m_{H_2O}$ is related to the mole fraction, $X^m_{H_2O}$, by the following expression:

$$a^m_{H_2O} = 0.25k \exp(6.25 - 2667/T)(X^m_{H_2O} - 0.5) \quad (9)$$

The value of the constant, k, as given by Burnham (1979), varies between 1.6 and 1.2 in the pressure range 10–20 kbar. Thus, the mole fraction of H_2O in this melt is 0.65–0.7.

In order to relate this value of $X^m_{H_2O}$ to the minimum water content of a CO_2-free, water-undersaturated andesitic melt in equilibrium with spinel peridotite, the effect of dissolved CO_2 on the silicate melt in the experiments must be considered. The data of Mysen et al. (1975) for CO_2 solubility in water- and CO_2-saturated andesite melt as a function of $CO_2/(CO_2 + H_2O)$ indicate that $X^m_{CO_2}$ is about 0.1. It was suggested above that with CO_2/H_2O about 1.2 in the melt, NBO/T is not affected. By correcting for the CO_2 content in this way, the $X^m_{H_2O}$ of water-undersaturated, CO_2-free andesitic melt is about 0.5—0.55. This mole fraction corresponds to 5—7 wt.% H_2O in the melt.

It is concluded, therefore, that andesitic melt may be formed by partial melting of spinel peridotite in the upper mantle if the water content of the system is sufficient to provide 5—7 wt.% H_2O to the melt. For 10% melting of the source region in the mantle, for example, this region must contain approximately 0.5 wt.% H_2O. It is noted that this water content corresponds closely to that suggested by Tatsumi (1982) on the basis of liquidus phase equilibria of Japanese andesite.

Expressions such as that in eq. 6 indicate the solubility mechanisms of H_2O in siliceous, highly aluminous silicate melts. Five wt.% H_2O corresponds to about 50 mole% H_2O. Solution of this much water in andesitic magma will therefore result in an increase in NBO/T of this melt from about 0.4 for the anhydrous composition to slightly higher than 1.0 for the hydrous andesite melt. Thus, hydrous andesite melt in equilibrium with four-phase peridotite in the upper mantle is less polymerized than volatile-free tholeiite, for example (NBO/T of tholeiite melt is between 0.8 and 0.9).

As a result of the complexing of OH with metal cations in the melt, hydrous andesite melt in equilibrium with four-phase peridotite in the mantle has little or no Al^{3+} in tetrahedral coordination. Inasmuch as Al + Si constitutes more than 95% of the tetrahedrally coordinated cations in andesitic magma, the solution of H_2O results in the network consisting almost only of Si^{4+}. Recalculated to an anhydrous basis, the stoichiometry of such a melt resembles that of a disilicate (Table III). From the bulk composition of andesite and data such as those in Figs. 2 and 3, one may estimate the anionic structure of anhydrous and hydrous andesitic magma.

TABLE III

Anionic constitution of hydrous andesite (7 wt.% H_2O; NBO/T = 1.2) recalculated to an anhydrous base

Three-dimensional units with no Al^{3+}	40%
Sheet units with no Al^{3+}	20%
Chain units with no Al^{3+}	30%
Monomers with no Al^{3+}	10%

Bulk composition: average Cenozoic andesite of Chayes (1969).

The enhanced abundance of anionic units with lower degree of polymerization in hydrous than in anhydrous andesite helps to explain the enhanced stability of more depolymerized liquidus minerals in an andesite system as compared with the anhydrous melt. Eggler (1972) showed, for example, that the plagioclase stability field is diminished relative to that of pyroxene as H_2O is added to Paricutin andesite. The occurrence of SiO_4^{4-} structural units in the hydrous melt leads to an appreciation of the possible stability of olivine in equilibrium with hydrous andesite melt at high pressure and temperature.

RELATIONSHIPS BETWEEN CRYSTAL-LIQUID EQUILIBRIA AND WATER CONTENT OF SILICATE MELT

Partition coefficients ($D_M^{\text{xtal-liq}} = C_M^{\text{xtal}}/C_M^{\text{liq}}$) of transition metals (e.g., M = Ni, Co, Fe, Mn) between silicate melts and a peridotite mineral such as olivine depend on the bulk composition of the melt (Watson, 1977; Hart and Davis, 1978; Takahashi, 1978). Mysen and Virgo (1980c) found that in the NBO/T range of natural magma and at constant temperature, D_{Ni} and D_{Mn} are linear functions of NBO/T of the melt. The data of Takahashi (1978) were obtained at NBO/T greater than about 1.5 in the temperature range 1200—1450°C. These data also show a dependence on NBO/T, although the relationship is not linear. The lack of linearity may be at least partly due to the variable temperature in the NBO/T range of the liquids of his experiments.

Because solution of water in silicate melt results in a rapid increase in NBO/T (e.g., eq. 6), the values of crystal-liquid partition coefficients would depend on the water content of the magma. At constant pressure and temperature, data such as those of Mysen and Virgo (1980c), Watson (1977), Takahashi (1978), and Hart and Davis (1978) may be combined with the determined water solubility mechanisms to formulate relationships between NBO/T, water content of magma, and the value of crystal-liquid partition coefficients. Examples with Mn^{2+} and Ni^{2+} for melts with NBO/$T < 1.0$ are (Mysen and Virgo, 1980b):

$$D_{\text{Ni}}^{\text{ol-liq}} = 19.2 - (15.7\text{NBO}/T + 0.23X_{\text{H}_2\text{O}}^m) \text{ at } 1500°\text{C} \quad (10)$$

and

$$D_{\text{Mn}}^{\text{ol-liq}} = 1.70 - (0.88\text{NBO}/T + 0.013X_{\text{H}_2\text{O}}^m) \text{ at } 1450°\text{C} \quad (11)$$

In eqs. 10 and 11, $X_{\text{H}_2\text{O}}^m$ is the mole fraction of water in the melt. Olivine is pure forsterite. By adjusting for differences in temperature between experimental data and that of hydrous melting of peridotite (~60% increase/100°C; Hart and Davis, 1978) and the effect of pressure on $D_{\text{Ni}}^{\text{ol-liq}}$ for melt with NBO/T ~ 0.4 (Mysen and Kushiro, 1979), values of $D_{\text{Ni}}^{\text{ol-liq}}$ for hydrous (~7wt.% H_2O) andesite melt in the upper mantle may be estimated. Application of ol-opx and opx-cpx partitioning data for Ni (Mysen, 1978) results in values of crystal-liquid nickel partition coefficients for hydrous andesite melt and several important minerals in peridotite upper mantle (Table IV). Similar estimates may be obtained for other geochemically important trace elements (e.g., rare earth elements) after their temperature and pressure dependence have been determined.

TABLE IV

Calculated nickel partition coefficients with hydrous andesite melt

Mineral	$D_{Ni}^{mineral\text{-}melt}$
Olivine	15.4
Clinopyroxene	6.2
Orthopyroxene	1.0
Garnet	3.1

Melt composition: average andesite of Chayes (1969). Water content of melt: 7 wt.%. Temperature: 1000°C. Pressure: 20 kbar. Anhydrous NBO/T: 0.4.

The partition coefficient, $(Fe/Mg)^{ol}/(Fe/Mg)^{liq}$, is often used to determine whether natural magma may have been in equilibrium with peridotite in the upper mantle without significant fractional crystallization subsequent to initial partial melting. The data of Takahashi (1978) indicate that the value of this partition coefficient decreases with increasing degree of polymerization (decreasing NBO/T) of the melt. Although his data do not cover the range of NBO/T of most natural magma, linear extrapolation into the NBO/T range of magma indicates that $(Fe/Mg)^{ol}/(Fe/Mg)^{liq}$ decreases by about 25% in the NBO/T range between 1.0 and 0.4. This difference in NBO/T corresponds to the difference between typical anhydrous andesite and the same andesitic melt with about 5 wt.% H_2O in solution. It is inferred, therefore, that the Mg/(Mg + Fe) of anhydrous andesite melt in equilibrium with peridotite at a given pressure, temperature, and fugacity of oxygen is lower than for the same andesite with 5 wt.% H_2O in solution. Iron-magnesium olivine-melt partition coefficients obtained at 1 atm may not be applied to hydrous melting of peridotite in the upper mantle before corrections for water content of the magma have been made.

ACKNOWLEDGEMENTS

Critical reviews by E.-R. Neumann, T.N. Irvine, D.H. Eggler and H.S. Yoder, Jr., are appreciated.

REFERENCES

Boettcher, A.L., 1973. Volcanism and orogenic belts — the origin of andesite. Tectonophysics, 17: 231—244.

Brawer, S.A., 1975. Theory of the vibrational spectra of some network and molecular glasses. Phys. Rev. B, 11: 3173—3194.

Brawer, S.A. and White, W.B., 1975. Raman spectroscopic investigation of the structure of silicate glasses. I. The binary silicate glasses. J. Chem. Phys., 63: 2421—2432.

Brawer, S.A. and White, W.B., 1977. Raman spectroscopic investigation of the structure of silicate glasses. II. Soda-alkaline earth-alumina ternary and quaternary glasses. J. Non-Cryst. Solids, 23: 261—278.

Burnham, C.W., 1975. Thermodynamics of melting in experimental silicate-volatile systems. Geochim. Cosmochim. Acta, 39: 1077—1084.

Burnham, C.W., 1979. The importance of volatile constituents. In: H.S. Yoder, Jr. (Editor), The Evolution of the Igneous Rocks — Fiftieth Anniversary Perspectives. Princeton University Press, Princeton, N.J., pp. 439—482.

Burnham, C.W., 1981. The nature of multicomponent alumino-silicate melts. In: D. Rickard and F.E. Wickman (Editors), Chemistry and Geochemistry at High Temperatures and High Pressures. Phys. Chem. Earth, 13/14: 197—227.

Burnham, C.W. and Davis, N.F., 1971. The role of H_2O in silicate melts. I. *P-V-T* relations in the system $NaAlSi_3O_8$-H_2O to 10 kilobars and 1000°C. Am. J. Sci., 270: 54—79.

Burnham, C.W. and Davis, N.F., 1974. The role of H_2O in silicate melts. II. Thermodynamic and phase relations in the system $NaAlSi_3O_8$-H_2O to 10 kilobars, 700°C to 1100°C. Am. J. Sci., 274: 902—940.

Carmichael, I.S.E., Turner, F.J. and Verhoogen, J., 1974. Igneous Petrology. McGraw-Hill, New York, N.Y., 739 pp.

Chayes, F., 1969. The chemical composition of Cenozoic andesite. In: A.R. McBirney (Editor), Proceedings of the Andesite. Conference. Oreg., Dept. Geol. Miner. Ind., Bull., 65: 1—11.

Chayes, F., 1975. The average composition of the commoner Cenozoic volcanic rocks. Carnegie Inst. Washington, Yearb., 74: 547—549.

Eggler, D.H., 1972. Water-saturated and water-undersaturated melting relations in a Paricutin andesite and an estimate of water content in natural magma. Contrib. Mineral. Petrol., 34: 261—271.

Eggler, D.H., 1974. Application of a portion of the system $CaAl_2Si_2O_8$-$NaAlSi_3O_8$-SiO_2-MgO-Fe-O-H_2O-CO_2 to genesis of the calc-alkaline suite. Am. J. Sci., 274: 297—315.

Eggler, D.H. and Burnham, C.W., 1973. Crystallization and fractionation trends in the system andesite-H_2O-CO_2 at pressures to 10 kb. Geol. Soc. Am. Bull., 84: 2517—2532.

Eggler, D.H. and Kadik, A.A., 1979. The system $NaAlSi_3O_8$-H_2O-CO_2. I. Compositional and thermodynamic relations of liquids and vapors coexisting with albite. Am. Mineral., 64: 1036—1049.

Eggler, D.H. and Rosenhauer, M., 1978. Carbon dioxide in silicate melts. II. Solubilities of CO_2 and H_2O in $CaMgSi_2O_6$ (diopside) liquids and vapors to 40 kb. Am. J. Sci., 278: 64—94.

Etchepare, J., 1972. Study by Raman spectroscopy of crystalline and glassy diopside. In: R.W. Douglas and B. Ellis (Editors), Glassy and Amorphous Materials. Wiley, New York, N.Y., pp. 337—346.

Fox, K.E., Furukawa, T. and White, W.B., 1982. Transition metal ions in silicate melts. II. Iron in sodium silicate glasses. Phys. Chem. Glasses, 23: 169—178.

Freund, F., 1981. Mechanisms of water and carbon dioxide solubility in oxides and silicates and the role of O^-. Contrib. Mineral. Petrol., 76: 474—482.

Fukuyama, H. and Navrotsky, A., 1980. Enthalpy of annealing of albite glass quenched from 1450°C at 10, 15 and 25 kilobars (abstract). EOS, Trans. Am. Geophys. Union, 61: 1147.

Furukawa, T. and White, W.B., 1979. Structure and crystallization of glasses in the $Li_2Si_2O_5$-TiO_2 system by Raman spectroscopy. Phys. Chem. Glasses, 20: 69—80.

Furukawa, T. and White, W.B., 1980. Vibrational spectra and glass structure. J. Non-Cryst. Solids, 38/39: 87—92.

Furukawa, T., Brawer, S.A. and White, W.B., 1978. The structure of lead silicate glasses determined by vibrational spectroscopy. J. Mater. Sci., 13: 268—282.

Furukawa, T., Fox, K.E. and White, W.B., 1981. The structure of silicate glasses. III. Raman intensities and structural units in sodium silicate glasses. J. Chem. Phys., 75: 3226—3237.

Galeener, F.L., 1979. Band limits and the vibrational spectra of tetrahedral glasses. Phys. Rev. B., 19: 4292—4298.

Gaskell, P.H., 1975. Construction of a model for amorphous tetrahedral materials using ordered units. Philos. Mag., 32: 211—229.

Gaskell, P.H. and Tarrant, I.D., 1980. Refinement of a random network model for vitreous silicon dioxide. Philos. Mag., B42: 265—286.

Hart, S.R. and Davis, K.E., 1978. Nickel partitioning between olivine and silicate melt. Earth Planet. Sci. Lett., 40: 203—220.

Hodges, F.N., 1973. Solubility of H_2O in forsterite melt at 20 kbar. Carnegie Inst. Washington, Yearb., 72: 495—497.

Kushiro, I., 1972. Effect of water on the composition of magmas formed at high pressures. J. Petrol., 13: 311—334.

Kushiro, I., 1974. Melting of hydrous upper mantle and possible generation of andesitic magma: an approach from synthetic systems. Earth Planet. Sci. Lett., 22: 294—299.

Kushiro, I., 1975. On the nature of silicate melt and its significance in magma genesis: regularities in the shift of liquidus boundaries involving olivine, orthopyroxene and silica minerals. Am. J. Sci., 275: 411—431.

Kushiro, I., 1976. Changes in the viscosity and structure of melt of $NaAlSi_2O_6$ composition at high pressures. J. Geophys. Res., 81: 6347—6350.

Kushiro, I., 1978a. Viscosity and structural changes of albite ($NaAlSi_3O_8$) melt at high pressure. Earth Planet. Sci. Lett., 41: 87—91.

Kushiro, I., 1978b. Density and viscosity of hydrous calc-alkalic andesite magma at high pressure. Carnegie Inst. Washington Yearb., 77: 675—678.

Kushiro, I., 1981. Viscosity change with pressure in the system CaO-Al_2O_3-SiO_2. Carnegie Inst. Washington, Yearb., 80: 339—341.

Kushiro, I., Syono, Y. and Akimoto, S.-I., 1968. Melting of a peridotite nodule at high pressures and high water pressures. J. Geophys. Res., 73: 6023—6029.

Kushiro, I., Shimizu, N., Nakamura, Y. and Akimoto, S.-I., 1972. Composition of coexisting liquid and solid phases formed upon melting of natural garnet and spinel lherzolites at high pressures: a preliminary report. Earth Planet. Sci. Lett., 14: 19—25.

Kushiro, I., Yoder, H.S., Jr., and Mysen, B.O., 1976. Viscosities of basalt and andesite melts at high pressures. J. Geophys. Res., 81: 6351—6356.

Lazarev, A.N., 1972. Vibrational Spectra and the Structure of Silicates. Consultants Bureau, New York, N.Y.

Liebau, F., 1981. The influence of cation properties on the conformation of silicate and phosphate anions. In: M. O'Keeffe and A. Navrotsky (Editors), Structure and Bonding in Crystals, Academic Press, London, pp. 97—132.

Long, D.A., 1977. Raman Spectroscopy. McGraw-Hill, New York, N.Y., 276 pp.

Lucovksy, G., 1979. Defect-controlled carrier transport in amorphous SiO_2. Philos. Mag., 39: 531—541.

Martin, R.M. and Galeener, F.L., 1981. Correlated excitations and Raman scattering in glasses. Phys. Rev. B., 23: 3071—3081.

Mitra, S.K., 1982. Molecular dynamics simulation of silicon dioxide glass. Philos. Mag. B, 45: 529—548.

Mysen, B.O., 1978. Experimental determination of nickel partition coefficients between liquid, pargasite and garnet peridotite minerals and concentration limits of behavior according to Henry's law at high pressure and temperature. Am. J. Sci., 278: 217—243.

Mysen, B.O. and Boettcher, A.L., 1975. Melting of a hydrous mantle. II. Geochemistry of crystals and liquids formed by anatexis of mantle peridotite at high pressures and high temperatures as a function of controlled activities of water, hydrogen and carbon dioxide. J. Petrol., 16: 549—590.

Mysen, B.O. and Kushiro, I., 1977. Compositional variations of coexisting phases with degree of melting of peridotite in the upper mantle. Am. Mineral., 62: 843—865.

Mysen, B.O. and Kushiro, I., 1979. The effect of pressure on the partitioning of nickel between olivine and aluminous silicate melt. Earth Planet. Sci. Lett., 42: 383—389.

Mysen, B.O. and Virgo, D., 1978. Influence of pressure, temperature and bulk composition on melt structures in the system $NaAlSi_2O_6$-$NaFe^{3+}Si_2O_6$. Am. J. Sci., 278: 1307—1322.

Mysen, B.O. and Virgo, D., 1980a. Solubility mechanisms of water in basalt melt at high pressures and temperatures: $NaCaAlSi_2O_7$-H_2O as a model. Am. Mineral., 65: 1176—1184.

Mysen, B.O. and Virgo, D., 1980b. Trace element partitioning and melt structure: an experimental study at 1 atm pressure. Geochim. Cosmochim. Acta, 44: 1917—1930.

Mysen, B.O. and Virgo, D., 1980c. Solubility mechanisms of carbon dioxide in silicate melts: a Raman spectroscopic study. Am. Mineral., 65: 885—899.

Mysen, B.O., Arculus, R.J. and Eggler, D.G., 1975. Solubility of carbon dioxide in natural nephelinite, tholeiite and andesite melts to 30 kbar pressure. Contrib. Mineral. Petrol., 53: 227—239.

Mysen, B.O., Virgo, D. and Scarfe, C.M., 1980a. Relations between the anionic structure and viscosity of silicate melts — a Raman spectroscopic study. Am. Mineral., 65: 690—710.

Mysen, B.O., Ryerson, F.J. and Virgo, D., 1980b. The influence of TiO_2 on the structure and derivative properties of silicate melts. Am. Mineral., 65: 1150—1165.

Mysen, B.O., Seifert, F.A. and Virgo, D., 1980c. Structure and redox equilibria of iron-bearing silicate melts. Am. Mineral., 65: 867—884.

Mysen, B.O., Virgo, D., Harrison, W.J. and Scarfe, C.M., 1980d. Solubility mechanisms of H_2O in silicate melts at high pressures and temperatures: a Raman spectroscopic study. Am. Mineral., 65: 900—914.

Mysen, B.O., Virgo, D. and Kushiro, I., 1981a. The structural role of aluminium in silicate melts — a Raman spectroscopic study at 1 atmosphere. Am. Mineral., 66: 678—701.

Mysen, B.O., Ryerson, F.J. and Virgo, D., 1981b. The structural role of phosphorus in silicate melts. Am. Mineral., 66: 106—117.

Mysen, B.O., Virgo, D. and Seifert, F.A., 1982. Structure of silicate melts: implications for chemical and physical properties of natural magma. Rev. Geophys., 20: 353—383.

O'Hara, M.J., 1965. Primary magmas and the origin of basalts. Scott. J. Geol., 1: 19—40.

Ryerson, F.J. and Hess, P.C., 1980. The role of P_2O_5 in silicate melts. Geochim. Cosmochim. Acta, 44: 611—625.

Seifert, F.A., Virgo, D. and Mysen, B.O., 1979. Melt structures and redox equilibria in the system Na_2O-FeO-Fe_2O_3-Al_2O_3-SiO_2. Carnegie Inst. Washington, Yearb., 78: 511—519.

Seifert, F.A., Mysen, B.O. and Virgo, D., 1981. Structure and properties of alumino-silicate melts with three-dimensional structures. Carnegie Inst. Washington Yearb., 80: 305—308.

Sen, N. and Thorpe, M.F., 1977. Phonons in AX_2 glasses: from molecular to band-like modes. Phys. Rev. B, 15: 4030—4038.

Sharma, S.K., Virgo, D. and Mysen, B.O., 1979. Raman study of the coordination of aluminum in jadeite melts as a function of pressure. Am. Mineral., 64: 779—788.

Shaw, H.R., 1963. Obsidian-H_2O viscosities at 1000 and 2000 bars in the temperature range 700° to 900°C. J. Geophys. Res., 68: 6337—6343.

Shimizu, N., 1974. Experimental study of the partitioning of K, Rb, Cs, Sr and Ba between clinopyroxene and liquid at high pressure. Geochim. Cosmochim. Acta, 38: 1789—1798.

Soules, T.F., 1979. A molecular dynamics calculation of the structure of sodium silicate glasses. J. Chem. Phys., 71: 4570—4578.

Stolen, R.H. and Walrafen, G.H., 1976. Water and its relation to broken bond defects in fused silica. J. Chem. Phys., 64: 2623—2631.

Takahashi, E., 1978. Partitioning of Ni^{2+}, Co^{2+}, Fe^{2+}, Mn^{2+}, and Mg^{2+} between olivine and silicate melt: compositional dependence of partition coefficients. Geochim. Cosmochim. Acta, 42: 1829—1845.

Tatsumi, Y., 1982. Origin of high-magnesian andesites in the Setouchi volcanic belt: II. Melting phase relations. Earth Planet. Sci. Lett., 60: 305—318.

Taylor, M. and Brown, G.E., 1979. Structure of mineral glasses. I. The feldspar glasses $NaAlSi_3O_8$, $KAlSi_3O_8$ and $CaAl_2Si_2O_8$. Geochim. Cosmochim. Acta, 43: 61—77.

Toop, G.W. and Samis, C.S., 1962. Some new ionic concepts of silicate slags. Can. Metall. Q., 1: 129—152.

Uys, J.M. and King, T.B., 1963. The effect of basicity on the solubility of water in silicate melts. Trans. Metall. Soc. AIME, 227: 492—500.

Velde, B. and Kushiro, I., 1976. Infrared spectra of high-pressure quenched silicate liquids. Carnegie Inst. Washington Yearb., 75: 618—621.

Verweij, H., 1979a. Raman study of the structure of alkali germanosilicate glasses. I. Sodium and potassium metagermanosilicate glasses. J. Non-Cryst. Solids, 33: 41—53.

Verweij, H., 1979b. Raman study of the structure of germanosilicate glasses. II. Lithium, sodium and potassium digermanosilicate glasses. J. Non-Cryst. Solids, 33: 55—69.

Verweij, H. and Konijnendijk, W.L., 1976. Structural units in K_2O-PbO-SiO_2 glasses by Raman spectroscopy. J. Am. Ceram. Soc., 59: 517—521.

Virgo, D., Mysen, B.O. and Kushiro, I., 1980. Anionic constitution of silicate melts quenched at 1 atm from Raman spectroscopy: implications for the structure of igneous melts. Science, 208: 1371—1373.

Visser, W. and Koster van Groos, A.F., 1979. Effect of pressure on the liquid immiscibility in the system K_2O-FeO-Al_2O_3-SiO_2-P_2O_5. Am. J. Sci., 279: 1160—1175.

Waff, H.S., 1975. Pressure-induced coordination changes in magmatic liquids. Geophys. Res. Lett., 2: 193—196.

Watson, E.B., 1977. Partitioning of manganese between forsterite and silicate liquids. Geochim. Cosmochim. Acta, 41: 1363—1374.

Wood, M.I. and Hess, P.C., 1980. The structural role of Al_2O_3 and TiO_2 in immiscible liquids in the system SiO_2-MgO-CaO-FeO-TiO_2-Al_2O_3. Contrib. Mineral. Petrol., 72: 319—328.

Woodcock, L.V., Angell, C.A. and Cheeseman, P., 1976. Molecular dynamics studies of the vitreous state: simple ionic systems and silica. J. Chem. Phys., 65: 1565—1577.

Yoder, H.S., Jr., 1969. Calcalkalic andesites: experimental data bearing on the origin of their assumed characteristics. In: A.R. McBirney (Editor), Proceedings of the Andesite Conference. Oregon Dept. Geol. Mineral Ind. Bull., 65: 77—89.

Journal of Volcanology and Geothermal Research, 18 (1983) 387—404
Elsevier Science Publishers B.V., Amsterdam — Printed in The Netherlands

FRACTIONATION PATHS OF ATKA (ALEUTIANS) HIGH-ALUMINA BASALTS: CONSTRAINTS FROM PHASE RELATIONS

DON R. BAKER and DAVID H. EGGLER

Geochemistry and Mineralogy Graduate Program, Department of Geosciences, The Pennsylvania State University, University Park, PA 16802 (U.S.A.)

(Received October 20, 1982)

ABSTRACT

Baker, D.R. and Eggler, D.H., 1983. Fractionation paths of Atka (Aleutians) high-alumina basalts: constraints from phase relations. In: S. Aramaki and I. Kushiro (Editors), Arc Volcanism. J. Volcanol. Geotherm. Res., 18: 387—404.

An Aleutian high-alumina basalt from the island of Atka at one atmosphere crystallizes plagioclase (1275°C) followed by olivine (1170°C) and clinopyroxene (1115°C). At oxygen fugacities along NNO, magnetite crystallizes below 1070°C, but its liquidus increases to at least 1175°C at an oxygen fugacity two log units above NNO. Phase relations at two kilobars pressure of melts containing small amounts of water are similar, although orthopyroxene and magnetite are observed to follow clinopyroxene. Amphibole crystallizes at near-liquidus temperatures only at water contents of melts approaching 4.5%. Amphibole assumes the liquidus in melts containing 5% water.

Anhydrous melts crystallize plagioclase to 19 kbar, where garnet and clinopyroxene assume the liquidus. Olivine yields to clinopyroxene as the highest-temperature subliquidus phase at about 9 kbar.

The array of compositions of basaltic Atka rocks, as displayed on appropriate pseudoternary projections, can be interpreted as a crystal fractionation path at moderate pressure (~8 kbar) and small melt-water contents. The interpreted fractionating minerals are olivine, clinopyroxene, plagioclase, and (probably) magnetite. (The actual phenocrysts in Atka basalts like AT-1, which lacks phenocrystic clinopyroxene, must have crystallized at pressure less than 8 kbar, however.) The compositions of two-pyroxene andesites from Atka can be interpreted to lie on a lower-pressure fractionation trend at melt water contents of 2—3%. Such water contents are consistent with the complete absence of amphibole in any Atka rocks and are suggestive that water contents of the basaltic magmas, if the basalts are parental to the andesites, were 1—2%.

INTRODUCTION

Volcanism in primitive intraoceanic arcs such as Scotia, Tonga, the Marianas, and the central Aleutians is not affected by underlying continental crust. Such crust has been shown to influence petrogenesis in evolved arcs (Coulon and Thorpe, 1981; Leeman, 1981). Although intraoceanic arcs are remote, field studies have become available in recent years. Stern (1979) and

0377-0273/83/$03.00

Dixon and Batiza (1979) found that high-alumina basalt is the parental magma to the andesitic rocks in the Mariana Islands, a parentage first proposed by Kuno (1968) on the basis of his extensive investigations of the more evolved Japanese arc. Perfit et al. (1980a) reached the same conclusion for a zoned pluton on the Aleutian island of Unalaska. Marsh (1982) proposed that high-alumina basalt is the parental magma for all central Aleutian andesites and dacites and that the high-alumina basalts represent primary magmas from subducted oceanic crust.

Because of the relative lack of experimental data on high-alumina basalts, phase relations of several such basalts from the central Aleutian island of Atka are being determined. In this paper phase relations (one atmosphere to 20 kbar) are reported for one basalt.

MATERIAL STUDIED

The high-alumina basalt investigated, AT-1, was donated by B.D. Marsh; its composition, mode, and norm appear in Table I. This rock, called an "andesitic basalt" by Marsh (1982), is slightly lower in magnesia and titania and higher in potash than classical high-alumina basalts of Kuno (1960).

AT-1, with 49.8% SiO_2, is among the more basic of the Atka basalts, which range in SiO_2 content down to 47% (Marsh, 1982). Its Mg-number, 47, is virtually identical to the average Mg-number of all basalts with less than 50% SiO_2 (B.D. Marsh, pers. commun., 1982).

TABLE I

Composition of the Atka basalt studied and of a calculated primary melt

	1	2	3	4		5	
SiO_2	49.89	49.80	48.9	Vesicles	0.37	Orthoclase	3.85
TiO_2	1.02	0.96	0.91	Groundmass	66.50	Albite	27.82
Al_2O_3	19.43	20.12	18.0	Plagioclase	29.25	Anorthite	30.03
Fe_2O_3	—	4.36	3.93	Olivine	3.00	Olivine	7.70
FeO	9.45	5.51	6.41	Magnetite	0.88	Hypersthene	12.32
MnO	0.28	0.18	—			Diopside	8.77
MgO	4.79	4.72	9.17			Apatite	0.38
CaO	9.17	9.66	8.68			Ilmenite	2.79
Na_2O	3.39	3.71	3.33			Magnetite	6.35
K_2O	0.76	0.78	0.67				
P_2O_5	—	0.23	—				
Total	98.18	100.03	100.00				

Explanations of columns: 1—average of 7 microprobe analyses of fused glass of sample AT-1, all iron as FeO, P_2O_5 not determined; 2—wet chemical analysis of AT-1; 3—An olivine-enriched primary melt calculated from 2 (see text); 4—mode of AT-1 in volume percent. The wet chemical and mode analyses are through the courtesy of B.D. Marsh; 5—CIPW norm of AT-1 in mole percent, based on column 2.

Chemical analyses of all Atka rocks (B.D. Marsh, pers. commun., 1981) are plotted in Fig. 1 in two projections modified after Walker et al. (1979) — a plagioclase (plag)-saturated olivine (ol)-diopside (di)-quartz + orthoclase (qz+or) projection and an ol-saturated di-plag-qz+or projection. At face value, the plots (Fig. 1a, c) suggest that AT-1 composition (Table I, column 1) is somewhat fractionated relative to compositions nearer the left ends of the arrays of points, and it may well be. It should be recognized, however, that the projected positions are quite sensitive to certain small differences in bulk composition, particularly the Na_2O contents of rocks. For instance, another chemical analysis of AT-1 (Table I, column 2), which is only slightly higher in Na_2O, plots much nearer to the left terminus of both arrays.

Marsh (1981) has interpreted the crystallization sequence of the more basic Atka lavas to be plagioclase, followed by olivine and magnetite, followed by clinopyroxene. The mode of AT-1 (Table I, column 4) is consistent with that sequence. Like most Atka lavas, the modal phenocryst content is rather high, suggesting the possibility of plagioclase accumulation. The REE analyses of rocks from the same locality and with almost identical chemistry do not have a positive europium anomaly (B.D. Marsh, pers. commun., 1981), however, indicating that accumulation of plagioclase was minimal.

EQUIPMENT AND METHODS

The rock was crushed with a hammer, then ground in a ball mill for ten minutes. Final grinding in a synthetic sapphire mortar or in the ball mill was continued until more than 85% of the grains were five microns or less in size; the maximum diameter of remaining grains was 30 microns. The rock was fired at 450°C in air for 24 hours and stored in a 110°C oven prior to use.

One-atmosphere experiments

The one-atmosphere experiments were performed in a gas mixing furnace; the oxygen fugacity was controlled to within 0.1 log unit of the desired value by mixing CO_2 and H_2 gases. Gases were purified before mixing by passing over hot copper turnings and Drierite, then measured via a glass manometer system (Darken and Gurry, 1945). The thermocouple was calibrated against the melting of gold (1064°C) and placed within 1.0 cm of samples. Samples of AT-1 were usually encased in $Ag_{50}Pd_{50}$ envelopes. In one experiment (Table II, column 4), AT-1 powder was pressed onto a platinum loop; to minimize iron loss from the sample the loop had previously been soaked in AT-1 melt for 70 hours at 1300°C (cf. Walker et al., 1979). Samples were suspended within the 1.1-cm-long isothermal zone of the furnace. Samples were quenched by lifting out of the furnace and blowing on them. Samples ceased glowing within a few seconds of removal from the hot zone.

A series of experiments at 1162°C was used to ascertain the time neces-

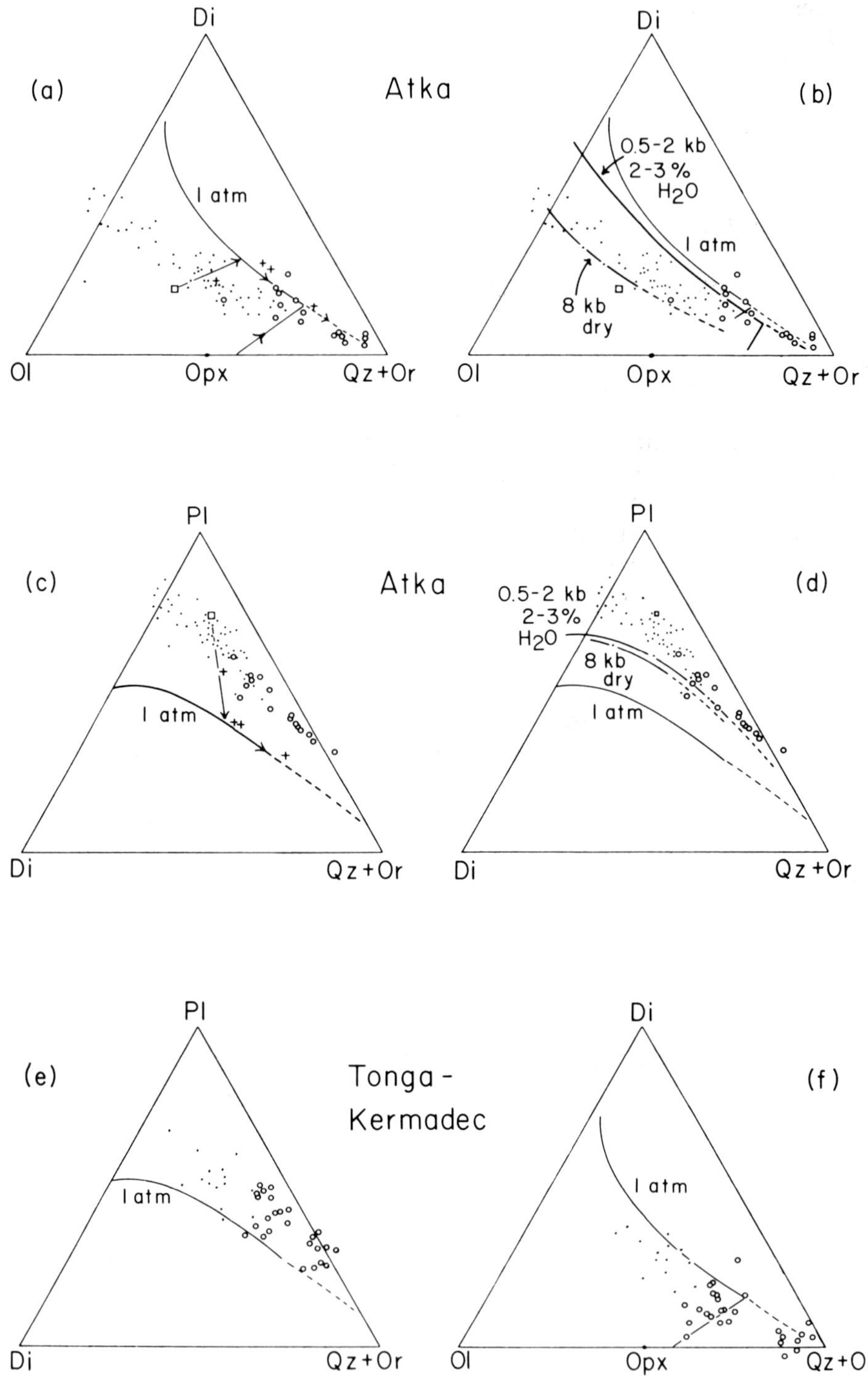
Atka
(a)
Di
1 atm
Ol
Opx
Qz+Or
(b)
Di
0.5-2 kb
2-3%
H2O
1 atm
8 kb
dry
Ol
Opx
Qz+Or
Atka
(c)
Pl
1 atm
Di
Qz+Or
(d)
Pl
0.5-2 kb
2-3%
H2O
8 kb
dry
1 atm
Di
Qz+Or
Tonga-
Kermadec
(e)
Pl
1 atm
Di
Qz+Or
(f)
Di
1 atm
Ol
Opx
Qz+Or

TABLE II

Compositions of phases from runs on AT-1 at 1 atmosphere pressure

	1	2	3	4	5	6	7	8
# Analyses	2	7	3	6	3	3	4	2
Temp. (°C)	1150	1135	1111	1100	1080	1080	1080	1080
$\log f_{O_2}$	−6.1	−8.34	−8.62	−8.80	−9.07	−9.07	−9.07	−9.07
Duration (hr)	12	24	24	60	37	37	37	37
SiO_2	52.98	51.97	53.16	53.27	55.51	35.39	51.38	54.05
TiO_2	1.52	1.55	2.38	2.62	3.26	0.41	0.88	0.09
Al_2O_3	15.07	15.10	12.94	12.91	13.49	0.20	1.78	26.04
FeO*	10.09	12.09	13.00	12.59	11.80	34.87	11.60	1.12
MnO	0.22	0.32	0.31	0.35	0.23	0.76	0.65	0.0
MgO	5.56	4.94	3.63	3.50	2.59	26.55	14.31	0.17
CaO	7.25	6.89	7.43	7.36	6.51	0.69	18.60	10.63
Na_2O	3.87	3.80	3.42	3.47	2.62	0.06	0.39	4.73
K_2O	1.22	1.37	1.90	1.69	2.48	0.0	0.03	0.47
Total	97.78	98.03	98.17	97.76	98.49	98.93	99.62	97.30
						Fo_{58}	$Wo_{39}En_{42}Fs_{19}$	An_{71}

Explanation of columns: 1—glass coexisting with olivine (ol), magnetite (mt) and plagioclase (plag) at oxygen fugacity NNO+2. 2—glass coexisting with ol and plag at NNO. 3—glass coexisting with ol, plag and clinopyroxene (cpx) at NNO. 4—glass coexisting with ol, plag and cpx at NNO; this experiment performed with a platinum loop. 5—glass coexisting with ol, plag and cpx at NNO. 6—olivine coexisting with glass in column 5. 7—clinopyroxene coexisting with glass in column 5. 8—plagioclase coexisting with glass in column 5.

Fig. 1. Pseudoternary phase diagrams showing liquidus boundary curves and rock compositions. For all compositions the projection procedure, modified after Walker et al. (1979) is: (1) Calculate ferric iron content of each analysis after Sack et al. (1980) for t = 1150 °C and f_{O_2} = NNO. (2) Convert oxide analysis to molar proportions. (3) Magnetite (mt) = Fe_2O_3; plagioclase (plag)= $Al_2O_3 + Na_2O - K_2O$; diopside (di)= $CaO - Al_2O_3 + Na_2O + K_2O$; olivine (ol) = $(FeO - Fe_2O_3 + MgO + MnO + Al_2O_3 - CaO - Na_2O - K_2O)/2$; quartz + orthoclase (qz+or) = $SiO_2 - 2K_2O - (Al_2O_3 + FeO - Fe_2O_3 + MnO + MgO + 3CaO + 11Na_2O + 3K_2O)/2$. (4) All compositions are projected from mt.
a and b. Compositions of Atka rocks and experimentally derived liquidus boundary curves plotted on mt-plag-saturated ol-di-qz+or projections. Dots: Atka rocks having olivine phenocrysts. Open circles: Atka rocks having orthopyroxene phenocyrsts. Open square: composition of AT-1 (anal. 1, Table I). Crosses: analyzed glasses from one-atmosphere, NNO, experiments on AT-1 in equilibrium with plag-ol or plag-ol-di. Solid lines: ol-di-plag and ol-opx-plag boundary curves. Dashed lines: opx-di-plag boundary curves. Curves at one atmosphere largely after Walker et al. (1979), except for ol-opx-plag on the ol-qz+or-plag join, from data of Eggler and Osborn (1982). For hydrous and high-pressure boundary curves see text.
c and d. Ol-mt-saturated plag-di-qz+or projections. Symbols same as a and b. Lines emanating from the open square in c and a indicate ol-plag fractionation paths.
e and f. Compositions of rocks from the Tonga-Kermadec arc (Ewart et al., 1977) in mt-ol-saturated plag-di-qz+or and mt-plag-saturated ol-di-qz+or projections.

sary for the partial melts of AT-1 to reach a constant composition. The duration of these experiments ranged from one hour to four days. Contents of TiO_2 and K_2O, which should act as incompatible elements, reach reasonably constant levels within one hour. At long durations, loss of iron is considerable — about 2.5 relative percent an hour. Duration of most one-atmosphere experiments on AT-1 varied between 12 and 24 hours, short enough to minimize iron loss and sufficiently long to allow the melt to reach a constant composition. Mass-balancing of some experiments (see below) indicates minimal iron loss.

Two-kilobar experiments

The two-kilobar experiments were carried out in an internally-heated pressure vessel (Holloway, 1971). Pressure was monitored by a manganin cell and is believed accurate to within 100 bars. Two-zone furnaces were used to control the temperature gradient along the capsule length. Two thermocouples, between which the capsules were situated, were calibrated against the one-atmosphere furnace thermocouples. In all experiments except those at 1050°C the thermal gradient was less than 2°C; during the 1050°C experiment the furnace cycled 5°C about the setpoint and created a thermal gradient of approximately 10°C. Oxygen fugacity was controlled in the vessel through the use of a AgPd membrane (Shaw, 1967). Samples were held at the temperatures and pressure of interest for 24 to 36 hours. Quenching was accomplished isobarically by simultaneously shutting off the power, pumping in cold gas, and rotating the vessel 90°. Although the vessel cooled to 700°C within a few minutes, quench crystals invariably formed.

Mixed CO_2-H_2O fluid phases were used to control the water content of the experimentally-produced melts (Eggler, 1972; Eggler and Burnham, 1973). The fluids were produced from various combinations of water, silver oxalate, and oxalic acid dihydrate. Mixtures of these compounds and AT-1 were sealed in 2-mm $Ag_{50}Pd_{50}$ capsules (except for experiments performed in the absence of a fluid phase — the capsules were loaded only with AT-1). Weights before and after welding indicated no volatile loss. The partial pressure of hydrogen in the vessel was controlled so that the oxygen fugacity of the water-saturated experiments was at the nickel-nickel oxide (NNO) buffer (Huebner and Sato, 1970). In mixed CO_2-H_2O fluids, the oxygen fugacity was lower than NNO; in a fluid of 50% H_2O and 50% CO_2, the oxygen fugacity is near the quartz-fayalite-magnetite buffer at two kilobars (Holloway and Burnham, 1972). In the absence of fluid, oxygen fugacity was not externally controlled.

After removal from the vessel, each capsule was centrifuged for ten minutes at 1500—1700 rpm on a 48-cm-diameter centrifuge. After weighing, the top of the capsule was punctured, allowing the CO_2 to be released; the capsules were reweighed and the observed weight loss attributed to the CO_2 content of the fluid. The capsules were then heated at 110°C for 15 minutes and cooled for ten minutes in a dessicator and weighed again. This step in the procedure was repeated a minimum of three times. The capsules would initially display a large weight loss as most of the water evaporated; subsequent losses in weight were small (approximately 0.5 mg) and defined a straight line that could be extrapolated back to time zero and used to calculate the total amount of water in the melt (Hamilton et al., 1964). This value is subtracted from the total weight of water originally present in the capsule to calculate the amount of water in the fluid phase. An alternative method of calculating the amount of water in the fluid phase is to use the initial weight loss upon heating. Both methods yielded values for the mole fraction of water in the fluid that were within three mole percent of each other. From the mole fraction of water in the fluid (X_w^{fl}), the weight percent of water in the melt can be calculated by a procedure similar to that of Eggler (1972). The mole fraction of water in the fluid is multiplied by the activity of pure water (Burnham et al., 1969) under identical conditions of pressure and temperature ($a_w^o{}_{P,T}$) to yield the activity of water in the melt (a_w^m):

$$a_w^m = a_w^{fl} = X_w^{fl}(a_w^o\ P,T) \tag{1}$$

where a_w^{fl} is the activity of water in the fluid. It is assumed in this calculation that H_2O mixes ideally in the fluid and that solution of CO_2 in the melt has little effect on phase relations. The activity of water in the melt is combined with a model of silicate-water interaction (Burnham, 1979) to calculate the actual H_2O content of the melt. Eggler (1972) has demonstrated for an andesitic composition that the calculated percent of water in the melt is equivalent to the weight percent of water as directly added.

Solid-media experiments

The 8—20-kbar experiments were performed in a single-stage solid-media apparatus (Boyd and England, 1960) using a floating piston technique. A negative 11% friction correction was applied to talc-pyrex assemblies (2.54-cm diam.) based on calibration from the decarbonation of magnesite in the presence of enstatite and forsterite (Newton and Sharp, 1975). The pressure was initially set 2 kbar over the run pressure, power was applied to the assembly to heat it to run temperature, and then the pressure was slowly dropped to run pressure. Pressure was maintained at ± 0.2 kbar. Temperatures were measured with $Pt\text{-}Pt_{90}Rh_{10}$ thermocouples and automatically controlled. No correction for the effect of pressure on the emf of the thermocouple was applied. Run temperatures are believed precise to ± 7°C.

Some sample materials were encased and welded in 4-mm $Ag_{60}Pd_{60}$ capsules (for experiments below 1290°C) or 2-mm $Pt_{95}Au_5$ capsules (above 1290°C). Other aliquots of material were loaded into graphite capsules that were welded inside Pt capsules. In all cases loaded capsules were dried for 1 hour at 110°C before welding. Both the metal-encased and graphite-encased materials were taken to final run conditions by two procedures in order to demonstrate reversibility of melting boundaries: some were heated directly to run temperature, whereas others were heated to a temperature above that of the boundary of interest for 1—6 hours and then dropped to a temperature below the boundary for 16—48 hours. Those heated directly to run temperature were run for 1 hour, although a few 8- and 10-kbar runs were up to 12 hours in duration.

The oxygen fugacity of the metal-encased and the graphite-encased runs is believed to have differed substantially. Minimum oxygen fugacity in the metal-encased runs would be set by the ferric—ferrous ratio of the bulk rock, presumably near NNO. Because some iron from the sample dissolves in metal, however, leaving excess O_2 in the sample, the oxygen fugacity could well have been much higher. In contrast, graphite capsules have been shown to reduce the oxygen fugacity of melts to values below the magnetite-wustite buffer (Thompson and Kushiro, 1972). Graphite capsules, also, of course, prevent loss of iron from the enclosed sample.

Analytical methods

The solid experimental products were removed from the capsules and examined optically. Selected one-atmosphere products were analyzed by the wavelength dispersive system of an automated ETEC electron microprobe. An accelerating potential of 15 kV and a specimen current of 0.012 microamperes were used for all analyses; the raw data were reduced by an on-line computer using the matrix correction procedures of Bence and Albee (1968) and of Albee and Ray (1970). Crystalline phases were analyzed with a two micron beam. The percentage of glass in the run products varied from 75% to 25%, and the size of the individual pools was 10 to 25 microns in diameter. Due to the small size of the pools, a two-micron beam diameter was used; although the possibility of sodium loss exists, it is probably slight, inasmuch as the measured sodium contents of superliquidus glasses did not differ in analyses using a two-micron and a twenty-micron beam diameter. All tabulated microprobe analyses (Tables I and II) are the average of multiple analyses. The number of analyses per phase varies from two to seven.

EXPERIMENTAL RESULTS

At one atmosphere, along the NNO buffer (Fig. 2), plagioclase is the liquidus phase at 1275°C. Not until 1170°C does the next crystalline phase, olivine, become stable. Clinopyroxene appears at 1115°C. Neither orthopyroxene nor magnetite was observed in any of the experiments along NNO. At two log units of oxygen fugacity above NNO, however, magnetite (3.28 wt.% TiO_2, as determined by microprobe analysis) crystallizes at 1175°C.

At two kilobar total pressure (Fig. 3), plagioclase is the liquidus phase ex-

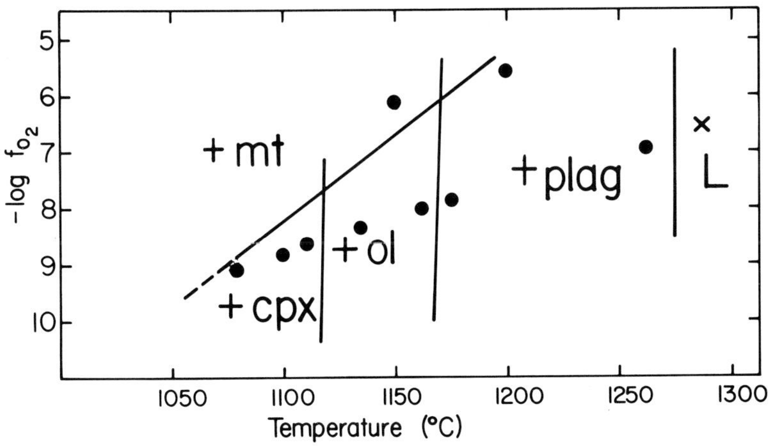

Fig. 2. Phase relations of high-alumina basalt AT-1 at one atmosphere pressure. The array of points at lower oxygen fugacity is along the NNO buffer. Labels denote the phase that crystallizes below each boundary. L-all liquid, plag-plagioclase, ol-olivine, cpx-clinopyroxene, mt-magnetite.

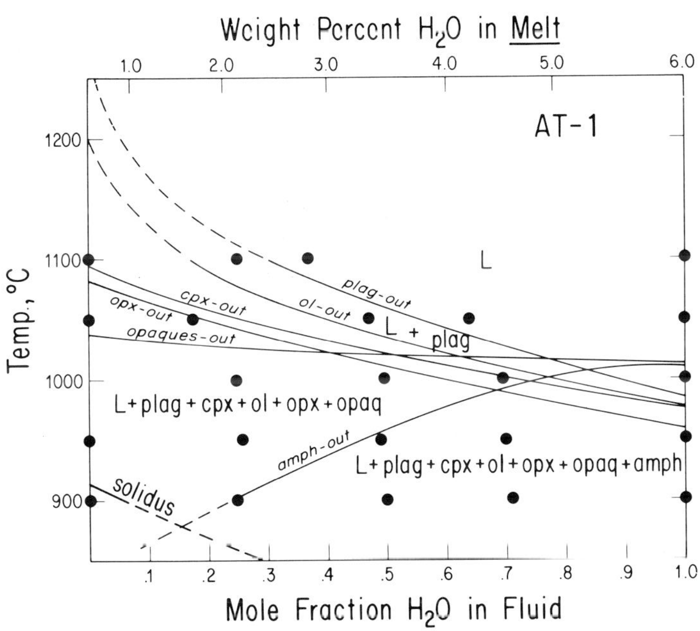

Fig. 3. Phase relations of AT-1 at two kilobars pressure as a function of temperature and the mole fraction of water in the coexisting CO_2-H_2O fluid. Experiments at $X_w^{fl} = 0$ were performed in the absence of fluid (see text). Abbreviations in addition to Fig. 2. opx-orthopyroxene, opaq-opaque, amph-amphibole.

cept at conditions near water-saturation. With small amounts of water in the melt, the subliquidus phase relations are similar to those at one atmosphere — olivine followed by clinopyroxene — although at two kilobars, orthopyroxene and an opaque phase were observed to follow clinopyroxene in the crystallization sequence. (Clinopyroxene was not observed in the run products of the 1100°C, 2 kbar, dry experiment, whereas clinopyroxene is stable to 1115°C at one atmosphere. The lack of clinopyroxene in the two kilobar experiment may be due to traces of water that depressed the clinopyroxene liquidus.) Amphibole melts near the liquidus with 4.5% water in the melt and at the liquidus with 5%. Olivine is present at temperatures within the stability fields of both orthopyroxene and amphibole; its modal abundance diminishes rapidly as the temperature is decreased, presumably due to a reaction relationship with liquid and amphibole (cf. Helz, 1973).

As detailed in a previous section, anhydrous experiments at 20 kilobars pressure were performed using four combinations of conditions — metal-encased and graphite-encased samples taken directly to temperature and taken to temperature from higher-temperature phase fields. As far as could be determined, the boundaries drawn in Fig. 4 are consistent with all runs. Such consistency implies that the boundaries are reversible and do not depend critically upon loss of iron from samples or upon variations in oxygen fugacity.

ANALYTICAL RESULTS

Electron microprobe analyses of the one-atmosphere experiments were performed to trace the liquid line of descent for AT-1 (Table II). Analyses of glasses all display low totals, presumably due to conductivity problems associated with vesicularity of the samples. In addition to the glass analyses in Table II, analyses of the phase coexisting with the glass at 1080°C are also presented; 1080°C was the maximum temperature at which clinopyroxene could be analyzed.

Analyses of glass and coexisting phases from selected experiments were mass-balanced with the bulk composition of AT-1 using a least-squares program (Holloway and Burnham, 1972). Residual sums were small, and the calculated modes were similar to optically estimated modes, indicating the validity of the analyzed compositions.

DISCUSSION

Historical perspective

High-alumina basalts in an arc setting were first mentioned by Powers (1932) in the Medicine Lake Highlands of California, but their significance was not suggested until the work of Kuno (1960); he determined that high-alumina basalt was a primary magma, not the product of the differentiation of a tholeiite or an alkalic basalt. Experiments were performed on the Warner high-alumina basalt at one atmosphere (sealed Pt capsules) and under water-saturated conditions at two, five, and ten kilobars by Yoder and Tilley (1962). It can be presumed that their experiments were conducted at high oxygen fugacity. They found plagioclase and a chrome spinel on the one-atmosphere liquidus at 1240°C, closely followed by olivine. Magnetite became the stable spinel at 1190°C and clinopyroxene appeared at 1160°C. Under two kilobars water pressure, olivine was the silicate liquidus phase at 1125°C (accompanied by magnetite, which was present at the highest temperatures studied at 2 kbar), with clinopyroxene and plagioclase crystallizing near 1075°C. Amphibole was stable below about 1010°C, and orthopyroxene was not observed in any of their experiments. Yoder and Tilley (1962) concluded that high-alumina basalts are not primary magmas, but may be derived by fractional crystallization of picritic parental magmas under hydrous conditions. Kuno (1966) subsequently modified his original hypothesis: high-alumina basalts may be produced from tholeiitic magmas, but only at high pressures. Kuno (1968) further argued that irrespective of the source of high-alumina basalts, they could produce calcalkaline andesites through fractional crystallization, and that the lack of iron enrichment in the more evolved rocks is due to the early crystallization of magnetite, as elegantly proposed by Osborn (1959, 1979).

Oxygen fugacity of the Atka rocks

Marsh (1982) determined from petrographic observations that magnetite appeared together with olivine in the crystallization sequence of the Atka high-alumina basalts. The phase relations of AT-1 at one atmosphere (Fig. 1) require an oxygen fugacity about two log units above NNO for the co-appearance of magnetite and olivine. Calculations by Marsh (1982), based primarily on ferrous/ferric contents of Atka rocks, also indicate conditions more oxidizing than NNO, although not as oxidizing as phase equilibria require.

Pressure of crystallization of Atka phenocrysts

Marsh (1976, 1980, 1982) has reasoned that the phenocrysts at Atka crystallized in the order plagioclase-olivine+magnetite-clinopyroxene and that crystallization occurred at pressures less than 5 kbar. Several magma chambers in other primitive island arcs have been interpreted to crystallize at no more than 6-kbar pressure (Dixon and Batiza, 1979; Stern, 1979; Perfit et al., 1980a). Phase equilibria are consistent with such shallow depths of crystallization. Crystallization of olivine before clinopyroxene from AT-1 magma requires pressure less than about 9 kbar under anhydrous conditions (Fig. 4), and probably considerably less than 9 kbar, inasmuch as AT-1 actually contains no clinopyroxene phenocrysts. Of course, the presence of H_2O in the

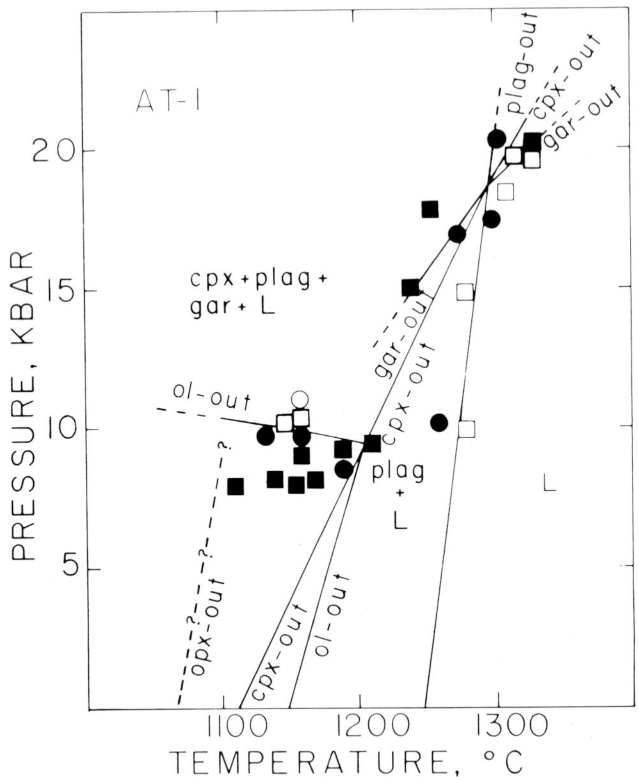

Fig. 4. Anhydrous phase relations of basalt AT-1. Phase-out notation indicates final melting temperature of indicated mineral. Squares: runs heated directly to temperature. Circles: run heated to a temperature above phase boundaries and then dropped. Filled symbols: graphite-encased sample. Open symbols: noble metal-encased sample. Four runs shown at ~10 kbar, 1175°C were actually all run at 10.0 kbar.

melt could increase that estimate of maximum pressure, because it raises the melting temperature of olivine relative to clinopyroxene in basalts (Nicholls and Ringwood, 1973). Such an effect requires a rather large amount of H_2O, however, which in turn would produce the offsetting effect of drastically lowering the plagioclase liquidus relative to the liquidi of the ferromagnesian minerals (e.g., Fig. 3).

Water content of Atka magmas

If it is assumed that phenocrysts in AT-1 crystallized at pressures less than about 5 kbar, Fig. 3 can be used to delimit the H_2O content of the magma. Phase relations in Fig. 3 were determined at 2 kbar, but the relative position of melting curves for basalts at 2 kbar are applicable to pressures of somewhat less than 1 kbar to at least 5 kbar (Holloway and Burnham, 1972; Eggler and Burnham, 1973).

Because the AT-1 phenocryst suite — plagioclase + olivine — does not include amphibole, H_2O content of the melt was certainly less than 5%. It may be argued, however, with less certitude, that the H_2O content was actually far less. Firstly, AT-1 and similar rocks contain no amphibole even as microphenocrysts or in the groundmass, indicating that subliquidus crystallization never entered the amphibole field. Secondly, in a section below, the array of compositions of Atka andesitic magmas is interpreted to have been saturated in plagioclase, orthopyroxene, and clinopyroxene at low pressure (2 kbar?) and 2—3% H_2O. On the assumption that the andesites originated by fractionation of anhydrous minerals from basaltic parents, the basaltic parental magmas must have contained less than 2—3% H_2O.

Low-pressure, anhydrous fractionation of Atka magmas

At low water contents, the sequence of crystallization of AT-1 is the same at one atmosphere and at two kilobars. The first phase in this sequence is plagioclase, followed by a mafic mineral at much lower temperatures (Figs. 2 and 3). *This sequence indicates that AT-1 liquid cannot have been produced by fractionation of a more mafic parent at low pressures, inasmuch as no mafic mineral is on the liquidus.*

The same point can be appreciated for the array of all Atka basalts by examining the plag-saturated di-ol-qz+or projection and the ol-saturated di-pl-qz+or projection, on which the one-atmosphere ol-di-plag and opx-di-plag boundary curves determined by Walker et al. (1979) are plotted (Fig. 1a, c). (Three points determined on the ol-di-plag boundary curve from experiments on AT-1 (Table II) nearly coincide with the curves of Walker et al. (1979) (Fig. 1a, c).) On both projections, the array of olivine-bearing basalts and basaltic andesites (points, Fig. 1a, c) is distinct from the one-atmosphere boundary curves.

The two arrays (Fig. 1a, c) correctly predict that all Atka basalts, like AT-1, would crystallize plagioclase and olivine before clinopyroxene at one atmosphere. For AT-1 bulk composition, fractionation of plagioclase and then olivine would move melts away from the olivine or plagioclase apex until they intersected (at about 1115°C) and then followed the ol-di-plag boundary curve. On both projections, the trend of liquids in equilibrium with olivine and plagioclase traverses the Atka trend, and therefore the more evolved high-alumina basalts (52—54% SiO_2) cannot be generated from a rock of composition similar to AT-1 at low pressures. Moreover, in detail the liquids in equilibrium with olivine and plagioclase do not replicate any of the more evolved basalts, principally because of the low alumina and high total iron and titania contents of the liquids. (An increase in oxygen fugacity to two log units higher than NNO stabilizes magnetite and should rapidly deplete the liquid in total iron, with only a slight change in the path of the liquid line of descent. Lower-temperature high-oxygen-fugacity runs have not been made, however.)

High-pressure and hydrous fractionation

Figures 1b and 1d display the plag-ol (or orthopyroxene [opx])-di boundary curves for melts containing 2—3% H_2O at pressure of 0.5—2 kbar and for anhydrous melts at 8-kbar pressure. The former curves have been estimated from liquidus and subliquidus phase assemblages in water-saturated and -undersaturated experiments on basalts, andesites and dacites by Yoder and Tilley (1962), Eggler (1972), Sekine et al. (1979), Graham (1981), and this study. The latter curves have been estimated from liquidus and subliquidus phase assemblages or from analyzed glass compositions in experiments at 6—12 kbar by Bender et al. (1978), Helz (1978), Thompson (1972), Kushiro and Thompson (1972), and this study (Fig. 4).

It is apparent, in Fig. 1b, d, that both total pressure and H_2O pressure displace the plag-ol(or opx)-di boundary curves in the same direction. To a first approximation, however, displacement in the plag-saturated projection (Fig. 1b) is principally due to total pressure, whereas displacement in the ol-saturated projection is due to a combination of pressure and H_2O content. On this less-than-perfect basis, the trend of Atka compositions may be interpreted to be due to a combination of relatively high pressure (~8 kbar?), at least for the basaltic rocks, and low to moderate H_2O content. The two-pyroxene andesites and dacites (open circles, Fig. 1) would appear to have fractionated at lower pressures, however, inasmuch as they plot closer to the one-atmosphere boundary curve in Fig. 1b. The diagrams are consistent with fractionation of andesite magma at pressures as low as 2 kbar and H_2O contents of 2—3%, whereas the dacites, which plot to the plagioclase-rich side of the "wet" boundary curve at highest values of qz+or (Fig. 1d), would appear to have contained even more H_2O. Higher H_2O contents of the dacites would certainly be expected from continued removal of anhydrous crystals.

Although magnetite has not been identified explicitly in the proposed fractionation sequence, its presence would be necessary to prevent the well-known Fenner or iron-enrichment trend, and fO_2 throughout the sequence would have to be maintained approximately parallel to a buffer such as NNO (e.g., Osborn, 1959; Eggler and Osborn, 1982). In reality, the projections used here (Fig. 1) avoid the magnetite problem by projecting all compositions from a calculated Fe_3O_4 melt content (see Fig. 1 caption).

Finally, it should be emphasized that the proposed fractionation sequence explains only the trend of *bulk compositions.* As discussed above, actual phenocrysts in the rocks must have crystallized at lower pressures.

Fractionation in other arcs

Analyses of basalts, andesites, and dacites from the Tonga-Kermadec arc are plotted on the same projections used above (Fig. 1e, f). The analyses form arrays similar to Atka arrays but displaced closer to the one-atmosphere boundary curves. If the arrays are interpreted as fractionation trends, it would appear that fractionation occurred at lower pressure (cf. Fig. 1f) than at Atka but still at low to moderate melt H_2O contents (cf. Fig. 1e).

Analyses from the South Sandwich Islands (Scotia arc; Baker, 1968) and the Marianas (Dixon and Batiza, 1979), not shown here, mirror the Tonga-Kermadec arrays much more closely than the Atka arrays.

Parental magma: speculation

Most petrologists since Bowen have looked to peridotite as the parental source rock for basaltic melts. Did Atka basalts have a peridotite parent, or, as Marsh (e.g., 1982) has proposed, an eclogitic parent in the form of subducted oceanic slab? This question revolves about myriad geochemical considerations beyond the scope of this paper. Within the scope of the paper, however, the question is really whether the left termini of the arrays in Fig. 1 represent the most primitive melts beneath Atka. For an eclogite source, the answer could well be yes. For a peridotite source, the answer is no — a primary melt would have to be modified by fractionation of olivine (in the simplest scenario) or olivine + clinopyroxene (Perfit et al., 1980b; Kay et al., 1982) to yield the low-Mg number Atka basalts (see also Uto, 1981).

To illustrate the magnitude of change involved in that fractionation, a primary melt (Mg number = 0.72) has been calculated for analysis 2 of AT-1 (Table I; primary melt = column 3) by adding olivine through use of equations 4, 6, 7, and 11 of Hanson and Langmuir (1978). The calculated liquid is a picritic high-alumina basalt with a SiO_2 content only nominally lower than AT-1. (It may be noted that a lower assumed Fe_2O_3 melt content increases the needed addition of olivine: the primary melt for column 1 contains 48.3% SiO_2, 15.0% Al_2O_3, and 14.9% MgO.)

CONCLUSIONS

Some classic high-alumina basalts may be low-pressure magmas derivative from more tholeiitic parents. Rock AT-1 and other basalts from Atka, however, cannot have been derived from more mafic parents at low pressures, because they have plagioclase on their liquidus at anhydrous and hydrous conditions. It is possible that AT-1 and similar Atka rocks have been derived from more mafic, damp parents by fractionation at moderate pressures (8 kbar?). In turn, they may have fractionated to andesitic liquids at moderate to low pressure, in the process concentrating the melt water contents to 2—3% by crystallization of anhydrous plagioclase, olivine, and clinopyroxene. Water contents throughout the basalt-andesite-dacite sequence never built to levels (>4.5—5 wt.%) sufficient that amphibole crystallized as a liquidus or subliquidus phase.

ACKNOWLEDGMENTS

The Editors of this volume on Arc Volcanism, Professors Aramaki and Kushiro, provided necessary and appreciated stimulus that saw this paper to completion. D.R. Baker would like to thank Professor D. Shimozuru and the organizing committee for the opportunity to present this work at the 1981 IAVCEI Symposium on Arc Volcanism. We greatly appreciate the cooperation of B.D. Marsh in supplying rock samples and unpublished analyses.

Research was supported by the Earth Sciences Section, National Science Foundation, Grant EAR 80-22262.

REFERENCES

Albee, A.L. and Ray, R., 1970. Correction factors for electron probe micro-analysis of silicates, oxides, carbonates, phosphates and sulphates. Anal. Chem., 42: 1408—1414.

Baker, P.E., 1968. Comparative volcanology and petrology of the Atlantic island arcs. Bull. Volcanol., 32: 189—206.

Bence, A.E. and Albee, A.L., 1968. Empirical correction factors for the electron microanalysis of silicates and oxides. J. Geol., 76: 382—403.

Bender, J.F., Hodges, F.N. and Bence, A.E., 1978. Petrogenesis of basalts from the project Famous area: experimental study from 0 to 15 kbars. Earth Planet. Sci. Lett., 41: 277—302.

Boyd, F.R. and England, J.L., 1960. Apparatus for phase equilibrium measurements at pressures up to 50 kilobar and temperatures up to 1750°C. J. Geophys. Res., 65: 741—748.

Burnham, C.W., 1979. The importance of volatile constituents. In: H.S. Yoder (Editor), The Evolution of Igneous Rocks, Fiftieth Anniversary Perspectives. Princeton University Press, Princeton, N.J., pp. 439—482.

Burnham, C.W., Holloway, J.R. and Davis, N.F., 1969. Thermodynamic properties of water to 1000°C and 10,000 bars. Geol. Soc. Am., Spec. Pap., 132, 96 pp.

Coulon, C. and Thorpe, R.S., 1981. Role of continental crust in petrogenesis of orogenic volcanic associations. Tectonophysics, 77: 79—94.

Darken, L.S. and Gurry, R.W., 1945. The system iron-oxygen. I. The wustite field and related equilibria. J. Am. Chem. Soc., 67: 1398—1412.

Dixon, T.H. and Batiza, R., 1979. Petrology and chemistry of recent lavas in the Northern Marianas: implications for the origin of island arc basalts. Contrib. Mineral. Petrol., 70: 167—181.

Eggler, D.H., 1972. Water-saturated and undersaturated melting relations in a Paricutin andesite and an estimate of water content in the natural magma. Contrib. Mineral. Petrol., 34: 261—271.

Eggler, D.H. and Burnham, C.W., 1973. Crystallization and fractionation in the system andesite-CO_2-H_2O at pressures to 10kbar. Geol. Soc. Am. Bull., 84: 2517—2532.

Eggler, D.H. and Osborn, E.F., 1982. Experimental studies of the system MgO-FeO-Fe_2O_3-$NaAlSi_3O_8$-$CaAl_2Si_2O_8$-SiO_2 — a model for subalkaline magmas. Am. J. Sci., 282: 1012—1041.

Ewart, A., Brothers, R.N. and Mateen, A., 1977. An outline of the geology and geochemistry, and the possible petrogenetic evolution of the volcanic rocks of the Tonga-Kermadec-New Zealand island arc. J. Volcanol. Geotherm. Res., 2: 205—250.

Graham, A.M., 1981. Melting relations of island arc lavas from Grenada, Lesser Antilles. In: C.E. Ford (Editor), Progress in Experimental Petrology. The Natural Environment Research Council Publications, Series D, 18: 126—132.

Hamilton, D.L., Burnham, C.W. and Osborn, E.F., 1964. The solubility of water and effects of oxygen fugacity and water content on crystallization in mafic magmas. J. Petrol., 5: 21—39.

Hanson, G.N. and Langmuir, C.H., 1978. Modelling of major elements in mantle-melt systems using trace element approaches. Geochim. Cosmochim. Acta, 42: 725—741.

Helz, R.T., 1973. Phase relations of basalts in their melting range at P_{H_2O} = 5 Kb. as a function of oxygen fugacity. J. Petrol., 14: 249—302.

Helz, R.T., 1978. The Petrogenesis of the Ice Harbor Member, Columbia Plateau, Washington: a Chemical and Experimental Study. Ph.D. Thesis, The Pennsylvania State University, 284 pp. (unpubl.)

Holloway, J.R., 1971. Internally heated pressure vessels. In: G.C. Ulmer (Editor), Research Techniques for High Pressures and High Temperatures. Springer, New York, N.Y., pp. 217—258.

Holloway, J.R. and Burnham, C.W., 1972. Melting relations of basalt with equilibrium water pressure less than total pressure. J. Petrol., 13: 1—29.

Huebner, J.S. and Sato, M., 1970. The oxygen fugacity-temperature relationships of manganese and nickel oxide buffers. Am. Mineral., 55: 934—952.

Kay, S.M., Kay, R.W. and Citron, G.P., 1982. Tectonic controls on tholeiitic and calc-alkaline magmatism in the Aleutian arc. J. Geophys. Res., 87: 4051—4072.

Kuno, H., 1960. High alumina basalt. J. Petrol., 1: 121—145.

Kuno, H., 1966. Lateral variation of basalt magma type across continental margins and island arcs. Bull. Volcanol., 29: 195—222.

Kuno, H., 1968. Origin of andesite and its bearing on the island arc structure. Bull. Volcanol., 32: 141—187.

Kushiro, I. and Thompson, R.N., 1972. Origin at some abyssal tholeiites from the Mid-Atlantic Ridge. Carnegie Inst. Washington, Yearb., 71: 403—406.

Leeman, W.P., 1981. The influence of crustal structure on compositions of subduction-related magmas (abstract) In: IAVCEI Symposium on Arc Volcanism, Aug. 1982, p. 203.

Marsh, B.D., 1976. Some Aleutian andesites: their nature and source. J. Geol., 84: 27—45.

Marsh, B.D., 1980. Geology and petrology of northern Atka, Aleutian Islands, Alaska (abstract). Geol. Soc. Am. Abstr. Programs, 12: 476.

Marsh, B.D., 1981. On the crystallinity, probability of occurrence, and rheology of lava and magma. Contrib. Mineral. Petrol., 78: 85—98.

Marsh, B.D., 1982. The Aleutians. In: R.S. Thorpe (Editor), Orogenic Andesites and Related Rocks. Wiley, New York, N.Y., pp. 99—113.

Newton, R.C. and Sharp, W.E., 1975. Stability of forsterite + CO_2 and its bearing on the role of CO_2 in the mantle. Earth Planet. Sci. Lett., 26: 239—244.

Nicholls, I.A. and Ringwood, A.E., 1973. Effects of water on olivine stability in tholeiites and the production of silica-saturated magmas in the island-arc environment. J. Geol., 81: 285—300.

Osborn, E.F., 1959. The role of oxygen pressure in the crystallization and differentiation of basaltic magma. Am. J. Sci., 257: 609—647.

Osborn, E.F., 1979. The reaction principle. In: H.S. Yoder (Editor), The Evolution of Igneous Rocks, Fiftieth Anniversary Perspectives. Princeton University Press, Princeton, N.J., pp. 133—170.

Perfit, M.R., Brueckner, H., Lawrence, J.R. and Kay, R.W., 1980a. Trace element and isotopic variation in a zoned pluton and associated volcanic rocks, Unalaska Island, Alaska: a model for fractionation in the Aleutian calcalkaline suite. Contrib. Mineral. Petrol., 73: 69—87.

Perfit, M.R., Gust, D.A., Bence, A.E., Arculus, R.J. and Taylor, S.R., 1980b. Chemical characteristics of island-arc basalts: implications for mantle sources. Chem. Geol., 30: 227—256.

Powers, H.A., 1932. The lavas of the Modoc Lava-bed quadrangle, California. Am. Mineral., 17: 253—294.

Sack, R.O., Carmichael, I.S.E., Rivers, M. and Ghiorso, M.S., 1980. Ferric-Ferrous equilibria in natural silicate liquids at 1 bar. Contrib. Mineral. Petrol., 75: 369—376.

Sekine, T., Katsura, T. and Aramaki, S., 1979. Water saturated phase relations of some andesites with application to the estimation of the initial temperature and water pressure at the time of eruption. Geochim. Cosmochim. Acta, 43: 1367—1376.

Shaw, H.R., 1967. Hydrogen osmosis in hydrothermal experiments. In: P.H. Abelson (Editor), Researches in Geochemistry. Wiley, New York, N.Y., pp. 521—541. 521—541.

Stern, R.J., 1979. On the origin of andesite in the Northern Mariana island arc: implications from Agrigan. Contrib. Mineral. Petrol., 68: 207—219.

Thompson, R.N., 1972. Melting behavior of two Snake River lavas at pressures up to 35 kb. Carnegie Inst. Washington, Yearb., 71: 407—410.

Thompson, R.N. and Kushiro, I., 1972. The oxygen fugacity within graphite capsules in piston-cylinder apparatus at high pressures. Carnegie Inst. Washington, Yearb., 71: 615—616.

Uto, K., 1981. Variation of Al_2O_3 Content in the Cenozoic Japanese Basalts (abstract) In: IAVCEI Symposium on Arc Volcanism, Aug. 1981, pp. 400—401.

Walker, D., Shibata, T. and DeLong, S.E., 1979. Abyssal tholeiites from the Oceanographer Fracture Zone. II-phase equilibria and mixing. Contrib. Mineral. Petrol., 70: 111—125.

Yoder, H.S., Jr. and Tilley, C.E., 1962. Origin of basalt magmas: an experimental study of natural and synthetic rock systems. J. Petrol., 3: 342—532.

Journal of Volcanology and Geothermal Research, 18 (1983) 405—433
Elsevier Science Publishers B.V., Amsterdam — Printed in The Netherlands

A JURASSIC—CRETACEOUS ISLAND ARC IN THE LADAKH-HIMALAYAS

VOLKER J. DIETRICH[1], WOLFGANG FRANK[2] and KASPAR HONEGGER[1]

[1] *Department of Earth Sciences, ETH-Zentrum, 8092 Zürich (Switzerland)*
[2] *Geologisches Institut, Universität Wien, Universitätsstr. 7, A-1010 Wien (Austria)*

(Received October 20, 1982)

ABSTRACT

Dietrich, V.J., Frank, W. and Honegger, K., 1983. A Jurassic—Cretaceous island arc in the Ladakh-Himalayas. In: S. Aramaki and I. Kushiro (Editors), Arc Volcanism. J. Volcanol. Geotherm. Res., 18: 405—433.

The Ladakh Mesozoic ophiolite belt (western Himalaya) contains a pile of volcanic thrust sheets (Dras unit) which differ significantly in structure and composition from the ophiolitic mélange zones. The Dras unit is composed of pillow lavas, doleritic sills, very irregular basaltic (?basaltic andesites) and dacitic flows intercalated with pyroclastics, volcanoclastic sediments and radiolarian cherts. According to fossil evidence, this volcanism must have been active between Upper Jurassic and Upper Cretaceous.

The presence of relict primary minerals, such as magnesiochromite, clinopyroxene, hastingsitic hornblende and Ti-magnetite as well as distinctive bulk chemistries, suggests that the volcanics belong to island arc tholeiite and to calc-alkaline rock series, typical of present island arcs in the Caribbean and Pacific.

Model calculations incorporating probed phenocryst phases indicate that in addition to olivine, clinopyroxene and plagioclase, amphibole and titanomagnetite are crucial fractionating phases in the development of the dacites from a primitive tholeiitic melt. The latter process must have taken place at about 1000°C and at moderate depth of 5—15 km within or underneath the island arc. Today, hornblende-bearing mafic cumulates appear in the vicinity of Kargil within and close to the Dras volcanics.

In a Sr-evolution diagram, the Dras volcanics have yielded a "pseudo-isochron" with a low initial $^{87}Sr/^{86}Sr$ ratio of 0.7035 ± 0.0003, which is in the same range as the mean of modern island arc volcanics. However, a geologically unrealistic age of ~263 m.y., is obtained from the slope of this isochron.

The upper mantle is regarded as the source material for the island arc tholeiitic magmas. Enrichment in K, Ba, Sr and LREE supports the involvement of components derived from dehydration or incipient melting of subducted Tethyan oceanic crust in the mantle.

INTRODUCTION

Ophiolitic rocks associated with Mesozoic oceanic sediments are one of the most dominant and characteristic rock suites along the Indus-Tsangpo suture zone (Gansser, 1964, 1977) which is part of the 5000-km-long Peri-

0377-0273/83/$03.00

Indian suture zone running from the Andaman Islands in the SE to Karachi in the SW. This major structural divide has been interpreted as the destructive plate margin of a Tethyan oceanic lithosphere between the Himalayan orogenic belts of the Gondwana continent (High-Himalaya), and the Eurasian continental margin marked by the Karakorum and Trans-Himalayan (Kangdese) batholiths (Frank et al., 1977; Gansser, 1980; Srikantia and Razdan, 1980; Prasad et al., 1980; Honegger et al., 1982; Trommsdorff et al., 1982).

Until recently, it was generally assumed that ophiolites (as defined by the 1972 Penrose Conference) were generated by spreading processes at mid-ocean ridges. On the basis of basalt geochemistry, Miyashiro (1973, 1974, 1975a and b) pointed out that ophiolites may also represent remnants of island arcs, or parts of seamounts, aseismic ridges and fracture zones. Further distinction between basalts from mid-ocean ridges and those from back arc basins has been made, using "immobile" discriminative trace elements such as Ti, Zr, Y, Nb, Ba, Ta, Hf and rare earth elements (Wood et al., 1979, 1980; Saunders et al., 1980; Hawkins, 1980). In contrast to the pure geochemical approach using bulk chemical analyses Kuno (1968) and Cameron et al. (1979, 1980) have demonstrated, that phenocryst mineralogy and their order of crystallization are also critical indicators in the distinction of basalts from different tectonic environments.

Mafic and ultramafic rocks from the Indus-Tsangpo suture zone in the Ladakh-Himalaya had been recognized already by Lydekker in 1883. De Terra (1935) applied the early ophiolite concept of Steinmann for the Indus ophiolites. Frank et al. (1977) described in more detail the geology and tectonics of Ladakh and separated the Dras volcanic unit and its volcano-sedimentary sequences from the serpentinite and gabbro-bearing mélange zones.

The purpose of this paper is to present a detailed petrographic study of parts of the Dras volcanics, and to demonstrate, on the basis of major- and trace-element geochemistry, as well as relict phenocryst mineral chemistry, their fractionation behavior as typical island arc magmas derived by partial melting from upper mantle material.

ANALYTICAL TECHNIQUES

Mineral compositions were analyzed using an ARL electron microprobe type SEMQ. Operating conditions and correction procedures have been described in Dietrich et al. (1981). Formula proportions and site occupancies in amphiboles and pyroxenes were calculated with procedures described by Laird and Albee (1981).

Bulk chemical X-ray fluorescence analyses were obtained on an automated Philips sequential spectrometer PW 1450, using fused glass beads to determine the major-element composition, and 10-g powder pellets to determine the trace-element abundances Nb, Zr, Y, Sr, U, Rb, Th, Pb, Ga, Zn, Cu, Ni, Co, Cr, V, Ce, Nd, Ba, La, Sc and S. A computer program was used to calculate background, interference, mass absorption and standard deviations

(Nisbet et al., 1979). Old and new USGS reference samples were used for calibration. The accuracy is ± 2–3% at ~1000 ppm, ± 5–10% at ~100 ppm and ± 10–20% at 10 ppm. Using only a Cr-tube, the detection limits are around 3 ppm for most of the trace elements.

REE abundances (Table V) of selected samples have been analysed by radiochemical neutron activation analyses (RNAA) at the EIR-Würenlingen, Switzerland (Bajo and Wyttenbach, 1980). BCR-1 prepared in the same manner as the unknowns yielded: La = 24.8 ± 0.5, Ce = 53.3 ± 1.0, Nd = 28.2 ± 0.5, Sm = 6.8 ± 0.14, Eu = 1.95 ± 0.02, Gd = 6.8 ± 0.2, Tb = 0.99 ± 0.03, Tm = 0.50 ± 0.015, Yb = 3.40 ± 0.07 and Lu = 0.53 ± 0.015.

GEOLOGICAL SETTING

A pile of volcanic thrust sheets of up to 15 km thickness constitutes a major geological unit (the Dras unit) within the Ladakh ophiolite belt (northwest Himalaya). This volcanic unit (Fig. 1) follows the Indus-Tsangpo suture zone over a distance of nearly 400 km from the Karakorum area (Pakistan) southeast towards Leh (Ladakh). Equivalent units are not yet known in the central and eastern parts of the Himalaya. To the west in Kohistan (Pakistan), corresponding units may exist (Bard et al., 1980; Qasim Jan, 1980).

The Dras unit has steep thrust contacts against the ophiolite mélange and the Indus molasse. It is difficult to estimate the original thickness of the suite due to intense deformation. Intercalated tectonic slices of mélange rock type (e.g. serpentinites, gabbros, meta-basalts and marbles) further demonstrate the internal complication of the Dras. Near the village of Dras (type locality), and in the gorge between Pashkyum and Mulbekh, this unit is composed of pillow lava, doleritic sills, very irregular basaltic and dacitic flows intercalated with pyroclastics, volcano-clastic sediments and radiolarian cherts. In the vicinity of Dras, the radiolarian cherts yield Upper Callovian to Tithonian ages. However, Albian to Cenomanian Orbitulina limestones found as inclusions in basaltic flows indicate that volcanism must have been active until the Upper Cretaceous.

In a few localities, such as south of Pashkyum, the lava flows reach a thickness of 15–20 m each. Pillow lavas occur at the base of the Dras unit (Juteau and Reibel, 1981), dykes and normal andesites are very rare. Towards the east, the volcanics are gradually replaced by an Upper Cretaceous flyschoid and siliceous volcano-sedimentary facies ("unité de Nindam" after Bassoullet et al., 1978 and 1980). The coarse-grained flyschoid layers consist entirely of basaltic and dacitic rock fragments in a matrix of devitrified siliceous glass (i.e. chlorite, sericite, pumpellyite, prehnite, albite, quartz and sphene) and subordinate carbonate.

In the vicinity of Kargil, the Dras volcanic unit is underlain by mafic and ultramafic cumulates (Fig. 1) and intruded by diorites and granodiorites, which yield Upper Cretaceous to Tertiary ages (Honegger et al., 1982) and which may belong to the large composite and multiple Trans-Himalayan batholiths. Field observations suggest a possible genetic relationship between the Dras volcanics and the underlying cumulates, the latter being explained as relict magma chambers within the remnant arc.

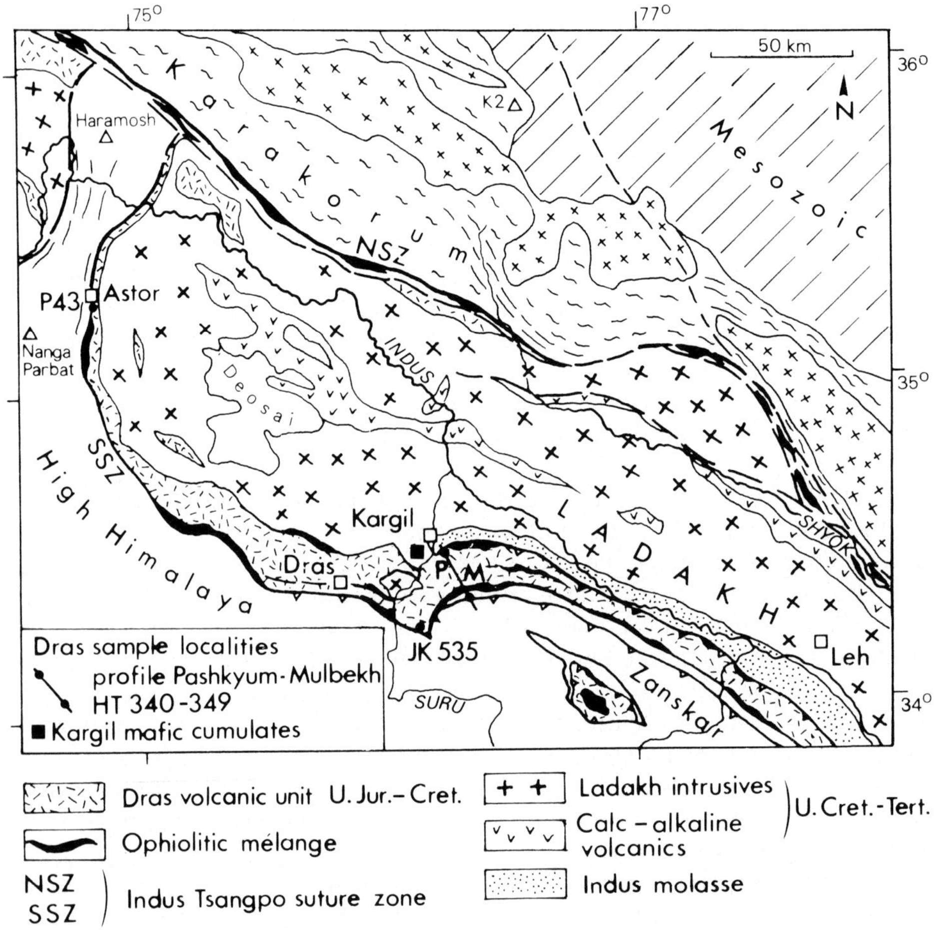

Fig. 1. Geological sketch map of Ladakh- and Northwest-Himalaya (after Gansser, 1964, 1980).

PETROGRAPHY AND MINERAL CHEMISTRY

The basaltic rocks are mostly porphyritic. Plagioclase transformed to albite, partly zoned clinopyroxene and also amphibole occur as macrophenocrysts. Relict microphenocrysts are magnesiochromites, Ti-magnetite and in one case (HT 344), quenched groundmass clinopyroxene. Serpentine pseudomorphs indicate the former presence of olivine or orthopyroxene (HT343).

The fine-grained groundmass contains the typical low-grade metamorphic mineral assemblage: chlorite, epidote, prehnite, pumpellyite and sphene ± quartz. Calcite is present in veins, amygdales and as tiny flakes in plagioclase (Table I).

Magnesiochromite is present as idiomorphic microphenocrysts (Table II and Fig. 2). Clinopyroxene is the most abundant phenocryst phase, often zoned rimwards to Fe-rich compositions (Table III). In the more evolved basalts, the clinopyroxene is progressively enriched in Ti and the Tschermaks-component (Al^{IV} and Ca, see Fig. 3). Using the eigenvector discrimination (not shown on a diagram) from Nisbet and Pearce (1978) the clinopyroxene compositions fall in the field of those from island arc basalts. Amphibole phenocrysts are less abundant than clinopyroxene in the basaltic

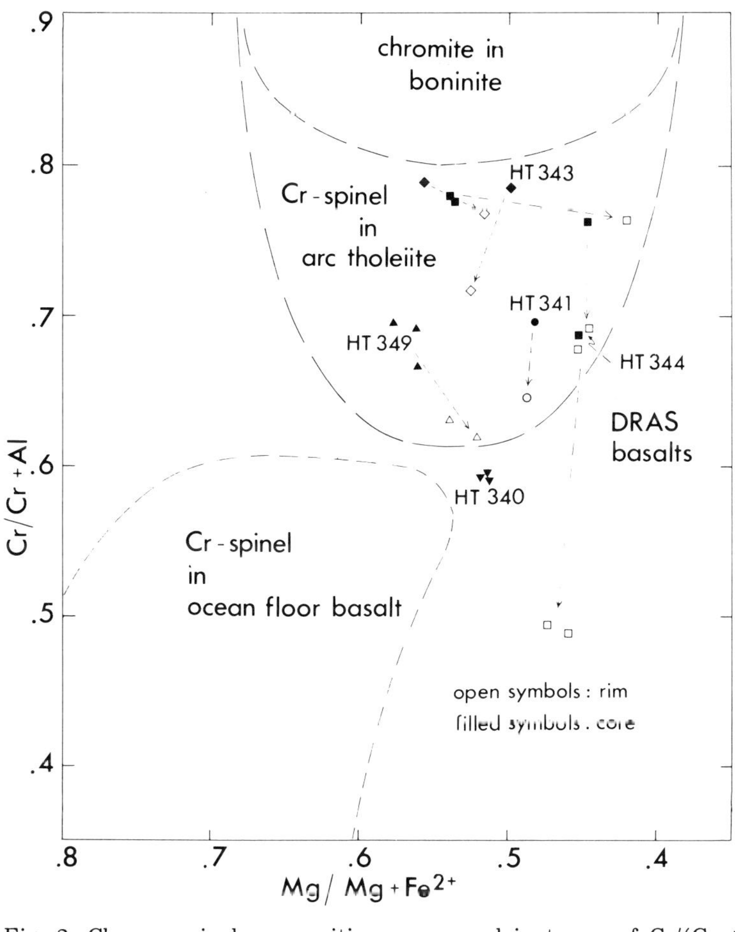

Fig. 2. Chromo-spinel compositions expressed in terms of Cr/(Cr + Al) and Mg/(Mg + Fe^{2+}) from Dras basalts. Fe^{2+} was calculated assuming stoichiometry. For comparison, spinel fields from boninites (Cameron et al., 1979), island arc tholeiites (Johnson and Arculus, 1978; Arculus, 1978) and from ocean floor basalts (Dietrich et al., 1981) are shown.

TABLE I

Mineralogy and textures of selected Dras volcanics from the Mulbekh-Pashkyum section (Fig. 1, HT 340—HT 349), JK 535 (Suru) and P 43 (Astor, Pakistan).
Due to metamorphic overprint it is impossible to give accurate modal compositions. (1—2%) = approx. phenocryst percentage of amphibole.
Abbreviations: Ti-mgt = titanomagnetite, plag = plagioclase, ol = olivine, cpx = clinopyroxene, opx = orthopyroxene, alb = albite, epd = epidote, chl = chlorite, pump = pumpellyite, preh = prehnite, tit = titanite, act = actinolite, py = pyrite.

Sample No.	Primary mineral-assemblage P = phenocryst MP = microphenocryst	Pseudomorphs	Metamorphic mineral assemblage	Texture, remarks
HT 344 basalt	P cpx ($En_{50-46}Wo_{45-44}Fs_{5-9}$) MP cpx ($En_{45}Wo_{41}Fs_{14}$) magnesiochromite	plag ol	alb, epd, chl, qtz, cct, leucoxene	sparsely phyric quenched groundmass cpx
HT 340 basalt	P cpx ($En_{50-45}Wo_{46-40}Fs_{5-15}$) MP cpx ($En_{46-45}Wo_{43-40}Fs_{13}$) P mg-hast.-hornblende (1—2%) magnesiochromite, Ti-mgt?	plag	alb, pump, qtz, chl, cct, tit, py	porphyritic to amygdular quartz-prehnite veins
HT 343 basalt	P cpx ($En_{47-46}Wo_{45-43}Fs_{7-11}$) MP cpx ($En_{45}Wo_{42}Fs_{13}$) magnesiochromite	plag ol opx?	alb, chl, prehn, pump, cct	porphyritic, partly fluidal amygdular
HT 341 basalt	P cpx ($En_{48-47}Wo_{45.5-45}Fs_{6-8}$) MP cpx ($En_{46}Wo_{45-40}Fs_{9-14}$) P mg-hastingsite (< 1%) magnesiochromite	plag	alb, chl, pump, epd	porphyritic, partly fluidal and amygdular quartz-calcite-prehnite veins
HT 349 basalt	P cpx ($En_{49}Wo_{46}Fs_{6}/En_{41}Wo_{46}Fs_{12}$) P mg-hastingsite (5—10%) magnesiochromite	plag	alb, chl, pump, epd, cct	brecciated porphyritic basalt, containing corroded green amphibole and xenoliths of dolerite and dacite
HT 348 dolerite	mg-hastingsite + mg-hast.-hornblende, Ti-mgt (20—25%)	plag	alb, chl, qtz, cct, prehnite	intersertal
JK 535 basalt	P brown amphibole (3—5%)	plag	alb, act, chl, pump, epd, tit, diopside?	porphyritic
HT 346 + HT 347 dacites	quartz	plag sanidine biotite	alb, cct, qtz, epd, sericite chl, tit	felsic partly porphyritic
P 43 basalt	P cpx ($En_{47}Wo_{43}Fs_{10}/En_{43.5}Wo_{45}Fs_{11.5}$) P mg-hastingsite (10—15%) ilmenite	plag	alb leucoxene	porphyritic

TABLE II

Microprobe analyses of Cr-spinel and Ti-magnetite from Dras basalts

Sample No: c = core r = rim	HT 344 c	r	c	r	HT 340	HT 343 c	r	c	r
SiO_2	0.08	0.08	0.06	0.08	0.09	0.09	0.09	0.08	0.09
TiO_2	0.54	0.52	0.31	0.42	0.69	0.52	0.81	0.62	1.03
Al_2O_3	9.91	10.22	13.26	21.79	17.21	9.81	10.49	10.01	12.77
Cr_2O_3	50.27	48.94	44.03	31.50	37.40	54.56	51.87	54.37	47.85
FeO_t	24.32	28.09	28.86	31.75	30.18	22.57	24.83	24.05	25.90
MgO	10.84	8.29	9.14	9.91	10.86	11.42	10.59	10.31	11.05
CaO	<0.01	<0.01	0.02	0.02	0.01	<0.01	<0.01	0.02	0.05
Total	95.96	96.15	95.68	95.47	96.44	98.98	98.69	97.46	98.73
Fe_2O_3	8.69	8.45	10.69	13.64	13.70	6.88	7.66	6.34	9.03
FeO	16.50	20.49	19.24	19.48	17.86	16.38	17.94	18.35	17.78
Cr/(Cr + Al)	0.773	0.763	0.690	0.492	0.593	0.789	0.768	0.785	0.715
$Mg/(Mg + Fe^{2+})$	0.539	0.419	0.458	0.476	0.520	0.554	0.513	0.500	0.526

	HT 341 c	r	HT 349 c	r		HT 348
SiO_2	0.10	0.10	0.13	0.13		0.11
TiO_2	0.52	0.58	0.38	0.52		4.85
Al_2O_3	13.37	15.97	14.23	17.42		1.70
Cr_2O_3	45.69	42.20	48.58	42.26		0.12
NiO			0.11	0.12	(MnO)	0.08
FeO_t	27.04	27.48	21.71	25.06		83.27
MgO	9.74	10.02	11.87	10.72		—
CaO	0.09	0.08	0.10	0.32		0.09
Total	96.55	96.43	97.18	96.55		90.24
Fe_2O_3	9.31	9.78	6.90	8.44		
FeO	18.67	18.68	15.50	17.47		
Cr/(Cr + Al)	0.696	0.639	0.696	0.619		
$Mg/(Mg + Fe^{2+})$	0.482	0.489	0.577	0.522		

rocks (Table I). Amphibole compositions (Fig. 4 and Table IVa) range from magnesio-hastingsite to the more iron- and silica-enriched magnesio-hastingsitic hornblende. It is interesting to note the high amounts of TiO_2 (2,55 to 1.4 wt.%) relative to the TiO_2 contents of the bulk rock composition (Table V) as well as Na_2O ranging from 2,65 to 2.32 wt.% and K_2O from 0.37 to 0.61 wt.%.

The primary mineral assemblage (Table I) as far as can be judged from relicts, is atypical for basalts generated in mid-ocean ridges, seamounts, aseismic ridges fracture zones and marginal basins (Cameron et al., 1979 and 1980; Wood et al., 1980). Spinel, clinopyroxene and in particular the amphibole compositions are very similar to those in alkalic and subalkalic basalts occurring in the Lesser Antilles island arc (Arculus, 1978) and the Aleutian island arc (Arculus et al., 1977).

TABLE III

Clinopyroxene compositions from Dras basalts

Sample No.	HT 344				P 43		HT 340				HT 343	
c = core r = rim	macrophenocryst		microphenocryst		macrophenocryst		macrophenocryst		microphenocryst		macrophenocryst	
	c	r	c	c	c	r	c	c	c	r	c	c-r
SiO_2	54.11	51.91	50.45	51.19	52.22	50.58	53.42	52.83	51.72	51.18	51.77	52.93
TiO_2	0.10	0.35	0.54	0.66	0.50	0.67	0.13	0.29	0.44	0.73	0.31	0.37
Al_2O_3	1.19	3.26	4.93	3.10	3.17	4.60	1.67	2.49	2.60	3.19	4.09	2.55
FeOt	3.12	5.71	6.90	8.80	6.21	7.11	3.04	6.57	8.57	8.12	4.87	6.09
MnO	n.d.	n.d.	n.d.	n.d.	n.d.	n.d.	n.d.	n.d.	n.d.	n.d.	n.d.	n.d.
MgO	17.70	15.55	14.70	14.73	15.51	14.00	17.00	16.20	15.23	14.67	15.47	15.55
CaO	22.84	22.25	21.65	20.35	21.42	22.03	23.61	20.91	20.32	21.15	23.11	22.01
Na_2O	0.15	0.24	0.19	0.24	0.33	0.37	0.13	0.16	0.28	0.25	0.15	0.20
NiO	<0.01	<0.01	<0.01	<0.01	<0.01	<0.01	0.05	<0.01	<0.01	0.04	<0.01	<0.01
Cr_2O_3	0.68	0.13	0.10	<0.01	0.18	<0.01	0.73	0.28	0.04	<0.01	0.18	0.06
Total	99.90	99.42	99.48	99.11	99.55	99.38	99.78	99.74	99.21	99.34	99.96	99.77
$Mg(Mg + Fe^{2+})$	0.91	0.85	0.81	0.76	0.82	0.80	0.92	0.82	0.78	0.78	0.86	0.82
En	50.1	47.1	46.6	44.9	46.6	44.4	49.0	47.7	46.0	44.9	47.3	45.9
Wo	44.9	44.4	42.7	40.8	43.0	44.4	46.5	41.4	40.8	42.5	45.4	44.1
Fs	5.0	8.6	10.7	14.3	10.5	11.2	4.5	10.9	13.2	12.7	7.4	10.0

TABLE III (*continued*)

	HT 343		HT 341				HT 349					
	phenocrysts		macrophenocryst		microphenocryst		macrophenocryst		macrophenocryst		microphenocryst	
	macro.r	mikro.c	c	r	c	c	c	r	c	r	c	r
SiO_2	50.88	50.39	51.95	52.80	49.70	51.81	52.63	47.42	51.78	51.53	49.69	49.92
TiO_2	0.57	0.68	0.29	0.30	0.54	0.38	0.23	0.79	0.24	0.38	0.63	0.47
Al_2O_3	4.51	4.50	3.38	2.60	4.49	1.96	2.98	7.11	3.39	4.03	5.95	4.70
FeO_t	6.87	8.83	4.48	4.86	7.08	8.78	4.00	8.15	4.01	4.25	5.77	8.26
MnO	n.d.	n.d.	0.11	0.12	0.13	0.34	n.d.	n.d.	n.d.	n.d.	n.d.	n.d.
MgO	14.75	14.02	15.59	15.95	14.10	15.30	16.39	11.93	15.81	15.41	14.18	12.90
CaO	21.72	21.37	22.93	22.51	22.50	20.00	23.15	22.94	22.94	22.98	22.86	22.77
Na_2O	0.26	0.30	0.17	0.17	0.20	0.28	0.17	0.24	0.22	0.20	0.28	0.30
NiO	<0.01	<0.01	n.d.	n.d.	n.d.	n.d.	<0.01	<0.01	0.04	0.06	0.04	<0.01
Cr_2O_3	0.14	<0.01	0.36	0.26	0.08	<0.01	0.16	<0.01	0.64	0.58	0.06	<0.01
Total	99.71	100.12	99.26	99.57	98.82	98.86	99.72	98.60	99.07	99.42	99.46	99.34
$Mg/(Mg + Fe^{2+} + Mn)$	0.81	0.78	0.86	0.85	0.83	0.77	0.89	0.79	0.89	0.87	0.86	0.78
En	46.3	44.8	46.9	47.0	45.7	45.9	48.6	42.2	48.2	47.3	46.9	42.0
Wo	43.1	42.5	45.5	44.8	45.1	40.4	45.6	46.5	45.7	45.6	45.4	46.2
Fs	10.6	12.7	7.5	8.2	9.2	13.7	5.8	11.2	6.0	7.0	7.7	11.8

End member percentages calculated subtracting $NaCrSi_2O_6$, $NaFe^{3+}Si_2O_6$, $NaAlSi_2O_6$, $CaTiAl_2O_6$, $CaCrAl_2O_6$, $CaFe^{3+}AlSiO_6$ and $CaAl_2SiO_6$ molecules.

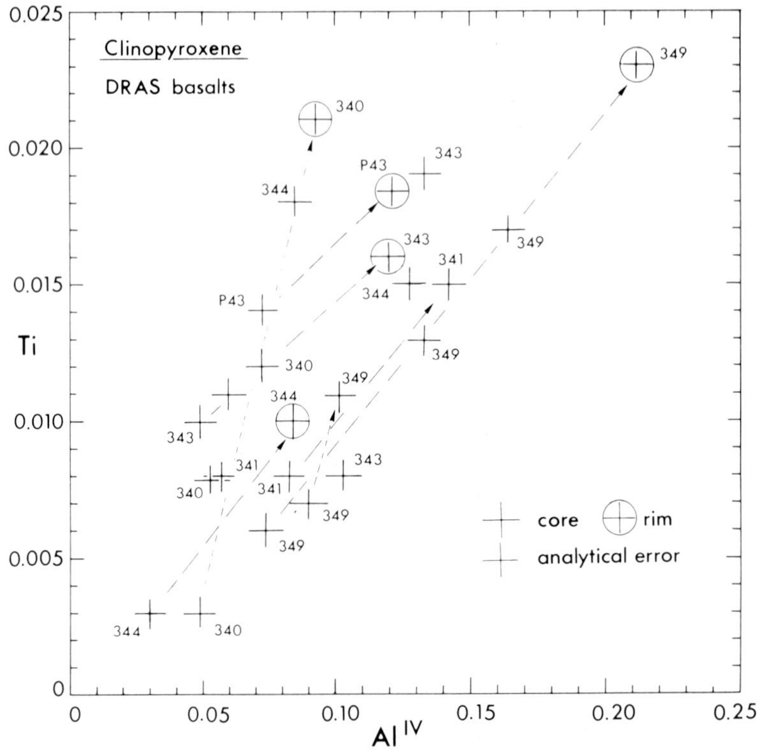

Fig. 3. Clinopyroxene compositions expressed in terms of Ti and Al^{IV} formula proportions. Linear positive correlations exist between Ti (as $CaTiAl_2O_6$ molecule) and Al^{IV} (mainly as the Tschermaks molecule $CaAl_2SiO_6$). The enrichment of these elements is consistent with the moderate iron enrichment (Table III) and marks the tholeiitic trend in island arc basalts.

Fig.4.a. Composition of calcic-amphiboles from Dras basalts and Kargil mafic cumulates shown in an alkali (A-site occupancies) versus silica variation diagram. In the Dras basalts, the increase of Na and K in the A-site is coupled with the increase of the Tschermaks molecule during fractionation (Fig. 4b, Al^{IV}), but with a decrease in silica. This process seems to be reversed in the mafic cumulates, which is in accordance with the proposed fractionation and accumulation model. Clearly visible is the overlap of amphibole composition between Dras basalts and rare relict core compositions from the Kargil cumulates. b. The same calcic-amphiboles as in Fig. 4a in a Ti versus Al^{IV} variation diagram. In general, the same positive linear correlation as in the clinopyroxene compositions (Fig. 3) is present. Two fractionation trends may be interpreted and shown with arrows in the diagram. The tholeiitic trend progressively leads to amphiboles highly enriched in Ti, whereas a possible calc-alkaline fractionation trend seems to lead to higher alkali and tschermakitic components.

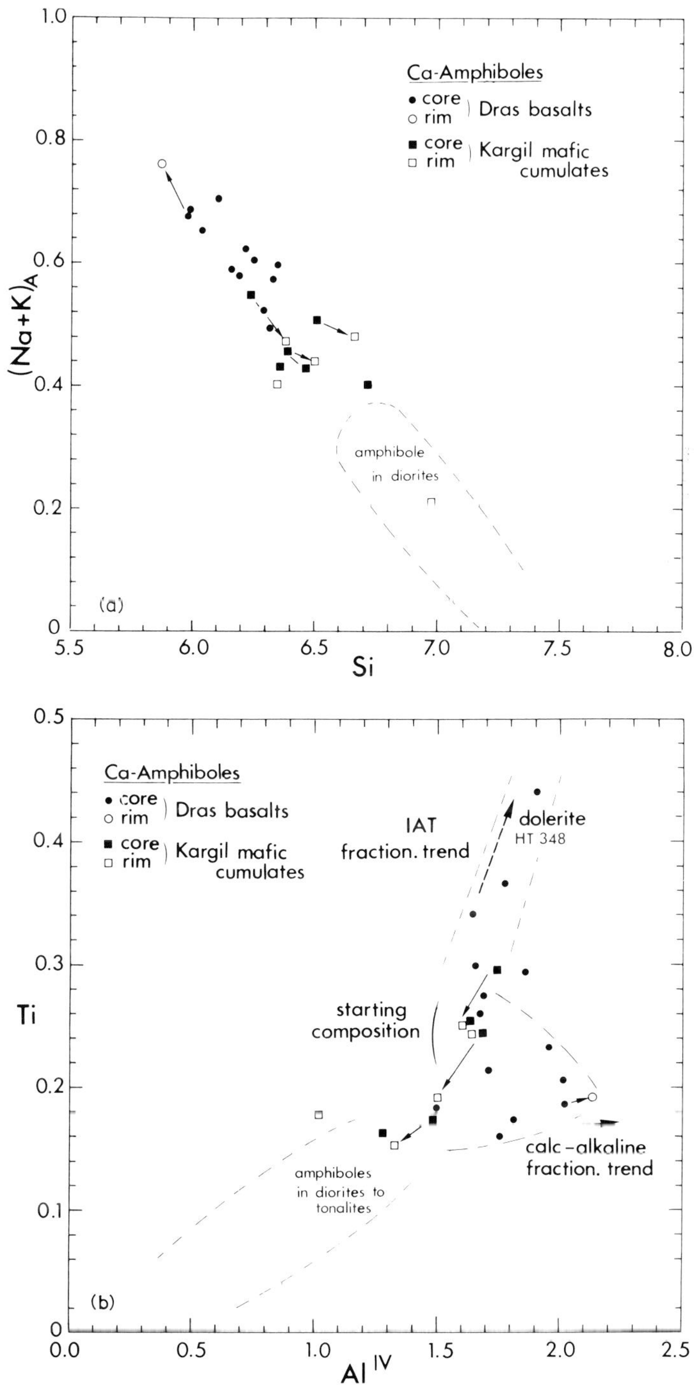
Ca-Amphiboles
core
rim
Dras basalts
core
rim
Kargil mafic cumulates
amphibole in diorites
(a)
(Na+K)A
Si
Ca-Amphiboles
core
rim
Dras basalts
core
rim
Kargil mafic cumulates
IAT fraction. trend
dolerite
HT 348
starting composition
calc-alkaline fraction. trend
amphiboles in diorites to tonalites
(b)
Ti
Al IV

TABLE IVa

Composition of amphiboles from the Dras basalts

Sample No: c = core r = rim	P 43 macrophenocrysts			HT 340 macrophenocrysts			HT 349 macrophenocrysts			HT 348 interstitial to subophitic			
	c	r	c	c	c	c	c	c	r	c	c	r	r
SiO_2	40.63	41.12	41.62	42.74	43.02	43.24	42.19	40.70	39.79	40.69	41.66	42.33	42.51
TiO_2	1.85	2.10	2.65	1.59	1.95	2.50	1.43	1.69	1.72	3.91	3.26	3.01	2.66
Al_2O_3	14.18	13.35	12.36	12.65	11.44	11.06	12.12	14.09	15.55	11.97	11.58	10.85	10.68
FeO_t	10.96	10.22	11.16	10.62	11.39	11.84	12.43	11.32	9.93	11.18	11.78	12.82	12.22
MgO	13.64	14.29	13.82	14.78	14.37	14.10	13.30	13.69	13.68	13.07	12.89	12.20	13.11
CaO	11.78	11.59	11.45	11.96	11.39	11.24	11.62	12.19	12.49	11.47	11.49	11.28	11.27
Na_2O	2.63	2.60	2.48	2.33	2.36	2.37	2.32	2.27	2.36	2.50	2.26	2.05	1.96
K_2O	0.39	0.48	0.37	0.38	0.38	0.36	0.50	0.61	0.61	0.64	0.64	0.95	0.93
Cr_2O_3	<0.02	0.12	<0.02	<0.02	<0.02	<0.02	<0.02	<0.02	0.04	<0.02	<0.02	<0.02	<0.02
MnO	n.d.	n.d.	n.d.	n.d.	n.d.	n.d.	n.d.	n.d.	n.d.	0.20	0.23	0.26	0.22
Total	96.06	95.88	95.90	97.05	96.30	96.71	95.92	96.55	96.17	95.65	95.79	95.72	95.56
Formula proportions (calculated to 24 oxygen, cations Ca-Na-K = 13 according to the procedure proposed by Laird and Albee, 1981)													
Si	5.987	6.048	6.140	6.189	6.289	6.311	6.247	5.981	5.867	6.10	6.22	6.36	6.35
Ti	0.205	0.232	0.294	0.173	0.214	0.274	0.160	0.186	0.191	0.44	0.37	0.34	0.30
Al(IV)	2.013	1.952	1.860	1,811	1.711	1.689	1.753	2.019	2.133	1.90	1.78	1.64	1.65
Al(VI)	0.451	0.364	0.289	0.350	0.261	0.214	0.363	0.421	0.571	0.21	0.26	0.28	0.22
Cr	—	0.015	—	—	—	—	—	—	0.005	—	—	—	—
Fe(3+)	0.611	0.625	0.583	0.678	0.716	0.676	0.621	0.625	0.438	0.28	0.33	0.27	0.47
Fe(2+)	0.739	0.633	0.795	0.607	0.676	0.769	0.919	0.767	0.786	1.13	1.14	1.34	1.05
Mg	2.995	3.132	3.039	3.192	3.133	3.066	2.937	3.000	3.008	2.92	2.87	2.73	2.92
Ca	1.858	1.827	1.811	1.856	1.783	1.756	1.844	1.920	1.974	1.84	1.84	1.82	1.80
Na(M4)	0.142	0.173	0.189	0.144	0.217	0.244	0.156	0.080	0.026	0.16	0.16	0.18	0.20
Na(A)	0.609	0.567	0.519	0.510	0.452	0.426	0.511	0.566	0.648	0.57	0.49	0.41	0.37
K	0.074	0.0741	0.069	0.070	0.071	0.068	0.094	0.114	0.114	0.12	0.12	0.18	0.18
$Mg/(Mg + Fe^{2+})$	0.802	0.832	0.793	0.840	0.823	0.799	0.762	0.796	0.793	0.72	0.71	0.67	0.72
Name	Mg-Ha	Mg-Ha	Mg-Ha	Mg-Ha	Mg-Ha-Hbl	Mg-Ha-Hbl	Mg-Ha	Mg-Ha	Mg-Ha	Mg-Ha	Mg-Ha	Mg-Ha-Hbl	Mg-Ha-Hbl

n.d. = not determined, Mg-Ha = magnesio-hastingsite, Mg-Ha-Hbl = magnesio-hastingsitic-hornblende.

TABLE IV(b)

Amphibole composition from the Kargil mafic cumulates

c = core r = rim	HT 125 Ol-Gabbro				HT 129 Gabbro-Norite			HT 124 Ol-Gabbro		
	c	r	c	Incl. in cpx	c	r		c	r	r
SiO_2	42.70	43.69	43.79	46.28	44.71	45.87	47.66	43.69	44.11	44.98
TiO_2	2.70	2.28	2.34	1.51	1.60	1.42	1.64	2.43	2.27	1.77
Al_2O_3	11.94	11.15	11.16	9.27	10.45	9.70	7.34	11.37	11.32	10.74
FeO	10.01	9.53	9.13	8.47	9.53	9.21	11.79	9.14	8.81	9.30
MnO	0.13	0.14	0.12	0.10	0.10	0.13	0.07	0.14	0.16	0.16
MgO	14.44	14.94	15.42	16.08	15.15	15.57	14.27	15.33	15.85	15.52
CaO	12.06	12.17	12.03	12.66	12.21	12.53	11.52	11.68	11.61	12.01
Na_2O	2.20	1.94	1.95	1.54	1.76	1.59	1.16	2.20	2.14	2.07
K_2O	0.21	0.12	0.07	0.03	0.50	0.45	0.41	0.06	0.09	0.04
Cr_2O_3	0.04	0.05	0.05	0.06	0.07	0.06	<0.01	0.03	<0.01	<0.01
Total	96.43	96.02	96.06	96.00	95.79	96.53	95.86	96.08	96.36	96.59
Si	6.25	6.38	6.36	6.72	6.51	6.67	6.98	6.34	6.35	6.49
Ti	0.30	0.25	0.26	0.16	0.18	0.16	0.18	0.26	0.25	0.19
Al(IV)	1.75	1.62	1.64	1.28	1.49	1.33	1.02	1.66	1.65	1.51
Al(VI)	0.31	0.31	0.27	0.31	0.32	0.33	0.25	0.29	0.27	0.32
Mn	0.02	0.02	0.01	0.01	0.01	0.02	0.01	0.02	0.02	0.02
Fe(3+)	0.40	0.42	0.54	0.26	0.37	0.25	0.39	0.57	0.69	0.49
Fe(2+)	0.82	0.75	0.57	0.77	0.80	0.87	1.06	0.54	0.37	0.63
Mg	3.15	3.26	3.34	3.48	3.31	3.37	3.12	3.32	3.40	3.34
Ca	1.89	1.91	1.87	1.97	1.92	1.95	1.81	1.82	1.79	1.86
Na(M4)	0.11	0.09	0.13	0.03	0.08	0.05	0.19	0.18	0.21	0.14
Na(A)	0.51	0.46	0.42	0.40	0.42	0.40	0.14	0.44	0.39	0.44
K	0.04	0.02	0.01	0.005	0.09	0.08	0.08	0.01	0.02	0.01
$Mg/(Mg + Fe^{2+})$	0.79	0.81	0.85	0.82	0.80	0.79	0.75	0.86	0.90	0.84
Name	Mg-Ha-Hbl		Tsch-Hbl	Mg-Hbl	Mg-Ha-Hbl		Mg-Hbl	Tschermakitic hornblende		

c = core, r = rim. Formula proportions (calculated to 24 oxygens, cations Ca-Na-K = 13)

TABLE V

Chemical composition of selected basalts and dacites from the Dras volcanics (Ladakh, India and Astor, Pakistan)

Sample No.	HT 344	HT 340	HT 343	HT 341	HT 349	HT 348	JK 535	HT 346	HT 347	P 43
SiO_2	49.15	48.69	45.81	47.62	45.38	44.86	47.25	59.04	63.67	42.22
TiO_2	0.61	0.64	0.69	0.78	0.99	1.32	0.78	0.39	0.33	1.65
Al_2O_3	13.07	14.12	15.22	15.81	17.65	19.28	18.15	15.52	16.37	14.00
Fe_2O_3	0.06	0.52	1.93	1.54	0.99	1.68	4.32	0.28	0.74	4.85
FeO	8.30	8.22	7.01	8.65	8.72	7.50	5.10	3.71	2.78	7.89
MnO	0.16	0.18	0.17	0.20	0.15	0.18	0.16	0.14	0.12	0.17
MgO	11.65	9.13	7.05	7.33	7.91	5.66	5.90	2.40	1.57	12.36
CaO	8.70	10.86	12.08	9.71	8.78	9.26	10.44	5.82	3.81	9.63
Na_2O	1.82	1.79	2.96	3.23	3.20	2.53	2.80	3.84	3.78	1.99
K_2O	0.83	0.40	0.05	0.28	1.13	1.09	1.30	1.49	1.91	0.25
P_2O_5	0.14	0.19	0.19	0.15	0.29	0.23	0.25	0.23	0.19	0.11
H_2O+	4.84	4.42	4.82	4.43	4.43	4.12	2.70	4.12	2.60	3.50
CO_2	0.85	0.62	1.67	0.33	0.17	2.15	0.25	1.63	2.01	0.40
Total	100.18	99.78	99.65	100.06	99.79	99.86	99.40	98.61	99.88	99.02
$Mg/(Mg + Fe^{2+})$	0.73	0.67	0.61	0.58	0.61	0.55	0.55	0.53	0.46	0.66
*Trace elements in ppm**										
Ba	74	63	31	51	128	237	188	248	425	87
Rb	13	5	<3	<3	25	28	34	40	49	<3
Sr	167	162	247	270	443	539	359	239	295	303
Y	12	16	13	17	20	24	13	8	10	13
Zr	31	46	35	43	51	52	52	125	138	50
V	235	215	249	315	280	392	328	76	48	748
Cr	636	329	218	156	246	92	70	3	4	171
Ni	198	118	57	43	87	24	32	<5	<5	41
Co	46	47	41	41	48	41	29	12	9	85
Cu	29	54	95	76	39	32	<3	3	3	120
Zn	54	66	58	144	62	67	35	58	74	88
Ga	11	13	13	14	15	16	9	8	11	12
Sc	36	34	36	38	38	57	34	7	4	83
S	460	617	136	808	58	77	<16	1369	<16	<16
La	3.5	5.4	5.26	n.d.	n.d.	12.5	12.6	21.9	26.5	5.8
Ce	9.5	12.80	13.49	n.d.	n.d.	26.4	29.5	44.00	46.9	13.2
Nd	7.17	9.12	8.82	n.d.	n.d.	16.05	17.9	18.6	17.7	8.8
Sm	2.22	2.52	2.62	n.d.	n.d.	4.21	4.00	3.33	3.15	2.92
Eu	0.72	0.82	0.86	n.d.	n.d.	1.23	1.11	0.88	0.77	0.98
Gd	2.48	2.65	2.73	n.d.	n.d.	3.79	3.26	2.64	2.20	2.84
Tb	0.40	0.42	0.42	n.d.	n.d.	0.61	0.48	0.39	0.35	0.55
Tm	n.d.	0.24	0.23	n.d.	n.d.	0.31	0.27	0.25	0.23	n.d.
Yb	1.47	1.64	1.54	n.d.	n.d.	2.22	1.78	1.80	1.92	1.72
Lu	0.24	0.26	0.25	n.d.	n.d.	0.34	0.29	0.30	0.33	0.25

*Nb < 2 ppm, U < 2 ppm, Th < 5 ppm, Pb < 6 ppm.

BULK CHEMISTRY

The bulk chemistry of the Dras volcanics also points to an island arc origin. The Dras volcanics can be divided into two major groups: basalts and dacites. Basaltic andesites and andesites have not been found. Most of the basaltic rocks are hypersthene and olivine normative while the dacites are quartz and hypersthene normative. The AFM diagram (Fig. 5) indicates more clearly that these rocks originated from a basaltic magma belonging either to island arc tholeiites (IAT) or to the calc-alkaline series. This is consistent with the weak to moderate iron enrichment and contrasts with the strong iron enrichment typical in mid-ocean ridge tholeiites (MORB).

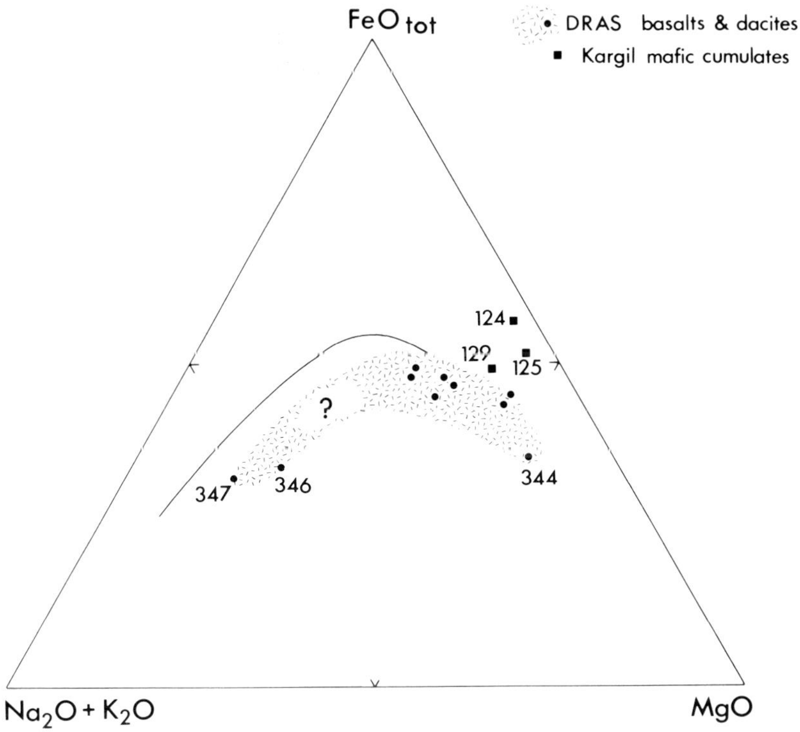

Fig. 5. AFM diagram showing the field of analysed Dras volcanics and the trend line of calc-alkaline series in the Izu-Hakone region in Japan (Miyashiro, 1975b). The points indicate the positions of the selected Dras basalts and dacites (Table V). Bulk analyses from cumulates are given in Honegger et al. (1982).

Chemically the Dras basalts are enriched in K, Ba, Rb, Sr and in the light rare earth elements, and are depleted in Ti, P, Zr, Y, Nb and in the heavy rare earth elements relative to MORB (Figs. 8 and 9).

Irregularities in the alkalies, Ca and Sr, are inferred to represent metasomatic exchange (spilitization and carbonatization) during alteration (Na_2O between 1.8 and 3.2 wt.%; K_2O between 0.05 and 1.3 wt.%). Ba and the rare earth element (REE) abundances do not seem to be affected much by alteration and metamorphic effects (for discussion see Hellman et al., 1979). The REE patterns (Fig. 9) of the different Dras volcanic rocks are very consistent, indicated by a small variation of La/Yb and La/Sm ratios, and coherent with their "immobile" trace element contents (such as Ti, Zr, Y, Nb, V, Cr and Ni after Pearce and Cann, 1973; Humphris and Thompson, 1978).

With the exception of the dolerite (HT 348) and the hornblende and clinopyroxene rich basalt P43 from Astor (Pakistan), TiO_2 ranges between 0.6 and 1.0 wt.%, which is approx. half the amount found in MORB (Sun et al., 1979; Wood et al., 1979, 1980; Dietrich et al., 1981). The P_2O_5/TiO_2 variation diagram (Fig. 6) shows the Dras basalts in the field of island arc tholeiites, clearly distinct from MORB type basalts outcropping in the mélange zones within or underneath the Dras units (Honegger et al., 1982). Using the Ti/Cr versus Ni discriminant method (Fig. 7), the Dras basalts also plot in the island arc tholeiite field.

The latter diagram shows close chemical similarity of the Dras volcanics and the mafic part (olivine gabbro, gabbro, norite etc.) of the Kargil intrusives. These intrusives may be interpreted as the cumulate counterpart to the overlying Dras volcanics.

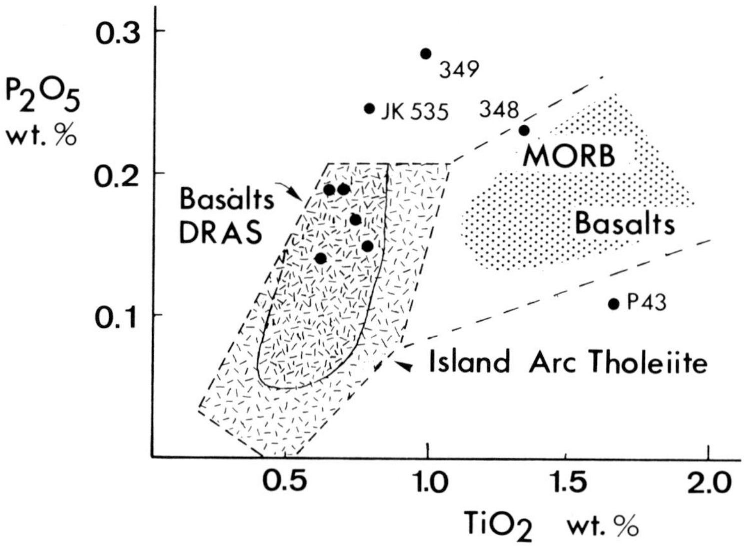

Fig. 6. P_2O_5/TiO_2 variation diagram. Ti and P are regarded as the two most stable elements during alteration and metamorphism of basaltic rocks. Field of analysed Dras basalts overlaps with the field for island arc tholeiites proposed by Hawkins (1980). Exceptions exist for the more evolved basalts (HT 349, JK 535, P 43) and the dolerite HT 348.

The log wt.% TiO_2/log ppm Zr variation diagram (Fig. 8) finally is used to compare the Dras volcanics, of probable island arc affinity, with volcanics from island arcs in the West Pacific. The Dras basalts as well as the basalts from the Mariana arc (Dietrich et al., 1978; Hawkins, 1980; comparative review in Wood et al., 1980) plot along the same TiO_2/Zr ratio (160). Fractionation of olivine, spinel, plagioclase and to some extent clinopyroxene does not change this ratio, as TiO_2 and Zr are progressively enriched during fractionation and overlap with the field of ocean floor basalts

(MORB). Boninites and marianites (Sharaskin et al., 1980) are also plotted in order to show the distinct difference between island arc tholeiites and these peculiar, magnesium- and silica-rich lavas.

It is interesting to recognize in the same diagram that the Dras dacites appear with TiO_2/Zr ratios < 30, although they are intercalated with the basalts and thus could bear a direct genetic relationship. Another fractionation process must therefore be involved if the dacites were generated from the basalts.

The occurrence of dacite flows (HT 346 and 347, Table V) intercalated with the basaltic flows also provides further evidence for an island arc environment of the Dras volcanics. The volume proportions between basalts and dacitic rocks in the areas of sampling are in the order of 8:2 (crude estimate).

Andesitic rocks have not been found, but may occur in other areas of the Dras unit.

The relatively low concentrations of Ba, Rb, Sr, Zr and Y in the Dras dacites are very similar to those in dacites from island arcs in the S.W. Pacific, Tonga-Kermadec and the lesser Antilles. In contrast, dacites from volcanoes within belts along continental margins (Ewart, 1979) have trace-element contents which are two to three times higher than those found in the Dras volcanics.

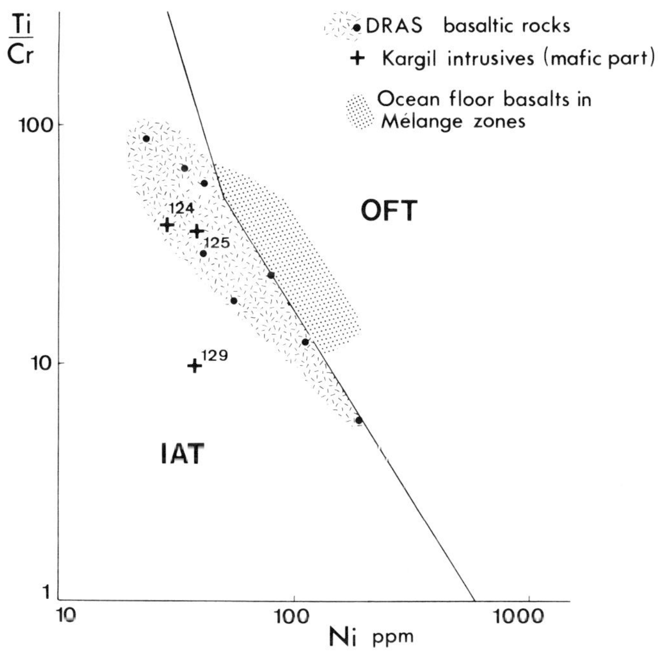

Fig. 7. Ti/Cr versus Ni diagram proposed by Beccaluva et al. (1979). This diagram discriminates between ocean floor tholeiites (OFT equivalent with MORB) and island arc tholeiites (IAT). Analyses of the Kargil mafic intrusives and ocean floor basalts in Honegger et al. (1982).

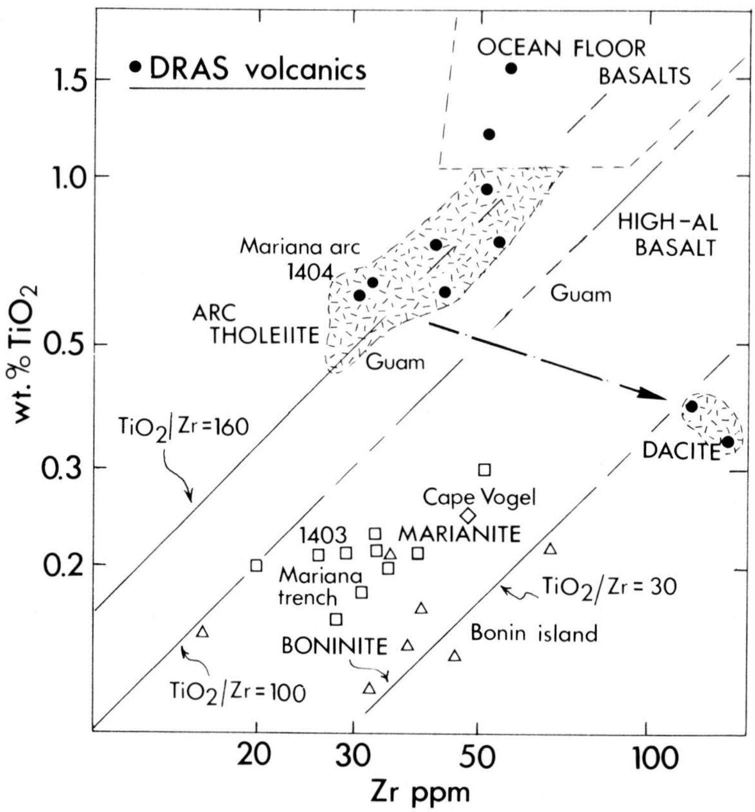

Fig. 8. Log TiO_2 (wt.%) versus log Zr (ppm) variation diagram. For comparison data from island arc basalts (Dietrich et al., 1978 and unpublished data; Johnson and Arculus, 1978), from boninites and marianites (Dietrich et al., 1978; Sharaskin et al., 1980) and from ocean floor basalts (Dietrich et al., 1981) are shown. The arrow between the Dras basalts (island arc tholeiites) and the Dras dacites should indicate the possible genetic relationship, discussed in the fractionation model (Table VI).

MAGMATIC DIFFERENTIATION

In terms of differentiation two extreme basaltic rock types can be recognized. HT 344, is a basalt that appears to be the most primitive rock, containing Mg-rich quenched groundmass clinopyroxene (Fig. 3) and magnesiochromite (Fig. 2). This sample yields the highest MgO and Cr_2O_3 contents, 100 Mg/Mg + Fe^{2+} = 72.8 assuming Fe_2O_3/FeO = 0.1, the lowest TiO_2 content and a flat REE pattern without an Europium anomaly (Fig. 9). Thus, this sample has similar major- and trace-element characteristics to the "hypersthene rock series" (Yajima et al., 1972). Because of its primitive composition we assign this rock as being close to the primary magma of the Dras volcanics. HT 348, a dolerite, is one of the most evolved basaltic rocks, containing the most abundant brown, Ti-rich amphibole and Ti-magnetite.

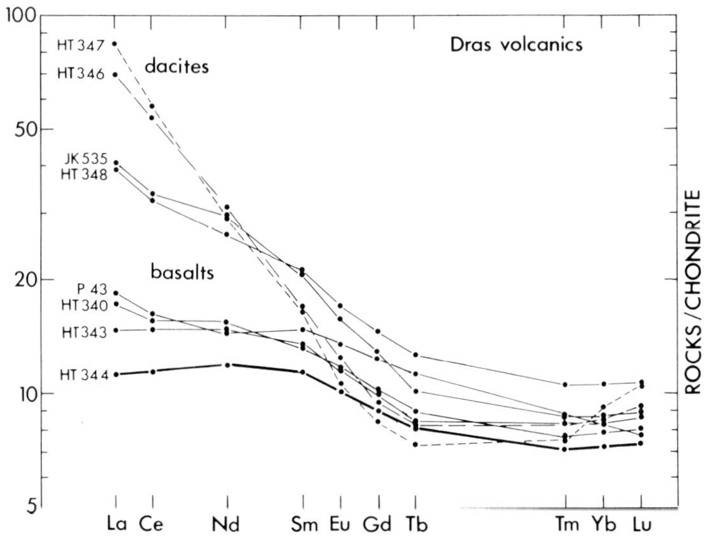

Fig. 9. Chondrite normalized (Leedy-chondrite/1.2; Masuda et al., 1973) rare earth patterns of Dras basalts and dacites. HT 344 basalt seems to have the most primitive pattern, although LREE patterns are 2—3 times higher than primitive normal MORB.

This sample has the lowest MgO and Cr_2O_3 contents, 100 mg/Mg + Fe^{2+} = 54.6 and the highest TiO_2, Rb, Ba, Sr, Zr, Y and LREE contents.

The basalt HT 349 and the dolerite HT 348 may have been derived from a primitive Mg, K, Ba and Sr-rich, basaltic liquid such as HT 344 by a rather simple fractionation process involving extraction of about 40—50% of olivine, spinel and clinopyroxene.

In order to derive the intercalated dacites from a primitive basaltic liquid, a fractionation process has to be envisaged involving as major fractionation components, phases which are capable of reducing Mg, Fe and Ti and increasing Si and Na contents in the residual liquid. Hastingsitic amphiboles and Ti-magnetite are the only possible phases. This fractionation process is already indicated in the appearance of the phenocryst phases in the basaltic rocks. Olivine, spinel, orthopyroxene (?), clinopyroxene and plagioclase have to be added in order to reduce Mg, Ca, Cr and Ni.

A mass-balance fractionation model has been calculated (Table VI) starting with the initial liquid composition of the primitive basalt HT 344 and separating amphibole (HT 340), clinopyroxene (HT 344), Ti-magnetite (HT 348), olivine and plagioclase using the probed compositions (Tables II—IV). The average olivine and plagioclase compositions (Table VI) had to be assumed due to their alteration in the Dras basalts. The model calculation is also based on the assumption that sufficient H_2O was present in the initial liquid to produce amphibole as a liguidus phase, an assumption supported by the appearance of amphibole phenocrysts in the different Dras basalts.

TABLE VI

Fractional crystallization model for Dras volcanics (Values in wt.%)

Model A (basalt to dacite)

wt.%	Basalt HT 344	-20% Amph (HT 340)	-17% Ol (Fo 75)	-10% Cpx (HT 344)	-10% Plag (An 85)	-2% Ti-Mgt (HT 348)	Residual liquid	Model A dacite	Actual composition HT 346, anhydrous
SiO_2	52.5	8.5	6.47	5.2	4.7	—	27.63	65.7	65.00
TiO_2	0.65	0.34	—	0.034	—	0.097	0.18	0.42	0.43
Al_2O_3	14.0	2.5	—	0.27	3.32	0.03	7.88	18.7	17.10
FeO*	8.9	2.2	3.77	0.54	0.04	1.66	0.69	1.64	4.40
MgO	12.5	2.8	6.43	1.6	—	—	1.67	3.97	2.64
CaO	8.15	2.3	—	2.2	1.72	—	1.93	4.58	4.12
Na_2O	1.95	0.46	—	0.02	0.18	—	1.29	3.07	4.23
K_2O	0.90	0.076	—	0.033	0.01	—	0.78	1.85	1.64
total	100.0						42.05	99.93	99.56

FeO*: total iron expressed as FeO. HT 344: assumed parental melt, recalculated anhydrous and calcite-free.

Model B (dacite to rhyolite)

	Model Dacite B	-10% Amph (HT 349)	-10% Plag (An 52)	-2% Mgt	Residual liquid	Model rhyolite
SiO_2	65.5	4.05	5.54	—	55.91	71.2
TiO_2	0.24	0.174	—	—	0.066	0.08
Al_2O_3	17.7	1.36	2.81	—	13.53	17.2
FeO*	4.05	1.25	—	1.66	1.14	1.45
MgO	2.91	1.28	—	—	1.63	2.07
CaO	4.9	1.19	1.1	—	2.61	3.32
Na_2O	2.68	0.23	0.54	—	1.91	2.43
K_2O	1.79	0.06	0.02	—	1.71	2.18
total	99.77				78.5	99.93

Model B dacite = Basalt (HT 344) - [25% Amph (HT 340), 15% Ol (Fo_{84}), 7% Cpx (HT 344), 10% Plag (An 80), 2% Ti-Mgt (HT 348)]

Two model dacite compositions A and B, similar to those analyzed in the Dras volcanics, were derived extracting total volumes of 59% (model A and B) accumulating minerals, such as amphibole, olivine, clinopyroxene, plagioclase and Ti-magnetite from HT 344. Continued fractionation of 10% plagioclase (An_{52}), 10% amphibole and 2% magnetite could theoretically produce rhyolitic lavas from the dacitic liquid. Andesitic rocks have been disregarded in the fractionation model, for the reason, that these rocks do not occur between the basalts and dacites in the sampling areas; a puzzling fact, which might be explained by the silica immiscibility effect in the magmas (Irvine, 1975). Similar fractionation models involving the cumulate phases amphibole, plagioclase, pyroxene, magnetite ± olivine and least-squares testing have been developed for basalt-dacite suites in the Lesser Antilles arc (Arculus and Wills, 1980).

To test the validity of the theoretical fractionation model the residual cumulates can be checked. These rocks occur in the vicinity of Kargil as rather irregularly distributed hornblende-bearing olivine gabbros, gabbros and norites (Fig. 1). Because of their wide range in modal and bulk compositions (Honegger et al., 1982) only the mineral compositions can be compared with those obtained from the Dras volcanics.

Olivine ranges in the cumulates from Fo_{72} to Fo_{77}, plagioclase from An_{80} to An_{91}, clinopyroxene from $En_{47}Wo_{47}Fs_6$ to $En_{44}Wo_{46}Fs_{10}$ and Ti-magnetite contains 0.45 to 1.6 wt.% TiO_2. In addition orthopyroxene (e.g. bronzite in HT 129, $En_{77\text{-}76}Wo_{<0.2}Fs_{23\text{-}24}$) occurs in some rocks.

The brown amphiboles are partly zoned and are generally high in TiO_2 (1.4 to 2.7 wt.%). In two diagrams the amphibole compositions from the Dras basalts and the Kargil mafic intrusives are compared (Fig. 4a, b). Relict amphibole core compositions from the cumulates overlap with those from the basalts (only in two cases). The major distinction between the two sets of amphiboles appears in the alkali, Si and Al (IV) contents (Fig. 4a, b). The basaltic amphiboles (magnesiohastingsitic hornblende to magnesio-hastingsite after the classification of Leake, 1978) have Na_2O of 2.3 to 2.6 wt.% and K_2O of 0.37 of 0.61%. Thus, they may have crystallized under equilibrium conditions with the basaltic liquids. In contrast the majority of amphiboles from the cumulate mafic intrusives have smaller amounts of the A-site occupancies (Na_2O: 1.3 to 2.3 wt.%, Table IVb) and K_2O as well as slightly higher Si contents (Fig. 4a). Thus, these amphiboles plot in the fields of tschermakitic hornblende and magnesio hornblende, possibly indicating re-equilibration with the neighbouring cumulate phases (i.e. the amphiboles of HT 129 with orthopyroxene). As result of the accumulation of these amphiboles the residual liquid must become silica- and sodium-enriched, buffered by quartz, plagioclase and Ti- magnetite crystallization, and tends towards dacitic composition. Only a few relict amphibole cores in the cumulates have hastingsitic compositions indicative of the original compositions involved in the fractionation process from primitive basaltic liquid.

Very little is known about the stability field of hastingsitic amphiboles in basaltic systems. Allen et al., (1975) and Allen and Boettcher (1978) synthesized similar amphiboles from andesitic and quartz-tholeiitic compositions near the vapor-saturated liquidus at 920°C and pressures between 10 and 18 kbar. The upper stability field of these amphiboles is limited by the appearance of garnet at 18 kbar, but the lower stability field is not yet known, but could be extrapolated down to pressures of a few kbar. The latter possibility is suggested by the occurrence of the mafic cumulates within the Dras volcanic thrust pile as well as by the subsolidus stability field of amphibole coexisting with forsterite, anorthite, clino- and orthopyroxene at low pressures (Obata and Thompson, 1981).

Cawthorn and O'Hara (1976) suggested on the basis of phase relationships in the system CaO-MgO-Al_2O_3-SiO_2-Na_2O-H_2O, that amphibole is an important near-liquidus phase in basic and intermediate magma compositions over a range of 2—10 kbar water pressure. Thus amphibole has been considered to play an important role in the differentiation of calc-alkaline magmas.

In this respect we assume that amphibole fractionation might have occurred at a moderate depth of about 5—15 km in the Dras island arc environment.

ISOTOPE CHEMISTRY AND ORIGIN OF THE MAGMAS

Rb-Sr isotopic investigations were carried out on the same samples from the gorge between Pashkyum and Mulbekh. When plotted on a Rb-Sr-evolution diagram (Fig. 10) 13 out of 15 samples are arranged along a trend line with a slope corresponding to 263 m.y. Evidently, this number cannot be related to the true age of the Dras volcanism. Based on geological and stratigraphical observations, the Dras volcanics must have been erupted over a time span of about 30 to 40 m.y., from Upper Jurassic to Middle Cretaceous. Thus a scattered field of data points should be expected in the Sr-diagram. A remarkable feature is the arrangement of the data points as a "pseudo-isochron" with an initial $^{87}Sr/^{86}Sr$ ratio of 0.7035, which is in the same range as the mean of modern island arc volcanics (Faure, 1977; Stern, 1979; Meijer and Reagan, 1981). We regard the "pseudo-isochron" as a characteristic magmatic feature excluding postmagmatic seawater alteration and regional metamorphism.

In addition, the Rb-Sr results (Honegger et al., 1982) obtained from large samples (15—25 kg) of the Kargil mafic intrusives have also been shown in the same diagram. Of interest here is the overlap of the data from both rock suites but the lack of isotopic homogenization in the mafic cumulates compared with the volcanics. The low initial Sr-ratios of both rock suites point to very similar source materials in the mantle, and no significant contamination by sialic crust.

Similar, apparently meaningless, Sr-pseudo-isochrons and cluster of data points have been reported from several island arc volcanics in the Pacific

(Kay et al., 1978), from the Italian volcanics (Hawkesworth and Vollmer, 1979) and from the Japanese islands (Zashu et al., 1980). Kay et al. (1978) interpreted the Sr and Pb isotopic data from the Aleutian island arc as the result of mixtures of continent-derived sediments with melts from underthrust oceanic crust and overlying mantle.

A simple two-component mixing model of source material cannot account for the Dras "pseudo-isochron" because of the lack of correlation of $^{87}Sr/^{86}Sr$ versus $1/Sr$. However, a multistage mixing of different components, e.g. different parts and ages of oceanic crust and mantle could more readily explain the "pseudo-isochron". Another interpretation of the age figure of 263 m.y. could be, that it results from an older homogenization event in the source, oceanic crust or mantle; a possible mechanism producing "pseudo-isochrons" as discussed by Brooks et al. (1976).

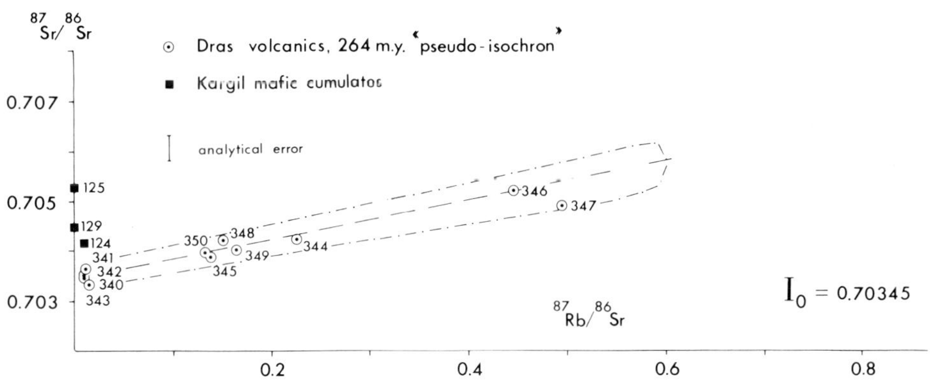

Fig. 10. Rb-Sr evolution diagram for whole rocks from Dras volcanics and the Kargil mafic cumulates (data from Honegger et al., 1982).

CONCLUSION

The distinctive bulk chemistry (e.g. low Ti-contents but high K, Ba and LREE contents) as well as the mineral chemistry of the phenocrysts in the basaltic rocks suggest that the mafic Dras volcanics can be classified as island arc volcanics, which have erupted in an island arc environment.

Petrographic observations show two different sequences of crystallization:

(1) In the island arc tholeiite series: olivine, magnesiochromite, clinopyroxene and plagioclase crystallized first, followed by hastingsite, magnesio-hastingsitic hornblende, and titanomagnetite.

(2) The crucial fractionating minerals in the development of the calc-alkaline dacitic rocks were certainly titanomagnetite and hornblende. Both phases reduce Ti starting from a relative primitive tholeiitic, water-bearing magma. As soon as these minerals began to crystallize and to accumulate (together with plagioclase and clinopyroxene), the residual melt composition

could change from the tholeiitic composition to the silica- and alkali-enriched calc-alkaline magma composition. With increasing silica activity, relatively Si- and Fe-rich amphibole started crystallizing, probably together with orthopyroxene, phases which have been found in the mafic cumulates of the Kargil intrusives within and near the Dras volcanics.

The latter process must have taken place at approx. 950 to 1000°C at a depth of 5—15 km within or underneath the arc. Unfortunately calculations of eruption temperatures using two-pyroxene geothermometry is not possible due to metamorphic overprinting of orthopyroxene.

The fractionation process forming the Dras basalts and dacites probably started from a simple primitive tholeiitic composition and does not necessarily require different magma batches generated by different partial melting processes in the mantle.

According to Kushiro (1972) slightly quartz-normative basalts as well as calc-alkalic andesites and dacites could be generated from wet peridotite. Qtz-normative island arc tholeiites were formed experimentally by melting wet spinel lehrzolite at or above 1200°C and at pressures <15 kbar (Takahashi et al., 1981). Tatsumi and Kushiro (1981) could experimentally demonstrate that even 1 wt.% H_2O in the mantle is sufficient to generate primary high-mg-andesites under similar P and T conditions, if the degree of partial melting is high enough.

The discussion on the magma source can be also based on the Rb-Sr isotopic data (yielding a very low $^{87}Sr/^{86}Sr$ initial of 0.7034) as well as on the geochemistry (e.g. REE distribution). It appears that the Dras volcanics as well as the Kargil intrusive complex had very similar magma sources in the upper mantle, both rock suites probably derived by partial melting of lherzolitic mantle material. The U-shaped REE patterns do not reflect original garnet in the mantle source, but indicate more likely amphibole fractionation.

The most primitive basalt from our collection (HT 344) has a relatively flat REE pattern (Fig. 9). LREE are enriched by the factor of 2—3 relative to primitive mid-ocean ridge basalts (Dietrich et al., 1981). This seems to be an original feature and is consistent with the higher contents of K, Ba and Sr, although some alteration effects have to be considered. This, together with the appearance of amphibole at middle stages of fractionation, suggests the involvement of components in the magma derived from dehydration or incipient melting of subducted Tethyan oceanic crust. Such a mechanism is shown on Fig. 11.

Contamination by remnants of sialic crust can be excluded because of low $^{207}Pb/^{206}Pb$ ratios of 0.046—0.050, obtained from Upper Cretaceous granodiorites in the Kargil intrusive complex (Honegger et al., 1982).

The Dras island arc volcanism probably started in Upper Jurassic time, producing pillow lavas, and massive flows in deep sea as well as in shallow-water environments. Dacitic flows alternating with island arc basalts are common. The upper age limit of the Dras volcanics is uncertain but probably

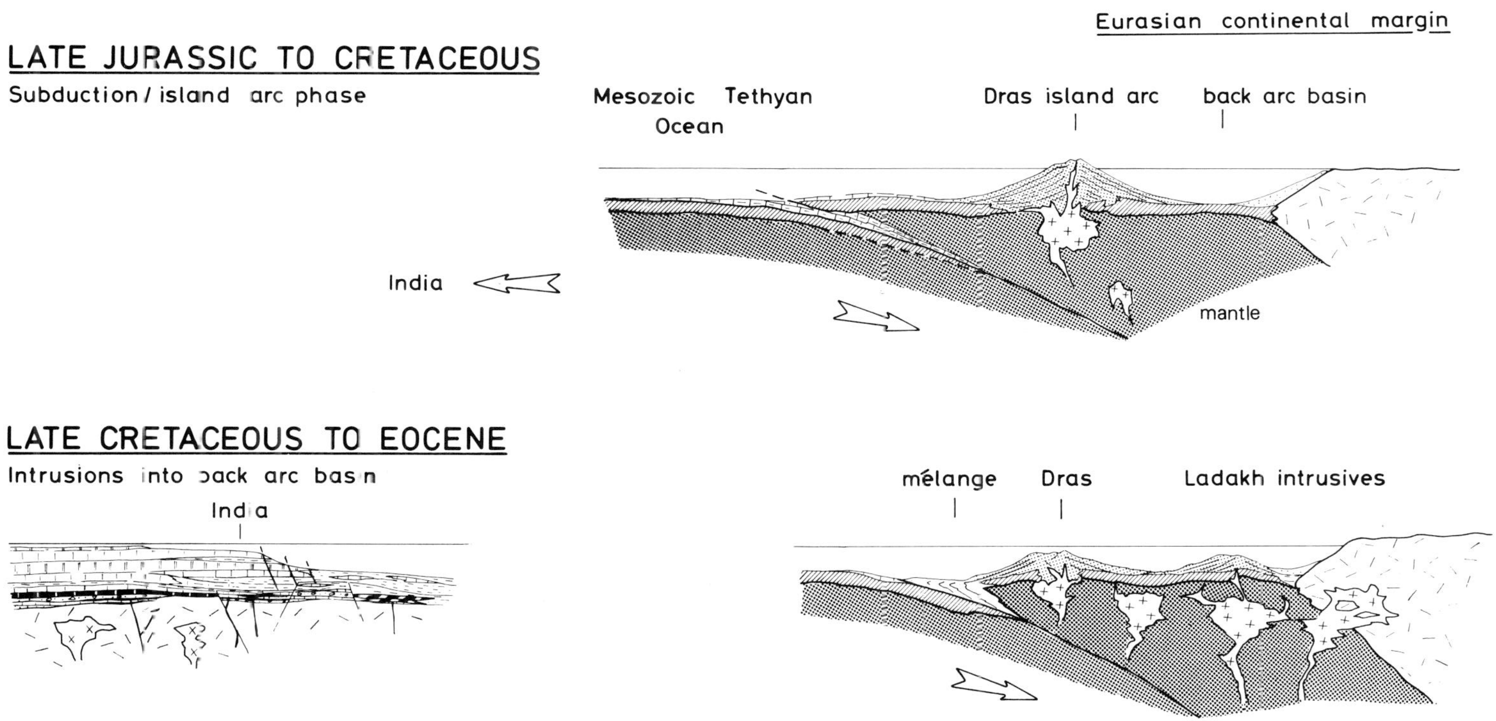

Fig. 11. Dras island arc evolution during Late Jurassic and Late Cretaceous to Eocene (partly after Trommsdorff et al., 1982).

post Cenomanian, and it is interesting to note, that although tectonic contacts exist in this thrust belt, the Dras volcanics pass into the shallow water volcano-detrital Nindam formation, which yields Upper Cretaceous ages. This observation suggests that in terms of sedimentation, morphology and time span, a typical back arc flysch apron existed similar to those known from present island arcs in the Pacific.

Hornblende-bearing basalts and basaltic andesites seem to be more abundant in island arcs bordering continental crust or in destructive continental margins than in oceanic island arcs (Ewart, 1976; Meijer and Reagan, 1981). Volcanics comparable to those in the Dras unit have been reported by Arculus et al. (1977) from the eastern Aleutian islands, that is in a transitional tectonic environment between oceanic and continental island arcs and by Arculus (1978) from the Lesser Antilles island arc. The Caribbean tectonic environment between the mature island arc of the Lesser Antilles and the Aves Ridge, which is partly built up of granodiorites and granites (Fox et al., 1971) seems to be a close actualistic equivalent to the Mesozoic—Tertiary Himalayan-Dras situation.

ACKNOWLEDGEMENTS

We thank A. Gansser, V. Trommsdorff (Zürich) and G. Fuchs (Geologische Bundesanstalt, Vienna) for many critical discussions concerning Ladakh geology and petrology. J. Sommerauer (Zürich) provided much help on the electron microprobe. The XRF analyses have been made with the assistance of A. Esenwein from the Eidgenössische Materialprüfungsanstalt (Dübendorf, Zürich). Neutron activation analyses of the rare earth elements were carried out by S. Bajo and A. Wyttenbach at the Eidgenössisches Institut für Reaktorforschung (Würenlingen, Switzerland).

We also thank S. Harley (Zürich) and R.J. Arculus (Canberra) for a thorough review which greatly improved the manuscript.

REFERENCES

Allen, J.C. and Boettcher, A.L., 1978. Amphiboles in andesites and basalt. II Stability as a function of P-T-f_{H_2O}-f_{O_2}. Am. Mineral., 63: 1074—1087.

Allen, J.C., Boettcher, A.L. and Marland, G., 1975. Amphiboles in andesite and basalt: I. Stability as a function of P-T-f_{O_2}. Am. Mineral., 60: 1069—1085.

Arculus, R.J., 1978. Mineralogy and petrology of Grenada, Lesser Antilles island arc. Contrib. Mineral. Petrol., 65: 413—424.

Arculus, R.J. and Wills, K.J.A., 1980. The petrology of plutonic rocks and inclusions from the Lesser Antilles Island arc. J. Petrol., 21: 743—799.

Arculus, R.J., Delong, S.E., Kay, R.W., Brooks, C. and Sun, S.S., 1977. The alkalic rock suite of Bogoslof Island, eastern Aleutian arc, Alaska. J. Geol., 85: 177—186.

Bajo, S. and Wyttenbach, A., 1980. Extraction liquide-liquide des terres rares et application à leur dosage dans les silicates par activation neutronique. EIR Techn. Mitt. (Intern. rep.). TM-44-80-2, 20 pp.

Bard, J.P., Maluski, H., Matte, Ph. and Proust, F., 1980. The Kohistan sequence: Crust and mantle of an obducted island arc. Proc. Int. Comm. on Geodynamics, Group 6, Peshawar, Nov. 1979, Spec. Iss., Geol. Bull. Univ. of Peshawar, 13: 87—94.

Bassoullet, J.P., Colchen, M., Marcoux, J. and Mascle, G., 1978. Une transversale de la zone de l'Indus Khalsi à Phothaksar, Himalaya du Ladakh. C.R. Acad. Sci. Paris, 286: 563—566.

Bassoullet, P., Colchen, M., Juteau, T., Marcoux, J. and Mascle, G., 1980. L'édifice de nappes du Zanskar (Ladakh, Himalaya). C.R. Acad. Sci. Paris, Sér. D, 290: 389—392.

Beccaluva, L., Ohnenstetter, D. and Ohnenstetter, M., 1979. Geochemical discrimination between ocean floor and island arc tholeiites. Application to some ophiolites. Ofioliti 4 (1): 67—72.

Brooks, C., Hart, S.R., Hofmann, A. and James, D.E., 1976. Rb/Sr mantle isochrons from oceanic regions. Earth Planet. Sci. Lett., 32: 51—61.

Cameron, W.E., Nisbet, E.G. and Dietrich, V.J., 1980. Petrographic dissimilarities between ophiolitic and ocean-floor basalts. Proceedings Internat. Ophiolite Symp. Nicosia, Cyprus, 1979, pp. 182—192.

Cameron, W.E., Nisbet, E.G. and Dietrich, V.J., 1979. Boninites, komatiites and ophiolitic basalts. Nature, 280: 550—553.

Cawthorn, R.G. and O'Hara, M.J., 1976. Amphibole fractionation in calc-alkaline magma genesis. Am. J. Sci., 276: 309—329.

De Terra, H., 1935. Yale North India Expedition. Article II, Geological studies in the Northwest Himalaya between the Kashmir and Indus valleys. Mem. Conn. Acad. Arts Sci., 8: 18—76.

Dietrich, V.J., Emmermann, R., Oberhaensli, R. and Puchelt, H., 1978. Geochemistry of basaltic and gabbroic rocks from the West Mariana Basin and the Mariana Trench. Earth Planet. Sci. Lett., 39: 127—144.

Dietrich, V.J., Gansser, A., Sommerauer, J. and Cameron, W.E., 1981. Palaeogene komatiites from Gorgona Island, East Pacific — A primary magma for ocean floor basalts? Geochem. J., 15: 141—161.

Ewart, A., 1976. Mineralogy and chemistry of modern orogenic lavas. Some statistics and implications. Earth Planet. Sci. Lett., 31: 417—432.

Ewart, A., 1979. A review of the mineralogy and chemistry of Tertiary-recent dacitic, latitic, rhyolitic, and related salic volcanic rocks. In: F. Barker (Editor), Trondhjemites, Dacites, and Related Rocks. Elsevier, Amsterdam-Oxford-New York, pp. 13—121.

Faure, G., 1977. Principles of Isotope Geology. John Wiley, New York, N.Y., 464 pp.

Fox, P.J., Schreiber, E. and Heezen, B.C., 1971. The geology of the Caribbean crust: Tertiary sediments, granitic and basic rocks from the Avis Ridge. Tectonophysics, 12: 89—109.

Frank, W., Gansser, A. and Trommsdorff, V., 1977. Geological observations in the Ladakh area (Himalayas), A preliminary report. Schweiz. Mineral. Petrogr. Mitt., 57: 89—113.

Gansser, A., 1964. Geology of the Himalayas. Interscience Publishers, John Wiley and Sons Ltd., London, 289 pp.

Gansser, A., 1977. The great suture zone between Himalaya and Tibet — A preliminary account. Himalaya - Sci. de la Terre. Coll. Intern. 1976. Centre Nat. Rech. Sci. Paris, 268: 181—192.

Gansser, A., 1979. Ophiolitic belts of the Himalayan and Tibetan region. In: Internat. Atlas of Ophiolites, 1 : 2.500 000. Geol. Soc. Am. Map and Chart Series MC-33.

Gansser, A., 1980. The division between Himalaya and Karakorum. Proc. Intern. Comm. on Geodynamics, Group 6, Peshawar Nov. 1979. Spec. Issue, Geol. Bull. Univ. of Peshawar, 13: 9—22.

Hawkesworth, C.J. and Vollmer, R., 1979. Crustal contamination versus enriched mantle: 143Nd/144Nd and 87Sr/86Sr evidence from the Italian volcanics. Contrib. Mineral. Petrol., 69: 151—165.

Hawkins, J.W., 1980. Petrology of back-arc basins and island arcs: Their possible role in the origin of ophiolites. Proc. Int. Ophiolite Symp. Ophiolites, Cyprus, 1979, pp. 244—254.

Hellman, P.L., Smith, R.E. and Henderson, P., 1979. The mobility of the rare earth elements: Evidence and implications from selected terrains affected by burial metamorphism. Contrib. Mineral. Petrol., 71: pp. 23—44.

Honegger, K., Dietrich, V., Frank, W., Gansser, A., Thoeni, M. and Tromsdorff, V., 1982. Magmatism and metamorphism in the Ladakh-Himalayas. Earth Planet. Sci. Lett., 60: 253—292.

Humphris, S.E. and Thompson, G., 1978. Trace-element mobility during hydrothermal alteration of oceanic basalts. Geochim. Cosmochim. Acta, 42: 127—136.

Irvine, T.N., 1975. The silica immiscibility effect in magmas. Carnegie Inst. Washington, Yearb., 74: 484—492.

Johnson, R.W. and Arculus, R.J., 1978. Volcanic rocks of the Witu Islands, Papua New Guinea. The origin of magmas above the deepest part of the New Britain Benioff zone. Bull. Volcanol., 41: 609—655.

Juteau, T. and Reibel, G., 1981. Indus suture: petrographic description of the pillow-lavas of the Dras formation, and of the ophiolites of the Spongtang-Photaksar nappe. Inst. meeting EUG, Strasbourg, 1981, in: Terra cognita, Spec. Iss., p. 19 (abstr.).

Kay, R.W., Sun, S.S. and Lee-Hu, C.-H., 1978. Pb and Sr isotopes in volcanic rocks from the Aleutian Islands and Pribilof Islands, Alaska. Geochim. Cosmochim. Acta, 42: 263—273.

Kuno, H., 1968. Differentiation of basalt magmas. In: H. Hess and A. Poldervaardt (Editors), Basalts, Vol. 2. Interscience Publishers, New York, London, Sydney, pp. 623—688.

Kushiro, I., 1972. Effect of water on the composition of magmas formed at high pressures. J. Petrol., 13: 311—334.

Laird, J. and Albee, A.L., 1981. High-pressure metamorphism in mafic schist from northern Vermont. Am. J. Sci., 281: 97—126.

Leake, B.E., 1978. Nomenclature of amphiboles. Am. Mineral., 63: 1023—1052.

Lydekker, R., 1883. The geology of Kashmir and Chamba territories and the British districts of Khagan. Mem. Geol. Surv. Ind., 22: 31—34.

Masuda, A., Nakamura, N. and Tanaka, T., 1973. Fine structures of mutually normalized rare-earth patterns of chondrites. Geochim. Cosmochim. Acta, 37: 239—248.

Meijer, A. and Reagan, M., 1981. Petrology and geochemistry of the Island of Sarigan in the Marian Arc; calc—alkaline volcanism in an oceanic setting. Contrib. Mineral. Petrol., 77: 337—354.

Miyashiro, A., 1973. The Troodos ophiolitic complex was probably formed in an island arc. Earth Planet. Sci. Lett., 19: 218—224.

Miyashiro, A., 1974. Volcanic rock series in island arcs and active continental margins. Am. J. Sci., 274: 321—355.

Miyashiro, A., 1975a. Island arc volcanic rock series: a critical review. Petrology, 1: 177—187.

Miyashiro, A., 1975b. Volcanic rock series and tectonic setting. Annu. Rev. Earth Planet. Sci., 3: 251—269.

Nisbet, E.G. and Pearce, J.A., 1977. Clinopyroxene composition in mafic lavas from different tectonic settings. Contrib. Mineral. Petrol., 63: 149—160.

Nisbet, E.G., Dietrich, V.J. and Esenwein, A., 1979. Routine trace element determination in silicate minerals and rocks by X-ray fluorescence. Fortschr. Miner., 57 (2): 264—279.

Pearce, J.A. and Cann, J.R., 1973. Tectonic setting of basic volcanic rocks determined using trace element analyses. Earth Planet. Sci. Lett., 19: 290—300.

Obata, M. and Thompson, A.B., 1981. Amphibole and chlorite in mafic and ultramafic rocks in the lower crust and upper mantle. Contrib. Mineral. Petrol., 77: 74—81.

Prasad, U., Singh, B., Venkatraman, K. and Ghosh, D.B., 1980. Petrochemical characteristics of basic and ultrabasic rocks of Dras area, "Indus ophiolite belt", Ladakh, India. Proc. Int. Ophiolite Symp. Cyprus 1979, pp. 444—453.

Qasim Jan, M., 1980. Petrology of the obducted mafic and ultramafic metamorphites from the southern part of the Kohistan island arc sequence. Proc. Intern. Comm. on Geodynamics, Group 6, Peshawar Nov. 1979, Spec. Issue Geol. Bull. Univ. of Peshawar, 13: 95—108.

Saunders, A.D., Tarney, J., Marsh, N.G. and Wood, D.A., 1980. Ophiolites as ocean crust and marginal basin crust: A geochemical approach. Proc. Int. Ophiolite Symp. Cyprus 1979, pp. 193—204.

Sharaskin, A.Ya., Dobretsov, N.L. and Sobolev, N.V., 1980. Marianites: The clinoenstatite bearing pillow-lavas associated with the ophiolite assemblage of Mariana trench. Proc. Int. Ophiolite Symp. Cyprus 1979, pp. 473—479.

Srikantia, S.V. and Razdan, M.L., 1980. The ophiolite-sedimentary belt of the Indus tectonic zone of the Ladakh Himalaya. Its stratigraphic and tectonic significance. Proc. Int. Ophiolite Symp. Cyprus 1979, pp. 430—441.

Stern, R.J., 1979. Strontium isotopes in island arcs and marginal basins from the Circum-Pacific: regional variations and implications for magma genesis. Geol. Soc. Am., Abstr. Programs, 11: 524.

Sun, S.-S., Nesbitt, R.W. and Sharaskin, A.Y., 1979. Geochemical characteristics of mid-ocean ridge basalts. Earth Planet. Sci. Lett., 44: 119—138.

Takahashi, E., Kushiro, I. and Tatsumi, Y., 1981. Melting of a peridotite at high pressures and its bearing on island arc magmas. Abstr., 1981 IAVCEI Symposium Arc Volcanism Aug. 28—Sept. 9, 1981, Tokyo and Hakone, p. 365.

Tatsumi, Y. and Kushiro, I., 1981. Origin of high magnesian andesite. Abstract, 1981 IAVCEI Symposium Arc volcanism Aug. 28—Sept. 9, 1981, Tokyo and Hakone, p. 375.

Trommsdorff, V., Dietrich, V. and Honegger, K., 1982. The Indus Suture Zone: Paleotectonic and igneous evolution in the Ladakh-Himalayas. In: K.J. Hsii (Editor), Mountain Building Processes. Academic Press, London, pp. 213—219.

Wood, D.A., Joron, J.L. and Treuil, M., 1979. A re-appraisal of the use of trace elements to classify and discriminate between magma series and different tectonic settings. Earth Planet. Sci. Lett., 45: 326—336.

Wood, D.A., Joron, J.L., Marsh, J.G., Tarney, J. and Treuil, M., 1980. Major- and trace-element variations in basalts from the North Phillipine Sea drilled during Deep Sea Drilling Project Leg 58: A comparative study of back-arc basin basalts with lava series from Japan and mid-ocean ridges. Init. Rep. DSDP Project 58. U.S. Government Printing Office, Washington D.C., pp. 873—894.

Yajima, T., Higuchi, H. and Nagasawa, H., 1972. Variation of rare earth concentrations in pigeonitic and hypersthenic rock series from Izu-Hakone Region, Japan. Contrib. Mineral. Petrol., 35: 235—244.

Zashu, S., Kaneoka, I. and Aoki, K.-I., 1980. Sr-Isotope study of mafic and ultramafic inclusions from Itinome-gata, Japan. Geochem. J., 14: 123—128.

Journal of Volcanology and Geothermal Research, 18 (1983) 435—447
Elsevier Science Publishers B.V., Amsterdam — Printed in The Netherlands

ON THE LATERAL VARIATIONS IN CHEMICAL COMPOSITION AND VOLUME OF QUATERNARY VOLCANIC ROCKS ACROSS JAPANESE ARCS

IKUO KUSHIRO

Geological Institute, University of Tokyo, Tokyo 113 (Japan)

(Received October 20, 1982)

ABSTRACT

Kushiro, I., 1983. On the lateral variations in chemical composition and volume of Quaternary volcanic rocks across Japanese arcs. In: S. Aramaki and I. Kushiro (Editors), Arc Volcanism. J. Volcanol. Geotherm. Res., 18: 435—447.

The Fe/Mg+Fe) ratios (X_{Fe}) of the Quaternary basalts (SiO_2 < 53 wt.%) in the Japanese arcs were examined. The X_{Fe} of relatively magnesian basalts decreases from the volcanic front toward the Japan Sea across the arcs. Based on the partition coefficient of Mg-Fe^{2+} between olivine and liquid, it is suggested that all the basalts near the volcanic front, which are mostly tholeiitic basalts, are significantly fractionated, whereas many basalts near the Japan Sea, which are mostly alkali basalts, are little fractionated. The K_2O content in the primary basalt magmas increases toward the Japan Sea. Combining the X_{Fe} and K_2O data, it is suggested that relatively large amounts of tholeiitic magmas are produced near the volcanic front, but they fractionate during their ascent, whereas smaller amounts of alkali basalt magmas are formed near the Japan Sea, but they can ascend with less fractionation. The density of primary tholeiite magma is significantly larger than that of primary alkali basalt magmas. It is most likely that primary tholeiite magmas cannot ascend beyond the upper crust and would fractionate to produce less dense tholeiitic magmas near the volcanic front, whereas primary alkali basalt magmas can ascend through the upper crust without fractionation, as far as buoyancy is the principal ascending force. In the Japanese arcs, the stress field may be less compressional near the Japan Sea than near the volcanic front, so that magmas can ascend more rapidly in the latter region than in the former. These two factors may be responsible for the above mentioned chemical variations of basalt magmas across the arcs. The variation in volume of the Quaternary volcanic rocks across the arcs can be explained by the presence of a melt-rich zone above but nearly parallel to the subducted slab.

INTRODUCTION

One of the major problems in the genesis of volcanic rocks in island arcs is the lateral variations of their chemical composition and mineralogy across the arcs. The volcanic rocks in the Japanese Islands, which form typical island arcs, show such variations and have been studied in detail.

0377-0273/83/$03.00

The spatial variations in mineralogy of the Quaternary volcanic rocks in the Japanese Islands were first summarized by Koto (1916), who showed that pyroxene andesites occur mainly in northeast Japan, whereas hornblende and biotite bearing andesites as well as bronzite andesite occur mainly in the southwest Japan. Tomita (1935) studied alkali rocks in and around the Japan Sea and showed that they are different in chemical composition and mineralogy from non-alkalic rocks in the Pacific side of the Japanese Islands. Kuno (1954a) extended the earlier works on the mineralogical variations of the Quaternary volcanic rocks in the Japanese Islands and showed the distribution of volcanoes which bear rocks with biotite, those with hornblende, and those with pyroxenes but no hydrous minerals. Later Kuno (1959, 1960) demonstrated the variations in chemical compositions of basalts across the Japanese arcs; tholeiitic basalts occur in the trench-side zone, high-alumina basalts in the intermediate zone, and alkali basalts in the Japan Sea-side zone. Sugimura (1960) also showed a continuous variation in chemical composition of volcanic rocks across the Japan arcs. He introduced Θ index (SiO_2 - $47\times(Na_2O+K_2O)$ /Al_2O_3) and showed that the Θ index decreases from the volcanic front toward the Japan Sea.

Kuno (1959, 1960) considered that the above chemical variations of basalts are due to the variations in chemical composition of basalt magmas produced by partial melting of peridotites along the deep seismic zone which becomes deeper toward the Japan Sea. His assumption that basalt magmas produced at depths change their chemical compositions little during their ascent to the surface was criticized by O'Hara (1965), who argued that most basalt magmas are products of fractional crystallization at shallow levels. Fukuyama (1978) pointed out that most basalts erupted in the Japanese arcs cannot be in equilibrium with the upper mantle peridotites based on the Fe^{2+}-Mg and Ni-Mg partitioning between olivine and liquid at 1 atm, determined by Roeder and Emslie (1970) and Fukuyama (1978), respectively. Experimental data on the Fe^{2+}-Mg partitioning between olivine and liquid at pressures to 9 kbar given by Green and Ringwood (1967) are roughly the same as those at 1 atm. Recently Takahashi et al. (1981) showed that the coefficient at pressures up to 25 kbar is essentially the same as that at 1 atm by Roeder and Emslie (1970), at least under anhydrous conditions. Based on these data, we can use the Fe^{2+}/Mg of un-oxidized basalts as a criterion for whether or not basalt magmas erupted in the Japanese arcs can be in equilibrium with the upper mantle peridotites.

In the present study, the variations in the Fe*/(Mg+Fe*) ratios (total iron as Fe*) of the Quaternary basalts in the Japanese arcs have been examined and the cause of the variations is discussed.

Another problem to be discussed is variation in the amount of volcanic rocks across the island arcs. Sugimura et al. (1963) showed that the amount of Quaternary volcanic rocks in the Japanese Islands decreases sharply from the volcanic front toward the Japan Sea.

VARIATIONS IN X_{Fe} OF BASALTS ACROSS JAPANESE ARCS

The Fe*/(Mg+Fe*) ratios (X_{Fe} hereafter), where Fe* represents total iron, of the Quaternary basalts ($SiO_2 < 53$ wt.%) in the Japanese Islands have been examined. In surveying the analyses of basalts, those with relatively low Fe/Mg have been carefully looked for. Fig. 1 shows the X_{Fe} of the basalts plotted against distance from the volcanic front (VF). It is shown in the figure that basalts near the volcanic front, which are mostly tholeiitic basalts, have relatively large X_{Fe}. There are many analyses of basalts which occur near the volcanic front, yet there is no basalt with X_{Fe} smaller than 0.42. On the other hand, many basalts near the Japan Sea, which are mostly alkali basalts, have relatively small X_{Fe}. In these plots, basalts with abundant olivine phenocrysts which are most likely of cumulate origin have been excluded, because

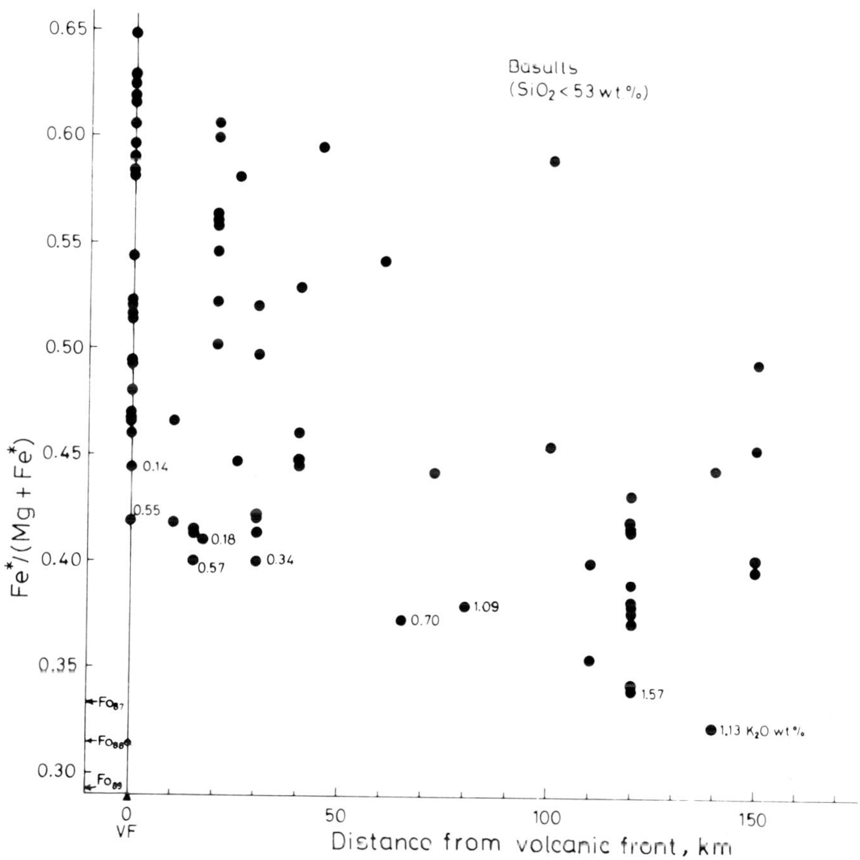

Fig. 1. Fe*/(Mg + Fe*) ratios of relatively magnesian basalts in the Japanese arcs plotted against distance from the volcanic front (VF). Total iron as Fe*. Numbers are wt.% of K_2O in the most magnesian basalts in the respective regions. Many of the data used in this figure are based on the compilation by Ono (1962), Kuno (1962) and Aoki (1978).

their X_{Fe} does not represent those of the magmas. Whether olivine is cumulate or not was checked by the X_{Fe} of bulk basalt composition and that of the core of olivine phenocrysts. In these plots, total iron is taken as FeO. However, part of iron in these basalts is ferric. If 10 wt.% of the total iron is Fe_2O_3, as in the least oxidized basalts in the Japanese Islands, and is excluded from the total iron, all the points shift down by about 0.03. It is also found that the systematic along-arc variations of X_{Fe} are not observed.

Olivine in most peridotite inclusions in basaltic rocks in the Japanese Islands, such as those in Ichinomegata and Takashima, is more magnesian than Fo_{88} (e.g., Kuno and Aoki, 1970; Aoki, 1971). Magmas with X_{Fe} less than 0.32 can be in equilibrium with olivine of $> Fo_{88}$, as judged from the partition coefficient for Mg-Fe between olivine and basaltic liquid by Roeder and Emslie (1970) at 1 atm and Takahashi et al. (1981) at higher pressures. Assuming that all the basalts in the Japanese Islands have been derived from peridotitic upper mantle containing olivine more magnesian than Fo_{88}, even the most magnesian basalts near the volcanic front must have been significantly fractionated, whereas those near the Japan Sea are not or little fractionated. This argument is supported by the frequent occurrence of peridotite inclusions in the basalts near the Japan Sea and the absence of such inclusions in those near the volcanic front. Presence of peridotite inclusions indicates that mafic minerals were not separated significantly from magmas during their ascent from the upper mantle to the surface.

Contamination of magmas by crustal materials such as granitic and gabbroic rocks and amphibolite, which are often found as inclusions in volcanic rocks, cannot produce the above variation of X_{Fe} of magnesian basalts.

The K_2O content of basalts generally increases with increasing X_{Fe} in a given zone. This may be a consequence of fractional crystallization of basalt magmas, although the patterns of fractional crystallization (e.g., degree of iron enrichment) are not always the same. However, when the magnesian basalts of different zones are compared, an opposite relation is obtained. The K_2O contents in the relatively magnesian basalts are shown in Fig. 1. The K_2O content generally increases with increasing distance from the volcanic front. This has been known since the work by Kuno (1959) and Sugimura (1960). Therefore, as far as the relatively magnesian basalts are concerned, those with larger X_{Fe} contain smaller amounts of K_2O. Primary magmas near the volcanic front should contain less K_2O than the fractionated basalts, because the K_2O increases by fractional crystallization. The above relation should, therefore, be more pronounced when the primary magmas are compared.

There is no positive evidence that the K_2O content in the upper mantle systematically increases toward the Japan Sea. Higher K_2O in the primary magmas in the Japan Sea than in those near the volcanic front suggests, therefore, that the primary basaltic magmas are formed by higher degree of partial melting near the volcanic front, whereas they are formed by smaller degree of partial melting near the Japan Sea. This possibility has been suggested by several investigators (e.g., Miyashiro, 1974; Sakuyama, 1979).

Now we have a difficult problem; that is, near the volcanic front, larger amounts of primary tholeiitic magmas are formed, but they cannot ascend to the surface without fractional crystallization, while near the Japan Sea, smaller amounts of primary alkali basalt magmas are formed, but they can ascend to the surface without significant fractionation. In addition, the depth of segregation of primary alkali basalt magmas from the upper mantle is probably deeper than that of primary tholeiite magmas, as discussed later.

CAUSE OF THE VARIATION OF X_{Fe} IN BASALTS ACROSS JAPANESE ARCS

The above-mentioned problem may be explained, if the primary magmas formed or segregated in the upper mantle near the volcanic front cannot ascend rapidly so that they fractionate during their ascent, whereas those formed near the Japan Sea can ascend rapidly so that they have little chance to fractionate. Because the X_{Fe} of the magnesian basalts decreases from the volcanic front toward the Japan Sea, magmas may ascend more and more rapidly toward the Japan Sea.

The structures of the crust and the upper mantle in the Japanese Islands have been studied by several seismologists (e.g., Matsuzawa, 1959; Ludwig et al., 1966; Yoshii, 1979). According to their results, there is no systematic change in the velocity structure of the crust and the upper mantle across the northern Honshu arc. It is unlikely, therefore, that the structures of the crust and the upper mantle are the cause of differences in the rate of ascent of magmas. In the present paper, following two factors that might lead to such a difference are considered.

One factor is the variation of the stress regime across the arc. If the lower crust and the upper mantle closer to the trench are under compressional stress, while those near the back-arc basin are under tensional stress, magmas formed in the upper mantle may be unable to ascend more rapidly near the volcanic front than near the back-arc basin. Under tensional stress, cracks would be more easily formed, through which magmas can rise as far as the amount of magma is enough to fill the cracks, or magmas can make fractures more easily immediately above the ascending magmas. A possible model of the arc—back-arc basin system with such stress patterns has been suggested by Toksöz and Hsui (1978). This model may be applied to the Japanese arcs with the Japan Sea as a back-arc basin, although the spreading of Japan Sea is believed to have been inactive during the Quaternary. Nakamura and Uyeda (1980) suggested on the basis of the presence of 'monogenetic volcanoes' near the Japan Sea that the stress field was less compressional on the Japan Sea side than on the trench side of the northeast Honshu arc at least several thousands years ago. 'Monogenetic volcanoes' are formed by single eruptions and their vents are often scattered within a restricted area, forming a volcanic field. Such volcanic fields are correlated to the tensional stress field by Nakamura and Uyeda. Quaternary 'monogenetic volcanoes' are also present in the Japan Sea side of southwest Honshu.

The other factor is the density variations of magmas across the Japanese arcs. This possibility has been examined by measuring densities of basaltic melts at high pressures. For the purpose of the present arguments, the densities of primary basalt magmas or close to them in the Japanese arcs must be compared. One of the most magnesian alkali basalts from Oki-Dogo Island in the Japan Sea (Uchimizu, 1966) and a magnesian K-rich alkali basalt from Abu, Sanin district (Koyaguchi, 1982) were chosen for the measurements (Table I, Nos. 1 and 2). The chemical composition of the primary tholeiite magma was obtained by the following way; olivine was successively added to a tholeiitic basalt containing only olivine as phenocrysts, keeping equilibrium between the bulk composition and adding olivine in each step, until the bulk composition becomes in equilibrium with olivine of Fo_{88} or Fo_{89}.

TABLE I

Chemical compositions of basalts used for the density measurements

	1	2	3	4
SiO_2	46.88	50.30	47.19	49.70
TiO_2	1.94	1.41	0.44	0.74
Al_2O_3	15.20	15.75	15.72	14.97
FeO*	10.12	7.87	10.38	10.56
MnO	0.15	0.13	0.10	—
MgO	10.11	9.17	12.78	13.03
CaO	9.62	7.47	11.60	9.00
Na_2O	3.04	1.81	1.53	1.56
K_2O	1.42	3.63	0.25	0.28
P_2O_5	0.36	0.53	0.01	—
Total	99.29	98.26	100.00	99.84

1. Alkali olivine basalt from Oki-Dogo in the Japan Sea (Uchimizu, 1966).
2. K-rich alkali olivine basalt from Abu, San-in district (Koyaguchi, 1982).
3. A mixture of a magnesian tholeiite from Izu (Kuno, 1954b) and olivine of Fo_{91}.
4. Estimated composition of primary tholeiite in the northeast Japan arc (Tatsumi, unpubl. data).

* Total iron as FeO.

This procedure, applied by Tatsumi and Ishizaka (1982), is based on the assumption that the basalt was formed by fractional crystallization involving separation of olivine alone. Similar method was applied by Rhodes (1981) to estimate primary magmas. One of the most magnesian tholeiite in the Izu region (Kuno, 1954b) containing only olivine as phenocrysts is chosen for this calculation. The composition of the primary tholeiite thus estimated is of olivine tholeiite. The starting material for the measurements was a glass made from a mixture of the above tholeiite and olivine of Fo_{91} (88:12 by weight) (Table I, No. 3) which is nearly identical in composition to the estimated primary tholeiite. The composition of another primary tholeiite magma in the northeast Honshu estimated by Tatsumi (unpubl. data) by the same procedure (Table I, No. 4) was also tested.

The density measurements were made with the falling-sphere method by Kushiro et al. (1976) and Fujii and Kushiro (1977) with olivine (Fo_{90}) and BN as spheres. Details of the density measurements of these basaltic melts at high pressures are not given here because of limitation of this article, but they are given in a separate report (Kushiro, 1982).

The results of the measurements are shown in Fig. 2. The density of the primary tholeiite melt ($Fe_2O_3/(FeO+Fe_2O_3)$ = 0.73 just after the run) increases from 2.65 g/cm^3 at 1 atm to 2.78 g/cm^3 at 10 kbar. Another primary tholeiite melt ($Fe_2O_3/(FeO+Fe_2O_3)$ = 0.36) increases from 2.66 g/cm^3 to 2.79 g/cm^3 from 1 atm to 10 kbar. The density of alkali basalt melt from Oki-Dogo ($Fe_2O_3/(FeO+Fe_2O_3)$ = 0.39) increases from about 2.62 g/cm^3 at

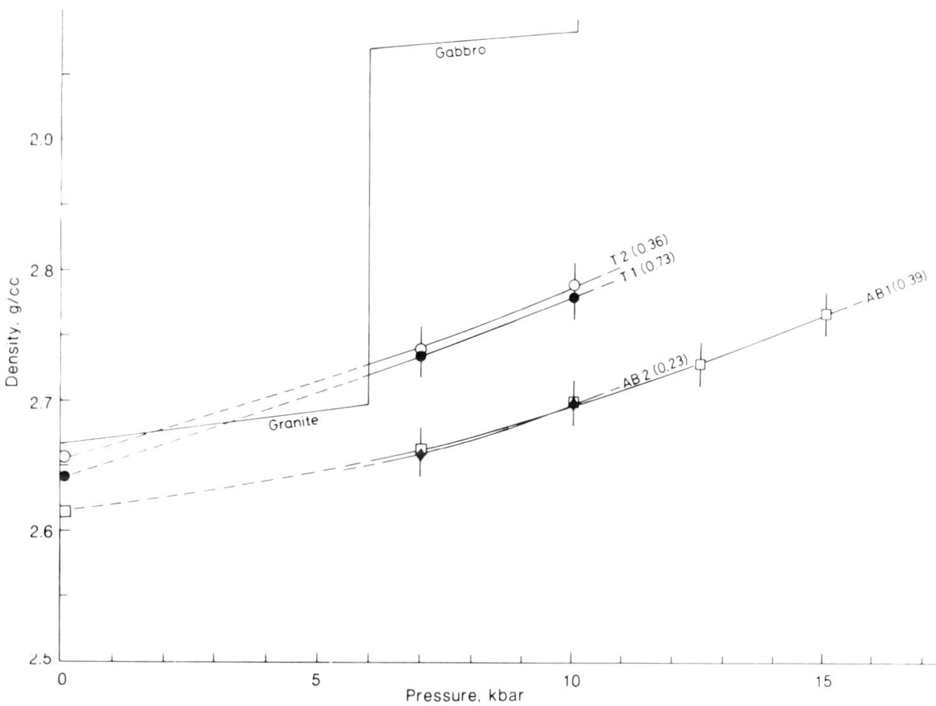

Fig. 2. Changes of density with pressure of melts of four different basaltic compositions. T1, primary tholeiite composition in the Izu region (No. 3, Table I); T2, primary tholeiite composition in the northeast Japan arc (Tatsumi, unpubl. data) (No. 4, Table I); AB1, alkali olivine basalt from Oki Dogo Island (Uchimizu, 1966) (No. 1, Table I), K-rich alkali olivine basalt from Abu district (Koyaguchi, 1982) (No. 2, Table I). Numbers in parentheses are Fe_2O_3/FeO ratios of the glass quenched after the measurements, determined by the conventional wet chemical analysis method. The density profile in the northeast Japan arc is based on the densities of average granites and gabbros given by Daly et al. (1966) and the compressibility data of these rocks given by Birch (1966).

1 atm to 2.70 g/cm^3 at 10 kbar. The K-rich alkali olivine basalt melt (Fe_2O_3/($FeO+Fe_2O_3$) = 0.23) shows a very similar density variation. The 1-atm densities of these melts were calculated by the method of Bottinga and Weill (1970). Temperatures of measurement were about 100°C higher than the liquidi of these basalts. However, the density change of basalt melt within 100°C interval is less than 0.01 g/cm^3 (Murase and McBirney, 1973; Fujii and Kushiro, 1977) and does not affect the following arguments. It is clear from the figure that the density of the primary tholeiite melt is larger than that of alkali basalt melt by 0.07 — 0.08 g/cm^3 at pressures between 5 and 10 kbar. The densities of these melts with the same Fe_2O_3/FeO were calculated using the partial molar volumes of Fe_2O_3 and FeO with the assumption that of alkali basalt melt by 0.07--0.08 g/cm^3 at pressures between 5 and ference in density between the primary tholeiite and alkali olivine basalt melts is larger when the melts with the same Fe_2O_3/FeO ratios are compared. The difference would be due to the differences in amount mostly of alkalies and partly of MgO and FeO. The partial molar volumes of Na_2O and K_2O are relatively large compared with those of mafic oxides (Bottinga and Weill, 1970). The alkali basalts contain about 2.6 wt.% more K_2O and Na_2O (about 25 wt.% normative ab and or) than the primary tholeiite.

A possible density profile of the crust and the upper mantle in the north Honshu arc is shown in Fig. 2 on the assumption that the lower crust consists of gabbro and the upper crust mostly of granites. It is shown that the primary tholeiite melt is denser than the average granites at pressures greater than about 2 kbar (density data from Daly et al., 1966 and compressibility data from Birch, 1966), whereas alkali basalt melts are less dense than the granitic rocks, even if the Fe_2O_3/($FeO+Fe_2O_3$) of the melts are less than 0.1. The alkali basalts near the Japan Sea frequently contain amphibole, indicating that the alkali basalt magmas contained some H_2O and other volatile components such as CO_2 and F, which would further reduce the density of the magmas. The partial molar volume of H_2O in silicate melt is relatively large. For example, addition of 1 wt.% H_2O reduces the density of albite melt by about 0.05 g/cm^3 at pressures between 5 and 7 kbar (Burnham and Davis, 1969). On the other hand, tholeiites near the volcanic front do not contain any primary hydrous minerals, and plagioclase crystallizes in the early stage of crystallization, indicating that the H_2O content in the primary tholeiite magma must be small (Saykuyama, 1983).

Assuming that buoyancy is the principal ascending force of magma, the primary tholeiite magmas cannot ascend beyond the upper crust consisting mainly of granite and metamorphosed sedimentary rocks and it would fractionate to produce less dense magmas in the lower to middle parts of the crust. The possibility of such a density filtering has been suggested by several investigators (e.g., Gill, 1973; Neumann, 1980; Stolper and Walker, 1980; and Coulon and Thorpe, 1981). On the other hand, alkali basalt magmas without significant amounts of mafic phenocrysts can ascend continuously to the surface through the granitic layer.

The ascent of magma in the shallow levels depends not only on the density difference between the magmas and the surrounding rocks (i.e., buoyancy), but also on other factors such as the amount of volatiles in the magma. Even heavy magmas could erupt by vesiculation caused by high vapor pressure in relatively shallow magma reservoirs. So the buoyancy argument may not be valid for the final stage, immediately prior to eruption.

In conclusion, magmas near the Japan Sea contain larger amounts of alkalies and volatile components than those near the volcanic front, and their resultant lower density may facilitate more rapid ascent of magmas near the Japan Sea. In addition, it is likely that the stress field is less compressional near the Japan Sea than near the volcanic front. Combination of these two factors may account for differences in the lateral variations of the X_{Fe} or the degree of fractionation of basalt magmas across the Japanese arcs.

CAUSE OF VOLCANIC FRONT AND VARIATION IN VOLUME OF VOLCANIC ROCKS ACROSS THE JAPANESE ARCS

Next problem to be discussed is the existence of the volcanic front (Sugimura, 1960) and the variation in the volume of erupted materials across the Japanese arcs. The presence of a sharp volcanic front would probably mean that the region of magma generation is restricted to the west side (continental side) of the volcanic front.

There are two possibilities on the generation of magmas only in the west side of the volcanic front. One possibility is that flux (mainly H_2O), which causes the melting of the wedge mantle, is supplied from the subducted oceanic lithosphere only in the west side of the volcanic front. The other possibility is that the flux is supplied to the wedge mantle in both sides of the volcanic front, but the temperature of the wedge mantle exceeds the solidus of the mantle material only in the west side of the volcanic front. Dehydration of the subducted oceanic lithosphere has been discussed by many investigators (e.g., Wyllie, 1973; Ringwood, 1974; Fyfe and McBirney, 1975; Anderson et al., 1976; Delany and Helgeson, 1978). According to these studies, dehydration of the oceanic lithosphere, especially oceanic crust, would take place at depths between 40 and 150 km almost continuously, and it seems unlikely that the volcanic front (~ 100 km depth) is the place of initiation of dehydration of the oceanic lithosphere, although there is a possibility that dehydration may take place more efficiently near the volcanic front than elsewhere. Moreover, all the H_2O from the subducted lithosphere may not ascend vertically, but part of it might migrate along the upper surface of the subducted slab for a certain distance. This could also cause the movement of H_2O into the wedge mantle in the non-volcanic zone.

The volume of the Quaternary volcanic materials in the Japanese arcs was estimated by Sugimura et al. (1963). They showed that the volume is largest near the volcanic front and sharply decreases toward the Japan Sea. The fact that larger amounts of volcanic materials are erupted near the volcanic front would indicate that the larger amounts of magmas are formed near the volcanic front.

The temperature distribution beneath the Japanese arcs or the island arcs in general has been estimated by several geophysicists (e.g., Hasebe et al., 1970; Toksöz et al., 1971; Griggs, 1972; Fujii and Kurita, 1978; Hsui and Toksöz, 1979). In all the models, the temperature in the asthenosphere of the mantle wedge increases from the volcanic front toward the Japan Sea. Thus, magmas are formed most abundantly in the coldest part of the mantle wedge in the volcanic zone. To explain this paradox and asymmetry, migration of magma toward the volcanic front is considered here. Okada (1978) has inferred a melt-rich zone just above but parallel to the subducted slab beneath the volcanic zone in the north Honshu. In such a tilted melt-rich zone, melt would migrate upward, and segregate from the upper end of this zone (i.e., beneath the volcanic front). Marsh (1979) has demonstrated experimentally that the largest amount of melt would segregate from the upper end of a tilted melt zone. Probably, the upper end of the melt-rich zone is defined by the wet solidus of peridotite, as discussed above. Magma may segregate from the melt-rich zone as a magma batch or a diapir of crystal-melt mush (Kushiro, 1981, Sakuyama and Kushiro, 1981). Figure 3 is a schematic picture showing the migration of magma in a possible melt-rich

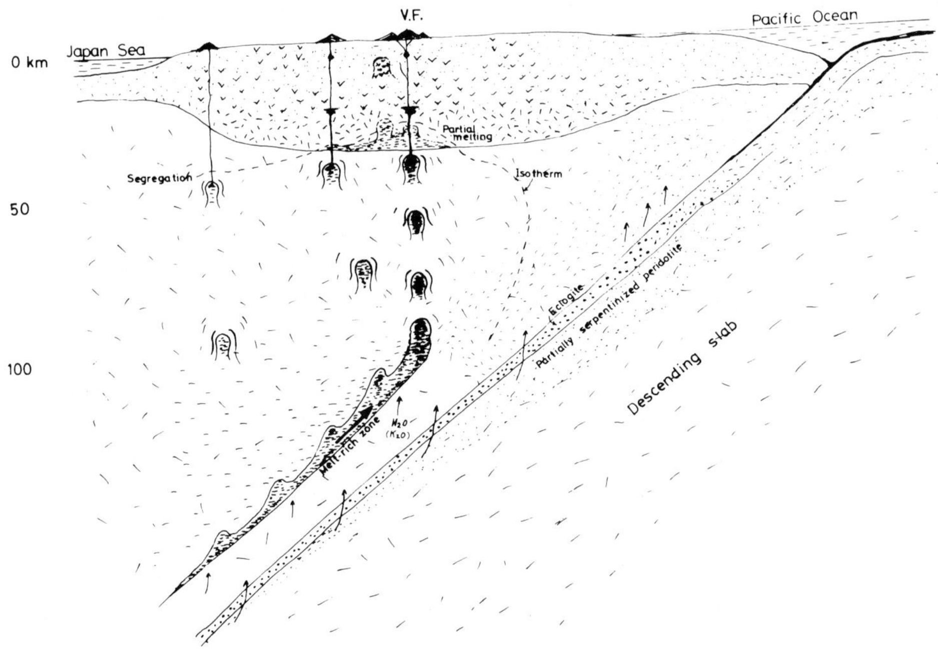

Fig. 3. A schematic picture illustrating migration of melt in a melt-rich layer beneath the volcanic zone in the Japanese arcs, as well as ascent of magmas or diapirs in the wedge mantle. Partial melting of the lower crust to produce rhyolitic magmas is suggested near the volcanic front (VF).

zone above the subduction zone. Above the melt-rich zone, only a small amount of melt would be present, because most of H_2O supplied from the subducted slab would be trapped in the melt-rich zine, although the temperature would be above the wet solidus of peridotite (but below the dry solidus). The magmas or diapirs ascend through the higher temperature region to shallow levels in the wedge mantle, where magmas would segregate.

The depth of segregation of primary magmas would become deeper from the volcanic front toward the Japan Sea. Recent experimental studies by Takahashi (1980) indicate that an alkali basalt from Oki-Dogo Island could be in equilibrium with upper mantle lherzolite at about 15 kbar and 1275°C under dry conditions. In the presence of a small amount of H_2O, the depth of equilibration would become slightly deeper. On the other hand, a magnesian high-alumina basalt from southern Kyushu could be in equilibrium with upper mantle lherzolite at about 10 kbar and 1250°C under dry conditions (Uto and Tatsumi, unpubl. data), and would be slightly deeper in the presence of a small amount of H_2O. Primary tholeiite magmas could be in equilibrium with the upper mantle peridotite also at about 11 kbar under anhydrous conditions (Tatsumi, unpubl. data). As mentioned before, the degree of partial melting would become smaller from the volcanic front toward the Japan Sea. This may partly be due to the increase of the depth of segregation of magma toward the Japan Sea (Kushiro, 1980). Less compressional stress in the upper mantle near the Japan Sea, as mentioned above might cause segregation of magmas at deeper levels, but no entirely satisfactory explanation of the phenomenon is currently available.

ACKNOWLEDGEMENTS

The author thanks Drs. F. Chayes, H. Fukuyama, K. Nakamura and M. Sakuyama for their critical reading of the manuscript.

REFERENCES

Anderson, R.N., Uyeda, S. and Miyashiro, S., 1976. Geophysical and geochemical constraints of converging plate boundaries. Part I: Dehydration in the downgoing slab. J.R. Astron. Soc., 44: 337—357.

Aoki, K., 1971. Petrology of mafic inclusions from Itinome-gata, Japan. Contrib. Mineral. Petrol., 30: 314—331.

Aoki, K., 1978. Petrographical characters of Quaternary volcanic rocks in the Japanese Islands (in Japanese). In: I. Kushiro and S. Aramaki (Editors), Earth Sciences, 3. Material Sciences of the Earth II. Iwanami-Shoten, Tokyo, pp. 153—170.

Birch, F., 1966. Compressibility; elastic constants. In: S.P. Clark (Editor), Handbook of Physical Constants. Geol. Soc. Am., Mem., pp. 97—173.

Bottinga, Y. and Weill, D.F., 1970. Densities of liquid silicate systems calculated from partial molar volumes of oxide components. Am. J. Sci., 269: 169—182.

Burnham, C.W. and Davis, N.F., 1969. Partial molar volume of water in albite melts (abstr.). Trans. Am. Geophys. Union, 50: 338.

Coulon, C. and Thorpe, R.S., 1981. Role of continental crust in petrogenesis of orogenic volcanic associations. Tectonophysics, 77: 79—93.

Daly, R.A., Manger, G.E. and Clark, S.P., 1966. Density of rocks. In: S.P. Clark (Editor), Handbook of Physical Constant. Geol. Soc. Am., Mem., pp. 19—26.

Delany, J.M. and Helgeson, H.C., 1978. Calculation of the thermodynamic consequences of dehydration in subducting oceanic crust to 100 kb and 800°. Am. J. Sci., 278: 638—686.

Fujii, T. and Kushiro, I., 1977. Density, viscosity, and compressibility of basaltic liquid at high pressures. Carnegie Inst. Washington, Yearb., 76: 419—424.

Fukuyama, H. 1978. Petrological Study of Ni, Rb and Sr in Magmas with Special Reference to the Mantle-Magma Equilibrium. Ph.D. Thesis, Univ. of Tokyo.

Fyfe, W.S. and McBirney, A.R., 1975. Subduction and the structure of andesitic volcanic belts. Am. J. Sci., 275-A: 285—297.

Gill, R.C.O., 1973. Mechanism for the salic magma bias of continental alkaline provinces. Nature, 242: 41—42.

Green, D.H. and Ringwood, A.E., 1967. The genesis of basaltic magmas. Contrib. Mineral. Petrol., 15: 103—190.

Griggs, D.T., 1972. The sinking lithosphere and the focal mechanism of deep earthquakes. In: E.C. Robertson (Editor), The Nature of the Solid Earth. McGraw-Hill, New York, N.Y., 667 pp.

Hasebe, K., Fujii, N. and Uyeda, S., 1970. Thermal processes under island arcs. Tectonophysics, 10: 335—355.

Hsui, A.T. and Toksöz, M.N., 1979. The evolution of thermal structures beneath a subduction zone. Tectonophysics, 60: 43—60.

Koto, B., 1916. On the volcanoes of Japan, I, II, III, IV and V. J. Geol. Soc. Tokyo, 23: 1—13, 17—28, 29—56, 77—94 and 95—127.

Koyaguchi, T., 1982. Geology and Petrology of Abu District. M.Sc. Thesis, Univ. of Tokyo.

Kuno, H., 1954a. Volcanoes and Volcanic Rocks (in Japanese). Iwanami-Shoten, Tokyo, 255 pp.

Kuno, H., 1954b. Geology and petrology of Omuro-yama volcano group, North Izu. J. Fac. Sci., Univ. Tokyo, Sec.II, 9: 241—265.

Kuno, H., 1959. Origin of Cenozoic petrographic provinces of Japan and surrounding areas. Bull. Volcanol., Sec.II, 20: 37—76.

Kuno, H., 1960. High-alumina basalt. J. Petrol., 1: 125—145.

Kuno, H., 1962. Catalogue of the Active Volcanoes of the World including Solfatara Fields; Part 11, Japan, Taiwan and Marianas. Int. Assoc. Volcanology.

Kuno, H. and Aoki, K., 1970. Chemistry of ultramafic nodules and their bearing on the origin of basaltic magmas. Phys. Earth Planet. Inter., 3: 273—301.

Kushiro, I., 1980. Changes with pressure of degree of partial melting and K_2O content of liquids in the system Mg_2SiO_4 - $KAlSiO_4$ - SiO_2. Carnegie Inst. Washington, Yearb., 79: 267—271.

Kushiro, I. 1981. Origin of magmas in island arcs: A discussion based on experimental petrology (in Japanese). J. Geol. Soc. Jpn., 87: 769—780.

Kushiro, I., 1982. Density of tholeiite and alkali basalt magmas at high pressures. Carnegie Inst. Washington, Yearb., 81: 305—309.

Kushiro, I., Yoder, H.S. and Mysen, B.O., 1976. Viscosities of basalt and andesite melts at high pressures. J. Geophys. Res., 81: 6351—6356.

Ludwig, W.J., Ewing, J.I., Ewing, M., Murauchi, S., Den, N., Asano, S., Hotta, H., Hayakawa, M., Ichimiya, K. and Noguchi, I., 1966. Sediments and structure of the Japan trench. J. Geophys. Res., 71: 2121—2137.

Marsh, B.D., 1979. Island arc development: some observations, experiments, and speculations. J. Geol., 87: 687—713.

Matsuzawa, T., 1959. On the crustal structure in northeast Japan by explosion seismic observations. Bull. Earthq. Res. Inst., 37: 123—154.

Miyashiro, A., 1974. Volcanic rock series in island arc and active continental margins. Am. J. Sci., 274: 321—355.

Murase, T. and McBirney, A.R., 1973. Properties of some common igneous rocks and their melts at high temperatures. Geol. Soc. Am. Bull., 84: 3563—3592.

Nakamura, K. and Uyeda, S., 1980. Stress gradient in arc—back arc regions and plate subduction. J. Geophys. Res., 85: 6419—6428.

Neumann, E-R, 1980. Petrogenesis of the Oslo region larvikites and associated rocks. J. Petrol., 21: 499—531.

O'Hara, M.J., 1965. Primary magmas and the origin of basalts. Scott. J. Geol., 1: 10—40.

Okada, H., 1978. Fine structure of the descending lithosphere as revealed by 'ScSp' phase. Int. Geodyn. Conf. Tokyo (abstr.), pp 120—121.

Ono, K., 1962. Chemical composition of volcanic rocks in Japan. Geol. Surv. Japan, Publ.

Rhodes, J.M., 1981. Characteristics of primary basaltic magmas. In: Basaltic Volcanism on the Terrestrial Planets. Pergamon Press, New York, N.Y., pp 409—432.

Ringwood, A.E., 1974. The petrological evolution of island arc system. J. Geol. Soc. London, 130: 183—204.

Roeder, P.L. and Emslie, R.F., 1970. Olivine—liquid equilibrium. Contrib. Mineral. Petrol., 29: 275—289.

Sakuyama, M., 1979. Lateral variation of H_2O contents in Quaternary magmas of northeastern Japan. Earth Planet. Sci. Lett., 43: 103—111.

Sakuyama, M., 1983. Petrology of arc volcanic rocks and their origin by mantle diapir. In: S. Aramaki and I. Kushiro (Editors), Arc Volcanism. J. Volcanol. Geotherm. Res., 18: 297—320.

Sakuyama, M. and Kushiro, I., 1981. Subduction zone and volcanic zone (in Japanese). Kagaku, 51: 499—507.

Stolper, E. and Walker, D., 1980. Melt density and average composition of basalt. Contrib Mineral. Petrol., 74: 7—12.

Sugimura, A., 1960. Zonal arrangements of some geophysical and petrological features in Japan and its environs. J. Fac. Sci., Univ. Tokyo, Ser.II, 12: 133—153.

Sugimura, A., Matsuda, T., Chinzei, K. and Nakamura, K., 1963. Quantitative distribution of late Cenozoic volcanic materials in Japan. Bull. Volcanol., Ser.II, 26: 125—140.

Takahashi, E., 1980. Melting relations of an alkali-olivine basalt to 30 kbar, and their bearing on the origin of alkali basalt magmas. Carnegie Inst. Washington, Yearb., 79: 271—276.

Takahashi, E., Kushiro, I. and Tatsumi, Y., 1981. Melting of a peridotite at high pressures and its bearing on island arc magmas. Abstracts, IAVCEI Symposium Arc Volcanism, Tokyo, pp. 365—366.

Tatsumi, Y. and Ishizaka, K., 1982. Magnesian andesite and basalt from Shodo-Shima islands, southwest Japan, and their bearing on the genesis of calc-alkaline andesites. Lithos, 15: 161—172.

Toksöz, M.N. and Hsui, A.T., 1978. Numerical studies of back-arc convection and the formation of marginal basins. Tectonophysics, 50: 177—196.

Toksöz, M.N., Minear, J.W. and Julian, B.R., 1971. Temperature field and geophysical effects of a downgoing slab. J. Geophys. Res., 76: 1113—1138.

Tomita, T., 1935. On the chemical compositions of the Cenozoic alkaline suite of the Circum-Japan Sea region. J. Shanghai Sci. Inst., Sec.2, 1: 227—306.

Uchimizu, M., 1966. Geology and petrology of alkali rocks from Dogo, Oki Islands. J. Fac. Sci., Univ. Tokyo, Sec.II, 16: 85—159.

Wyllie, P.J., 1973. Experimental petrology and global tectonics: A review. Tectonophysics, 17: 189—209.

Yoshii, T., 1979. A detailed cross-section of the deep seismic zone beneath northeastern Honshu, Japan. Tectonophysics, 55: 349—360.

Journal of Volcanology and Geothermal Research, 18 (1983) 449—459
Elsevier Science Publishers B.V., Amsterdam — Printed in The Netherlands

PLEISTOCENE VOLCANIC ROCKS FROM THE FLY-HIGHLANDS PROVINCE OF WESTERN PAPUA NEW GUINEA: A NOTE ON NEW Sr AND Nd ISOTOPIC DATA AND THEIR PETROGENETIC IMPLICATIONS

P.J. HAMILTON[1], R.W. JOHNSON[2], D.E. MACKENZIE[2], and R.K. O'NIONS[1]

[1] *Department of Earth Sciences, University of Cambridge, Downing Street, Cambridge CB2 3EQ (United Kingdom)*

[2] *Bureau of Mineral Resources, P.O. Box 378, Canberra City, A.C.T. 2601 (Australia)*

(Received October 20, 1982)

ABSTRACT

Hamilton, P.J., Johnson, R.W., Mackenzie, D.E. and O'Nions, R.K., 1983. Pleistocene volcanic rocks from the Fly-Highlands province of western Papua New Guinea: a note on new Sr and Nd isotopic data and their petrogenetic implications. In: S. Aramaki and I. Kushiro (Editors), Arc Volcanism. J. Volcanol. Geotherm. Res., 18: 449—459.

RB-Sr and Sm-Nd isotopic and trace-element-abundance values have been determined for 15 mafic and intermediate rocks from six Pleistocene volcanic centres of the Fly-Highlands province. $^{87}Sr/^{86}Sr$ and ϵ_{Nd} values range from 0.70362 to 0.70540, and +1.9 to +5.9, respectively. These new data can be accounted for by contamination of mantle-derived magmas by the continental crust through which the magmas have risen. They do not, however, preclude derivation of some of the Sr and Nd from subducted crust, nor are they inconsistent with Sr and Nd enrichments having taken place by means of mantle metasomatic events. Nevertheless, there is no Benioff zone beneath the Fly-Highlands province (although there is geological evidence for Cretaceous subduction). A preferred interpretation is that uncontaminated, mantle-derived magmas are related to the Pliocene crustal uplift that caused the development of the highlands and which formed in response to a mid-Tertiary continent/island-arc collision.

INTRODUCTION

The provenance of the major-, minor- and trace-element constituents of subduction-related volcanic rocks, although a subject of vigorous research, remains contentious. Possible contributory sources include underthrust basaltic oceanic crust (altered in different degrees) and associated sediments, the wedge of mantle above the downgoing slab, and the crust of the arc-trench system itself. These sources are potentially distinguishable by means of their different element ratios and isotopic compositions. A combined Nd and Sr isotopic approach is adopted in this paper in an attempt to assess the

0377-0273/83/$03.00

relative roles of potential source materials for mafic and intermediate volcanic rocks from the Fly-Highlands province of western Papua New Guinea (Fig. 1).

PREVIOUS WORK

Details of the volcanic geology, tectonics, and geochemistry of the Fly-Highlands province are reported elsewhere (Mackenzie, 1976; Johnson et al., 1978a; Johnson, 1979; Mackenzie and Johnson, in press), and only a brief description is required here. Mainly Pleistocene volcanic rocks overlie up to 10 km of Mesozoic and Tertiary sediments which rest on 20—30 km of Palaeozoic continental crust near the northern edge of the Australian craton (Fig. 1). A wide range of Pleistocene volcanic rocks has been identified — from alkali basalt, basanite, and tholeiitic basalt (olivine and quartz normative) to trachybasalt and andesite (mainly low-Si). All analysed samples have high concentrations of several incompatible elements relative to those for ocean-floor and many island-arc volcanic rocks. Crustal inclusions are found in some Fly-Highlands lava flows, and particularly in those of more fractionated (low Mg-no.) composition.

The magmas of the different eruptive centres do not appear to be genetically related to one another in any simple way. Rather, each eruptive centre is characterised by different phenocryst assemblages, proportions of rock types, chemical differentiation trends, and Sr-isotope compositions (Page

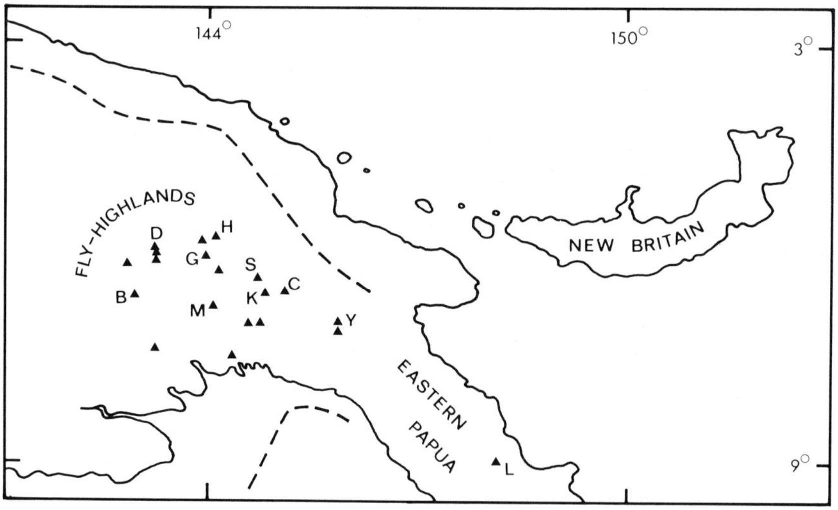

Fig. 1. Distribution of Pleistocene volcanoes of the Fly-Highlands province, Papua New Guinea. Dashed line refers to postulated edge of Palaeozoic continental crust. Letters represent volcanic centres (triangles) referred to in this study: *B*=Bosavi; *C*=Crater Mountain; *D*=Doma Peaks; *G*=Giluwe; *H*=Hagen; *K*=Karimui; *L*=Lamington; *M*=Murray; *S*=Suaru; *Y*=Yelia.

and Johnson, 1974; Mackenzie, 1976). Some general geochemical features of the province can, however, be recognised. For example, incompatible elements in the more mafic rocks generally increase northwards towards the edge of the Palaeozoic crust (Mackenzie, 1976). Further, a slight negative correlation between Sr contents and $^{87}Sr/^{86}Sr$ values is evident for the samples from Mounts Hagen, Suaru, and Karimui, and Crater Mountain measured by Page and Johson (1974). However, Mackenzie (1980) observed no such correlation where additional data (including unpublished Sr-isotope ratios supplied by R.W. Page) from Mounts Hagen, Giluwe, Murray, Bosavi, Doma Peaks, and several other centres in the province were included in the plot. Johnson et al., (1978a, fig. 9) also noted that Rb/Sr and $^{87}Sr/^{86}Sr$ values are positively correlated for the most mafic samples from volcanoes in the north of the province, nearest a zone of maximum Pliocene uplift.

RESULTS

Samples and methods

Fifteen fresh rock samples (Table I) from six Fly-Highlands volcanoes (Fig. 1) have been analysed for Nd and Sr-isotope ratios at the University of Cambridge. The selected samples cover a broad spectrum of compositions from basanite to high-Si andesite (Table I), and most have been analysed for rare-earth elements (REE; Mackenzie and Johnson, in press). Some were also used previously for the Sr-isotopic assessment that was included in the study by Johnson et al. (1978a).

The Nd and Sr-isotope ratios (Table II) were measured using an automated Vacuum Generators Micromass 54E mass-spectrometer using techniques previously described by Hooker et al. (1981). Rb, Sr, Sm, and Nd concentrations (Table II) were analysed by isotope dilution (Cambridge) and X-ray-fluorescence spectrometry (Department of Geology, Australian National University, Canberra) and, where duplicated, are in agreement within the limits of analytical error.

Trace elements

Mackenzie and Chappell (1972) and Mackenzie (1976) demonstrated that the abundances of several incompatible trace elements (e.g. Rb, Sr, Sm, Nd; Table I) are in general greater than those in rocks containing similar amounts of SiO_2 from many island arcs, such as the Aleutians (e.g. McCulloch and Perfit, 1981), the Japanese Islands (e.g. Nohda and Wasserburg, 1981), and New Britain (e.g. DePaolo and Johnson, 1979). However, some island-arc volcanic rocks — for example, from northern Java (Whitford et al., 1981) and the Aeolian Islands (Keller, 1974) — and some continental-margin rocks, such as those from parts of the Andes (Francis et al., 1977; Hawkesworth et al., 1979a), have similar or even greater abundances of these elements.

TABLE I

Fifteen Fly-Highlands volcanic-rock samples

Volcano and sample number[a]	Rock name[b]	Mg-no.[c]
Mount Hagen		
1741	high-K hy-trachybasalt	77
0600A	high-K olivine-tholeiitic basalt	69
1607	low-K, low-Si andesite	68
1786	low-K, high-Si andesite	50
Mount Giluwe		
32	high-K hy-trachybasalt	65
56	high-K, low-Si andesite	69
79	high-K, low-Si andesite	70
35	high-K, low-Si andesite	53
Doma Peaks		
0012	high-K, low-Si andesite	64
0212	high-K, high-Si andesite	56
Mount Bosavi		
0002D	alkali basalt	73
0004	olivine-tholeiitic basalt	65
Mount Murray		
5030E	olivine-tholeiitic basalt	66
5030G	high-K, low-Si andesite	45
Mount Suaru		
2108	basanite	63

[a]Sample numbers are last 4 of 5 digits and letters of BMR registered numbers, except for the Mount Giluwe samples.
[b]Rock nomenclature based on normative scheme of Johnson et al. (1978b). Fe_2O_3 values of whole-rock analyses set = 0.2 (total-Fe-as Fe_2O_3).
[c]Mg.no. = 100 Mg/Mg+Fe^{2+}), where Fe^{2+} derives from oxidation state used in (b).

Chondrite-normalised REE patterns are light-REE enriched; many samples have chondrite-normalised La/Yb values in the range 5—15 (e.g. Johnson et al., 1978a, Fig. 9). Plagioclase, which concentrates Eu, is a common phenocryst in the samples, but Eu anomalies are absent or small (Eu/Eu* between 0.9 and 1.1), suggesting either that Eu was present mainly as Eu^{3+}, or that plagioclase fractionation and accumulation was limited, or both. However, Sr/Nd values are relatively high, and chondrite-normalised trace-element diagrams are characterised by positive Sr (and Ba) anomalies.

Sr isotopes

$^{87}Sr/^{86}Sr$ values for the 15 Fly-Highlands samples range from 0.70362 ± 6 for basanite 2108 from Mount Suaru, to 0.70540 ± 6 for olivine-tholeiitic-

TABLE II

Rb-Sr and Sr-Nd data

	Rb[a]	Sr	Sm	Nd	Sr/Nd	$^{87}Rb/^{86}Sr$	$^{147}Sm/^{144}Nd$	$^{87}Sr/^{86}Sr \pm 2\sigma_m$	$\epsilon_{Nd} \pm 2\sigma_m$[b]
Mt Hagen									
1741	70.4(74)	561(560)	3.32(3.3)	14.1(14)	40	0.363	0.142	0.70428 ± 5	3.9 ± 0.4
0600A	67.2(69)	622(616)	3.57(3.9)	15.0(16)	41	0.312	0.144	0.70422 ± 5	3.1 ± 0.6
1607	29.4(30)	642(610)	3.67(3.7)	17.3(15)	37	0.132	0.128	0.70427 ± 6	3.1 ± 0.4
1786	(64)	(934)	(3.4)	(19)	49	0.198	0.108	0.70432 ± 7	3.4 ± 0.4
Mt Giluwe									
32	81.9(86)	715(720)	3.75(4.1)	15.7(17)	46	0.331	0.144	0.70414 ± 5	2.5 ± 0.3
56	62.4(66)	721(715)	3.88(4.2)	19.0(20)	38	0.250	0.123	0.70422 ± 5	2.8 ± 0.3
79	101 (108)	707(712)	3.57	17.4	41	0.413	0.124	0.70423 ± 5	2.6 ± 0.5
35	(82)	(970)	(4.1)	(19)	51	0.245	0.130	0.70423 ± 8	4.5 ± 0.3
Doma Peaks									
0012	(69)	(889)	(5.0)	(26)	34	0.224	0.116	0.70464 ± 4	4.2 ± 0.4
0212	(92)	(917)	(4.5)	(34)	27	0.290	0.080	0.70441 ± 4	3.8 ± 0.4
Mt Bosavi									
0002D	(29.0)	(673)	(5.0)	(22)	31	0.121	0.137	0.70470 ± 3	3.8 ± 0.5
0004	(39.5)	(689)	(4.7)	(27)	26	0.166	0.105	0.70540 ± 6	1.9 ± 0.4
Mt Murray									
5030E	(38.0)	(636)	(3.8)	(16)	40	0.173	0.144	0.70432 ± 5	3.1 ± 0.5
5030G	(81)	(787)	(5.7)	(35)	22	0.298	0.098	0.70462 ± 7	2.0 ± 0.5
Mt Suaru									
2108	52.2	1358	6.00	30.2	45	0.111	0.120	0.70362 ± 6	5.9 ± 0.5

[a] Concentrations given in ppm. Values in parenthesis obtained by X-ray-fluorescence spectrometry. Other values obtained by isotope dilution (±0.5% precision).

[b] $\epsilon_{Nd} = [[(^{143}Nd/^{144}Nd) / 0.512638] - 1] \times 10^4$

Average values for isotopic standards are currently $^{143}Nd/^{144}Nd = 0.512664 \pm 0.000007$ (2σ) for USGS standard rock BCR-1 and $^{87}Sr/^{86}Sr = 0.71025 \pm 0.00002$ for NBS987 $SrCO_3$.

basalt 0004 from Mount Bosavi (Table II). Sr contents and $^{87}Sr/^{86}Sr$ values are similar to those reported by Page and Johnson (1974) and are similarly negatively correlated. The correlation is, however, crude, and if the data for samples 2108 and 0004 are eliminated the remaining 13 sample points form only a cluster (Fig. 2).

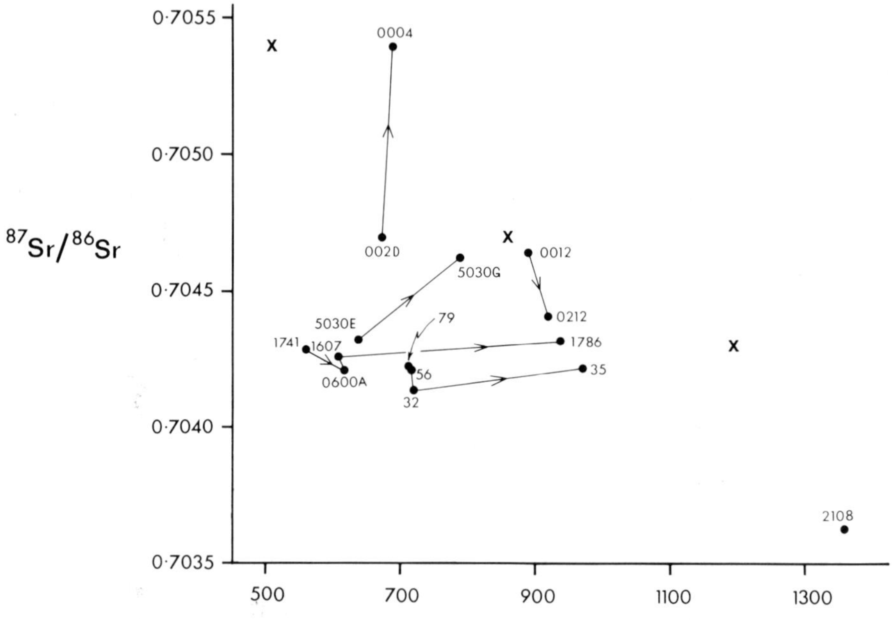

Fig. 2. $^{87}Sr/^{86}Sr$ versus Sr values for Fly-Highlands samples (dots). Data for eastern Papua samples (crosses) are from Page and Johnson (1974). Arrows in this diagram and in Fig. 3 connect data points for rocks from the same volcano in order of decreasing Mg-no. (Table I), but the tie lines do not necessarily represent liquid lines of descent (simple crystal fractionation) or linear contamination trends.

Sr contents and $^{87}Sr/^{86}Sr$ values for three eastern Papua volcanic rocks (Page and Johnson, 1974) are also negatively correlated, and the $^{87}Sr/^{86}Sr$ values are similar to those of the Fly-Highlands samples (Fig. 2). This similarity in Sr-isotope geochemistry also extends to the positive correlation between Rb/Sr and $^{87}Sr/^{86}Sr$ values for mafic samples (Johnson et al., 1978a, fig. 9).

Nd isotopes

$^{143}Nd/^{144}Nd$ values expressed as deviations from the assumed bulk-Earth value in parts per 10^4 (ϵ_{Nd}) range from +5.9 for sample 2108 to +1.9 for sample 0004. The Nd and Sr-isotopic ratios are negatively correlated — a feature found in many other rock suites (Figs. 3 and 4) — but, again, the correlation is non-existent if data for samples 2108 and 0004 are excluded. There is no correlation between ϵ_{Nd} and $^{147}Sm/^{144}Nd$ values.

ϵ_{Nd} and $^{87}Sr/^{86}Sr$ values for about half of the Fly-Highlands samples fall within the mantle array defined by oceanic basalts, whereas those for the other half fall on the high-$^{87}Sr/^{86}Sr$ side of the array, apparently trending to-

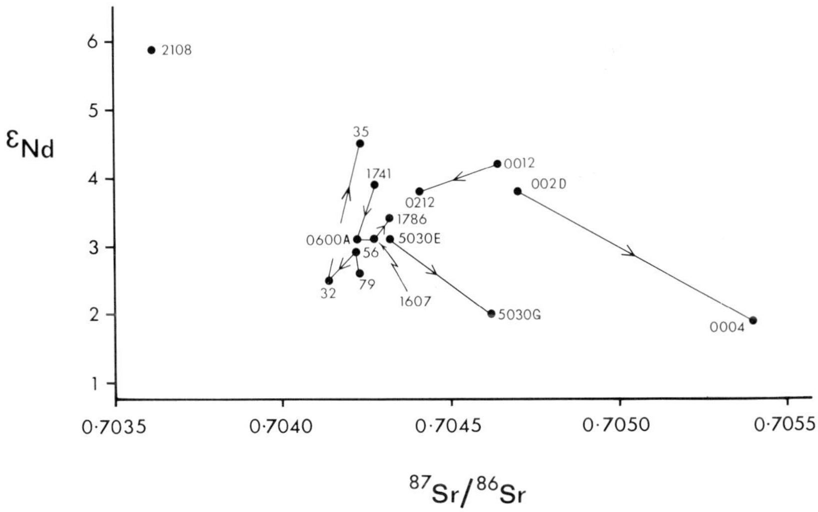

Fig. 3. ϵ_{Nd} versus $^{87}Sr/^{86}Sr$ values for Fly-Highlands samples (dots). See Fig. 2 for note on significance of arrows.

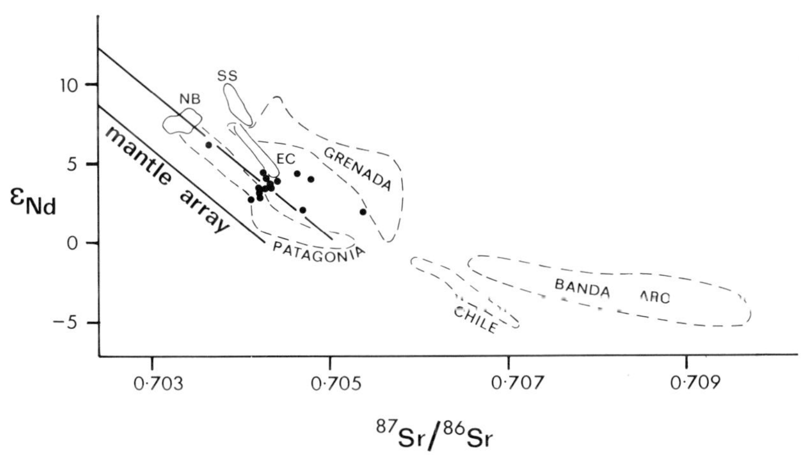

Fig. 4. ϵ_{Nd} versus $^{87}Sr/^{86}Sr$ values for Fly-Highlands samples (Fig.3) compared with data fields for other arc-trench-type areas. Small fields NB, SS, and EC refer to New Britain, South Sandwich Islands, and Ecuador, respectively. Data are from Whitford et al. (1981), DePaolo and Johnson (1979), Hawkesworth et al. (1977, 1979a, b), Hawkesworth and Powell (1980), and Francis et al. (1977).

wards upper continental crust compositions. The ranges of Nd and Sr-isotope compositions are similar to those of some Andean volcanic rocks (Fig. 4).

DISCUSSION

Volcanic rocks from several subduction-related suites have been shown to have Nd and Sr-isotope values that plot to the right of the ϵ_{Nd}-$^{87}Sr/^{86}Sr$ mantle array (Fig. 4). For example, Hawkesworth et al. (1977, 1979a) showed that South Sandwich Islands and Andean rocks have compositions that fall on the high-$^{87}Sr/^{86}Sr$ side of the array. Some authors (e.g. DePaolo and Wasserburg, 1977; Hawkesworth et al., 1977) have identified such shifts as the contribution of continent-derived, seawater Sr (high $^{87}Sr/^{86}Sr$) to the magmas. Further, in areas such as the Lesser Antilles (Hawkesworth et al., 1979b; Hawkesworth and Powell, 1980), the Banda arc (Whitford et al., 1981), and Japan (Nohda and Wasserburg, 1981), where an accompanying shift in ϵ_{Nd} is observed, contamination by subducted crust or sediments (or both) has been proposed as the most likely explanation. However, in many instances the mechanism whereby such material was incorporated — by means of subduction of oceanic crust and sediments, or contamination by the crust of the arc itself — is not immediately apparent.

Sr/Nd values for Fly-Highlands rocks are high, ranging from 22 to 51 (Table II) and, in common with volcanic rocks from New Britain and other island arcs, are higher than in oceanic basalts (DePaolo and Johnson, 1979). However, high Sr/Nd values are also characteristic of some continental flood lavas, basaltic lower crustal materials, and intra-continental peridotite nodules (e.g. Arculus and Johnson, 1981). Thus, although the high values may relate to a component derived from subducted crust, an additional possibility to be considered in the case of the high Sr/Nd values of Fly-Highlands rocks is contamination of magmas by the underlying continental crust, or redistribution in the upper mantle of Sr relative to Nd by metasomatic processes unrelated to subduction.

There is no current intermediate-depth seismicity that defines a Benioff zone beneath the Fly-Highlands province (except in the extreme east beneath Mount Yelia, which, however, occupies an anomalous tectonic setting relative to that of the volcanoes farther west; e.g. Johnson, 1979). Thus, there is considerable doubt that the Fly-Highlands volcanism is directly related to late Cainozoic subduction (Mackenzie and Chappell, 1972; Johnson et al., 1978a). However, because Cretaceous arc-trench-type rocks are known in the Highlands, Johnson et al. (1978a) proposed that a subduction-related enrichment event about this time may have chemically modified magma source regions beneath the Fly-Highlands area, and that magma generation from these enriched sources was related to crustal uplift in the Pliocene caused by a mid-Tertiary continent/island-arc collision (Johnson and Jaques, 1980). The positive correlation between Rb/Sr and $^{87}Sr/^{86}Sr$ mentioned above for mafic Fly-Highlands samples was regarded as having possible temporal significance because the 100 ± 80 m.y. 'age' of the pseudoisochron was late Mesozoic to Paleogene. However, the error for the age is large, and no comparable chronological information is available from any other decay scheme (there is no correlation between the ϵ_{Nd} and $^{147}Sm/^{144}Nd$ values given in Table II), so critical evaluation of this interpretation is difficult. Further, the Rb-Sr data for Mounts Murray and Bosavi do not accord with those defining the pseudoisochron.

The simplest, but not necessarily the only, explanation for the Rb-Sr and Nd-isotopic and trace-element data presented here is that magmas unrelated to subduction of oceanic crust have interacted with the crust underlying the Fly-Highlands province. This explanation is to some extent supported by the occurrence of crustal inclusions in the volcanic rocks, although the petrographic evidence for reaction and digestion is not particularly strong. Arculus et al. (1983, this volume) provide mineralogical evidence that supports a concept of crustal contamination of the andesitic and trachybasaltic magmas of Mount Lamington, eastern Papua (Fig. 1).

The inverse relationships between Sr and $^{87}Sr/^{86}Sr$ (Fig. 2) and ϵ_{Nd} and $^{87}Sr/^{86}Sr$ (Figs. 3 and 4) are so poor that they cannot be taken as evidence for two-component (mantle-crust) mixing. These poor correlations may indicate that neither the uncontaminated magmas nor the crustal contaminants were chemically or isotopically homogeneous. Note, in particular, in Fig. 2 (and Tables I and II) that a group of lower-Sr and lower-$^{87}Sr/^{86}Sr$ values corresponds to relatively unfractionated, mostly mafic rocks from Mounts Hagen, Giluwe, and Murray, whereas a group of higher values corresponds to andesites from these same centres, as well as from Doma Peaks, and to basalts from Mount Bosavi (the Mount Suaru basanite could belong to either group). Single-volcano tie-lines, from high to low-Mg-no. rocks, in Fig. 2 have zero or positive slopes, and are evidence that if the differences in $^{87}Sr/^{86}Sr$ or Sr, or both, are because of crustal contamination, then the contaminating material must have been chemically and isotopically heterogeneous. The contaminant beneath Mounts Hagen and Giluwe could have been young crust that has similar Sr-isotopic values to those of the magmas. Magmas of Mounts Murray and Bosavi, on the other hand, may have been contaminated by older, upper crust that is more and less enriched in radiogenic Sr and Nd, respectively (see, especially, Fig. 3).

Crustal contamination appears to have contributed to the incompatible-element-enriched composition of the Fly-Highlands rocks, but the crust is unlikely to have been the only source of these elements. This is because the group of more fractionated (low Mg-no.) rocks identified above as the more contaminated (higher Sr, generally higher $^{87}Sr/^{86}Sr$), have abundances for some incompatible elements that are lower than those of the first group from the same volcano. Further, all of the rocks that plot within the mantle array in Fig. 4 are incompatible-element enriched. These include mafic, near primary samples whose bulk compositions preclude digestion of large proportions of differentiated crustal material; they do not resemble magmas that stagnated in reservoirs for long periods during which fractionation and contamination could take place.

CONCLUSIONS

Differences in Nd and Sr isotopic values and element abundances for 15 Fly-Highlands rock samples are consistent with contamination of mantle-derived, fractionated magmas by the continental crust that underlies this Pleistocene volcanic province. Both the pristine magmas and the crustal contaminants were probably isotopically and chemically heterogeneous. There is no Benioff zone beneath the Fly-Highlands area, and contemporaneous subduction of oceanic lithosphere appears to have played no part in the generation of the magmas, although an earlier period of subduction (perhaps in the Cretaceous) could have chemically modified upper-mantle source regions beneath the province. The Nd and Sr data are not inconsistent with the view that some Nd and Sr was derived from a subduction zone, but the simplest interpretation is contamination of magmas by the crust that underlies the Fly-Highlands province itself. The uncontaminated magmas are considered to have formed in response to a continent-arc collision that began in the mid-Tertiary, and to be related to the Pliocene crustal uplift that caused the development of the highlands.

ACKNOWLEDGEMENTS

This research was supported by a grant from the National Environment Research Council and the Royal Society of London. The paper is Department of Earth Sciences (University of Cambridge) Contribution No. ES297. R.W.J. and D.E.M. publish with the permission of the Director of the Bureau of Mineral Resources, and gratefully acknowledge the cooperation and interest of Dr B.W. Chappell who allowed access to the X-ray-fluorescence spectrometry facilities.

REFERENCES

Arculus, R.J. and Johnson, R.W., 1981. Island-arc magma sources: a geochemical assessment of the roles of slab-derived components and crustal contamination. Geochem. J., 15: 109—133.

Arculus, R.J., Johnson, R.W., Chappell, B.W., McKee, C.O. and Sakai, H., 1983. Ophiolite-contaminated andesites, trachybasalts, and cognate inclusions of Mount Lamington, Papua New Guinea: anhydrite-amphibole-bearing lavas and the 1951 cumulodome. In: S. Aramaki and I. Kushiro (Editors), Arc Volcanism. J. Volcanol. Geotherm. Res., 18: 215—247.

DePaolo, D.P. and Johnson, R.W., 1979. Magma genesis in the New Britain island arc: constraints from Nd and Sr isotopes and trace-element patterns. Contrib. Mineral. Petrol., 70: 367—379.

DePaolo, D.P. and Wasserburg, G.J., 1977. The sources of island arcs as indicated by Nd and Sr isotopic studies. Geophys. Res. Lett., 4: 465—474.

Francis, P.W., Moorbath, S. and Thorpe, R.S., 1977. Strontium isotope data for recent andesites in Ecuador and North Chile. Earth Planet. Sci. Lett., 37: 197—202.

Hawkesworth, C.J. and Powell, M., 1980. Magma genesis in the Lesser Antilles island arc. Earth Planet. Sci. Lett., 51: 297—308.

Hawkesworth, C.J., O'Nions, R.K., Pankhurst, R.J., Hamilton, P.J. and Evensen, N.M., 1977. A geochemical study of island arc and back arc tholeiites from the Scotia Sea. Earth Planet. Sci. Lett., 36: 253—262.

Hawkesworth, C.J., Norry, M.J., Roddick, J.C., Baker, P.E., Francis, P.W. and Thorpe, R.S., 1979a. $^{143}Nd/^{144}Nd$, $^{87}Sr/^{86}Sr$, and incompatible element variations in calc-alkaline andesites and plateau lavas from South America. Earth Planet. Sci. Lett., 42: 45—57.

Hawkesworth, C.J., O'Nions, R.K. and Arculus, R.J., 1979b. Nd and Sr isotope geochemistry of island arc volcanics, Grenada, Lesser Antilles. Earth Planet. Sci. Lett., 45: 237—248.

Hooker, P.J., Hamilton, P.J. and O'Nions, R.K., 1981. An estimate of the Nd isotope composition of Iapetus seawater from ca 490 Ma metalliferous sediments. Earth Planet. Sci. Lett., 56: 180—188.

Johnson, R.W., 1979. Geotectonics and volcanism in Papua New Guinea: a review of the late Cainozoic. BMR J. Aust. Geol. Geophys., 4: 181—207.

Johnson, R.W. and Jaques, A.L., 1980. Continent-arc collision and reversal of arc polarity: new interpretations from a critical area. Tectonophysics, 63: 111—124.

Johnson, R.W., Mackenzie, D.E. and Smith, I.E.M., 1978a. Delayed partial melting of subduction modified mantle in Papua New Guinea. Tectonophysics, 46: 197—216.

Johnson, R.W., Mackenzie, D.E. and Smith, I.E.M., 1978b. Volcanic rock associations at convergent plate boundaries: reappraisal of the concept using case histories from Papua New Guinea. Geol. Soc. Am. Bull., 89: 96—106.

Keller, J., 1974. Petrology of some volcanic rock series of the Aeolian Arc, Southern Tyrrhenian Sea: calc-alkaline and shoshonitic associations. Contrib. Mineral. Petrol., 46: 29—47.

Mackenzie, D.E., 1976. Nature and origin of late Cainozoic volcanoes in western Papua New Guinea. In: R.W. Johnson (Editor), Volcanism in Australasia. Elsevier, Amsterdam, 221—238.

Mackenzie, D.E., 1980. Volcanoes of the Papua New Guinea Highlands: Volcanism on a Continental Margin and its Relationship to Subduction. PhD Thesis, University of Melbourne (unpubl.).

Mackenzie, D.E. and Chappell, B.W., 1972. Shoshonitic and calc-alkaline lavas from the Highlands of Papua New Guinea. Contrib. Mineral. Petrol., 35: 50—62.

Mackenzie, D.E. and Johnson, R.W., in press. Pleistocene volcanoes of the western Papua New Guinea highlands: morphology, geology, petrography, and modal and chemical analyses. Bur. Miner. Res. Aust. Rep. 146.

McCulloch, M.T. and Perfit, M.R., 1981. $^{143}Nd/^{144}Nd$, $^{87}Sr/^{86}Sr$ and trace element constraints on the petrogenesis of Aleutian island arc magmas. Earth Planet. Sci. Lett., 56: 167—179.

Nohda, S. and Wasserburg, G.J., 1981. Nd and Sr isotopic study of volcanic rocks from Japan. Earth Planet. Sci. Lett., 52: 264—276.

Page, R.W. and Johnson, R.W., 1974. Strontium isotope ratios of Quaternary volcanic rocks from Papua New Guinea. Lithos, 7: 91—100.

Whitford, D.J., White, W.M. and Jezek, P.A., 1981. Neodymium isotopic composition of Quaternary island arc lavas from Indonesia. Geochim. Cosmochim. Acta, 45: 989—995.

Journal of Volcanology and Geothermal Research, 18 (1983) 461—482
Elsevier Science Publishers B.V., Amsterdam — Printed in The Netherlands

TRACE-ELEMENT AND ISOTOPIC CONSTRAINTS ON THE SOURCE OF MAGMAS IN THE ACTIVE VOLCANO AND MARIANA ISLAND ARCS, WESTERN PACIFIC

ROBERT JAMES STERN and EMI ITO

Programs in Geosciences, The University of Texas at Dallas, Richardson, TX 75080 (U.S.A.)

Department of Geology and Geophysics, University of Minnesota, Minneapolis, MN 55455 (U.S.A.)

(Received October 20, 1982)

ABSTRACT

Stern, R.J. and Ito, E., 1983. Trace-element and isotopic constraints on the source of magmas in the active Volcano and Mariana island arcs, Western Pacific. In: S. Aramaki and I. Kushiro (Editors), Arc Volcanism. J. Volcanol. Geotherm. Res., 18: 461—482.

Analytical results of the relative and absolute abundance of LIL-incompatible trace elements (K, Rb, Cs, Sr, and Ba) and isotopic compositions ($^{18}O/^{16}O$, $^{87}Sr/^{86}Sr$, and $^{143}Nd/^{144}Nd$) are summarized for fresh samples from active and dormant volcanoes of the Volcano and Mariana island arcs. The presence of thickened oceanic crust ($T \sim 15$—20 km) beneath the arc indicates that while hybridization processes resulting in the modification of primitive magmas by anatectic mixing at shallow crustal levels cannot be neglected, the extent and effects of these processes on this arc's magmas are minimized. All components of the subducted plate disappear at the trench. This observation is used to reconstruct the composition of the crust in the Wadati-Benioff zone by estimating proportions of various lithologies in the crust of the subducted plate coupled with analyses from DSDP sites. Over 90% of the mass of the subducted crust consists of basaltic Layers II and III. Sediments and seamounts, containing the bulk of the incompatible elements, make up the rest. Bulk Western Pacific seafloor has $^{87}Sr/^{86}Sr \sim 0.7032$, $\delta^{18}O \sim +7.2$, K/Rb ~ 510, K/Ba ~ 46, and K/Cs ~ 13,500. Consideration of trace-element data and combined $\delta^{18}O$ - $^{87}Sr/^{86}Sr$ systematics limits the participation of sediments in magmagenesis to less than 1%, in accord with the earlier results of Pb-isotopic studies. Combined $^{143}Nd/^{144}Nd$ - $^{87}Sr/^{86}Sr$ data indicate little, if any, involvement of altered basaltic seafloor in magmagenesis. Perhaps more important than mean isotopic and LIL-element ratios is the restricted range for lavas from along over 1000 km of this arc. Mixtures of mantle with either the subducted crust or derivative fluids should result in strong heterogeneities in the sources of individual volcanoes along the arc. Such heterogeneities would be due to: (1) gross variations of crustal materials supplied to the subduction zone; and (2) lesser efficiency of mixing processes accompanying induced convection between arc segments (parallel to the arc) as compared to that perpendicular to the arc. The absence of these heterogeneities indicates that either some process exists for the efficient mixing of mantle and subducted material parallel to the arc or that subducted materials play a negligible role in the generation of Mariana-Volcano arc melts.

U.T.D. Contribution Number 421.

0377-0273/83/$03.00

Consideration of plausible sources in the mantle indicates that (1) an unmodified MORB-like mantle cannot have generated the observed trace-element and isotopic composition of this arc's magmas, while (2) a mantle similar to that which has produced alkali-olivine basalts (AOB) of north Pacific "hot spot" chains is indistinguishable in many respects from the source of these arc lavas.

INTRODUCTION

The origin and evolution of magmas at convergent plate boundaries are among the most intriguing of all petrologic problems. Recycling of crustal materials back into the Earth's mantle is accepted as a critical corollary of plate tectonic axioms, and appears to be the only mechanism capable of impeding the irreversible differentiation of the earth (Hofmann and White, 1980). It follows that our understanding of the evolution of the Earth's crust and mantle depends on how well we know especially the following two aspects of the subduction process: (1) Over what portion of the earth's history has such activity been important? and (2) What are the relative proportion and absolute masses of material returned to the mantle compared with that distilled back into the crust as a result of arc magmatism? The first of these questions is beyond the scope of this paper. Results of our isotopic and geochemical studies in the western Pacific, however, bear on the second.

In spite of two decades of field, petrologic, and geochemical investigations into problems of arc magmagenesis, none of the present generalized models have proven entirely acceptable. Considering the wide range in the composition of arc volcanic rocks, this is to be expected. Following the complicated interactions resulting from the differing thermal regimes, tectonic styles, and compositional variations present in subducted crusts and overlying mantle wedges of the Earth's convergent plate boundaries, this should result in a similar range in the processes and products of melting. Recognition of the potential independence of many dynamic processes among arcs thus suggests that the study of each arc should be based on the evidence from that arc alone instead of attempting to apply conclusions reached from studies of other arcs. The story that will emerge after integrating such studies will doubtless be complex, but should more accurately reflect reality.

It is in this spirit that we outline trace element and isotopic constraints on magmagenesis in two related arc systems, the Mariana and Volcano arcs. We will argue in the next section that these arcs are built on primitive crust and that melts rising through this crust should not be contaminated. Concentration of the major elements and "compatible" trace elements (i.e. trace elements with $D^{xl/liq.} \geqslant 0.1$) will vary in response to crystal-liquid equilibrium in the melt zone as well as re-equilibration during ascent and storage in magma chambers; unless the ascent rate approaches the adiabat, the abundance of these elements in eruptive rocks will reveal little about the composition of the source. Concentrations of the large ion lithophile (LIL) incompatible elements ($D^{xl/liq.} < .1$; e.g. K, Rb, Cs, Ba, and to a lesser extent, Sr) will also vary as a function of the fractionation history of the magmas. However, insofar as mica and amphibole are not residual, the *ratios* of LIL incompatible elements K/Rb, K/Ba, K/Cs, etc., should change little during moderate degrees of fractionation. These phases should be entirely consumed during moderate degrees of melting and are very rare as phenocrysts in Volcano and Mariana arc lavas. Isotopic compositions of especially the heavier elements Sr, Nd, and Pb also should not change during fractionation, and the isotopic composition of O (expressed as $\delta^{18}O$)[1] should also be unaffected at magmatic temperatures. A recent paper by Kyser et al. (1981) suggests that oxygen may fractionate at magmatic temperatures, but we consider their evidence to be inconclusive. Thus we contend that LIL incompatible element ratios K/Rb, K/Cs, and K/Ba as well as isotopic ratios of O, Sr, Nd, and Pb in Mariana-Volcano arc lavas should be very similar to that of the source. In the succeeding section, we argue that we can outline the incompatible element and isotopic characteristics of sources in the subducted crust and mantle wedge. By comparing these with the isotopic and trace-element compositions in fresh arc lavas, we will attempt to help resolve the source of melts in this system.

GEOLOGIC SETTING: THE ROLE OF THE CRUST IN THE OVER-RIDING PLATE

The Mariana and Volcano arcs are southern parts of the volcanic chain that stretches over 2000 km from Tokyo to Guam (Fig. 1). The Wadati-Benioff zone dips 45° in the north to depths of ~ 500 km. Beneath the Marianas, the seismic zone dips ~ 30° to 150 km, then plunges vertically to over 600 km. Beneath the Volcano arc seismicity is diffuse and a well-defined Wadati-Benioff zone does not exist. Seismicity beneath the Volcano arc is concentrated at depths shallower than 100 km (Katsumata and Sykes, 1969). This chain is built on oceanic crust as a result of the change in plate motion of the Pacific plate ca. 40 m.y. ago which converted a transform fault connecting segments of the Kula Pacific Ridge into the site of west-dipping subduction (Uyeda and Miyashiro, 1974). Subsequent back-arc rifting led to the development of the Parece Vela Basin in the Upper Oligocene and Lower Miocene and the Mariana Trough since the Pliocene (Crawford et al., 1981).

The evolution of the Mariana-Volcano arc system thus reconstructed indicates it is built on Lower Tertiary oceanic crust. Geophysical surveys indicate this crust has since been thickened to 18—22 km (Murauchi et al., 1968; Sager, 1980). This is an increase of 12—16 km over the thickness of normal oceanic crust (~ 5—6 km). Assuming thickening occurred continuously over the arc's history, this represents a thickening rate of 300—400 m/million years (~ 0.01 mm/yr). Since periods of little igneous activity in the arc are inferred for much of the Oligocene, Upper Miocene, and Pliocene (Crawford et al., 1981), this estimate of thickening rate is a minimum. Thickening may

[1] $\delta^{18}O = (R/R_{SMOW} - 1) \times 1000; R = {}^{18}O/{}^{16}O$; SMOW = standard mean ocean water

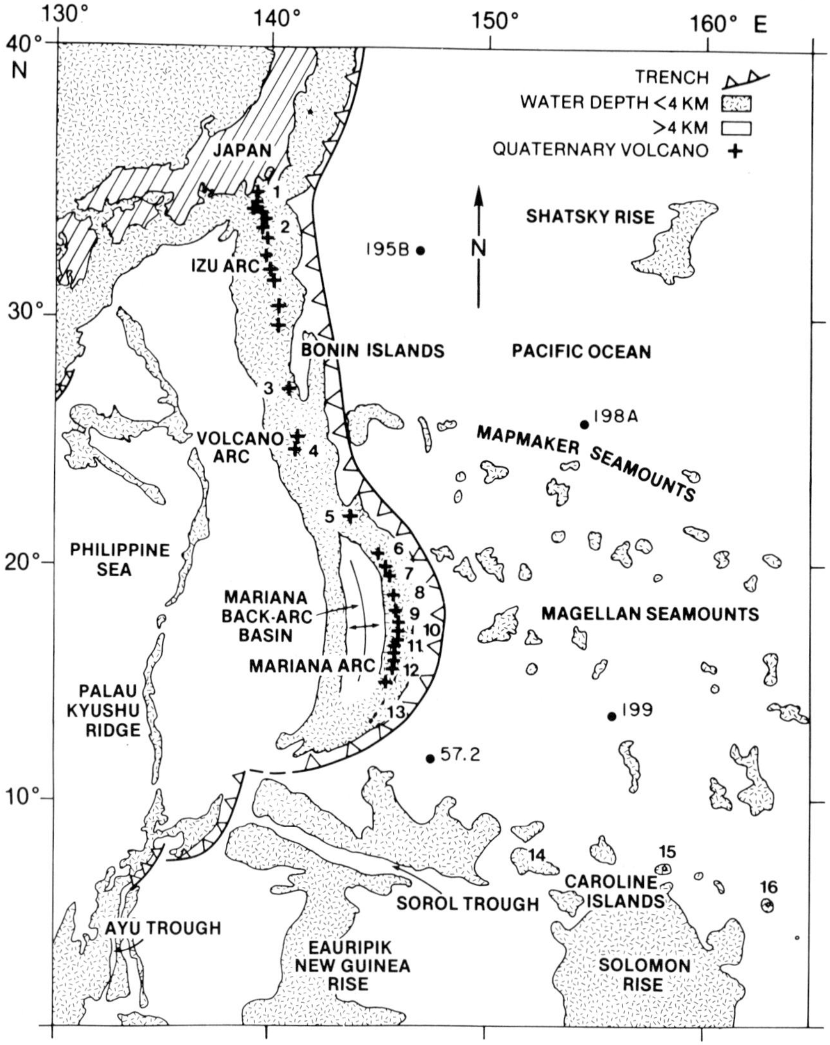

Fig. 1. Map of the western Pacific, modified after the chart of Chase et al. (1971). In addition to the location of major bathymetric features, volcanic islands studied in the course of this investigation are identified by numbers: *1* = O-Shima; *2* = Myojin Sho; *3* = Nishino-Shima; *4* = Iwo Jima; *5* = Fukujin (seamount); *6* = Uracas; *7* = Asuncion; *8* = Agrigan; *9* = Pagan; *10* = Guguan; *11* = Sarigan; *12* = Esmeralda Bank (seamount); *13* = Mariana Frontal Arc; *14* = Truk; *15* = Ponape; *16* = Kusaie (Kosrae). Also shown are the locations of Deep Sea Drilling Project sites 57.2, 195B, 198A and 199.

have been accomplished by the development of layered igneous bodies within and below the crust as well as outpourings of lava (Isshiki, 1963; Stern, 1979).

The fact that the Mariana-Volcano arc, one of the youngest and most primitive of all arcs, nevertheless sits on crust which has tripled or quadrupled in thickness since the early Tertiary limits the confidence with which we can resolve whether the melts of this system were derived from subducted crust or mantle. Some lavas of arcs built on thickened crust elsewhere (Andean-type margins) are contaminated by anatectic crustal melts (e.g., Tilton and Barreiro, 1980). We expect that while the possibility of similar processes occurring beneath the Mariana-Volcano arc cannot be precluded, attendant compositional changes in ascending melts should be minor. This conclusion results from the following considerations: (1) the thickened crust, composed of Lower Tertiary oceanic crust, anhydrous sub-volcanic gabbroic bodies, and arc lavas, will be relatively refractory; (2) the density of this crust is about ~ 2.9 g/cm^3 (Sager, 1980) compared to ~2.6 g/cm^3 for basaltic melts (Bottinga and Weill, 1970). Under these conditions, stagnation of basaltic melts in the crust should be limited to the development of shallow magma chambers beneath the larger edifices, leaving little opportunity for chemical exchange between melt and crust; and (3) the chemical and isotopic similarity between basaltic melts rising through the crust and the crust itself should insure that what chemical exchange does occur will not severely affect derivative melts. In summary, while the Mariana-Volcano arc is one of the most promising of all arcs for "seeing through" the effects of crustal contamination suspected in other arcs, such contamination cannot be dismissed. We can only say that of all arc systems, the extent and effects of contamination should be minimized in this system.

COMPOSITION AND ROLE OF SUBDUCTED CRUST AND SEDIMENTS

The Volcano and Mariana arcs are a promising system in which to study problems of arc magmagenesis, not only because of their primitive crustal setting, but also because it appears that everything that arrives at the trench disappears down it. Samplings of the inner wall of the Mariana Trench have recovered peridotites, gabbros, and basalts as well as more exotic quartz diorites, dacites, and boninites (Dietrich et al., 1978; Anonymous, 1978; Bloomer and Hawkins, 1980). Offscraped pelagic sediments have not been recovered; we conclude that the western Pacific seafloor and pelagic sediment cover are carried intact into the subduction zone.

The crust arriving at the Mariana Trench is the oldest seafloor in the Pacific (Early Jurassic; Larson and Chase, 1972; Hilde et al., 1977). In contrast to arcs such as the Aleutians and Cascades where subducted sediments have an arc provenance and are, therefore, compositionally similar to the arc lavas, sediments subducted beneath the Mariana-Volcano arc are dominantly pelagic. The thickness of the sedimentary section is uncertain because DSDP drill holes have not reached basement. Sub-bottom penetration in DSDP Leg 20 holes approached 500 m, while reflection profiles in the region suggest basement lies 800 m or more beneath the seafloor (Davies, 1973; Zachariadis, 1973). We assume a conservative sediment thickness of 500 m in subsequent calculations. The lithologic proportions of this is estimated from DSDP Leg 6 and 20 summaries (Fischer and Heezen, 1971; Heezen et al., 1973); this composite section is composed of 140 m Tertiary volcanogenic sediment, 217 m Mesozoic and Lower Tertiary chalk, 120 m Mesozoic abyssal clay, and 23 m Mesozoic chert.

In addition to basaltic crust and pelagic sediments, seamounts are also being supplied to the trench. Since alkalic rocks have not been recovered from the inner wall of the Mariana Trench, it appears these are also being subducted. The volume of 95 seamounts with edifice heights of at least 350 m were calculated for the region east of the Mariana-Volcano arc, between 10° and 25° N and between 150° and 165° E, from the bathymetric chart of Chase et al. (1971). The volumes, assumed densities, and masses of the three components of western Pacific crust are listed in Table I.

TABLE I

Nature of the crust being subducted beneath the Mariana-Volcano arc

	Volume ($\times 10^5$ km^3)	Density (g/cm^3)	Mass ($\times 10^{18}$ kg)	% of total subducted mass
Basaltic seafloor[1]	82.5	3.0	24.8	91.8
Sediments[2]	7.5	1.52	1.14[3]	4.2
Seamounts[4]	3.64	3.0	1.09	4.0

[1] Thickness = 5.5 km.
[2] Thickness = 500 m.
[3] Mass of sediments calculated using 140 m Tertiary volcanogenic sediment ($\rho \sim 1.2$ g/cm^3), 217 m Mesozoic and Lower Tertiary chalk ($\rho \sim 1.6$ g/cm^3), 120 m Mesozoic abyssal clay ($\rho \sim 1.6$ g/cm^3), and 23 m Mesozoic chert ($\rho \sim 2.3$ g/cm^3).
[4] Volume based on point-counting bathymetric highsbetween 10—25° N and 150—160°E (Chase et al., 1971).

Trace-element, $^{87}Sr/^{86}Sr$, and $\delta^{18}O$ analyses for 10 representative samples of western Pacific sediments are reported in Table II. Using the assumed lithologic proportions already presented and densities listed in Table I, the bulk composition of subducted western Pacific sediments was determined (Table III). More difficult to assess is the bulk composition of the seamounts and basaltic crust. Very little is known about the seamounts; in Table III these are assumed to be composed of typical ocean island alkali basalt (Engel et al., 1965). Basaltic basement beneath the sediments has not been recovered, but for modelling purposes is assumed to consist of 10% weathered basalt and 90% fresh MORB (Hart et al., 1974; Hart, 1975). Using these values, the bulk composition of trace elements and Sr- and O-isotopes in the subducted western Pacific seafloor has been reconstructed in Table III.

Comparison of LIL incompatible element ratios between these subducted components and the fields occupied by Mariana-Volcano arc lavas is shown in Fig. 2. The following relationships between subducted seafloor and the arc lavas are apparent:

TABLE II

Trace-element and isotopic composition of western Pacific sediments

	199 1-2 92-93	199 10-2 38-39	199 11-1 29-31	198A 2-2 103-104	198A 3-3 77-78	198A CC	195B 3-1 124-125	195B 3-1 129-130	59.2 4-1 109-111	59.2 1-3 99-100
K (ppm)	8763	3836	7476	31,218	30,867	2589	511	2624	16,715	39,156
Rb	14.9	10.1	10.2	152.7	124.2	8.4	1.2	9.0	30.5	21.2
Sr	832	1035	381	142.7	192.7	21.5	0.53	561.2	221.6	216.6
Cs	0.49	0.51	0.65	11.2	10.1	0.63	0.06	0.49	1.20	0.98
Ba	580	678	160.2	380.0	364.3	30.3	2.90	1288	107.3	183.5
$^{87}Sr/^{86}Sr$	0.70841 ± 5	0.70763 ± 5	0.70807 ± 6	0.71499 ± 3	0.71230 ± 8	0.71056 ± 4	—	0.70758 ± 8	0.70554 ± 8	0.70526 ± 9
$\delta^{18}O$ (‰)	30.1	28.4	29.5	—	—	—	25.7	27.4	—	—
% loss (110°C)	—	1.3	3.6	28.1	27.5	3.3	1.3	0.9	64.3	70.2
K/Rb	588	330	733	204	249	308	426	292	548	1847
K/Cs	17,884	7522	11,502	2787	3056	4110	8517	5355	13,929	39,955
K/Ba	15.1	5.7	46.7	82.2	84.7	85.4	176.2	2.04	155.8	213.4
Rb/Sr	0.018	0.010	0.027	1.070	0.645	0.391	2.26	0.016	0.138	0.098
Ba/Sr	0.697	0.655	0.420	2.66	1.89	1.41	5.47	2.30	0.484	0.847

Lithologies: 199 1-2 92-93: Lower Tertiary foram ooze; 199 10-2 38-39: Paleocene nanno chalk; 199 11-1 29-31: Paleocene limestone; 198A 2-2 103-104: Late Cretaceous clay; 198A 3-3 77-78: Late Cretaceous clay; 198A CC: Late Cretaceous chert: 195B 3-1 124-125: Lower Cretaceous chert; 195B 3-1 129-130: Lower Cretaceous chalk; 59.2 4-1 109-111: Early Miocene tuff; 59.2 1-3 99-100; Early Miocene tuff.
Trace elements analyzed by isotope dilution. $^{87}Sr/^{86}Sr$ normalized to $^{86}Sr/^{88}Sr = 0.1194$ and $^{87}Sr/^{86}Sr = 0.70800$.
Oxygen isotope data based on $\delta^{18}O$ for NBS-28 quartz = 9.50‰.

TABLE III

Composition of subducted western Pacific seafloor

	Sediments[1]	Seamounts[2]	Basaltic crust[3]	Bulk western Pacific seafloor
K (ppm)	10205	14027	1265	2160
Rb	32.3	33	1.7	4.2
Sr	361	815	125.6	164
Cs	2.3	0.35	0.06	0.16
Ba	382	498	11.5	47
$^{87}Sr/^{86}Sr$	0.7083	0.7034	0.7027	0.7032
$\delta^{18}O$ (‰)	24.5[4]	6.0[5]	6.5[6]	7.2
K/Rb	316	425	744	514
K/Ba	27	28	110	46
K/Cs	4440	40,000	21,000	13,500
Rb/Sr	0.09	0.04	0.01	0.03
Ba/Sr	1.06	0.61	0.09	0.29

[1] Bulk western Pacific sediment composition based on weighted average of each of the 4 major lithologies in Table II.

$$X^i = \sum_{j=1}^{4} \frac{C_j^i\,(1-L_j)\,\rho_j\,T_j}{\bar{\rho}\;500}$$

x^i = mean concentration of element "i" in subducted sediment
C_j^i = concentration of element "i" in lithology "j"
L = fraction of weight lost during drying
ρ = bulk density
T = thickness of sediments in meters
j = Chalk, chert, clay, volcanic ash

[2] Elemental abundances for alkali basalt (Engel et al., 1965); $^{87}Sr/^{86}Sr$ is for northern Pacific ocean islands (Hedge, 1978). Cs from K/Cs ~ 40,000 (Hofmann et al., 1978).

[3] Hypothetical basaltic crust composed of 10% weathered basalt (AD5-18 intermediate; Hart et al., 1974) and 90% fresh "Average MORB" (Hart, 1975).

[4] $\delta^{18}O$ for entire sedimentary column after Garlick and Dymond (1970), Savin and Epstein (1970), and Lawrence et al. (1975), as well as data from Table II.

[5] Oceanic island alkalic lavas (Taylor, 1968)

[6] Muehlenbachs and Clayton, 1972.

(1) Since the fields occupied by sediments, basaltic crust, bulk western Pacific seafloor, or MORB-type mantle do not correspond to that defined by Mariana-Volcano arc lavas, none of these alone can be the source of the arc magmas.

(2) Melting of subducted seamounts alone, or seamounts plus a small amount of western Pacific basaltic crust is allowed.

(3) Melting of 10—20% sediments and 80—90% basaltic crust, with or without seamounts, is allowed.

(4) Mixing and melting of 2—5% sediment and 95—98% MORB-type mantle is allowed.

(5) Melting of an ocean-island or "hot-spot" type mantle is allowed.

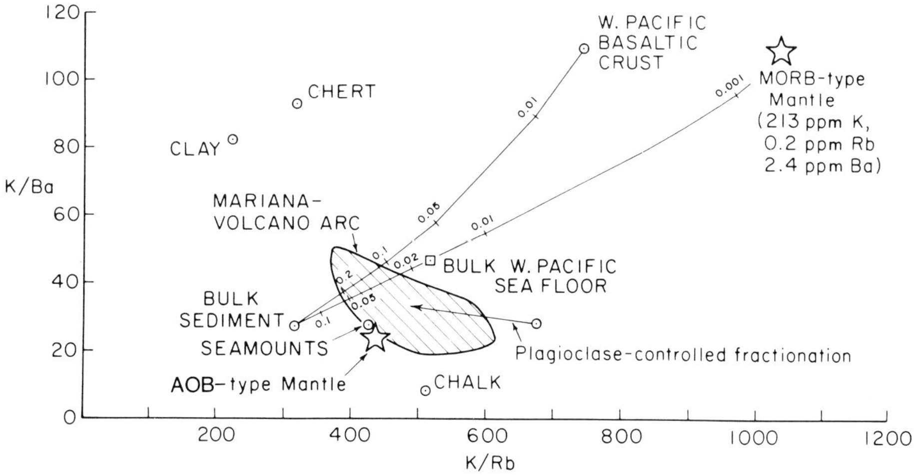

Fig. 2. Incompatible element ratios of Mariana-Volcano arc lavas and possible sources in the mantle and subducted seafloor. Fresh rocks from this arc have a limited range in K/Rb (350—600) and K/Ba (20—50). This contrasts with much greater ranges in mean ratios for pelagic and wind-blown sediments of the western Pacific. Bulk sediment (Table II) has a K/Rb ~ 320 and K/Ba ~ 27. Mixtures of bulk sediment with western Pacific basaltic crust intersect the field of Mariana-Volcano arc lavas at 10—20% sediment, a much higher proportion than is permitted on the basis of Pb isotopic data (Meijer, 1976). The compositions of sources expected in the overlying mantle wedge are indicated by stars. Values for the MORB-type mantle were determined by assuming that (1) MORB = 20% partial melt of the mantle: and (2) K, Rb, and Ba are totally partitioned into the melt. Thus, the MORB-type mantle has 0.2× the concentration of K, Rb, and Ba of the "Average MORB" reported by Hart (1975). Such a mantle clearly cannot be the source of these arc magmas, although mixtures with 2—5% sediment do intersect the field for arc lavas. Finally, an AOB-type mantle similar to that responsible for the generation of the average oceanic alkali basalt of Engel et al. (1965) is very similar to the source of Mariana-Volcano arc lavas.

Combined use of oxygen and strontium isotopes is also useful for evaluating the role of the subducted crust. Subducted sediments have $\delta^{18}O \sim 25‰$ and $^{87}Sr/^{86}Sr \sim 0.708$ while the bulk subducted western Pacific seafloor should have $\delta^{18}O \sim 7.2‰$ and $^{87}Sr/^{86}Sr \sim 0.7033$ (Table III). Both of these are different from Mariana-Volcano arc lavas ($^{87}Sr/^{86}Sr$ = 0.7033—0.7034; $\delta^{18}O$ = 5.5—6.8‰; Ito and Stern, 1981; Douthitt and Dixon, 1981) While we cannot rule out mixing of bulk seafloor with the mantle to produce the arc values, the data limits the participation of sediments in the mixing or melting process to 1% or less (Fig. 3). The limited range of $\delta^{18}O$ and $^{87}Sr/^{86}Sr$ values exhibited by Mariana-Volcano arc lavas (1.3‰ total variation in $\delta^{18}O$, ±0.04% variation in $^{87}Sr/^{86}Sr$) contrasts strongly with the Banda arc (±3.6‰ total variation in $\delta^{18}O$, ±0.35% variation in $^{87}Sr/^{86}Sr$), where compelling arguments favoring the mixing of sediments and mantle have been presented (Magaritz et al., 1978). On this basis, it is clear that suggestions 3 and 4 of the previous paragraph are wrong. Consideration of

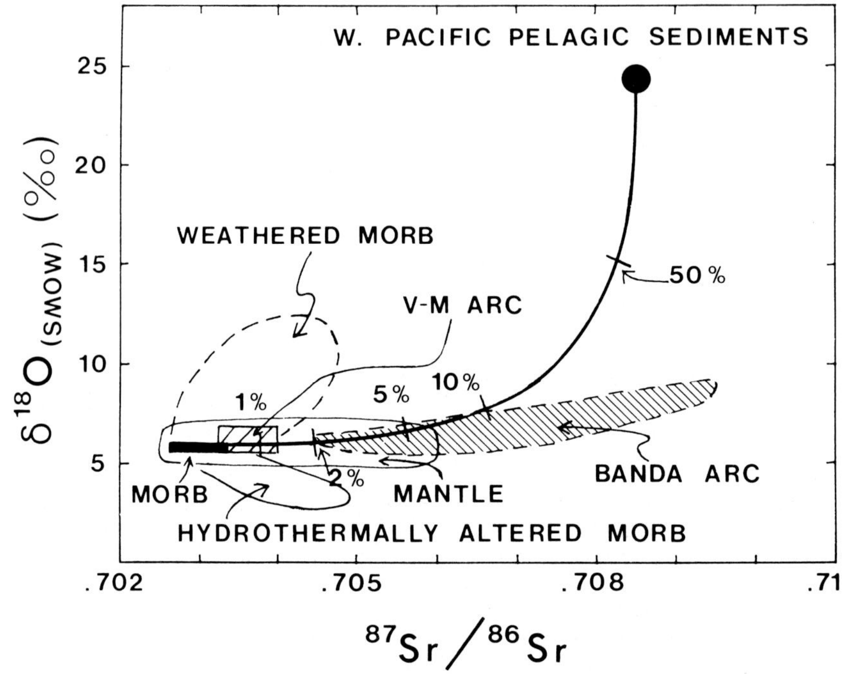

Fig. 3. Plot of the isotopic composition of O and Sr in the Volcano-Mariana arc (V-M; box with diagonal lines; data from Ito and Stern, 1981) and possible sources in the subducted crust and mantle. The composition of the bulk western Pacific pelagic sediments is from Table III. The range of values expected from weathered and hydrothermally altered MORB is also shown. Also shown is the wide range of values reported for the Banda arc (Magaritz et al., 1978) where Whitford and coworkers have made a strong case that subducted components can be isotopically identified in analyzed lavas. Note that the range of isotopic compositions for the Banda arc, ~ 500 km long, is about an order of magnitude greater than that of the ~ 1000 km long Mariana-Volcano arc. For comparison, the range of $\delta^{18}O$ and $^{87}Sr/^{86}Sr$ reported for MORB-like basalts of the Mariana Trough (back-arc basin) and Ayu Trough are also reported (solid black field) as well as the range of $\delta^{18}O$ and $^{87}Sr/^{86}Sr$ for mantle-derived AOB or "hot spot" basalts (field labelled "mantle", Kyser and O'Neil, 1978; Duncan and Compston, 1976). Also shown is a mixing line between sediments and MORB-like mantle. Note that no more than 1% sediment can be involved in magmagenesis beneath this arc.

the $\delta^{18}O$ - $^{87}Sr/^{86}Sr$ data coupled with arguments based on Pb-isotopic analyses of Mariana arc lavas and western Pacific sediments (Meijer, 1976) limits sediment involvement in magmagenesis to 1% or less.

The fact that sediment involvement in Mariana-Volcano arc magmagenesis is in all cases 1% or less suggests it may not contribute at all. The composition of the bulk subducted western Pacific seafloor listed in Table III is highly idealized. In reality, the relative proportions of the ratio of sediments to seamounts, sediment lithologies, as well as the thickness of sediments delivered to the subduction zone must vary enormously along the strike of the arc. Note, for example, that in Fig. 1 the Mapmaker and Magellan Seamounts represent probable "hot-spot" chains now arriving at the trench in the northern parts of the Volcano and Mariana arcs, respectively. The proportion of clays to chalks within the pelagic sedimentary succession depends on the depth and thus on the age of the crust when it passed beneath the equatorial zone of high biogenic accumulation rates (Heezen et al., 1973). It is impossible at present to do more than point out that the bulk composition of subducted crust plus sediments plus seamounts varies greatly, both in time and in space. If mixing of this material with the mantle controls the composition of arc melts, these mixes should also be heterogeneous. While mixing processes in the overlying mantle wedge should be relatively efficient as a result of induced mantle convection (Toksöz and Bird, 1977), flow lines should be normal to the strike of the arc. Convective mixing of subducted crust, sediments, and derivative fluids with the mantle should thus be much less efficient along strike of the arc than for portions of the mantle perpendicular to the arc. Insofar as these processes are responsible for the generation of arc melts, the composition of lavas from volcanoes along the arc should also show considerable variation in the incompatible element trace element ratios and isotopic compositions. Such source region heterogeneity is not characteristic of the Mariana-Volcano arc. This is indicated in Fig. 4 which shows variations in $^{87}Sr/^{86}Sr$ for the various subducted sources as well as data for the Mariana-Volcano arc. Clearly the source of these arc melts is remarkably homogeneous, especially considering these data come from islands as much as 1000 km apart.

COMPOSITION AND ROLE OF THE MANTLE WEDGE

The recognition that the source of Mariana-Volcano arc melts is isotopically homogeneous and that sediment involvement must be less than 1% suggests an important role for the mantle wedge. Unfortunately no mantle xenoliths have been recovered from this arc, so we have no direct evidence bearing on the composition of the mantle wedge beneath the Mariana-Volcano arc. Instead, we will assume this mantle is similar to that of the suboceanic mantle elsewhere beneath the Pacific, and proceed from this to a discussion of whether or not such a mantle could be responsible for Mariana-Volcano arc magmagenesis.

The suboceanic mantle can be grossly subdivided into two reservoirs on the basis of isotopic and trace element composition. One is responsible for the generation of mid-ocean ridge basalts, the MORB-type mantle. This has been strongly depleted in the incompatible elements and light REE over much of the Earth's history; consequently it is characterized by low $^{87}Sr/^{86}Sr$ (<0.7030) and high $^{143}Nd/^{144}Nd$ (>0.5131*, e.g. White, 1979). Very

*$^{143}Nd/^{144}Nd$ values discussed here have been renormalized to BCR-1 = 0.51271 and UCSD Nd = 0.51196.

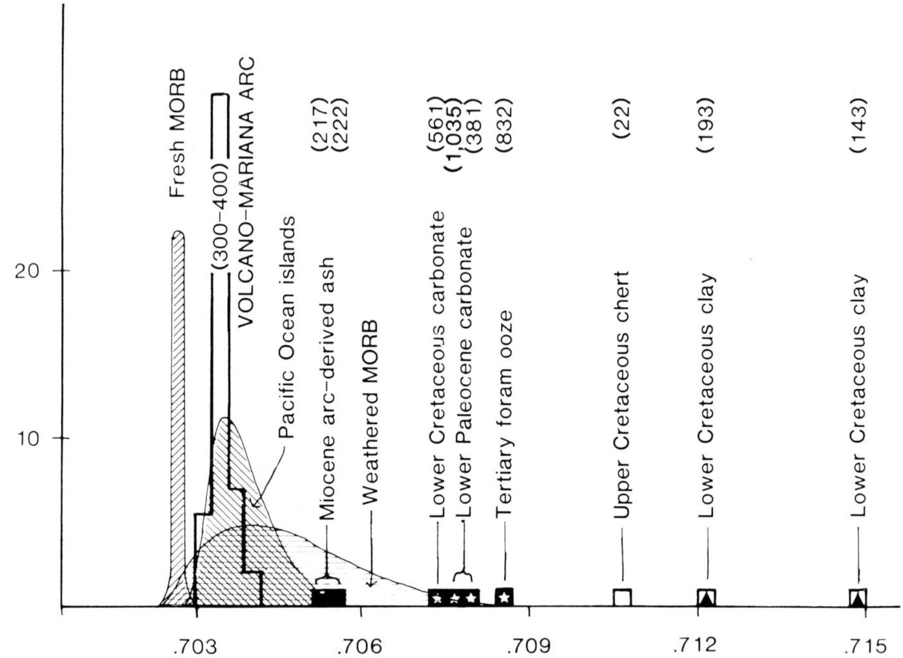

Fig. 4. Isotopic composition of Mariana-Volcano arc lavas and plausible sources, showing the homogeneity of $^{87}Sr/^{86}Sr$ in the source of its melts. Figures in parentheses indicate the concentration of Sr in ppm. Data for sediments are from Table II; sources for the arc are listed in Fig. 5. The distribution of $^{87}Sr/^{86}Sr$ in fresh MORB, weathered MORB, and Pacific Ocean islands is idealized from the voluminous literature on the subject. Note that the homogeneity in the composition of the arc melts seems more consistent with an origin in the mantle than with a source involving the participation of extremely heterogeneous crustal components.

high K/Rb (~ 1000) and K/Ba (~ 100) are also diagnostic. Insofar as similar lavas are erupted along 66,000 km of submarine ridges, the MORB-type mantle is probably the most homogeneous of all mantle reservoirs. The other obvious sub-oceanic mantle reservoir is that responsible for generating "hot-spot" or alkali olivine basalt island chains, hereafter referred to as AOB-type mantle. Although this source is strongly heterogeneous beneath the Atlantic, south Pacific, and Indian Oceans, it appears to be relatively homogeneous beneath the north Pacific (Hedge, 1978). The relative disposition of MORB- and AOB-type reservoirs in the mantle under this region is not known, but both sources are present beneath the western Pacific as shown by the alkalic rocks of the Caroline Islands (AOB-type mantle; Yagi, 1960, Stern, 1981; Mattey, 1981) and MORB-like basalts of the Mariana Trough and Ayu Trough (Hart et al., 1972; Fornari et al., 1979; Ito and Stern, 1981).

Consideration of LIL incompatible element ratios (Fig.2) suggests that MORB-type mantle alone cannot generate the Mariana-Volcano arc lavas.

This conclusion is substantiated by consideration of the isotopic systematics, since Mariana-Volcano arc lavas have significantly more radiogenic $^{87}Sr/^{86}Sr$ (~0.7034) and less radiogenic $^{143}Nd/^{144}Nd$ (~0.5130) as well as slightly heavier O ($\delta^{18}O \sim 6$) than do MORB. These data indicate that none of the characteristics expected from the melting of a MORB-type mantle are reflected in the diagnostic isotopic and trace element ratios of Mariana-Volcano arc lavas.

Melting of an AOB-type mantle to generate the arc lavas has many attractive features. Such a reservoir has the same incompatible element ratios as the Mariana-Volcano arc lavas (Fig. 2). Oxygen and strontium isotope ratios for AOB encompass the field occupied by Mariana-Volcano arc lavas (note field labelled "Mantle" in Fig. 3). Recognition of the gross similarity of the geochemical source characteristics of the arc and AOB has lead to comparative geochemical and isotopic studies between the lavas of this arc system and those of nearby north Pacific ocean islands. The details of that investigation are reported elsewhere (Stern, 1981) but the main results bear directly on the problem being discussed here.

A comparison of $^{87}Sr/^{86}Sr$ and incompatible element ratios of the rocks of the Mariana-Volcano arc and those of Kilauea and Caroline "Hot Spots" is shown in Fig. 5. Inspection of the distribution of data in this figure shows that the mean and range of especially $^{87}Sr/^{86}Sr$, K/Rb, and K/Ba between arc and hot spots are remarkably similar. K/Cs for the arc is often lower than for the hot spots, possibly indicating the transport of Cs from subducted crust up into the mantle, but it is noteworthy that K/Cs in fresh Caroline Islands basalts is often low as well. Available oxygen isotope data is also quite similar between these arcs and hot spots. Kilauea tholeiites have $\delta^{18}O$ between + 4.9 and + 6.2 with a mean of + 5.5 (Kyser and O'Neil, 1978). Basalts from Ponape, in the Caroline Islands, have $\delta^{18}O$ ranging from + 5.4 to + 7.1, with a mean $\delta^{18}O$ of + 6.3 (Douthitt and Dixon, 1981). Mariana-Volcano arc lavas analyzed by Ito and Stern (1981) have $\delta^{18}O$ between + 5.5 and + 6.8 with a mean $\delta^{18}O$ of + 6.0. Douthitt and Dixon (1981) report a similar range in $\delta^{18}O$ of + 5.2 to + 6.7 and mean $\delta^{18}O$ of + 5.8 for submarine and subaerial Mariana arc volcanoes.

Combined $^{87}Sr/^{86}Sr$ - $^{143}Nd/^{144}Nd$ isotopic systematics are valuable for evaluating the possibility that subducted seafloor participates in arc magmagenesis (e.g., DePaolo and Wasserburg, 1977; Hawkesworth et al., 1977, DePaolo and Johnson, 1979). A comparison of Nd- and Sr- isotopic composition for the Mariana and Volcano arcs with those expected of plausible sources in the mantle and subducted crust was outlined by Stern (1981) and is graphically presented in Fig. 6. Clearly, none of the sources in the subducted slab (with the exception of seamounts) alone could generate the arc isotopic compositions. Instead, the field for the Mariana-Volcano arcs coincides with that of the Caroline and Kilauea "Hot Spots" and lies entirely within the field defined for AOB-type mantle.

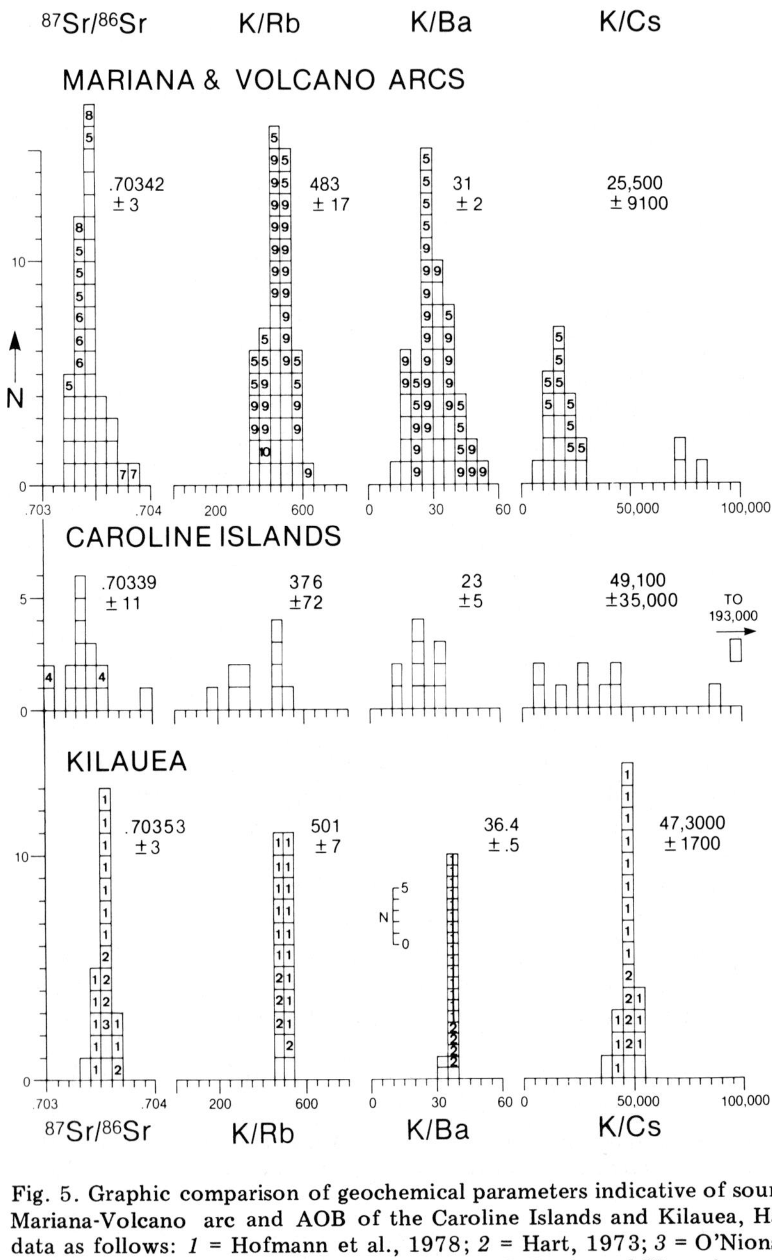

Fig. 5. Graphic comparison of geochemical parameters indicative of source compositions, Mariana-Volcano arc and AOB of the Caroline Islands and Kilauea, Hawaii. Sources of data as follows: *1* = Hofmann et al., 1978; *2* = Hart, 1973; *3* = O'Nions et al., 1977; *4* = Hedge, 1978; *5* = Dixon and Batiza, 1979; *6* = DePaolo and Wasserburg, 1977; *7* = Meijer, 1976; *8* = Stern, 1979; *9* = Chow et al., 1980. Boxes left empty represent analyses performed using the solid-source mass spectrometer facilities at DTM and are reported in Stern and Bibee (1980), Stern (1981), Stern (1982), Ito and Stern (1981) and Dixon and Stern (in press). All $^{87}Sr/^{86}Sr$ data have been renormalized to E + A $SrCO_3$ = 0.70800 or NBS 987 = 0.71022.

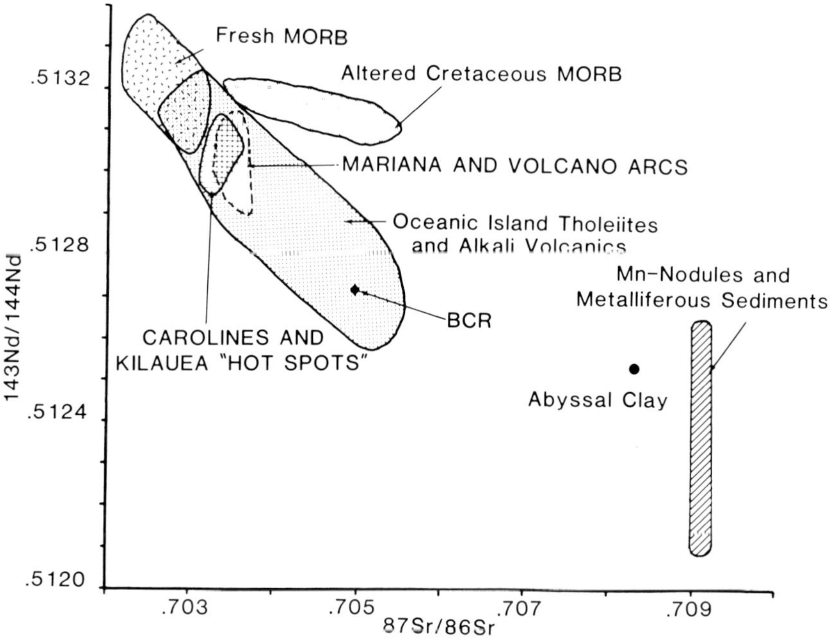

Fig. 6. Plot of $^{143}Nd/^{144}Nd$ vs. $^{87}Sr/^{86}Sr$ for volcanic rocks of the active Mariana-Volcano arc as well as plausible sources in the mantle and subducted crust. Isotopic data for Arcs are from DePaolo and Wasserburg (1977), Stern and Bibee (1980), Stern (1981), and Dixon and Stern (in press). Sources in the subducted slab include altered Mesozoic MORB (McCulloch et al., 1980; Jahn et al., 1980) and sediments represented by abyssal clay, Mn-nodules, and metalliferous sediments (McCulloch and Wasserburg, 1978; O'Nions et al., 1978). Mantle sources include those of fresh MORB and AOB (DePaolo and Wasserburg, 1976; O'Nions et al., 1977; White and Hofmann, 1978; White, 1979; Hawkesworth et al., 1979; Zindler et al., 1979; Cohen et al., 1980). Note the AOB reservoir in the north Pacific, as represented by basalts from the Caroline Islands (Stern, 1981) and Kilauea (O'Nions et al., 1977) defines a more restricted range than that of global AOB and that the field defined by this source coincides with that of the Mariana and Volcano arcs. All data have been renormalized to $^{87}Sr/^{86}Sr$ E + A = 0.70800 and/or NBS 987 = 0.71022 and $^{143}Nd/^{144}Nd$ BCR-1 = 0.51271.

DISCUSSION AND CONCLUSIONS

The suggestion that the volcanic rocks of some arcs contain the isotopic and trace-element signature of an AOB-type mantle is not a new one. Sinha and Hart (1972) first noted that the Pb-isotopic and trace-element characteristics of Tongan volcanics ". . . do not differ in any important way from the volcanic rocks of oceanic islands." Meijer (1976) noted that the Pb-isotopic compositions of Mariana active arc volcanic rocks correspond to those of more alkalic north Pacific sea-floor basalts. Church and Tilton (1973) noted that the Pb- and Sr-isotopic compositions of Cascade lavas were very similar to that of oceanic alkali basalt. Church (1976) subsequently retracted this observation as a result of new analytical data which plotted above the array defined by basalts from the Juan de Fuca and Gorda Ridges and associated seamounts. Lead from other oceanic island basalts such as Hawaii (Tatsumoto, 1966) or Gough and Ascension (Gast, 1969) nevertheless encompasses an isotopic composition that includes the Cascade data; for this reason, the original observation of Church and Tilton (1973) still seems valid.

Development of combined $^{143}Nd/^{144}Nd$ - $^{87}Sr/^{86}Sr$ techniques has also failed to settle the issue. While earlier studies of basalts and andesites from active Mariana arc (DePaolo and Wasserburg, 1977) and Bismarck arc volcanoes (DePaolo and Johnson, 1979) indicate that the isotopic composition of these magmas falls within the field defined by ocean island basalts, arc basalts from outside the circum-Pacific do not (Hawkesworth et al., 1977; Hawkesworth and Powell, 1980). The earlier suggestion of Armstrong (1971) that subducted sediments are involved in magmagenesis in the Lesser Antilles appears substantiated, and the recent Pb-isotope studies of South Sandwich arc volcanic rocks (Barreiro et al., 1981) indicate a similar involvement of sediments there. Perhaps the most convincing argument for subducted sediment involvement in arc magmagenesis comes from the Banda Arc where Whitford and his co-workers (Whitford et al., 1981; Whitford and Jezek, 1979; Magaritz et al., 1978) have assembled a compelling story based on trace element and isotope geochemistry.

Recent analytical studies of arc volcanic rocks show that concentrations of high-field strength trace cations (HFSTC) such as Ti, Th, U, Zr, Nb, Hf, Ta, and Y as well as the Rare Earth Element (REE) are generally lower than in MORB and are *much* lower than in AOB (see reviews by Perfit et al., 1980, and Arculus and Johnson, 1981). These elements are generally treated as incompatible elements in MORB and AOB. This depletion of HFSTC and REE coupled with LIL incompatible element abundances that are comparable to AOB leads to the very high ratios of LIL/HFSTC and LIL/REE characteristic of arc magmas. Trace element studies of Mariana lavas show such high ratios are also found there (Dixon and Batiza, 1979).

The high LIL/HFSTC and LIL/REE ratios observed in arc lavas must be the result of either melting of a unique mantle source tapped only beneath arcs or is the result of chemical decoupling during evolution of sources or melts. While there is no evidence to rule out the first possibility, neither is there any evidence to support it; consideration of Occum's Razor indicates we should not pursue this possibility further at present. We conclude that chemical decoupling of LIL and HFSTC incompatible elements must have occurred.

Proponents of source modification argue that LIL elements may be released during dehydration of the subducted crust into the overlying mantle wedge (Best, 1975). Many models also call on the mantle wedge thus affected to have been previously and severely depleted in all incompatibles, possibly during an earlier melting event (e.g., Kay, 1980). Melting of this partially recharged material would generate the observed high LIL/HFSTC ratios. We object to models calling on such an origin for Mariana-Volcano arc magmas for the following reasons:

(1) The age and nature of the prior depletion event is speculative. There is no reason to expect that the source of arc melts should commonly involve material that would otherwise be *least* expected to melt. Neither can the depletion event be the result of early arc volcanism, since high LIL/HFSTC and LIL/REE ratios are found at all stages of arc volcanism.

(2) There is evidence that the source of Mariana-Volcano magmas are not depleted in the non—LIL basaltic fraction. High abundances of CaO and Al_2O_3 characterize these and other arc magmas, and plagioclase in these basalts is unusually rich in anorthite (Dixon and Batiza, 1979; Stern, 1979). These elements should largely reside in the mantle in garnet and clinopyroxene, minerals attacked in any significant melting event. While elevated calcium and aluminum abundances in Mariana-Volcano arc lavas often reflect plagioclase accumulation, aphyric lavas are not depleted in these elements. Neither can the anorthitic plagioclase be in equilibrium with a CaO-, Al_2O_3-poor melt. How HFSTC could have been stripped from this source without affecting CaO and Al_2O_3 is difficult to imagine, nor does it seem likely that CaO and Al_2O_3 were introduced from subducted crust without also adding the HFSTC.

(3) We cannot imagine the physical processes accompanying mantle metasomatism beneath the Volcano-Mariano arc that could have generated the observed regularity in LIL incompatible element ratios and isotopic compositions. As previously noted, the materials supplied to the subduction zone are strongly heterogeneous, especially the sediments and seamounts. Since flow regime set up by induced convection in this region should result in very inefficient, longitudinal mixing, the result of mantle metasomatism in this environment should be the generation of a very heterogeneous mantle. If this were the case, we should expect derivative melts to manifest major LIL and isotopic variation along strike of the arc. This is not observed in the Mariana-Volcano arc; instead LIL ratios and isotopic compositions in fresh samples from active portions of the arc are more constant than many AOB or "hot-spot" chains and are only slightly less homogeneous than MORB.

We prefer to explain the unusual and high ratios of LIL to HFSTC and REE in some arc magmas as resulting from the sequestering of the latter two groups into some mineral phase(s) stable during melting or subsequent fractionation beneath arcs but not during melting to form AOB or MORB. Our suggestion differs little from that of Gill (1976). We appreciate the fact that empirical or experimental evidence to support the presence of this phase is lacking at present. Nevertheless we believe the data outlined in this and other studies on the Mariana-Volcano arc compel the conclusion that these magmas come from a source remarkably similar to that responsible for at least some of the chemical features of AOB elsewhere in the north Pacific Basin.

In general, we conclude that many of the intra-oceanic island arcs of the circum-Pacific reflect a source remarkably similar in many respects to that of AOB-type mantle, and that other primitive arcs from outside the circum-

Pacific do not. These results seem to lend further support to the conclusion of Stern (1982) that the similarity of regional variation in $^{87}Sr/^{86}Sr$ between hot spots basalts and circum-Pacific intra-oceanic arcs suggests a similar mantle source. We do not know the reason for this striking difference in the processes of magma-genesis beneath otherwise similar arc systems. Perhaps the apparent mechanical decoupling of subducted slab from mantle wedge which is common in western Pacific arcs (Uyeda and Kanamori, 1979) is accompanied by a chemical decoupling which inhibits the migration of LIL-enriched fluids from the slab into the overlying mantle wedge.

ACKNOWLEDGEMENTS

Much of this work was accomplished while we were postdoctoral fellows at the Department of Terrestrial Magnetism, Carnegie Institution of Washington. The use of DTM facilities and assistance of DTM staff in the course of this and related geochemical studies was greatly appreciated, as was the assistance of T.C. Hoering of the Geophysical Laboratory in maintaining the stable isotope mass spectrometers. Samples contributed by J. Ossaka of the Tokyo Institute of Technology, G. Corwin of the U.S.G.S., and P. Fryer of Hawaii Institute of Geophysics, and their assistance is greatly appreciated.

REFERENCES

Anonymous, 1978. Near the Philippines, Leg 60 ends in Guam. Geotimes, 23: 19—22.

Arculus, R.J. and Johnson, R.W., 1981. Island-arc magma sources: A geochemical assessment of the roles of slab-derived components and crustal contamination. Geochem. J., 15: 109—133.

Armstrong, R.L., 1971. Isotopic and chemical constraints on models of magma genesis in volcanic arcs. Earth Planet. Sci. Lett., 12: 137—142.

Barreiro, B.A., Dixon, T.H. and Tilton, G.R., 1981. Pb isotope variation in the Marianas and South Sandwich Island Arcs. Geol. Soc. Am., Abstr. with Programs, 12: 43.

Best, M.G., 1975. Migration of hydrous fluids in the upper mantle and potassium variation in calc-alkalic rocks. Geology, 3: 429—432.

Bloomer, S. and Hawkins, J., 1980. Arc-derived plutonic and volcanic rocks from the Mariana Trench Wall. Trans. Am. Geophys. Union, 61: 1143.

Bottinga, Y. and Weill, D.F., 1970. Densities of liquid silicate systems calculated from partial molar volumes of oxide components. Am. J. Sci., 269: 169—182.

Chase, T.E., Menard, H.W. and Mammerickx, J., 1971. Bathymetry of the North Pacific (Chart No. 6). Institute of Marine Resources, Technical Report Series TR-11.

Chow, T.J., Stern, R.J. and Dixon, T.H., 1980. Relative and absolute concentrations of K, Rb, Sr, and Ba in recent lavas from the Northern Mariana Island Arc. Chem. Geol., 28: 111—121.

Church, S.E., 1976. The Cascade Mountains revisited. A re-evaluation in light of new lead isotopic data. Earth Planet. Sci. Lett., 29: 175—188.

Church, S.E. and Tilton, G.R., 1973. Lead and Strontium isotopic studies in the Cascade Mountains: Bearing on andesite genesis. Geol. Soc. Am. Bull., 84: 431—454.

Cohen, R.S., Evensen, N.M., Hamilton, P.J. and O'Nions, R.K., 1980. U-Pb, Sm-Nd, and Rb-Sr systematics of mid-ocean ridge basalt glasses. Nature, 283: 149—153.

Crawford, A.J., Beccaluva, L. and Serri, G., 1981. Tectono-magmatic evolution of the West Philippine-Mariana region and the origin of boninites. Earth Planet. Sci. Lett., 54: 346—356.

Davies, T.A., 1973. A seismic reflection profile between the Bonin Trench and 160° E. In: B.C. Heezen et al. (Editors), Initial Reports of the Deep Sea Drilling Project, Leg 20. U.S. Government Printing Office, Washington, D.C., pp. 505—521.

DePaolo, D.J. and Johnson, R.W., 1979. Magma genesis in the New Britain island-arc: Constraints from Nd and Sr isotopes and trace-element patterns. Contrib. Mineral. Petrol., 70: 367—379.

DePaolo, D.J. and Wasserburg, G.J., 1976. Inferences about magma sources and mantle structure from variations of $^{143}Nd/^{144}Nd$. Geophys. Res. Lett., 3: 743—746.

DePaolo, D.J. and Wasserburg, G.J., 1977. The sources of island arcs as indicated by Nd and Sr isotopic studies. Geophys. Res. Lett., 4: 465—468.

Dietrich, V., Emmermann, R., Oberhansli, R. and Puchelt, H., 1978. Geochemistry of basaltic and gabbroic rocks from the West Mariana Basin and the Mariana Trench. Earth Planet. Sci. Lett., 39: 127—144.

Dixon, T.H. and Batiza, R., 1979. Petrology and chemistry of recent lavas in the Northern Marianas: Implications for the origin of island arc basalts. Contrib. Mineral. Petrol., 70: 167—181.

Dixon, T.H. and Stern, R.J., in press. Petrology and chemistry of submarine volcanoes in the Southern Mariana Arc. Geol. Soc. Am. Bull.

Douthitt, C. and Dixon, T.H., 1981. Oxygen and strontium isotopic ratios of volcanic rocks from the Mariana Island Arc. Trans. Am. Geophys. Union, 62: 1901.

Duncan, R.A. and Compston, W., 1976. Sr-isotopic evidence for an old mantle source region for French Polynesian volcanism. Geology, 4: 728—732.

Engel, A.E.J., Engel, C.G. and Havens, R.G., 1965. Chemical characteristics of oceanic basalts and the upper mantle. Geol. Soc. Am. Bull., 76: 719—734.

Fornari, D.J., Weissel, J.K., Perfit, M.R. and Anderson, R.N., 1979. Petrochemistry of the Sorol and Ayu Troughs: Implications for crustal accretion at the northern and western boundaries of the Caroline Plate. Earth Planet. Sci. Lett., 45: 1—15.

Fischer, A.G. and Heezen, B.C., 1971. Introduction. In: A.G. Fischer et al. (Editors), Initial Reports of the Deep Sea Drilling Project, Leg 6. U.S. Government Printing Office, Washington, D.C., pp. 3—16.

Gast, P.W., 1969. The isotopic composition of lead from St. Helena and Ascension Islands. Earth Planet. Sci. Lett., 5: 353—359.

Garlick, G.D. and Dymond, J.R., 1970. Oxygen isotope exchange between volcanic materials and ocean water. Geol. Soc. Am. Bull., 81: 2137—2142.

Gill, J.B., 1976. Evolution of the mantle: Geochemical evidence from alkali basalt: Comment. Geology, 4: 625—626.

Hart, S.R., 1973. Submarine basalts from Kilauea Rift, Hawaii: Nondependence of trace element composition on extrusion depth. Earth Planet. Sci. Lett., 20: 201—203.

Hart, S.R., 1975. Deep Sea Drilling Project Leg 34 basalts and ocean ridge basalts: A geochemical comparison. Carnegie Inst. Wash. Ybk., 74: 218—224.

Hart, S.R. ,Glassley, W.E. and Karig, D.E., 1972. Basalts and sea floor spreading behind the Mariana island arc. Earth Planet. Sci. Lett., 15: 12—18.

Hart, S.R., Erlank, A.J. and Kable, E.J.D., 1974. Sea floor basalt alteration: Some chemical and Sr isotopic effects. Contrib. Mineral. Petrol., 44: 219—230.

Hawkesworth, C.J. and Powell, M., 1980. Magma genesis in the Lesser Antilles Island Arc. Earth Planet. Sci. Lett., 51: 297—308.

Hawkesworth, C.J., O'Nions, R.K., Pankhurst, R.J., Hamilton, P.J. and Evensen, N.M., 1977. A geochemical study of island-arc and back-arc tholeiites from the Scotia Sea. Earth Planet. Sci. Lett., 36: 253—262.

Hawkesworth, C.J., Norry, M.J., Roddick, J.C. and Vollmer, R., 1979. $^{143}Nd/^{144}Nd$ and $^{87}Sr/^{86}Sr$ ratios from the Azores and their significance in LIL-element enriched mantle. Nature, 280: 28—31.

Hedge, C.E., 1978. Strontium isotopes in basalts from the Pacific Ocean basin: Earth Planet. Sci. Lett., 38: 88—94.

Heezen, B.C., MacGregor, I.D., Foreman, H.P., Forristall, G., Hekel, H., Hesse, R., Hoskins, R.H., Jones, E.J.W., Kaneps, A., Krasheninnikov, V.A., Okada, H. and Ruef, M.H., 1973. The post Jurassic sedimentary sequence on the Pacific Plate; A kinematic interpretation of dichronous deposits, In: B.C. Heezen, I.D. MacGregor et al., Initial Reports of the Deep Sea Drilling Project, Leg 20. U.S. Government Printing Office, Washington, D.C., pp. 725—738.

Hilde, T.W.C., Uyeda, S. and Kroenke, L., 1977. Evolution of the Western Pacific and its margin. Tectonophysics, 38: 145—165.

Hofmann, A.W. and White, W.M., 1980. The role of subducted oceanic crust in mantle evolution. Carnegie Inst. Washington, Yearb., 79: 477—483.

Hofmann, A.W., Wright, T.L. and Feigenson, M., 1978. Trace element and Sr isotope abundances in recent lavas from Kilauea, Hawaii. Carnegie Inst. Washington, Yearb., 77: 590—596.

Isshiki, N., 1963. Petrology of Hachijojima volcano group, Seven Izu Islands, Japan. J. Fac. Sci., Univ. Tokyo, Sec. II, 15: 91—134.

Ito, E. and Stern, R.J., 1981. Oxygen- and strontium-isotope investigations on the origin of volcanism in the Izu—Volcano—Mariana Island Arc. Carnegie Inst. Washington Yearb., 80: 449—454.

Jahn, B.-M., Bernard-Griffiths, J., Charlot, R., Cornichet, J. and Vidal, F., 1980. Nd and Sr isotopic compositions and REE abundances of Cretaceous MORB (Holes 417D and 418A, Legs 51, 52, and 53). Earth Planet. Sci. Lett., 48: 171—184.

Katsumata, M. and Sykes, L.R., 1969. Seismicity and tectonics of the Western Pacific: Izu—Mariana—Caroline and Ryukyu—Taiwan Regions. J. Geophys. Res., 74: 5923—5948.

Kay, R.W., 1980. Volcanic arc magmas: Implications of a melting-mixing model for element recycling in the crust-upper mantle system. J. Geol., 88: 497—522.

Kyser, T.K. and O'Neil, J.R., 1978. Oxygen isotope relations among oceanic tholeiites, alkali basalts, and ultramafic nodules. In: Short Papers of the Fourth International Conference of Geochronology, Cosmochronology, Isotope Geology. U.S. Geol. Surv., Open-File Rep., 78-701; 237—240.

Kyser, T.K., O'Neil, J.R., and Carmichael, I.S.E., 1981. Oxygen isotope thermometry of basic lavas and mantle nodules. Contrib. Mineral. Petrol., 77: 11—23.

Larson, R.L. and Chase, C.G., 1972. Late Mesozoic evolution of the Western Pacific Ocean. Geol. Soc. Am. Bull., 83: 3627—3644.

Lawerence, J.R., Gieskes, J.M. and Broecker, W.S., 1975. Oxygen isotope and cation composition of DSDP pore waters and the alteration of Layer II basalts. Earth Planet. Sci. Lett., 27: 1—10.

Magaritz, M., Whitford, D.J. and James, D.E., 1978. Oxygen isotopes and the origin of high $^{87}Sr/^{86}Sr$ andesites. Earth Planet. Sci. Lett., 40: 220—230.

Mattey, D.P., 1981. Secular geochemical variation in lavas from Truk, Ponape, and Kusaie, Eastern Caroline Islands. Trans. Am. Geophys. Union, 62: 1068.

McCulloch, M.T. and Wasserburg, G.J., 1978. Sm-Nd and Rb-Sr chronology of continental crust formation. Science, 200: 1003—1011.

McCulloch, M.T., Gregory, R.T., Wasserburg, G.J. and Taylor, H.P., Jr., 1980. A neodymium, strontium and oxygen isotopic study of the Cretaceous Samail Ophiolite and implications for the petrogenesis and seawater-hydrothermal alteration of oceanic crust. Earth Planet. Sci. Lett., 46: 201—211.

Meijer, A., 1976. Pb and Sr isotopic data bearing on the origin of volcanic rocks from the Mariana island-arc system. Geol. Soc. Am. Bull., 87: 1358—1369.

Muehlenbachs, K. and Clayton, R.N., 1972. Oxygen isotope geochemistry of submarine greenstones. Can. J. Earth Sci., 9: 471—478.

Murauchi, S., Den, N., Asano, S., Hotta, H., Yoshii, T., Asanuma, T., Hagiwara, K., Ichikawa, K., Sato, T., Ludwig, W.J., Ewing, J.I., Edgar, N.T. and Houtz, R.E., 1968. Crustal structure of the Philippine Sea. J. Geophys. Res., 73: 3143—3171.

O'Nions, R.K., Hamilton, P.J. and Evenson, N.M., 1977. Variations in $^{143}Nd/^{144}Nd$ and $^{87}Sr/^{86}Sr$ in oceanic basalts. Earth Planet. Sci. Lett. 34: 13—22.

O'Nions, R.K., Carter, S.R., Cohen, R.S., Evensen, N.M. and Hamilton, P.J., 1978. Pb, Nd, and Sr isotopes in oceanic ferromanganese deposits and ocean floor basalts. Nature, 273: 435—438.

Perfit, M.R., Gust, D.A., Bence, A.E., Arculus, R.J. and Taylor, S.R., 1980. Chemical characteristics of island-arc basalts: Implications for mantle sources. Chem. Geol., 30, 227—256.

Sager, W.W., 1980. Mariana-Arc structure inferred from gravity and seismic data. J. Geophys. Res., 85: 5382—5388.

Savin, S.M. and Epstein, S., 1970. The oxygen and hydrogen isotope geochemistry of ocean sediments and shales. Geochim. Cosmochim. Acta, 34: 43—63.

Sinha, A.K. and Hart, S.R., 1972. A geochemical test of the subduction hypothesis for generation of island arc magmas. Carnegie Inst. Washington, Yearb., 71: 309—312.

Stern, R.J., 1979. On the origin of andesite in the Northern Mariana Island arc: Implications from Agrigan. Contrib. Mineral. Petrol., 68: 207—219.

Stern, R.J., 1981. A common mantle source for Western Pacific island arc and "hot spot" magmas- Implications for layering in the upper mantle. Carnegie Inst. Washington, Yearb., 80: 455—461.

Stern, R.J., 1982. Strontium isotopes from circum-Pacific intra-oceanic island arcs and marginal basins: Regional variations and implications for magmagenesis. Geol. Soc. Am. Bull., 93: 477—486.

Stern, R.J. and Bibee, L.D., 1980. Esmeralda Bank: Geochemistry of an active submarine volcano in the Mariana Island Arc and its implication for magmagenesis in island arcs. Carnegie Inst. Washington, Yearb., 79: 465—472.

Tatsumoto, M., 1966. Isotopic composition of lead in volcanic rocks from Hawaii, Iwo Jima, and Japan. J. Geophys. Res., 71: 1721—1733.

Taylor, H.P., Jr., 1968. The oxygen isotope geochemistry of igneous rocks. Contrib. Mineral. Petrol., 19: 1—71.

Tilton, G.R. and Barreiro, B.A., 1980. Origin of lead in Andean calc-alkaline lavas, southern Peru. Science, 210: 1245—1247.

Toksöz, M.N. and Bird, P., 1977. Formation and evolution of marginal basins and continental plateaus. In: M. Talwani and W.C. Pitmann III (Editors), Island Arcs, Deep Sea Trenches, and Back-Arc Basins. Am. Geophys. Union, Washington, D.C., pp. 379—393.

Uyeda, S. and Kanamori, H., 1979. Back-arc opening and the mode of subduction. J. Geophys. Res., 84: 1049—1061.

Uyeda, S. and Miyashiro, A., 1974. Plate tectonics and the Japanese Island: A synthesis. Geol. Soc. Am. Bull., 85: 1159—1170.

White, W.M., 1979. Geochemistry of basalts from the FAMOUS area: A re-examination. Carnegie Inst. Washington, Yearb., 78: 325—330.

White, W.M. and Hofmann, A.W., 1978. Geochemistry of the Galapagos Islands: Implications for mantle dynamics and evolution. Carnegie Inst. Washington, Yearb., 77: 596—606.

Whitford, D.J. and Jezek, P.A., 1979. Origin of late-Cenozoic lavas from the Banda Arc, Indonesia: Trace element and Sr isotope evidence. Contrib. Mineral. Petrol., 68: 141—150.

Whitford, D.J., White, W.M. and Jezek, P.A., 1981. Neodymium isotopic composition of Quaternary island arc lavas from Indonesia. Geochim. Cosmochim. Acta, 45: 989—995.

Yagi, K., 1960. Petrochemistry of the alkalic rocks of the Ponape Island, western Pacific Ocean. Rept. 21st Int. Geol. Cong., Pt. 13: 108—122.

Zachariadis, R.G., 1973. Site survey report: Kana Keoki sites 3 and 4, Honolulu to Ponape, 1971. In: B.C. Heezen et al. (Editors), Initial Reports of the Deep Sea Drilling Project, Leg 20. U.S. Government Printing Office, Washington, D.C., pp. 487—503.

Zindler, A., Hart, S.R., Frey, F.A. and Jakobsson, S.P., 1979. Nd and Sr isotope ratios and rare earth element abundances in Reykjanes Peninsula basalts: Evidence for mantle heterogeneity beneath Iceland. Earth Planet. Sci. Lett., 45: 249—262.

Journal of Volcanology and Geothermal Research, 18 (1983) 483—510
Elsevier Science Publishers B.V., Amsterdam — Printed in The Netherlands

OXYGEN AND STRONTIUM ISOTOPIC CHARACTERISTICS OF CALC-ALKALIC VOLCANIC ROCKS FROM THE CENTRAL AND WESTERN JAPAN ARCS: EVALUATION OF CONTRIBUTION OF CRUSTAL COMPONENTS TO THE MAGMAS

YUKIHIRO MATSUHISA and HAJIME KURASAWA

Geological Survey of Japan, 1-1-3 Higashi, Yatabe, Ibaraki 305 (Japan)

(Received October 20, 1982)

ABSTRACT

Matsuhisa, Y. and Kurasawa, H., 1983. Oxygen and strontium isotopic characteristics of calc-alkalic volcanic rocks from the central and western Japan arcs: evaluation of contribution of crustal components to the magmas. In: S. Aramaki and I. Kushiro (Editors), Arc Volcanism. J. Volcanol. Geotherm. Res., 18: 483—510.

$^{18}O/^{16}O$ and $^{87}Sr/^{86}Sr$ ratios were determined for Quaternary calc-alkalic volcanic rocks from six volcanic rock suites in the central and western Japan arcs. The $\delta^{18}O$ values relative to SMOW and $^{87}Sr/^{86}Sr$ ratios range from +6.3 to +9.9‰ and 0.70357 to 0.70684, respectively. Both the O- and Sr-isotopic compositions are higher than those for island-arc primitive magmas and their differentiates. The isotopic compositions of the calc-alkalic rocks cannot be derived by a simple fractional crystallization of the primitive magmas. On the other hand, the ^{18}O- and ^{87}Sr-enrichment is confined to the rock suites located in well-developed island arcs having thick continental-type crust with low or negative Bouguer anomalies. Involvement of ^{18}O- and ^{87}Sr-rich crustal material in the magma formation is suggested.

The isotopic compositions vary remarkably within individual rock suites as well as from volcano to volcano. The data points in $\delta^{18}O$ vs. $^{87}Sr/^{86}Sr$ plot accord with a mixing model between primitive magmas and crustal material of dioritic composition on an average, assuming their comparative Sr contents. The primitive magmas involved could not be low-Sr tholeiites, but magmas more or less enriched in incompatible elements including Sr, which correspond to high-alkali tholeiites or alkali basalts and their evolved magmas. The nature of the primitive magmas seems to change from tholeiitic to more alkalic with progressing island-arc evolution.

Mixing of crust-derived melts is more plausible than assimilation of solid-rocks for involving 20 to 30% crustal material in the magmas along simple mixing curves. Isotopic variations between the rock suites are ascribed to variable Sr concentration ratio of the end-members, variable isotopic compositions of crustal material or variable mixing ratio of the end-members. Extremely high-$\delta^{18}O$ rocks with moderate increase in $^{87}Sr/^{86}Sr$ ratio suggest another mixing process in shallower magma chambers between andesite magmas and metasedimentary rocks having high $\delta^{18}O$ and $^{87}Sr/^{86}Sr$ values but low Sr content. Subsequent fractional crystallization of once-derived magmas would be the prominent process for the rock suites showing gradual increase in ^{18}O up to 1‰ with uniform $^{87}Sr/^{86}Sr$ ratios.

0377-0273/83/$03.00

INTRODUCTION

Arc volcanism is characterized by prevalence of calc-alkalic andesites as well as island-arc tholeiites. Generally speaking, the main rocks in immature island arcs built up directly on the sea-floor are basalts and basaltic andesites of tholeiitic rock series, whereas those in well-developed island arcs with thick, continental-type crust are andesites and dacites of tholeiitic and calc-alkalic rock series (Miyashiro, 1974). Calc-alkalic rocks show, as a whole, more evolved features than do tholeiitic or alkalic rocks which have more primitive basaltic members. Various models have been proposed to interpret the genesis of the calc-alkalic andesites, which include differentiation of basaltic magma, mixing of basaltic magma with crustal materials, direct melting of the mantle wedge under hydrous condition, and melting of subducted ocean crust (Gill, 1981).

Controversy has been held on possible involvement of subducted ocean crust in the generation of island-arc magmas on the basis of trace-element or isotopic data. In case of calc-alkalic andesites occurring in well-developed island arcs, however, it seems desirable to evaluate possible modification of magmas by reaction with crustal material during their ascent before relating their characteristics to ultimate source region.

O and Sr isotopes would be a good indicator of crust contamination or assimilation because of high enrichment in ^{18}O and ^{87}Sr of crustal rocks relative to primary magmas (Taylor, 1980; James, 1981). In well-developed arc systems and continental margins, many authors suggested that $^{87}Sr/^{86}Sr$ ratios of primary magmas of calc-alkalic rocks, and partly tholeiitic rocks as well, have been modified by mixing with ^{87}Sr-rich crustal material (e.g. Ewart and Stipp, 1968; Leeman and Manton, 1971; Francis et al., 1977, 1980; Briqueu and Lancelot, 1979; Leeman, 1981). On the basis of O-Sr isotopic correlation, involvement of crustal material in subduction zones has been discussed for the Andean andesites (James, 1978, 1981) and the Banda arc rock suite (Magaritz et al., 1978).

The Japanese island arcs, where different types or stages of island-arc evolution are observed, would be one of the best fields for studying arc volcanism. Tatsumoto and Knight (1969) interpreted U-Th-Pb isotopic systematics of the primitive basalts across the Japanese island arcs as due to vertical differentiation of the upper mantle about 3.6 Ga ago or mixing of mantle material with subducted ocean crust including pelagic sediments. They ascribed enrichment in radiogenic Pb isotopes of calc-alkalic rocks to crustal contamination. Hedge and Knight (1969) pointed out the less radiogenic tendency of Pb and Sr isotopes toward continent side across northern Honshu. Contribution of subducted slab material to the magma sources as well as crustal contamination across the arc were also discussed on the basis of Nd-Sr isotopic correlation (Nohda and Wasserburg, 1981).

Masuda and Aoki (1979) discussed trace-element correlation including Cr, Ni, Th and REE for the volcanic rocks from northeast Japan to conclude generation of independent primary magmas of tholeiitic and calc-alkalic by different degrees of partial melting of peridotitic upper mantle and subsequent fractional crystallization. Matsuhisa (1979) distinguished two types of andesites of the East Japan island arcs on their O isotopic ratios. That is, andesites and dacites from immature arcs lie on differentiation trend of island-arc tholeiite which could be directly originated from the upper mantle, whereas, andesites from well-developed arcs are relatively enriched in ^{18}O, suggesting differentiation at lower temperatures, mixing of ^{18}O-rich crustal material, or involvement of subducted ocean crust.

In the present study, Quaternary calc-alkalic rock suites from central and western parts of the Japanese island arcs where continental-type crust is well developed were analyzed for O and Sr isotopes. The results will be discussed in view of mixing models in O-Sr isotopic correlation.

GEOLOGIC SETTING

The Japanese islands are grouped into two major arc-systems on the basis of their geologic and geophysical characteristics: the East Japan island arcs and the West Japan island arcs (Sugimura and Uyeda, 1973). Development of trench and dipping of the Benioff zone toward continent are clearly observed in the East Japan island arcs, but not so clear in the West Japan island arcs (Fig. 1). The areas of four main islands have a well-developed continental-type crust, consisting of succession of sedimentary and igneous rocks of Paleozoic to Tertiary. The crustal thickness increases toward continent (westward) in northern Honshu and reaches the maximum (> 36 km) in central Honshu (Sugimura and Uyeda, 1973).

Six volcanoes consisting of calc-alkalic rocks suites, Yatsugataké and Kiso-Ontaké from central Japan, and Daisen, Aonoyama, Unzen and Aso from western Japan were selected for oxygen and strontium isotopic analyses. Ages of their volcanic activity are dated to be from 0.3 Ma to present. Their localities are shown in Fig. 1 together with the volcanoes in the East Japan island arcs studied by Matsuhisa (1979). From the view point of arc structure, Yatsugataké and Kiso-Ontaké may belong to the East Japan island arcs. K_2O content increases from Yatsugataké (0.9%) toward Kiso-Ontaké (1.5%) at SiO_2=55%. The four other volcanoes seem to belong to a common arc system; the West Japan island arcs. However, the arc structure is not so clearly reflected in chemical characteristics of the rocks. For example, K_2O content in Aso (3.3 to 3.9%), which is located on the volcanic front, is much higher than that of an inner volcano, Unzen (2.3 to 2.5%) at SiO_2=61%. This observation may relate to crustal thickness as suggested by Ui and Aramaki (1978).

Yatsugataké is a volcano group which extends about 50 km long in NS direction at the convergence of two major tectonic features: the Median Tectonic Line and the Fossa Magna (Kawachi, 1974, 1977). There is a gap in chemical composition between southern and northern groups. The southern

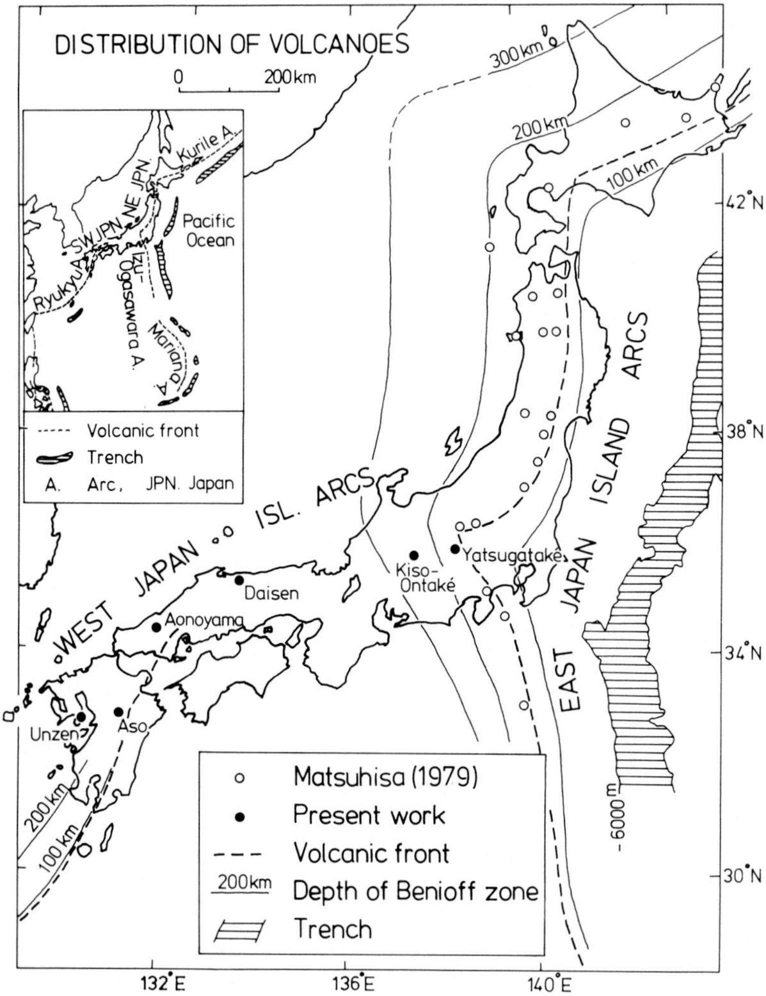

Fig. 1. The Japanese island arcs and the distribution of the volcanoes for the present study (solid circles) and Matsuhisa (1979) (open circles).

group consists of olivine-two-pyroxene andesites with SiO_2 content of 53 to 54% , whereas the northern group consists of hornblende-two-pyroxene and dacites with SiO_2 content of 59 to 63% (Kawachi, 1974, 1977). The Yatsugataké volcanic rocks contain xenoliths of a large variety of basement rocks (Kawachi, 1977).

The Kiso-Intaké volcano in central Honshu consists of a calc-alkalic rock suite of augite-olivine basalts, hornblende-two-pyroxene andesites and hornblende-hypersthene rhyolites. The volcanic activity is divided into four stages with a long quiescent period between the Stages I and II (Kobayashi et al., 1975). The volcano is built up on the basement of Paleozoic formations consisting of slate, sandstone and chert, and Cretaceous rhyolite welded-tuffs and granitoids.

The Daisen and Aonoyama volcanoes, both located in western Honshu, consist of biotite-bearing hornblende andesites, with a very narrow range of chemical composition (SiO_2 = 61—64%), which erupted on the basement of Cretaceous granitoids (Ota, 1962; Kurasawa, 1965). Murakami (1975) reported several types of granulitic inclusions in the volcanic rocks from western Honshu.

The Unzen volcano located in northwestern Kyushu consists of biotite-bearing pyroxene-hornblende andesites and dacites. Biotite-bearing is characteristic of the volcanoes distributed along the Sea of Japan, including Daisen, Aonoyama and Unzen. Kurasawa and Takahashi (1965) distinguished six stages of volcanic activity at Unzen, stressing discontinuity by tectonic movement between the Stages III and IV. They described that rocks in the Stages I to III are characterized by pyroxenes, whereas rocks erupted after the tectonic movement have higher amounts of biotite and hornblende.

The Aso volcano in central Kyushu erupted large-scale pyroclastic flows four times and formed a large caldera, 24 by 18 km in diameter, which was succeeded by central cone activity (Ono and Watanabe, 1981). The pre-caldera rocks are predominantly felsic with some intermediate rocks and minor volume of basic rocks, while the commonest rock type in the post-caldera stage is basaltic andesite. Most samples for the pre-caldera stage were collected from essential lenses and pumices in pyroclastic deposits with the exception of lavas of the A2/1 stage.

ANALYTICAL PROCEDURES

Identical splits of powdered samples were used for both O and Sr isotopic analyses. SiO_2 and K_2O (recalculated as K in Table I) contents were taken from published data except for the Aso volcano (unpublished data by K. Ono).

Oxygen was extracted from rocks by reaction with bromine pentafluoride, then quantitatively converted to carbon dioxide using heated carbon for mass-spectroscopic analysis (Matsuhisa et al., 1971). The isotopic data are reported in the conventional $\delta^{18}O$ notation defined as follows:

$$\delta^{18}O = \left[\frac{(^{18}O/^{16}O)_{sample}}{(^{18}O/^{16}O)_{standard}} - 1\right] \times 1000\ (^0/_{00})$$

where standard is Standard Mean Ocean Water (SMOW) (Craig, 1961). Average $\delta^{18}O$ value of $+9.3^0/_{00}$ was obtained for the quartz reference sample NBS-28 on this scale. Experimental error is $\pm 0.1^0/_{00}$.

Strontium isotopic compositions were measured on a VG Isomass 54E double collector mass spectrometer with on-line computer facilities using a single tantalum filament. All measurements have been normalized to $^{86}Sr/^{88}Sr$=0.1194 and presented as $(^{87}Sr/^{86}Sr)n$ in Table I. Average $^{87}Sr/^{86}Sr$ ratio of 0.70805±2 were obtained for the Eimer and Amend standard $SrCO_3$.

Rb and Sr contents were determined by X-ray fluorescence with an uncertainty of ±2 to 4%.

RESULTS AND POST-ERUPTIVE ALTERATION

Isotopic results are presented in Table I together with SiO_2, K, Rb and Sr concentrations. The total variation ranges of $\delta^{18}O$ values and $^{87}Sr/^{86}Sr$ ratios are +6.3 to +9.9‰ and 0.70357 to 0.70684, respectively. Sr content ranges from 220 to 730 ppm with the exception of extremely high values of around 1000 ppm for the Aonoyama andesites. K/Rb ratios are mostly in the range of 200 to 350 with higher ratios of 350 to 750 for the Yatsugataké and Aonoyama rocks. Rb/Sr ratios vary from 0.02 to 0.12 for the Yatsugataké rock suite to 0.10 to 0.58 for the Aso rock suite.

Before considering isotopic variation in magmas it is necessary to examine the possible effects of weathering processes to isotopic composition. Weathering processes may increase both H_2O content and $\delta^{18}O$ value of rocks because of formation of clay minerals at low temperatures (Muehlenbachs and Clayton, 1972; Lawrence and Taylor, 1972). The lavas from Kiso-Ontaké, Yatsugataké and Unzen have total water contents lower than 2% with no correlation to $\delta^{18}O$ values, showing no evidence of alteration by weathering (Fig. 2a). At the Aso volcano no correlation between $\delta^{18}O$ value and water content is found in case of lavas, essential lenses and scoriae, all of which have water content less than 2%. However, pumice samples have water contents of 3 to 6% with high $\delta^{18}O$ values as a whole. Tie-lines in Fig. 2b connecting pumice and rock samples which have similar chemical composition in the same eruption stage suggest that the ^{18}O-enrichment has resulted from hydration. On the other hand, no remarkable change in $^{87}Sr/^{86}Sr$ ratio is observed for the corresponding sample pairs (Table I). Therefore, it is inferred that subsequent post-eruptive hydration of pumice has effected a change in the $\delta^{18}O$ values but not altered the $^{87}Sr/^{86}Sr$ ratios. The oxygen isotopic data for the pumice samples will not be included in the following discussion.

ISOTOPIC VARIATION IN INDIVIDUAL VOLCANOES

Figures 3 and 4 show $\delta^{18}O$ and $^{87}Sr/^{86}Sr$ variations in order of eruption in the volcanoes of Yatsugataké, Kiso-Ontaké, Unzen and Aso. Time scale and volume of products of eruptions are not specified. The isotopic compositions vary remarkably within single volcano as well as from volcano to volcano. It is noticed that rocks having discordantly high $\delta^{18}O$ values or $^{87}Sr/^{86}Sr$ ratios appear at intervals in a sequence of eruptions, although high $\delta^{18}O$ values are not always accompanied with high $^{87}Sr/^{87}Sr$ ratios. At Unzen and Aso, eruptions in earlier stages as a whole have higher $\delta^{18}O$ values and $^{87}Sr/^{86}Sr$ ratios.

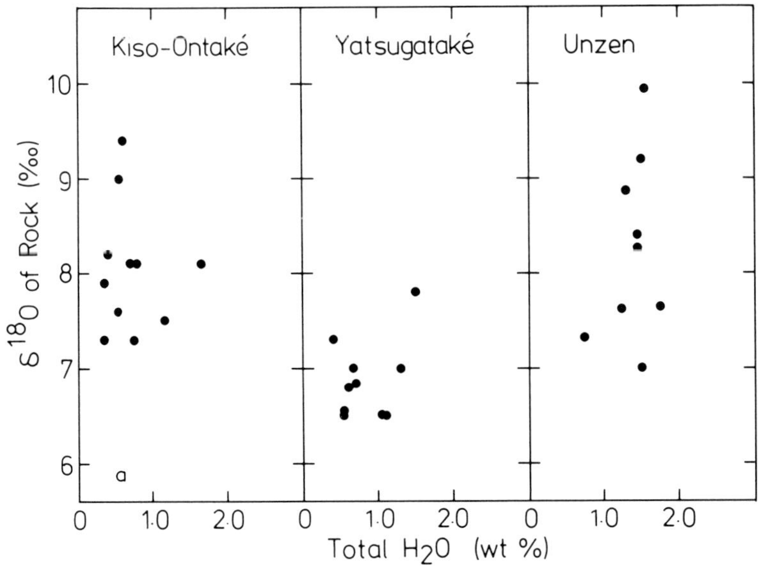

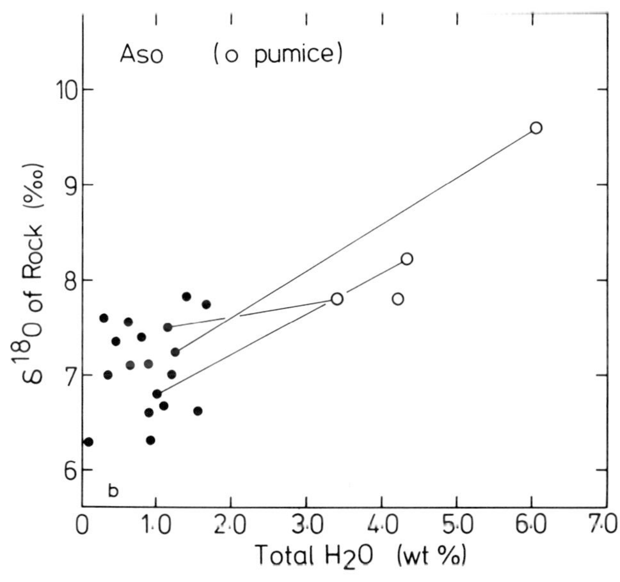

Fig. 2a. Plot of $\delta^{18}O$ vs. weight percent total water of the volcanic rocks from Kiso-Ontaké, Yatsugataké and Unzen. b. Plot of $\delta^{18}O$ vs. weight percent total water of the Aso volcanic rocks. Data points for pumice and rock samples with similar chemical compositions in the same stage are connected by tie-lines.

TABLE I

Chemical and isotopic data for the volcanic rocks from the central and western Japan arcs

Sample No.	Rock type	SiO_2 (wt.%)	K (%)	Rb (ppm)	Sr (ppm)	$(^{87}Sr/^{86}Sr)n$	$\delta^{18}O_{SMOW}$ (‰)
1. Yatsugataké							
766	Augite-olivine basalt	48.9	0.46	6	633	0.70372	—
933	Augite-olivine andesite	50.9	1.02	25	521	0.70388	—
11	Two-pyroxene andesite	52.9	0.96	22	725	0.70379	+6.8
240	ditto	53.0	0.90	16	636	0.70369	+7.0
894	Olivine-two-pyroxene andesite	54.4	0.79	17	719	0.70380	+6.5
886	ditto	54.0	0.90	23	634	0.70411	+7.3
897	ditto	52.7	0.90	21	581	0.70449	+6.5
1054	Hornblende-two-pyroxene dacite	62.6	1.83	30	260	0.70388	+6.8
1110	ditto	62.8	1.95	56	484	0.70490	+7.0
396	Hornblende-two-pyroxene andesite	58.9	1.34	37	540	0.70382	+7.8
984	ditto	61.7	1.73	55	577	0.70375	+6.5
988	ditto	60.2	1.33	30	629	0.70387	+6.6
2. Kiso-Ontaké							
1	Olivine-augite andesite	51.2	1.21	34	569	0.70526	+7.5*
2	Two-pyroxene-hornblende rhyolite	65.4	2.78	123	537	0.70533	—
3	Augite-olivine basalt	50.8	1.15	31	638	0.70569	+7.3*
4	Two-pyroxene-hornblende andesite	63.3	2.06	84	728	0.70684	+8.2*
5	Two-pyroxene andesite	59.6	2.81	139	573	0.70523	+8.1*
6	Hornblende-two-pyroxene andesite	59.1	2.26	114	602	0.70515	+8.1
7	Hypersthene rhyolite	69.3	2.46	94	462	0.70602	+9.4*
8	Two-pyroxene andesite welded tuff	61.7	2.21	80	554	0.70598	+7.6
9	Two-pyroxene andesite	62.2	2.14	87	536	0.70564	+9.0*
10	ditto	55.4	1.29	41	671	0.70537	+7.3
11	ditto	57.9	1.63	57	558	0.70672	+7.9
12	Hornblende-two-pyroxene andesite	57.9	1.76	65	599	0.70562	+8.1*
3. Daisen							
DS-1	Porphyritic dike	—	—	—	—	—	+7.3
DS-4	Pyroxene andesite	61.5	1.73	44	670	0.70495	+6.9
DS-5A	Biotite-hornblende andesite	63.9	—	—	—	—	+7.0
DS-5B	ditto	63.9	—	—	—	0.70522	+7.1
DS-6	ditto	62.3	—	—	—	0.70491	+7.2
4. Aonoyama							
58112502	Hornblende andesite	61.5	1.40	19	1073	0.70357	+7.1
58112505	ditto	61.8	1.61	43	952	0.70389	+7.3

5. Unzen							
57120223	Hypersthene-hornblende andesite	61.7	2.08	72	376	0.70481	+8.9
57113024	Two-pyroxene-biotite-hornblende andesite	60.2	2.09	92	350	0.70509	+9.9
57120201A	ditto	59.4	2.00	69	432	0.70457	+8.4
57120125	Biotite-hornblende andesite	61.2	2.00	72	358	0.70481	+8.3
57120102	Two-pyroxene-biotite-hornblende andesite	61.5	1.92	62	422	0.70427	+7.3
57120228B	ditto	55.3	1.67	51	416	0.70428	+9.2
57120301	Hypersthene-biotite-hornblende dacite	62.4	2.24	85	348	0.70452	+7.6
57120206	Biotite-hornblende dacite	59.4	2.00	67	335	0.70462	+7.3
57120227	Two-pyroxene-biotite-hornblende dacite	60.5	2.18	71	405	0.70458	+7.6
57112903	Olivine-biotite-hornblende andesite	55.8	1.53	46	366	0.70430	+7.0
57113002	ditto	55.6	1.53	47	349	0.70425	—
57113003	Two-pyroxene-biotite-hornblende dacite	63.8	2.00	74	310	0.70470	+7.6
6a Aso (pre-caldera)							
64AS86-I-1	Rhyolite lens	66.1	4.00	129	221	0.70449	+7.6
68WF121A1	Andesite lava	54.8	1.48	47	486	0.70415	+7.1
67TD469	Andesite lava	61.6	3.27	91	341	0.70417	+7.8
67WF80G	Andesite lens	61.0	3.22	101	361	0.70426	+7.4
68AS67A	Andesite pumice	61.4	2.82	95	318	0.70424	+7.8
68TD588TL	Rhyolite pumice	65.9	3.61	116	235	0.70422	+9.6
62AS15B	Rhyolite lens	65.2	3.89	127	240	0.70420	+7.2
61AS25	Andesite scoria	57.8	2.50	84	418	0.70417	+6.6
64TM2	Rhyolite tuff	67.9	3.85	133	243	0.70420	+7.5
67AS318-2	Rhyolite-pumice	68.5	3.86	134	235	0.70416	+7.8
61TD12-OB	Andesite lens	63.3	3.13	105	380	0.70413	+7.0
61AS22B	Andesite scoria	56.0	1.97	60	567	0.70411	+6.7
68ME501A3	Rhyolite pumice	68.8	3.22	115	279	0.70418	+8.2
64AS142A2	Rhyolite lens	67.7	3.37	113	319	0.70420	+6.8
KC-611	Andesite scoria	51.2	1.26	—	—	0.70413	+6.3
6b Aso (central cones)							
69WF195A1	Rhyolite lava	71.8	4.24	146	126	0.70438	+7.1
80AS561A	Dacite lava	67.6	3.31	—	—	0.70434	+7.0
80AS567A	Andesite lava	57.1	1.82	—	—	0.70438	+7.6
KWTNO5	Andesite lava	61.1	3.09	—	—	0.70418	+7.4
80AS596	Dacite lava	63.7	2.88	—	—	0.70431	+7.7
80AS587V	Basalt lava	51.1	1.08	—	—	0.70403	+6.3
KWASO4	Basalt lava	51.7	1.22	—	—	—	+6.7
AS741015	Andesite scoria	52.9	1.57	—	—	0.70410	+6.6

*Matsuhisa et al. (1973).

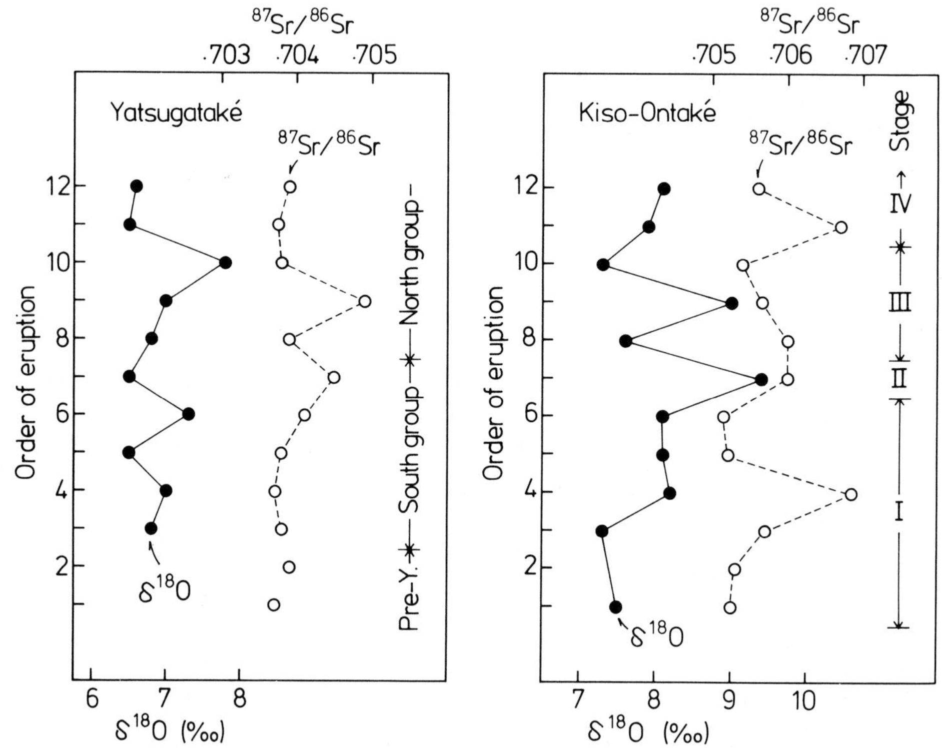

Fig. 3. $\delta^{18}O$ and $^{87}Sr/^{86}Sr$ variations in order of eruptions in the Yatsugataké and Kiso-Ontaké volcanoes. "Pre-Y." stands for the pre-Yatsugataké eruptions.

If magmatic processes controlled by fractional crystallization or melting of a homogeneous reservoir are dominant, a smooth increase in ^{18}O (up to 1‰) would be expected at least within a single stage of volcanic activity (Matsuhisa, 1979). Such magmatic processes produce little isotopic fractionation of strontium, so that $^{87}Sr/^{86}Sr$ ratios would be virtually constant throughout a rock suite derived from a common magma. The magnitude of isotopic variations observed in Figs. 3 and 4 is much greater than that expected in a magmatic process of a single origin. Assimilation of ^{18}O-rich or ^{87}Sr-rich material into magmas will be demonstrated in $\delta^{18}O$ vs. SiO_2 plot (Figs. 5 to 7) and in $^{87}Sr/^{86}Sr$ vs. Sr plot (Fig. 8).

$\delta^{18}O$ vs. SiO_2 plot

Matsuhisa (1979) showed that island-arc tholeiites from Hachijo-jima have $\delta^{18}O$ values of around +5.7‰ identical to that for mantle-derived basalts from other oceanic regions. The $\delta^{18}O$ values of andesites and dacites from Hachijo-jima (+6.0 to +6.7‰ identical to that for mantle-derived basalts the tholeiitic magma. Ito and Stern (1981) reported similar $\delta^{18}O$-range for tholeiitic rocks from the Mariana island arc.

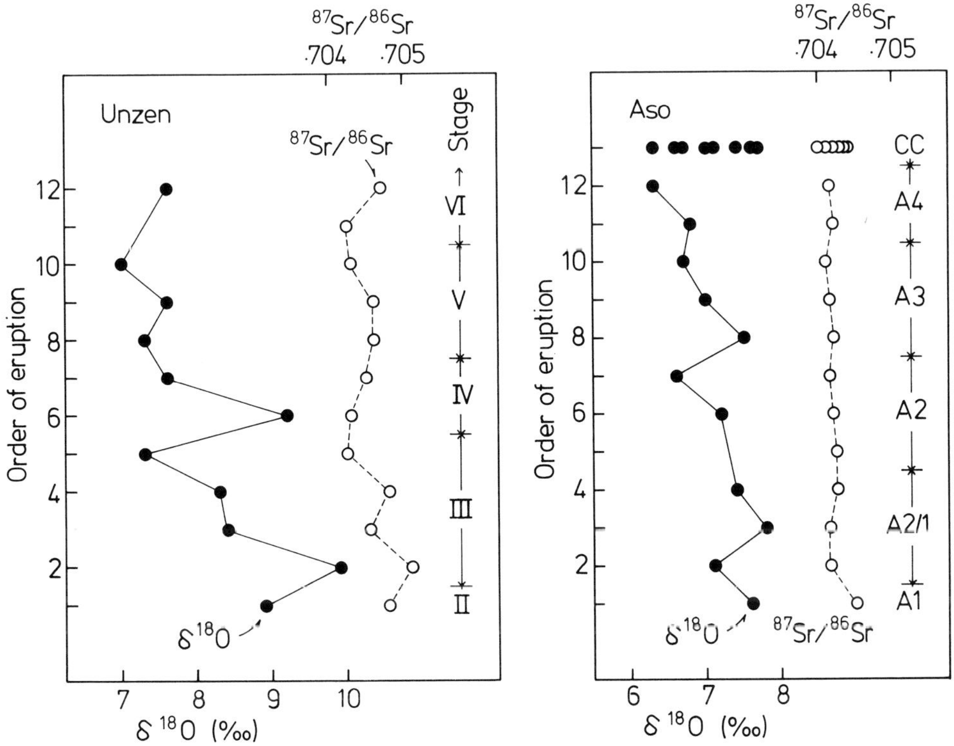

Fig. 4. $\delta^{18}O$ and $^{87}Sr/^{86}Sr$ variations in order of eruptions in the Unzen and Aso volcanoes. At the Aso volcano CC stands for the central cones, while the eruptions from A1 through A4 are of the pre-caldera stages.

Figures 5 to 7 demonstrate that the calc-alkalic rocks from the central and western Japan arcs are evidently enriched in ^{18}O, by 0.5 to 3.5‰, relative to the island-arc tholeiite rock suite.

The Kiso-Ontaké rock suite has $\delta^{18}O$ values ranging from +7.3 to +9.4‰. Rocks from the Stages I, III and IV form two variation trends in the plot, being parallel to and enriched in ^{18}O by 1.0 to 1.5‰ relative to the differentiation trend of island-arc tholeiite magma (Fig. 5). A rhyolite of the Stage II and an andesite of the Stage III have $\delta^{18}O$ values of +9.4 and +9.0‰ respectively, distinctly higher than the main trends. In the Stage III at Unzen, successive eruptions with virtually constant chemical composition show a large variation of $\delta^{18}O$ values gradually decreasing from +9.9 to +7.3‰ with order of eruption. The main variation trend excepting for the anomalously high $\delta^{18}O$ rocks of the Unzen volcano is again parallel to and enriched in ^{18}O by 1‰ than the differentiation trend of island-arc tholeiite magma (Fig. 6).

The Aso volcanic rocks have $\delta^{18}O$ values in the range of +6.3 to +7.8‰. The variation range is much smaller than those of the above two volcanoes. Two or three different trends each of which corresponds to a stage of erup-

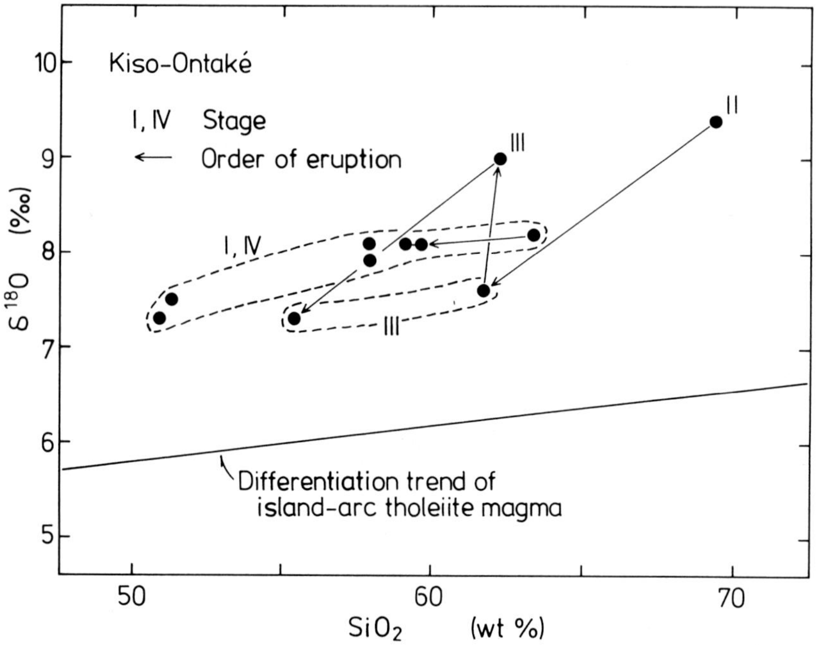

Fig. 5. Plot of $\delta^{18}O$ vs. SiO_2 content for the volcanic rocks from Kiso-Ontaké. The differentiation trend of island-arc tholeiite magma induced from the Hachijo-jimo rock suite (Matsuhisa, 1979) is given for comparison.

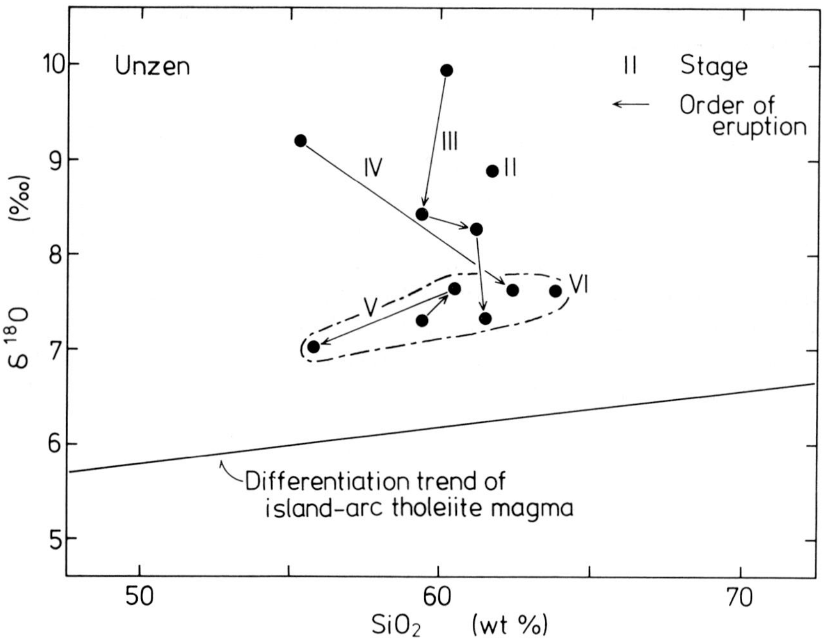

Fig. 6. Plot of $\delta^{18}O$ vs. SiO_2 content for the volcanic rocks from Unzen. The main variation trend is enclosed by a broken line.

tion are identified for the Aso volcanic suite in the diagram (Fig. 7). The trends show enrichment in ^{18}O by 0.5 to 1.5‰ compared to the variation trend of island-arc tholeiite magma. The products of central cones scatter over the composition range of the pre-caldera rocks. The lavas of the Stage A2/1 together with some of the central cone products form a variation trend about 1‰ higher than others.

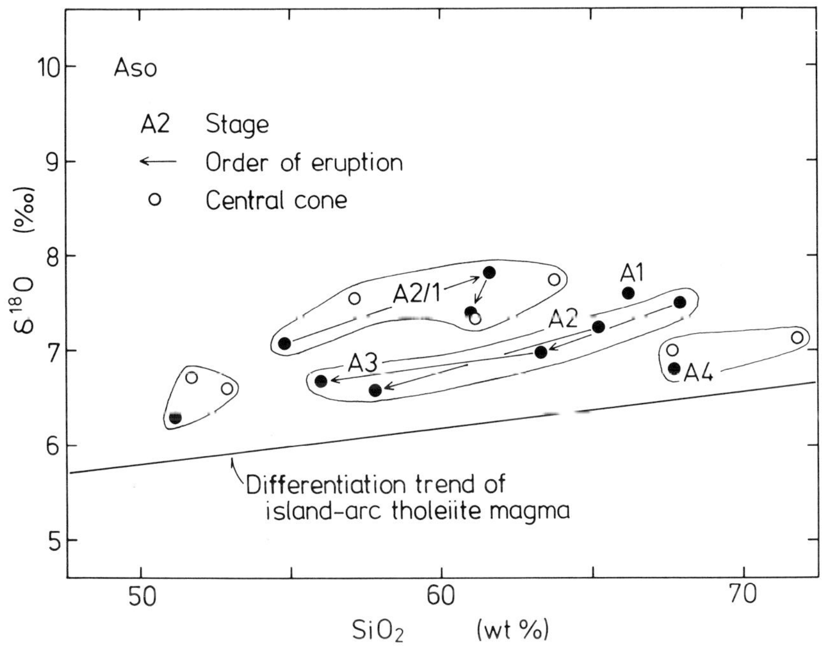

Fig. 7. Plot of $\delta^{18}O$ vs. SiO_2 content for the volcanic rocks from Aso. The pre-caldera and post-caldera eruptions are shown by solid and open circles, respectively.

Since the calc-alkalic rocks studied are porphyritic more or less, isotopic and chemical compositions of whole-rocks do not necessarily represent those of magmas. Each variation trend would represent a suite of rocks which have variable amount of phenocrysts with similar composition of melt (matrix), or rocks derived from a common magma by fractional crystallization. Anyhow, parallel variation trends in the plot cannot be derived from one another by simple fractional crystallization. Instead, they may suggest derivation of magmas from isotopically heterogeneous sources. However, mixing of an isotopically homogeneous primary magma with ^{18}O-rich material in varied proportions and subsequent fractional crystallization could also produce a set of parallel variation trends.

It is noticed that different variation trends occur within a single volcano as well as among volcanoes in separate regions. At Kiso-Ontake and Unzen, large spread of $\delta^{18}O$ values is recognized even within a single stage. It is not plausible that magmas are derived from isolated, isotopically heterogeneous sources in limited time and space. Variable degree of assimilation or variable $\delta^{18}O$ values of assimilated material seemingly produced the $\delta^{18}O$ variation.

$^{87}Sr/^{86}Sr$ vs. Sr plot

Figure 8 shows variation in $^{87}Sr/^{86}Sr$ ratio of the rocks against Sr content. The composition ranges are also shown for island-arc tholeiites (IATH) from the Izu arc (Pushkar, 1968; Kurasawa, 1979; Notsu, 1981b; Hirano, 1982),

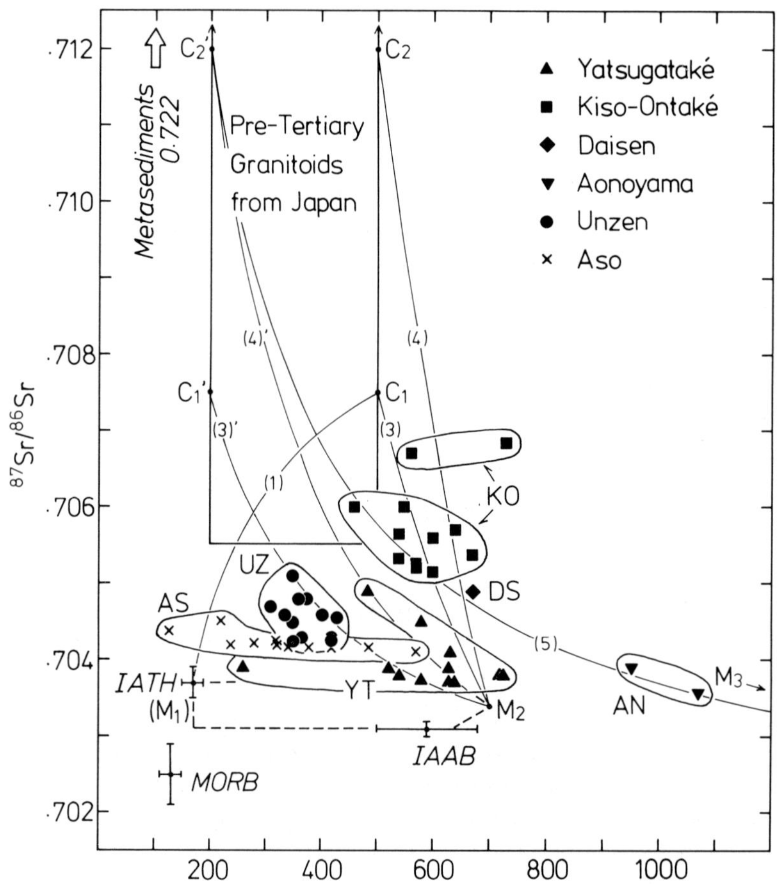

Fig. 8. Plot of $^{87}Sr/^{86}Sr$ vs. Sr content for the calc-alkalic volcanic rocks from the central and western Japan arcs. The composition ranges are also shown for island-arc tholeiites (IATH), island-arc alkalic basalts (IAAB) and pre-Tertiary granitoids and metasediments from Japan, and mid-ocean ridge basalts (MORB). Possible composition range of the primitive magmas in the Japanese island-arcs are indicated by the area enclosed by a broken line. Mixing curves (1) to (5) and end-members of mixing M_1, M_2, M_3, $C_1(C_1')$ and C_2 (C_2') correspond to those in Fig. 9.

for island-arc alkalic basalts (IAAB) from Oshima-Oshima (Hedge and Knight, 1969; Katsui et al., 1978; Notsu, 1981a) and Irozaki (Kurasawa, 1979), and for MORB (Dasch et al., 1973; Hart et al., 1974; O'Nions et al., 1977). The large rectangle in higher $^{87}Sr/^{86}Sr$ region represents variable composition range for Pre-Tertiary granitoids including gabbroic rocks from central and western Japan (Shigeno and Yamaguchi, 1976; Shibata and Ishihara, 1979a,b).

It is remarkable that the calc-alkalic rocks show a large spread of $^{87}Sr/^{86}Sr$ ratios ranging from 0.7036 to 0.7068, distributing between the primitive basalts and the Pre-Tertiary granitoids. The Yatsugataké volcanic rocks have two distinct trends in the diagram: (a) virtually constant $^{87}Sr/^{86}Sr$ ratios of 0.7037 to 0.7039 with variable Sr content and (b) gradual increase in $^{87}Sr/^{86}Sr$ ratio up to 0.7049 with decreasing in Sr content. The two trends converge into a point with $^{87}Sr/^{86}Sr \approx 0.7038$ and ppm Sr $\approx$ 700. The Kiso-Ontaké volcanic rocks have the highest and highly variable $^{87}Sr/^{86}Sr$ ratios of 0.7052 to 0.7068. A general tendency of increase in $^{87}Sr/^{86}Sr$ ratio with decreasing in Sr content is observed. The Unzen volcanic rocks have $^{87}Sr/^{86}Sr$ ratios of 0.7042 to 0.7051 with a relatively narrow range of Sr content (310 to 430 ppm), yet an overall negative correlation between $^{87}Sr/^{86}Sr$ ratio and Sr content is recognized. $^{87}Sr/^{86}Sr$ ratios of the Aso volcanic rocks fall in narrow range of 0.7040 to 0.7045 with a large variation of Sr content. The Aonoyama andesites have the highest Sr content of around 1100 ppm with a low $^{87}Sr/^{86}Sr$ ratio of 0.7036.

A rock suite produced by melting and fractional crystallization of a common source material would be distributed along a horizontal line in the diagram depending upon Sr partition between minerals and melt, and degree of crystallization. Alternatively, contamination by radiogenic Sr or combined contamination-fractional crystallization would produce an increase or decrease in $^{87}Sr/^{86}Sr$ ratio with Sr content depending upon $^{87}Sr/^{86}Sr$ ratio and Sr content of contaminating material (Briqueu and Lancelot, 1979).

Virtually uniform $^{87}Sr/^{86}Sr$ ratios with variable Sr content at Yatsugataké and Aso suggest fractional crystallization with no remarkable, if any, contamination by ^{87}Sr-rich material. Gradual decrease in Sr content indicates removal of plagioclase-dominant cumulates from the magmas (Higuchi and Nagasawa, 1969; Philpotts and Schnetzler, 1970). Different amounts of plagioclase phenocrysts may also give apparently variable Sr content of the rocks. At Kiso-Ontaké, however, correlation between Sr and plagioclase-phenocryst contents is not observed (Kobayashi et al., 1975).

Negative correlation of $^{87}Sr/^{86}Sr$ vs. Sr observed in a part of Yatsugataké, Kiso-Ontaké, Unzen and probably in Aonoyama suggests assimilation of ^{87}Sr-rich but Sr-poor material relative to the magmas. Since the data points for these suites are rather scattered, the assimilation process cannot involve a simple binary mixing.

Isotopic variation among volcanoes strongly suggests assimilation of ^{87}Sr-rich material at different degrees, or origin of the magmas in isotopically heterogeneous sources. It is observed that the calc-alkalic andesites have Sr contents (normally 200 to 700 ppm) much higher than those of primitive basalts, especially of tholeiites (150 to 190 ppm). If assimilation is assumed, mixing is required between a ^{87}Sr-poor, Sr-rich magma and ^{87}Sr-rich material with variable Sr content. Mixing curves presented in Fig. 8 will be referred in the following discussion.

DISCUSSION

Both O and Sr isotopic data suggest involvement of ^{18}O- and ^{87}Sr-rich crustal material in the calc-alkalic magmas. The large spread in isotopic compositions of the calc-alkalic rocks within a volcano as well as among volcanoes may be due to variable degree of assimilation or variable isotopic compositions of assimilated material. Possible rock types for assimilation and possible composition of primary magmas will be discussed on the basis of mixing models in $\delta^{18}O$-$^{87}Sr/^{86}Sr$ correlation.

Primitive magmas in island arcs

It is well known that the most primitive island-arc magmas observed are tholeiitic on the volcanic front, and become alkalic toward the continent side (Kuno, 1960; Dickinson and Hatherton, 1967). Calc-alkalic rocks typically are distributed throughout island arcs including tholeiitic and alkalic provinces. Although primitive magmas directly related to calc-alkalic rocks have not been identified unambiguously, they may be in composition range between the two extreme members of the most primitive magmas across island arcs: low-alkali tholeiite on ocean side and alkali olivine basalt on continent side.

Tholeiites from primitive arcs have uniform $\delta^{18}O$ values of +5.7 ± 0.2‰ (Muehlenbachs and Hoering, 1972; Matsuhisa et al., 1973; Matsuhisa, 1979; Ito and Stern, 1981) which are identical to the values of primitive basalts from other oceanic regions including MOR basalts. According to Matsuhisa (1979), alkali olivine basalts from Oshima-Oshima located in the Sea of Japan west of Hokkaido have the same $\delta^{18}O$ value of +5.7‰. Therefore, we could safely assume that the most primitive basaltic magmas in the Japanese island arcs have $\delta^{18}O$ value of +5.7 ± 0.2‰ regardless of rock types.

On the other hand, it has been observed that $^{87}Sr/^{86}Sr$ ratios of the most primitive magmas decrease from 0.7037 for tholeiites on ocean side to 0.7031 for alkali olivine basalts on continent side of the Japanese island arcs (Pushkar, 1968; Hedge and Knight, 1969; Kurasawa, 1979; Notsu, 1981a,b). Variation in $^{87}Sr/^{86}Sr$ ratio is also found within tholeiites along the volcanic front, reaching the lowest value of 0.7034 at the Mariana arc (Notsu, 1981b; Ito and Stern, 1981; Stern, 1981); they are more or less enriched in ^{87}Sr relative to fresh MOR basalts ($^{87}Sr/^{86}Sr$ = 0.7022 to 0.7029) (Dasch et al., 1973; Hart et al., 1974; O'Nions et al., 1977).

It seems still ambiguous whether Sr isotopic variation in the primitive island-arc magmas is due to involvement of subducted lithosphere or isotopic heterogeneity in the mantle wedge. Altered submarine basalts are enriched in ^{87}Sr up to $^{87}Sr/^{86}Sr$ ratio of at least 0.704 by reaction with seawater (Dasch et al., 1973; Hart et al., 1974). On the basis of Nd-Sr isotopic correlations, DePaolo and Wasserburg (1977) and Hawkesworth and his co-workers (Hawkesworth et al., 1979; Hawkesworth and Powell, 1980) suggested input of seawater-modified Sr from the subducted ocean lithosphere in the generation of island-arc magmas. However, Stern (1981) questioned that possibility by demonstrating that the rocks from the primitive Mariana arc lie in the "mantle array" on the Nd—Sr isotopic diagram. Arculus and Johnson (1981) also questioned addition of ^{87}Sr-rich material from subducted ocean lithosphere to island-arc magmas because of no positive correlation between the magnitude of Sr anomalies and $^{87}Sr/^{86}Sr$ ratios in most island-arc rock suites.

Sr content of the primitive magmas varies from 150 to 190 ppm for tholeiites to 500 to 680 ppm for alkali olivine basalts across the Japanese island arcs (Pushkar, 1968; Hedge and Knight, 1969; Kurasawa, 1979; Hirano, 1982). Because Sr would behave as an incompatible element in the mantle at depths where plagioclase is not stable, partial melting of the subducted ocean crust which has 140 ppm Sr (Dasch et al., 1973; Hart et al., 1974), as much as that of the island-arc tholeiites, may produce a magma highly enriched in Sr. Island-arc tholeiite magma with 170 ± 20 ppm Sr cannot be derived by direct melting of the subducted ocean crust except for total melting or extreme dilution by Sr-poor mantle-derived magma.

If the primitive magmas are produced by partial melting of the mantle wedge, the variation in Sr content across island arcs could be produced by variable degree of melting of mantle material, or variable partition coefficient of Sr (solid/melt), both being smaller in source regions for alkali olivine basalt (Gast, 1968; Philpotts and Schnetzler, 1970). Sr-poor tholeiite magmas can be produced by larger degree of partial melting of the mantle wedge. This assumption accords with the observation that a large quantity of volcanic products (as much as 75% of the total across island arcs) is asymmetrically distributed on the volcanic front where tholeiitic rocks are predominant (Sugimura and Uyeda, 1973, fig. 121).

Crustal rocks

Possible rocks to be assimilated include pre-Tertiary granitoids and metasedimentary rocks composing the basements of the volcanoes. Lower crust may consist of basic members of old intrusive rocks or buried previously erupted volcanic rocks.

Cretaceous granitoids, which are among the most dominant rock types in the basement of central and western Japan, have $\delta^{18}O$ values ranging from +7 to +12‰ (Matsuhisa et al., 1972; Honma and Sakai, 1975; Honma,

1978). The composition range differs with rock series of granitoids, being higher for ilmenite-series granitoids associated with metamorphic rocks (Matsuhisa et al., 1972; Ishihara and Matsuhisa, 1977).

Cretaceous granitoids including gabbroic rocks from Japan have a large variation of $^{87}Sr/^{86}Sr$ ratios ranging from 0.7055 to 0.728 (present value) (Shigeno and Yamaguchi, 1976; Shibata and Ishihara, 1979a,b). Extremely high $^{87}Sr/^{86}Sr$ ratios occur only in relatively rare high Rb/Sr rock types, and probably should be eliminated in estimating average composition range. Shibata and Ishihara (1979b) pointed out that regional variation of $^{87}Sr/^{86}Sr$ ratios (initial ratios in their case) are positively related to crustal thickness. The ranges of the present values of crustal $^{87}Sr/^{86}Sr$ ratio are 0.707 to 0.722 for central Japan, and 0.7055 to 0.7095 for western Japan; Sr contents range from 200 to 500 ppm depending upon rock types.

Nohda (1973) reported $^{87}Sr/^{86}Sr$ ratios ranging from 0.7164 to 0.7549 for the Yatsushiro granite and gneiss which have been considered to be a member of deep-seated basement rocks of the Japanese islands. These rocks give a whole-rock isochron age of 352 Ma with an initial $^{87}Sr/^{86}Sr$ ratio of 0.7037. $^{87}Sr/^{86}Sr$ ratios of 0.707 to 0.716 were measured for Mesozoic to Paleozoic basement rocks from some other areas in Japan (Shibata et al., 1972).

Metasedimentary rocks from the Ryoke metamorphic belt have $\delta^{18}O$ values and $^{87}Sr/^{86}Sr$ ratios ranging from +12 to +18$^0/_{00}$ (Honma and Sakai, 1975; Ishihara and Matsuhisa, 1977) and from 0.720 to 0.729 (Ueno et al., 1969; Shigeno and Yamaguchi, 1976), respectively. Sr contents of silicic and argillaceous metamorphic rocks from the Yanai district, western Honshu, are as low as 12 to 108 ppm (Shigeno and Yamaguchi, 1976).

Mixing models in $\delta^{18}O$ - $^{87}Sr/^{86}Sr$ correlation

Figure 9 shows $\delta^{18}O$ vs. $^{87}Sr/^{86}Sr$ plot for the calc-alkalic rocks from the central and western Japan arcs. Composition ranges of the granitoids including gabbroic rocks are shown by two large rectangles for central and western Japan. The small rectangle with M in lower left of the diagram represents possible composition range of the island-arc primitive magmas. The composition range of MORB is also shown for comparison.

The least differentiated and contaminated rocks having the lowest $\delta^{18}O$ and $^{87}Sr/^{86}Sr$ values in each volcano are shown by solid symbols. They are distributed on a linear trend between the primitive magmas and the crustal material except for Aonoyama, suggesting that isotopic variations among the rock suites might have been derived by mixing of the two end-members in variable proportions. Within individual volcanoes two types of variation trends are recognized: (a) $\delta^{18}O$-increase with increasing $^{87}Sr/^{86}Sr$ being concordant to the main variation trend among the rock suites; and (b) $\delta^{18}O$-increase with virtually constant or very small increase in $^{87}Sr/^{86}Sr$ ratio.

Because oxygen contents of silicate rocks are more or less identical regardless of rock types (they vary only from 45 to 50 wt.%), change in $\delta^{18}O$ value

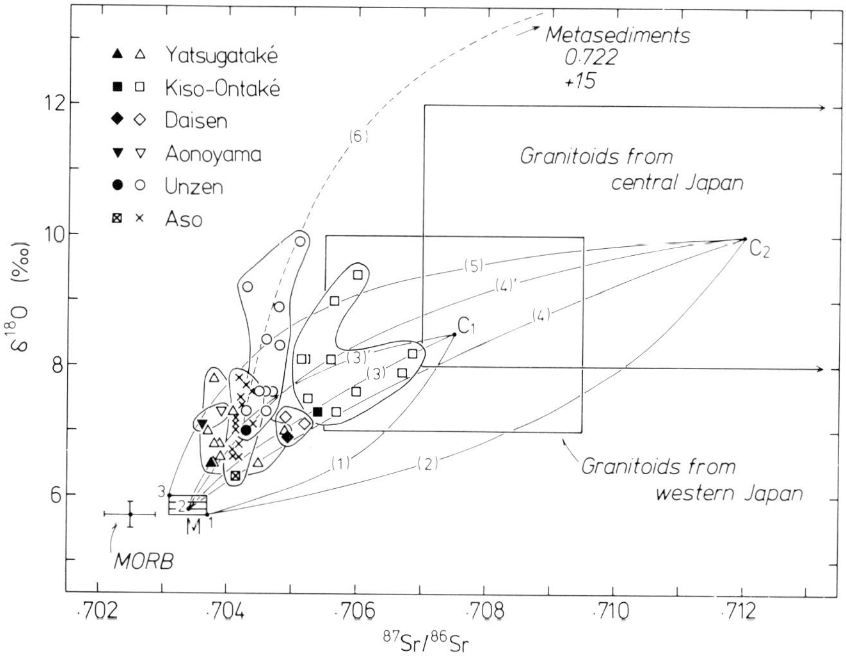

Fig. 9. Plot of $\delta^{18}O$ vs. $^{87}Sr/^{86}Sr$ for the calc-alkalic volcanic rocks from the central and western Japan arcs. The composition range assumed for the primitive magmas is given by the small rectangle with *M*1, *2* and *3* in lower left of the diagram. The composition ranges of the granitoids from western and central Japan are shown with their average compositions of C_1 and C_2, respectively. Mixing curves (1) to (6) are calculated for different combinations of end-members (See text for the details).

by mixing of rocks and magmas would be proportional to the mixing ratio of the end-members. Whereas, magnitude of change in $^{87}Sr/^{86}Sr$ ratio by mixing strongly depends on Sr contents of the end-members which vary widely in different rock types. The general tendency of linear increase in $\delta^{18}O$ with increasing $^{87}Sr/^{86}Sr$ ratio among the rock suites strongly suggests mixing of two end-members having similar Sr contents. On the other hand, the $\delta^{18}O$-increase with virtually constant $^{87}Sr/^{86}Sr$ ratio observed within a single volcano may suggest small ratio of Sr content of assimilated material to that of magma, or lack of assimilation.

Since the calc-alkalic andesites have Sr content in the range of 300 to 700 ppm, the end-members of mixing should have similar Sr content to produce a linear mixing in the diagram, or initial magma should have higher Sr content to produce a mixing curve convex upward. If the degree of melting in the source is smaller for calc-alkalic magmas than for tholeiitic magmas (Masuda and Aoki, 1979), Sr content as well as other incompatible elements of magma may increase. However, it is supposed that 500 to 680 ppm Sr reported for alkali olivine basalts on continent side would suggest possible upper limit of Sr content of the primitive magmas in the Japanese island arcs.

Another possible mechanism to concentrate Sr in magma is fractional crystallization in plagioclase-absent condition. Plagioclase does not occur as a liquidus phase of andesitic melt at 15 kb under H_2O-saturated conditions (Kushiro, 1974). If plagioclase does not occur dominantly down to 10 kbar (Green and Ringwood, 1967; Wyllie, 1979), fractional crystallization of magma ascending into thick crust might be initiated in plagioclase-absent or plagioclase-deficient condition (Ui and Aramaki, 1978). Table II shows model composition of primitive tholeiitic and alkalic magmas in island arcs and their evolved magmas assuming fractional crystallization with Sr partition coefficient (solid/liquid) of 10^{-2}. If hornblende appears as a dominant cumulate phase, Sr partition coefficient between solid and melt would become closer to 10^{-1} (Gast, 1968; Philpotts and Schnetzler, 1970). If primitive magma is alkalic more or less, 20% fractional crystallization can produce a magma having Sr content as high as 700 ppm with no remarkable change in SiO_2 content and $\delta^{18}O$. The composition range for the primitive magmas (M) in Fig. 9 is taken from Table II. The upper limit of $\delta^{18}O$ value corresponds to 50% crystallization of the initial magmas.

Model compositions of crust in the Japanese island arcs are given in Table III, which were calculated on the basis of the composition ranges of the pre-Tertiary granitoids and metasedimentary rocks. DePaolo and Wasserburg (1979) adopted average compositions of $^{87}Sr/^{86}Sr = 0.722$ and ppm Sr = 350 for upper crust, and $^{87}Sr/^{86}Sr = 0.7105$ and ppm Sr = 570 for lower crust.

TABLE II

Model composition of primitive and evolved magmas in plagioclase-absent condition

f*	Tholeiitic magma			Alkalic magma		
	$\delta^{18}O$	$^{87}Sr/^{86}Sr$	Sr (ppm)	$\delta^{18}O$	$^{87}Sr/^{86}Sr$	Sr (ppm)
1	5.7	0.7037	170	5.7	0.7031	700
0.8	5.8	0.7037	210	5.8	0.7031	870
0.6	5.9	0.7037	280	5.9	0.7031	1160
0.5	6.0	0.7037	420	6.0	0.7031	1390

f* = residual fraction of magma.
$\alpha(^{18}O/^{16}O) = 0.9995$, $D(Sr) = 10^{-2}$.

TABLE III

Model composition of crust in the Japanese island arcs

	$\delta^{18}O$	$^{87}Sr/^{86}Sr$	Sr (ppm)
Granitoids in Western Japan (C_1)	+8.5	0.7075	200—500
Granitoids in Central Japan (C_2)	+10.0	0.712	200—500
Metasediments	+15.0	0.722	50—150

The $^{87}Sr/^{86}Sr$ ratios in our model are a little lower than their estimation, probably reflecting younger age of the Japanese basements than the continental crust.

Mixing lines for several different combinations of end-members are shown in Fig. 9. They are reproduced in Fig. 8 together with end-member compositions.

Assimilation of crustal material C_1 or C_2 by the primitive tholeiite magma (M_1) produces the mixing curves (1) and (2) convex downward in $\delta^{18}O$ vs. $^{87}Sr/^{86}Sr$ plot. It is clear from the data points in both Figs. 8 and 9 that the calc-alkalic rocks from the central and western Japan arcs cannot be derived by this mixing model. In order to derive linear mixing for tholeiitic magma, assumption of crystallization more than 70% of the initial magma at depth and subsequent assimilation of crustal material is required. Neither fractional crystallization nor assimilation by the primitive tholeiite magma is plausible for derivation of the calc-alkalic rocks. Magaritz et al. (1978) concluded source contamination by subducted continental material for andesites from the Banda arc on the basis of their lying on a downward-convex curve in the $\delta^{18}O$ vs. $^{87}Sr/^{86}Sr$ plot. James (1981) also assumed the same source contamination model for the Andean andesites. This would not be the case for the calc-alkalic rocks from the central and western Japan arcs.

If a magma derived by small-degree partial melting in the source or evolved in plagioclase-absent condition is assumed to have $\delta^{18}O = +5.8‰$, $^{87}Sr/^{86}Sr = 0.7034$ and ppm Sr = 700 (M_2 in Figs. 8 and 9), mixing with crustal material C_1 or C_2 produces curves slightly convex upward in $\delta^{18}O$ vs. $^{87}Sr/^{86}Sr$ plot depending upon Sr content of crustal material: 500 ppm Sr of C_1 and C_2 for the curves (3) and (4), 200 ppm Sr of C_1 and C_2 for the curves (3)′ and (4)′, respectively. One extreme case is given by the curve (5), where mixing is assumed between a highly evolved alkalic magma (M_3: $\delta^{18}O = +6.0‰$, $^{87}Sr/^{86}Sr = 0.7031$ and ppm Sr = 1400) and Sr-poor (200 ppm) crustal material C_2.

The least differentiated and contaminated rocks are distributed on the curves (3), (3)′, (4) or (4)′ except for the Aonoyama andesites. That is, the calc-alkalic magmas can be produced by mixing of crustal material and a Sr-rich magma of mantle origin derived by small-degree partial melting or fractional crystallization in plagioclase-absent condition. Isotopic variation among the calc-alkalic magmas would depend upon Sr concentration ratio of the end-members or isotopic compositions of assimilated material. Since the Unzen rocks lie on a mixing curve more strongly convex upward [(3)′ or (4)′], the ratio of Sr in the assimilated material to that in the magma seems to be smaller than that for other volcanoes including Kiso-Ontaké, Daisen and Aso. This accords with lower Sr contents of the Unzen rocks in Fig. 8. Andesites from Aonoyama which have the peculiar composition of low $^{87}Sr/^{86}Sr$ ratio with high Sr content and $\delta^{18}O$ value may have been produced by assimilation of Sr-poor crustal rocks by a magma extremely enriched in Sr as suggested by the curve (5) in Figs. 8 and 9.

A Daisen andesite and some of the Kiso-Ontaké rocks seem to lie on the curve (5) in $^{87}Sr/^{86}Sr$ vs. Sr plot (Fig. 8) as well, but they do not in $\delta^{18}O$ vs. $^{87}Sr/^{86}Sr$ plot (Fig. 9). The inconsistency in the two plots suggests that they are not products of the mixing applicable to the Aonoyama andesites except for the two data points with high $\delta^{18}O$ values of the Kiso-Ontaké rocks.

A part of isotopic variations within individual volcanoes, especially of high $^{87}Sr/^{86}Sr$ ratios in Yatsugataké and Kiso-Ontaké, seem to satisfy the mixing curves (3), (3)′, (4) or (4)′. The same mixing process that worked for producing isotopic variations among the different rock suites may partly have produced the isotopic variation within individual volcanoes as well. It is difficult to distinguish multiple stages of mixing in this case because the curvature of the mixing curves is small.

However, another trend within individual volcanos, that is, $\delta^{18}O$-increase with virtually constant or very small increase in $^{87}Sr/^{86}Sr$ ratio, does not fit the mixing models discussed above. Instead, this trend could be produced by fractional crystallization of magma or assimilation of ^{18}O- and ^{87}Sr-rich but Sr-poor material. It has been suggested on the basis of $\delta^{18}O$ vs. SiO_2 plot (Figs. 5 to 7) that some part of ^{18}O-enrichment (up to $1‰$) could be ascribed to fractional crystallization if it occurs with a constant $^{87}Sr/^{86}Sr$ ratio. Some magmas once formed by mixing of more primitive magmas and crustal material may have been fractionated without subsequent remarkable contamination. Broken curve (6) in Fig. 9 shows the mixing trend between an andesite magma with $\delta^{18}O = +7.0‰$, $^{87}Sr/^{86}Sr = 0.70425$ and ppm Sr = 600, and a metasedimentary rock with $\delta^{18}O = +15‰$, $^{87}Sr/^{86}Sr = 0.722$ and ppm Sr = 50. The data points with extremely high $\delta^{18}O$ values (higher than $+8.0‰$) from the Unzen volcano are distributed along this curve. Therefore, multistage assimilation involving granitoids and metasediments is proposed for the Unzen andesites and probably for the high-$\delta^{18}O$ rocks from Kiso-Ontaké.

Estimation of the proportion of assimilated material to magma is quite unreliable because of large composition range of crustal material. However, if simple mixing is assumed on the mixing curves in Fig. 9, 20 to 30% crustal material is required to produce the calc-alkalic magmas. Assimilation of country rocks into a magma must involve simultaneous crystallization because of heat balance (Taylor, 1980). He demonstrated that combined assimilation-fractional crystallization could markedly modify a simple mixing curve in $\delta^{18}O$ vs. $^{87}Sr/^{86}Sr$ plot toward ^{87}Sr-rich direction especially in later differentiates. Because the main trend of the calc-alkalic rocks in the present study, deduced from the least differentiated and assimilated rocks, seems to lie on simple mixing curves (Fig. 9), mixing might have taken place between primary magmas and crust-derived melts instead of rock-assimilation.

Relation with arc structure

If assimilation of crustal material takes place, thickness and composition of crust through which magma ascends likely would affect isotopic compositions of the magma. Figure 10 shows correlation between $\delta^{18}O$ values of the volcanic rock suites normalized to at SiO_2 = 60 and Bouguer anomalies of the areas where volcanoes are located. The rock suites from the East Japan island arcs in Matsuhisa (1979) are shown together. Bouguer anomalies, which are assumed to roughly correspond to crustal thickness, were taken from Tomoda (1973). A large spread in $\delta^{18}O$ to higher values is recognized in the areas with Bouguer anomalies lower than 60 mgal.

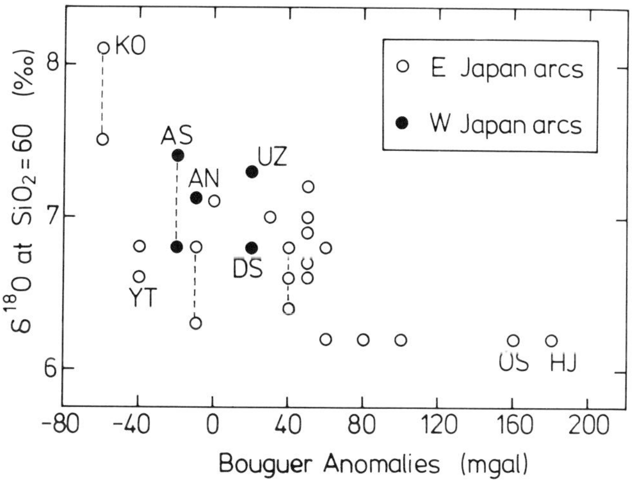

Fig. 10. $\delta^{18}O$ variation vs. Bouguer anomalies for the volcanic rock suites from the Japanese island arcs. The O isotopic data include the present study and Matsuhisa (1979). The $\delta^{18}O$ values were normalized at SiO_2 =60 on the variation trends of individual rock suites. Bouguer anomalies were taken from Tomoda (1973). Abbreviations: *YT* = Yatsugatakė, *KO* = Kiso-Ontakė, *DS* = Daisen, *AN* = Aonoyama, *UZ* = Unzen, *AS* = Aso, *OS* = Oshima-Oshima, *HJ* = Hachijo-jima.

Correlation between the lowest $^{87}Sr/^{86}Sr$ ratio of each rock suite in the present study and Bouguer anomalies is shown in Fig. 11. The data for the Hachijo-jima tholeiites (Notsu, 1981b) and the Oshima-Oshima alkali olivine basalts (Hedge and Knight, 1969; Katsui et al., 1978; Notsu, 1981a) are shown for comparison. Again, a large spread in $^{87}Sr/^{86}Sr$ ratio to higher values is recognized in the areas with Bouguer anomalies lower than 20 mgal. Notsu (1981a) pointed out the increase in $^{87}Sr/^{86}Sr$ ratio along the volcanic front of the East Japan arc toward central Honshu where the crust becomes thickest.

Although simple negative correlation between the isotopic compositions and Bouguer anomalies is not detected, the general tendency for high $\delta^{18}O$ and $^{87}Sr/^{86}Sr$ values to be associated with low Bouguer anomalies suggests that enrichment in ^{18}O and ^{87}Sr of the magmas is related to thick continental-type crust. Thicker crust is favorable to melting of crust at depth, fractionation of the primitive magmas in plagioclase-unstable condition and grea-

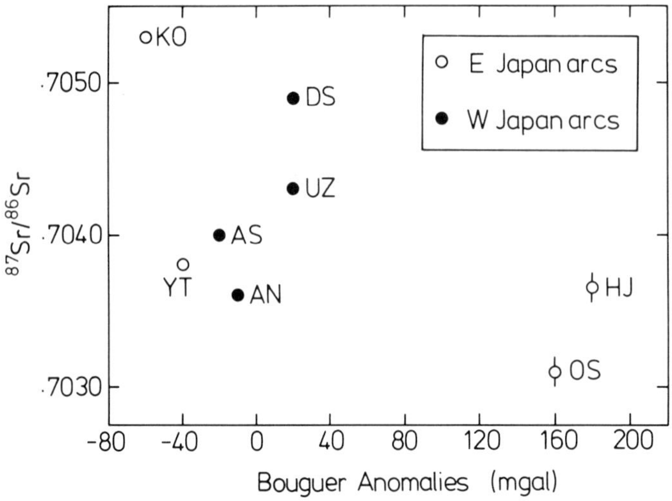

Fig. 11. $^{87}Sr/^{86}Sr$ variation vs. Bouguer anomalies for the volcanic rock suites from the central and western Japan arcs. The data from literatures for the Oshima—Oshima and Hachijo-jima rock suites are shown together for comparison. The lowest $^{87}Sr/^{86}Sr$ ratio was taken for each rock suite. The abbreviations are the same as in Fig. 10.

ter probabilities of mixing of crustal material during magma ascent. The present results accord with the Nd-Sr isotopic correlation by Nohda and Wasserburg (1981) that suggests a correlation between increase in crustal component of magmas and thickness of old continental crust across central Japan.

CONCLUDING REMARKS

Involvement of crustal material is strongly suggested on the basis of O and Sr isotopic data for the calc-alkalic volcanic rocks from the central and western Japan arcs where thick continental-type crust is developed. Data points in $\delta^{18}O$ vs. $^{87}Sr/^{86}Sr$ plot accord with mixing models between island-arc primitive magmas and crustal material having dioritic composition on an average. Any type of the primitive magmas could potentially be the end member of the mixing. If tholeiitic magmas are involved, the rocks derived by the mixing would lie on a convex downward curve as demonstrated by Magaritz et al. (1978) for the Banda arc. However, the calc-alkalic rocks from the central and western Japan arcs are distributed along the mixing curves convex upward in the diagram, suggesting high Sr content with low $^{87}Sr/^{86}Sr$ ratio and $\delta^{18}O$ value for the primitive magmas. Neither fractional crystallization nor assimilation by the primitive theoleiite magma is plausible for derivation of these calc-alkalic rocks. While, high-alkali tholeiite or alkali basalt magmas enriched in incompatible elements are more likely as the primitive magmas to derive them by mixing. The same mixing models seem to be applicable to the calc-alkalic andesites from the Peruvian Andes which have $\delta^{18}O$ values of +7.2 to +8.6 and $^{87}Sr/^{86}Sr$ ratios of 0.7054 to 0.7075 (Magaritz et al., 1978) and those from southern Chile which have $\delta^{18}O$ values of +5.9 to +8.3 and $^{87}Sr/^{86}Sr$ ratios of 0.705 to 0.707 (Muehlenbachs and Stern, 1980). In mature island arcs, the primitive magmas, which used to be tholeiite-dominant in earlier stages of island-arc evolution, seem to become more alkalic and have greater probabilities of crustal contamination during their ascent.

Isotopic differences between eruptions as well as between rock suites are ascribed to variable Sr-concentration ratio of the end-members, variable isotopic compositions of assimilated material, or variable mixing ratio of the end-members. Extremely high-$\delta^{18}O$ rocks with moderate increase in $^{87}Sr/^{86}Sr$ ratio (e.g. a part of the Unzen andesites) suggest another mixing process between andesite magmas and metasedimentary rocks in shallower magma chambers. Fractional crystallization of magmas once derived by mixing is also inferred for the rock suites showing gradual increase in ^{18}O up to 1‰ with virtually uniform $^{87}Sr/^{86}Sr$ ratios.

ACKNOWLEDGEMENTS

We gratefully acknowledge Dr. S. Kawachi of Hokkaido University, Dr. T. Kobayashi of Toyama University and Dr. K. Ono of the Geological Survey of Japan for generously offering the samples used in this study. We thank Dr. S. Aramaki of Tokyo University for his encouragement to submit this paper to this special issue.

REFERENCES

Arculus, R.J. and Johnson, R.W., 1981. Island-arc magma sources: a geochemical assessment of the roles of slab-derived components and crustal contamination. Geochem. J., 15: 109—133.

Briqueu, L. and Lancelot, J.R., 1979. Rb-Sr systematics and crustal contamination models for calc-alkaline igneous rocks. Earth Planet. Sci. Lett., 43: 385—396.

Craig, H., 1961. Standard for reporting concentrations of deuterium and oxygen-18 in natural waters. Science, 133: 1833—1834.

Dasch, E.J., Hedge, C.E. and Dymond, J., 1973. Effect of sea water interaction on strontium isotope composition of deep-sea basalts. Earth Planet. Sci. Lett., 19: 177—183.

DePaolo, D.J. and Wasserburg, G.J., 1977. The sources of island arcs as indicated by Nd and Sr isotopic studies. Geophys. Res. Lett., 4: 465—468.

DePaolo, D.J. and Wasserburg, G.J., 1979. Petrogenetic mixing models and Nd-Sr isotopic patterns. Geochim. Cosmochim. Acta, 43: 615—627.

Dickinson, W.R. and Hatherton, T., 1967. Andesitic volcanism and seismicity around the Pacific. Science, 157: 801—803.

Ewart, A. and Stipp, J.J., 1968. Petrogenesis of the volcanic rocks of the Central North Island, New Zealand, as indicated by a study of Sr^{87}/Sr^{86} ratios, and Sr, Rb, K, U and Th abundances. Geochim. Cosmochim. Acta, 32: 699—736.

Francis, P.W., Moorbath, S. and Thorpe, R.S., 1977. Strontium isotope data for recent andesites in Ecuador and north Chile. Earth Planet. Sci. Lett., 37: 197—202.

Francis, P.W., Thorpe, R.S., Moorbath, S., Kretzschmar, G.A. and Hammil, M., 1980. Strontium isotope evidence for crustal contamination of calc-alkaline volcanic rocks from Cerro Galan, northwest Argentina. Earth Planet. Sci. Lett., 48: 257—267.

Gast, P.W., 1968. Trace element fractionation and the origin of tholeiitic and alkaline magma types. Geochim. Cosmochim. Acta, 32: 1056—1086.

Gill, J.B., 1981. Orogenic Andesites and Plate Tectonics. Springer-Verlag, Berlin-Heidelberg, 390 pp.

Green, D.H. and Ringwood, A.E., 1967. An experimental investigation of the gabbro to eclogite transformation and its petrological applications. Geochim. Cosmochim. Acta, 31: 767—833.

Hart, S.R., Erlank, A.J. and Kable, E.J.D., 1974. Sea floor basalt alteration: some chemical and Sr isotopic effects. Contrib. Mineral. Petrol., 44: 219—230.

Hawkesworth, C.J. and Powell, M., 1980. Magma genesis in the Lesser Antilles island arc. Earth Planet. Sci. Lett., 51: 297—308.

Hawkesworth, C.J., Norry, M.J., Roddick, J.C. Baker, P.E., Francis, P.W. and Thorpe, R.S., 1979. $^{143}Nd/^{144}Nd$, $^{87}Sr/^{86}Sr$, and incompatible element variations in calc-alkaline andesites and plateau lavas from south America. Earth Planet. Sci. Lett., 42: 45—57.

Hedge, C.E. and Knight, R.J., 1969. Lead and strontium isotopes in volcanic rocks from northern Honshu, Japan. Geochem. J., 3: 15--24.

Higuchi, H. and Nagasawa, H., 1969. Partition of trace elements between rock-forming minerals and the host volcanic rocks. Earth Planet. Sci. Lett., 7: 281—287.

Hirano, M., 1982. Arc Volcanism in Izu-islands, Japan, based on Sr/Ca-Ba/Ca diagram. Ph.D. Thesis, Univ. Tsukuba.

Honma, H., 1978. $^{18}O/^{16}O$ and D/H ratios of the Cretaceous granitic rocks of the Chugoku belt, southwest Japan. Short Papers, 4th Int. Conf. Geochron. Cosmochron. and Isotope Geol., 1978, U.S. Geol. Surv., Open-File Rept., 78-701: 187—189.

Honma, H. and Sakai, H., 1975. Oxygen isotope study of metamorphic and granitic rocks of the Yanai district in the Ryoke belt, Japan. Contrib. Mineral. Petrol., 52: 107—120.

Ishihara, S. and Matsuhisa, Y., 1977. The magnetite-series and ilmenite-series granitoids and their $^{18}O/^{16}O$ isotope ratios (in Japanese). Abstr. 84th Annu. Meet. Geol. Soc. Japan.

Ito, E. and Stern, R.J., 1981. Oxygen and strontium isotopic investigations on the origin of volcanism in the Izu-Volcano-Mariana island arc. Carnegie Inst. Washington Yearb., 80: 449—455.

James, D.E., 1978. Origin of high $^{87}Sr/^{86}Sr$ ratios in central Andean calc-alkaline lavas. Short Papers, 4th Int. Conf. Geochron. Cosmochron. and Isotope Geol., U.S. Geol. Surv., Open-File Rep., 78-701: 199—201.

James, D.E., 1981. The combined use of oxygen and radiogenic isotopes as indicators of crustal contamination. Annu. Rev. Earth Planet. Sci., 9: 311—344.

Katsui, Y., Oba, Y., Ando, S., Nishimura, S., Masuda, Y., Kurasawa, H. and Fujimaki, H., 1978. Petrochemistry of the Quaternary volcanic rocks of Hokkaido, north Japan. J. Fac. Sci., Hokkaido Univ., Ser. IV, 18: 449—484.

Kawachi, S., 1974. Geology of the Tateshinayama District (in Japanese with English abstract). Quadrangle Series, Nagano (36), No. 24, Geol. Surv. Jpn., 119 pp.

Kawachi, S., 1977. Geology of the Yatsugatake District (in Japanese with English abstract). Quadrangle Series, Tokyo (8), No. 35, Geol. Surv. Jpn., 92 pp.

Kobayashi, T., Ohmori, E. and Ohmori, T., 1975. Petrochemistry of Ontake Volcano. Bull. Geol. Surv. Jpn., 26: 497—512.

Kuno, H., 1960. High-alumina basalt. J. Petrol., 1: 121—145.

Kurasawa, H., 1965. Petrology and chemistry of Quaternary volcanic rocks in the western part of Chugoku Provinces, southwest Japan. Bull. Geol. Surv. Jpn., 16: 217—226.

Kurasawa, H., 1979. Isotopic composition of Strontium in volcanic rocks from Fuji, Hakone, and Izu Area, central Japan (in Japanese with English abstract). Bull. Volcanol. Soc. Jpn., 24: 135—152.

Kurasawa, H. and Takahashi, K., 1965. Petrology and chemistry of rocks of the Unzen volcano and its basement, Kyushu, southwest Japan. Bull. Geol. Surv. Jpn., 16: 258—274.

Kushiro, I., 1974. Melting of hydrous upper mantle and possible generation of andesitic magma: an approach from synthetic systems. Earth Planet. Sci. Lett., 22: 294—299.

Lawrence, J.R. and Taylor, H.P. Jr., 1972. Hydrogen and oxygen isotope systematics in weathering profiles. Geochim. Cosmochim. Acta, 36: 1377—1393.

Leeman, W.P., 1981. The influence of crustal structure on compositions of subduction-related magmas. Abstracts, 1981 IAVCEI symposium, Arc Volcanism, Tokyo, p. 203.

Leeman, W.P. and Manton, W.I., 1971. Strontium isotopic composition of basaltic lavas from the Snake River Plain, southern Idaho. Earth Planet. Sci. Lett., 11: 420—434.

Magaritz, M., Whitford, D.J. and James, D.E., 1978. Oxygen isotopes and the origin of high $^{87}Sr/^{86}Sr$ andesites. Earth Planet. Sci. Lett., 40: 220—230.

Masuda, T. and Aoki, K., 1979. Trace element variations in the volcanic rocks from the Nasu zone, northeast Japan. Earth Planet. Sci. Lett., 44: 139—149.

Matsuhisa, Y., 1979. Oxygen isotopic compositions of volcanic rocks from the East Japan island arcs and their bearing on petrogenesis. J. Volcanol. Geotherm. Res., 5. 271—296.

Matsuhisa, Y., Honma, H., Matsubaya, O. and Sakai, H., 1972. Oxygen isotopic study of the Cretaceous granitic rocks in Japan. Contrib. Mineral. Petrol., 37: 65—74.

Matsuhisa, Y., Matsubaya, O. and Sakai, H., 1971. BrF_3-technique for the oxygen isotopic analysis of silicates and water. Mass Spectroscopy, 19: 124—133.

Matsuhisa, Y., Matsubaya, O. and Sakai, H., 1973. Oxygen isotope variations in magmatic differentiation processes of the volcanic rocks in Japan. Contrib. Mineral. Petrol., 39: 277—288.

Miyashiro, A., 1974. Volcanic rock series in island arcs and active continental margins. Am. J. Sci., 274: 321—355.

Muehlenbachs, K. and Clayton, R., 1972. Oxygen isotope studies of fresh and weathered submarine basalts. Can. J. Earth Sci., 9: 172—184.

Muehlenbachs, K. and Hoering, T.C., 1972. Oxygen isotope geochemistry of some rocks from island arcs. Carnegie Inst. Washington Yearb., 71: 545—547.

Muehlenbachs, K. and Stern, C.R., 1980. $\delta^{18}O$ variation among calc-alkaline volcanic rocks from S. Chile: Implications for the effects of ridge subduction on the source region of orogenic volcanism. EOS, Trans. Am. Geophys., 61: 401.

Murakami, N., 1975. High-grade metamorphic inclusions in Cenozoic volcanic rocks from west Sanin, southwest Japan. Japan Assoc. Mineral. Petrol. Econ. Geol., 70: 424—439.

Nohda, S., 1973. Rb-Sr dating of the Yatsushiro granite and gneiss, Kyushu, Japan. Earth Planet. Sci. Lett., 20: 140—144.

Nohda, S. and Wasserburg, G.J., 1981. Nd and Sr isotopic study of volcanic rocks from Japan. Earth Planet. Sci. Lett., 52: 264—276.

Notsu, K., 1981a. Strontium isotope composition in volcanic rocks from Japan. Abstracts, 1981 IAVCEI Symposium, Arc Volcanism, Tokyo, pp. 265—266.

Notsu, K., 1981b. $^{87}Sr/^{86}Sr$ ratios of arc volcanic rocks (in Japanese). Abstr., 1981 Annu. Meet. Geochem. Soc. Jpn., pp. 24—25.

O'Nions, R.K., Hamilton, P.J. and Evensen, N.M., 1977. Variations in $^{143}Nd/^{144}Nd$ and $^{87}Sr/^{86}Sr$ ratios in oceanic basalts. Earth Planet. Sci. Lett., 34: 13—22.

Ono, K. and Watanabe, K., 1981. Geology of Aso Volcano. Abstr., 1981 IAVCEI Symposium, Arc Volcanism, Tokyo, p. 280.

Ota, R., 1962. Akasaki and Daisen (in Japanese with English abstract). Explanatory Text of the Geological Map of Japan, Okayama (9), No. 19, Geol. Surv. Jpn., 41 pp.

Philpotts, J.A. and Schnetzler, C.C., 1970. Phenocryst-matrix partition coefficients for K, Rb, Sr, and Ba, with applications to anorthosite and basalt genesis. Geochim. Cosmochim. Acta, 34: 307—322.

Pushkar, P., 1968. Strontium isotope ratios in volcanic rocks of three island arc areas. J. Geophys. Res., 73: 2701—2714.

Shibata, K. and Ishihara, S., 1979a. Rb-Sr whole-rock and K-Ar mineral ages of granitic rocks in Japan. Geochem. J., 13: 113—119.

Shibata, K. and Ishihara, S., 1979b. Initial $^{87}Sr/^{86}Sr$ ratios of plutonic rocks from Japan. Contrib. Mineral. Petrol., 70: 381—390.

Shibata, K., Wanless, R.K., Kano, H., Yoshida, T., Nozawa, T., Igi, S. and Konishi, K., 1972. Rb-Sr ages of several so-called basement rocks in the Japanese islands (in Japanese with English abstract). Bull. Geol. Surv. Jpn., 23: 505—510.

Shigeno, H. and Yamaguchi, M., 1976. A Rb-Sr isotopic study of metamorphism and plutonism in the Ryoke belt, Yanai district, Japan (in Japanese with English abstract). J. Geol. Soc. Jpn., 82: 687—698.

Stern, R.J., 1981. A common mantle source for Western Pacific island arc and "hot spot" magmas — implications for layering in the upper mantle. Carnegie Inst. Washington Yearb., 80: 455—462.

Sugimura, A. and Uyeda, S., 1973. Island Arcs: Japan and its Environs. Elsevier, Amsterdam, 247 pp.

Tatsumoto, M. and Knight, R.J., 1969. Isotopic composition of lead in volcanic rocks from central Honshu — with regard to basalt genesis. Geochem. J., 3: 53—86.

Taylor, H.P., Jr., 1980. The effects of assimilation of country rocks by magmas on $^{18}O/^{16}O$ and $^{87}Sr/^{86}Sr$ systematics in igneous rocks. Earth Planet. Sci. Lett., 47: 243—254.

Tomoda, Y., 1973. Free Air and Bouguer Gravity Anomalies in and around Japan. Univ. Tokyo Press, Tokyo (map).

Ueno, N., Ozima, M. and Ono, A., 1969. Geochronology of the Ryoke metamorphism-- Rb-Sr, K-Ar isotopic investigations of the metamorphic rocks in the Ryoke metamorphic belt. Geochem. J., 3: 35—43.

Ui, T. and Aramaki, S., 1978. Relationship between chemical composition of Japanese island-arc volcanic rocks and gravimetric data. Tectonophysics, 45: 249—259.

Wyllie, P.J., 1979. Petrogenesis and the physics of the earth. In: H.S. Yoder, Jr. (Editor), The Evolution of the Igneous Rocks. Princeton Univ. Press, Princeton, N.J., pp. 483—520.

Journal of Volcanology and Geothermal Research, 18 (1983) 511—529
Elsevier Science Publishers B.V., Amsterdam — Printed in The Netherlands

GENESIS OF BASALT MAGMAS AND THEIR DERIVATIVES UNDER THE IZU ISLANDS, JAPAN, INFERRED FROM Sr/Ca—Ba/Ca SYSTEMATICS

NAOKI ONUMA[1], MASATAKA HIRANO[2] and NAOKI ISSHIKI[3]

[1] *Department of Earth Sciences, Ibaraki University, Mito, Ibaraki (Japan)*
[2] *Institute of Chemistry, University of Tsukuba, Niihari, Ibaraki (Japan)*
[3] *Geological Survey of Japan, Tsukuba, Ibaraki (Japan)*

(Received October 20, 1982)

ABSTRACT

Onuma, N., Hirano, M. and Isshiki, N., 1983. Genesis of basalt magmas and their derivatives under the Izu Islands, Japan, inferred from Sr/Ca—Ba/Ca systematics. In: S. Aramaki and I. Kushiro (Editors), Arc Volcanism. J. Volcanol. Geotherm .Res., 18: 511—529.

The Sr/Ca—Ba/Ca systematics defined for a series of volcanic rocks provided by volcanoes of the Izu Islands, Japan, have cast a new light on the origin and evolution of basalt magmas and their derivatives:

(1) The mantle material in the source region of primary basalt magmas beneath the Izu Islands shows a chondritic value of Sr/Ca and Ba/Ca ratios.

(2) Both the tholeiite magma and the high-alumina/calc-alkali basalt magma are primary with higher degrees (15—20% for the former) and lower degrees (8—11% for the latter) of partial melting of a common mantle material.

(3) The primary basalt magmas evolve independently via crystal fractionation process in respective magma chambers at shallower depths each providing a series of andesite and dacite magmas corresponding to respective primary basalt magmas.

(4) The crystal fractionation process in magma chamber is controlled mainly by plagioclase and clinopyroxene crystallization in terms of the alkaline earth elements. The plagioclase/clinopyroxene ratio decreases during crystal fractionation process. The chemical environments of magma chambers are similar to each other in the tholeiite series and in the high-alumina basalt/calc-alkali rock series.

(5) The end products provided by the crystal fractionation process lie within Bowen's petrogeny's residua system, making a thin, silicic crust under the volcanic islands near the Izu Peninsula. The calc-alkali rhyolites in these islands are derived from the thin silicic crust via melting process by the heat of intruded primary basalt magmas.

(6) The regional distribution of degree of partial melting indicates variations from 15 to 20% along the volcanic front and from 8 to 11% in the region behind it. The fact suggests that an interaction between the mantle wedge under the Philippine Sea Plate and the subducting slab of the Pacific Plate beneath the Izu Islands is different from place to place, with respect to temperature distribution and/or water supply from the subducting slab.

0377-0273/83/$03.00

INTRODUCTION

Quaternary volcanoes of the Izu Islands which are located at the northernmost position of the Izu-Mariana island arc have been formed on the Philippine Sea Plate induced by subduction of the Pacific Plate under the Philippine Sea Plate. As shown in Fig. 1, the volcanoes of the Izu Islands are classified into two groups: one is volcanoes situated on the volcanic front parallel to the Izu-Ogasawara Trench (from north to south, Ōshima-Miyakejima-Mikurajima-Hachijōjima) and the other is volcanoes situated behind the volcanic front and oblique to the Izu-Ogasawara Trench (from northeast to southwest, Ōshima vs. Toshima-Udonejima-Niijima-Shikinejima-Kōzushima; Miyakejima vs. Ōnoharajima; Mikurajima vs. Inambajima). Detailed studies on geology and petrology in the region have been carried out by one of the authors (N.I.) during past 20 years (Isshiki, 1958, 1960, 1963, 1964a, b, 1977, 1978, 1980, 1981; Isshiki et al., 1963). According to these studies, the petrology of the volcanic rocks in the Izu Islands can be summarized as follows. The volcanic rocks constituting the volcanoes on the volcanic front are dominantly basalts with subordinate andesites and rare dacites. These volcanic rocks form a distinct chemical group characterized by high excess silica and low alkalis, and belong to the tholeiite series. On the other hand, the volcanic rocks constituting the volcanoes behind the volcanic front are high-alumina basalt, and calc-alkali basalts, andesites and rhyolites. The basalts and andesites from the volcanoes on the volcanic front contain subcalcic clinopyroxene in groundmass and the dacites have porphyritic ferroaugite and fayalite in glassy groundmass. The andesites from the volcanoes behind the volcanic front are characterized by the presence of orthopyroxene and clinopyroxene in groundmass and the rhyolites in the same area contain porphyritic hypersthene, cummingtonite, hornblende and biotite in glassy or cryptocrystalline groundmass.

This is the first comprehensive study of trace-element geochemistry of the volcanic rocks in the Izu Islands region. The problem to be addressed by our geochemical approach are: (1) the nature of the mantle material situated under the Izu Islands region; (2) the nature of the partial melting process in the mantle; (3) the nature of crystal fractionation processes which operated within the magma chamber; and (4) the interaction between primary magmas and granitic crust. The problems could be elucidated by a new geochemical diagram named "Sr/Ca—Ba/Ca diagram" or the SB diagram for short (Onuma, 1980, 1981; Onuma and Hirano, 1981). The SB diagram enables us to visualize both the partial melting process of the mantle material and the crystal fractionation process in magma chamber.

Therefore, we have determined the alkaline earth elements (Ca, Sr and Ba) of more than one hundred volcanic rocks from ten volcanic islands in the Izu Islands region. Ca, Sr and Ba contents in the samples have been

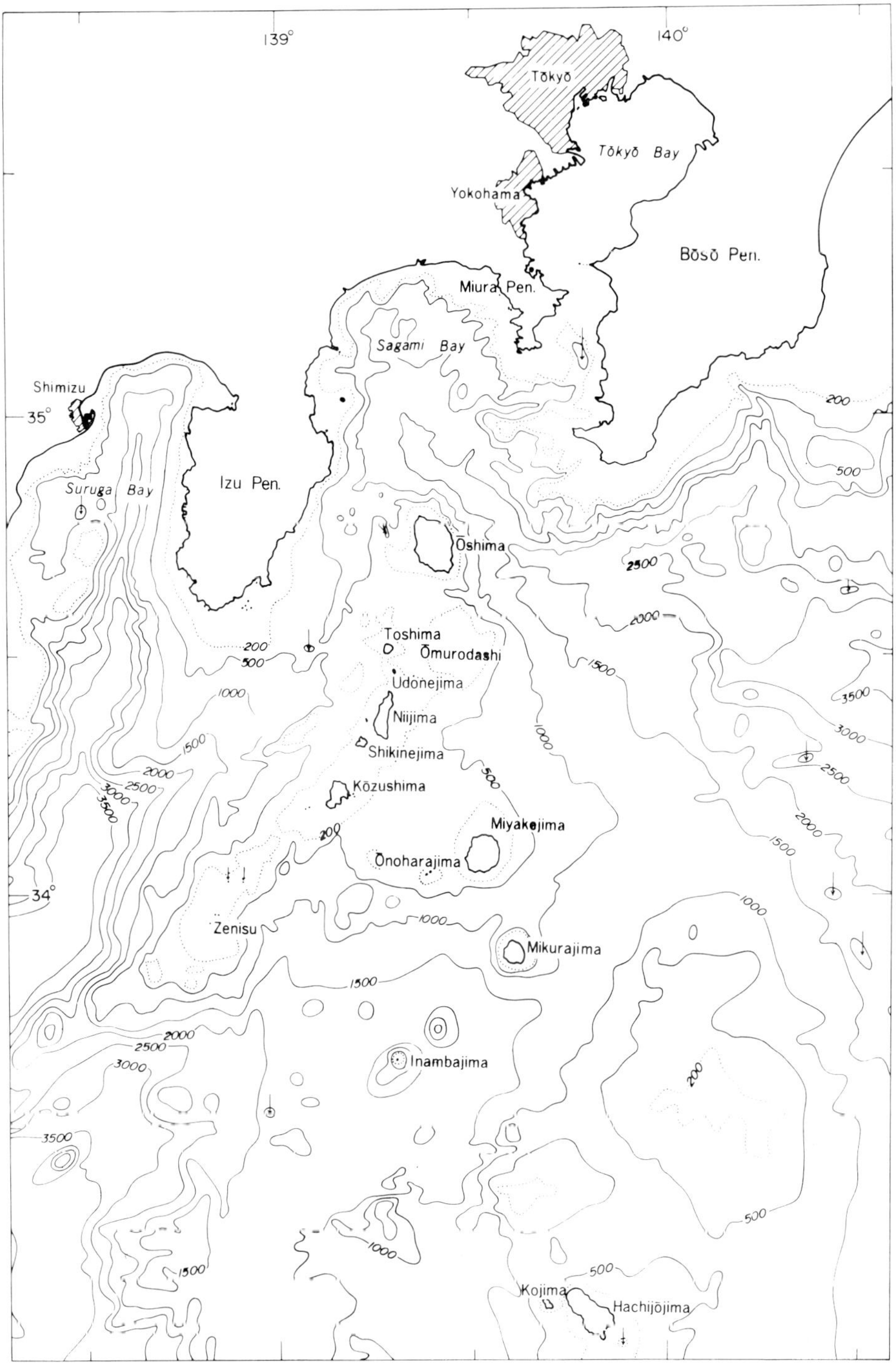

Fig. 1. Locality map of the Izu Islands region based on Ocean Sounding Charts G 1406 and G 1506 published by Maritime Safety Agency, Japan, in 1974.

determined simultaneously by an inductively coupled plasma — optical emission spectrometry (Hirano et al., 1980). The geological setting of each volcano, petrographical description of the samples and presentation of the data obtained is being published elsewhere (Onuma et al., 1981b, Hirano et al., 1982, Isshiki et al., 1982, Onuma et al., 1983).

In this paper, we will discuss regional problems related to the genesis of basalt magmas and their derivatives under the Izu Islands, Japan, based on Sr/Ca—Ba/Ca systematics. The main topics are: (1) the nature of the mantle materials and genesis of the two different primary basalt magmas under the Izu Islands region; (2) a comparison of basalt-andesite-dacite of the tholeiite series, and high-alumina basalt and basalt-andesite-rhyolite of the calc-alkali rock series; and (3) the regional distribution of degree of partial melting of the mantle materials beneath the Izu Islands region.

Sr/Ca—Ba/Ca DIAGRAM FOR ELUCIDATION OF MAGMA GENESIS

In this section, the principles of SB diagram for elucidating the origin and evolution of magmas will be introduced (Onuma, 1980, 1981; Onuma et al., 1981a, b). As is well known, the large divalent triads, Ca^{2+} (100pm), Sr^{2+} (117pm) and Ba^{2+} (136pm), show different behavior in igneous processes such as partial melting and crystal fractionation, controlled by crystal structure of mineral phases in silicate melt (Onuma et al., 1968; Higuchi and Nagasawa, 1969; Jensen, 1973; Matsui et al., 1977). Therefore, a three-element plot or Sr/Ca—Ba/Ca diagram is considered to be a sensitive geochemical indicator for analyzing magma genesis.

Distribution of Ca, Sr and Ba in a mantle material

First, let us consider the distribution of the three elements (Ca, Sr and Ba) within a primitive mantle material such as phlogopite-bearing garnet peridotite. This is composed of olivine, orthopyroxene, clinopyroxene and garnet as major phases and phlogopite and apatite as minor phases. The major phase olivine does not accept the three large ions because of crystal structure control. The large ion, Ca, is contained mainly in clinopyroxene and garnet and in small amounts in orthopyroxene and phlogopite and apatite as minor phases. The larger cations, Sr and Ba, do not enter into the major phases (olivine, orthopyroxene, clinopyroxene and garnet) also because of crystal structure control. The larger cations are contained mainly in phlogopite and apatite as minor phases.

SB systematics established by partial melting

When the mantle material is gradually heated, partial melting occurs. A low degree of partial melting produces a large enrichment of Sr and Ba in the melt, since the accessory minerals such as phlogopite and apatite are among the first to enter the melt. The initial Sr/Ca and Ba/Ca ratios of the melt are the largest ones. Increasing degree of partial melting dilutes Sr and Ba contents in the melt with addition of Sr- and Ba-poor major phases. Ca content of the melt increases with addition of Ca from garnet and clinopyroxene when these minerals begin to melt, so that Sr/Ca and Ba/Ca ratios of the melt decrease with constant Sr/Ba ratio. The melt generated by complete melting of the mantle material shows the same ratios as those of the mantle material.

Partial melting of the mantle and the corresponding variation of Sr/Ca and Ba/Ca ratios in the generated melts are shown schematically in Fig. 2. As shown in Fig. 2, a series of melts derived from the mantle material with different degree of partial melting form a Sr/Ca—Ba/Ca systematics (SB systematics) which is represented by a line with a slope of 45° through the mantle material on the log-log SB diagram. The slope of 45° on the diagram implies addition of Ca to or subtraction of Ca from a melt with a constant Sr/Ba ratio. We call the line a "partial melting line". The residual solids left during partial melting have low Sr/Ca and Ba/Ca ratios and fall on the line below the mantle material.

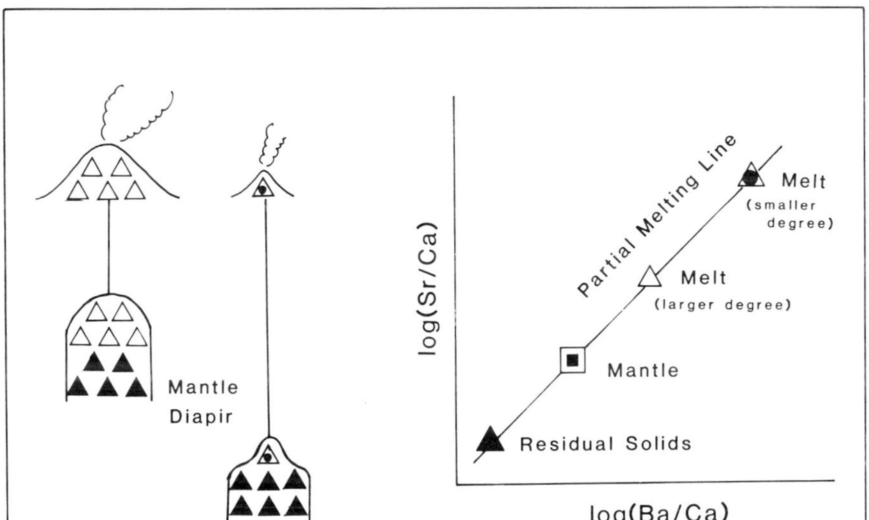

Fig. 2. A schematic presentation of a relationship between magma genesis by partial melting in mantle diapirs and Sr/Ca and Ba/Ca ratios of generated primitive melts. The primitive melt (△) with larger degree of partial melting shows smaller Sr/Ca and Ba/Ca ratios (△), while the primitive melt (△) with smaller degree shows larger Sr/Ca and Ba/Ca ratios (▲). A series of the primitive melts (primary basalt magmas) with different degrees, if available, forms a line (or SB systematics) with a slope of 45° through mantle (▣) and residual solids (▲). The line is called "partial melting line", since the SB systematics is established by partial melting process operated in mantle diapirs.

Once generated, melt or primary magma is separated from the residual solids and ascends to the surface. During the ascent olivine crystals precipitate from the primary magma. However, Sr/Ca and Ba/Ca ratios of the magma do not change, since olivine does not accept such larger cations. Therefore, both ratios of the primary magma are frozen. In other words, Sr/Ca and Ba/Ca ratios of the primary magma correspond to degree of partial melting, unless the primary magma is influenced by subsequent crystal fractionation process in "magma chamber" at a shallow depth.

SB systematics established by crystal fractionation

When the primary magma generated in the mantle ascends to a shallow depth and reaches to "magma chamber", crystal fractionation occurs. Crystallization of olivine, orthopyroxene, magnetite and ilmenite in the magma chamber does not change either Sr/Ca or Ba/Ca ratios of the primary magma, since those minerals do not accept the three cations. However, crystallization of plagioclase, clinopyroxene and hornblende does change Sr/Ca and Ba/Ca ratios of the primary magma and form evolved melts.

In the case of plagioclase crystallization, Sr/Ca ratio in the evolved melt hardly changes, while Ba/Ca ratio in the evolved melt increases greatly, since plagioclase accepts both Ca and Sr with similar degree but excludes Ba. In the case of clinopyroxene (hornblende) crystallization, both Sr/Ca and Ba/Ca ratios in the evolved melt increase. Increase of Ba/Ca ratio is much greater than that of Sr/Ca ratio, since Ca enters, while Sr is hard to enter and Ba does not enter into clinopyroxene (hornblende).

The crystal fractionation process in the magma chamber and the corresponding variation of Sr/Ca and Ba/Ca ratios in evolved melts are schematically shown in Fig. 3. A series of evolved melts derived from the primary magma via plagioclase and clinopyroxene (hornblende) crystallization in the magma chamber form another SB systematics which is a curve with a gentle slope through the primary magma on the log-log SB diagram. On the SB systematics, basalts and their derivatives provided by a volcano should be lined up in order of the primary magma-basalt-andesite-dacite. In this sense, we call the line "crystal fractionation line". The slope of the line is controlled by plagioclase/clinopyroxene ratio precipitated in the magma chamber. The larger Pl/Cpx ratio corresponds to the flatter slope and the smaller Pl/Cpx ratio corresponds to the steeper slope. The cumulates separated from the primary magma via crystal fractionation in the magma chamber show similar Sr/Ca ratio and lower Ba/Ca ratio compared to the primary magma so that they fall on certain positions extrapolated from the crystal fractionation line through the primary magma.

The "crystal fractionation line" in Fig. 3 might be called "mixing line" under certain conditions. For example, a similar SB systematics can be established by a basalt-andesite-dacite series which is derived from the

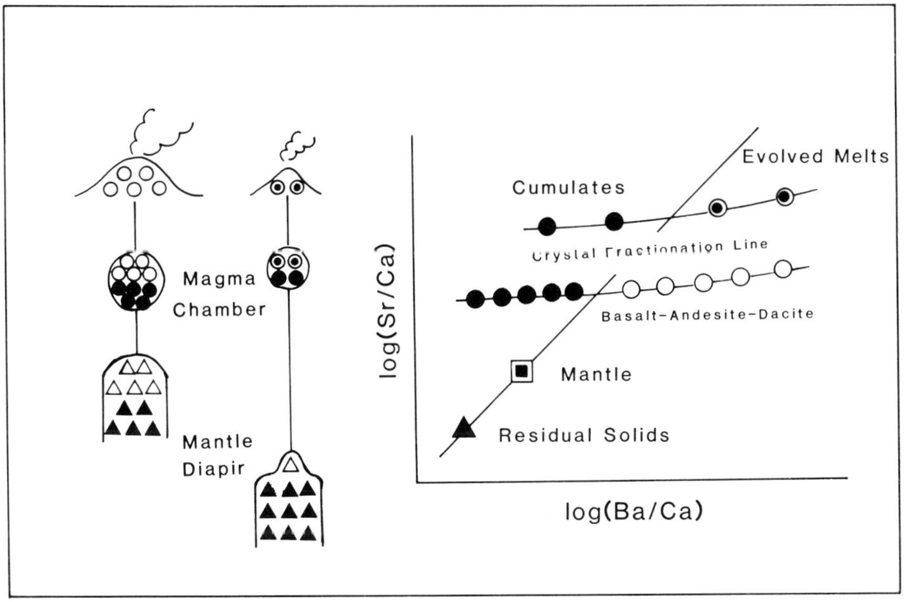

Fig. 3. A schematic presentation of a relationship between evolution of the primitive melts by crystal fractionation process in magma chambers at shallower depths and Sr/Ca and Ba/Ca ratios of evolved melts. The primitive melts (△) generated in mantle diapirs ascend to magma chambers and change their Sr/Ca and Ba/Ca ratios by plagioclase/clinopyroxene fractional crystallization. The evolved melts (○, ⊙) provide different series of SB systematics which are represented by respective derivatives. These lines with gentle slopes established by series of volcanic rocks (basalt-andesite-dacite) and corresponding cumulates (●) are called "crystal fractionation lines".

mixing process of a primary magma and the crust with granitic composition. Therefore, we must be careful in interpreting a given systematics from a volcano situated on the continental crust.

NATURE OF THE MANTLE AND ORIGIN OF THE TWO DIFFERENT PRIMARY BASALT MAGMAS UNDER THE IZU ISLANDS REGION

As pointed out in the previous section, an SB systematics corresponding to a partial melting line could be defined by a series of primary magmas derived from the mantle with different degrees of partial melting. However, we do not know Sr/Ca and Ba/Ca ratios of the mantle material situated under the Izu Islands region, and we cannot draw the partial melting line with a slope of 45° through the mantle material in the SB diagram.

A clue to solve the problem has come up from a study of SB systematics in the differentiated meteorites (Onuma and Hirano, 1981). The source material for eucritic meteorites was considered to be chondritic meteorites in terms of Sr/Ca and Ba/Ca ratios. The chondritic meteorites, as is well known, show non-volatile elemental abundances representative of the solar system. The non-volatile elements such as Ca, Sr and Ba are expected not to be fractionated during the accretion process of planetary materials on a growing planet, and subsequent metal-silicate fractionation processes within the planet. Therefore, we have decided to employ Sr/Ca and Ba/Ca ratios of chondritic meteorites as those of a hypothetical mantle material and to draw a line with a slope of 45° through chondritic meteorites as a hypothetical partial melting line (Onuma, 1981; Onuma et al., 1981a, b). If the primary basalt magmas with different degrees of partial melting of the mantle material could be available from the Izu Islands region, and if the data from the primary basalts would fall on the hypothetical partial melting line through chondritic meteorites, we could say something about nature of the mantle material and partial melting process operated within the mantle under the Izu Islands region.

Fig. 4 shows a hypothetical partial melting line with a slope of 45° through chondrites and an SB systematics defined by a series of the most primary basalts from volcanoes in the Izu Islands region. The data are taken from Tera et al. (1970) for chondrites and Onuma et al. (1981a, b, 1982) and Hirano et al. (1982) for rocks of the tholeiite series in volcanoes situated on the volcanic front (Ōshima, Miyakejima, Mikurajima and Hachijōjima) and Isshiki et al. (1982) for those of the high-alumina basalt/calc-alkali rock series in volcanoes behind the volcanic front (Toshima, Udonejima and Niijima), respectively.

Figure 4 strongly suggests that the mantle material situated under the

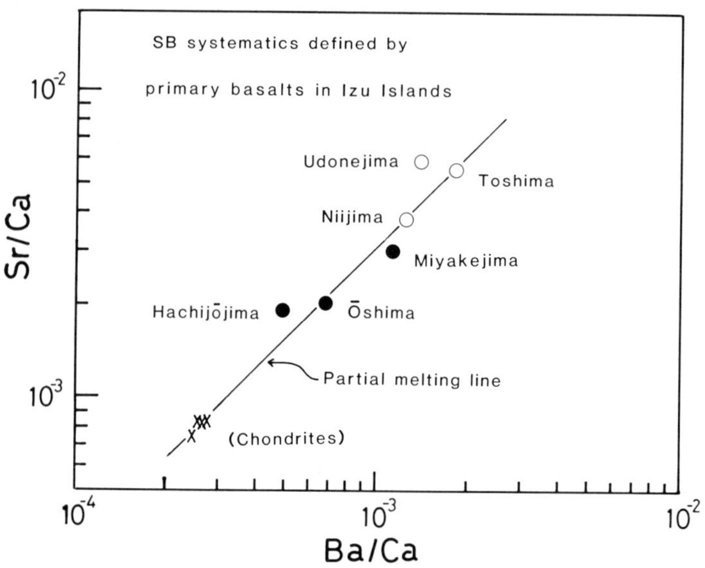

Fig. 4. SB systematics established by primary basalts in the Izu Islands region. A partial melting line is drawn through chondrites with slope of 45°. • = primary basalts of the tholeiite series; ○ = primary basalts of high-alumina basalt and calc-alkali rock series.

Izu Islands region has a chondritic composition in terms of Sr/Ca and Ba/Ca ratios, and at least the two different primary basalt magmas are derived from the common mantle material, since the primary basalts from the volcanoes in this region fall on the hypothetical partial melting line. The tholeiite magmas (●) generated along the volcanic front parallel to the Izu-Ogasawara Trench show higher degrees of partial melting of the mantle material, because the primary basalts have smaller Sr/Ca and Ba/Ca ratios as shown in Fig. 4. On the other hand, the high-alumina basalt/calc-alkali magmas (○) generated behind the volcanic front show lower degrees of partial melting of the mantle material, since the primary basalts have larger Sr/Ca and Ba/Ca ratios as shown in Fig. 4.

EVOLUTION OF ROCKS OF THE THOLEIITE SERIES ERUPTED ON THE VOLCANIC FRONT PARALLEL TO THE IZU-OGASAWARA TRENCH

In the previous section, we discussed the origin of two different primary magmas generated by partial melting in the mantle at depths. In this section, we will discuss the evolution of the tholeiite magma in relation to crystal fractionation process in a "magma chamber" at a shallower depth.

Figure 5 shows various trends of SB systematics defined by basalt-andesite-dacite of the tholeiite series which was erupted from the volcanoes

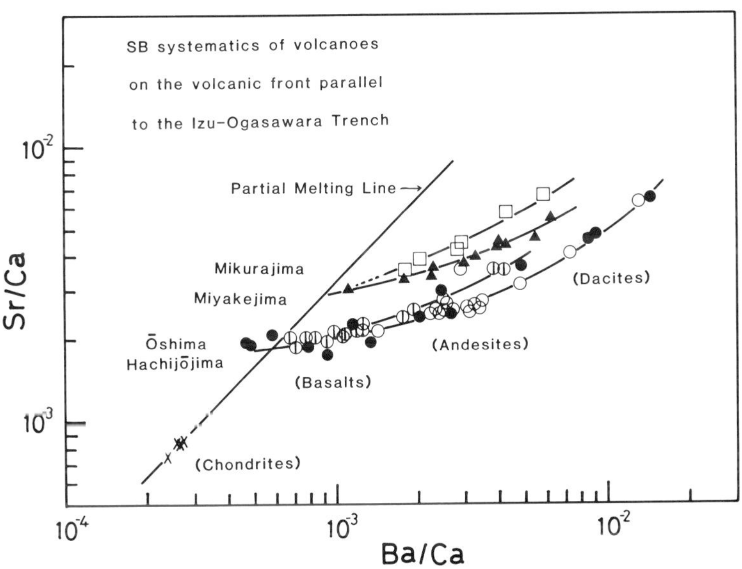

Fig. 5. Various SB systematics established by volcanic rocks from volcanoes situated on the volcanic front parallel to the Izu Ogasawara Trench. The volcanic rocks belong to the tholeiite series. ⦶ = the older volcanoes in Ōshima; ○ = the younger volcano in Ōshima; ● = Higashiyama volcano in Hachijōjima; ▲ = Miyakejima volcano; □ = Mikurajima volcano.

of the volcanic front parallel to the Izu-Ogasawara Trench. The volcanic islands lined from north to south are Ōshima, Miyakejima, Mikurajima and Hachijōjima, as shown in Fig. 1. The data are taken from Onuma et al. (1981b) for Ōshima, Hirano et al. (1982) for Miyakejima and Mikurajima, and Onuma et al. (1983) for Hachijōjima, respectively.

As shown in Fig. 5, the tholeiite series provided by the stratovolcanoes form various curves, starting from the partial melting line and growing upward from basalts through andesites to dacites in order. We call the curve the SB systematics established by crystal fractionation process in "magma chambers" at shallower depths under the volcanoes. The slope of the SB systematics is considered to be controlled by plagioclase/clinopyroxene ratio crystallizing in "magma chamber". The concave upward shape of the SB systematics indicates that plagioclase crystallization prevails at the early stage and clinopyroxene crystallization prevails at the later stage of the crystal fractionation process in "magma chamber".

In Ōshima Island, there are four volcanoes called Okata, Gyōjanoiwaya, Fudeshima and Ōshima volcanoes. In the SB diagram, the three older volcanoes [Okata, Gyōjanoiwaya and Fudeshima volcanoes (⦶)] give the same SB systematics, while the younger volcano [Ōshima volcano (○)] shows a different SB systematics. The two different trends of SB systematics branching off from a common precursor suggest that the primary basalt magmas were the same in terms of degrees of partial melting of the mantle material, but crystal fractionation processes in the "magma chamber" were different in terms of crystallized plagioclase/clinopyroxene ratios. The plagioclase/clinopyroxene ratio for the three older volcanoes is smaller than that for the younger volcano (Onuma et al., 1981b).

Miyakejima volcano (▲) and Mikurajima volcano (□) in the SB diagram show two different trends of SB systematics, branching off from a common precursor. The two trends of SB systematics suggest that the primary basalt magmas generated in the mantle under the two different volcanic islands were the same in terms of degrees of partial melting, while the environment of the two magma chambers at shallower depths were different in terms of plagioclase/clinopyroxene crystallization (Hirano et al., 1982).

Higashiyama volcano in Hachijōjima Island provided basalt-andesite-dacite of the tholeiite series. The volcanic rocks define an SB systematics (●) which is quite similar to the SB systematics of Ōshima Island (⦶, ○), as shown in Fig. 5. The similarity suggests that partial melting process operated in the mantle as well as crystal fractionation process carried out in the magma chambers were quite similar under the two different islands (Onuma et al., 1983).

The tholeiitic basalts erupted from the volcanoes on the volcanic front parallel to the Izu-Ogasawara Trench are primary basalt magmas with higher degrees of partial melting of the primitive mantle material with chondritic Sr/Ca and Ba/Ca ratios. The degrees of partial melting vary 1.3 fold within the tholeiitic basalts. The andesites and dacites of the tholeiite series are derived from the primary basalt magmas via crystal fractionation process in "magma chambers" at shallower depths. The crystal fractionation process in the "magma chambers" under the volcanoes are slightly different from each other, in terms of plagioclase/clinopyroxene crystallization.

COMPARISON OF THE THOLEIITE SERIES AND THE HIGH-ALUMINA BASALT/ CALC-ALKALI ROCK SERIES

In this section, we will present comparison of SB systematics defined by the tholeiite series and the high-alumina basalt/calc-alkali rock series. As shown in Fig. 1, there are several volcanic islands running in echelon from the volcanic front toward the southwest, such as Oshima vs. Toshima-Udonejima-Niijima-Shikinejima-Kōzushima, Miyakejima vs. Ōnoharajima, and Mikurajima vs. Inambajima. The volcanoes situated behind the volcanic front provide rocks of the high-alumina basalt/calc-alkali rock series in contrast with rocks of the tholeiite series erupted from the volcanoes on the volcanic front.

Figure 6 shows a comparison of basalt-andesite-dacite of the tholeiite series from Ōshima vs. basalt-andesite of the high-alumina basalt/calc-alkali rock series from Toshima and Niijima. The data are taken from Onuma

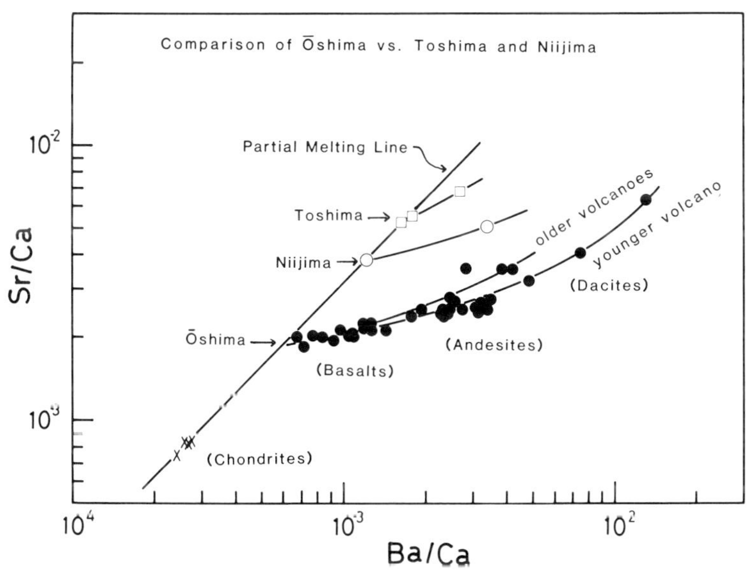

Fig. 6. Comparison of SB systematics established by the tholeiite series and high-alumina basalt/calc-alkali rock series Ōshima vs. Toshima and Niijima system. • = Ōshima (tholeiite series); ○ = Niijima (high-alumina basalt/calc-alkali rock series); □ = Toshima (high-alumina basalt/calc-alkali rock series).

et al. (1981b) for Ōshima (•), and Isshiki et al. (1982) for Toshima (□) and Niijima (○), respectively.

Figure 6 suggests that major difference between the two rock series is caused by difference in the degrees of partial melting of a common mantle material. The degree of the tholeiite series from Ōshima is larger than that of the high-alumina basalt/calc-alkali rock series from Toshima and Niijima. The basalts from Toshima show the smallest degree in the Izu Islands region. Minor differences are also seen in the slopes of SB systematics which might reflect differences in chemical environments of "magma chambers". The parallelisms in the SB systematics between Niijima and the younger volcano (Ōshima volcano) in Oshima, and between Toshima and the older volcanoes (Okata, Gyōjanoiwaya and Fudeshima volcanoes) in Ōshima suggest that crystal fractionation processes which occurred in "magma chambers" are similar to each other in terms of plagioclase and clinopyroxene crystallization.

Figure 7 shows a comparison of basalt-andesite of the tholeiite series from Miyakejima (•) and cognate inclusion with basaltic composition and andesite of the calc-alkali rock series from Ōnoharajima (○). The data are taken from Hirano et al. (1982). Figure 7 suggests that major differences in the two rock series, again, correspond to the different degrees of partial melting of a common mantle material and there is no difference in crystal fractionation process in "magma chambers" between them, as indicated by the two parallel series of SB systematics.

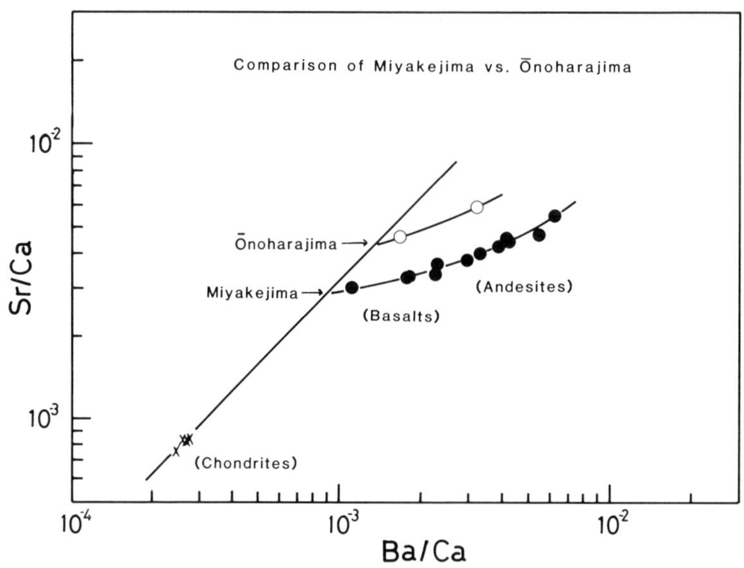

Fig. 7. Comparison of SB systematics established by tholeiite series and calc-alkali rock series in Miyakejima vs. Ōnoharajima system. • = Miyakejima volcano (tholeiite series); ○ = Ōnoharajima volcano (calc-alkali rock series).

The comparative study reveals that the different primary basalt magmas might originate in the mantle by partial melting of a common mantle material with chondritic Sr/Ca and Ba/Ca ratios. The rocks of the tholeiite series might be derived from primary basalt magmas with larger degrees of partial melting via crystal fractionation process in "magma chambers" at shallower depths, while the rocks of the high-alumina basalt/calc-alkali rock series might be derived from primary basalt magmas with smaller degrees of partial melting through crystal fractionation process in "magma chamber". There seems no essential difference in crystal fractionation process in "magma chambers" between the different series in terms of plagioclase and clinopyroxene crystallization.

PETROGENY'S RESIDUA SYSTEM AND GENESIS OF CALC-ALKALI RHYOLITES

During this investigation, we have noticed two new trends revealed in the SB diagram: One is convergence of the various series of SB systematics of the volcanoes in the Izu Islands into a silicic rock and the other is an unexpected SB systematics defined by a series of rhyolites from Niijima and Kōzushima. Figure 8 shows the new findings. The various series of SB systematics defined both by the tholeiite series and the high-alumina basalt/calc-alkali rock series expressed by solid lines are transferred from Figs. 5, 6 and 7 and the data of calc-alkali rhyolites from Niijima and Kōzushima are taken from Isshiki et al. (1982).

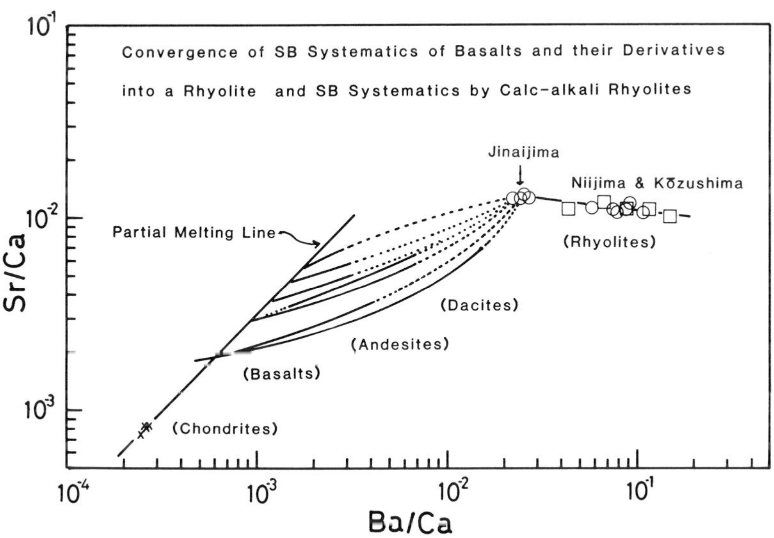

Fig. 8. Convergence of various SB systematics established by tholeiite series and high-alumina basalt/calc-alkali rock series into a rhyolite (Jinaijima) and a different SB systematics established by calc-alkali rhyolites from Niijama and Kōzushima. ○ = calc-alkali rhyolites from Niijima; □ = calc-alkali rhyolites from Kōzushima.

The extrapolation lines (the dashed lines in Fig. 8) of the SB systematics, though large extrapolations, seem to converge into a point represented by the rhyolite of Jinaijima which is a small island located west of Niijima. The other rhyolites from Niijima and also from Kōzushima form a quite different SB systematics with a slightly negative slope, and Jinaijima type rhyolite is situated at the left end of the SB systematics.

The convergence of the various trends of SB systematics defined by the tholeiite series and the high-alumina basalt/calc-alkali rock series into a silicic rock with composition represented by Jinaijima type rhyolite, suggests that the silicic rock might be an end product derived from crystal fractionation process operated in "magma chambers" at shallower depths under the volcanoes. The various primary basalt magmas with different degrees of partial melting of the mantle material are thought to evolve through the paths of basalt-andesite-dacite magmas into the silicic magma with composition of Jinaijima type rhyolite, regardless of the differences of their initial compositions. The convergent point may lie within the low-temperature trough in the equilibrium $NaAlSiO_4$ - $KAlSiO_4$ - SiO_2 system or in the Bowen's "petrogeny's residua system" (Bowen, 1937). Crystal fractionation of different primary basalt magmas leads always to a residual magma enriched in alkali aluminium silicate.

The unexpected SB systematics defined by rhyolites from Niijima and Kōzushima in Fig. 8 may be explained by the following two mechanisms: (1) partial melting of a silicic rock with composition of Jinaijima type rhyolite leaving plagioclase as residual solid; or (2) total melting of the silicic rock followed by plagioclase crystallization. Removal of plagioclase from a given silicate melt causes great increase of Ba/Ca ratio with constant Sr/Ca ratio or slight decrease of the ratio in the evolved silicate melts. The rhyolite from Jinaijima may be a silicate melt formed, by total melting, from an end product of crystal fractionation process while the other rhyolite magmas of Niijima and Kōzushima may be melts evolved from an original one with composition of Jinaijima type rhyolite by removal of plagioclase.

Therefore, we imagine that there may exist a thin silicic crust with composition of Jinaijima type rhyolite seated at a shallow depth under Niijima and Kōzushima. If the proposed model that the rhyolites are derived by melting from the thin silicic crust is correct, the question on the mechanism of the crustal melting must be answered. A clue to the problem is found in the cognate inclusions in the rhyolites from these islands.

The cognate inclusions in the rhyolites from Niijima give very similar compositions to that of a basalt of the high-alumina basalt series from this island, as shown in Fig. 9. The data are taken from Isshiki et al.(1982). The striking similarity suggests that the cognate inclusion represents a patch of the high-alumina basalt magma which may have intruded below the thin silicic crust under Niijima and melted the crust and provided the rhyolitic magmas. The rhyolite volcanism seen at Shikinejima and Kōzushima in the Izu Islands region may be explained by the same mechanism.

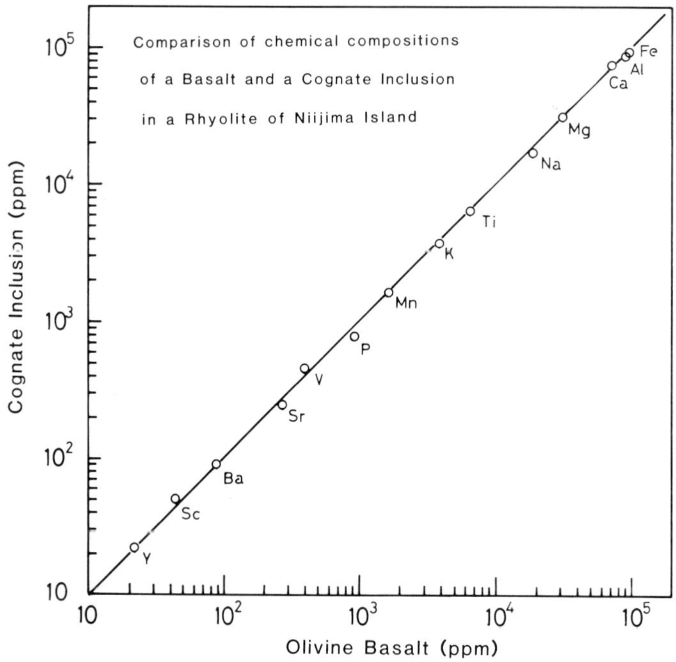

Fig. 9. Comparison of chemical compositions of a basalt and a cognate inclusion in a rhyolite of Niijima.

REGIONAL DISTRIBUTION OF DEGREES OF PARTIAL MELTING IN THE MANTLE BENEATH THE IZU ISLANDS

As schematically shown in Fig. 3, one of the most important pieces of information which can be deduced from the SB diagram is "SB index" as a measure of degree of partial melting of the mantle material. The "SB index" is defined by an intersection of the partial melting line and the crystal fractionation line established by a series of volcanic rocks from a given volcano. The "SB indices" have been calibrated tentatively as "degrees" of partial melting, assuming a simple melting model of a garnet peridotite (Onuma and Hirano, 1983). Fig. 10 shows the estimated "degrees" of partial melting in the mantle beneath the Izu Islands region.

The primary basalt magmas in the Izu Islands region may have been generated by an interaction between the mantle wedge under the Philippine Sea Plate and a subducting slab of the Pacific Plate. As shown in Fig. 10, generally speaking, the "degrees" of partial melting in the mantle under the volcanic front are larger than those in the mantle behind the volcanic front.

However, a detailed inspection reveals that there is no simple regularity between the "degrees" of partial melting and the distances from the plate boundary. For example, in the volcanic islands on the volcanic front, the

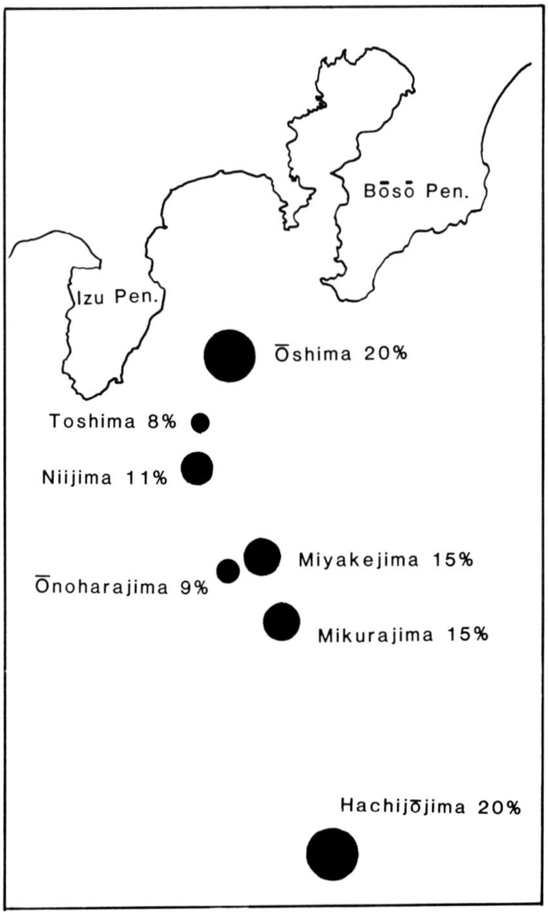

Fig. 10. Regional distribution of "degree" of partial melting of the mantle material beneath the Izu Islands region. The figures for "degree" are tentatively determined by a batch melting model and are meaningful only in relative sense.

degree decreases from Ōshima (20%) to Miyakejima (15%), holds constant from Miyakejima to Mikurajima (15%), and increases from Mikurajima to Hachijōjima (20%). Thus, the "degrees" are not constant but variable in the volcanoes on the front. The similar variation of the "degrees" can be seen in the volcanic islands in echelon array from the volcanic front toward the southwest. The "degree" decreases from Ōshima (20%) to Toshima (8%) and then increases from Toshima to Niijima (11%). In the system of Miyakejima (15%) vs. Ōnoharajima (9%), the "degree" decreases toward the southwest.

Therefore, we may conclude that the interaction between the mantle wedge under the Philippine Sea Plate and the Pacific Plate is different from place to place beneath the Izu Islands region.

The "degree" of partial melting of the mantle material shows (primary basalt magma/mantle material) ratio in a mantle diapir, or simply (liquid/ mantle) ratio. The (liquid/mantle) ratio is considered to be controlled mainly by temperature and water content in the diapir. Higher temperature may produce a diapir with larger (liquid/mantle) ratio at constant water content. Higher water content may also provide a diapir with larger (liquid/mantle) ratio by decreasing melting point of the mantle material. Pressure effect on (liquid/mantle) ratio may be conceivable when the movement of an ascending diapir in the mantle is considered.

The regional variation of the (liquid/mantle) ratio shown in Fig. 10 suggests that: (1) if the (liquid/mantle) ratio is mainly controlled by temperature, the temperature distribution in the mantle beneath the Izu Islands region may be not a smooth one, but a rough one in the local scale; and (2) if the (liquid/mantle) ratio is mainly controlled by water supplied from the subducting slab to the mantle wedge, the amounts of water may be not a constant one but a variable one from place to place in the mantle beneath the Izu Islands region.

It is hoped that, as a next step, new thermometers, barometers as well as hygrometers which are applicable to analyzing the nature of mantle diapir may be developed to understand chemical process occurring under the island arc regions.

CONCLUSIONS

The conclusions obtained in this investigation are summarized as follows:

(1) The SB systematics defined by a series of volcanic rocks of a given volcano reveals characteristics of the volcano with respect to partial melting process in "mantle" and crystal fractionation process in "magma chamber" at a shallower depth.

(2) The mantle material in a source region of primary basalt magmas beneath the Izu Islands has a chondritic Sr/Ba ratio.

(3) The two different magmas, the tholeiite and the high-alumina/calc-alkali basalt magmas generated under the Izu Islands, are considered to be primary basalt magmas with larger degrees (15—20%) for the former and smaller degrees (8—11%) for the latter of partial melting of a common mantle material.

(4) The parental basalt magmas evolve independently via crystal fractionation process in respective "magma chambers" and provide a series of andesite and dacite magmas corresponding to the respective basalt magmas.

(5) The crystal fractionation process in terms of Sr/Ca and Ba/Ca ratios in a "magma chamber" is controlled mainly by plagioclase and clinopyroxene crystallization.

(6) The physicochemical conditions of the "magma chambers" are,

more or less, similar both for the tholeiite series and the high-alumina basalt/calc-alkali rock series of the Izu Islands.

(7) End products of the crystal fractionation process of the primary basalt magmas lie within petrogeny's residua system. The end product presumably constituting a thin crust under the volcanic islands near the Izu Peninsula have a silicic composition similar to that of the rhyolite now found in Jinaijima. The majority of the calc-alkali rhyolites from the volcanic islands are derived from the thin silicic crust via melting process by the heat of basalt magmas intruded under the crust.

(8) Regional distribution of the "degrees" of partial melting of a common mantle material beneath the Izu Islands indicates variations from 15 to 20% along the volcanic front and from 8 to 11% in the region behind it.

(9) An interaction between the mantle wedge under the Philippine Sea Plate and the subducting slab of the Pacific Plate beneath the Izu Islands is considered to be different from place to place with respect to temperature distribution and/or water supply from the subducting slab. However, we have not yet drawn a definitive conclusion about a wirepuller who controls the "degrees" of partial melting of the mantle material.

ACKNOWLEDGEMENTS

During the course of this investigation, we have been provided with very helpful suggestions and comments on magma genesis from many persons. Profs. I. Kushiro and K. Aoki gave us general information about partial melting process of the mantle materials. One to one correspondence between a volcano and mantle diapir has been claimed by Dr. M. Sakuyama. Dr. K. Notsu gave us encouragements through this work. Constructive comments on the manuscript have been provided by Profs. S. Uyeda and Y. Matsui. Our thanks go to these persons. One of us (N.O.) was supported in part by Grant-in-Aid for Scientific Research, No. 504112 and No. 56110005 from the Ministry of Education, Science and Culture, Japan.

REFERENCES

Bowen, N.L., 1937. Recent high-temperature research on silicates and significance in igneous geology. Am. J. Sci., Fifth Ser., 33: 1—21.

Higuchi, H. and Nagasawa, H., 1969. Partition of trace elements between rock-forming minerals and the host volcanic rocks. Earth Planet. Sci. Lett., 7: 281—287.

Hirano, M., Notsu, K. and Onuma, N., 1980. Rapid simultaneous 17 elements analysis of some Yamato meteorites by ICP-OES. Mem. Natl. Inst. Polar Res., Spec. Iss., 17: 152—158.

Hirano, M., Isshiki, N. and Onuma, N., 1982. Sr/Ca—Ba/Ca systematics in Miyakejima, Ōnoharajima, Mikurajima and Inambajima volcanoes, the Izu Islands, Japan. Geochem. J., 16: 79—87.

Isshiki, N., 1958. Petrology of plutonic cognate ejecta from Nishi-yama volcano, Hachijō-jima, the Seven Izu Islands, Japan. Jpn. J. Geol. Geogr., 26: 55—74.

Isshiki, N., 1960. The geological sheet map "Miyake-jima", scale 1 : 50,000, and its explanatory text. Geol. Surv. Jpn., 85 pp. (in Japanese with English abstract).

Isshiki, N., 1963. Petrology of Hachijō-jima Volcano Group, Seven Izu Islands, Japan. J. Fac. Sci., Univ. Tokyo, Sec. II, 15: 91—134.

Isshiki, N., 1964a. Mode of eruption of Miyake-jima volcano in historic times. Bull. Volcanol., 27: 29—48.

Isshiki, N., 1964b. Ō-shima volcano. Guidebook for Excursion 2, Geol. Surv. Jpn., 24 pp.

Isshiki, N., 1977. Activity of Miyake-jima volcano during the last 3,000 years (abstract). Bull. Volcanol. Soc. Jpn., 25: 307 (in Japanese).

Isshiki, N., 1978. Geology of the Toshima district. Quadrangle Series, scale 1:50,000. Geol. Surv. Jpn., 34 pp. (in Japanese with English abstract, 2pp.).

Isshiki, N., 1980. Geology of the Mikurajima, Inambajima and Zenisu Districts. Quadrangle Series, scale 1:50,000. Geol. Surv. Jpn., 35 pp. (in Japanese with English abstract, 3pp.).

Isshiki, N., 1981. Petrology and chemistry of Quaternary volcanoes of the Izu-Volcano-Mariana arc. Abstr., 1981 IAVCEI Symposium, Arc Volcanism, Tokyo and Hakone, p. 148.

Isshiki, N., Nakamura, K., Hayakawa, M., Hirasawa, K., Yukutake, T., Arai, Y. and Iwasaki, B., 1963. Structure of caldera of Ōshima volcano, Izu, as revealed by drilling. Bull. Volcanol. Soc. Jpn., Second Series, 8: 61—106 (in Japanese with English abstract).

Isshiki, N., Onuma, N. and Hirano, M., 1982. Sr/Ca—Ba/Ca systematics of Quaternary volcanoes in Toshima, Udonejima, Niijima and Kōzushima, the Izu Islands, Japan. Geochem. J., 16: 119—127.

Jensen, B.B., 1973. Patterns of trace element partitioning. Geochim. Cosmochim. Acta, 37: 2227—2242.

Matsui, Y., Onuma, N., Nagasawa, H., Higuchi, H. and Banno, S., 1977. Crystal structure control in trace element partition between crystal and magma. Bull. Soc. Fr. Minéral. Cristallogr., 100: 315—324.

Onuma, N., 1980. Sr-Ba systematics on meteorites, the moon and the earth. Solar System Science Symposium II. Tokyo, The Institute of Space and Astronautical Science, pp. 1—6 (in Japanese).

Onuma, N., 1981. Sr/Ca—Ba/Ca systematics on eucrite parent body, the moon and the earth. Proc. 14th ISAS Lunar Planet. Symp. Tokyo, The Institute of Space and Astronautical Science, pp. 334—340.

Onuma, N. and Hirano, M., 1981. Sr/Ca—Ba/Ca systematics in Antarctic Ca-rich achondrites and their origins. Mem. Natl. Inst. Polar Res., Spec. Iss., 20: 202—210.

Onuma, N. and Hirano, M., 1983. SB index vs. degree of partial melting of mantle. In prep.

Onuma, N., Higushi, H., Wakita, H. and Nagasawa, H., 1968. Trace element partition between two pyroxenes and the host lava. Earth Planet. Sci. Lett., 5: 47—51.

Onuma, N., Hirano, M. and Isshiki, N., 1981a. Sr/Ca—Ba/Ca systematics in Quaternary volcanoes of Izu Islands, Japan. Abstr. 1981 IAVCEI Symposium, Arc Volcanism, Tokyo and Hakone, pp. 281—282.

Onuma, N., Hirano, M. and Isshiki, N., 1981b. Sr/Ca—Ba/Ca systematics in four volcanoes of Ōshima, Izu Islands, Japan. Geochem. J., 15: 315—324.

Onuma, N., Hirano, M. and Isshiki, N., 1983. Sr/Ca—Ba/Ca systematics of four volcanoes of Hachijōjima, Izu Islands, Japan. In prep.

Tera, F., Eugster, O., Burnett, D.S. and Wasserburg, G.J., 1970. Comparative study of Li, Na, K, Rb, Cs, Ca, Sr, Ba abundances in achondrites and Apollo 11 Lunar Samples. Proc. Apollo 11 Lunar Sci. Conf., 2: 1637—1657.

Journal of Volcanology and Geothermal Research, 18 (1983) 531—548
Elsevier Science Publishers B.V., Amsterdam — Printed in The Netherlands

STRONTIUM ISOTOPE COMPOSITION IN VOLCANIC ROCKS FROM THE NORTHEAST JAPAN ARC

KENJI NOTSU

Institute of Chemistry, University of Tsukuba, Sakura-mura, Ibaraki 305 (Japan)

(Received October 20, 1982)

ABSTRACT

Notsu, K., 1983. Strontium isotope composition in volcanic rocks from the Northeast Japan arc. In: S. Aramaki and I. Kushiro (Editors), Arc Volcanism. J. Volcanol. Geotherm. Res., 18: 531—548.

In the Northeast Japan arc, a number of Quaternary volcanoes form a long, narrow belt, parallel to the Japan Trench. $^{87}Sr/^{86}Sr$ ratios were determined in 52 specimens of volcanic rocks from 27 volcanoes in the Northeast Japan arc area. The results reveal that the ratios change systematically in space. Decreasing $^{87}Sr/^{86}Sr$ ratios across the arc were confirmed over a wide area of Northeast Japan. In the same direction, increases in both Rb and Sr contents were also found. The regular trends are considered to be a strong constraint for elucidation of subduction-originated magma genesis at the Eurasia plate vs. Pacific plate boundary. In the northern region of the Northeast Japan arc, $^{87}Sr/^{86}Sr$ ratios in volcanic rocks along the volcanic front were almost constant (0.7038—0.7045) and slightly higher than those from the Izu-Ogasawara arc (0.7032—0.7038). This suggests that "interactions" between the Eurasia plate and the Pacific plate, and those between the Philippine Sea plate and the Pacific plate are slightly different. The southern region of the Northeast Japan arc, where the direction of the volcanic front bends from southward to westward, showed anomalously high $^{87}Sr/^{86}Sr$ ratios, reaching to 0.7077. This region coincides with the triple junction of the Eurasia, Pacific and Philippine Sea plates, suggesting "anomalous interaction" at the triple junction.

INTRODUCTION

Volcanic activity is intense along island arcs at subduction zones of plate margins. In eastern Japan, there are a number of Quaternary volcanoes, making a long, narrow belt, from Hokkaido via northeastern and central Honshu to Izu and Ogasawara islands, parallel to the Japan and Izu-Ogasawara Trenches. In the northern part of this volcanic belt, which is named the Northeast Japan arc, the Pacific plate subducts beneath the Eurasia plate, while in the southern part, the Izu-Ogasawara arc, the Pacific plate goes beneath the Philippine Sea plate. The Northeast Japan arc is jointed to the Kurile arc to the northeast and the Izu-Ogasawara arc is jointed to the Mariana arc to the south.

In the Northeast Japan arc, distribution of volcanoes is not uniform in space. The population of volcanoes is abundant on the eastern side facing the Japan Trench and decreases from east to west. A sharp boundary, which is called "volcanic front" (Sugimura, 1960), is observed at the eastern edge of the volcanic belt. Physical properties of subduction have been investigated most precisely on this arc. A visualized feature of subducting oceanic slab was first reported by Hasegawa et al. (1978), who discovered the double-planed structure of the deep seismic zone over a wide area, more than 300 km × 200 km, through the determination of microearthquake hypocenters.

Corresponding with the subducting feature of the oceanic slab, various kinds of lateral variations in petrology, chemical and isotopic compositions of arc volcanic rocks have been discussed since Kuno's (1966) work. He pointed out that three types of basalts, namely tholeiite, high-alumina basalt and alkali-olivine basalt, occur zonally from east to west along the island arc. Before Kuno's work, Kawano et al. (1961a) noticed that alkali contents in volcanic rocks from the Northeast Japan arc increased from east (tholeiite series) to west (calc-alkali rock series) based on chemical compositions of 180 rocks. Increase in K_2O contents, when normalized to SiO_2 contents, away from the volcanic front was commonly observed in many island arc systems on earth (Dickinson, 1968) and Ui and Aramaki (1978) summarized the K_2O zoning on Japanese island arcs in connection with the Bouguer anomaly. Recent works on volcanic rocks from the Northeast Japan arc have shown that some trace-elements contents, such as Ba, Th, U, REE and F contents, also increased along with K_2O content from east to west (Masuda, 1979; Ishikawa et al., 1980). Chondrite-normalized REE patterns also changed. The inclination of the pattern for lighter REE was flat in volcanic rocks from the volcanic front side and was gradually steeper away from it (Fujimaki and Kurasawa, 1980; Fujitani and Masuda, 1981). Furthermore, Sakuyama (1979) pointed out that H_2O contents in arc magma increased from east to west in the Northeast Japan arc.

Isotopic studies were also done in this region. Hedge and Knight (1969) measured Sr and Pb isotope compositions in volcanic rocks from a 40°N traverse cross-section across the Northeast Japan arc, showing that both $^{87}Sr/^{86}Sr$ ratios and radiogenic components of Pb isotopes decreased away from the volcanic front. Nohda and Wasserburg (1981) determined Sr and Nd isotope compositions in volcanic rocks from the same 40°N traverse cross section, confirming the previous Sr data and revealed that they fell near the "mantle array" (DePaolo and Wasserburg, 1979) on the Nd and Sr isotope diagrams.

Some models about the origin of arc magma have been proposed to compromise these lateral variations in volcanic rocks, underground structure obtained from geophysical data, such as gravity, earthquake, heat flow, etc., and the underground condition estimated from phase petrological investigations, but a unique model has not been reported yet. In order to construct a realistic model, isotopic data, especially $^{87}Sr/^{86}Sr$ ratios, in volcanic rocks are essential. However, in the case of the Northeast Japan arc, $^{87}Sr/^{86}Sr$ ratios have been previously determined only for a limited few volcanoes. Except the traverse works of $^{87}Sr/^{86}Sr$ ratios by Hedge and Knight (1969) and Nohda and Wasserburg (1981), we can enumerate: Oshima-oshima (Hedge and Knight, 1969), Oshima-oshima and Oshima-kojima (Katsui et al., 1978), Osoreyama (Togashi, 1981), Ichinomegata (Kaneoka et al., 1978), Akita-komagatake (Kurasawa, 1979), Towada (Kurasawa, 1972), Asama, Eboshi and Myoko (Nohda and Wasserburg, 1981) and Iizuna, Kurohime, Myoko and Yakeyama (Ishizaka et al., 1977). Even if all these data are compiled, the spatial distribution of $^{87}Sr/^{86}Sr$ ratios in volcanic rocks from the Northeast Japan arc remains unclear.

The purpose of this study is to present a comprehensive set of $^{87}Sr/^{86}Sr$ ratios of volcanic rocks from many Quaternary volcanoes situated in a wide area of the Northeast Japan arc. From this approach, we can expect to clarify the regional variations of $^{87}Sr/^{86}Sr$ ratios, both in across-arc and along-arc directions. This approach is considered to be essential to understand the "interaction" between the Eurasia plate and the Pacific plate in the northeastern Honshu area, as well as among the Eurasia plate, the Pacific plate and the Philippine Sea plate at the central Honshu area (Kanto-district) where there is a unique triple junction.

SAMPLES

Rock samples analyzed in this work were taken from 28 representative volcanoes on the Northeast Japan arc, which were selected with regard to their spatial distribution. Localities of volcanoes investigated are shown in Fig. 1, together with those of other Quaternary volcanoes in the Northeast Japan arc. For an individual volcano, usually two to four representative rock samples were analyzed. These rock samples were selected considering the rock types or chemical compositions of rocks and their erupting stages.

Rock samples from some volcanoes were newly collected by the present author with the aid of other persons, referring to the previous publications about geology and volcanology of them. Rock samples of other volcanoes were provided by several persons. Comments on each volcano are listed in Table I.

ANALYTICAL METHODS

Strontium isotope ratios were determined using a VG-Micromass MM-30 double-collector-type mass spectrometer with on line computer facilities, after the Sr in the rock samples was separated by a conventional cation exchange technique. Standard strontium isotope reagent, NBS 987, was analyzed with every five rock samples to check reproducibilities of the isotope measurements. The measured range of $^{87}Sr/^{86}Sr$ ratios of NBS 987 was 0.71028 to 0.71033 during the period of this study. As the statistical error

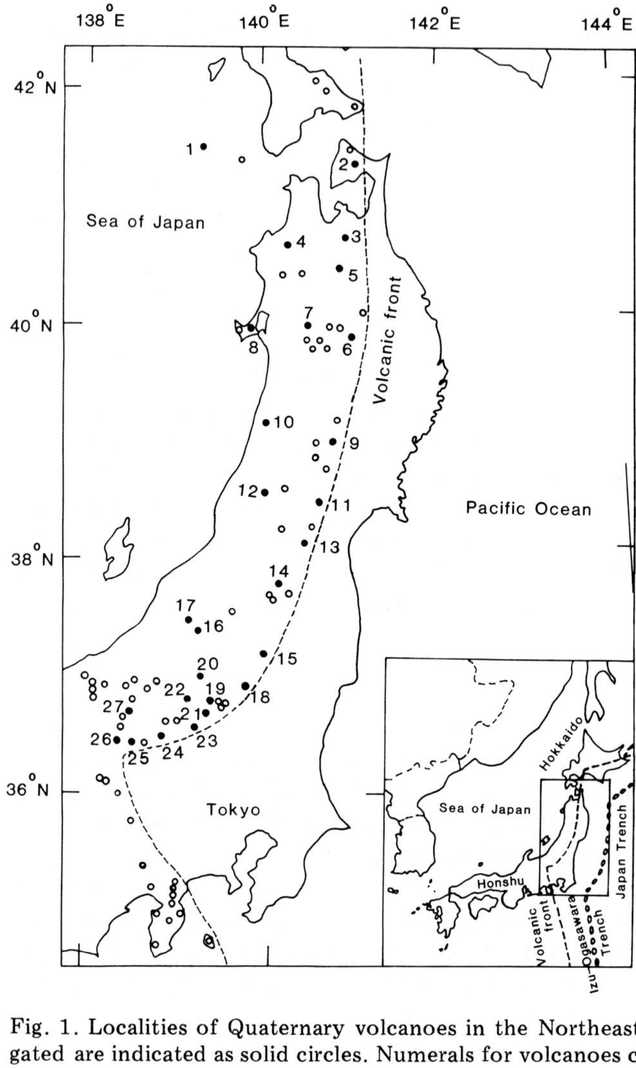

Fig. 1. Localities of Quaternary volcanoes in the Northeast Japan arc. Volcanoes investigated are indicated as solid circles. Numerals for volcanoes correspond to those in Table I.

of each isotope analysis was usually less than 0.00002 (2σ value of standard deviation in $^{87}Sr/^{86}Sr$ ratios), which was less than the reproducibility range (<0.00005), the uncertainty associated with $^{87}Sr/^{86}Sr$ ratio analysis of rock samples was estimated less than 0.00005.

Concentrations of Rb and Sr were measured by (wave length dispersive) X-ray fluorescence spectrometry on a few grams of pressed powder disk samples. Matrix effects were corrected using background radiation as an internal standard (Murad, 1973). Uncertainties in Rb and Sr determinations were within ±5% for concentrations of more than 10 ppm. For 6 samples, Rb and Sr were determined by energy-dispersive X-ray fluorescence spectrometry. The latter XRF technique enabled us to determine Rb and Sr in about 5 mg of powdered rock sample as precise as the former conventional one, although the detection limit was several times higher.

RESULTS

Analytical results are presented in Table I. In this study, $^{87}Sr/^{86}Sr$ ratios for several rocks from the same volcano were found to agree with one another usually within 0.0003, regardless of rock types and/or their erupting stages. Though the difference of 0.0003 in $^{87}Sr/^{86}Sr$ ratio is meaningfully greater than the uncertainty of each isotope analysis of this work, this level of difference does not seem significant in comparison with the $^{87}Sr/^{86}Sr$ ratio variations among volcanoes. It is, of course, important to discuss the minute difference in $^{87}Sr/^{86}Sr$ ratios in several rocks from one volcano, with regard to the evolution of magma composition in the magma chamber of the volcano. It is a very interesting problem whether or not there is any difference in $^{87}Sr/^{86}Sr$ ratio among basalt, andesite and dacite from a given volcano and/or between associated tholeiite and calc-alkaline series rocks and/or among rocks from successive volcanic stages throughout the life of the volcano. In this work, however, we dared to neglect this problem to make clear the wide-scale variation of the $^{87}Sr/^{86}Sr$ ratio related to a global tectonic phenomenon, such as subduction of the Pacific plate beneath the Eurasia plate. Throughout this paper, thereafter, the variation in $^{87}Sr/^{86}Sr$ ratios in several volcanic rocks from one volcano is not treated, but it will be discussed elsewhere in a separate paper.

In Fig. 2 the spatial distribution of $^{87}Sr/^{86}Sr$ ratios in Quaternary volcanic rocks from the Northeast Japan arc is given. In this figure, $^{87}Sr/^{86}Sr$ ratios are classified into 5 groups: less than 0.7032, 0.7032—0.7038, 0.7038—0.7045, 0.7045—0.7055 and more than 0.7055. This figure also includes previously measured results by other investigators for the volcanoes which were not studied in this work. They are identified by circles with small alphabetical symbols corresponding to their references. $^{87}Sr/^{86}Sr$ ratios in volcanic

TABLE I

Analytical results of volcanic rocks from the Northeast Japan arc

No.*	Name of volcano	Sample No.	Rock type	Rb (ppm)	Sr (ppm)	$^{87}Rb/^{86}Sr$	$^{87}Sr/^{86}Sr$
1	Oshima-oshima	OO2A	Andesite	94	637	0.430	0.70311
		OO3-2	Basalt	58	573	0.295	0.70313
2	Osoreyama	OSO04	Andesite	3.7	215	0.046	0.70406
		OSO07	Dacite	18	218	0.240	0.70419
		OSO13	Andesite	11	221	0.145	0.70388
3	Hakkoda	SA64090102	Andesite	25	266	0.274	0.70405

TABLE I (*continued*)

No.*	Name of volcano	Sample No.	Rock type	Rb (ppm)	Sr (ppm)	$^{87}Rb/^{86}Sr$	$^{87}Sr/^{86}Sr$
4	Iwaki	Iw.797121	Andesite	25	326	0.223	0.70407
		Iw.797122	Andesite	27	299	0.263	0.70406
		Iw.797123	Andesite	28	304	0.268	0.70409
		Iw.7910134	Andesite	21	324	0.189	0.70437
5	Towada	TOW01	Andesite	7.4	300	0.072	0.70421
		TOW06	Andesite	11	289	0.111	0.70415
6	Iwate	IWT13-2	Basalt	2.2	280	0.023	0.70440
		IWT19	Andesite	4.8	250	0.056	0.70432
7	Moriyoshi	MOR7-1	Andesite	17	403	0.123	0.70379
		MOR7-13	Dacite	59	289	0.594	0.70403
		MOR7-14	Andesite	25	416	0.175	0.70395
8	Kampu	KMP01	Andesite	79	668	0.344	0.70343
		KMP03	Andesite	78	651	0.349	0.70356
9	Kurikoma	KUR03	Andesite	31	206	0.439	0.70416
		KUR04-01	Andesite	28	217	0.376	0.70425
10	Chokai	CHO01	Andesite	69	386	0.520	0.70352
		TU7461502	Andesite	68	396	0.500	0.70347
11	Funagata	FUN05	Andesite	18	250	0.210	0.70407
		FUN07-01	Basalt	3.1	169	0.053	0.70425
12	Gassan	GAS13	Andesite	63	413	0.444	0.70352
		GAS14	Andesite	64	382	0.488	0.70347
13	Zao	ZAO09	Andesite	8.0	334	0.070	0.70414
		ZAO10	Basalt	9.7	364	0.078	0.70384
14	Azuma	AZ01	Andesite	45	219	0.598	0.70471
		AZ02	Andesite	63	201	0.912	0.70485
15	Nasu	52414	Andesite	34	250	0.396	0.70500
		52714	Basalt	7.7	265	0.095	0.70489
16	Asakusa	A7	Andesite	42**	588**	0.208	0.70391
		A203	Andesite	50**	588**	0.248	0.70425
17	Sumon	KC30	Andesite	65**	420**	0.451	0.70437
		KC57	Andesite	52**	383**	0.395	0.70414
18	Takahara	SA62101315	Dacite	72	209	1.003	0.70588
19	Nikkoshirane	ET78080405	Dacite	60	255	0.685	0.70593
20	Hiuchi	HIU04	Andesite	45	357	0.367	0.70538
21	Sukai	SA7808050	Andesite	38	313	0.353	0.70639
22	Hotaka	TY7782612	Andesite	22	290	0.221	0.70563
		TY78101408	Dacite	59	270	0.636	0.70577
23	Akagi	AK399-7	Andesite	9.4	368	0.074	0.70759
		AK528-1	Dacite	68	308	0.643	0.70704
		AK838-1	Andesite	9.6	359	0.078	0.70775
24	Haruna	OO661023-6	Dacite	--***	264**	—	0.70488
		OO681127-19	Basalt	--***	319**	—	0.70544

25	Asama	AS5292	Andesite	38	294	0.376	0.70417
		SA77103006	Andesite	37	289	0.373	0.70416
26	Eboshi	SA77112506	Andesite	8.1	410	0.058	0.70404
27	Kusatsushirane	Ns1	Andesite	48	278	0.503	0.70396

*This number corresponds to that in Fig. 1.
**Energy dispersive XRF data (see the text).
***Not determined.

1. Oshima-oshima: provided by Dr. Y. Nakamura, Univ. Tokyo. OO2A and OO3-2 are Nishiyama upper lava and parasitic crater lava of central cone, respectively.
2. Osoreyama: collected by the author, referring to Togashi (1977). OSO04, OSO07 and OSO13 are Kamabuseyama lava, Shojiyama lava and Tsurugiyama lava, respectively.
3. Hakkoda: provided by Prof. S. Aramaki, Univ. Tokyo.
4. Iwaki: provided by Dr. F. Uemura, Geological Survey of Japan. Iw.797121 was collected from somma, Iw.797122 and Iw.797123 were from central cone and Iw.7910134 was from upper stream of Ushironagane-zawa.
5. Towada: collected by the author, referring to Taniguchi (1972). TOW01 and TOW06 are 2nd stage lavas from Nakayama peninsula.
6. Iwate: collected by the author, referring to Onuma (1962). IWT13-02 IWT19 are lavas of Yakushidake cone and Yakehashiri, respectively.
7. Moriyoshi: provided by Dr. M. Sakuyama, Univ. Tokyo. MOR7-1 was collected from summit of central cone and MOR7-13 and MOR7-14 were from southern flank.
8. Kampu: provided by Dr. M. Sakuyama.
9. Kurikoma: collected by the author. KUR03 and KUR04-01 are lavas of Kurikomayama and Tsurugisan, respectively.
10. Chokai: CHO01 (Shinzan lava erupted in 1801AD) was provided by Prof. S. Aramaki and TU7461502 was provided by Dr. T. Ui, Kobe Univ.
11. Funagata: collected by the author. FUN05 and FUN07-01 are lavas of Izumigatake and Kurohanayama, respectively.
12. Gassan: collected by the author, referring to Shibahashi (personal communication, 1981). AS13 is Gassan central cone lava and GAS14 is Shinsen-ike lava.
13. Zao: collected by the author. ZAO09 and ZAO10 are lavas of Fubosan.
14. Azuma: provided by Dr. M. Sakuyama. AZ01 and AZ02 are lavas of Higashiazumasan and Issaikyosan, respectively.
15. Nasu: provided by Prof. S. Aramaki. 52414 is upper Chausudake lava and 52714 is Numahara lava.
16. Asakusa: provided by Prof. K. Chihara, Niigata Univ.
17. Sumon: provided by Prof. K. Chihara. KC30 is lava of Eboshidake and KC57 is Kitaone lava.
18. Takahara: provided by Prof. S. Aramaki. SA62101315 is pumice block in pumice flow deposit.
19. Nikkoshirane: provided by Prof. S. Aramaki.
20. Hiuchi: provided by Prof. S. Aramaki.
21. Sukai: provided by Prof. S. Aramaki.
22. Hotaka: provided by Mr. T. Yamaguchi, Univ. Tokyo. TY7782612 and TY78101408 correspond to Ap-1 and Tg-2 in Yamaguchi's (1981) paper.
23. Akagi: provided by Prof. I. Moriya, Kanazawa Univ.
24. Haruna: provided by Dr. O. Oshima, Univ. Tokyo. OO661023-6 is Futatsudake pumice flow and OO681127-19 is main cone lava.
25. Asama: provided by Prof. S. Aramaki. AS5292 is Shimonobutai lava and SA-77103006 is Onioshidashi lava (Aramaki, 1963).
26. Eboshi: provided by Prof. S. Aramaki. SA77112506 is Takamine lava flow.
27. Kusatsushirane: provided by Mr. Y. Hayakawa, Univ. Tokyo. Ns1 is Ishizu lava.

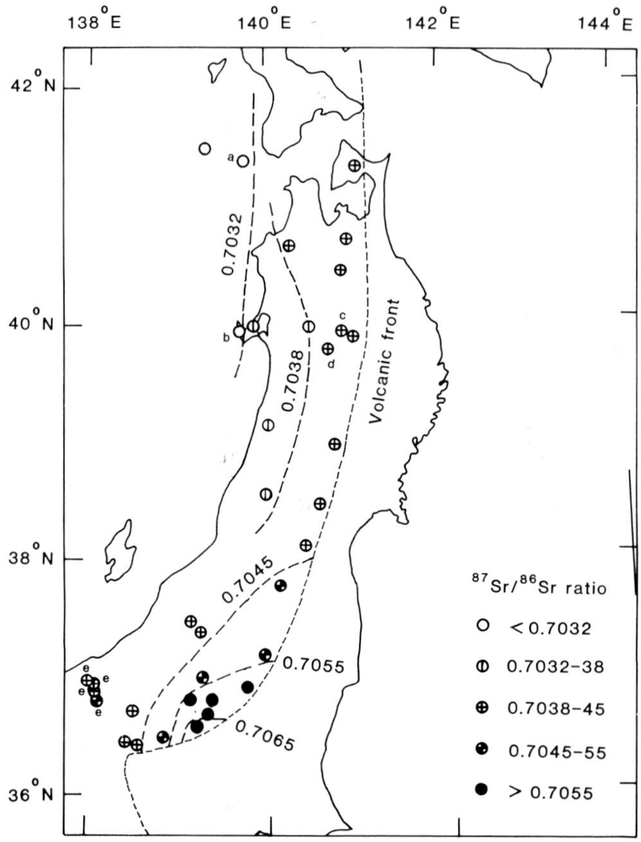

Fig. 2. Spatial distribution of $^{87}Sr/^{86}Sr$ ratios in Quaternary volcanic rocks from the Northeast Japan arc. Circles without alphabetical symbols were analyzed in this work. Others are: *a*, Katsui et al. (1978); *b*, Kaneoka et al. (1978); *c*, Nohda and Wasserburg (1981); *d*, Kurasawa (1979); and *e*, Ishizaka et al. (1977).

rocks from some volcanoes were determined both in this work and in the previous ones, being found to agree with each other according to the 5 groups of classification (Oshima-oshima, Iwate and Kampu by Hedge and Knight (1969), Oshima-oshima by Katsui et al. (1978), Osoreyama by Togashi (1981) and Moriyoshi, Kampu, Asama and Eboshi by Nohda and Wasserburg (1981)). Figure 2 demonstrates that the spatial distribution of $^{87}Sr/^{86}Sr$ ratios in volcanic rocks from the Northeast Japan arc is significantly characteristic and that the ratios change systematically in both across-arc and along-arc directions.

Spatial distributions in Rb and Sr concentrations in andesites and basalts are shown in Figs. 3 and 4, from which lateral variations in both elements are clearly observed. Rb and Sr contents in both andesites and basalts increase in across-arc direction with increasing depth of the Wadati-Benioff zone.

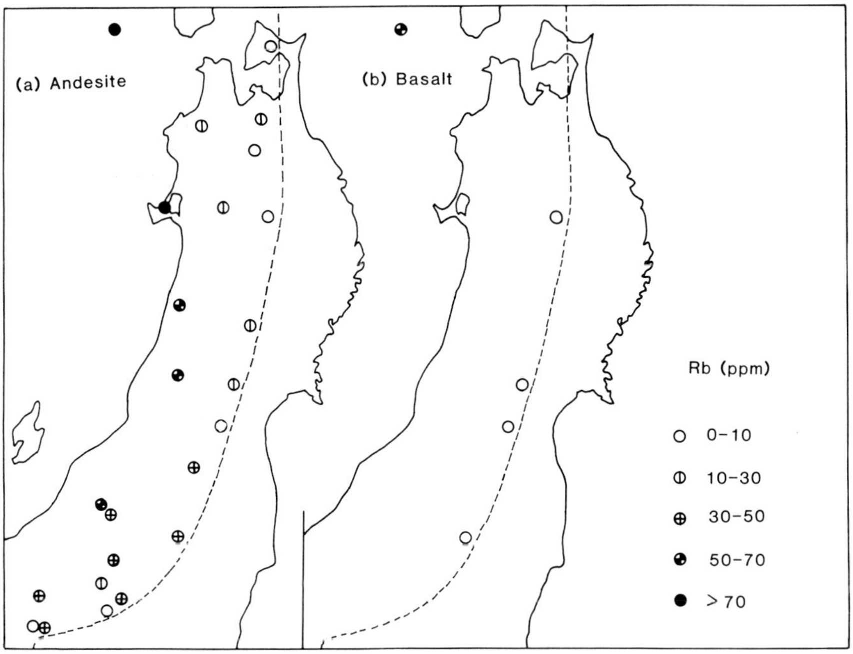

Fig. 3. Spatial distribution of Rb concentrations in Quaternary volcanic rocks from the Northeast Japan arc.

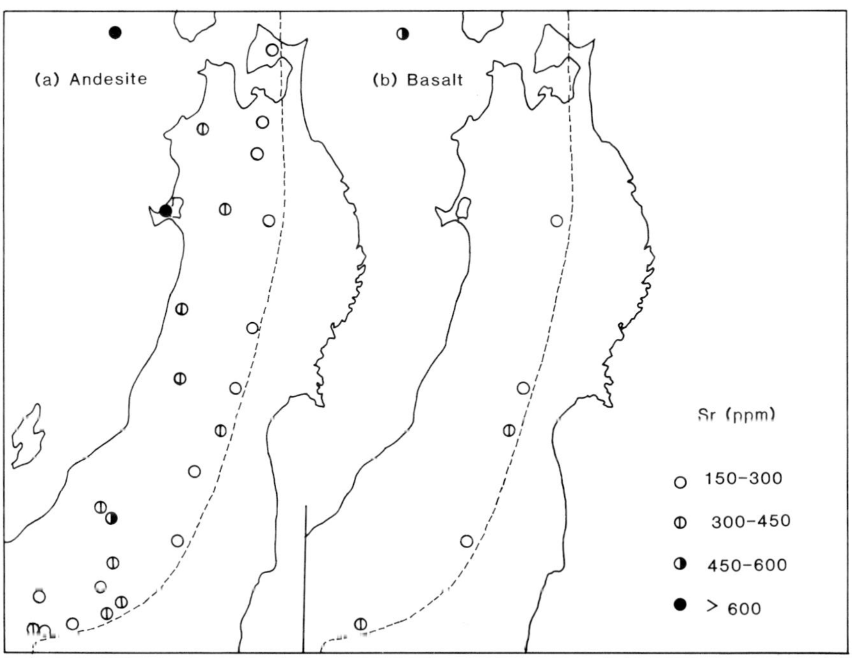

Fig. 4. Spatial distribution of Sr concentrations in Quaternary volcanic rocks from the Northeast Japan arc.

VARIATION OF $^{87}Sr/^{86}Sr$ RATIOS IN ACROSS-ARC DIRECTION

Figure 2 indicates a decrease in $^{87}Sr/^{86}Sr$ ratios of volcanic rocks from the volcanic front side to the back arc one, in other words, with increasing depth of the Wadati-Benioff zone, is observed in a wide area of the Northeast Japan arc. This observation expands the earlier finding by Hedge and Knight (1969), confirming that zones with similar $^{87}Sr/^{86}Sr$ ratios spread from north to south parallel to the Japan Trench. Figure 5 shows decreasing gradients of $^{87}Sr/^{86}Sr$ ratios against the distance from the volcanic front. From this figure, the gradients are found to be roughly equal for all volcanoes on the back arc side, except the case of Iwaki volcano, in which the $^{87}Sr/^{86}Sr$ ratio is similar to that from the volcanic front side. It has been pointed out that alkali, especially K_2O, contents in volcanic rocks from Iwaki volcano are rather similar to those from the volcanoes in the volcanic front side (Kawano et al. 1961b).

$^{87}Sr/^{86}Sr$ ratio variation away from the volcanic front, or with increasing depth of the Wadati-Benioff zone, has been observed from other island arc

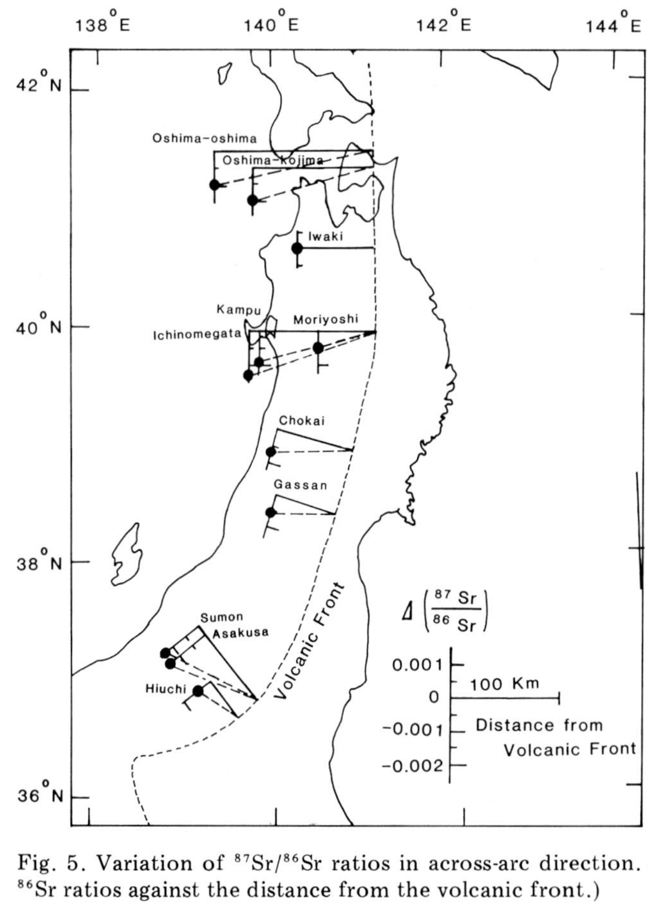

Fig. 5. Variation of $^{87}Sr/^{86}Sr$ ratios in across-arc direction. (Decreasing gradients of $^{87}Sr/^{86}Sr$ ratios against the distance from the volcanic front.)

systems. In the Scotia arc, tholeiites from the South Sandwich Islands (volcanic front side) had significantly higher $^{87}Sr/^{86}Sr$ ratios than those from Scotia Sea Rise (back-arc side) (Hawkesworth et al., 1977). This tendency was similar to that observed in the Northeast Japan arc. Light REE enrichment in the back-arc side was commonly observed in both the Scotia arc and the Northeast Japan arc (Fujitani and Masuda, 1981). But, in the Scotia arc, $^{143}Nd/^{144}Nd$ ratios were indistinguishable in rocks from both the back-arc side and the volcanic front side, unlike in the Northeast Japan arc (Nohda and Wasserburg, 1981). In the Sunda arc, $^{87}Sr/^{86}Sr$ ratios in normal calc-alkaline and tholeiitic rocks increased with increasing depth of the Wadati-Benioff zone (Whitford, 1975). This tendency is opposite in sense to that observed in the Northeast Japan arc.

In the Northeast Japan arc, many kinds of lateral variations in petrology and geochemistry of volcanic rocks have been observed, as was mentioned before. Lateral variation in K_2O, Ba, Th, U, REE and F contents and chondrite-normalized REE patterns in volcanic rocks and estimated H_2O contents in magmas strongly suggest the degree of partial melting of mantle material is small in the back arc side, growing larger toward the volcanic front side, because these elements have a tendency to concentrate in a liquid phase during the partial melting process. The smaller the degree of partial melting, the higher are the concentrations of these elements in partially melted liquid. Sakuyama (1979) explained the relation between decrease of the H_2O contents in the magma and increase of the degree of partial melting. Variation of REE patterns was also understood in terms of the degree of partial melting. Lateral variations of Rb and Sr contents, which have already been reported in other volcanic arcs (Hutchison, 1976) and were newly observed in the Northeast Japan arc by this work, are compatible with the above-mentioned model, because the behaviors of Rb and Sr during partial melting processes resemble those of K_2O, etc. According to Kushiro (1981), the degree of partial melting is closely related to the depth of primary magma generation, based on recent results from experimental petrology. As lateral variation in petrology of volcanic rocks has been explained in terms of the differences in the depth of primary magma generation; the depth increased from the volcanic front side to the back-arc side (Kuno, 1966), "the degree of partial melting" model becomes consistent with "the depth of primary magma" model. Next, we will discuss the strontium isotope data with relation to the subduction model mentioned above.

Suppose that $^{87}Sr/^{86}Sr$ ratios in the melted regions in mantle are almost identical, the following two processes are speculated to explain the observed variation in $^{87}Sr/^{86}Sr$ ratios in volcanic rocks. One process is that $^{87}Sr/^{86}Sr$ ratios in initially melting material and later melting material are different under an unequilibrated partial melting process. This means that a few phases of the mantle material melt in the back arc side and that more phases, including the first, melt in the volcanic front side under the condition that isotope exchange processes do not take place between the melted phase and the residual solid phase. If so, an apparent isochronic relation, analogous to a mantle isochron (Brooks et al., 1976), is expected on a($^{87}Sr/^{86}Sr$—$^{87}Rb/^{86}Sr$) correlation diagram from arc volcanic rocks with different $^{87}Sr/^{86}Sr$ ratios. But it is not the case for the Northeast Japan arc, based on data in this work. The other process is that some materials with different $^{87}Sr/^{86}Sr$ ratios are mixed with the originally melting material of either or both sides. As volcanic rocks from the back-arc side have slightly higher $^{87}Sr/^{86}Sr$ ratios than normal-type MORB (Sun et al., 1979) and those from the volcanic front side have even higher $^{87}Sr/^{86}Sr$ ratios, we must explain the slight difference in $^{87}Sr/^{86}Sr$ ratios between in the back-arc side rocks and in MORB as well as lateral variation of $^{87}Sr/^{86}Sr$ ratios in an across-arc direction. As to the latter problem, it is reasonable to suppose that original material of volcanic rocks from the volcanic front side may be more contaminated with materials with a higher $^{87}Sr/^{86}Sr$ ratio. Candidates of the contaminating materials are: (1) crustal material; (2) subducting slab suffering from sea water alteration, and (3) marine sediments accompanying the subducting slab.

The thickness of the crust under the volcanoes concerned is almost constant in an across-arc direction except under the Oshima-oshima volcano (Yoshii, 1972) and the $^{87}Sr/^{86}Sr$ ratios of crustal materials, which are represented by those of plutonic rocks near the volcanoes concerned, do not change significantly and seem rather higher in the back-arc side (Shibata and Ishihara, 1979). These findings suggest that it is not easy to explain the $^{87}Sr/^{86}Sr$ ratio difference in across-arc direction by contamination of crustal materials. Kobayashi and Hori (1979) proposed that stress-field controlled crustal contamination might change the $^{87}Sr/^{86}Sr$ ratios in volcanic rocks, taking account of Nakamura's (1977) idea that upward penetratability of original magma was dependent on the regional stress field. He suggested that, where magma ascended easily, crustal contamination might be less and resulting in lower $^{87}Sr/^{86}Sr$ ratios in volcanic rocks. As, at present Northeast Japan, the stress field is less compressional on the Japan Sea Coast (back-arc side) than on the trench side (Nakamura and Uyeda, 1980), the observed lateral variation of $^{87}Sr/^{86}Sr$ ratios is consistent with Kobayashi and Hori's (1979) proposal. Although crustal material may not be the best candidate to explain the lateral variation of $^{87}Sr/^{86}Sr$ ratios, it may remain a promising candidate to explain the slight increase of $^{87}Sr/^{86}Sr$ ratios in volcanic rocks from the back arc side in comparison with those in MORB, because the lowest $^{87}Sr/^{86}Sr$ ratio was obtained from Oshima-oshima volcano which is situated on the thinnest crust in the studied region.

DePaolo and Wasserburg (1977) pointed out that sea water contributed to island arc magma in some regions through Nd and Sr isotope studies of arc volcanic rocks. This supports the second candidate, subducting slab suffering from sea water alteration, as the contaminating material. However, Nd and Sr isotope studies of volcanic rocks on the 40° N across-arc traverse in Northeast Japan arc showed no evidence of sea water involvement (Nohda and Wasserburg, 1981).

The third candidate, oceanic sediment accompanying the subducting slab, is prospective. It seems reasonable that the contribution of the subducting slab component becomes smaller away from the volcanic front, because the upper surface of the subducting slab is deeper toward the back-arc side. However, the mechanical process of mixing of the mantle wedge material with the subducting oceanic slab component and forming the island arc magma is, at present, not well known. Possible pictures are: strontium in the sediment is removed with water emanating from the surface of the subducting slab to the mantle wedge or melt zone observed in the lowest portion of the mantle wedge (Okada, 1979) expands in the surface of the subducting slab. Chemical aspects of mixing, or "interaction" process can be related to geophysical properties of subduction, the degree of coupling of two plates and/or evolutional stage of subduction (Uyeda and Kanamori, 1979).

VARIATION OF $^{87}Sr/^{86}Sr$ RATIOS IN ALONG-ARC DIRECTION

Figure 6 shows the variation of $^{87}Sr/^{86}Sr$ ratios in along-arc direction. This figure also includes the data of volcanoes located to the south of Asama volcano by Kurasawa (1979) and Nohda and Wasserburg (1981). Volcanoes along the volcanic front are divided into two groups by Asama volcano as a boundary position, so far as $^{87}Sr/^{86}Sr$ ratios are concerned. Volcanoes northward from Asama volcano have $^{87}Sr/^{86}Sr$ ratios of higher than 0.7038, while southern volcanoes have those from 0.7032 to 0.7038. Asama volcano is situated at the boundary point of $^{87}Sr/^{86}Sr$ ratio variation and actually stands where the direction of the volcanic front is twisted. In the region northward of Asama volcano, $^{87}Sr/^{86}Sr$ ratios are constant with values of 0.7038 to

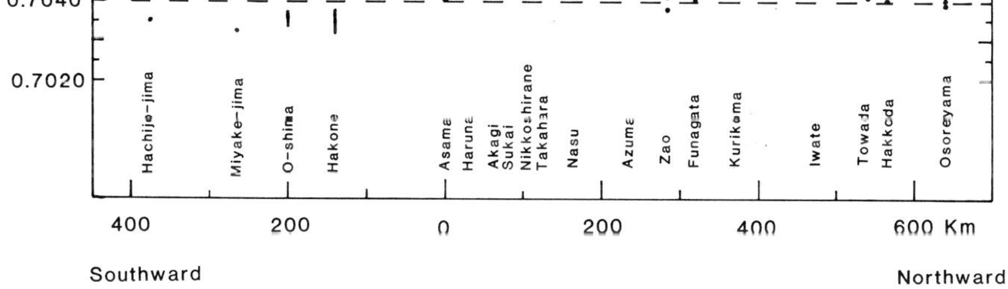

Fig. 6. Variation of $^{87}Sr/^{86}Sr$ ratios of volcanic front volcanoes in the along-arc direction. Data of Hakone and O-shima are from Kurasawa (1979) and Miyake-jima and Hachijo-jima from Nohda and Wasserburg (1981).

0.7045 between Osoreyama and Zao volcanoes (in Northeastern Honshu area), where the directions of the volcanic front and the Japan Trench are almost parallel. On the other hand, between Azuma and Haruna volcanoes (in central Honshu area), $^{87}Sr/^{86}Sr$ ratios are significantly higher than those in volcanoes in Northeastern Honshu area, with the maximum value at Akagi volcano. In this area, the volcanic front curves from southward to westward and parallelism between the volcanic front and the Japan Trench is lost. Further going to Asama, Eboshi and Kusatsushirane volcanoes along the volcanic front, $^{87}Sr/^{86}Sr$ ratios decrease to the level of Northeastern Honshu area.

Anomalously high $^{87}Sr/^{86}Sr$ ratios around Akagi volcano are suggested to have relation to the bending feature of the volcanic front. Similar anomalously high $^{87}Sr/^{86}Sr$ ratios were reported in other island arc systems on earth. In the Banda arc, $^{87}Sr/^{86}Sr$ ratio was maximum at Serua island (Whitford and Jezek, 1979), and in the Lesser Antilles arc at St. Lucia island (Smith et al., 1980). These three cases mentioned above have the following common features. The volcano with the highest $^{87}Sr/^{86}Sr$ ratio stands on nearly the central position along a bending arc and have similar maximum values of $^{87}Sr/^{86}Sr$ ratios (the Northeast Japan arc: 0.7077, the Banda arc: 0.7095 and the Lesser Antilles arc: 0.7092). In the case of the Banda arc, Whitford and Jezek (1979) explained the high $^{87}Sr/^{86}Sr$ ratio as the result of mixing a crustal component of subducted lithosphere with a mantle derived component. In the Lesser Antilles arc case, Pushkar et al. (1973) interpreted the high $^{87}Sr/^{86}Sr$ ratio as due to mixing of marine sediment with mantle material. Although it is possible to explain high $^{87}Sr/^{86}Sr$ ratios around Akagi volcano in the Northeast Japan arc as a result of these types of contamination, we have no other chemical data, at present, to support this hypothesis. The possibility that magmas around Akagi volcano are generated by remelting of crustal material cannot be excluded, until new evidence about the genesis of magma in this area has been presented.

Akagi and its neighboring volcanoes are situated in a very unique environment in terms of subduction-related volcanism, unlike other volcanoes with anomalously high $^{87}Sr/^{86}Sr$ ratios on bending arcs, such as the Banda and Lesser Antilles arcs. Under the central Honshu area around Akagi volcano, three plates meet (triple junction). The Pacific plate and the Philippine Sea plate subduct beneath the Eurasia plate, from east to west and from south to north, respectively. Nakamura and Shimazaki (1981) showed the descending features of both the Pacific and Philippine Sea plates under the central Honshu area (Kanto district) through the determination of microearthquake hypocenters, showing that the head of the Philippine Sea plate reached to the close south of Akagi volcano under the ground. If the Philippine Sea plate extends under Akagi volcano, the depth of the plate surface is extrapolated to about 100 km. Around Akagi volcano, the depth of the Pacific plate surface is between 100 and 140 km. This situation of a triple junction under the Kanto district is unique in the world and may be related to the anomalously high $^{87}Sr/^{86}Sr$ ratios in volcanic rocks. Further systematic investigations of not only $^{87}Sr/^{86}Sr$ ratios but also other isotope ratios and chemical compositions in volcanic rocks from this area are necessary.

Considering that high $^{87}Sr/^{86}Sr$ ratios in volcanic rocks from the central Honshu area are due to a peculiar geographical or tectonic situation, normal $^{87}Sr/^{86}Sr$ ratios in volcanic rocks along the volcanic front of the Northeast Japan arc, where the Pacific plate subducts beneath the Eurasia plate, are 0.7038—0.7045, which are obtained from volcanoes in the Northeastern Honshu area. These values are slightly higher than those from volcanic rocks in the Izu-Ogasawara arc, where the Pacific plate thrusts under the Philippine Sea plate. Physical properties of the two subduction systems are different, as shown schematically in Fig. 7. The dip angle of the Wadati-Benioff zone is sharper and the thickness of the crust over the mantle wedge is thinner in the Izu-Ogasawara arc than in the Northeast Japan arc. These differences may affect $^{87}Sr/^{86}Sr$ ratios in volcanic rocks from both arcs.

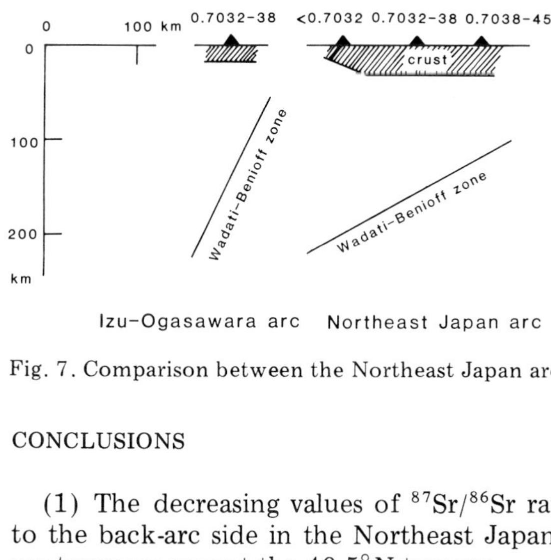

Fig. 7. Comparison between the Northeast Japan arc and the Izu-Ogasawara arc.

CONCLUSIONS

(1) The decreasing values of $^{87}Sr/^{86}Sr$ ratios from the volcanic front side to the back-arc side in the Northeast Japan arc are confirmed for all across-arc traverses except the 40.5° N traverse.

(2) $^{87}Sr/^{86}Sr$ ratios in the along-arc direction are constant (0.7038—0.7045) over 350 km in northern region of the Northeast Japan arc. While, in southern region of this arc (central Honshu), $^{87}Sr/^{86}Sr$ ratios change with an anomalously high maximum value reaching to 0.7077.

(3) The anomalously high $^{87}Sr/^{86}Sr$ ratio area coincides with the triple junction of the Eurasia plate, the Philippine Sea plate and the Pacific plate, suggesting "anomalous interaction" at the triple junction.

(4) There is a systematic difference in $^{87}Sr/^{86}Sr$ ratios between the Northeast Japan arc and the Izu-Ogasawara arc. $^{87}Sr/^{86}Sr$ ratios of the former (0.7038—0.7045) are higher than those of the latter (0.7032—0.7038), suggesting that "interactions" between the Eurasia plate and the Pacific plate,

and those between the Philippine Sea plate and the Pacific plate are slightly different.

ACKNOWLEDGEMENTS

I wish to thank to Prof. N. Onuma for continuous discussion and for critical reading of the manuscript. I am also grateful to Prof. S. Aramaki for encouragement and discussion during the course of this work as well as for donating rock samples. Constructive comments have been provided by Drs. K. Aoki, N. Fujii, N. Isshiki, I. Kaneoka, Y. Kobayashi, H. Mabuchi, M. Sakuyama and H. Wakita. My thanks go to these persons. Rock samples were donated by Drs. S. Aramaki, M. Sakuyama, Y. Nakamura, F. Uemura, T. Ui, K. Chihara, T. Yamaguchi, I. Moriya, O. Oshima and Y. Hayakawa and field assistance was provided by Drs. M. Sakuyama (in Towada, Iwate, Funagata and Zao volcanoes), H. Yurimoto (Osoreyama), Y. Nakamura (Kurikoma) and F. Masuda (Gassan). I also wish to thank these persons. Dr. M. Sakuyama kindly gave me information about some rock samples. XRF facilities were provided by Geophysical Institute, University of Tokyo and Rigaku Industrial Corporation. This work was partially supported by a grant from the Ministry of Education, Science and Culture, Japan (No. 574212).

REFERENCES

Aramaki, S., 1963. Geology of Asama volcano. J. Fac. Sci. Univ. Tokyo, Sect. II, 14: 229—443.

Brooks, C., Hart, R.S., Hofmann, A. and James, D.E., 1976. Rb-Sr mantle isochrons from oceanic regions. Earth Planet. Sci. Lett., 32: 51—61.

DePaolo, D.J. and Wasserburg, G.J., 1977. The sources of island arcs as indicated by Nd and Sr isotopic studies. Geophys. Res. Lett., 4: 465—468.

DePaolo, D.J. and Wasserburg, G.J., 1979. Petrogenetic mixing models and Nd-Sr isotopic pattern. Geochim. Cosmochim. Acta, 43: 615—627.

Dickinson, W.R., 1968. Circum-Pacific andesite types. J. Geophys. Res., 73: 2261—2269.

Fujimaki, H. and Kurasawa, H., 1980. Lateral variation of REE pattern of basaltic magma across the Japan arc. J. Jpn. Assoc. Miner. Petrogr. Econ. Geol., 75: 313—322.

Fujitani, T. and Masuda, A., 1981. Light REE inclination and distance from volcanic front; a case of volcanic rocks in Northeastern Japan. Geochem. J., 15: 269—281.

Hasegawa, A., Umino, N. and Takagi, A., 1978. Double-planed structure of the deep seismic zone in the northeastern Japan arc. Tectonophysics., 47: 43—58.

Hawkesworth, C.J., O'nions, R.K., Pankhurst, R.J., Hamilton, P.J. and Evensen, N.M., 1977. A geochemical study of island-arc and back-arc tholeiites from the Scotia Sea. Earth Planet. Sci. Lett., 36: 253—262.

Hedge, C.E. and Knight, R.J., 1969. Lead and strontium isotopes in volcanic rocks from northeastern Honshu, Japan. Geochem. J., 3: 15—24.

Hutchison, C.S., 1976. Indonesian active volcanic arc: K, Sr and Rb variation with depth to the Benioff zone. Geology, 4: 497—408.

Ishikawa, K., Kanisawa, S. and Aoki, K., 1980. Content and behavior of fluorine in Japanese Quaternary volcanic rocks and petrogenetic application. J. Volcanol. Geotherm. Res., 8: 161—175.

Ishizaka, K., Yanagi, T. and Hayatsu, K., 1977. A strontium isotopic study of the volcanic rocks of Myoko volcano group, central Japan. Contrib. Mineral. Petrol., 63: 295—307.

Kaneoka, I., Matsuda, J., Zashu, S., Takahashi, E. and Aoki, K., 1978. Ar and Sr isotopes of mantle-derived rocks from the Japanese islands. Bull. Volcanol., 41: 424—433.

Katsui, Y., Oba, Y., Ando, S., Nishimura, S., Masuda, Y., Kurasawa, H. and Fujimaki, H., 1978. Petrochemistry of the Quaternary volcanic rocks of Hokkaido, North Japan. J. Fac. Sci. Hokkaido Univ. Ser. IV, 18: 449—484.

Kawano, Y., Yagi, K. and Aoki, K., 1961a. Petrography and petrochemistry of the volcanic rocks of Quaternary volcanoes of northeastern Japan. Sci. Rep. Tohoku Univ. Ser. III, 7: 1—46.

Kawano, Y., Aoki, K. and Kadowaki, K., 1961b. Petrology of Iwaki volcano. J. Jpn. Assoc. Miner. Petrogr. Econ. Geol., 46: 101—110 (in Japanese with English abstract).

Kobayashi, Y. and Hori, K., 1979. A relationship between a tectonic stress and $^{87}Sr/^{86}Sr$ ratios of calc-alkaline rocks in the Sunda and Banda arcs. Bull. Volcanol. Soc. Jpn., 24: 83—84 (in Japanese with English abstract).

Kuno, H., 1966. Lateral variation of basalt magma type across continental margins and island arcs. Bull. Volcanol., 24: 195—222.

Kurasawa, H., 1972. Isotopic composition of strontium in volcanic rocks from Towada volcano, northeast Japan. Bull. Volcanol. Soc. Jpn., 17: 158—159 (abstract in Japanese).

Kurasawa, H., 1979. Isotopic composition of strontium in volcanic rocks from Fuji, Hakone and Izu area, central Japan. Bull. Volcanol. Soc. Jpn., 24: 135—152 (in Japanese with English abstract).

Kushiro, I., 1981. Origin of magmas in island arcs: a discussion based on experimental petrology. J. Geol. Soc. Jpn., 87: 769—780 (in Japanese with English abstract).

Masuda, Y., 1979. Lateral variation of trace element contents in Quaternary volcanic rocks across Northeast Japan. Bull. Univ. Osaka Pref. Ser. A, 28: 105—125.

Murad, E., 1973. Determination of trace elements in unfused rock and mineral samples by X-ray fluorescence. Anal. Chim. Acta, 67: 37—53.

Nakamura, K., 1977. Volcanoes as possible indicators of tectonic stress orientation — principle and proposal. J. Volcanol. Geotherm. Res., 2: 1—16.

Nakamura, K. and Uyeda, S., 1980. Stress gradient in arc—back arc regions and plate subduction. J. Geophys. Res., 85: 6419—6428.

Nakamura, K. and Shimazaki, K., 1981. Sagami and Suruga troughs and plate subduction. Kagaku, 51: 490—498 (in Japanese).

Nohda, S. and Wasserburg, G.J., 1981. Nd and Sr isotopic study of volcanic rocks from Japan. Earth Planet. Sci. Lett., 52: 264—276.

Okada, T., 1979. New evidence of the discontinuous structure of the descending lithosphere as revealed by ScSp phase. J. Phys. Earth, 27: 53—64 (suppl.).

Onuma, K., 1962. Petrography and petrochemistry of the rocks from Iwate volcano, northeastern Japan. J. Jpn. Assoc. Miner. Petrogr. Econ. Geol., 47: 192—204.

Pushkar, P., Steuber, A.M., Tomblin, J.F. and Julian, G.M., 1973. Strontium isotopic ratios in volcanic rocks from St. Vincent and St. Lucia, Lesser Antilles. J. Geophys. Res., 78: 1279—1287.

Sakuyama, M., 1979. Lateral variation of H_2O contents in Quaternary magmas of northeastern Japan. Earth Planet. Sci. Lett., 43: 103—111.

Shibata, K. and Ishihara, S., 1979. Initial $^{87}Sr/^{86}Sr$ ratios of plutonic rocks from Japan. Contrib. Mineral. Petrol., 70: 381—390.

Smith, A.L., Roobol, M.J. and Gunn, B.M., 1980. The Lesser Antilles — a discussion of island arc magmatism. Bull. Volcanol., 43: 287—302.

Sugimura, A., 1960. Zonal arrangements of some geophysical and petrological feature in Japan and its environs. J. Fac. Sci. Univ. Tokyo, Sect. II, 12: 133—153.

Sun, S.-S., Nesbitt, R.W. and Sharaskin, A.Y., 1979. Geochemical characteristics of mid-ocean ridge basalts. Earth Planet. Sci. Lett., 44: 119—138.
Taniguchi, H., 1972. Petrological study on Towada volcano. J. Jpn. Assoc. Miner. Petrogr. Econ. Geol., 67: 128—138 (in Japanese with English abstract).
Togashi, S., 1977. Petrology of Osore-yama volcano, Japan. J. Jpn. Assoc. Miner. Petrogr. Econ. Geol., 72: 45—60 (in Japanese with English abstract).
Togashi, S., 1981. Petrology and geochemistry of andesite and dacite from Osoreyama volcano, northern Honshu, Japan. Abstr. 1981 IAVCEI Symposium, Arc Volcanism. Tokyo, pp. 381—382.
Ui, T. and Aramaki, S., 1978. Relationship between chemical composition of Japanese island-arc volcanic rocks and gravimetric data. Tectonophysics, 45: 249—259.
Uyeda, S. and Kanamori, H., 1979. Back-arc opening and the mode of subduction. J. Geophys. Res., 84: 1049—1061.
Whitford, D.J., 1975. Strontium isotopic studies of the volcanic rocks of the Sunda arc, Indonesia, and their petrogenetic implications. Geochim. Cosmochim. Acta, 39: 1287—1302.
Whitford, D.J. and Jezek, P.A., 1979. Origin of late-Cenozoic lavas from Banda arc, Indonesia: Trace element and Sr isotope evidence. Contrib. Mineral. Petrol., 68: 141—150.
Yamaguchi, T., 1981. Geology of Hotaka volcano. J. Geol. Soc. Jpn., 87: 823—832 (in Japanese with English abstract).
Yoshii, T., 1972. Terrestrial heat flow and features of the upper mantle beneath the Pacific and the Sea of Japan. J. Phys. Earth, 20: 271—285.

Journal of Volcanology and Geothermal Research, 18 (1983) 549—560
Elsevier Science Publishers B.V., Amsterdam — Printed in The Netherlands

ALKALI MAPPING OF THE JAPANESE QUATERNARY VOLCANIC ROCKS

S. ARAMAKI and T. UI

Earthquake Research Institute, University of Tokyo, Bunkyo-ku, Tokyo 113 (Japan)
Department of Earth Sciences, Faculty of Science, Kobe University, Nada-ku, Kobe 657 (Japan)

(Received October 20, 1982)

ABSTRACT

Aramaki, S. and Ui, T., 1983. Alkali mapping of the Japanese Quaternary volcanic rocks. In: S. Aramaki and I. Kushiro (Editors), Arc Volcanism. J. Volcanol. Geotherm. Res., 18: 549—560.

Of ca. 200 Quaternary volcanic centers in the Japanese islands, ca. 120 are characterized by K_2O and Na_2O concentrations of their ejecta normalized to SiO_2 = 60%. K_2O values give a geographically well-defined trend of regularly increasing away from the volcanic front. Na_2O values show no geographical trend. Relationships between the SiO_2-normalized K_2O and the depth of the deep seismic zone (K-*h* relation) are different for different arcs. The Southwest Japan arc shows distinctly higher K_2O and less regular geographic trend than the others probably due to crustal contamination. In the normal segments of the Northeastern Japan belt, K_2O increases ca. 2.5 to 3.0% per 100 km depth of the seismic zone. Groups of volcanic centers near the junction between the Northeast Honshu and Izu-Mariana arcs give characteristically lower K_2O while Rishiri volcano far behind the arc junction between the Kurile and Honshu arcs also gives a low K_2O value.

INTRODUCTION

The across-arc change of magma chemistry related to subduction is one of the most firmly established features of arc volcanism. Patterns of regular and monotonous changes in abundances of K and related LIL elements etc. or changes of various ratios with the spatial configuration of the subduction zone give important constraints on the models of the arc magma generation although such regular patterns or combination of them have not yet led to any unequivocal model (Gill, 1981). This paper summarizes an effort of characterizing spatial distribution of K and Na concentrations in the magmas of the Quaternary volcanic centers in the Japanese islands using published data. The risk is high in being misled to erroneous conclusions because of heterogeneous quality of available data and sampling biases. However, large amount of available data and the lack of systematic effort so far made to collect and process them properly justify this kind of statistical approach.

0377-0273/83/$03.00

DATA PROCESSING

About 1600 high-quality chemical analyses used in this report were selected from various publications and checked individually against the possible misprints and other errors (for details see Aramaki and Ui, 1978a) Of ca. 200 Quaternary volcanic centers in the Japanese islands, only about half of them carry more than 3 analyses. To characterize the levels of K_2O or Na_2O contents of the lavas and pyroclastics erupted from those volcanic centers, indices such as K_{55}, K_{60} etc. as proposed by Hatherton and Dickinson (1968) might be useful. In order to save those centers with only a few analyses, the following modification was made to get the equivalent of Dickinson's K-value.

An inspection of all the available K_2O-SiO_2 plots in Harker diagrams for Japanese Quaternary centers gives a standard set of K_2O-SiO_2 variation trend lines as shown in Fig. 1. This standard set of trend lines is expressed by an equation:

$$Z \times (0.365 \times SiO_2 \text{ (wt.\%)} - 1.0) - 3.65 \times K_2O \text{ (wt.\%)} = 2.46$$

where Z is a constant for any particular "standard trend line". This set of "standard trend lines" match closely that given by Gill (1981, figs. 1,2) used in his definition of andesites. Using this equation, the K-value can be computed from any single chemical analysis. For example, an equivalent of the K_{60} value [hereafter referred to as $K_2O(SiO_2 = 60)$] is obtained by the equation

$$K_2O(SiO_2 = 60) = \frac{1.19 \times K_2O + 0.804}{0.0365 \times SiO_2 - 1.0} - 0.674$$

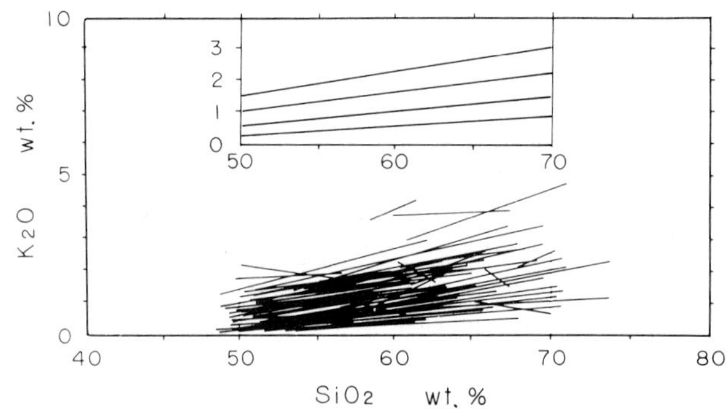

Fig. 1. K_2O-SiO_2 variation diagram of the Japanese Quaternary volcanic centers. Straight lines are linear regression lines for the individual centers. Length of a line indicates the SiO_2 range of the available analyses for a particular volcanic center. Four straight lines in the top-center represent a family of "standard trend lines" for the Japanese Quaternary volcanic rocks. They are expressed by an equation: $Z \times (0.0365 \times SiO_2 \text{ (wt.\%)} - 1.0) - 3.65 \times K_2O \text{ (wt.\%)} = 2.46$ where Z is a constant for any particular "standard trend line". Thus, a K_2O value normalized against a certain value of SiO_2, say 60%, may be computed for any single analysis if such a set of "standard trend lines" is assumed.

An analysis with K_2O = 1.34% and SiO_2 = 50.7% will give the $K_2O(SiO_2 =60)$ value of 2.23%, if the volcanic suite follows the "standard trend". Thus, these modified K-values can be assigned to individual volcanic centers even with a single analysis. The difference between Dickinson's K_{60} values derived from the K_2O-SiO_2 variation diagrams using many points and the averages of $K_2O(SiO_2 = 60)$ computed from individual analyses is small. The difference is less than 0.2% (K_2O) for 87% of the samples and less than 4% (K_2O) for 97%. When computing K_{60} and $K_2O(SiO_2 = 60)$ values, analyses with 50% > SiO_2 > 70% are excluded. This is because more than 90% of the analyses fall in this range (Aramaki and Ui, 1978a) and the linear expression of the "standard trend lines" best fits within this range. The principal mode of frequency is at 57% SiO_2 so that standardization at SiO_2 = 60% would be most appropriate.

Using the $K_2O(SiO_2 = 60)$ values, about 120 volcanic centers can be characterized by their level of K_2O contents, thus doubling the number of centers available with the Dickinson's K_{60} values (Ui and Aramaki, 1978). The result is plotted in Fig. 2. The numbers indicate $K_2O(SiO_2 = 60) \times 100$ in weight percent. Figure 3 is a similar plot for $Na_2O(SiO_2 = 60)$ values. The "standard trend lines" for Na_2O-SiO_2 plots turned out to be a set of parallel straight lines in the range 50% < SiO_2 < 70%. The $Na_2O(SiO_2 = 60)$ values are given by the equation:

$$Na_2O(SiO_2 = 60) = Na_2O - 0.0825 \times (SiO_2 - 60.0)$$

GEOGRAPHICAL VARIATION TREND OF $K_2O(SiO_2 = 60)$ AND $Na_2O(SiO_2 = 60)$ VALUES

Figure 2 gives a more detailed picture of zonal K_2O variation than the previously published maps of K_{55} and K_{60} distribution (Ui and Aramaki, 1978). The regularity is such that only 10 out of 170 data points fall on the wrong side of the contour lines (Fig. 2). The regular zonation is more pronounced in the East Japan volcanic belt (Izu-Mariana-Northeast Honshu (Japan)-Kurile-Kamchatka arcs combined, as defined by Sugimura, 1965; see inset map in Fig. 3) than in the Southwest Japan volcanic belt (or arc). Along the volcanic front of the East Japan volcanic belt, gentle but coherent ups and downs of $K_2O(SiO_2 = 60)$ values are observed. There are highs in central Hokkaido, central (around 39°N lat.), and southern (37°N—36°N lat.) Northeast Honshu arc and south of the junction of the Northeast Honshu and Izu-Mariana arcs. Along the front values for the Southwest Japan volcanic belt are distinctly higher than those for the Eastern belt.

In strong contrast, the $Na_2O(SiO_2 = 60)$ values do not show any appreciable spatial trend (Fig. 3). No regular increase away from the volcanic front or systematic variation along the front can be observed.

CORRELATION BETWEEN THE K_2O LEVELS AND THE DEPTHS OF THE DEEP SEISMIC ZONE

Figure 4 shows the relation between the depth of the deep seismic zones (hereafter called Wadati-Benioff zone) and the $K_2O(SiO_2 = 60)$ values. The depth of the Wadati-Benioff zone is obtained from the contoured map published by Utsu (1974; tabulated in Aramaki and Ui, 1978b). Utsu's contours give smoothed values of the vertical distance from the surface to the center of the dipping seismic zone. Each point characterizes the level of K of a single volcanic center. The grouping of the points according to volcanic arcs is shown in Fig. 5. The so-called K-*h* relation (Dickinson, 1975, etc.) differs from arc to arc (Fig. 5); K is higher in the Southwest Japan arc (SWJ) and lower in the Izu-Mariana arc (IM) than in the Northeast Honshu arc (NEH). K is higher and more variable in the Hokkaido part of the Kurile-Kamchatka arc (HKK) than in NEH. The depth of the Wadati-Benioff zone (*h*) is deeper (160—200 km) in IM than in the other three arcs (generally 120—160 km except near the junction between IM and NEH where *h* reaches 180 km). For the same value of $K_2O(SiO_2 = 60)$, *h* is the deepest in IM, intermediate in NEH and shallower in SWJ and HKK. This order is not directly correlated with either the speed of subduction or the slope of the Wadati-Benioff zone.

DISCUSSION

Although the plotting such as Fig. 5 including all points may give an impression that the correlation between K and *h* is very poor, the straight-line correlation is more or less well defined when a limited length of the segment of the volcanic arc is taken. In Fig. 5, most points fall inside the envelopes designated for respective arcs. The envelopes for IM, NEH and HKK generally give a slope of 2.5 to 3.0 K_2O % per 100 km of *h* (K-value increases this amount when depth of the Wadati-Benioff zone is increased by 100 km). This favorably compares with similar values for the Indonesian arcs (2.85, computed from data given in Hutchison, 1976) and "Indo-Pacific" arcs (2.8, from Ninkovich and Hays, 1972). It is much higher than those given by Nielson and Stoiber (1973) for various arcs (less than 1.0, see Fig. 5), by Hatherton and Dickinson (1968) for New Zealand (1.4) and by Dickinson (1975, Fig. 2) for world-wide composite plotting (1.2). The difference is probably due to the way of sampling the chemical and geophysical data. In the East Japan volcanic belt, for example, a narrow segment across northern Honshu of the NEH arc (including Iwate, Hachimantai and Kampu volcanoes) gives a steep slope in the K-*h* plot while the segment across the junction of IM and NEH arcs gives a gentle K-*h* slope. The volcanic centers in the latter segment are marked as M (Myoko, Iizuna, Kurohime etc.), N (Ontake, Tateyama, Hakusan, etc.) and H (Haruna, Akagi, Asama etc.) in Fig. 5. In other words, these centers erupt magmas with much lower K than those in northern

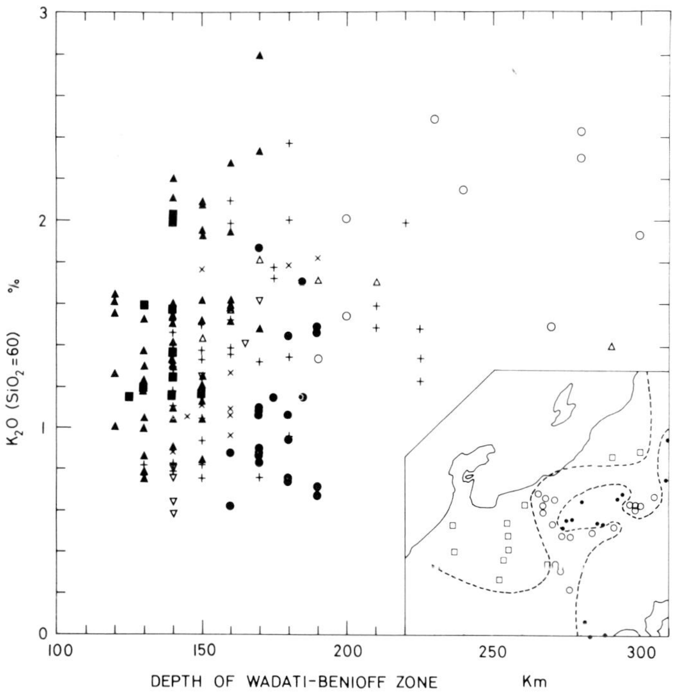

Fig. 4. Diagram showing the relationship between the SiO_2-normalized K_2O content [$K_2O(SiO_2 = 60)$] of the Japanese Quaternary volcanic centers and the depth of the deep seismic zone (Wadati-Benioff zone). Symbols are volcanic centers of the Northeast Honshu arc + (<50 km from the volcanic front) and × (>50 km), those of the Izu-Mariana arc • (< 50 km) and ○(>50 km), those of the Southwest Japan arc ▪ (all <50 km), and those of the Kurile arc (Hokkaido part only) ▲ (<50 km) and △ (>50 km). Inset is a map of the junction between the Izu-Mariana and Northeast Honshu arcs showing zonal distribution of phenocryst assemblage of volcanic rocks (after Sakuyama, 1977a, b). Symbols indicate position of vents which erupted magmas with phenocrysts of: (1) pyroxenes and others but no amphibole and biotite (•); (2) amphibole and others but no biotite (○); and (3) biotite and others (□).

Honshu for the same h. The shape of the K-contours in this segment as shown in Fig. 2 may give a false impression that N and M groups produce high-K magmas. However, the isobaths of the Wadati-Benioff zone make much less sharp bend at the arc junction than the volcanic front and the zone of volcanoes do thus giving deeper h values for the centers near the junction. The low K-h slopes such as those computed by Nielson and Stoiber (1973), which are reproduced in Fig. 5, may be the result of inclusion of such anomalous centers. A point with R is for Rishiri volcano off the northern tip of Hokkaido (Fig. 2). It should be noted that all the anomalous (low-K) centers are located behind the arc junctions.

The inset map in Fig. 4 shows the combination of mafic phenocryst minerals in the lavas (after Sakuyama, 1977a, b). Solid circles indicate vol-

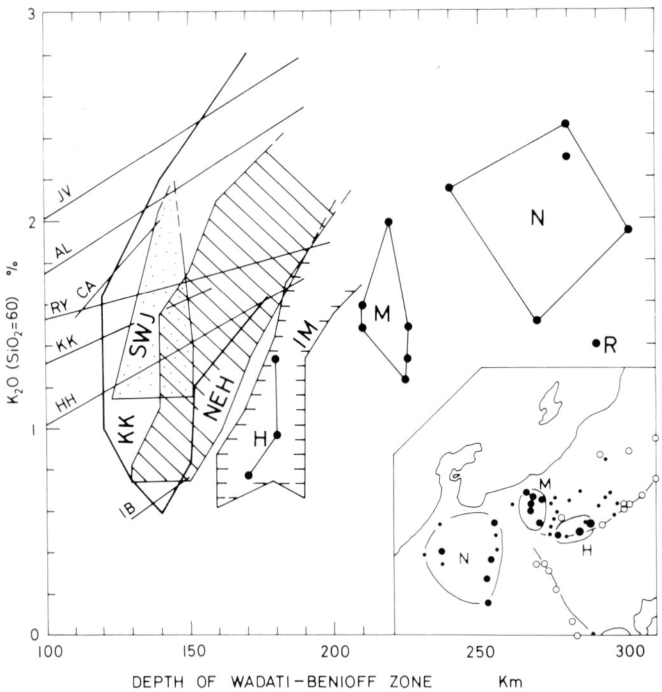

Fig. 5. Diagram similar to Fig. 4 showing relationship between $K_2O(SiO_2 = 60)$ values and the depth of the Wadati-Benioff zone. Envelopes for individual arcs are indicated: *NEH* = Northeast Honshu arc; *IM* = Izu-Mariana arc; *SWJ* = Southwest Japan arc; and *KK* = Kurile-Kamchatka arc (Hokkaido part only). Solid circles are the centers falling outside the general envelopes of the arcs with definitely lower $K_2O(SiO_2 = 60)$ values for the corresponding depths of the Wadati-Benioff Zone. *N*, *M* and *H* refer to the groups indicated in the inset map. Straight lines are the regression lines given by Nielson and Stoiber (1973): *JV* = Java; *AL* = Aleutians; *CA* = Central America; *KK* = Kurile-Kamchatka; *HH* = Honshu-Hokkaido; and *IB* = Izu-Bonin arcs. Inset map shows the distribution of volcanic centers near the junction between the NEH and IM arcs. Large solid circles correspond to those in the main diagram. Open circles are the centers with the "normal" level of $K_2O(SiO_2 = 60)$ values falling inside the envelopes of NEH or IM. Small solid circles are centers with no available analyses.

canic centers without biotite and hornblende phenocrysts in any of the ejecta, open circles those with some hornblende but no biotite phenocrysts and open squares those with biotite phenocrysts in some of the ejecta. The boundary lines between these three zones run parallel with the volcanic front in southern Hokkaido and northern Honshu (in Sakuyama's figures, 1977a, b), but near the junction with the IM arc, there is some regional anomaly as shown in Fig. 4. In the NEH arc, Notsu (1983, fig. 2) showed a regular decrease in $^{87}Sr/^{86}Sr$ ratios of the lavas away from the front except one anomaly with very high ($^{87}Sr/^{86}Sr > 0.706$) ratios. The anomaly lies just east of the junction between the NEH and IM arcs and roughly coincides with the anomaly shown in Fig. 4. Whether these anomalous features are just coincidences or are produced by some common causes such as interactions between the subducted plate segments at the junction where a geometry of the high-angle bend of the surface plate boundaries requires bending, crum pling or doubling of the downgoing slabs (Aoki, 1974) may be a matter yet to be shown. However, these features are very conspicuous and may offer important constraints on the mechanism of magma generation along the subduction zones.

The K-*h* trend in the SWJ arc is more complex and difficult to interprete. There is a gap in the distribution of Quaternary volcanic centers in central Kyushu (between Aso and Kirishima). The segment south of this appears to give generally lower and possibly more regular $K_2O(SiO_2 = 60)$ values than the northern segment. The northern segment includes Aso caldera complex where Sr-isotope, REE and other trace-element concentrations strongly indicate various degrees of crustal contamination in different rocks. In general, the effect of crustal contamination is thought to be stronger in southwestern Japan than in the northeastern part (Kurasawa, 1968, 1975; Shuto and Kagami, 1975) which probably affects higher and irregular K values.

Only about half of the Quaternary volcanic centers have been characterized in this report because of the lack of available analyses. Available data are very much heterogeneous in quality as well as in quantity. It is apparent that a more systematic way of collection of analyses is necessary in order to characterize chemical features of the Japanese Quaternary volcanism in relation to time and space. Such a systematic approach is now under way to build a chemical data base. Similarly, overall configuration of the Wadati-Benioff zone will be defined more precisely by the mass of new seismic data now rapidly accumulating.

ACKNOWLEDGEMENTS

We thank Prof. T. Maruyama, Earthquake Research Institute, for the advice on data processing and Mrs. K. Kudo for assistance in data gathering and preparation of the manuscript. Prof. I. Kushiro critically read the manuscript to greatly improve the contents.

REFERENCES

Aoki, H., 1974. Plate tectonics of arc-junction at central Japan. J. Phys. Earth, 22: 141—161.

Aramaki, S. and Ui, T., 1978a. Major element frequency distribution of the Japanese Quaternary volcanic rocks. Bull. Volcanol., 41: 390—407.

Aramaki, S. and Ui, T., 1978b. List of geodynamic parameters of Quaternary volcanoes of Japan, Mariana, Kurile, and Kamchatka. Contrib. Geodynamics Project Japan, No. 78-2.

Dickinson, W.R., 1968. Circum-Pacific andesite types. J. Geophys. Res., 73: 2261—2269.

Dickinson, W.R., 1975. Potash-depth (K-h) relations in continental margin and intra-oceanic magmatic arcs. Geology, 3: 53—56.

Gill, J., 1981. Orogenic Andesites and Plate tectonics. Springer-Verlag, Berlin, 390 pp.

Hatherton, T. and Dickinson, W.R., 1968. Andesitic volcanism and seismicity in New Zealand. J. Geophys. Res., 73: 4615—4619.

Hutchison, C.S., 1976. Indonesian active volcanic arc: K, Sr, and Rb variation with depth to the Benioff zone. Geology, 4: 407—408.

Kurasawa, H., 1968. Isotopic composition of lead and concentrations of uranium, thorium and lead in volcanic rocks from Dogo of the Oki Islands, Japan. Geochem. J., 2: 11—28.

Kurasawa, H., 1975. Isotope geology of volcanic rocks. Bull. Volcanol. Soc. Jpn, Ser. 2, 20: 307—317 (in Japanese with English abstract).

Nielson, D.R. and Stoiber, R.E., 1973. Relationship of Potassium content in andesitic lavas and depth to the seismic zone. J. Geophys. Res., 78: 6887—6892.

Ninkovich, D. and Hays, J.D., 1972. Mediterranean island arcs and origin of high potash volcanoes. Earth Planet. Sci. Lett., 16: 331—345.

Notsu, K., 1983. Strontium isotope composition in volcanic rocks from Northeast Japan arc. In: S. Aramaki and I. Kushiro (Editors), Arc Volcanism. J. Volcanol. Geotherm. Res., 18: 531—548.

Sakuyama, M., 1977a. Lateral variation on phenocryst assemblages in volcanic rocks of the Japanese islands. Nature, 269: 134.

Sakuyama, M., 1977b. Lateral variation of H_2O contents in Quaternary magma of northeastern Japan. Bull. Volcanol. Soc. Jpn., Ser. 2, 22: 263—271 (in Japanese with English abstract).

Shuto, K. and Kagami, H., 1975. Regional variations in the Sr isotopic composition in the Cenozoic tholeiitic-calc-alkaline volcanic rocks from the intra- and circum-Pacific region. Earth Science (J. Assoc. Geol. Collab. Jpn; Chikyu Kagaku), 29: 75—86 (in Japanese with English abstract).

Sugimura, A., 1965. Distribution of volcanoes and seismicity of the mantle in Japan. Bull. Volcanol. Soc. Jpn., Ser. 2, 10: 37—58 (in Japanese with English abstract).

Ui, T. and Aramaki, S., 1978. Relationships between chemical composition of Japanese island-arc volcanic rocks and gravimetric data. Tectonophysics, 45: 249—259.

Utsu, T., 1974. Epicenter distribution around Japan. Kagaku, 44: 739—746 (in Japanese with English abstract).

Journal of Volcanology and Geothermal Research, 18 (1983) 561—588
Elsevier Science Publishers B.V., Amsterdam — Printed in The Netherlands

THE INFLUENCE OF CRUSTAL STRUCTURE ON COMPOSITIONS OF SUBDUCTION-RELATED MAGMAS

WILLIAM P. LEEMAN

Department of Geology, Rice University, Houston, TX 77001 (U.S.A.)

(Received October 20, 1982)

ABSTRACT

Leeman, W.P., 1983. The influence of crustal structure on compositions of subduction-related magmas. In: S. Aramaki and I. Kushiro (Editors), Arc Volcanism. J. Volcanol. Geotherm. Res., 18: 561—588.

A survey of Sr isotopic ratios and other compositional features of subduction-related magma suites reveals significant correlations between these averaged parameters and characteristics of the underlying crust (i.e., thickness, composition, and age). These observations lead to the conclusion that crust and(or) mantle rocks in the hanging walls of subduction zones are involved in modification of primary mafic magmas (typically basalt or basaltic andesite). It is proposed that mafic magmas will stagnate within the crust or uppermost mantle where they may differentiate and react with wall rocks. The extent to which such processes manifest themselves will depend upon details of the local crustal structure. In particular, the composition and age of the crust will strongly influence such parameters as Sr, Nd and Pb isotopic compositions. Such data strongly indicate the involvement of crustal rocks in locales underlain by old sialic crust (e.g., central Andes). Depending upon the level of magma stagnation and evolution within the crust, different trends in isotopic composition may result. These isotopic trends may be enhanced by partial melting of the wall rocks to produce relatively silicic anatectic magmas, and locally they may reflect subduction of continental sediments. Interpretation of the isotopic data may be more ambiguous in locales underlain by younger and more mafic continental crust (Cascades, E Eleutians) and those underlain by oceanic crust owing to the similarity in isotopic composition of primary magmas and the latter crustal materials. Yet some degree of crustal involvement in magmatic evolution seems highly probable even in these more primitive terranes. Consequently, most island arc magmas, and especially those more evolved than basalt, are probably not primary in the sense that they do not represent direct melts of the upper mantle. Studies of arc volcanic rocks may yield misleading conclusions concerning processes of magma generation related to subduction unless evolutionary processes are defined and their effects considered. It appears that modern volcanic arcs provide a poor analog for models of early crustal development because the modern mantle-derived magmatic components are more mafic in composition than average continental crust.

INTRODUCTION

Despite the considerable effort invested in studies of subduction-related tectonism and magmatism, many questions remain concerning how volcanic arc magmas are generated and subsequently evolve prior to eruption. Increasing evidence supports the view that individual volcanic arcs and subduction systems differ in significant ways. Accordingly, it is perhaps unrealistic to expect any general model to explain the characteristics of all arcs. Yet the diverse "personalities" of volcanic arcs provide important clues about the nature and relative importance of processes and mechanisms involved in arc magmatism.

With respect to proportions of volcanic rock types and details of their geochemistry, volcanic arcs situated entirely on oceanic crust differ systematically from those situated on continental or transitional crust (Miyashiro, 1974; Dickinson, 1975; Ewart et al., 1977). These sorts of distinctions (as discussed below) logically can be attributed to spatial variations in: (1) processes of magma genesis; (2) processes of magma evolution during its ascent, or possibly a combination of both. The evidence summarized herein, in my opinion, largely favors (2) and can be rationalized on the basis of differing crustal (and possibly lithospheric mantle) structure and composition from one arc to another. This view leads to perhaps somewhat unconventional implications concerning the significance of most andesitic magmas and their role in crustal evolution.

GENERAL MODELS FOR ARC VOLCANISM

All the volcanic arcs discussed here are associated with seismically defined subduction zones (except perhaps the Cascades) where oceanic lithosphere is thrust beneath a wedge of asthenospheric mantle which is overlain by oceanic, continental, or transitional (relative to the former two) lithosphere. It is useful to consider arc magmatism in terms of a mass-balance approach, with a number of reservoirs from which magmas could potentially be extracted. These include: (1) the slab of subducted oceanic crust and lithosphere (possibly including part of its sediment veneer); and (2) the overlying asthenospheric wedge which is considered to be a convecting medium (cf. Marsh, 1979b). A primary magma source within the overriding lithosphere or accretionary prism seems unlikely, in general, because regional heat flow (hence, geothermal gradient) is too low to produce mafic magmas at such shallow depths (cf., Watenabe et al., 1977). An obvious, though uncommon, exception to this view is shallow magmatism related to subduction of a spreading center (cf., DeLong and Fox, 1977; Dunkley and Hughes, 1981).

Worldwide compositional variations in the subducted material are to be expected due to differences in the degree of alteration of oceanic crust (which varies with age of the seafloor), and the thickness and composition of its sediment veneer, part of which may be subducted (Gill, 1981). Pb-isotopic compositions of volcanic rocks from some Pacific and Atlantic arcs suggest that the respective magmas may contain up to a few percent of Pb derived from the corresponding pelagic sediments (Tatsumoto, 1969; Armstrong and Cooper, 1971; Kay et al., 1978; Karig and Kay, 1981). Combined Sr and Nd-isotopic data further indicate that subducted oceanic crust (more or less equilibrated with seawater) may serve as a viable source for significant components of these elements in arc magmas, either by direct-melting or via ill-defined fluid-transfer processes associated with dehydration of this reservoir (e.g., Hawkesworth and Powell, 1980; Kay, 1980; Thorpe et al., 1981; Arculus and Johnson, 1981). Unfortunately, it has not been convincingly demonstrated that relatively buoyant low-strength sediments or tectonic slivers of crustal basement rocks can be subducted to depths where they may reasonably be involved in magma genesis (cf., Gill, 1981).

A number of authors (e.g., Perfit et al., 1980; Arculus and Johnson, 1981) have summarized arguments for derivation of arc magmas from the asthenospheric wedge above subduction zones. Thermal conditions in this region are likely more conducive to generation of high-temperature basaltic magmas (cf., Marsh, 1979a), and convection there could conceivably replenish magmatic components that would otherwise tend toward depletion in long-lived arcs (e.g., Japan, Andes, Cascades). Isotopic data mentioned above would imply that such asthenospheric sources must either be regionally varied in composition or else be in partial communication with slab or sediment reservoirs, perhaps via transfer of fluids or magmatic components.

A major barrier to resolving the sources of arc magmas is our incomplete understanding of processes by which such magmas are modified during their ascent. Isotopic data, in particular, clearly reveal that magmas from many arcs have either formed by fusion of heterogeneous sources, or else have behaved as open systems with respect to wall rock contamination, magma mixing or other processes. This paper analyzes comparative data for most recently active volcanic arcs in an attempt to focus on this problem and to distinguish primary from secondary features of arc magmas.

COMPARISON OF VOLCANIC ARCS

Numerous authors (Jakes and White, 1971, 1972; Miyashiro, 1974; Dickinson, 1975; Molnar and Atwater, 1978; Uyeda and Kanamori, 1979; Yokokura, 1981; and Gill, 1981) summarized many physical and compositional features of subduction-related volcanic arcs. Geologic and tectonic characteristics of arcs or arc segments are clearly correlated with such features as plate convergence rates, geometry of subduction (or seismic) zones, age of subducted oceanic crust (hence, lithospheric thickness and depth of seismic zone), nature of the overriding lithospheric plate (oceanic vs. continental), duration of subduction, and so on. Young, immature intra-oceanic arcs are commonly characterized by relatively high proportions of mafic volcanic rocks, typically of tholeiitic (i.e., pigeonitic series) affinities. More mature island and continent margin arcs are usually underlain by thicker crust, and are characterized by greater proportions of silicic volcanic rocks. In such arcs, tholeiitic, calc-alkalic (i.e., hypersthenic) and alkalic series rocks are more or less well-developed. Compositional features of arc volcanic rocks are poorly correlated with most of the physical or geometrical parameters mentioned above (Coulon and Thorpe, 1981; Gill, 1981). However, there appears to be some correlation between composition of arc volcanic rocks and crustal thickness (Dickinson, 1975; Leeman, 1980; Coulon and Thorpe, 1981; Karig and Kay, 1981). This correlation is manifest most clearly in the relative proportions of mafic and silicic volcanic rocks, and their trace-element and isotopic compositions.

The Andes (Pichler and Zeil, 1972; Klerkx et al., 1977; Thorpe and Francis, 1979; Thorpe et al., 1981) and Tonga—Kermadec—New Zealand (Ewart et al., 1977) provide two striking examples. In each case, subducted oceanic lithosphere differs little along strike with respect to consumption rate, age, and presumably composition. Major differences along strike include subduction geometry, volume and provenance of arc-trench sediments, composition of oceanic sediment, and thickness, composition, and age of the overriding crust. In both New Zealand and the Andes, volcanic rocks have trace-element and isotopic compositions (Pb, Sr, Nd) that seemingly reflect a continental component for each of these elements (Armstrong and Cooper, 1969; Oversby and Ewart, 1972; Magaritz et al., 1978; Thorpe and Francis, 1979; Tilton and Barreiro, 1980; James, 1981; Perfit et al., 1981). For the Andes, there is little geologic evidence for subduction of significant quantities of continent-derived sediments (Thorpe et al., 1981). On the other hand, Armstrong (1981) convincingly argues that appreciable pelagic sediment may be subducted at this and other subduction zones. If it is assumed that there is little difference in the volume and composition of pelagic sediment subducted along strike, then along strike differences in magma compositions plausibly reflect differences in magma—crust interactions. Similar arguments could apply to New Zealand. The Tonga—Kermadec arcs are far removed from a continental sediment source, rest on oceanic lithosphere and are underlain by thin crust; their volcanic rocks are more primitive in composition than those of New Zealand. Clearly, geochemical differences along this arc system cannot be attributed to differences in the slab reservoir, and appear to be related to the presence of continental crust beneath New Zealand or, possibly, subduction of sediments derived therefrom.

Many other volcanic arcs display along strike geochemical differences that coincide with changes in either the structure of sub-arc crust (i.e., age and composition), differences in sediment provenance and composition, or differences in the nature of the subducted lithosphere. Obvious examples of the latter include continent—continent, continent—arc, or oceanic plateau—arc collision zones (Bird et al., 1975; Hamilton, 1979;

Ben-Avraham et al., 1981). Some specific examples illustrate these factors. Compiled compositional data for volcanic arcs are summarized in Tables I and II.

The Aleutian—Alaska arc is situated on Oligocene or older oceanic crust to the west and a thick continental crust east of the Bering shelf that includes a variety of Paleogene to Mesozoic plutonic and sedimentary rocks (Burk, 1965). Among Quaternary volcanic rocks from the Alaska peninsula segment, silicic rocks (andesite - dacite - rhyolite) predominate volumetrically, whereas such rocks are subordinate further west (Table I). Available data suggest that Sr isotopic ratios differ little from west to east and interpolated K_2O values at 57.5% SiO_2 ($K_{57.5}$) tend to be lower in the east. Such trends seem contrary to expectation if arc magmas interact with crustal rocks. However, detailed Sr isotopic studies of the Edgecumbe volcanic field (in SE Alaska) strongly indicate contamination by crustal material or mixing of anatectic crustal melts with mantle-derived magmas (Myers and Marsh, 1981; Kosco, 1981).

The Indonesian arcs lie in a tectonically complex region whose geology has been summarized by Hamilton (1979). The western (Sunda) arc is situated in part on continental crustal fragments of Cretaceous age, whereas the eastern (Banda) arc appears to be established mainly on oceanic crust. Isotopic data (Sr, O, Nd, Pb) strongly indicate incorporation of sialic crustal material in the Indonesian magmas, and most strikingly in those from the Banda arc (Whitford, 1975; Magaritz et al., 1978; Whitford and Jezek, 1979; Morris and Hart, 1980; Whitford et al., 1981). In this area, it appears that crustal rocks or sediments derived from the subducting Australian plate are somehow involved in magma genesis or evolution.

The Japanese island arcs are subdivided into four distinct structures including the Izu—Mariana, southwest Japan, northeast Japan and Hokkaido—Kurile zones (Miyashiro, 1974; Aramaki and Ui, 1978). The Izu-Mariana arc lies on oceanic crust and consists dominantly of mafic tholeiite series lavas which have primitive Sr, Nd, and O isotopic compositions and trace-element contents (cf., Stern, 1981; Ito and Stern, 1981). These authors point out the compositional similarity between Mariana basalts and those from oceanic islands, such as Hawaii and the Carolines, in support of their origin more or less directly by partial fusion of similar mantle sources. However, Pb isotopic data for Mariana lavas are consistent with incorporation of at least small amounts of subducted pelagic sediment (Meijer, 1976). Regardless of the exact origin of these magmas, the lavas studied are fairly uniform in isotopic composition (see Table II), and their major and trace-element compositional variance can be explained in terms of fractional crystallization of observed phenocryst assemblages (Stern, 1979; Dixon and Batiza, 1979).

The other Japanese arcs are underlain by more mature crust, at least in part of continental character (Sugimura and Uyeda, 1973). The crust beneath the major Japanese islands averages about 30 km in thickness

TABLE I

Some vital statistics for subduction-related volcanic arcs (late Cenozoic except as noted)

Volcanic arcs	No. Anal.	% ADR	K_2O (57.5)	FeO/MgO (57.5)	t_c (km)	Rate (cm/yr)
I. CONTINENTAL MARGIN ARCS						
A. *With mainland volcanoes*						
Alaska—E Aleutian	255[a]	73	1.2	1.8	35—40	5.7
Cascades: a. All	490	68	1.3	1.7	25—35	~ 2.0
b. Mt. Hood	190[b]	67	1.2	1.9		
c. Mt. Mazama	90[b]	70	1.3	1.5		
Central America	337	30	1.5	2.3	30—40	7.9*
Kamchatka	330	46	1.3	2.1	30—40[h]	8.5
Andes: a. Northern	54	78	1.5	2.5	50—60	9.0
b. Central	200	84	2.2	1.6	50—70	9.1
c. Southern	66	65	1.4	2.0	30—40	10.4
Mexico (western and central)	261[c]	67	1.5	1.4	40	5.6
Turkey—NW Iran	139	75	2.3	2.2	~50	~ 3.0
N America Cordillera (Oligocene):						
a. San Juan	335[d]	88 (89)[m]	2.8	—	~50[d]	—
b. Mogollon—Datil	665[e]	84	2.5	—	40—50	—
B. *With island or peninsular volcanoes*						
Aegean	168	63	2.4	1.7	25—35[i]	2-4*
II. INTRA-OCEANIC ARCS						
A. *With detached continental basement*						
Sunda (Sumatra—Java)	262	40	1.7	2.3	25—35	6.4
SW Japan (Kyushu)	541[f]	53 (87)	1.5	2.1	25—30[j]	5.6
NE Japan (Honshu)	595[f]	64 (63)	1.0	2.2	25—35[j]	9.0
Hokkaido	125	60 (51)	1.1	2.6	25—30[j]	6.9
New Zealand	234[g]	78	1.5	1.7	35	3.4*
Papua	78	52	2.1	1.6	20—35	9.0
B. *Stationary arc without active back-arc spreading*						
W Aleutian	131[a]	47	1.6	2.5	15—25	4.1
Banda—Bali	81	50	1.8	2.2	~20	~ 6.6
Kurile	217	62	1.2	2.2	15—30	9.0
Eolian	142	46	2.4	2.3	~20—25	~ 1.0
Lesser Antilles: a. All	777	53	0.9	2.2	~30—35	~ 2.0
b. St. Kitts	173	64	0.6	2.5		
c. Dominica	280	81	0.9	2.4		
d. Grenada	264	41	1.2	1.7		
C. *Migratory arc with active back-arc spreading*						
Izu—Mariana	420[f]	25 (30)	1.0	4.2	15—18	3.7*
Tonga	108[g]	48	0.6	3.1	~12	9.1*
Kermadec	82[g]	22	0.4	3.0	~18	6.9*
S Sandwich (Scotia)	34	36 (< 32)	0.7	4.1	15[k]	9.0*
Melanesia: a. Bismarck (New Britain)	175	57	0.9	2.3	20—40	9.2
b. S Vanuatu (New Hebrides)	146	8	2.2	2.9	15—20[l]	9.6
c. Solomon	63	16	2.3	2.2	15—20	10.0

Key: % ADR = percent of all analyses with >57% SiO_2; $K_2O_{57.5}$ and $FeO/MgO_{57.5}$ are interpolated values at 57.5 % SiO_2 (Gill, 1981); t_c = estimated crustal thickness; rate = estimated net convergence rate normal to each arc (resultant of normal components of motion of both subducted and overrriding plates in most cases). Asterisk indicates overriding plate is retreating from trench, and some form of back-arc extension is recognized.

Data sources: Major-element analyses are from compilation of Gill (1981) except: (a) Kosco, 1981; (b) Leeman, unpubl.; (c) Pal et al., 1978; (d) Lipman, 1975; (e) Bornhorst, 1980; (f) Aramaki and Ui, 1978; (g) Ewart et al., 1977; (m) figures in parentheses indicate volumetric estimates (see text).
Crustal thickness estimates are from compilations of Gill (1981), Soller et al., 1981, except: (h) Yermakov, 1981; (i) Le Pichon and Angelier, 1981; (j) Sugimura and Uyeda, 1973; (k) Baker, 1968; (l) Dickinson, 1975, Appendix.
Convergence rate estimates are from Yokokura (1981), Molnar and Atwater (1978), and Gill (1981).

TABLE II

Compilation of $^{87}Sr/^{86}Sr$ data for volcanic arcs

Volcanic arcs	No. Anal.	Range	Mean	References
I. CONTINENTAL MARGIN ARCS				
Alaska (SE) (b,a,d,r)*	8	0.7030—0.7044	0.7035	1— 2
Cascades (b,a,d,r)				3— 8
N California	49	0.7027—0.7043	0.7035	
Oregon	40	0.7027—0.7047	0.7038	
Washington	11	0.7026—0.7041	0.7036	
Central America (b,a,d)	40	0.7031—0.7046	0.7040	9—11
Kamchatka—Kurile	12	0.7031—0.7036	0.7033	11
Andes				13—22, 46
Ecuador—Columbia (b,a)	35	0.7038—0.7045	0.7043	
Central Peru (a,d)	4	0.7042—0.7051	0.7046	
S Peru (Barroso—Arequipa) (a,d,r)	37	0.7054—0.7081	0.7069	
N Chile—NW Argentina—Bolivia (b,a,d,r)	73	0.7050—0.7115	0.7071	
S Central Chile (b,a)	6	0.7038—0.7042	0.7040	
Patagonia (b,aob)	15	0.7034—0.7053	0.7043	
Mexico (W—Central) (b,a,d)	49	0.7032—0.7045	0.7039	23—25
Turkey—NW Iran (a,d)		0.7036—0.7065	~0.705	26—27
N. America Cordillera (Oligocene)				
San Juan (b,a,d,r)	27	0.7050—0.7075	0.7059	28
Mogollon—Datil (b,a,d,r)	47	0.7039—0.7114	0.7076	29
Sierra Madre Occidental (d,r)	27	0.7043—0.7074	0.7051	30—31
Aegean (b,a,d,r,l)	24	0.7037—0.7134	0.7065	32—33
II. INTRA-OCEANIC ARCS				
A. *With detached continental basement*				
Sunda (b,a,d,r)	108	0.7039—0.7066	0.7047	34—37
Japan (b,a,d,r)				38—44
Kyushu (SW)	—	0.7037—0.7095	>0.705	
Honshu (NE)	78	0.7026—0.7058	0.7040	
Hokkaido	46	0.7026—0.7058	0.7037	
New Zealand (b,a,r)	16	0.7041—0.7062	0.7052	45
Papua	9	0.7037—0.7054	0.7044	50
N Sardinia (b,a,r)	9	0.7054—0.7089	0.7066	52—53
B. *Stationary arc without active back-arc spreading*				
Aleutian (b,a)	24	0.7028—0.7037	0.7032	54
Banda (b,a,d,r)	35	0.7042—0.7095	0.7064	34—35, 37
Eolian (b,a,d,r,l)	33	0.7030—0.7065	0.7050	55—56

(continued)

TABLE II (continued)

Volcanic arcs	No. Anal.	Range	Mean	References
Lesser Antilles				57—63, 9
St. Kitts (b,a,d)	14	0.7036—0.7040	0.7037	
St. Lucia (b,a,d)	10	0.7035—0.7092	0.7063	
St. Vincent (b,a,d)	17	0.7039—0.7043	0.7041	
Dominica (b,a,d,r)	23	0.7041—0.7068	0.7045	
Carriacou (b,a)	5	0.7049—0.7054	0.7052	
Grenada (b,a)	52	0.7039—0.7058	0.7049	
C. Migratory arc with active back-arc spreading				
Izu—Mariana (b,a,d)	54	0.7032—0.7043	0.7036	9, 47, 64—68
Tonga—Kermadec (b,a,d)	19	0.7033—0.7057	0.7038	69—70
S. Sandwich (b,a,d)	24	0.7031—0.7042	0.7038	71—72
Melanesia				
Bismarck (b,a,d,r)	45	0.7032—0.7041	0.7036	47—51
Vanuatu (b,aob,a)	—	—	0.7024	73
Fiji (b,a,d,r)	44	0.7032—0.7053	0.7040	74—77
Solomon (b,a)	13	0.7030—0.7043	0.7037	75

*Letters indicate rock types analyzed: basalt (b), alkali-olivine basalt (aob), andesite (a), dacite (d), rhyodacite—rhyolite (r), latite (l).

Abbreviated references: *1*. Kosco (1981) J. Geol. 89, 459; *2*. Myers and Marsh (1981) CMP 77, 272; *3*. Church and Tilton (1973) GSAB 84, 431; *4*. Hedge et al. (1970) EPSL 8, 434; *5*. Peterman et al. (1970) GSAB 81, 311; *6*. Mertzman (1979) CMP 70, 81; *7*. Mertzman and Williams (1981) GCA 45, 1463; *8*. Leeman (1980; and unpubl.) EOS 61, 68; *9*. Pushkar (1968) JGR 73, 2701; *10*. Thorpe et al. (1979) Nature 277, 44; *11*. Rose et al. (1977) J. Geol. 85. 63; *12*. Hedge and Gorshkov (1978) Dokl. Akad. Nauk SSSR 233, 163; *13*. Klerkx et al. (1977) JVGR 2, 48; *14*. James et al. (1976) GSAB 87, 592; *15*. Noble et al. (1975) Geology 3, 501; *16*. Pichler and Zeil (1972) BV 35, 424; *17*. Thorpe et al. (1979) EPSL 42, 359; *18*. Francis et al. (1977) EPSL 37, 197; *19*. Briqueu and Lancelot (1979) EPSL 43, 385; *20*. Hawkesworth et al. (1979) EPSL 42, 45; *21*. James (1980) CIW 79, 452; *22*. Jaramillo - Mejia (1980) Ph.d. (Univ. Houston); *23*. Whitford and Bloomfield (1976) CIW 75, 207; *24*. Moorbath et al. (1978) Nature 271, 437; *25*. Verma (1981) IAVCEI, 404; *26*. Lambert et al. (1974) J. Geol. 82, 419; *27*. Gulen (1980) EOS 61, 412; *28*. Lipman et al. (1978) GSAB 89, 59; *29*. Bornhorst (1980) Ph.D. (Univ. New Mexico); *30*. Lanphere et al. (1980) Nature 286, 594; *31*. Moll (1981) JGR 86, 10321; *32*. Pe and Gledhill (1975) Lithos 8, 209; *33*. Pe (1975) Chem. Geol. 15, 53; *34*. Whitford (1975) GCA 39, 1287; *35*. Whitford and Jezek (1979) CMP 68, 141; *36*. Leo et al. (1980) JVGR 7, 139; *37*. Whitford (1978) CIW 77, 613; *38*. Hedge and Knight (1969) Geochem. J. 3, 15; *39*. Ishizaka et al. (1977) CMP 63, 245; *40*. Kurasawa and Fujimaki (1977) BVSJ 22, 249; *41*. Nohda and Wasserburg (1981) EPSL 52, 264; *42*. Kurasawa (1979) BVJS 24, 135; *43*. Shuto (1974) Sci. Rep. Tokyo Univ. Educ. 12, 75; *44*. Notsu (1981) IAVCEI, 265; *45*. Ewart and Stipp (1968) GCA 32, 699; *46*. Francis et al. (1980) EPSL 48, 257; *47*. DePaolo and Wasserburg (1977) Geophys. Res. Lett. 4, 465; *48*. Peterman et al. (1970) GSAB 81, 39; *49*.

Peterman and Heming (1974) GSAB 85, 1265; *50*. Page and Johnson (1974) Lithos 7, 91; *51*. DePaolo and Johnson (1979) CMP 70, 367; *52*. Dupuy et al. (1974) GCA 38, 1287; *53*. Beckinsale et al. (1981) IAVCEI, 36; pers. commun.; *54*. Kay et al. (1978) GCA 42, 263; *55*. Barberi et al. (1974) EPSL 21, 269; *56*. Klerkx et al. (1974) EPSL 23, 297; *57*. Hedge and Lewis (1971) CMP 32, 39; *58*. Donnelly et al. (1971) GSA Mem. 130, 181; *59*. Pushkar et al. (1973) JGR 78, 1279; *60*. Hawkesworth and Powell (1980) EPSL 51, 297; *61*. Hawkesworth et al. (1979) EPSL 45, 237; *62*. Stipp and Nagle (1978) EOS 59, 404; *63*. Smith et al. (1980) BV 43, 287; *64*. Meijer (1976) GSAB 87, 1358; *65*. Stern (1979) CMP 68, 207; *66*. Dixon and Batiza (1979) CMP 70, 167; *67*. Ito and Stern (1981) CIW 80, 449; *68*. Matsuda et al. (1977) Tectonophys. 37, 141; *69*. Ewart et al. (1973) J. Petrol. 14, 429; *70*. Ewart (1980, pers. commun.); *71*. Gledhill and Baker (1973) EPSL 19, 309; *72*. Hawkesworth et al. (1977) EPSL 36, 253; Lancelot et al. (1978) USGS Open-File Rept. 78-701, 240; *74*. Gill (1970) CMP 27, 174; *75*. Gill and Compston (1973) in Western Pacific Island Arcs, Marginal Seas, Geochemistry (ed. P. Coleman), 543; *76*. Gill (1976) GSAB 87, 1384; *77*. Gill (1976) Geology 4, 123.

Key to abbreviated publications: BV = Bull. Volcanol.; BVSJ = Bull. Volcanol. Soc. Jpn.; CMP = Contrib. Mineral. Petrol.; CIW = Carnegie Inst. Wash. Yearb.; EPSL = Earth Planet. Sci. Lett.; GCA = Geochim. Cosmochim. Acta; GSAB = Geol. Soc. Am. Bull.; JGR = J. Geophys. Res.; JVGR = J. Volcanol. Geotherm. Res.

but is at least 35 km thick in southern Honshu where Paleozoic (and possibly Precambrian) metamorphic and igneous rocks occur (cf., Nohda and Wasserburg, 1981). The long-wavelength gravity field differs over southern and northern Japan (Ui and Aramaki, 1978) and, coupled with the similarity in crustal thickness in these areas, indicates that the crust of southern and southwest Japan is of lower density (greater sialic component) than that beneath northeastern Japan. Detailed estimates of the volume of Quaternary volcanic rock types (Aramaki and Ui, 1978) show that the proportion of andesite - dacite - rhyolite eruptives is much higher in southwest Japan where sialic crust is more voluminous. Sr isotopic ratios in these rocks show a similar correlation with crystal type. Volcanic rocks in southwestern Japan have $^{87}Sr/^{86}Sr$ ratios typically higher than 0.705, whereas those from northeastern Japan typically have ratios less than 0.704; anomalously high ratios are observed in the region just northeast of Tokyo where crustal thickness reaches a maximum (cf., Shuto, 1974; Kurasawa, 1979; Notsu, 1981). Limited data for Nd and O isotopic compositions reveal negative and positive correlations, respectively, with Sr isotope ratios in many cases (Matsuhisa, 1979, 1981; Nohda and Wasserburg, 1981; Nakamura, 1981). These data are consistent with magma—crust interactions, although inhomogeneities in the source regions could also be responsible for the observations. Limited Pb isotopic data have been interpreted as indicating the involvement of a small pelagic sediment component in Japanese magmas (Tatsumoto, 1969), but the same data could be explained by the alternatives stated above (cf., Matsuhisa, 1977), particularly if old crust or mantle rocks are involved. As with the

Aleutians and Andes, there is little reason to suspect significant differences in the composition of subducted material along strike for these arcs. Thus, differences in the proportions of volcanic rock types and their isotopic compositions seemingly reflect processes unrelated to the subduction process itself. The contrast between the Izu—Mariana and other Japanese arc volcanic rocks is obviously related to differences in the structure, composition, and age of the underlying crust.

Another striking example of such a relation is provided by the Aegean arc, although data there are relatively sparse. The Aegean region has been progressively distended southeastward since late Miocene time as a result of collision between the Arabian and Eurasian plates in the vicinity of Turkey and northeast Iran (LePichon and Angelier, 1979, 1981). Much of the Aegean Sea is characterized by easterly extension, whereas the Aegean arc is overriding a portion of the African plate (or at least a thick sedimentary sequence derived therefrom). Available data show that the proportions of Aegean arc volcanic rock types vary along strike with crustal thickness (cf., Gill, 1981; LePichon and Angelier, 1981), as does Sr isotopic composition (Pe, 1975; Pe and Gledhill, 1975). In Greece, where about 35 km of continental crust is present, intermediate to silicic lavas are dominant and have high $^{87}Sr/^{86}Sr$ ratios (>0.706). In the central portion of the arc, where crustal thickness is about 25 km, both of these parameters decrease. However, both these parameters and also K_2O/Na_2O ratios are higher than for intraoceanic arcs (Marianas, Aleutians, etc.), presumably due to partial involvement of the subducted Mediterranean sediments in magma genesis.

In the continent—continent collision zone of southwest Turkey and northwest Iran, the crust is anomalously thick and arc volcanics apparently show some compositional similarities to those of the Andean arc — namely, an abundance of intermediate and silicic rock types, many of which display elevated Sr isotope ratios and K_2O contents (Tables I and II). In this area, it is believed that some portion of the Arabian plate (late Precambrian continental crust) has been subducted (Bird et al., 1975).

The Lesser Antilles provide a well-documented example of just how complicated an intra-oceanic arc may be. The active arc (Pliocene to Recent) displays significant compositional differences along strike. From north to south, dominant magma types range from tholeiitic to calcalkalic to alkalic, with widely varied proportions of rock types, and $^{87}Sr/^{86}Sr$ ratios generally increase (Brown et al., 1977; Smith et al., 1980; Rea and Baker, 1980; Hawkesworth and Powell, 1981). These spatial variations have been attributed in part to several factors including: (a) differing depths of magma genesis and degrees of melting of a uniform source; (b) compositional variations in source rocks; (c) increased component of subducted continent-derived sediment closer to South America; and (d) westward migration of the northern part of the arc in Miocene (?) time, so that this portion is more immature than the central and southern parts. The available data do not allow a clear distinction between the relative importance of these alternatives. There is debate concerning the geometry of subduction and whether it even occurs in the southern part of the arc. Unfortunately many of the active volcanic arcs in the world are even less thoroughly characterized with respect to petrochemistry of the volcanic rocks, geophysical studies of crustal structures, seismicity, etc., and geologic development and geochronology.

These few examples reveal that volcanic arcs differ widely in terms of their tectonic setting and the nature of the crust on which they are situated. Following Miyashiro (1974) and Dickinson (1975), volcanic arcs may be classified according to such characteristics. The most primitive intraoceanic arcs have relatively thin mafic crust and an abundance of tholeiitic mafic volcanic rocks ($<57\%$ SiO_2) characterized by low K_2O contents and isotopic compositions of Sr, Nd, O, and Pb that resemble or approach those of ocean floor and some oceanic island basalts. On the other extreme are continental margin arcs situated on variably mature continental crust. Volcanic rocks from such arcs are more diverse (tholeiitic, calcalkalic, or alkalic series), often more evolved (intermediate-silicic compositions abundant or dominant) and typically have higher K_2O contents. Isotopic compositions of Sr, Nd, and Pb commonly trend toward values typical of sialic crustal materials. Such isotopic trends are most pronounced where the underlying crust is old or highly evolved in composition (cratonic vs. accreted oceanic or magmatic arc terranes). "Transitional" arcs, containing detached slivers of old continental crust, but separated from nearby continent masses by marginal seas, commonly display intermediate characteristics. The above comparison also points out the apparent connection between magma compositions and subduction of sialic detritus (Banda and Aegean arcs) or continental crust (Turkey–Iran and probably Himalayan arcs).

Uyeda and Kanamori (1979) further classify volcanic arcs in terms of tectonic style related to absolute velocities of the subducting and overriding plates. Those arcs in which the overriding plate retreats from the subduction zone are characterized by the presence of an actively spreading back-arc basin (e.g., Marianas, Tonga-Kermadex, Vanuatu, etc.), whereas those in which the overriding arc converges on, or is stationary with respect to, the subduction zone are characterized by compressional tectonics (e.g., Andes). In the former case, the volcanic arc is often characterized by a predominance of relatively primitive mafic magmas, particularly in the back-arc region. Such configurations are not restricted to intraoceanic arcs. The southern Cascades may represent early stages of development of a back-arc ensialic basin due to clockwise rotation of the arc (Magill and Cox, 1981; Oldow and Leeman, in prep.). Central America apparently represents more advanced stages of this configuration as dominantly mafic (basalt, basaltic andesite) Quaternary volcanoes there are situated in an extensional graben (cf., Pichler and Weyl, 1973). In contrast, earlier

Neogene volcanism there was more typical of an Andean style arc with abundant rhyolite ignimbrites, particularly in Guatemala and Honduras where Precambrian (?) basement is present; these rocks have more radiogenic Sr than the Quaternary volcanic rocks and appear to contain a significant component of crustal material (Pushkar et al., 1972; Pichler and Weyl, 1973). Note that data for the earlier rocks are not included in Tables I and II, so the values tabulated for Central America are not entirely representative of the entire arc. Back-arc and even intra-arc spreading is significant for some intraoceanic arcs. One example is Vanuatu (New Hebrides) where a relatively recent intra-arc basin has developed. Volcanic rocks in this area are dominantly basaltic (some mildly alkalic) and may not be entirely representative of subduction-related magmatism (cf., Colley and Warden, 1974). These examples suffice to illustrate that the nature of magmatism associated with a presumed subduction complex may evolve significantly with time as local tectonic regimes change in response to plate interactions.

Despite the many complexities alluded to, the limitations of the available data for virtually all volcanic arcs, and the published caveats against generalized petrogenetic models to account for such arcs (e.g., Arculus and Johnson, 1978), it is nevertheless instructive to examine the influence of crustal structure on the nature of their volcanic products.

A MODEL FOR MAGMA—CRUST INTERACTION

The apparent influence of crustal structure and composition on the nature of arc volcanic rocks seems well established, although of debatable significance judging from the extensive literature. Consideration of geologic, tectonic and geochemical data for some thirty volcanic arcs (Table I) provides the basis for a model of magma-crust interaction (Leeman, 1980). In this model, it is first assumed that, aside from differences in the degree of alteration of oceanic crust and in the proportion and composition of sediments subducted, magmatic arcs develop in similar ways. Young primitive intraoceanic arcs must represent a relevant end member, and in fact such arcs apparently differ little with respect to proportions of rock types (dominantly basaltic) or their geochemical characteristics. Isotopic data suggest that in these cases the source rocks could be asthenospheric mantle (wedge) with minor components of slab (including small amounts of pelagic sediment)-derived material (Perfit et al., 1980; Arculus and Johnson, 1981; Gill, 1981). It is further assumed that primitive arc magmas (basaltic) ascend mainly because of density contrast with mantle wall rocks. Beneath intraoceanic arcs, such magmas would ascend efficiently as diapiric blobs (Marsh, 1979b) to depths at which the magma - wall rock density contrast reached zero. For primitive basaltic magmas ascending beneath oceanic crust, isostatic equilibrium would be attained at shallow depths within a few kilometers of the surface, whereas beneath continental crust equilibrium would likely be attained near the crust—mantle interface or within the lower crust if it comprised granulite facies metamorphic rocks (Fig. 1). A major consequence of the density - filter effect of the crust (cf., Gill, 1973) is that primitive magmas stagnate at greater depths beneath continental relative to oceanic crust.

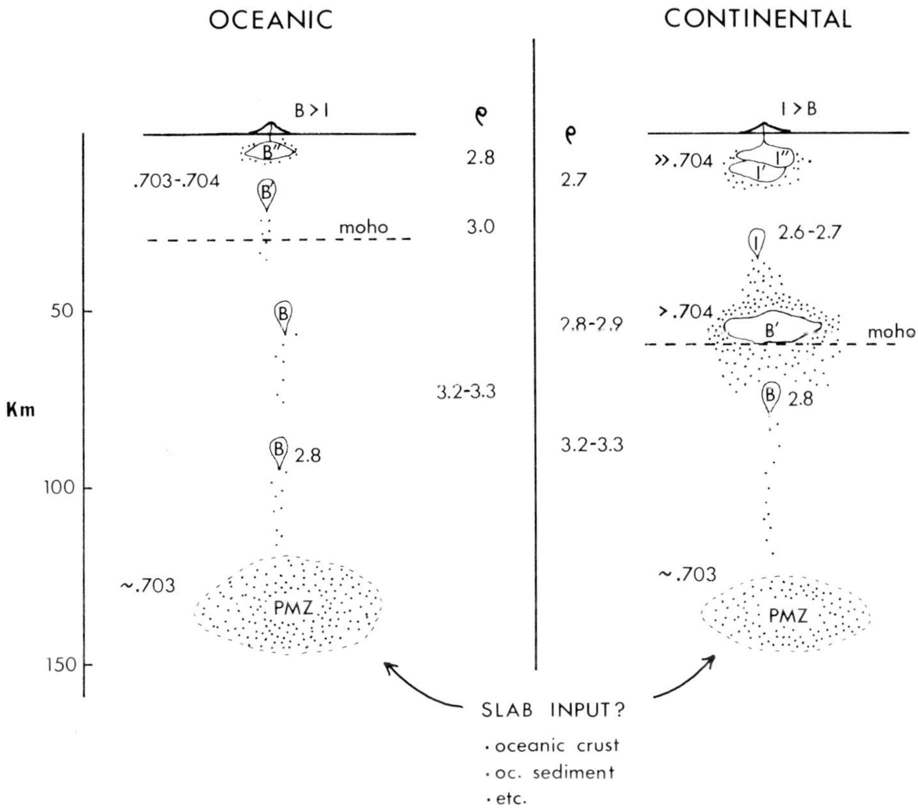

Fig. 1. Schematic model for magma ascent under two extreme cases: beneath oceanic or continental crust. Primitive basaltic magma (*B*) is generated in partial melt zones (*PMZ*) from similar source materials (perhaps mantle wedge with possible slab input) having $^{87}Sr/^{86}Sr$ ratios near 0.703. Basaltic magmas (with density [ρ] near 2.8 g/cm^3) rise through a denser mantle (ρ = 3.2—3.3) to levels where they reach isostatic equilibrium in shallow, mafic oceanic crust with low $^{87}Sr/^{86}Sr$ or in deep, high-grade metamorphic continental crust with slightly elevated $^{87}Sr/^{86}Sr$ (> 0.704). Upon stagnation, differentiation yields evolved magmas of tholeiite series (*B'*, *B''*, etc.) within oceanic island arc crust, and of calcalkaline series (*I*, or intermediate SiO_2) within continental crust. In the latter case, intermediate differentiates (ρ = 2.6—2.7) further ascend to shallow crustal levels where they stagnate and further evolve (*I'*, *I''*, etc.) and interact with wallrocks having elevated $^{87}Sr/^{86}Sr$ (>>0.704, depending on age of crustal rocks). Eruptive rocks would be dominantly basaltic (*B* > *I*) in oceanic island arcs, whereas intermediate rocks would dominate (*I* > *B*) in continental margin arcs.

In continental settings primitive magmas must evolve toward intermediate to silicic compositions, with corresponding reduction in density, before they could ascend to shallower levels or erupt at the surface. If these magmas are hydrous, as is commonly assumed, high-pressure crystallization would include amphibole (and perhaps biotite in more silicic magmas) as a significant near-liquidus phase. The significance of amphibole fractionation in producing calcalkalic magma series has been stressed by many authors (e.g., Eggler and Burnham, 1973; Cawthorn and O'Hara, 1976; Allen and Boettcher, 1978). Furthermore, petrographic evidence indicates that amphibole is indeed an important fractionated phase in many volcanic arcs. Anderson (1980) discusses evidence for amphibole fractionation from basaltic magmas, and numerous authors describe amphibole-bearing cognate xenoliths from both calcalkalic and alkalic series lavas; mineral geobarometers indicate that those from the Lesser Antilles represent cumulate assemblages derived from mafic magma chambers at depths of 10 to 30 km (Powell, 1978; Takahashi, 1978; Arculus and Wills, 1980; Conrad and Kay, 1980).

In oceanic settings primitive magmas will stagnate at shallower levels where amphibole is not a stable liquidus phase, and crystal fractionation will be dominated by anhydrous phase assemblages including one or more of the following: plagioclase, olivine, orthopyroxene, augite, and magnetite (so-called POAM fractionation of Gill, 1981). Accordingly, differentiation would produce tholeiitic series magmas which are characteristic of immature island arcs, but also are represented in more evolved arcs such as northeast Japan. As a given arc evolves with time, repeated influx of magmas must cause the crust to thicken, and because low-density, intermediate to silicic magmas would concentrate in the shallow crust, the depth of stagnation of primitive basaltic magmas must increase. In this way, immature island arc crust may evolve toward mature continental crust, given sufficient time and magma influx. Correspondingly, magma types erupted at the surface would initially comprise tholeiitic series lavas (mafic lavas dominant), and with time calcalkalic series lavas (intermediate to silicic lavas dominant) would predominate. In detail, this simple progression may not always be observed. For example, changes in tectonic conditions could result in intra-arc extension which would favor more efficient ascent of primitive magmas (Eichelberger, 1978a).

A very important corollary to this model is the thermal effect of magma influx into continental crust. As noted by Hodge (1974) and Patchett (1980), the repeated influx of hot basaltic magma into the crust could result in extensive anatexis of wall rocks. If the wall rocks are close to their solidus temperature to begin with (as is conceivably the situation where magma influx has proceeded for long times), the volume of anatectic melt produced could approach that of the mafic magma body that is emplaced. This process could result in genesis of voluminous silicic magma directly from crustal sources as is envisaged for mid-Tertiary Cordilleran volcanism in North and Central America, and more recently in the central Andes. Alternatively, mantle-derived magmas could assimilate significant quantities of such anatectic melts while simultaneously undergoing crystal fractionation (cf., Taylor, 1980; DePaolo, 1981). The effects of crustal assimilation or anatexis on isotopic compositions of erupted magmas could be slight or significant, depending on the age and composition of the crust. In old cratonic terranes (e.g., Andes, southwest Japan, North America Cordillera), the erupted magmas could have elevated $^{87}Sr/^{86}Sr$ and lower $^{143}Nd/^{144}Nd$ than those observed in immature intraoceanic arcs. Pb isotopic ratios could be more or less radiogenic, depending on whether the magmas interacted with upper or lower crustal rocks. In younger accretionary continental terranes (e.g., southwest Alaska, Cascades, western Mexico), comprising fragments of Phanerozoic oceanic or magmatic arc tectonic slivers, the crustal rocks and subduction-derived magmas may differ only little in isotopic composition. Hence, volcanic arc magmas could interact significantly with such crust and still retain isotopic compositions close to those of immature intraoceanic arcs (cf., Kosco, 1981; Myers and Marsh, 1981).

These points are illustrated in Fig. 2 which contrasts Sr and Pb isotopic trends for several volcanic suites either situated on old thick crust or for which subduction of craton-derived sediment has been inferred. Negative correlations of the isotopic ratios for San Juan basaltic andesites (Lipman et al., 1978) and Peruvian andesites and dacites (Tilton and Barriero, 1980) indicate that these lavas contain significant components of Sr and Pb from aged high-grade metamorphic rocks (James, 1981). Such rocks (granulites) are uniquely characterized by present-day $^{87}Sr/^{86}Sr$ ratios only slightly higher than modern oceanic volcanics, and their $^{206}Pb/^{204}Pb$ ratios are distinctly less radiogenic (cf., Doe, 1970). San Juan rhyolites presumably evolved in shallow crustal reservoirs where they interacted with old upper crustal rocks containing both radiogenic Sr and Pb (Lipman et al., 1978). The Banda volcanics (basalt to dacite) clearly reflect incorporation of radiogenic Sr and Pb from a similar upper crustal source which is thought to be sediments derived from the Archean craton of western Australia (Whitford et al., 1977; Morris and Hart, 1980). As an interesting comparison, two late Cenozoic magma suites from Mt. Taylor volcano, New Mexico (Leeman and Lipman, 1982), are shown in Fig. 2. Alkalic lavas (alkali olivine basalt to trachyte) some containing ultramafic xenoliths, presumably are derived more directly from mantle depths; they display no correlation between Sr and Pb isotopic compositions. Calcalkalic lavas (basalt to dacite) display a negative correlation similar to that for the San Juan basaltic andesites; presumably these magmas evolved in deep crustal magma chambers where they assimilated Sr and Pb from granulitic wall rocks. One rhyolite from this volcano has Pb and Sr isotopic ratios like those in San Juan rhyolites. Some of these suites have $^{207}Pb/^{204}Pb$ ratios significantly higher than those observed in pelagic sediments or ocean

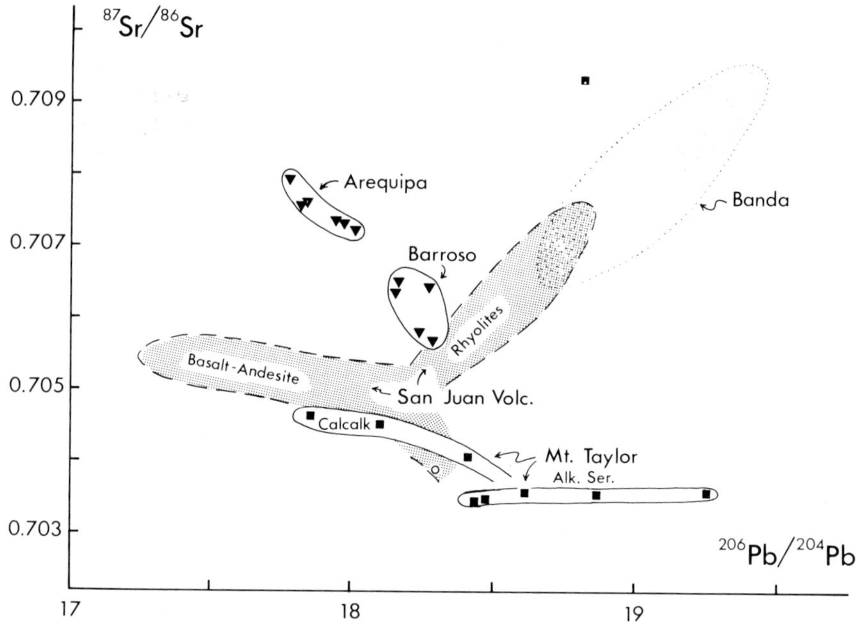

Fig. 2. Sr and Pb isotopic trends for volcanic suites from areas underlain by Precambrian crust (Mt. Taylor, New Mexico; San Juan Volcanics, Colorado; Arequipa and Barroso Volcanics, southern Peru), or involving subduction of sediments derived largely from Precambrian shield (Banda arc, Indonesia). Solid squares denote rocks from Mt. Taylor; open circle indicates a relatively primitive basaltic andesite from the San Juan Volcanics. See text for discussion and references.

floor and oceanic island volcanic rocks. Incorporation of old sialic rocks, or sediments derived therefrom, more plausibly explain isotopic trends like those in Fig. 2. Subduction of such materials seems required for the Banda arc, but in the other examples contamination of mantle-derived magmas or anatexis of old crustal rocks seems more plausible.

It is not argued that a crustal-interaction model alone can explain compositional variations in all arcs. Compositional inhomogeneities in the slab or wedge reservoirs also may contribute to compositional differences within or between volcanic arcs. However, consideration of hydrostatic principles does suggest that silicic arc lavas (andesite - dacite - rhyolite) generally represent evolved rather than primitive magmas. High-Mg andesites (boninites) may possibly be exceptional, but they are thought to form under special circumstances (e.g., fusion of water-rich subducted slab; Kushiro and Sato, 1978; Tatsumi and Ishizaka, 1981). With this rare exception, primary subduction-related magmas seem to be basaltic. If this is the general case, modern island arcs do not appear to be very close analogs for early crustal growth models as proposed by Taylor (1977).

For additional discussion of magma—crust interactions the reader is referred to papers by Eichelberger (1978b), Taylor (1980), DePaolo (1981) and Coulon and Thorpe (1981).

EVALUATION OF CRUSTAL INFLUENCE

A survey of island arc characteristics reported in Tables I and II allows a quantitative evaluation of the possible relations between crustal structure, tectonic factors, subduction geometry and kinematics, etc. and compositional features of arc volcanic rocks. In terms of the magma—crust interaction model just described, it is to be expected that the volumetric proportions of erupted rock types should be strongly correlated with crustal thickness and composition (i.e., the capacity of the crust to regulate magma ascent and thereby control conditions of crystallization). A detailed survey of crustal structure beneath volcanic arcs is beyond the scope of this paper; in any event, data on crustal structures are too limited at present to warrant such a survey (cf., Gill, 1981). Estimates of crustal thickness are available for portions of most volcanic arcs, but are of varied quality. No attempt is made here to weight these estimates according to their quality. Reported ranges in crustal thickness reflect both intra-arc variance and uncertainties in measurements; the median values are used in all computations and plots. Volumetric estimates have been reported for Japanese (Aramaki and Ui, 1978), Lesser Antilles (Baker, 1968; Brown et al., 1977), South Sandwich (Baker, 1968), and Cascades (McBirney, 1978) arcs and for the San Juan volcanics (Lipman, 1975). These estimates were made using different methods and assumptions, and cannot be directly compared; they are informative and indicate that the proportions of rock types do change significantly with time in some arcs. Because reliable volumetric estimates for rock types generally are not available, the proportions of rock types has been approximated by the number of chemical analyses within specific ranges of SiO_2 contents as follows: basalts (<53% SiO_2), basaltic andesites (53—57%), andesites (57—63%), dacites and rhyolites (>63%). Frequency distributions of these analyses are given for twenty-nine arc systems in Fig. 3, and the percentage of analyses with greater than 57% SiO_2 (andesite - dacite - rhyolite or ADR) is given in Table I. It is clear that this approach is subject to sampling biases, as is evident for Central America and Vanuatu as discussed earlier. However, it yields reasonable agreement with volumetric estimates for the Japan (except for southwest Japan) and South Sandwich arcs and the San Juan volcanic field. This method is certainly the most objective one for the present purposes. As with all approaches that average large numbers of data, there is no convenient means to express within-arc variance in composition, and in the present form the tabulation does not facilitate comparison of other elemental components. Despite these limitations, several important relations are apparent.

For example, K_2O contents may differ considerably between (or with-

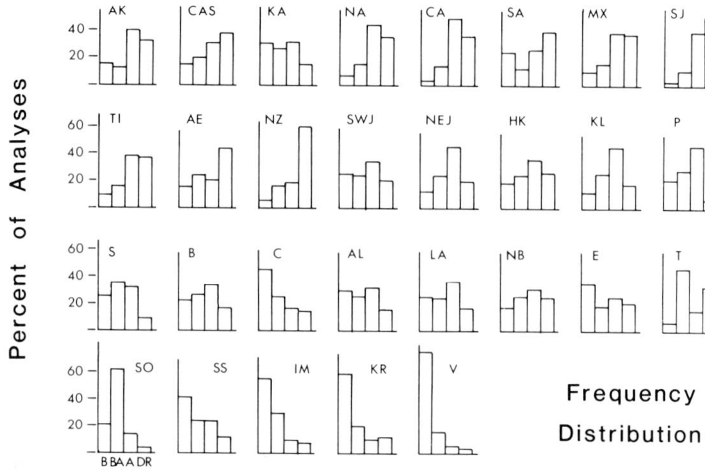

Fig. 3. Histograms of frequency distribution of 29 volcanic arcs. Class intervals correspond to B (<53% SiO), BA (53—57%), A (57—63%), DR (>63%). Arcs with continental basement in whole or part include: Alaska (*AK*), Cascades (*CAS*), Kamchatka (*KA*), Central America (*C*), N Andes (*NA*), Central Andes (*CA*), S Andes (*SA*), Mexico (*MX*), San Juan, Colorado (*SJ*), Turkey—Iran (*TI*), Aegean (*AE*), New Zealand (*NZ*), SW Japan (*SWJ*), NE Japan (*NEJ*), Hokkaido (*HK*), Papua (*P*), Sunda (*S*). Those presumably underlain by oceanic basement include: Kurile (*KL*), Banda (*B*), Aleutian (*AL*), Lesser Antilles (*LA*), New Britain (*NB*), Eolian (*E*), Tonga (*T*), Kermadec (*KR*), Solomon (*SO*), S Sandwich (*SS*), Izu—Mariana (*IM*), and Vanuatu (*V*). The Mogollon—Datil (*MD*) field (Oligocene) resembles the contemporaneous San Juan volcanics and is not shown. See text for further comments.

in) arcs. Mean K_2O values interpolated from SiO_2 variation diagrams at 57.5% SiO_2 range from 0.4 to 2.9, and although arcs with continental basement have higher K_2O values as a group compared with intraoceanic arcs, almost the entire range is observed in the latter arcs. Mean FeO/MgO ratios, also interpolated at 57.5% SiO_2 are systematically lower in continental margin arcs, and reflect a predominance of calcalkalic volcanic rocks (cf., Miyashiro, 1974). Neither of these averaged parameters (K_2O, FeO/MgO) displays a correlation with percent ADR. However, mean K_2O/Na_2O ratios computed for basaltic andesites and andesites both show significant correlations with percent ADR (Fig. 4). This ratio is considered to be a better index of differentiation than K_2O alone because it is not so strongly dependent on the composition or degree of melting of the magma source rocks. Yet K_2O is enriched relative to Na_2O in residual liquids during fractionation of clinopyroxene, amphibole, and plagioclase-bearing assemblages. The K_2O/Na_2O ratio displays a progressive increase with increasing proportion of ADR, and is highest for continental margin arcs; note that this relation is observed in rocks having restricted and comparable ranges in SiO_2 contents, but very different crustal structures.

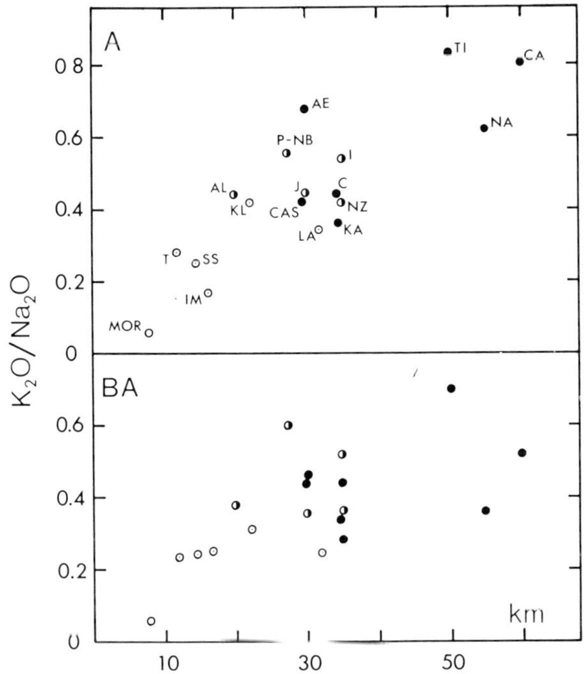

Fig. 4. Plot of estimated crustal thickness vs. K_2O/Na_2O ratios in basaltic andesites (BA) and andesites (A) based on values compiled by Sugisaki (1976) and Brown et al. (1977). Each point is labeled as in Fig. 3, except Indonesia (*I*), Alaska—Aleutian (*AL*) and Japan (*J*) arcs are combined. Values for mid-ocean ridge basalts (*MOR*) are shown for comparison. Symbols represent Group I arcs (●), Group II arcs with continental basement in part (◑), and Group II arcs with oceanic basement (○).

Figure 5 shows the relation between crustal thickness and percent ADR for the volcanic arcs or arc segments listed in Table I. The degree of correlation between these parameters is indicated by several least-squares regressions tabulated in Table III; logarithmic functions best fit the data, and, interestingly, intersect the abscissa at a reasonable thickness for normal oceanic crust. There is little difference between regression functions including volcanic arcs with or without continental basement; although correlation coefficients differ somewhat, all indicate a significant correlation at a high level of confidence. A clear relation exists between crustal thickness and proportion of evolved rock types as is predicted by the magma-crust interaction model.

Figure 6 shows the relation between crustal thickness and Sr isotopic composition based on the data in Table II. The lowest observed $^{87}Sr/^{86}Sr$ ratios (most primitive magmas) display a positive correlation with crustal thickness and this ratio is systematically higher for volcanic arcs with continental basement or for arcs where continental sediments are subducted. This relation is improved if old lower precision data for the Cascades is

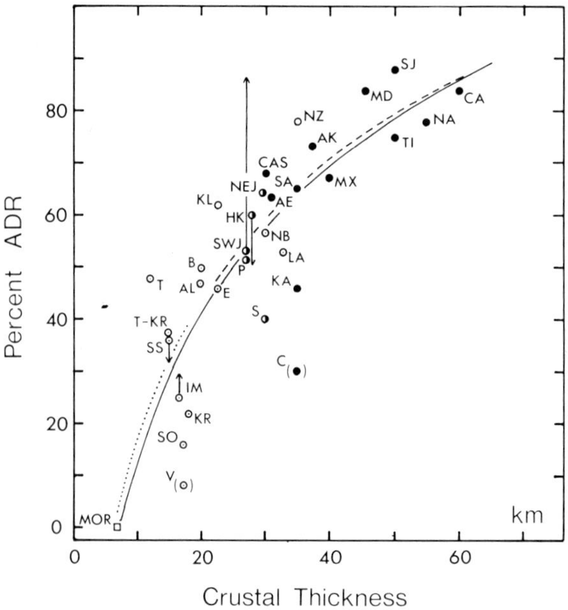

Fig. 5. Plot of estimated crustal thickness (median of ranges given in Table I) vs. percent ADR (andesite-dacite-rhyolite as defined in text). Solid curve is the least-squares logarithmic fit to all data, except Central Amercia, as given by fit 2 in Table III. Dashed and dotted curves represent fits 4 and 5, respectively. Symbols are keyed as in Fig. 3. Points of vertical arrows indicate volumetric estimates of percent ADR from Aramaki and Ui (1978) and Baker (1968). Points for Central America and Vanuatu may be non-representative for reasons discussed in the text.

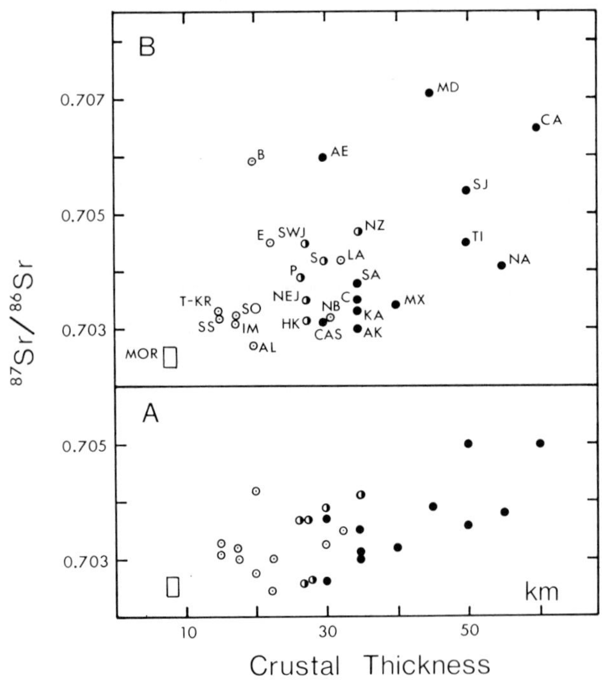

TABLE III

Least-squares fits to crustal thickness vs. percent ADR

Data set	No. Pts.	Fit[a]	Slope	Intercept	R
1. All arcs in Table I	30	lin	1.30	14.6	0.765
		log	39.34	−77.1	0.765
2. All arcs except Central America	29	lin	1.33	14.7	0.802
		log	40.85	−81.0	0.811
3. Excluding Central America, Vanuatu; Tonga—Kermadec combined	27	lin	1.24	18.8	0.829
		log	40.27	−78.2	0.854
4. All arcs with continental basement except Central America	17	lin	0.97	30.2	0.736
		log	39.90	−76.7	0.748
5. All arcs on oceanic basement except Vanuatu; Tonga—Kermadec combined	10	lin	1.59	9.2	0.664
		log	37.24	−69.5	0.691

[a]Slope (S) and intercept (I) parameters are given for linear ($y = Sx + I$) and logarithmic ($y = S \ln X + I$) fits; y = percent ADR, x = crustal thickness in km (median values of ranges given in Table I). The correlation coefficient (R) is significant for all fits at the 98% confidence level or better. Logarithmic fits are intuitively more realistic in that they project to zero percent ADR for standard oceanic crustal thickness (~7 km).

excluded (minimum $^{87}Sr/^{86}Sr$ best estimate is about 0.7029). Mean $^{87}Sr/^{86}Sr$ ratios depend to some extent on the number of radiogenic samples analyzed; these values scatter considerably, but generally show a positive correlation with crustal thickness and presence of continental basement as is predicted by the model. However, these correlations are surprisingly strong considering that the age of continental crust and its composition differ considerably for these arcs.

Nd and Pb isotopic analyses are too sparse at present to warrant similar treatment, but in general they too show expected correlations with crustal thickness as indicated for Pb in Fig. 2. Nd isotopic ratios for Izu-Mariana (Stern, 1981), Aleutians (Langmuir and Goldstein, 1981), South Sandwich (Hawkesworth et al., 1977), Lesser Antilles (Hawkesworth et al., 1979; Hawkesworth and Powell, 1980), and New Britain (DePaolo and Johnson, 1979) resemble those for mid-ocean ridge tholeiites and some ocean island basalts, whereas Andes (cf., James, 1981), Indonesia (Whitford et al., 1981),

Fig. 6. Plot of estimated crustal thickness vs. initial $^{87}Sr/^{86}Sr$ ratios in volcanic rocks: (A) Lowest observed ratios, usually in lowest SiO_2 rocks; (B) Mean of all observed ratios (all rock types). Symbols are the same as in Fig. 3. Ratios typical of normal (N-type) mid-ocean ridge basalts are indicated for comparison.

Japan (Nohda and Wasserburg 1981) and New Zealand (Perfit et al., 1981) volcanic rocks have Nd isotopic ratios that trend toward those of old continental materials.

Finally, the tabulated compositional features of volcanic arc magmas show little or no correlation with convergence rates or geometric parameters related to subduction (cf., Coulon and Thorpe, 1981; Gill, 1981), although such factors appear to influence the tectonic and geologic evolution of the arcs.

CONCLUSIONS

Proportions of rock types and their major-element and isotopic compositions display strong correlations with crustal structure that support the view that volcanic arc magmas interact with crustal wall rocks during their ascent and evolution. Some of the erupted magmas in continental margin arcs may represent crustal melts. In some arcs (Banda, New Zealand, Aegean) a crustal signature may be imparted to the magmas by subduction of continent-derived sediments. Dominantly mafic magmas with more primitive isotopic compositions are characteristic of immature intraoceanic arcs. Even these magmas may have evolved substantially, prior to eruption, via crystallization and wall rock interactions. However, the mantle and crustal wall rocks in these settings are so similar in composition to the subduction-related magmas that contamination may be hard to detect. By analogy with immature arcs, it is suggested that basaltic magmas are equally important in continental margin arcs. But because they cannot ascend efficiently through low density continental crust they erupt relatively infrequently unless extensional tectonics prevail. Instead, they evolve by protracted crystallization at lower crustal depths and provide heat to generate significant volumes of crustal melts which may either be assimilated by the mantle-derived magmas or ascend as distinct magma bodies. Of course, similar processes also may operate to some extent in the upper mantle portion of the lithosphere. Wall rock interactions at this level may be hard to detect, and could be significant in extent, if the mantle rocks differ little in isotopic composition from the ascending primitive magmas. However, if this is the case, it would be difficult to account for the crustal isotopic signature in continental margin arc volcanics by mantle processes alone.

As a final comment, it is stressed that the treatment in this paper is based on generalized concepts and averaged data that may not be entirely representative. The model and discussions presented are intended to focus attention on the probability that volcanic arc magmas are seldom primary; rather, most have evolved to some extent during ascent. Until the evolutionary processes are addressed and evaluated adequately, it will be impossible to arrive at unambiguous models of magma genesis and mantle evolution.

The experience of compiling the data used in this paper has revealed the rather haphazard and incomplete characterization of most volcanic arcs. Clearly, much more and better data are needed for crustal structure, and more systematic sampling, analysis, and description of the volcanic rocks and their geologic relations should be a goal for the future. Finally, we know very little about physical processes attending subduction, magma extraction and ascent, etc. and chemical processes involved in element distributions between silicate melts, crystalline phases, and fluids. These aspects are amenable to experimental and theoretical evaluation.

ACKNOWLEDGEMENTS

The work presented here has been supported by National Science Foundation Grant EAR79-19998. My data compilation was nearly completed when publications by Coulon and Thorpe (1981) and Gill (1981) appeared. I subsequently utilized much of the data tabulated in Gill's Appendix in order to reduce the number of references cited and to ensure that my own biases in the compilation were minimized. C. Coulon and S. Aramaki kindly provided me with some of their compiled material. I thank A. Walters for help with manuscript preparation. Finally, I thank I. Kushiro and S. Aramaki for their encouragement to submit this paper, and their patience when it arrived past their deadline.

REFERENCES

Allen, J.C. and Boettcher, A.L., 1978. Amphiboles in andesite and basalt II. Stability as a function of P-T-f_{H_2O}-f_{O_2}. Am. Mineral., 63: 1074—1087.

Anderson, A.T., Jr., 1980. Significance of hornblende in calc-alkaline andesites and basalts. Am. Mineral., 65: 837—851.

Aramaki, S. and Ui, T., 1978. Major element frequency distribution of Japanese Quaternary volcanic rocks. Bull. Volcanol., 41: 390—407.

Aramaki, S. and Ui, T., 1983. Alkali mapping of the Japanese Quaternary volcanic rocks. In: S. Aramaki and I. Kushiro (Editors), Arc Volcanism. J. Volcanol. Geotherm. Res., 18:549—560.

Arculus, R.J., 1981. Island arc magmatism in relation to the evolution of the crust and mantle. Tectonophysics, 75: 113—133.

Arculus, R.J. and Johnson, R.W., 1978. Criticism of generalized models for the magmatic evolution of arc trench systems. Earth Planet. Sci. Lett., 39: 118—126.

Arculus, R.J. and Johnson, R.W., 1981. Island-arc magma sources: a geochemical assessment of the roles of slab-derived components and crustal contamination. Geochem. J., 15: 109—133.

Arculus, R.J. and Wills, K.J.A., 1980. The petrology of plutonic blocks and inclusions from the Lesser Antilles island arc. J. Petrol., 21: 743—799.

Armstrong, R.L., 1981. Radiogenic isotopes: the case for crustal recycling on a near-steady-state no-continental-growth Earth. Philos. Trans. R. Soc. London, Ser. A, 301: 433—472.

Armstrong, R.L. and Cooper, J.A., 1971. Lead isotopes in island arcs. Bull. Volcanol., 35: 27—63.

Baker, P.E., 1968. Comparative volcanology and petrology of the Atlantic island-arcs. Bull. Volcanol., 32: 189—206.

Ben-Avraham, Z., Nur, A., Jones, B. and Cox, A., 1981. Continental accretion: from oceanic plateaus to allochthonous terranes. Science, 213: 47—54.

Bird, P., Toksöz, M.N. and Sleep, N.H., 1975. Thermal and mechanical models of continent-continent convergence zones. Geophys. Res., 80: 4405—4416.

Bornhorst, T.J., 1980. Major and Trace Element Geochemistry and Mineralogy of Upper Eocene to Quaternary Volcanic Rocks of the Mogollon-Datil Volcanic Field, Southwestern New Mexico. Ph.D. Thesis, Univ. New Mexico, 1108 pp.

Brown, G.M., Holland, J.G., Sigurdsson, H., Tomlin, J.F. and Arculus, R.J., 1977. Geochemistry of the Lesser Antilles volcanic island. Geochim. Cosmochim. Acta, 41: 785—801.

Burk, C.A., 1965. Geology of the Alaska Peninsula - Island Arc and Continental Margin. Geol. Soc. Am. Mem. 99.

Cawthorn, R.G. and O'Hara, M.J., 1976. Amphibole fractionation in calc-alkaline magma genesis. Am. J. Sci., 276: 309—329.

Colley, H. and Warden, A.J., 1974. Petrology of the New Hebrides. Geol. Soc. Am. Bull., 85: 1635—1646.

Conrad, W.K. and Kay, R.W., 1980. Petrology and significance of olivine and amphibole bearing xenoliths in Aleutian andesites. EOS, Trans. Am. Geophys. Union, 61: 400.

Coulon, C. and Thorpe, R.S., 1981. Role of continental crust in petrogenesis of orogenic volcanic associations. Tectonophysics, 77: 79—93.

DeLong, S.E. and Fox, P.J., 1977. Geological consequences of ridge subduction. In: M. Talwani and W.C. Pitman, III (Editors), Island Arcs, Deep Sea Trenches and Back-Arc Basins. Am. Geophys. Union, Maurice Ewing Series, 1: 221—228.

DePaolo, D.J., 1981. Trace element and isotopic effects of combined wallrock assimilation and fractional crystallization. Earth Planet. Sci. Lett., 53: 189—202.

DePaolo, D.J. and Johnson, R.W., 1979. Magma genesis in the New Britain island arc: Constraints from Nd and Sr isotopes and trace element patterns. Contrib. Mineral. Petrol., 70: 367—379.

Dickinson, W.R., 1975. Potash-depth (K-h) relations in continental margin and intra-oceanic magmatic arcs. Geology, 3: 53—56.

Dixon, T.H. and Batiza, R., 1979. Petrology and chemistry of recent lavas in the northern Marianas: Implications for the origin of island arc basalts. Contrib. Mineral. Petrol., 70: 167—181.

Doe, B.R., 1970. Lead Isotopes. Springer-Verlag, New York , N.Y., 137 pp.

Dunkley, P.N. and Hughes, G.W., 1981. Volcanism and the evolution of the ensimatic Solomon islands arc. Abstr. 1981 IAVCEI Symposium Arc Volcanism, Tokyo, pp. 76—77.

Eggler, D.H. and Burnham, C.W., 1973. Crystallization and fractionation trends in the system andesite - H_2O - CO_2 - O_2 at pressures to 10 kb. Geol. Soc. Am. Bull., 84: 2517—2532.

Eichelberger, J.C., 1978a. Andesitic volcanism and crustal evolution. Nature, 275: 21—27.

Eichelberger, J.C., 1978b. Andesites in island arcs and continental margins: Relationship to crustal evolution. Bull. Volcanol., 41: 480—500.

Ewart, A., Brothers, R.N. and Mateen, A., 1977. An outline of the geology and geochemistry, and the possible petrogenetic evolution of the volcanic rocks of the Tonga-Kermadec-New Zealand island arc. J. Volcanol. Geotherm. Res., 2: 205—250.

Gill, J.B., 1981. Orogenic Andesites and Plate Tectonics. Springer-Verlag, New York, N.Y, 385 pp.

Gill, R.C.O., 1973. Mechanism for the salic magma bias of continental alkaline provinces. Natura Phys. Sci., 242: 41—42.

Hamilton, W., 1979. Tectonics of the Indonesian region. U.S. Geol. Surv. Prof. Pap., 1078.

Hawkesworth, C.J. and Powell, M., 1980. Magma genesis in the Lesser Antilles island arc. Earth Planet. Sci. Lett., 51: 297—308.

Hawkesworth, C.J., O'Nions, R.K., Pankhurst, R.J., Hamilton, P.J. and Evensen, N.M., 1977. A geochemical study of island-arc and back-arc tholeiites from the Scotia Sea. Earth Planet. Sci. Lett., 36: 253—262.

Hawkesworth, C.J., O'Nions, R.K. and Arculus, R.J., 1979. Nd and Sr isotope geochemistry of island arc volcanics, Grenada, Lesser Antilles. Earth Planet. Sci. Lett., 45: 237—248.

Hodge, D.S., 1974. Thermal model for origin of granitic batholiths. Nature, 251: 297—299.

Ito, E. and Stern, R.J., 1981. Oxygen and strontium isotopic investigations on the origin of volcanism in the Izu-Volcano-Mariana arc. Carnegie Inst. Washington Yearb., 80: 449—455.

Jakes, P. and White, A.J.R., 1971. Composition of island arcs and continental growth. Earth Planet. Sci. Lett., 12: 224—230.

Jakes, P. and White, A.J.R., 1972. Major and trace element abundances in volcanic rocks of orogenic areas. Geol. Soc. Am. Bull., 83: 29—40.

James, D.E., 1981. The combined use of oxygen and radiogenic isotopes as indicators of crustal contamination. Annu. Rev. Earth Planet. Sci., 9: 311—344.

Karig, D.E. and Kay, R.W., 1981. Fate of sediments on the descending plate of convergent margins. Philos. Trans. R. Soc. London, Ser. A, 301: 233—251.

Kay, R.W., 1980. Volcanic arc magmas: implications of a melting-mixing model for element recycling in the crust-upper mantle system. J. Geol., 88: 497—522.

Kay, R.W., Sun, S.-S. and Lee-Hu, C.N., 1978. Pb and Sr isotopes in volcanic rocks form the Aleutian Islands and Pribolof Islands, Alaska. Geochim. Cosmochim. Acta, 42: 263—273.

Klerkx, J., Deutsch, S., Pichler, H. and Zeil, W., 1977. Strontium isotopic composition and trace element data bearing on the origin of Cenozoic volcanic rocks of the central and southern Andes. J. Volcanol. Geotherm. Res., 2: 49—71.

Kosco, D.G., 1981. The Mt. Edgecumbe volcanic field, Alaska: an example of tholeiitic and calc-alkaline volcanism. J. Geol., 89: 459—477.

Kurasawa, 1979. Isotopic composition of strontium in volcanic rocks from Fuji, Hakone, and Izu area, central Japan. Bull. Volcanol. Soc. Jpn., 24: 135—152.

Kushiro, I. and Sato, H., 1978. Origin of some calc-alkalic andesites in the Japanese islands. Bull. Volcanol., 41: 576—585.

Langmuir, C. and Goldstein, S., 1981. Constraints on arc petrogenesis from the Aleutians and Central America. Abstr. 1981 IAVCEI Symposium Arc Volcanism, Tokyo, pp. 198—199.

Leeman, W.P., 1980. Influence of crustal structure on compositions of subduction-related magmas. Geol. Soc. Am., Abstr. with Programs, 12: 470.

Leeman, W.P. and Lipman, P.W., 1982. Pb and Sr isotopic studies of Mt. Taylor volcano, New Mexico. EOS. Trans. Am. Geophys. Union, 63: 1143.

LePichon, X. and Angelier, J., 1979. The Hellenic arc and trench system: A key to the neotectonic evolution of the eastern Mediterranean area. Tectonophysics, 60: 1—42.

LePichon, X. and ANgeller, J., 1981. The Aegean Sea. Philos. Trans. R. Soc. London, Ser. A, 300: 357—372.

Lipman, P.W., 1975. Evolution of the Platoro Caldera complex and related volcanic rocks, southeastern San Juan Mountains, Colorado. U.S. Geol. Surv. Prof. Pap. 852, 128 pp.

Lipman, P.W., Doe, B.R., Hedge, C.E. and Steven, T.A., 1978. Petrologic evolution of the San Juan volcanic field, southwestern Colorado: Pb and Sr isotope evidence. Geol. Soc. Am. Bull., 89: 59—82.

Magaritz, M., Whitford, D.J. and James, D.E., 1978. Oxygen isotopes and the origin of high $^{87}Sr/^{86}Sr$ andesites. Earth Planet. Sci. Lett., 40: 220—230.

Magill, J. and Cox, A., 1981. Post-Oligocene tectonic rotation of the Oregon Western Cascade Range and the Klamath Mountains. Geology, 9: 127—131.

Marsh, B.D., 1979a. Island arc development: Some observations, experiments and speculations. J. Geol., 87: 687—713.

Marsh, B.D., 1979b. Island-arc volcanism. Am. Sci., 67: 161—172.

Matsuhisa, Y., 1977. Isotope geochemistry of an island arc traverse. Geochem. J., 11: 107—109.

Matsuhisa, Y., 1979. Oxygen isotopic compositions of volcanic rocks from the east Japan island arcs and their bearing on petrogenesis. J. Volcanol. Geotherm. Res., 5: 271—296.

Matsuhisa, Y., 1981. Oxygen isotopic characteristics of island-arc andesite. Abstr. 1981 IAVCEI Symposium Arc Volcanism, Tokyo, pp. 221—222.

McBirney, A.R., 1978. Volcanic evolution of the Cascade Range. Annu. Rev. Earth Planet. Sci., 6: 437—456.

Meijer, A., 1976. Pb and Sr isotopic data bearing on the origin of volcanic rocks from the Mariana Island arc system. Geol. Soc. Am. Bull., 87: 1358—1369.

Miyashiro, A., 1974. Volcanic rock series in island arcs and active continental margins. Am. J. Sci., 274: 321—355.

Molnar, P. and Atwater, T., 1978. Interarc spreading and cordilleran tectonics as alternates related to the age of subducted oceanic lithosphere. Earth Planet. Sci. Lett., 41: 330—340.

Morris, J.D. and Hart, S.R., 1980. Lead isotope geochemistry of the Banda arc. EOS, Trans. Am. Geophys. Union, 61: 1157.

Myers, J.D. and Marsh, B.D., 1981. Geology and petrogenesis of the Edgecumbe volcanic field, SE Alaska: the interaction of basalt and sialic crust. Contrib. Mineral. Petrol., 77: 272—287.

Nakamura, N., 1981. Nd and Sr isotopic variations in Quaternary volcanic rocks from northeast Japan. Abstr. 1981 IAVCEI Symposium Arc Volcanism, Tokyo, pp. 252—253.

Nohda, S. and Wasserburg, G.J., 1981. Nd and Sr isotopic study of volcanic rocks from Japan. Earth Planet. Sci. Lett., 52: 264—276.

Notsu, K., 1981. Strontium isotope composition in volcanic rocks from Japan. Abstr. 1981 IAVCEI Symposium Arc Volcanism, Tokyo, pp. 265—266.

Oversby, V.M. and Ewart, A., 1972. Lead isotopic compositions of Tonga-Kermadec volcanics and their petrogenetic significance. Contrib. Mineral. Petrol., 37: 181—210.

Pal, S., Lopez-M., Perez-R., J. and Terrell, D.J., 1978. Magma characterization of the Mexican Volcano Belt (Mexico). Bull. Volcanol., 41: 379—389.

Patchett, P.J., 1980. Thermal effects of basalt on continental crust and crustal contamination of magmas. Nature, 283: 559—561.

Perfit, M.R., McCulloch, M.T. and Froude, D., 1981. Sr- and Nd-isotopic variations in volcanic and plutonic rocks from the Aleutian islands and Taupo volcanic zone, New Zealand: Implications for island arc magma genesis. Abstr. 1981 IAVCEI Symposium Arc Volcanism, Tokyo, pp. 292—293.

Perfit, M.R., Gust, D.A., Bence, A.E., Arculus, R.J. and Taylor, S.R., 1980. Chemical characteristics of island-arc basalts: Implications for mantle sources. Chem. Geol., 30: 227—256.

Pe, G.G., 1975. Strontium isotope ratios in volcanic rocks from the northwestern part of the Hellenic arc. Chem. Geol., 15: 53—60.

Pe, G.G. and Gledhill, A., 1975. Strontium isotope ratios in volcanic rocks from the southeastern part of the Hellenic arc. Lithos, 8: 209—214.

Pichler, H. and Zeil, W., 1972. The Cenozoic rhyolite - andesite association of the Chilean andesites. Bull. Volcanol., 35: 1—29.

Pichler, H. and Weyl, R., 1973. Petrochemical aspects of Central American magmatism. Geol. Rundsch., 62: 357—396.

Powell, M., 1978. Crystallization conditions of low-pressure cumulate nodules from the Lesser Antilles island arc. Earth Planet. Sci. Lett., 39: 162—172.

Pushkar, P., McBirney, A.R. and Kudo, A.M., 1972. The isotopic composition of strontium in Central American ignimbrites. Bull. Volcanol., 35: 265—294.

Rea, W.J. and Baker, P.E., 1980. The geochemical characteristics and conditions of petrogenesis of the volcanic rocks of the northern Lesser Antilles — a review. Bull. Volcanol., 43: 325—336.

Shuto, 1974. Strontium isotopic study of the Tertiary acid volcanic rocks from the southern part of northeast Japan, Sci. Rept. Tokyo Univ. Education, 12: 75—140.

Smith, A.L., Roobol, M.J. and Gunn, B.M., 1980. The Lesser Antilles - A discussion of the island arc magmatism. Bull. Volcanol., 43: 287—302.

Soller, D.R., Ray, R.D. and Brown, R.D., 1981. A Global Crustal Thickness Map. Phoenix Corp., McLean, Virginia.

Stern, R.J., 1979. On the origin of andesite in the northern Mariana island arc: implications from Agrigan. Contrib. Mineral. Petrol., 68: 207—219.

Stern, R.J., 1981. A common mantle source for western Pacific island arc and "hotspot" magmas: Implications for layering in the upper mantle. Carnegie Inst. Washington Yearb., 80: 455—462.

Sugimura, A. and Uyeda, S., 1973. Island Arcs: Japan and its Environs. Elsevier, New York, N.Y., 247 pp.

Sugisaki, R., 1976. Chemical characteristics of volcanic rocks: relation of plate movements. Lithos, 9: 17—30.

Takahashi, E., 1978. Petrologic model of the crust and upper mantle of the Japanese island arcs. Bull. Volcanol., 41: 529—547.

Tatsumi, Y. and Ishizaka, K., 1981. Existence of andesitic primary magma: An example from southwest Japan. Earth Planet. Sci. Lett., 53: 124—130.

Tatsumoto, M., 1969. Lead isotopes in volcanic rocks and possible ocean floor thrusting beneath island arcs. Earth Planet. Sci. Lett., 6: 369—376.

Taylor, H.P., Jr., 1980. The effects of assimilation of country rocks by magmas on $^{18}O/^{16}O$ and $^{87}Sr/^{86}Sr$ systematics in igneous rocks. Earth Planet. Sci. Lett., 47: 243—254.

Taylor, S.R., 1977. Island-arc models and the composition of the continental crust. In: M. Talwani and W.C. Pitman, III (Editors), Island Arcs, Deep Sea Trenches and Back-Arc Basins. Am. Geophys. Union, Maurice Ewing Series, 1: 325—335.

Thorpe, R.S. and Francis, P.W., 1979. Variations in Andean andesite compositions and their petrogenetic significance. Tectonophysics, 57: 53—70.

Thorpe, R.S., Francis, P.W. and Harmon, R.S., 1981. Andean andesites and crustal growth. Philos. Trans. R. Soc. London, Ser. A, 301: 305—320.

Tilton, G.R. and Barreiro, B.A., 1980. Origin of lead in Andean calcalkalic lavas, southern Peru. Science, 210: 1245—1247.

Ui, T. and Aramaki, S., 1978. Relationship between chemical composition of Japanese Island-arc volcanic rocks and gravimetric data. Tectonophysics, 45: 249—259.

Uyeda, S. and Kanamori, H., 1979. Back-arc opening and the mode of subduction. J. Geophys. Res., 84: 1049—1061.

Watenabe, T., Langseth, M.G. and Anderson, R.N., 1977. Heat flow in back-arc basins of the Western Pacific. In: M. Talwani and W.C. Pitman, III (Editors), Island Arcs, Deep Sea Trenches and Back-Arc Basins. Am. Geophys. Union, Maurice Ewing Series, 1: 137—162.

Whitford, D.J., 1975. Strontium isotopic studies of the volcanic rocks of the Sunda arc, Indonesia, and their petrogenetic implications. Geochim. Cosmochim. Acta, 39: 1287—1302.

Whitford, D.J. and Jezek, P.A., 1979. Origin of late Cenozoic lavas from the Banda arc, Indonesia: Trace element and Sr isotope evidence. Contrib. Mineral. Petrol., 68: 141—150.

Whitford, D.J., Compston, W., Nicholls, I.A. and Abbott, M.J., 1977. Geochemistry of late Cenozoic lavas from eastern Indonesia: Role of subducted sediments in petrogenesis. Geology, 5: 571—575.

Whitford, D.J., White, W.M. and Jezek, P.A., 1981. Neodymium isotopic composition of Quaternary island arc lavas from Indonesia. Geochim. Cosmochim. Acta, 45: 989—995.

Yermakov, V.A., 1981. The role of volcanism in the transformation of the Earth's crust at continental margins with reference to Kamchatka. Tectonophysics, 77: 95—132.

Yokokura, T., 1981. On subduction of dip angles. Tectonophysics, 77: 63—77.

Journal of Volcanology and Geothermal Research, 18 (1983) 589—631
Elsevier Science Publishers B.V., Amsterdam — Printed in The Netherlands

ISOTOPE GEOCHEMISTRY AND ORIGIN OF CALC-ALKALINE LAVAS FROM A CALEDONIAN CONTINENTAL MARGIN VOLCANIC ARC

M.F. THIRLWALL

Department of Earth Sciences, University of Leeds, Leeds (United Kingdom)

(Received October 20, 1982)

ABSTRACT

Thirlwall, M.F., 1983. Isotope geochemistry and origin of calc-alkaline lavas from a Caledonian continental margin volcanic arc. In: S. Aramaki and I. Kushiro (Editors), Arc Volcanism. J. Volcanol. Geotherm. Res., 18: 589—631.

Analyses for major and trace elements, including REE, and Sr, Nd and Pb isotopes are reported from a suite of Siluro-Devonian lavas from Fife, Scotland. The rocks form part of a major calc-alkaline igneous province developed on the Scottish continental margin above a WNW-dipping subduction zone. Within the small area (ca. 15 km^2) considered, rock types range from primitive basalts and andesites (high Mg, Ni and Cr) to lavas more typical of modern calc-alkaline suites with less than 30 ppm Ni and Cr. There is a marked silica gap between these rocks (< 62%) and the rare rhyolites (> 74%), yet the latter can be generated by fractional crystallization from the more mafic lavas. In contrast, variation in incompatible element concentrations and ratios in the mafic lavas can not be generated by fractional crystallization processes. Increasing SiO_2 is accompanied by increasing Rb, K, Pb, U and Ba relative to Sr and high field strength elements, increasing LREE enrichment and increasing ϵ_{Sr} calculated at 410 Ma, and by decreasing HREE, Eu/Eu*, Sm/Nd and ϵ_{Nd} (410). ϵ_{Nd} and ϵ_{Sr} are roughly anticorrelated and have more radiogenic compositions than the mantle array, in common with data reported elsewhere from this part of the arc. The correlation extrapolates up to cross the mantle array within the composition field of the contemporary MORB source, and extrapolates down towards the probable compositional range of Lower Palaeozoic greywackes, which may form the uppermost 8 km of the crust, or may be supplied to the source by subduction. One sample, however, lies within the mantle array, and closely resembles lavas from northwestern parts of the arc, where a mantle source with mild time-integrated Rb/Sr and LREE enrichment has been inferred. The lavas have relatively high initial $^{207}Pb/^{204}Pb$ for their $^{206}Pb/^{204}Pb$, a feature which has been interpreted elsewhere as the result of incorporation of a sediment component into arc magmas. The systematic changes with increasing SiO_2 in isotopic and chemical parameters can be explained by mixing of a greywacke-derived component with depleted mantle. The various possible mixing mechanisms are discussed, and it is considered most likely that mixing occurred in the mantle source through greywacke subduction. The bulk of the Rb, K, Ba and Pb in the lavas is probably recycled from the crust, whereas less than some 40% of the Sr and Nd is recycled. The calc-alkaline chemical trends are solely a function of mixing with the sediment component.

0377-0273/83/$03.00

INTRODUCTION

Most of the igneous and tectonic activity in Britain in the Lower Palaeozoic can be linked to plate margin processes occurring during progressive closure of a NE—SW-trending ocean, Iapetus (Dewey, 1971; Wright, 1976). The final stages of closure in the Silurian and ?early Devonian were accompanied by the development of an accretionary prism (Leggett et al., 1979) and major calc-alkaline igneous activity on the northern continent in Scotland. The igneous activity comprises a suite of diorite-granite plutons, the Newer granite suite, which range in age from 435 to 392 Ma (Halliday et al., 1979), and a basalt-andesite-rhyolite volcanic suite associated with the continental sediments known as the Old Red Sandstone (ORS).

The Old Red Sandstone volcanic rocks crop out throughout Scotland between the Iapetus suture and the SW Highlands (Fig. 1). Thirlwall (1979, 1981) has described the petrochemistry of these rocks and has shown that they are typical of continental margin calc-alkaline volcanics. Somewhat unusual is the occurrence of basalts and andesites rich in Mg, Ni and Cr interbedded with more normal low-Mg calc-alkaline lavas, but these may be matched with high-Mg andesites in Japan and elsewhere (Tatsumi and Ishizaka, 1981). Many of these high-Mg lavas must have incompatible element concentrations little different from those of primary magmas. Thirlwall (1981) demonstrated that concentrations of Sr, Ba, K, P, LREE and the ratio La/Y in primitive lavas north of the Southern Uplands accretionary prism all increased markedly to the NW, away from the Iapetus suture. The geographic change was defined by three principal regions: the S and N Midland Valley and the SW Highlands (Fig. 1). This spatial chemical variation was compared with that described from many modern arcs, and a tectonic model for the final stages of Iapetus closure was proposed, based on the relatively precise location of the subduction zone that could be inferred from the chemical data.

As the spatial variation is defined by primitive lavas it is possible to constrain closely its genesis. Thirlwall (1981) demonstrated that variation in the extent of fractional crystallization or in the nature of the fractionating assemblage could not produce the spatial variation. The lack of spatial variation in Rb, Zr and Ti is not consistent with higher concentrations in the NW being the result of a smaller degree of partial melting, nor can the lack of spatial variation in Rb be explained by a contribution from subducted material increasing northwestward (Thirlwall, 1982a).

Sr and Nd isotope data have been reported for a wide selection of primitive ORS lavas by Thirlwall (1982a). Initial ratios and ϵ-parameters were calculated at 410 Ma, the best estimate for the age of ORS volcanism N of the accretionary prism. $^{87}Sr/^{86}Sr_{410}$ is everywhere low (< 0.705) and shows no spatial variation, but in contrast ϵ_{Nd} (410) is mostly lower in rocks from the SW Highlands than in those from the Midland Valley. The NW increase in some incompatible element concentrations associated with the decrease in ϵ_{Nd}, and especially a strong Sm-Nd pseudoisochron relationship, can not be explained by any process of crustal contamination. Thirlwall (1982a) concluded that a deep mantle source with mild time-integrated LREE enrichment was necessary to explain the SW Highlands data, and that the spatial chemical variation at least in part reflected the derivation of magmas from deeper, less LREE-depleted mantle at greater distances from the Iapetus suture.

Although Midland Valley samples are colinear with SW Highland samples on the Sm-Nd pseudoisochron and show considerable overlap, the two regions are completely separated on the ϵ_{Nd}-ϵ_{Sr} diagram. Samples from the SW Highlands lie within or slightly below the mantle array and contrast with Midland Valley samples which show a marked negative correlation, parallel to but offset to more radiogenic compositions than the mantle array. Thirlwall (1982a) tentatively ascribed this difference to a greater contribution from subducted lithosphere in the Midland Valley but noted that this was not easily reconciled with lower Sr/Ce than in the SW Highlands (cf. Hawkesworth and Powell, 1980).

Important problems which remain are the precise nature of the slab derived component and the relationships between interbedded high-Mg andesites and low-Mg basaltic rocks. It is possible that the unusual isotope geochemical relationships discussed by Thirlwall (1982a) are in part an artefact of considering together rocks from a large region. The present contribution reports chemical and isotope data from lavas in a very small area (ca. 15 km^2) of the eastern Midland Valley, at the mouth of the Firth of Tay, near Dundee (Fig. 1). This area was chosen because of the good coastal exposure, the wide range of rock types available and the relative abundance of petrographically fresh samples (Thirlwall, 1979).

The Old Red Sandstone lavas near Dundee were last described by Geikie (1902) and Harris (1928), although K-Ar ages of 400 ± 5 Ma (recalculated to modern constants) were reported by Evans et al. (1971) and a new edition of the Institute of Geological Sciences 1:50 000 sheet 49 (Arbroath) has recently been published. The rocks form part of the southern limb of the Sidlaw Anticline and dip at 10—20° to the SE. The base of the ORS sequence is not seen and the basement on which the ORS lavas rest is uncertain, but it probably includes Lower Palaeozoic sediments comparable to those of the accretionary prism further south, perhaps together with late Pre-Cambrian metasediments (Dalradian) and/or a ?Pre-Cambrian granulite facies complex (Graham and Upton, 1978). The oldest units in the present area comprise a highly altered rhyolite flow (OC117, Fig. 1) and petrographically similar, but much fresher, rhyolite boulders in a conglomerate (OC128, 126). This is overlain by some 1500 ft (500 m) (Geikie, 1902) of basalts and andesites with an intrusion of felsite (OC111). Lavas comprise almost all the Lower ORS sequence south of the Tay, but thin quickly to the north, where several poorly exposed lavas are intercalated with sandstones. A large suite of mafic intrusions, probably mainly sills and plugs, is present in the Dundee area (e.g. OC141). The ORS volcanics are overlain unconformably by sandstones and shales of the Upper Old Red Sandstone (Upper Devonian—Carboniferous) on both sides of the Tay.

AGE OF THE VOLCANICS

The Lower ORS has customarily been regarded as Lower Devonian in age on the basis of fossil fish and plant spores. Thirlwall (1981) pointed our that the K-Ar ages of Evans et al. (1971) were more consistent with a late Silurian age for the volcanic sequences. This is made more likely by the probability that the K-Ar ages are somewhat younger than crystallization ages because of K and Ar mobility during deuteric alteration. Biotite and plagioclase have therefore been separated by conventional magnetic techniques from OC128A, a rhyolite boulder from the lowest conglomerate seen in the present area, and from OC72, a small olivine-diorite body intruded higher in the sequence at Glenfarg, about 20 miles (32 km) SW of

TABLE I

Rb-Sr age determinations

OC128A	Rb	Sr	$^{87}Rb/^{86}Sr$	$^{87}Sr/^{86}Sr$
Whole rock	81.5	161.7	1.459	0.71388 ± 10
Plagioclase	14.87	612.7	0.0702	0.70462 ± 4
Biotite	513.0	36.81	41.25	0.94298 ± 8
3-point age: 429 ± 57 (MSWD = 144)				
Bi-Pl age: 406.5 ± 5.6			$^{87}Sr/^{86}Sr_{init.}$ = 0.70421 ± 4	
OC72	Rb	Sr	$^{87}Rb/^{86}Sr$	$^{87}Sr/^{86}Sr$
Whole rock	38.6*	647*	0.1725	0.70477 ± 3
Plagioclase	26.50	755.4	0.1015	0.70444 ± 2
Biotite	341.0	43.75	22.83	0.83619 ± 10
3-point age: 406 ± 20 (MSWD = 13)				
Bi-WR age: 407.3 ± 5.6			$^{87}Sr/^{86}Sr_{init.}$ = 0.70377 ± 4	

* = XRF analyses. Errors are 2σ, and have been multiplied by $\sqrt{MSWD}$ for the 3-point regressions. The high MSWD is a function of alteration of groundmass glass in OC128A, and of sericitization of feldspar in OC72 (see text). Analyses of the standards NBS607 and GSP1 gave the following results:

	Rb	Sr	$^{87}Sr/^{86}Sr$
NBS607	525.2	65.21	1.19792 ± 5
NBS607	523.1	64.90	1.19785 ± 18
NBS607*	523.9	65.5	1.20039 ± 20
GSP1	253.3	232.3	0.76839 ± 13

* = recommended.

the present area. Rb—Sr isotope analyses are reported for the separates and the whole rock powders in Table I. For both samples the 3-point isochron yields high MSWD, which is considered to be the result of alteration. Biotite and plagioclase are completely fresh in OC128A, but lie in an oxidised felsic groundmass, and the age of 407 ± 6 Ma calculated from this mineral pair is considered the most reliable. It is conceivable that the biotite has undergone some ^{87}Sr-loss to the groundmass, in which case this would represent a minimum age. OC72 is a very fresh rock containing fresh olivine, and it is felt that the slightly higher initial $^{87}Sr/^{86}Sr$ of the plagioclase (0.70385) is the result of minor sericitization. No significant difference is obtained between a bi-pl and a bi-WR age, however, and 407 ± 6 Ma is taken as the best estimate of the age of intrusion. These dates are within error of the 410 Ma age assumed for the ORS lavas by Thirlwall (1982a).

Errors on these Rb-Sr ages are quoted on the basis of 0.7% error on $^{87}Rb/^{86}Sr$. While this error may be appropriate on the absolute age, the relative error on Leeds biotite ages is substantially less, and probably of the order of ± 3 Ma at 400. The age of 407 Ma can not be high in the lower Devonian, and must represent late Silurian or possibly earliest Devonian (cf. timescale of McKerrow et al., 1980).

SAMPLING AND ANALYTICAL TECHNIQUES

Sample locations of the analysed Fife ORS lavas are given in Table II and most are shown on Fig. 1. Four samples from the area further west than Fig. 1 have been isotopically analysed (OC102, 103, 104 and 152), of which data from the last two are presented by Thirlwall (1982a). Exposure is much more restricted in this region, but chemical data from all samples collected east of Glenfarg are presented in Appendix 2, and are used in some of the diagrams. All samples were collected from massive, non-vesicular portions of lava flows.

Major- and trace-element analyses were performed using the Philips PW1450/20 X-ray fluorescence spectrometer of the Grant Institute of Geology, University of Edinburgh. XRF techniques used were summarized by Thirlwall (1981) and were described in detail by Thirlwall (1979). The techniques were devised to give very high precision and accuracy: 1 standard deviation XRF reproducibility for the analyses reported here are typically SiO_2 0.17%, Al_2O_3 0.06%, Fe_2O_3 0.05%, MgO 0.03%, CaO 0.02%, Na_2O 0.05%, K_2O 0.005%, TiO_2 0.003%, MnO 0.008%, P_2O_5 0.002%, LOI 0.05%, Ni 0.4 ppm, Cr 2 ppm, V 3, Sc 0.3, Cu 0.6, Zn 0.6, Sr 2.2, Rb 0.5, Zr 0.8, Nb 0.4, Ba 4.3, Th 1.5, La 1.0, Ce 1.2, Nd 0.9 and Y 0.4 ppm, and it is estimated that analyses are nearly this accurate. Confirmation of the accuracy is provided by the agreement between XRF and isotope dilution (ID) La, Ce and Nd mostly to within the 2sd XRF reproducibility (Table II), as the LREE are among the most difficult elements to analyze by XRF of those reported here. Pb was analysed using similar techniques on the Philips

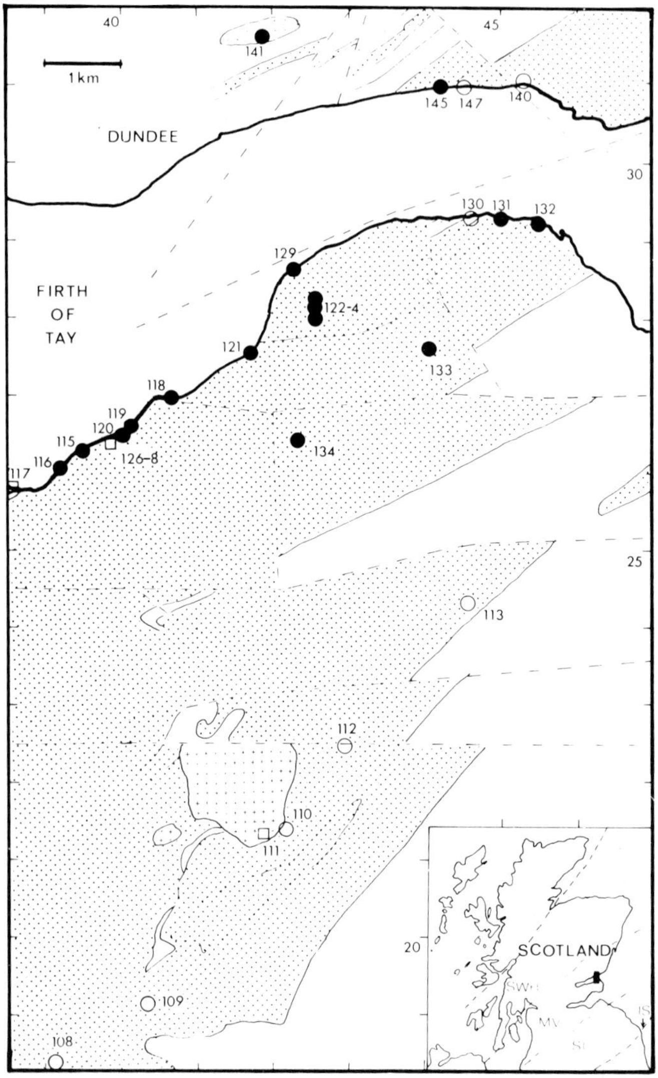

Fig. 1. Locality map of the lavas studied, shown as a black rectangle on the inset map of Scotland. Abbreviations: *SWH* = Southwest Highlands; *MV* = Midland Valley; *SU* = Southern Uplands (accretionary prism); *IS* = Iapetus suture. Distribution of rock types and faults from Institute of Geological Sciences 1:50 000 sheet 49 (Arbroath). Stippled ornament = lava flows; crosses = minor intrusions. • = high-Cr lavas, ○ = low-Cr lavas (Fig. 3), □ = rhyolites.

PW1400 XRF of the Department of Earth Sciences, Leeds University. Peak and background count times of about 25 min were used, giving precision of ca. ±0.2 ppm (twice standard error). Calibration was based on 15 samples analysed by isotope dilution ranging from picrite-basalt to rhyolite, including 6 samples from Table IV. The XRF and ID Pb determinations agree within analytical error.

Mineral analyses were performed using the Cambridge Instruments electron microprobe of the Grant Institute of Geology, University of Edinburgh. All phases were analysed using wavelength dispersive methods, except for the titanomagnetites which were analyzed using a Link Systems energy dispersive system.

REE were determined by isotope dilution using the highly precise method of Thirlwall (1982b), and all were determined on aliquots from the solution used for Nd isotope analysis. Complete solution at aliquotting was demonstrated by centrifugation. Sr and Nd isotope analyses were performed using the techniques of Thirlwall (1982a) using the VG-Micromass 30 and Isomass 54E mass spectrometers of the Department of Earth Sciences, University of Leeds.

Pb was separated in HBr on anion exchange columns, and was purified by anodic electrodeposition and loaded with silica gel and H_3PO_4 on single Re filaments. Blank levels for Pb were uniformly less than 4 ng, and are insignificant. Samples analyzed for isotopic composition were also analyzed for U, and six for Pb, by isotope dilution using ^{208}Pb and ^{235}U spikes. Pb was analyzed at $> 1100°C$ on the VG-Micromass 30 at Leeds, while U was analyzed using triple Re filaments on the Micromass 30. Data for the Pb standard SRM981 are reported in Table IV, and correspond to mass fractionation of about 0.06% per mass unit, a correction which has been applied to the sample analyses.

PETROGRAPHY AND MINERAL CHEMISTRY

Almost all of the ORS lava flows sampled in the Fife area are porphyritic. One or more of olivine, orthopyroxene, calcic clinopyroxene and plagioclase are present as phenocrysts (Table II), sometimes together with titanomagnetite. Hornblende is absent, and biotite and apatite are restricted to the infrequent rhyolites and acid intrusions. A variety of groundmass textures are present, from ophitic in some of the basalts to fresh brown glass in some andesites (OC98, 123, 145). All rocks show some signs of deuteric alteration: olivine is rarely fresh, but pseudomorphed by iddingsite, and orthopyroxene is often replaced by pale green chlorite. All samples analyzed for Sr isotopes have less than 2% loss on ignition, representing a normal volatile content for fresh volcanic rocks. In these samples, the slight alteration has probably had little effect on K, Rb, Sr concentrations and the Sr isotopes, for there are several systematic relationships between these and less mobile chemical parameters. The three samples with a fresh glass groundmass have anomalously low K and Rb (e.g. Fig. 4); this is thought to be a product of alkali leaching during hydration of the glass, as similar effects are seen in glassy andesites of the Cheviot Hills (Thirlwall, 1979). Pb is well correlated with SiO_2 (Fig. 4), except for two otherwise fresh samples (OC102, 119) which have anomalously high Pb, and U shows a very good correlation with Rb: Rb/U increases systematically from 19 to 23 over a range of 19 to 75 ppm Rb. This strongly suggests that U and Pb have not been modified by alteration in many rocks, and that the calculated initial Pb isotope ratios are meaningful, except for OC119.

Mean analyses of phenocryst minerals are presented in Appendix 1. Plagioclase is labradorite-andesine in the basalts and andesites, but compositional variation between zones of a single crystal is often as large as the total variation in mean compositions. The compositions of mafic phenocrysts from Fife lavas are plotted on Fig. 2, where they are distinguished by the Cr content of the host lava (Fig. 3). Olivine is only fresh in OC152 (bulk-rock analysis in Thirlwall, 1982a). It is magnesian (up to Fo_{86}) and normally zoned. The predicted olivine in equilibrium with a liquid of composition OC152 (assuming $Fe_2O_3/(Fe_2O_3 + FeO)$ of 0.2) is Fo_{91} (Roeder and Emslie, 1970), and it is thought that the anomalously Mg-rich nature of this sample is due to olivine accumulation (ca. 15% phenocrysts). Pyroxenes are Mg-rich, orthopyroxene being bronzite-hypersthene and clinopyroxene low-Na, -Ti, -Al diopsidic augite. Cr contents are often high. In general, the more Mg-rich pyroxenes occur in the Cr-rich lavas (Fig. 2), but some of the most siliceous lavas have very Mg-rich pyroxenes (e.g. OC123, chemically very

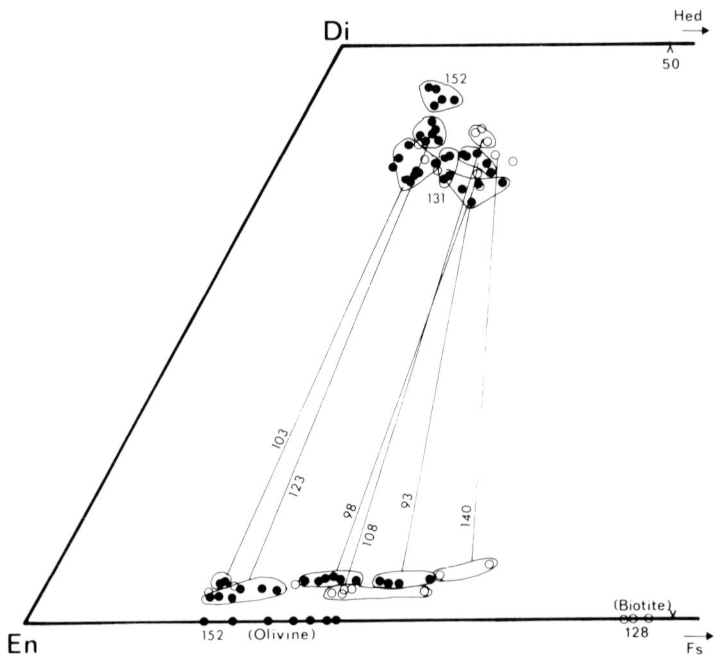

Fig. 2. Compositions of mafic phenocrysts in terms of atomic Mg (En), Fe (Fs) and Ca. Solid symbols for high-Cr lavas (Fig. 3). Tie-lines are between fields of coexisting pyroxenes.

similar to OC124). This can not be explained by titanomagnetite separation causing higher Mg/Fe in residual liquids, as the pyroxenes are also Cr-rich. Instead, the pyroxenes must have crystallised from primitive magmas. Coexisting pyroxene pairs from Cr-poor lavas (OC108, 140) indicate somewhat lower temperatures than in Cr-rich lavas (Fig. 2), consistent with the Cr-poor lavas being the products of more extensive high-level fractional crystallization. The 2-pyroxene temperature, calculated by the method of Wood and Banno (1973), broadly decreases with decreasing clinopyroxene and bulk rock Cr from 1150°C in OC103 to 980°C in OC140.

MAJOR- AND TRACE-ELEMENT CHEMISTRY

General features

Major- and trace-element analyses are presented in Table II for those samples for which isotope data are reported, and additional data are given in Appendix 2. The samples range from 50 to 79% SiO_2, but there is a pronounced silica gap between the dominant basalts and andesites with < 62% and the rare rhyolites and acid intrusions with > 74% (Fig. 3). The

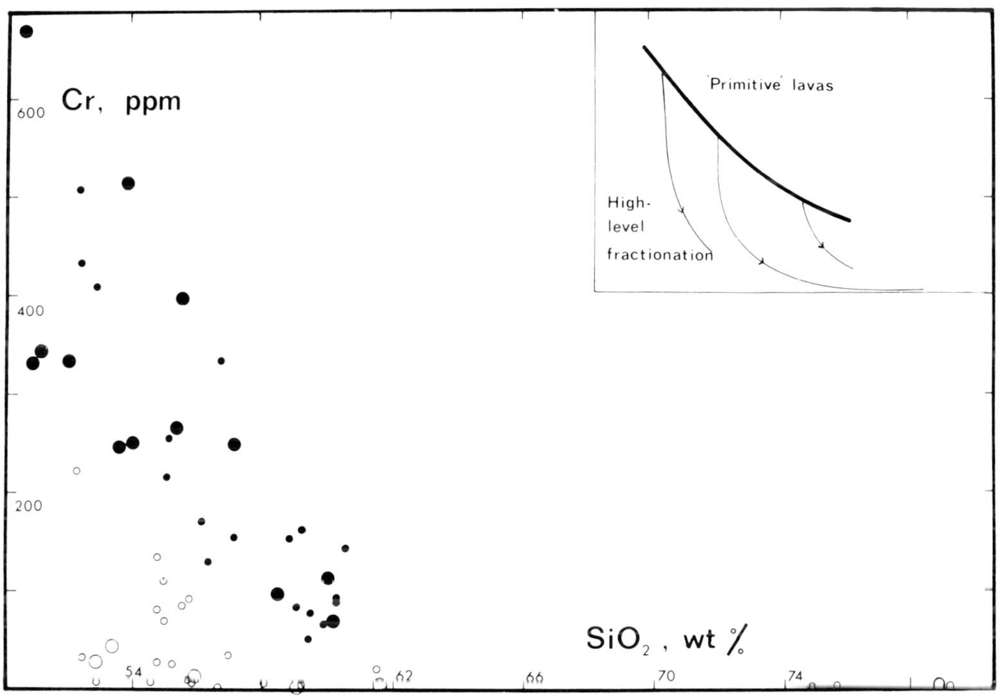

Fig. 3. Cr-SiO_2 variation diagram for all analyzed ORS lavas E of Glenfarg, defining high-Cr (relatively primitive) and low-Cr (fractionated) lavas. Large symbols are samples for which isotope data are presented.

TABLE II

Major- and trace-element compositions of Old Red Sandstone lavas

	OC102	OC103	OC109	OC110	OC112	OC115	OC116	OC118
SiO_2	50.84	55.55	61.56	53.33	55.90	53.80	54.00	51.97
Al_2O_3	17.51	15.73	17.27	18.58	18.71	16.81	16.86	17.19
Fe_2O_3	10.45	6.72	5.48	7.69	7.43	6.79	7.57	7.76
MgO	6.24	7.10	2.32	5.18	2.76	7.20	6.25	7.67
CaO	8.09	7.25	4.71	8.85	6.90	8.26	7.80	7.86
Na_2O	3.69	3.66	4.13	3.62	4.55	3.71	3.95	3.83
K_2O	1.032	1.916	2.616	1.011	1.644	1.415	1.398	1.019
TiO_2	1.510	1.158	1.013	1.163	1.542	1.277	1.413	1.657
MnO	0.080	0.085	0.099	0.132	0.110	0.125	0.142	0.132
P_2O_5	0.241	0.306	0.232	0.161	0.338	0.273	0.334	0.300
LOI	1.60	0.97	0.84	0.64	0.59	1.17	1.04	1.92
Ni	183	193	8	23	14	219	154	127
Cr	330	398	7	44	14	513	250	333
V	199	154	115	197	162	179	147	188
Sc	29.9	22.6	12.6	26.3	16.9	29.5	19.9	31.2
Cu	49	28	10	18	25	20	39	16
Zn	68	71	68	68	67	63	61	90
Pb	18.7	8.7	19.6	8.6	12.9	11.5	7.3	5.5
Zr	161	236	232	117	244	186	208	190
Nb	6.7	9.0	13.6	6.7	11.4	8.4	11.7	9.4
Ba	286	512	640	269	400	399	456	250
Th	3	5	12	3	6	4	3	n.d.
Nd"	20	28	30	17	31	23	27	24
Y	27.4	23.2	24.3	22.8	32.2	23.6	25.6	26.5
La	18"	25"	35.7	13.3	26.4	21.1	26.4	17"
Ce	37"	55"	69.8	30.1	58.1	46.3	55.2	42"
Nd			29.44	17.40	29.52	24.05	26.38	
Sm	—	—	5.472	4.093	6.215	5.071	5.327	—
Eu	—	—	1.468	1.346	1.889	1.567	1.623	—
Gd	—	—	4.75	4.25	5.99	4.74	4.98	—
Dy	—	—	4.15	3.90	5.40	3.94	4.28	—
Er	—	—	2.299	2.247	3.099	2.129	2.342	—
Yb	—	—	2.144	2.055	2.873	1.878	2.092	—
Pheno-crysts	pl-ol	pl-opx -cpx	pl-opx cpx-mt	pl-cpx -ol	pl-ol -cpx	ol-pl	pl	ol-pl
"%"	25	10	20	10	20	10	20	5
Grid ref.	NO358 252	NO363 254	NO404 192	NO422 214	NO430 225	NO393 262	NO389 258	NO407 270

Oxides in wt.%, trace elements in ppm. n.d. = not detected. Fe_2O_3 = total iron as Fe_2O_3, LOI = loss on ignition, " = La, Ce, Nd by XRF. OC141 is from a minor intrusion. Phenocrysts are listed in approximate order of abundance. Abbreviations: ol = olivine, cpx = calcic clinopyroxene, opx = orthopyroxene, pl = plagioclase, bi = biotite, ap = apatite, mt = titanomagnetite. Phases underlined are present as pseudomorphs only. "%" refers to the approximate volume percent phenocrysts.

OC119	OC124	OC128	OC129	OC131	OC133	OC140	OC141	OC147
51.10	60.15	78.86	55.32	59.99	57.12	52.80	58.45	59.11
17.03	16.52	12.02	17.43	16.18	15.91	17.94	16.50	17.88
9.92	6.12	0.76	6.78	6.82	7.18	8.97	6.53	6.20
6.50	3.55	0.16	5.30	2.83	5.84	4.95	4.22	2.90
7.64	5.89	0.45	7.15	5.16	6.71	7.66	5.81	4.72
4.02	4.11	4.57	4.24	4.13	3.65	4.16	4.17	4.78
1.124	2.114	3.128	1.570	2.440	1.674	1.233	2.204	2.356
1.601	1.058	0.109	1.389	1.225	1.078	1.593	1.142	1.278
0.103	0.113	0.038	0.089	0.061	0.115	0.149	0.091	0.160
0.296	0.257	0.044	0.258	0.391	0.245	0.281	0.268	0.300
1.46	0.88	0.51	1.09	0.79	0.82	0.61	0.96	1.00
139	31	4	103	83	127	29	46	7
342	69	4	266	113	249	28	96	4
193	123	8	147	144	125	178	113	133
32.1	17.8	0.6	23.0	18.3	19.1	21.7	17.5	15.3
27	18	5	22	19	22	21	23	9
76	64	13	64	73	67	89	65	88
19.8	15.1	22.5	10.9	23.1	10.7	10.1	13.2	20.1
184	224	94	197	231	209	209	220	229
10.6	14.0	16.7	11.9	14.8	14.7	11.1	13.4	14.2
285	552	791	356	844	389	367	620	616
2	12	10	7	7	9	2	10	11
23	33	21	25	37	27	27	29	33
30.5	23.4	21.3	25.2	20.3	24.2	30.2	24.5	25.7
19.1	35.9	18.9	26.4	44.7	31.8	22.9	33.1	32"
43.3	71.1	41.8	53.6	85.1	62.5	51.4	66.1	69"
24.03	30.15	17.56	24.94	35.80	27.22	27.39	29.12	
5.455	5.547	3.731	5.179	6.173	5.248	5.916	5.552	—
1.778	1.499	0.793	1.575	1.699	1.480	1.849	1.570	—
5.57	4.77	3.24	5.01	4.95	4.74	5.79	4.89	—
5.01	3.95	3.27	4.41	—	4.12	5.26	4.11	—
2.794	2.180	1.987	2.464	2.006	2.332	3.018	2.266	—
2.489	2.000	2.089	2.230	1.766	2.152	2.774	2.061	—
ol-pl	opx cpx-ol	pl-bi mt-ap	ol-mt	pl-opx ol-cpx	ol-pl	pl-ol cpx	ol-pl	pl
7	5	5	5	25	10	10	1	2
NO402	NO426	NO397	NO424	NO451	NO441	NO454	NO429	NO446
268	283	264	287	293	276	311	329	310

rocks are mostly quartz-normative, although some of the most magnesian have a little normative olivine. The rhyolites and some altered andesites are peraluminous, but this is probably a function of alkali loss during alteration. In common with other samples from the Midland Valley, they are typically calc-alkaline to high-K calc-alkaline (Thirlwall, 1981). Variation diagrams against SiO_2 show wide scatter (e.g. Figs. 3 and 4), although some elements show general decreases (Fe, Ca, Sc and V) or general increases (K, Rb, Ba, Pb, Th, La) with increasing SiO_2 (Fig. 4 and Thirlwall, 1979). Variation diagrams against MgO do not improve the scatter, although Ni and Cr show general decreases with decreasing MgO. Many samples have high Mg, Ni and Cr, and these include many andesites (Fig. 3). An arbitrary division into relatively low and high Cr samples is made on Fig. 3, and this will be used later to assess effects of fractional crystallization. Thirlwall (1981) inferred that ORS lavas with > 100 ppm Ni and > 150 ppm Cr were primitive, and represented primary magmas which had undergone less than 25% fractional crystallization of mafic minerals. Basaltic lavas with > 100 ppm Ni could be parental to the low-Cr, -Ni basalts and basic andesites, but fractional crystallization of the observed phenocryst phases does not provide sufficient silica enrichment to generate high-Cr, -Ni andesites from basaltic parents. These andesites, similarly, could not be parental to the low-Cr basalts. High-level fractionation of observed phenocryst phases could generate the low-Cr lavas from a wide variety of high-Cr parents produced at depth (Fig. 3), and could explain some of the scatter on variation diagrams.

Ni correlates well with Cr, suggesting that their concentrations are principally governed by fractional crystallization or accumulation of mafic phases, but not by accumulation of olivine alone. For example, accumulation of only 5% olivine with 0.22% NiO (OC152) would result in Ni increase of ca. 100 ppm, but little change in Cr. As most of the high-Ni, high-Cr lavas analyzed contain only olivine and plagioclase phenocrysts, and as many are only sparsely porphyritic (Table II), their primitive nature can not be explained by phenocryst accumulation. In OC152, olivine is associated with Cr-rich augite and magnetite phenocrysts, accumulation of which can account for the high Ni, Cr, Mg and low SiO_2 of this sample. Some primitive andesites have Ni-, Cr-rich bronzite and augite phenocrysts (analyses in appendices); in OC103 and 131 these are greatly subordinate to plagioclase phenocrysts, and if they were wholly accumulative could not account for more than 50% of the whole-rock Cr or more than 20% of the whole-rock Ni. This is not the case for OC124, but six similar samples have been analyzed, most from different flows, and these all have comparably high Ni and Cr for 60% SiO_2 (OC122, 123, 134—136, 138, analyses in Appendix 2). It is thought unlikely that all seven samples would have accumulated pyroxenes to the same extent. Further, the high Mg/Fe, Ni and Cr of the pyroxenes suggest crystallization from a highly primitive liquid. The primitive andesites must, therefore, represent liquid compositions which can not be generated by fractional crystallization of the observed phenocryst phases from the associated primitive basalts.

Low-Cr mafic lavas

Trends on variation diagrams plotted against SiO_2 become more clearly defined when only the high-Cr samples as defined by Fig. 3 are considered. These rocks show markedly higher Mg and Ni than low-Cr samples, higher Ca, Sc and lower Al, Fe, Na and Y for a given SiO_2 content (e.g. Fig. 4).

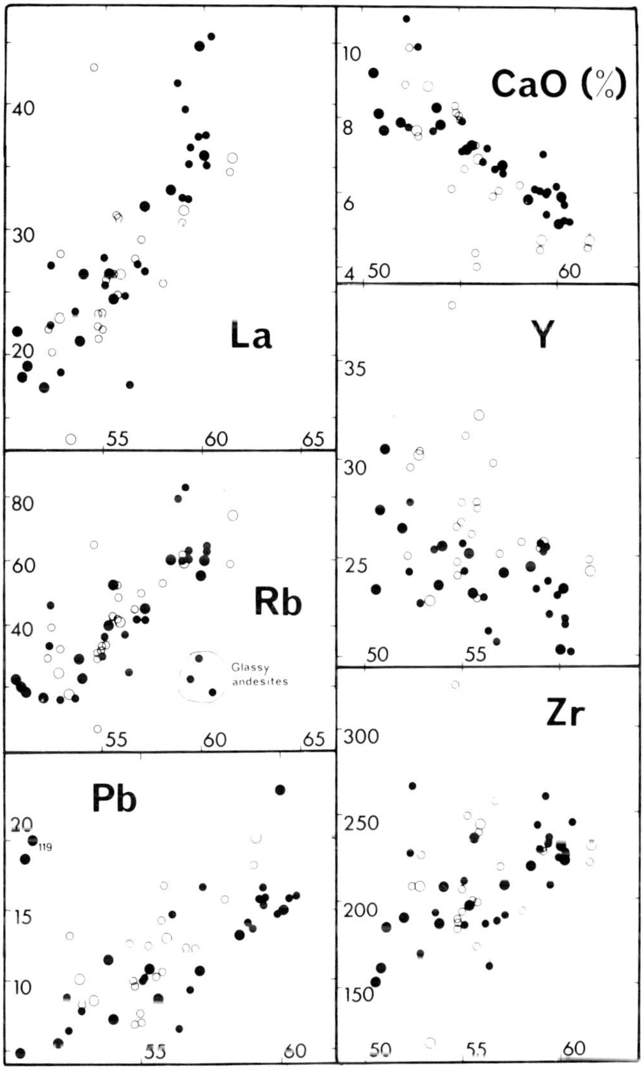

Fig. 4. Variation diagrams against SiO_2 for all analyzed lavas E of Glenfarg. Symbols as Fig. 3.

In the high-Cr samples, Mg, Ca, Sc, V and Ni decrease linearly with increasing SiO_2 while Al remains essentially constant. These chemical differences between high- and low-Cr samples are consistent with derivation of the latter from the former by some 20—30% fractionation of olivine and augite. As these, with plagioclase, are the low-pressure phenocryst phases it is probable that this was a high-level fractional crystallization process operating on a wide range of relatively primitive basalts and andesites produced at depth (cf. Fig. 3). Isotopic data presented later confirm that the low-Cr samples can be so produced.

Such a process provides little SiO_2 enrichment in the derivative low-Cr magmas, and is unable to change incompatible element ratios significantly. For example, La/Y correlates well with SiO_2 in all samples, with low-Cr lavas offset to slightly higher SiO_2 because of minor SiO_2 enrichment during the high-level olivine-augite fractionation (Fig. 5). Similarly, OC152 is offset to lower SiO_2 by olivine accumulation. Another example is given by Fig. 6, where both high- and low-Cr samples lie on the same tight correlations of Eu/Eu* with Rb/Sr and SiO_2. Again, OC152 is offset to lower silica.

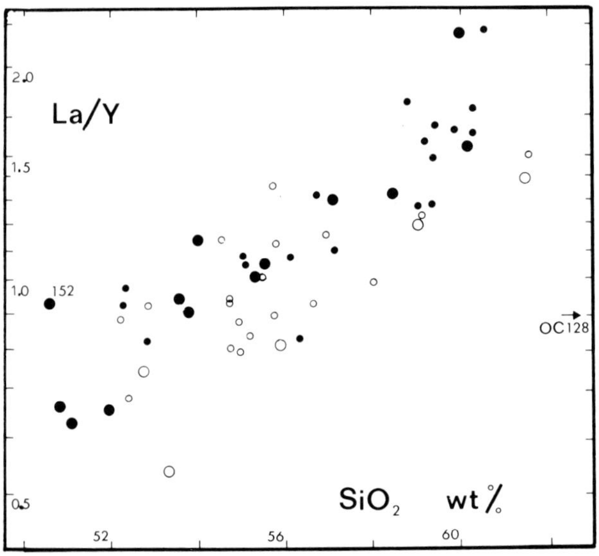

Fig. 5. La/Y vs. SiO_2 for all analyzed ORS lavas E of Glenfarg. Symbols as Fig. 3. The relatively higher SiO_2 of many low-Cr samples probably reflects high-level fractional crystallization.

Acid magmas

The rhyolites and acid intrusions are highly depleted in Fe, Mg, Ca, Ti, P, Ni, Cr, V and Sc, and moderately depleted in Zn and Al relative to the mafic lavas. OC128 has a pronounced negative Eu anomaly (Eu/Eu* = 0.7). These characteristics are indicative of extensive fractional crystallization

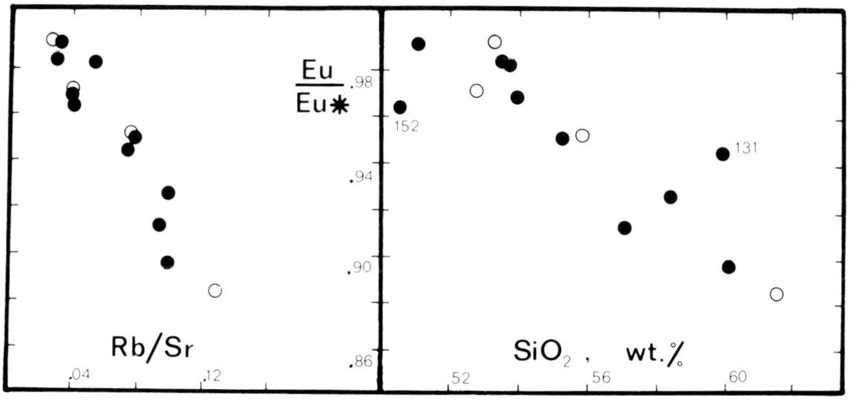

Fig. 6. Relationships of Eu anomaly to Rb/Sr and SiO_2. Symbols as Fig. 3. Eu/Eu* = (Eu/0.077) (Sm × Gd/0.203 × 0.276)$^{-1/2}$ OC131 has anomalously high Eu/Eu* and low Rb/Sr because it is transitional to SW Highland lavas.

of plagioclase, pyroxenes, titanomagnetite and apatite. Isotopic data presented later suggest that mafic lavas OC115 or 140 might represent parental magmas to OC128, and at least 50% crystallization is indicated by the increases in K, Rb, Ba, Pb and Nb concentrations. It is possible that biotite fractionation has occurred, and that therefore no element can be regarded as incompatible. The lower Zr/Nb (ca. 6.0) and Zr requires that the fractionating assemblage contains some 0.05% zircon, while the P depletion requires some 1% apatite. The presence of apatite, together with general increase in D_{REE} with increasing magmatic SiO_2, can explain the lower REE concentrations in OC128 and the stronger relative depletion in Gd and Dy, so that Yb > Er. The amount of zircon removed is insufficient to give rise to HREE depletion (even with D_{Yb}^{Zircon} = 300, Nagasawa, 1970).

High-Cr lavas

Many element concentrations in the relatively primitive lavas vary in rough linearity with SiO_2 (Fig. 4). In principal, this chemical variation could arise through fractional crystallization of a higher-pressure mineral assemblage than seen as phenocrysts, by varying degrees of partial melting of a common source or by some mixing process.

The hypothesis of high-pressure fractional crystallization can best be tested by using the chemistry of andesites such as OC124 and basalts such as OC118/119 to place constraints on the nature of the fractionating assemblage. The element which shows greatest increase over the silica range is Rb, with a 230% increase to OC124. OC124 would therefore have to represent at least 70% crystallization of OC118/119 liquid on a Rayleigh fractionation model (D_{Rb}^{bulk} = 0). Sr is essentially constant between 500 and 700 ppm in all samples, but shows a slight (10%) increase from OC118/119 to OC124. D_{Sr}^{bulk} would therefore be close to 1, implying that plagioclase would constitute about half the fractionating assemblage ($D_{Sr}^{plag} = 2$). This would be consistent with the good negative correlations of Eu/Eu* with Rb/Sr and SiO_2 (Fig. 6), which would imply crystallization in a mildly reducing environment.

The strong correlation between La/Y and SiO_2 (Fig. 5) implies marked steepening of the chondrite-normalized REE pattern with increasing SiO_2 (Y/Dy = 5.9 ± 0.1 in analyzed lavas). Taking $D_{La}^{Gt} = D_{La}^{Cpx} = 0.08$, $D_{Y}^{Cpx} = 0.66$ and $D_{Y}^{Gt} = 15$ (Thirlwall, unpubl. data) and assuming Rayleigh fractionation, the La/Y increase from OC118/119 to OC124 would require some 80% augite or 6% garnet fractionation. The figure of 80% augite fractionation is far in excess of that permitted by Ni and Cr concentrations in the andesites, and amphibole is equally unsuitable as an alternative.

In addition to Rb/Sr and La/Y, there is wide variation in other incompatible element ratios. Although La correlates well with SiO_2 (Fig. 4), and shows a large increase between OC118/119 and OC124, Nd is poorly correlated and only increases by 25%. Garnet does not provide sufficient separation between Nd and La to cause this difference. Zr and Nb show very small increases with SiO_2, and are only poorly correlated with each other. As distribution coefficients for these elements are closely similar it is very difficult to change the Zr/Nb ratio by limited fractional crystallization (Pearce and Norry, 1979). P_2O_5 is poorly correlated with SiO_2. This variation in incompatible element ratios requires the high-P phase assemblage to include several accessory minerals, such as apatite and a Zr-bearing phase.

In order to explain the trace-element chemistry of the high-Cr lavas by fractional crystallization it is therefore necessary to propose a high-P liquidus assemblage of ca. 50% plagioclase, 6% garnet, various accessory phases with olivine and pyroxenes to make up the remainder. Such a fractionating assemblage can not generate the SiO_2 enrichment of the andesites, for it does not contain sufficient low-SiO_2 phases. Hornblende and magnetite have been suggested as providing the silica enrichment in calc-alkaline suites (Osborn, 1962; Cawthorn and O'Hara, 1976); use of appropriate partition coefficients for Ni, Cr and V (Irving, 1978; Shervais, 1982) shows that in both cases silica enrichment from 50 to 60% would be accompanied by decline in Ni, Cr and, in the reducing conditions implied by the Eu data V concentrations, to an order of magnitude less than those observed in the primitive andesites. Further, the phase assemblage required by the incompatible trace-element data is very implausible, and most of the phases involved require to be totally resorbed at depth, as there is no petrographic evidence of garnet, apatite or Zr-bearing accessory phases. Phenocryst plagioclase is absent from several rocks with low Eu/Eu* and where present is often subordinate to high-level olivine and pyroxenes. It is concluded that the high-Cr andesites can not be produced by fractional crystallization of the basalts, as their silica and incompatible element enrichment can not be achieved without major depletion in Ni and Cr, and the wide variation in incompatible element ratios can not be produced without invoking implausible accessory phases.

It is conceivable that the high-Cr andesites were generated by fractional crystallization of basaltic magmas not seen as lavas. It is equally difficult to develop the SiO_2 enrichment without Ni, Cr and V depletion, however, and this mechanism can not explain the continuous chemical variation from primitive basalts to andesites. The latter objection also applies to derivation of the andesites by fractional crystallization of a primary andesite magma (e.g. Tatsumi and Ishizaka, 1981) produced by hydrous melting of the mantle.

The high Ni, Cr and Mg/Fe of the basalts requires, as observed by Thirlwall (1982a), that these were derived from a mantle source, followed by some 10—20% olivine fractionation during ascent. The primitive andesites also can not have been in equilibrium with mantle olivine, but even if they were derived from a primary andesite magma, this could not be generated from the same mantle source as the basalts. The mantle source for the primary andesite would have to be much more hydrous, and would have to show marked enrichment in Rb relative to elements such as Nd and Zr, unless these were retained in residual phases during melting. It is clear that the high-Cr lavas can not be generated by variable degrees of partial melting of a common source, even if olivine fractionation during ascent is permitted.

The chemical features of the high-Cr lavas can not be explained by mixing between the basalts and the rare acid magmas. The rhyolite OC128 is less enriched in many incompatible elements than some of the andesites and does not fall anywhere near the correlations between incompatible elements or ratios and SiO_2 in the high-Cr lavas (e.g. Fig. 5). The other acid magmas are similar to OC128.

Concentrations of several incompatible elements in representative analyzed lavas, normalized to estimated mantle (Hawkesworth et al., 1979), are plotted in Fig. 7. Thirlwall (1982a) discussed the behaviour of mean primitive ORS lavas (> 100 ppm Ni) from the three main regions of the ORS province using a similar diagram. Rocks from all three regions show enrichment in hydrophile elements (K, Rb, Ba) relative to REE and high field strength elements (HFSE — Zr, Nb, Ti) similar to that known from most modern arc lavas (e.g. Hawkesworth et al., 1979; Hawkesworth and Powell, 1980). The following important points may be noted from Fig. 7:

(1) The Fife lavas are not enriched in Sr relative to Nd and HFSE. This lack of enrichment was noted in the mean Midland Valley primitive lavas by Thirlwall (1982a): the relatively constant Sr/Nd and Sr/Zr through a wide range of SiO_2 and Zr implies that this is not a result of plagioclase fractionation.

(2) They are strongly enriched in K, Rb and also Ba, Pb and U relative to Sr, Nd and HFSE.

(3) They are strongly depleted in Ti, Y and the HREE.

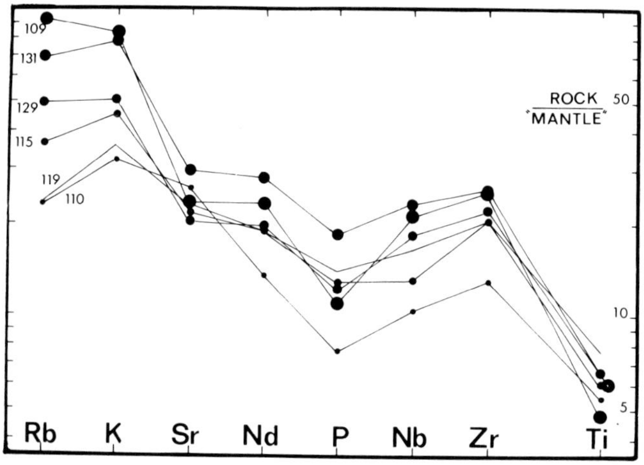

Fig. 7. Concentrations of incompatible trace elements of representative Fife ORS lavas normalized to an estimated mantle composition (after Hawkesworth et al., 1979: Rb = 0.8 ppm, K_2O = 0.0312%, Sr = 25 ppm, Nd = 1.26 ppm, P_2O_5 = 0.021%, Nb = 0.64 ppm, Zr = 8.9 ppm, TiO_2 = 0.21%). Increasing symbol size reflects increasing ϵ_{Sr} (410).

(4) Rb and K (with U, Pb and Ba) show the greatest range in concentrations, and except for the low-Cr basalt OC110, Sr, Nd and Zr concentrations have a very restricted range. The anomalously low REE and HFSE of OC110, in association with low Mg, Ni and Cr, require a parental magma with much lower concentrations of incompatible elements than any sample analyzed.

(5) The extent of enrichment in K, Rb, Pb and Ba relative to Sr, Nd and HFSE is correlated with increasing SiO_2 and La/Y, and decreasing Eu/Eu* and K/Rb.

These chemical data strongly suggest that the Fife lavas originated by mixing between a basaltic component and a siliceous component enriched in K, Rb, Pb, U, Ba and LREE. This was followed by shallow fractional crystallization of olivine and clinopyroxene to produce the low-Cr lavas, and further fractionation of plagioclase and minor phases to produce the acid magmas. The source of the siliceous component may best be identified using isotopic data.

Sr AND Nd ISOTOPE DATA

Sr and Nd isotope analyses are reported in Table III. Initial ratios are calculated as ϵ-parameters (DePaolo and Wasserburg, 1977) at 410 Ma to be consistent with the calculations of Thirlwall (1982a). An error of ± 20 Ma in the age results in trivial changes in ϵ_{Sr} and ϵ_{Nd}, however, as Rb/Sr and Sm/Nd are sufficiently close to UR and CHUR values respectively. ϵ_{Sr} shows a range from −7 to +14 ($^{87}Sr/^{86}Sr_{410}$ = 0.70352—0.70500) and ϵ_{Nd} from +5.2 to −1.5, ranges comparable to those reported by Thirlwall

TABLE III

Isotope data and Rb, Sr concentrations (XRF) for Old Red Sandstone lavas

	Rb	Sr	$^{87}Sr/^{86}Sr$	Sm/Nd	$^{143}Nd/^{144}Nd$	$^{148}Nd/^{144}Nd$	ϵNd	ϵSr
OC102	20.7	441	0.70468 ± 6*	—	—	—	—	−1.8
OC103	52.8	550	0.70592 ± 3*	—	—	—	—	+4.1
OC109	74.5	580	0.70707 ± 3	0.1859	0.512425 ± 8	0.241561 ± 8	+0.1	+12.6
OC110	18.6	658	0.70422 ± 2	0.2352	0.512729 ± 7	0.241588 ± 9	+4.4	−3.8
OC112	41.2	542	0.70532 ± 3	0.2105	0.512645 ± 5	0.241575 ± 6	+3.6	+0.3
OC115	29.3	543	0.70505 ± 4*	0.2109	0.512664 ± 7	0.241585 ± 10	+3.9	+1.8
OC116	23.5	584	0.70511 ± 1	0.2019	0.512485 ± 8	0.241563 ± 8	+0.7	+5.9
OC118	17.2	547	0.70405 ± 2	—	—	—	—	−7.0
OC119	18.9	573	0.70415 ± 2	0.2271	0.512756 ± 8	0.241567 ± 6	+5.2	−6.0
OC124	60.2	607	0.70664 ± 2	0.1840	0.512418 ± 11	0.241552 ± 6	0.0	+13.5
OC128	65.6	128	—	0.2125	0.512649 ± 8	0.241582 ± 8	+3.6	+3.0
OC129	39.8	510	0.70560 ± 5*	0.2077	0.512577 ± 6	0.241586 ± 6	+2.3	+3.9
OC131	55.5	745	0.70579 ± 3	0.1724	0.512322 ± 11	—	−1.5	+7.4
OC133	45.2	483	0.70594 ± 5*	0.1928	0.512556 ± 9	0.241595 ± 9	+2.4	+4.9
OC140	25.0	618	0.70440 ± 3	0.2160	0.512690 ± 6	0.241580 ± 8	+4.3	−4.1
OC141	60.5	610	0.70667 ± 2	0.1907	0.512455 ± 10	—	+0.5	+13.9
OC147	59.8	658	0.70570 ± 2	—	—	—	—	+2.1

ϵ-parameters are calculated at 410 Ma, using: $(^{143}Nd/^{144}Nd)_{CHUR,\ O} = 0.51264$, $(^{87}Sr/^{86}Sr)_{UR,\ O} = 0.7045$, $(^{147}Sm/^{144}Nd)_{CHUR} = 0.1936$, $(^{87}Rb/^{86}Sr)_{UR} = 0.0839$, $\lambda^{147}Sm = 6.54 \times 10^{-12}\ yr^{-1}$, $\lambda^{87}Rb = 1.42 \times 10^{-11}\ yr^{-1}$. Errors quoted on isotope analyses are 2 standard errors on the mean, and apply to the last digit(s) quoted. * = Sr analyzed on the MM30. Nd isotope ratios are normalized to $^{146}Nd/^{144}Nd = 0.7219$, $^{87}Sr/^{86}Sr$ to $^{86}Sr/^{88}Sr = 0.1194$.

(1982a) for Midland Valley lavas, but extended to lower ϵ_{Nd} and slightly higher ϵ_{Sr}. ϵ_{Sr} is negatively correlated with ϵ_{Nd}, but the correlation band is offset to more radiogenic compositions than that of modern mantle-derived volcanic rocks, the mantle array (Fig. 8). This offset is identical to that described by Thirlwall (1982a) from all Midland Valley primitive lavas, and is comparable to the offsets displayed by many modern arc suites (e.g. Hawkesworth et al., 1979; Hawkesworth and Powell, 1980). It is particularly striking that almost the whole isotopic variation of the Midland Valley lavas can be shown in an area of 15 km^2.

One analyzed sample (OC131) falls within the mantle array and is isotopically indistinguishable from SW Highlands samples (cf. Thirlwall, 1982a). This sample has the highest concentrations of Sr, Ba, P, LREE and the lowest HREE of any sample reported here, and thus closely corresponds in chemistry to many SW Highland lavas (e.g. L50, Thirlwall, 1982a). The occurrence of very rare SW Highland-type lavas in the Midland Valley implies that the mildly enriched mantle required to explain the isotope geochemistry of SW Highland lavas must also be available to supply magma in the Midland Valley. This further confirms the view that the mantle heterogeneity must be expressed as a vertical stratification, rather than as lateral heterogeneity.

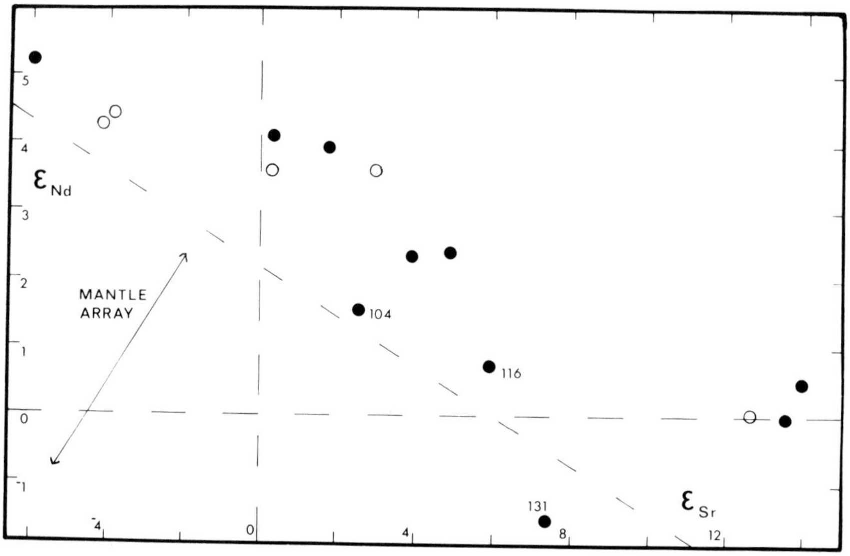

Fig. 8. ϵ_{Nd} (410) plotted against ϵ_{Sr} for Fife ORS lavas. Symbols as Fig. 3. The area beneath the diagonal line represents the region of modern mantle-derived volcanic rocks (the mantle array).

Two further samples (OC116 and 104) lie close to the mantle array; if these and OC131 are discounted the data form a general trend at a slight angle to the mantle array (Fig. 8). Such an inclined trend is much easier to account for than the apparent parallelism observed by Thirlwall (1982a). It may reflect mixing of Nd and Sr derived from a depleted mantle source with crustally derived Nd and Sr, perhaps from Lower Palaeozoic sediments either in the crust or subducted beneath the arc. Further mixing with Nd and Sr derived from a source similar to the SW Highlands enriched mantle might then account for the samples lying closest to the mantle array.

It is clear from Fig. 8 that the mafic rocks with low Cr and Ni are isotopically identical to high-Cr lavas. The rhyolite OC128 has ϵ_{Nd} identical to primitive lavas, and combined with the ϵ_{Sr} value from plagioclase separated from a neighbouring boulder (Table I) plots on the trend defined by more mafic lavas. These data confirm that the evolved lavas can be derived by fractional crystallization from the more primitive lavas.

The basalts and andesites plotted on Fig. 7 are distinguished by their ϵ_{Sr} values. It is clear that ϵ_{Sr} shows a progressive increase with Rb and K concentrations, and less clearly, also with Nd, Nb and Zr. No clear relationship is shown with Sr. A broad pseudoisochron relationship therefore exists between Rb/Sr and ϵ_{Sr}, whose slope corresponds to an "age" of 800 ± 90 Ma (2σ error multiplied by $\sqrt{MSWD}$). Better correlations exist between ϵ_{Sr} and Ba/Sr, SiO_2 and LREE enrichment (e.g. Fig. 9a). The last correlation is shown by samples analyzed throughout the north Midland Valley, and is strong evidence that ϵ_{Sr} has not been grossly modified by alteration. ϵ_{Nd}

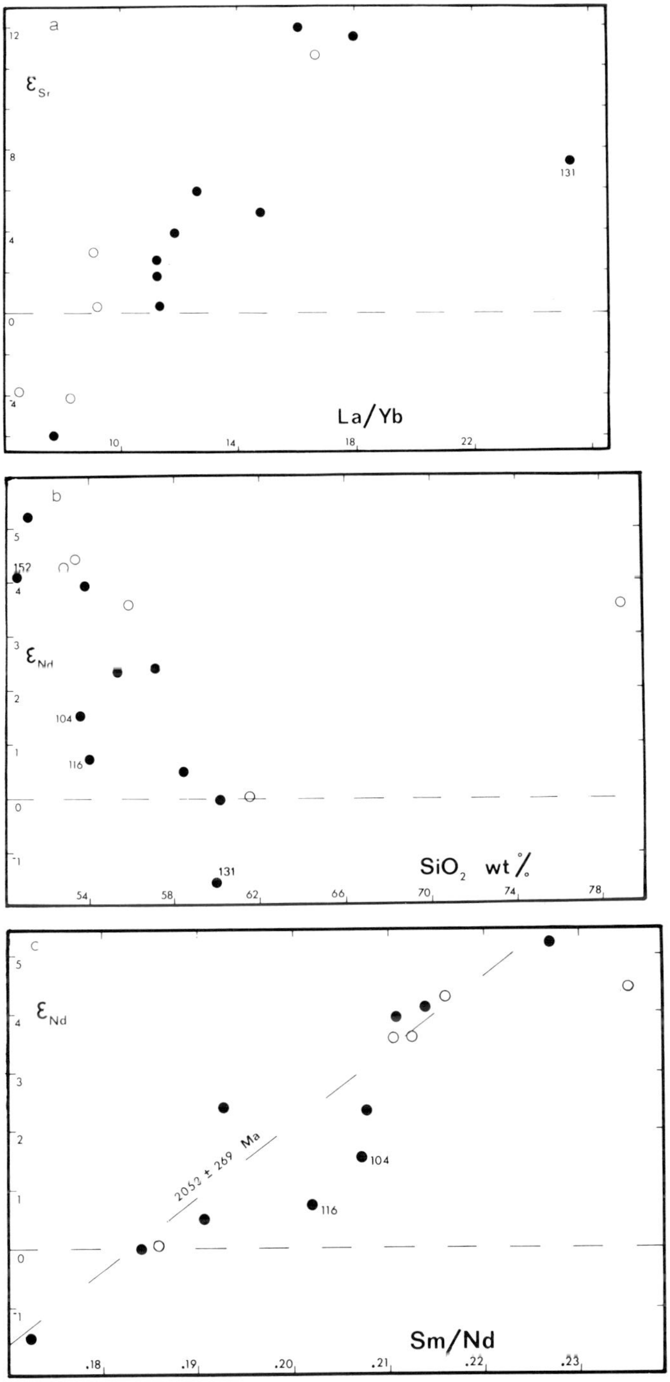

Fig. 9 a—c. Relationships between chemical and isotopic parameters in Fife ORS lavas. Note anomalous behaviour of OC116, 104 and 131. Symbols as Fig. 3.

similarly shows negative correlations with all the above parameters, for example with SiO_2 (Fig. 9b). The correlation with SiO_2 is much improved by omitting samples OC104 and 116 again, and allowing for the low SiO_2 in OC152 produced by olivine accumulation. ϵ_{Nd} forms a pseudoisochron relationship with Sm/Nd (Fig. 9c). Again omitting OC104 and 116, and also OC110, the low-Cr sample with anomalously low REE and HFSE concentrations, the slope corresponds to an "age" of 2053 ± 269 Ma, with initial ϵ_{Nd} (2050) of +17.6. This pseudoisochron is colinear with that obtained from all ORS lavas by Thirlwall (1982a).

The low-Cr samples all fall on the same correlations between isotopic and chemical parameters as the high-Cr samples, except for the high Rb/Sr of the rhyolite OC128 produced by plagioclase fractionation. The high-level fractionation process that generated the low-Cr lavas therefore modified neither incompatible element nor isotope ratios. This implies that the low-Cr lavas are a product of closed system fractional crystallization, with no concurrent wallrock assimilation.

The Sr and Nd isotope data for basalts and andesites, therefore, confirm the conclusion that the silica enrichment in this calc-alkaline suite is not a function of fractional crystallization. The primitive andesites clearly require a source isotopically distinct from that which yielded the associated basalts, and can not be explained by the hydrous mantle-melting hypothesis of Tatsumi and Ishizaka (1981). In summary, the Fife ORS basalts and andesites show progressive increase in ϵ_{Sr}, Rb/Sr, Rb/K, Ba/Sr, La/Y, Rb, U, Pb, K, Ba, La and less clearly Nd and Zr with increasing SiO_2, and progressive decrease in ϵ_{Nd}, Sm/Nd and Eu/Eu*. More siliceous samples are more offset to more radiogenic compositions than the mantle array. Sr concentrations show little change.

Pb ISOTOPIC DATA

Pb concentrations are high in the Fife ORS lavas, and unlike Sr and Nd, increase markedly with increasing SiO_2 (Fig. 4). Pb isotope analyses for eight samples are presented in Table IV, with U and Pb concentrations determined by isotope dilution. The initial Pb isotope ratios at 410 Ma are plotted in Fig. 10 and compared with data from MORB (Cohen et al., 1980; Dupré and Allègre, 1980) and for K-feldspar separates from Caledonian Newer Granites (Blaxland et al., 1979). The data are displaced to higher $^{207}Pb/^{204}Pb$ than modern MORB, but recalculation of the MORB field to 410 Ma assuming μ = 8 would place a few samples within the MORB field. This procedure makes several assumptions which are difficult to justify, however. There is a rough correlation between $^{207}Pb/^{204}Pb$ and $^{206}Pb/^{204}Pb$ with a slope steeper than that of MORB, but the spread of data is little greater than the errors introduced by mass fractionation and age correction. The data lie in the region between the Midland Valley and Southern Uplands granites of Blaxland et al. (1979) and show no trace

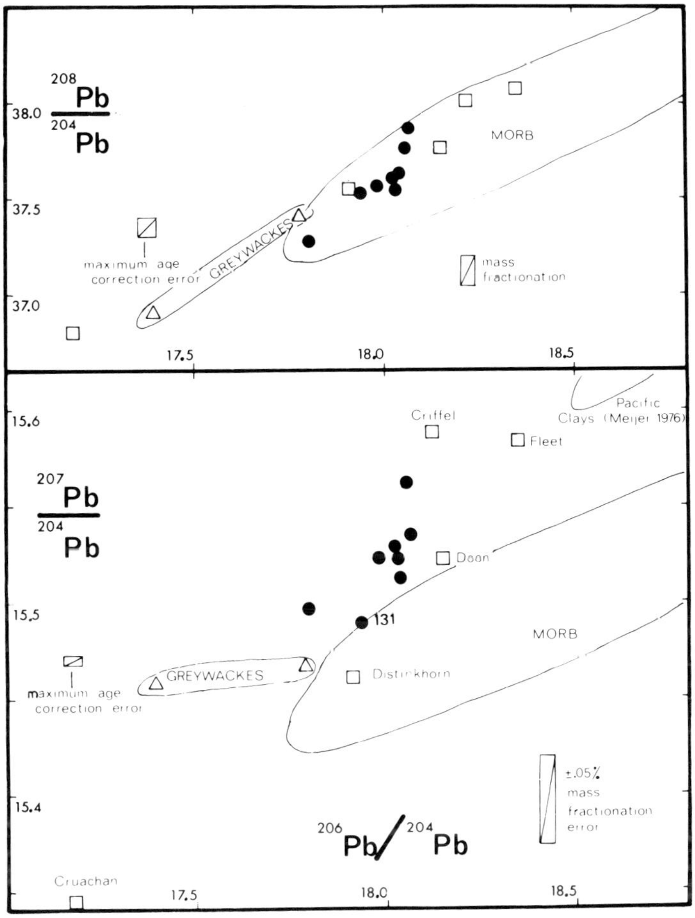

Fig. 10. Pb isotopic data for Fife ORS lavas (●) compared with some Caledonian Newer Granites (Blaxland et al., 1979, □), Southern Uplands greywackes (△), and MORB today (Cohen et al., 1980; Dupré and Allègre, 1980).

of the relatively unradiogenic Pb characteristic of Highland granites, considered by Blaxland et al. to be derived from Archaean granulite facies lower crust.

The small variation in Pb isotope ratios contrasts with the behaviour of Sr and Nd. It is probably significant that the basalt OC118 has the lowest ϵ_{Sr} and $^{206}Pb/^{204}Pb$, and that the high-ϵ_{Sr} andesite OC124 has the highest $^{207}Pb/^{204}Pb$, while the low $^{207}Pb/^{204}Pb$ of OC131 may relate to its transitional character to SW Highland compositions. The other samples are isotopically identical although OC119 has probably had Pb added during alteration (Fig. 4) which may explain its divergence from OC118. The relatively high $^{207}Pb/^{204}Pb$ of all samples and the steep $^{207}Pb/^{204}Pb$ vs. $^{206}Pb/^{204}Pb$ correlation are common features of modern arc volcanics and have been

TABLE IV

U-Pb isotope data

	$^{206}Pb/^{204}Pb$	$^{207}Pb/^{204}Pb$	$^{208}Pb/^{204}Pb$	U ppm	Pb ppm	$^{206}Pb^*/^{204}Pb$	$^{207}Pb^*/^{204}Pb$	$^{208}Pb^*/^{204}Pb$
OC109	18.720	15.567	38.50	3.24	19.5	18.02	15.53	37.60
OC112	18.626	15.555	38.32	1.86	13.0	18.03	15.52	37.55
OC118	18.559	15.538	38.25	1.00	5.5+	17.80	15.50	37.28
OC119	18.259	15.537	38.09	0.99	20.0	18.06	15.54	37.87
	18.276	15.559	38.17	—	19.7			
OC124	18.843	15.605	38.78	2.78	14.9	18.06	15.56	37.77
OC129	18.682	15.561	38.47	1.80	10.8	17.98	15.52	37.57
OC131	18.332	15.513	38.04	2.23	23.3	17.94	15.49	37.53
OC152	18.852	15.557	38.64	0.94	4.9+	18.04	15.51	37.61
SRM981	16.915 ±2 (2se)	15.462	36.605	(mean of five analyses)				

Pb isotope ratios have been corrected for average mass fractionation by normalizing to standard values for SRM981 ($^{206}Pb/^{204}Pb$ = 16.937, $^{207}Pb/^{204}Pb$ = 15.491, $^{208}Pb/^{204}Pb$ = 36.720). In-run errors are $< 0.06\%$. U and Pb concentrations by isotope dilution ($< 2\%$ error, 2σ), except + = XRF Pb ($\pm$ 0.3 ppm, 2σ).
*Pb isotope ratios estimated at 410 Ma using Th/U of 4.0. Th concentrations so estimated are within error of XRF determinations (compare Table II).

interpreted as a result of mixing Pb from the MORB-source with Pb from oceanic sediments (e.g. Sun, 1980). If the Pb isotope data for the Fife lavas is the product of mixing, then either the end-member Pb compositions are closely similar, or one has much larger Pb content than the other, and dominates the isotope composition of mixtures.

DISCUSSION — ORIGIN OF THE PRIMITIVE LAVAS

Depleted mantle

The isotope geochemistry of basalts and andesites from the Fife Old Red Sandstone requires the existence of several isotopically distinct source materials. It is possible that these represent extremes between which the source is continuously variable, but it is nevertheless useful to consider the nature of the source in terms of discrete end-members. One of the mixing end-members is required to have high ϵ_{Nd}, low ϵ_{Sr}, high Sm/Nd and Eu/Eu* $\cong$ 1, and is dominant in the basaltic rocks. Since many of the basalts are very primitive, this end-member must be derived from a mantle source, and the isotopic characteristics suggest that this has been depleted in Rb relative to Sr and in Nd relative to Sm for long periods of time. The basalts analyzed could either be direct melts of this source (i.e. pure depleted end-member) or they could merely represent the analyzed compositions least affected by the mixing process. That the basalts are mixtures is suggested by:

(a) The occurrence of a basalt with higher ϵ_{Nd} in the south Midland Valley (PE5, Thirlwall, 1982a): unfortunately, this sample may have altered Sr isotopes.

(b) Enrichment in Rb, K, U, Ba and Pb relative to Sr, Nd and HFSE is still present in the basalt with highest ϵ_{Nd}, but it is not as marked as in the andesites. This enrichment can not be generated by melting of normal mantle in the absence of Sr-, Nd- and HFSE-bearing residual phases.

(c) The basalts with highest ϵ_{Nd} are still offset to higher ϵ_{Sr} than the mantle array, though not to as great an extent as the andesites.

The composition of the depleted mantle end-member can not be reliably estimated until the mixing mechanism has been elucidated, for, as observed by DePaolo (1981), mixing trajectories do not always point to the end-members. However, it may here be noted that the correlation between ϵ_{Nd} and ϵ_{Sr} in Fig. 8 can be extended to intersect the centre of the mantle array at ϵ_{Nd} between +7.5 and +9.5, values typical of Lower Palaeozoic MORB (Jacobsen and Wasserburg, 1979; Thirlwall, unpubl. data). Similarly, the $^{207}Pb/^{204}Pb$ vs. $^{206}Pb/^{204}Pb$ correlation can be extended to intersect the probable compositional field of Lower Palaeozoic N-MORB. There is thus no need to infer a depleted end-member other than the MORB source or a melt derived from it.

An enriched mantle end-member?

The end-member with low ϵ_{Nd} and high ϵ_{Sr} could in principle be derived from a long-term enriched mantle source, or from a crustal source, with mixing occurring either during ascent of a magma derived from the depleted source (crustal contamination), or prior to melting, through recycling of crustal components in the subduction zone.

If the Nd-Sr isotopic variation were purely to reflect mixing between two mantle sources, then the enriched source would have to lie remote from the mantle array, or would have to have Sr/Nd much higher and ϵ_{Nd} much lower than any observed composition, in order to explain the ϵ_{Nd}-ϵ_{Sr} correlation (Fig. 8) as part of a convex upward hyperbola. However, the samples analyzed show a slight fall in Sr/Nd with decreasing ϵ_{Nd}, and most importantly, lavas with low ϵ_{Nd} are highly siliceous. Melts with 60—62% SiO_2 are unlikely to have been in equilibrium with peridotitic lithologies, and there is no intrinsic reason why a low-ϵ_{Nd} mantle source should give rise to more siliceous magmas. Further, an enriched mantle source would be expected to show equal enrichment of equally incompatible elements, and not the pronounced hydrophile element enrichment (K, Rb, Ba) shown by the Fife ORS lavas.

A mildly enriched source with $\epsilon_{Nd} \cong -2$ is however required to explain the isotope geochemistry of SW Highland lavas (Thirlwall, 1982a), but this lies mostly within the mantle array, and does not have the enrichment in Rb, K, U, Ba or Pb relative to Sr needed to explain the isotope geochemistry of the Fife andesites. However, as noted previously, one of the Fife andesites (OC131) lies within the mantle array and has high concentrations of Sr, Ba, P, and for 60% SiO_2, high K/Rb, La/Yb (Fig. 9a) and Eu/Eu* (Fig. 6) and appears to be transitional to SW Highland compositions. It is thought that OC131, and possibly the other two samples that lie close to the mantle array (OC104, 116) may represent mixing between the depleted mantle source and SW Highland enriched mantle, and that other samples may have some component from SW Highland type mantle.

The Sm/Nd pseudoisochron (Fig. 9c) is a feature common to all the ORS lavas: the regression defined solely by the Fife lavas analyzed here is identical within error in slope (ca. 2000 Ma "age") and intercept ("ϵ_{Nd} (2000)" = +17) to that defined by all ORS primitive lavas (Thirlwall, 1982a). It should be strongly emphasized that this can not represent an erupted mantle isochron, as the initial ratio would require the existence of extremely depleted mantle prior to 2000 Ma, for which there is no evidence from studies of Archaean rocks (e.g. McCulloch and Compston, 1981). Instead, the pseudoisochron must represent a mixing line: the colinearity with SW Highland compositions could suggest that Nd and REE concentrations are at least in part controlled by mixing between SW Highland enriched mantle and overlying depleted mantle.

Nature of the crustal component

The high SiO_2, Rb, K, U and Pb of the enriched end-member is therefore most likely to be crustally derived. Seismic refraction work across Scotland has shown that the Midland Valley crust is of normal thickness (ca. 35 km, Bamford, 1979), and is composed of a relatively thin (ca. 8 km) layer of sediments overlying metamorphic basement. The metamorphic basement is presumably that sampled by many Carboniferous diatremes, shown by Graham and Upton (1978) to contain many clasts from a K-, Rb-depleted granulite facies terrain. This is presumably also U-depleted in a similar fashion to the Lewisian granulites (Moorbath et al., 1969), and may have comparably unradiogenic Pb and relatively unradiogenic Sr. Unfortunately, few compositional data are available, but it is unlikely that a contribution from this source could provide the enrichment in K, Rb and Pb shown by the andesites without change in Pb isotopes to less radiogenic compositions. Indeed, the Pb isotopes of the andesites appear to be somewhat more radiogenic than in the basalts.

The thin layer of sediments beneath the Old Red Sandstone is probably comparable to the Lower Palaeozoic sediments of the Southern Uplands accretionary prism in bulk composition if not in structure. It is noteworthy, however, that no such sediment clasts have been retrieved from the Midland Valley Carboniferous diatremes, despite the wide range in Carboniferous and metamorphic clasts these contain (B.G.J. Upton, pers. commun., 1982). The Southern Uplands sediments range in age from Lower Ordovician to Wenlock and are preserved in a series of thrust-bound slices which show upward sequences from spilitic basalts with MORB affinities (Lambert et al., 1981) through radiolarian cherts and pelagic shales to greywackes (Leggett et al., 1979). The oldest slices, close to the Southern Uplands Fault, have basaltic basement and are dominated by greywackes, rich in mafic volcanic clasts, probably arc-derived, while in the younger slices decollement appears to have occurred at the pelagic shale horizon (Leggett et al., 1979) and the greywackes are richer in acid continental detritus (Halliday et al., 1980). Rb-Sr isotope data have been reported for a range of Southern Uplands greywackes by Halliday et al. (1980): ϵ_{Sr} (410) is low (mean of +9) in the Ordovician mafic clast greywackes but increases to high values (+70 or more) in the younger Ordovician and Silurian samples. Chemical and Pb isotope analyses of these samples are reported in Appendix 3, and mean incompatible element concentrations of the mafic and more acid types are shown in Fig. 11, normalized to the estimated mantle composition (cf. Fig. 7).

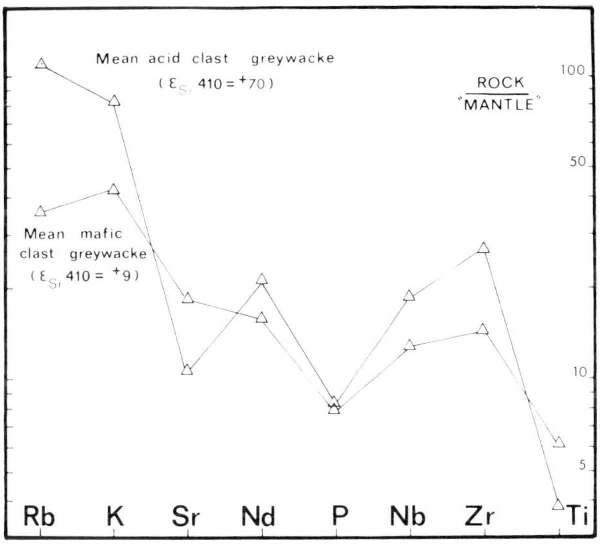

Fig. 11. Mean incompatible element concentrations of mafic (4 samples) and acid (7 samples) Southern Uplands greywackes, normalized to estimated mantle (cf. Fig. 7). ϵ_{Sr} data from Halliday et al. (1980).

It is immediately clear from comparing Fig. 11 with Fig. 7 that the analyzed sediments have the appropriate enrichment in K, Rb and also Ba and Pb relative to Sr, Nd and HFSE to generate the range in Rb/Sr, Rb/Nd etc. present in the lavas by mixing with a melt derived from a depleted mantle end-member. Additionally, they have marked depletion in

P and Ti which could explain their unenriched concentrations in the lavas. Although they are siliceous (59% and 65% SiO_2 in the mean mafic and acid greywackes, respectively), they have relatively high Ni and Cr concentrations (181 and 81 ppm Ni, respectively), and could thus produce the silica enrichment in the primitive andesites, by mixing with basalt. Since Sr/Nd of the sediments is similar to that of the lavas, and, presumably, the depleted mantle end-member, mixing lines on the ϵ_{Nd}-ϵ_{Sr} diagram would be straight. As Sr and Nd concentrations in the sediments are comparable to those of reasonable small partial melts of depleted mantle (?5% melts), Sr and Nd isotope ratios in the mixtures would not be dominated by either end-member, but show a wide spread. In contrast, K, Rb, Ba and Pb in the mixtures would be dominantly derived from the sediments, and thus their Pb isotopic composition would be essentially that of the sedimentary end-member.

The Sr isotope data of Halliday et al. (1980) suggest that the Sr isotopes of the andesites could be explained by derivation of all their Sr from mafic greywackes or more than 30% of their Sr from acid greywackes. Although Nd isotope data for the sediments are at present rather limited, both mafic [ϵ_{Nd} (410) ≅ +1, A.N. Halliday, pers. commun., 1982] and acid greywackes

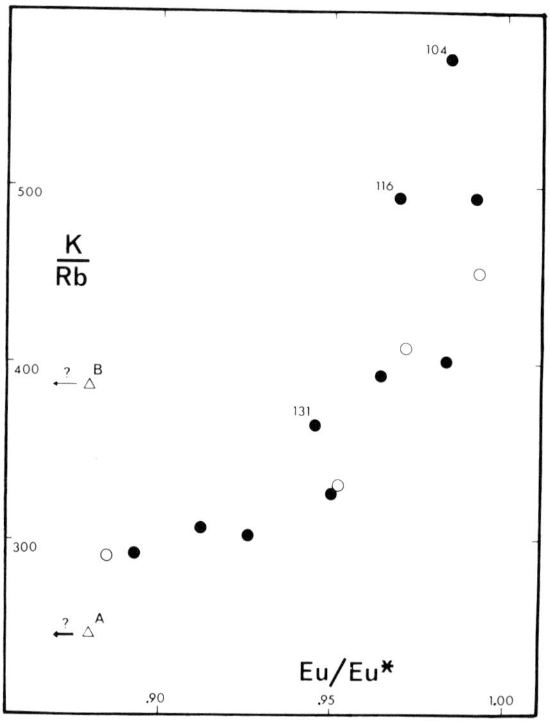

Fig. 12. Hyperbolic mixing relationship for K/Rb and Eu/Eu*. *A* and *B* are mean K/Rb of Southern Uplands acid and basic greywackes, respectively.

[ϵ_{Nd} (410) from −5 to −10, A.N. Halliday; A. Tindle and G. Rogers, pers. commun., 1982] lie on the extension of the Fife lava ϵ_{Nd}-ϵ_{Sr} correlation to more enriched compositions. The relatively tight clustering of the ORS data even at high ϵ_{Sr} suggests that an end-member with substantially higher ϵ_{Sr} than observed in the lavas i.e. acid greywacke, is the most plausible. Pb isotope data for two greywackes, one mafic, one acid, are compared with the Fife lavas in Fig. 10. Despite their terrigenous nature, the greywackes have less radiogenic Pb than the analyzed lavas, although this may reflect U-Pb alteration of the sediments, and the sediments may have a wider spread than those analyzed here. Pb in the lavas could be 100% derived from the sediments, but this can not be proved. The similarity between the lava Pb data and that from Southern Uplands granite feldspars (Blaxland et al., 1979) may again be noted (Fig. 10): the granites were suggested to be mixes of mantle- and sediment-derived material by Halliday et al. (1980) on the basis of Sr and O isotopic data, with greatest proportions of the sediment-derived component at Fleet and Criffel.

Hyperbolic mixing relationships will only be expected when comparison is made between isotope or element ratios of which the ratio of concentrations of the denominator elements is markedly different in the mixing end-members (Langmuir et al., 1978). An example of this is given in Fig. 12, where all samples except the three anomalously close to the mantle array fall on a well-defined hyperbola governed by the much higher Rb/REE ratio of the sediment component. Plagioclase fractionation is not needed to explain the Eu anomalies.

Mechanism of mixing

Four possible methods of incorporating a sedimentary component into a melt of depleted mantle (or vice versa) may be suggested.

(a) Bulk assimilation of sediment in the Midland Valley crust

The mechanism of bulk assimilation requires that the magmas be superheated prior to assimilation, otherwise fractional crystallization will occur to provide the heat necessary for assimilation (DePaolo, 1981). Concentrations of elements in the primitive lavas should lie on linear mixing trends between the two end-members. Pb and Sr concentrations in the andesites are higher than in the analyzed sediments, and the sediments only show a range in La/Y from 0.6 to 1.2. Bulk assimilation of these can not produce the marked increase in La/Y with increasing SiO_2, to values in excess of 1.5, shown by the lavas (Fig. 5).

(b) Assimilation-fractional crystallization

Assimilation associated with fractional crystallization (AFC) is a much more plausible mechanism than assimilation alone, as superheated magmas are relatively unusual. Such a mechanism could explain the higher Sr and Pb concentrations in the andesites if these were incompatible. As the assimilant has relatively high Ni and Cr, extensive fractional crystallization would not be manifested in fall of Ni, Cr and V to concentrations below those observed in the primitive andesites (DePaolo, 1981). The 6% garnet fractionation model suggested earlier is certainly feasible when associated with assimilation, but the 80% augite fractionation model would produce dramatic enrichment (ca. 4-fold) in Zr and Nb in the andesites, which is not observed. For the garnet fractionation model to be tenable, the AFC process must occur within the stability field of garnet in non-peraluminous basalt-basic andesite magma. The 8-km Lower Palaeozoic sediment layer in the Midland Valley (Bamford, 1979) is highly unlikely to be deep enough to encompass the garnet stability field. A further point which suggests that AFC processes are unlikely to have been operative is the behaviour of the lavas which can be shown chemically to have undergone fractional crystallization: the rhyolites and the low-Cr mafic lavas. These are isotopically indistinguishable from the primitive lavas and show no evidence for AFC processes at all. If the isotope geochemistry of the primitive lavas is to be explained by AFC processes, then these must be followed by fractional crystallization without assimilation occurring in shallower magma chambers.

(c) Incorporation of a sediment-derived melt

Contamination of a magma derived from depleted mantle by a melt produced from country rock sediment could explain the higher Sr and Pb in the lavas as their concentrations in the sediment melt would be a function of the residual mineralogy. In addition to providing enrichment in Rb, K, Pb, Ba and U, the sediment melt must also supply sufficient SiO_2 to the mixture to provide the increase from 50 to 60% observed in the andesites. As 77% is a maximum reasonable SiO_2 content for the sediment melt, the andesites would have to consist of at least 45% sediment melt. To generate the observed correlations with SiO_2, a sediment melt with 77% SiO_2 would have to have La/Y of about 5.6 (Fig. 5), ϵ_{Nd} of ca. −8, ϵ_{Sr} of ca. +50 and Eu/Eu* of about 0.7. While the latter three parameters are perfectly plausible it is difficult to see how a melt with La/Y as high as 5.6 can be generated from the sediments without residual garnet. Garnet is unlikely to remain residual at more than some 5% melting of a non-peraluminous quartz-rich sediment, and yet the andesites must contain at least 45 wt.% of this melt. Such a small percentage partial melt would have concentrations of many incompatible elements far too high to produce the andesite chemistry by 45% mixing with basalt. For example, if Rb were perfectly incompatible, the melt would contain some 1500 ppm Rb (mean ppm Rb in acid greywackes/fraction of melt formed) compared with the ca. 150 ppm Rb at 77% SiO_2 required to generate the SiO_2-Rb correlation (Fig. 4). To obtain only 150 ppm Rb in the melt, D_{Rb}^{bulk} would need to be about 0.4, with none of the Rb-bearing phases contributing to the melt. This seems most implausible.

Further, the 1500 ft (500 m) thick lava pile can be estimated to contain an average 20% sediment melt at 77% SiO_2: this is equivalent to total extraction of a 5% melt from a 2-km thick sedimentary pile. This is unlikely, since the Lower Palaeozoic layer only forms about the upper 8 km of the crust (Bamford, 1979), and there is no evidence of a melt component derived from the underlying metamorphic rocks, which are presumably hotter. The participating thickness of the sedimentary layer is also substantially reduced by the fact that fractional crystallization without melt incorporation must take place at higher levels in the layer, and that garnet is probably only stable at the base of the layer, if at all.

If the andesites were allowed to represent even greater % sedimentary melt component, the SiO_2 and La/Y of the melt could be reduced considerably, but correspondingly more melt must be available. An additional problem of incorporating such a large amount of sediment melt is that, since Ni and Cr levels in the melt would be low, their concentrations in the mixtures would be correspondingly low.

(d) Incorporation of sedimentary material into the source by subduction

Although it can not be conclusively proved, it seems unlikely that the sediment-derived component can be incorporated during ascent of mantle-derived magmas through the crust. The circumstantial evidence against this is principally the thinness of the Lower Palaeozoic sedimentary layer, and the fact that one major petrogenetic process, fractional crystallization to low-Cr lavas and rhyolite, has to take place in this layer anyway, thereby reducing the volume of sediment able to participate in a fundamentally different process.

The subduction zone environment permits recycling of crustal materials into the mantle source region of arc volcanics. Could the sedimentary component in the mixing relationships represent subducted material? Assuming a subduction rate of 2 cm yr^{-1} and a subduction angle of 45°, sediment supplied to the mantle beneath the N. Midland Valley would have taken some 10 Ma to travel from the oceanic trench. This corresponds to an age of the youngest sediments in an accretionary slice of 417 ± 6 Ma, perhaps mid-Silurian. The sedimentary component in the mixing relations is therefore likely to be of Silurian age, and not the Ordovician mafic clast greywackes collected by Halliday et al. (1980) from close to the Southern Uplands Fault. Silurian greywackes are a more appropriate end-member because of the relative coherence of the mixing relationships in the lavas at ϵ_{Sr} values comparable to those of the mafic clast greywackes. Decollement appears to have occurred at the pelagic shale horizon, however (Leggett et al., 1979), and greywackes may not have been subducted during the Lower Silurian. It is possible that the onset of ORS volcanism may reflect the first subduction of greywackes, and slowing in the rate of accretion, during the Wenlock (cf. van Breemen and Bluck, 1981).

The isotope geochemical effects of sediment subduction are difficult to constrain rigorously. It is likely that mantle close to the slab would be grossly modified by siliceous sediment-derived melt, and converted to pyroxenitic or, at lower temperatures, amphibolitic lithologies. These may rise as diapirs until their solidus is reached, when segregation of a primary andesite melt may occur. The isotope geochemistry of this primary andesite will depend on the original nature of the mantle above the slab, and on the melting relationships of the subducted sediments. Strong LREE enrichment in the primary andesite could thus be produced by residual garnet during the original sediment melting or during melting of the diapir, or by the existence of an LREE-enriched mantle region immediately above the subducted slab. The presence of the SW Highland mantle "signature" in one of the andesites analyzed here (OC131) suggests that magma generation occurred close to or within a boundary region between the mildly enriched SW Highlands mantle and overlying depleted mantle (Thirlwall, 1982a). The existence of SW Highland mantle, which we know can yield basaltic melts with low Sm/Nd and low ϵ_{Nd}, suggests that the behaviour of REE, except Eu, may be controlled by mixing with SW Highland mantle. Otherwise it is difficult to see why the ORS Sm/Nd pseudoisochron should be consistent throughout the ORS province.

The rising pyroxenitic diapirs will interact strongly with adjacent peridotitic mantle, especially after melt has begun to form. Small diapirs or the margins of large ones would be expected to interact most: they may either be mixed mechanically with the surrounding depleted mantle or they may incorporate a melt derived from it. Such modified peridotitic lithologies are expected to melt to yield the basaltic lavas. The isotopic composition of these will essentially be a function of the extent of mixing between the depleted mantle and the high SiO_2, low ϵ_{Nd}, high ϵ_{Sr} pyroxenitic diapir. Basaltic lavas with the isotopic signature of SW Highlands mantle are not produced, because only the mantle closest to the Benioff zone, and thus most influenced by the sediment-derived melt, is of this type. After segregation of an andesitic melt from the pyroxenitic diapir further interaction with overlying mantle is also highly probable, unless the magma path had developed an "insulating jacket" of pyroxenitic lithologies. Such a model can thus in principle produce the observed mixing relationships in the basalts and andesites. It is highly unlikely that pure end-member sources would ever be reflected in an erupted lava because the sediment-derived melt reacts immediately with peridotitic mantle, while the mantle alone is probably unable to melt until water and other components have been provided by the sedimentary melt.

SUMMARY

(1) The isotope geochemistry of relatively primitive basalts and andesites from the Fife Old Red Sandstone (late Silurian) is controlled by mixing between a component derived from depleted mantle and a component

derived from Lower Palaeozoic greywackes. Less primitive mafic lavas and rhyolites are produced by closed system fractional crystallization of these at high levels in the crust.

(2) Rb, K, Pb, U and Ba in the lavas are dominantly recycled from the crust, but their Sr and REE are largely mantle-derived. The relative lack of enrichment in Sr compared to many modern arcs is due to the subduction of low-Sr acid greywackes; enrichment of Sr in many modern arcs (e.g. the Lesser Antilles, Hawkesworth and Powell, 1980) may reflect subduction of carbonates. REE behaviour, except Eu, is probably governed by magma generation close to the boundary zone between deep mildly enriched mantle (Thirlwall, 1982a) and overlying depleted mantle.

(3) Mixing between components probably occurs close to the subducted oceanic lithosphere. Siliceous melts derived from subducted greywackes may convert the overlying mantle to pyroxenitic or amphibolitic bodies, which rise diapirically and undergo melting to produce "primary" andesite. Reaction between the diapirs or the primary andesite and surrounding depleted peridotite may produce the observed mixing relations in the lavas. Such a model is however not well constrained by the isotope geochemistry of the lavas or by phase relationships at the depth of interest.

ACKNOWLEDGEMENTS

Helpful criticism of the manuscript was provided by Drs M.J. Norry and I.W. Luff, while Dr R.A. Cliff and Mr P.G. Guise provided useful advice on techniques etc. The mineral separations were performed by Messrs. W. Wilkinson and T. Oddy. XRF and microprobe work were carried out at Edinburgh University during tenure of a NERC research studentship. Isotope and REE research at Leeds is funded by NERC and the Royal Society.

REFERENCES

Bamford, D., 1979. Seismic constraints on the deep geology of the Caledonides of northern Britain. In: A.L. Harris, C.H. Holland and B.E. Leake (Editors), The Caledonides of the British Isles — reviewed. Spec. Publ. Geol. Soc. London, 8: 93—96.

Blaxland, A.B., Aftalion, M. and van Breemen, O., 1979. Pb isotopic composition of feldspars from Scottish Caledonian granites, and the nature of the underlying crust. Scott. J. Geol., 15: 139—151.

Cawthorn, R.G. and O'Hara, M.J., 1976. Amphibole fractionation in calc-alkaline magma genesis. Am. J. Sci., 276: 309—329.

Cohen, R.S., Evensen, N.M., Hamilton, P.J. and O'Nions, R.K., 1980. U-Pb, Sm-Nd and Rb-Sr systematics of mid-ocean ridge basalt glasses. Nature, 283: 149—153.

DePaolo, D.J., 1981. Trace element and isotopic effects of combined wall-rock assimilation and fractional crystallization. Earth Planet. Sci. Lett., 53: 189—202.

DePaolo, D.J. and Wasserburg, G.J., 1977. The sources of island arcs as indicated by Nd and Sr isotopic studies. Geophys. Res. Lett., 4: 465—468.

Dewey, J.F., 1971. A model for the Lower Palaeozoic evolution of the southern margin of the early Caledonides of Scotland and Ireland. Scott. J. Geol., 7: 219—240.

Dupré, B. and Allègre, C.-J., 1980. Pb-Sr-Nd isotopic correlation and the chemistry of the North Atlantic mantle. Nature, 286: 17—22.

Evans, A.L., Mitchell, J.G., Embleton, B.J.J. and Creer, K.M., 1971. Radiometric age of the Devonian polar shift relative to Europe. Nature, Phys. Sci., 229: 50—51.

Geikie, A., 1902. The Geology of Eastern Fife. Mem. Geol. Surv. Great Britain (Scotland).

Graham, A.M. and Upton, B.G.J., 1978. Gneisses in diatremes, Scottish Midland Valley: petrology and tectonic implications. J. Geol. Soc. London, 135: 219—228.

Halliday, A.N., Aftalion, M., van Breemen, O. and Jocelyn, J., 1979. Petrogenetic significance of Rb-Sr and U-Pb isotopic systems in the 400 Ma old British Isles granitoids and their hosts. In: A.L. Harris, C.H. Holland and B.E. Leake (Editors), The Caledonides of the British Isles — reviewed. Spec. Publ. Geol. Soc. London, 8: 653—661.

Halliday, A.N., Stephens, W.E. and Harmon, R.S., 1980. Rb-Sr and O isotopic relationships in 3 zoned Caledonian granitic plutons, Southern Uplands, Scotland: evidence for varied sources and hybridization of magmas. J. Geol. Soc. London, 137: 329—348.

Harris, J.W., 1928. Notes on the extrusive igneous rocks of the Dundee district. Trans. Edinb. Geol. Soc., 12: 105—110.

Hawkesworth, C.J. and Powell, M., 1980. Magma genesis in the Lesser Antilles island arc. Earth Planet. Sci. Lett., 51: 297—308.

Hawkesworth, C.J., Norry, M.J., Roddick, J.C. and Baker, P.E., 1979. $^{143}Nd/^{144}Nd$, $^{87}Sr/^{86}Sr$ and incompatible element variations in calc-alkaline andesites and plateau lavas from South America. Earth Planet. Sci. Lett., 42: 45—57.

Irving, A.J., 1978. A review of experimental studies of crystal—liquid trace element partitioning. Geochim. Cosmochim. Acta, 42: 743—770.

Jacobsen, S.B. and Wasserburg, G.J., 1979. Nd and Sr isotopic study of the Bay of Islands ophiolite complex and the evolution of the source of mid-ocean ridge basalts. J. Geophys. Res., 84: 7429—7445.

Lambert, R.St.J., Holland, J.G. and Leggett, J.K., 1981. Petrology and tectonic setting of some Ordovician volcanic rocks from the Southern Uplands of Scotland. J. Geol. Soc. London, 138: 421—436.

Langmuir, C.H., Vocke, R.D.Jr., Hanson, G.N. and Hart, S.R., 1978. A general mixing equation with applications to Icelandic basalts. Earth Planet. Sci. Lett., 37: 380—392.

Leggett, J.K., McKerrow, W.S. and Eales, M.H., 1979. The Southern Uplands of Scotland: a Lower Palaeozoic accretionary prism. J. Geol. Soc. London, 136: 755—770.

McCulloch, M.T. and Compston, W., 1981. Sm-Nd age of Kambalda and Kanowna greenstones and heterogeneity in the Archaean mantle. Nature, 294: 322—327.

McKerrow, W.S., Lambert, R.St.J. and Chamberlain, V., 1980. The Ordovician, Silurian and Devonian timescales. Earth Planet. Sci. Lett., 51: 1—8.

Moorbath, S., Welke, H. and Gale, N.H., 1969. The significance of lead isotope studies in ancient, high grade, metamorphic basement complexes, as exemplified by the Lewisian rocks of northwest Scotland. Earth Planet. Sci. Lett., 6: 245—256.

Nagasawa, H., 1970. Rare earth concentrations in zircons and apatites and their host dacites and granites. Earth Planet. Sci. Lett., 9: 359—364.

Osborn, E.F., 1962. Reaction series for subalkaline igneous rocks based on different oxygen pressure conditions. Am. Mineral., 47: 211—226.

Pearce, J.A. and Norry, M.J., 1979. Petrogenetic implications of Ti, Zr, Y and Nb variations in volcanic rocks. Contrib. Mineral. Petrol., 69: 33—47.

Roeder, P.L. and Emslie, R.F., 1970. Olivine-liquid equilibrium. Contrib. Mineral. Petrol., 29: 275—289.

Shervais, J.W., 1982. Ti-V plots and the petrogenesis of modern and ophiolitic lavas. Earth Planet. Sci. Lett., 59: 101—118.

Sun, S.-S., 1980. Lead isotopic study of young volcanic rocks from mid-ocean ridges, oceanic islands and island arcs. Philos. Trans. R. Soc. London, Ser. A, 297: 409—445.

Tatsumi, Y. and Ishizaka, K., 1981. Existence of andesitic primary magma: an example from southwest Japan. Earth Planet. Sci. Lett., 53: 124—130.

Thirlwall, M.F., 1979. The Petrochemistry of the British Old Red Sandstone Volcanic Province. Ph.D. Thesis, Edinburgh University.

Thirlwall, M.F., 1981. Implications for Caledonian plate tectonic models of chemical data from volcanic rocks of the British Old Red Sandstone. J. Geol. Soc. London, 138: 123—138.

Thirlwall, M.F., 1982a. Systematic variation in chemistry and Nd-Sr isotopes across a Caledonian calc-alkaline volcanic arc: implications for source materials. Earth Planet. Sci. Lett., 58: 27—50.

Thirlwall, M.F., 1982b. A triple filament method for rapid and precise analysis of rare earth elements by isotope dilution. Chem. Geol., 35: 155—166.

Van Breemen, O. and Bluck, B.J., 1981. Episodic granite plutonism in the Scottish Caledonides. Nature, 291: 113—117.

Wood, B.J. and Banno, S., 1973. Garnet-orthopyroxene and orthopyroxene-clinopyroxene relationships in simple and complex systems. Contrib. Mineral. Petrol., 42: 109—124.

Wright, A.E., 1976. Alternating subduction direction and the evolution of the Atlantic Caledonides. Nature, 264: 156—160.

Appendices on following pages.

APPENDIX 1

Mean analyses of phenocryst minerals in Fife lavas

	OC152 Ol	OC84 Opx	OC93 Opx	OC98 Opx	OC103 Opx	OC108 Opx	OC123 Opx	OC140 Opx*
SiO_2	39.42	55.48	53.65	54.07	55.90	53.51	55.48	53.07
TiO_2	0.03	0.19	0.44	0.36	0.15	0.40	0.18	0.39
Al_2O_3	0.06	1.89	1.30	1.29	1.67	2.81	2.05	0.71
Cr_2O_3	0.05	0.44	0.09	0.15	0.37	0.05	0.66	0.03
FeO*	17.38	10.27	17.32	13.82	8.87	14.99	9.47	19.68
MnO	0.30	0.18	0.38	0.32	0.19	0.30	0.24	0.61
MgO	42.79	30.26	25.11	27.59	30.90	27.06	30.82	22.68
NiO	0.22	—	—	—	—	—	—	—
CaO	0.16	1.60	1.83	1.97	1.89	1.43	1.32	2.24
Total	100.41	100.32	100.12	99.59	99.95	100.54	100.23	99.41
En	(81.4)	81.4	69.5	75.1	83.0	74.1	83.1	64.2
Fs	(18.6)	15.5	26.9	21.1	13.4	23.0	14.3	31.3
Wo	—	3.1	3.6	3.8	3.6	2.8	2.6	4.6
N	5	3	4	5	3	4	5	2

* = ?xenocryst, *N* = number of core analyses in mean, Ol = olivine, Opx = orthopyroxene.

	OC84 Cpx	OC86 Cpx	OC93 Cpx	OC98 Cpx	OC103 Cpx	OC108 Cpx	OC123 Cpx	OC131 Cpx	OC140 Cpx	OC152 Cpx
SiO_2	51.04	49.33	51.56	51.28	52.01	51.91	52.24	51.90	52.06	50.51
TiO_2	0.82	1.15	0.80	0.72	0.63	0.74	0.58	0.80	0.93	0.89
Al_2O_3	3.69	6.34	2.18	2.36	3.63	2.76	2.50	2.15	1.80	3.86
Cr_2O_3	0.44	0.06	0.09	0.27	0.59	0.02	0.25	0.40	0.03	0.47
FeO*	7.81	6.53	8.57	9.35	6.24	8.72	6.38	7.87	9.80	5.93
MnO	0.18	0.14	0.20	0.27	0.18	0.27	0.18	0.23	0.35	0.15
MgO	15.72	15.24	15.98	15.94	17.15	15.05	16.50	16.45	14.94	15.67
CaO	19.25	19.48	19.43	18.33	18.94	20.42	20.42	19.24	19.48	21.55
Na_2O	0.49	0.59	0.35	0.55	0.47	0.42	0.35	0.38	0.44	0.30
Total	99.44	98.87	99.25	99.07	99.85	100.30	99.42	99.44	99.87	99.33
En	46.3	46.3	46.0	46.4	50.1	43.5	47.5	47.4	43.4	45.4
Fs	12.9	11.1	13.8	15.3	10.2	14.1	10.3	12.7	16.0	9.7
Wo	40.8	42.6	40.2	38.3	39.7	42.4	42.2	39.9	40.7	44.9
N	5	2	6	3	6	2	4	3	1	5

Cpx = clinopyroxene.

APPENDIX 1 (continued)

	OC84 Pl	OC86 Pl	OC93 Pl	OC98 Pl	OC103 Pl	OC108 Pl	OC128 Pl	OC131 Pl	OC140 Pl	OC152 Pl
SiO_2	53.81	53.41	54.71	56.30	53.90	52.72	64.54	54.38	51.81	50.30
TiO_2	0.08	0.04	0.08	0.10	0.08	0.06	0.01	0.07	0.06	0.05
Al_2O_3	28.78	27.98	27.73	25.78	28.43	29.53	21.80	27.40	29.52	30.03
FeO*	0.47	0.38	0.66	0.94	0.59	0.49	0.15	0.65	0.52	0.82
MgO	0.09	0.08	0.09	0.12	0.16	0.07	0.00	0.16	0.07	0.08
CaO	11.64	11.64	10.63	9.11	11.70	12.56	3.05	11.03	12.98	13.99
Na_2O	4.69	4.71	5.17	5.90	4.60	4.17	9.06	5.04	4.04	3.46
K_2O	0.34	0.26	0.52	0.63	0.38	0.22	0.98	0.43	0.26	0.26
Total	99.91	98.49	99.60	98.89	99.83	99.84	99.60	99.17	99.27	99.01
An	56.7	56.8	51.6	44.3	57.2	61.6	14.8	53.4	63.0	68.0
Ab	41.4	41.7	45.4	52.0	40.6	37.1	79.5	44.1	35.5	30.5
Or	2.0	1.5	3.0	3.7	2.2	1.3	5.7	2.5	1.5	1.5
N	5	6	5	4	4	4	5	3	6	5

Pl = plagioclase, *N* = number of core analyses in mean, FeO* = total iron as FeO, 1290 ppm Sr in OC152 Pl.

	OC98 Sp	OC128 Sp	OC152 Sp		OC128 Bi
SiO_2	0.00	0.46	0.31	SiO_2	36.04
TiO_2	15.20	2.30	8.66	TiO_2	4.10
Al_2O_3	2.00	2.70	2.90	Al_2O_3	14.57
Cr_2O_3	1.40	0.00	5.41	FeO*	18.97
Fe_2O_3	32.60	57.87	43.26	MnO	0.67
FeO	41.16	30.62	34.57	MgO	12.29
MnO	2.50	1.84	0.45	Na_2O	0.65
MgO	0.08	0.36	2.92	K_2O	8.63
ZnO	0.06	0.05	0.05	Total	95.93
CaO	0.14	0.00	0.07		
Total	95.15	96.21	98.60	N	3
N	4	5	6		

Sp = spinels, Bi = biotite. Spinel iron oxides recalculated assuming R_3O_4. Also 6560 ppm Ba in OC128 Bi.

APPENDIX 2

Chemical analyses of samples from east of Glenfarg for which isotope data have not been presented. Samples with lower numbers than OC108 are from further west than Fig. 1, others not on Fig. 1 are from further north

	OC82	OC83	OC84	OC85	OC86	OC90	OC91	OC92	OC93	OC94
SiO_2	54.73	55.75	55.50	52.22	55.20	52.84	52.30	52.40	58.79	59.20
Al_2O_3	18.15	17.18	17.34	17.76	18.27	16.23	16.09	18.75	15.72	15.55
Fe_2O_3*	7.85	7.68	7.51	8.63	8.08	8.46	7.25	7.23	6.49	6.83
MgO	4.32	4.54	4.80	5.23	3.71	6.92	6.84	3.23	4.83	3.74
CaO	8.32	7.26	7.17	8.87	6.61	9.90	10.66	9.88	6.08	7.03
Na_2O	3.68	3.89	3.63	3.76	4.38	3.35	3.16	4.47	3.37	3.13
K_2O	1.343	1.700	1.667	1.350	1.624	1.135	1.680	1.618	2.691	2.737
TiO_2	1.353	1.306	1.325	1.201	1.428	1.249	1.242	1.572	1.075	1.068
MnO	0.154	0.131	0.127	0.161	0.094	0.086	0.100	0.096	0.072	0.081
P_2O_5	0.283	0.288	0.291	0.292	0.359	0.234	0.294	0.300	0.291	0.291
LOI	0.74	0.78	0.76	1.06	0.76	2.75	3.53	3.10	1.39	2.76
Ni	52	63	62	103	19	210	296	41	90	89
Cr	28	92	85	222	26	409	507	33	154	163
V	187	150	138	191	117	167	184	200	112	110
Sc	26.5	20.5	19.1	25.9	16.0	25.9	26.2	21.9	17.1	18.6
Cu	90	30	27	30	17	32	37	28	37	26
Zn	75	65	68	67	64	73	93	127	66	61
Pb	10.4	10.6	10.3	—	12.5	7.9	8.9	13.2	14.1	15.8
Sr	627	521	541	566	628	597	674	709	735	661
Rb	31.3	42.1	43.1	29.9	34.0	16.8	33.6	39.5	79.6	83.2
Zr	187	199	200	188	249	169	228	209	244	261
Nb	9.7	12.3	11.6	7.0	12.7	7.7	7.6	7.6	13.3	14.0
Ba	390	400	412	422	425	330	519	446	796	725
Th	4	6	7	4	5	4	6	4	11	10
La"	23	25	26	22	26	19	22	20	42	40
Ce"	55	55	55	49	65	38	50	48	89	82
Nd"	26	28	26	25	30	18	28	27	39	37
Y	24.8	27.8	26.2	25.1	31.2	22.7	24.3	29.6	23.4	25.3
Pheno-	pl-ol	pl-cpx	pl-cpx	pl-opx	pl-ol	ol	pl-ol	ol-cpx	pl-opx	pl-opx
crysts	cpx	ol-opx	ol-opx	ol-cpx	cpx-?hb		cpx	pl	cpx-mt	cpx-mt
"%"	15	20	10	25	3	8	30	4	25	30
Grid	NO165	NO177	NO195	NO189	NO249	NO228	NO223	NO232	NO231	NO233
ref.	118	112	113	134	120	163	166	168	179	180

Abbreviations and symbols as Table II. OC111 is from an intrusion, the Lucklaw Hill Felsite, while OC107 may be intrusive.

OC95	OC96	OC98	OC100	OC101	OC105	OC106	OC107	OC108	OC111	OC113
56.63	52.36	60.55	56.35	54.73	55.80	61.47	55.73	58.03	74.81	52.84
18.22	15.77	15.87	16.78	16.70	18.41	17.28	18.52	17.61	14.56	18.64
7.79	8.36	5.79	7.29	7.64	7.04	5.10	7.02	6.62	1.22	9.41
2.81	8.19	4.52	5.13	5.73	4.71	2.94	4.62	2.93	0.38	3.51
5.91	7.73	5.21	7.17	8.14	4.01	4.53	4.38	6.19	0.91	7.49
4.54	3.14	3.90	4.09	3.70	4.31	4.29	3.18	4.36	2.40	3.99
1.936	1.829	1.973	1.186	0.911	3.005	2.351	4.454	2.035	6.08	1.598
1.240	1.405	1.045	1.238	1.304	1.560	0.996	1.266	1.358	0.128	1.669
0.118	0.149	0.158	0.098	0.109	0.052	0.115	0.075	0.101	0.042	0.074
0.439	0.370	0.358	0.251	0.281	0.365	0.267	0.253	0.251	0.047	0.344
1.13	2.53	1.49	0.59	1.07	1.90	1.15	3.71	0.98	2.01	1.19
6	207	79	56	99	23	22	15	12	4	17
3	433	145	130	135	8	23	12	8	4	8
112	166	108	136	155	164	107	183	162	3	166
14.5	24.0	12.9	20.0	22.7	20.5	13.5	21.7	17.0	1.2	19.5
11	26	23	26	35	15	19	16	20	3	16
62	74	71	66	70	120	65	84	69	24	82
12.4	6.5	16.0	6.6	6.9	16.8	17.0	14.3	15.8	14.7	8.3
615	544	850	558	610	537	623	500	656	102	665
45.2	46.4	19.0	25.3	7.7	48.8	59.2	52.6	53.2	127	32.6
258	266	245	162	183	240	222	173	194	115	226
13.4	9.3	15.1	8.7	11.3	13.1	12.5	9.4	13.4	20.3	16.0
511	517	959	335	371	461	645	619	595	1010	408
5	6	8	1	4	7	10	8	7	12	6
28	27	46	18	22	31	35	31	26	23	28
68	58	88	40	48	64	79	61	62	55	57
33	31	37	20	23	30	35	30	28	24	30
29.8	27.8	20.2	21.3	24.1	27.5	24.9	23.0	25.8	26.5	30.4
pl-mt	ol	pl-opx cpx-mt	pl-ol	ol-pl	ol	pl-opx cpx-mt	pl-opx	pl-opx cpx	pl-bi -mt	pl-ol
0	5	30	10	5	2	15	0	2	4	5
NO244	NO268	NO337	NO331	NO356	NO357	NO333	NO375	NO388	NO419	NO446
176	171	237	214	238	249	161	189	173	213	243

APPENDIX 2 (continued)

	OC117	OC120	OC121	OC122	OC123	OC126	OC130	OC132	OC134	OC135
SiO_2	79.18	57.15	56.13	60.27	59.86	75.59	59.03	55.10	60.26	59.38
Al_2O_3	13.89	17.53	17.41	16.55	16.44	13.43	17.97	16.36	16.92	16.94
Fe_2O_3	0.82	6.58	6.59	6.16	6.18	1.75	6.55	7.72	5.83	6.20
MgO	0.20	4.71	4.99	3.33	3.88	0.30	2.57	6.35	3.91	3.41
CaO	0.85	6.50	6.81	5.68	6.16	0.29	4.46	7.10	5.25	5.96
Na_2O	0.17	4.33	4.02	3.94	4.32	5.61	4.58	3.92	3.97	4.06
K_2O	4.725	1.798	1.741	2.269	1.468	3.067	2.431	1.467	2.189	2.178
TiO_2	0.130	1.096	1.113	1.047	1.061	0.120	1.275	1.303	1.076	1.155
MnO	0.069	0.120	0.120	0.072	0.145	0.028	0.110	0.089	0.063	0.104
P_2O_5	0.054	0.284	0.283	0.257	0.250	0.047	0.297	0.244	0.268	0.278
LOI	3.56	1.18	1.01	1.10	1.06	0.63	1.41	1.04	1.22	1.04
Ni	9	81	93	34	30	5	7	115	31	33
Cr	4	154	171	93	67	4	3	255	90	52
V	11	144	150	125	125	2	138	137	131	132
Sc	0.9	21.4	21.7	19.0	16.9	0.8	15.6	22.6	20.8	18.3
Cu	48	32	32	19	17	32	9	26	19	22
Zn	5	72	72	66	67	26	68	68	72	59
Pb	9.3	16.7	14.7	15.8	14.7	—	18.2	10.2	15.9	15.4
Sr	258	710	719	587	657	141	644	493	605	612
Rb	93	41.8	37.0	64.9	29.7	39.3	62.3	36.5	63.1	60.7
Zr	110	191	186	226	225	105	231	186	228	236
Nb	18.2	11.5	11.1	13.9	14.6	19.7	14.3	11.3	14.2	14.3
Ba	656	560	542	565	587	853	618	337	579	576
Th	13	4	6	11	10	13	10	7	9	10
La"	37	27	25	35	37	23	31	26	38	35
Ce"	73	55	50	73	72	53	69	53	74	76
Nd"	26	27	25	31	30	24	32	23	31	31
Y	17.0	24.2	23.0	21.9	23.1	21.9	25.6	24.3	21.6	23.8
Pheno-crysts	?pl-bi	ol-pl cpx	ol-pl cpx	opx cpx-ol	opx cpx-ol	pl-bi mt	pl-?cpx	ol-mt	opx cpx-ol	cpx-pl opx-ol
"%"	5	5	10	4	4	4	4	8	6	1
Grid ref.	NO385 258	NO400 265	NO418 276	NO426 280	NO426 282	NO397 264	NO447 293	NO454 293	NO424 265	NO477 333

OC122, 123, 134—136 and 138 have a few plagioclase microphenocrysts.
OC143, 150 and possibly 137 and 139 are from small intrusions. OC153 is sample S54857 in the Institute of Geological Sciences collection, Edinburgh, and has $^{87}Sr/^{86}Sr$ of 0.70567 ± 5, corresponding to ϵ_{Sr} (410) of +6.0.

OC136	OC137	OC138	OC139	OC143	OC145	OC148	OC149	OC150	OC153
59.03	54.95	59.35	54.97	56.94	59.44	54.76	55.04	54.54	56.74
16.60	17.31	16.68	17.31	17.23	16.19	17.38	16.80	18.11	15.56
6.38	7.41	6.78	7.80	7.06	6.11	7.79	7.47	8.40	7.25
3.62	4.65	3.66	4.37	4.16	4.75	4.41	5.29	3.36	6.65
6.05	8.06	5.41	7.94	6.04	6.02	8.13	7.91	6.09	6.61
4.26	3.97	3.79	3.94	4.49	3.94	3.86	3.65	4.17	3.63
2.189	1.431	2.229	1.373	1.962	1.595	1.344	1.655	2.387	1.808
1.153	1.491	1.149	1.410	1.309	1.119	1.428	1.503	1.664	1.100
0.080	0.142	0.053	0.130	0.121	0.157	0.095	0.109	0.067	0.105
0.272	0.302	0.274	0.267	0.297	0.231	0.264	0.332	0.578	0.276
0.98	0.66	1.14	0.56	1.07	1.19	0.89	1.15	1.01	0.86
43	42	35	30	41	60	31	111	14	164
83	110	53	69	36	78	81	215	8	334
116	158	100	139	125	121	157	169	125	125
18.3	23.4	18.4	19.4	17.9	16.2	20.3	22.2	15.2	18.9
20	30	19	27	26	33	24	32	32	36
65	61	73	61	65	90	61	69	72	66
13.7	7.7	16.6	7.0	12.3	15.9	9.6	10.0	12.6	9.4
590	611	617	567	585	640	599	641	592	573
60.4	32.0	63.5	33.5	50.2	23.2	29.6	30.4	65.3	42.0
229	207	233	194	220	209	189	212	326	188
15.0	11.2	14.2	10.2	13.0	14.2	10.0	14.4	17.9	11.7
559	399	575	410	518	584	399	461	552	584
11	4	12	6	6	14	3	6	10	8
33	23	32	22	29	37	21	28	43	27
73	51	68	51	62	79	49	60	92	61
30	26	29	25	29	32	25	28	46	26
25.7	26.8	25.5	27.8	25.2	22.1	26.6	25.7	37.8	20.7
opx	pl-ol	ol-opx	pl-ol	ol-pl	pl-opx	ol-pl	pl-ol	pl-ol	pl-opx
cpx-ol	cpx	cpx	cpx		cpx-mt				cpx-ol
5	25	8	20	1	35	5	15	1	20
NO485	NO475	NO504	NO489	NO392	NO443	NO513	NO556	NO522	NO265
338	355	355	370	315	310	375	386	417	174

APPENDIX 3

Chemical and Pb isotope analyses of Southern Uplands sediments

	1/2	1/3	1/4	1/5	3	4	5
SiO_2	59.89	58.51	58.70	59.14	66.20	69.86	61.53
Al_2O_3	13.07	13.73	13.29	12.71	14.83	11.79	13.62
Fe_2O_3*	9.31	9.45	8.86	9.22	5.45	4.52	6.54
MgO	6.98	6.37	6.22	7.53	3.72	2.93	4.67
CaO	4.16	5.28	5.20	4.79	2.00	4.32	7.88
Na_2O	2.85	3.13	3.41	2.70	2.20	2.46	1.75
K_2O	1.140	1.494	1.474	1.276	4.091	2.464	2.346
TiO_2	1.545	1.216	1.099	1.296	0.845	0.677	0.843
MnO	0.123	0.163	0.151	0.132	0.086	0.084	0.083
P_2O_5	0.133	0.200	0.188	0.146	0.197	0.168	0.171
LOI	5.30	3.90	2.70	3.95	3.35	4.43	9.15
Ni	199	157	157	211	79	63	73
Cr	596	456	421	686	201	161	165
V	292	283	246	270	108	97	141
Sc	39.1	31.4	28.2	34.7	17.9	14.7	20.4
Cu	22	28	25	27	7	14	18
Zn	61	94	78	82	73	61	87
Pb	4.7	10.2	11.5	8.1	6.7	17.7	9.0
Sr	244	595	616	393	256	251	161
Rb	28.1	31.2	29.5	27.3	135	73	96
Zr	146	129	123	121	306	262	222
Nb	9.3	8.1	8.1	7.7	14.5	11.3	12.7
Ba	188	603	631	267	1021	512	179
Th	3	10	10	5	10	8	14
La"	16	31	29	14	30	26	29
Ce"	29	59	58	33	73	57	63
Nd"	14	26	25	15	31	25	27
Y	25.6	28.1	25.2	24.1	30.0	24.5	28.7

	$^{206}Pb/^{204}Pb$	$^{207}Pb/^{204}Pb$	$^{208}Pb/^{204}Pb$	U ppm	$^{206}Pb^*/^{204}Pb$	$^{207}Pb^*/^{204}Pb$	$^{208}Pb^*/^{204}Pb$
1/3	18.737	15.519	38.59	2.31	17.79	15.47	37.40
9	18.852	15.538	38.95	1.83	17.40	15.46	36.90

Abbreviations and symbols as Tables II and IV. Localities and Sr isotope data in Halliday et al. (1979). 1/2—1/5 are mafic clast greywackes, the remainder are acid clast greywackes.

6	7	8	9
64.00	67.10	63.77	63.48
12.00	14.51	15.51	15.25
5.76	5.12	6.65	7.15
4.07	2.83	4.47	5.23
9.33	2.96	2.31	1.54
2.03	2.48	3.32	3.31
1.890	2.894	2.422	2.119
0.824	0.752	0.762	0.872
0.080	0.073	0.076	0.092
0.183	0.185	0.138	0.186
9.84	1.96	3.93	5.26
58	58	78	160
197	127	232	320
122	85	166	151
17.8	12.5	26.5	25.2
12	12	31	20
73	65	79	80
7.7	16.2	15.0	5.3
199	490	308	212
78	90	77	66
296	275	152	194
12.8	12.7	8.4	12.3
276	825	532	468
12	9	8	7
25	32	22	27
61	72	49	52
27	33	23	24
30.1	26.4	22.3	23.3

APPENDIX (Complementary tables of the article by T.E. Smith et al., pp. 249—278)

TABLE 1

Chemical analyses and structural formulae of plagioclases
Olivine-Pyroxene Gabbronorite Series

OLIVINE-CLINOPYROXENE SUB-SERIES

	MP-22		LP-106		C-16		C-36	
SiO_2	45.03		45.46		45.16		45.95	
Al_2O_3	33.89		34.43		34.59		35.16	
TiO_2	--		0.01		0.01		--	
Fe_2O_3	0.46		1.29		0.75		0.58	
MnO	0.02		0.02		0.02		0.01	
MgO			0.71		0.72		--	
CaO	18.35		17.63		18.12		18.06	
Na_2O	0.14		0.51		0.62		0.72	
K_2O	0.18		0.08		0.19		0.17	
P_2O_5	--		0.02		0.03		--	
Numbers of ions on the basis of 32(O)								
Si	8.455		8.368		8.324		8.408	
Al	7.501	16.02	7.472	16.00	7.514	15.94	7.582	16.07
Ti	--		0.001		0.001		--	
Fe^{3+}	0.065		0.179		0.105		0.080	
Mn	0.002		0.003		0.003		0.001	
Mg	--		0.194		0.197		--	
Ca	3.692	3.79	3.477	3.90	3.578	4.05	3.541	3.84
Na	0.052		0.182		0.222		0.256	
K	0.042		0.019		0.044		0.040	
P	--		0.003		0.005		--	
Ab	1.4		4.7		5.5		6.7	
An	97.5		94.9		93.4		92.3	
Or	1.1		0.5		1.1		1.0	
Name	Anorthite		Anorthite		Anorthite		Anorthite	
Trace Elements (ppm)								
Rb	2		--		2		1	
Sr	487		961		669		487	
Ba	15		13		25		12	

TABLE 1

Chemical analyses and structural formulae of plagioclases
Olivine-Pyroxene Gabbronorite Series

	OLIVINE-CLINOPYROXENE SUB-SERIES							
	MP-18		C-9-A		C-34		MP-59	
SiO_2	43.75		46.41		44.12		46.50	
Al_2O_3	35.82		34.73		33.65		29.85	
TiO_2	--		--		--		--	
Fe_2O_3	0.46		0.80		0.60		0.60	
MnO	0.01		0.02		0.01		0.01	
MgO	--		--		--		--	
CaO	18.28		18.04		18.02		18.06	
Na_2O	0.76		0.76		0.81		0.84	
K_2O	0.17		0.17		0.18		0.17	
P_2O_5	--		--		--		--	
Numbers of ions on the basis of 32(0)								
Si	8.150		8.471		8.373		8.924	
Al	7.864	16.08	7.472	16.05	7.525	15.98	6.752	15.76
Ti	--		--		--		--	
Fe^{3+}	0.065		0.110		0.086		0.087	
Mn	0.001		0.003		0.002		0.002	
Mg	--		--		--		--	
Ca	3.649	3.96	3.528	3.84	3.665	4.01	3.713	4.07
Na	0.273		0.269		0.297		0.311	
K	0.041		0.041		0.043		0.042	
P	--		--		--		--	
Ab	6.9		7.0		7.5		7.6	
An	92.1		92.0		91.4		91.3	
Or	1.0		1.0		1.1		1.1	
Name	Anorthite		Bytownite		Anorthite		Anorthite	
Trace Elements (ppm)								
Rb	--		1		1		--	
Sr	468		546		474		490	
Ba	11		11		15		12	

TABLE 1

Chemical analyses and structural formulae of plagioclases
Olivine-Pyroxene Gabbronorite Series

OLIVINE-CLINOPYROXENE SUB-SERIES

	MP-87		C-44		C-45		C-39	
SiO_2	44.98		46.40		52.93		54.08	
Al_2O_3	34.84		34.63		28.41		28.34	
TiO_2	--		0.02		0.02		0.01	
Fe_2O_3	0.48		0.30		0.60		0.47	
MnO	0.01		0.02		0.01		0.01	
MgO	--		--		--		0.12	
CaO	17.65		16.87		11.50		10.80	
Na_2O	1.28		1.30		5.04		5.4	
K_2O	0.19		0.17		0.49		0.34	
P_2O_5	0.01		--		0.05		0.09	
Numbers of ions on the basis of 32(0)								
Si	8.350		8.514		9.710		9.822	
Al	7.621	16.04	7.489	16.12	6.144	15.94	6.066	15.94
Ti	--		0.002		0.002		0.002	
Fe^{3+}	0.067		0.110		0.083		0.064	
Mn	0.002		0.003		0.002		0.002	
Mg	--		--		--		0.032	
Ca	3.510	4.02	3.36	3.87	2.261	4.18	2.101	4.13
Na	0.461		0.464		1.793		1.901	
K	0.044		0.039		0.114		0.078	
P	--		--		0.008		0.013	
Ab	11.5		12.0		43.0		46.0	
An	87.4		86.9		54.2		51.7	
Or	1.1		1.1		2.8		3.3	
Name	Bytownite		Bytownite		Labradorite		Labradorite	
Trace Elements (ppm)								
Rb	2		1		1		2	
Sr	546		552		549		489	
Ba	34		8		85		23	

TABLE 1

Chemical analyses and structural formulae of plagioclases
Olivine-Pyroxene Gabbronorite Series

	OLIVINE-CLINOPYROXENE SUB-SERIES					
	LP-134	LP-179	MP-31		CM-47	
SiO_2			42.30		46.22	
Al_2O_3			31.21		34.86	
TiO_2			0.01		0.08	
Fe_2O_3			0.66		0.49	
MnO			0.01		--	
MgO			0.43		0.61	
CaO			17.18		17.37	
Na_2O			1.25		1.35	
K_2O			0.17		0.11	
P_2O_5			--		--	
Numbers of ions on the basis of 32(0)						
Si			8.412		8.421	
Al			7.315	15.83	7.485	15.93
Ti			0.001		0.071	
Fe^{3+}			0.098		0.067	
Mn			0.002		--	
Mg			0.127		0.166	
Ca			3.660	4.31	3.391	4.06
Na			0.482		0.477	
K			0.043		0.026	
P			--		--	
Ab			11.2		11.7	
An	88.0	88.0	87.9		87.6	
Or			1.0		0.7	
Name	Bytownite	Bytownite	Bytownite		Bytownite	
Trace Element (ppm)						
Rb	1	2	1			
Sr	1075	934	553			
Ba	24	19	8			

TABLE 1

Chemical analyses and structural formulae of plagioclases
Olivine-Pyroxene Gabbronorite Series

AMPHIBOLE-ORTHOPYROXENE SUB-SERIES

	CM-60		C-53		CM-165		CM-152	
SiO_2	48.62		50.36		54.66		63.15	
Al_2O_3	32.39		30.86		28.66		22.09	
TiO_2	0.01		0.01		0.10		0.03	
Fe_2O_3	0.79		0.61		0.43		0.61	
MnO	--		0.01		--		0.01	
MgO	0.64		--		0.60		0.15	
CaO	15.22		13.24		11.54		8.45	
Na_2O	3.03		3.89		4.57		5.11	
K_2O	0.15		0.22		0.24		0.38	
P_2O_5	--		--		--		--	
Numbers of ions on the basis of 32(O)								
Si	8.848	15.912	9.257	16.04	9.801	15.923	11.192	15.89
Al	6.947		6.685		6.057		4.615	
Ti	0.013		0.002		0.014		0.004	
Fe^{3+}	0.108		0.084		0.058		0.082	
Mn	--	4.25	0.001	4.05	--	4.02	0.001	3.49
Mg	0.174		--		0.160		0.038	
Ca	2.968		2.608		2.217		1.604	
Na	1.069		1.386		1.589		1.756	
K	0.035		0.051		0.055		0.086	
P	--		--		--		--	
Ab	25.2		34.2		39.5		50.3	
An	73.9		64.4		59.1		47.0	
Or	0.9		1.3		1.4		2.5	
Name	Bytownite		Labradorite		Labradorite		Andesine	
Trace Elements (ppm)								
Rb			--					
Sr			734					
Ba			50					

TABLE 1

Chemical analyses and structural formulae of plagioclases
Olivine-Pyroxene Gabbronorite Series

OLIVINE-CLINOPYROXENE SUB-SERIES

	MP-86		CM-63A		CM-92		LP-107
SiO_2	45.96		46.07		45.72		
Al_2O_3	35.34		34.14		33.95		
TiO_2	--		0.01		0.01		
Fe_2O_3	0.58		0.85		1.09		
MnO	0.02		--		0.01		
MgO	--		0.44		0.44		
CaO	17.97		17.48		17.03		
Na_2O	0.93		1.19		0.82		
K_2O	0.17		0.01		0.08		
P_2O_5	0.01		--		0.01		
Numbers of ions on the basis of 32(O)							
Si	8.388	16.07	8.465	15.99	8.483	16.06	
Al	7.600		7.393		7.424		
Ti	--		0.009		0.001		
Fe^{3+}	0.080		0.128		0.152		
Mn	0.002	3.89	0.000	4.01	0.002	3.82	
Mg	--		0.121		0.122		
Ca	3.513		3.441		3.386		
Na	0.329		0.424		0.295		
K	0.041		0.022		0.019		
P	0.001		--		--		
Ab	8.4		10.6		7.7		
An	90.6		88.8		91.8		88.0
Or	1.1		0.6		0.5		
Name	Anorthite		Bytownite		Anorthite		Bytownite
Trace Elements (ppm)							
Rb	1		1		6		1
Sr	554		577		635		1167
Ba	24		22		21		25

TABLE 1

Chemical analyses and structural formulae of plagioclases
Olivine-Pyroxene Gabbronorite Series

AMPHIBOLE-ORTHOPYROXENE SUB-SERIES

	CM-68		CM-88		C-48		CM-119	
SiO_2	44.34		43.80		46.69		50.30	
Al_2O_3	32.01		31.31		32.91		31.44	
TiO_2	0.07		0.07		--		0.09	
Fe_2O_3	0.79		0.95		0.64		0.78	
MnO	--		--		0.02		--	
MgO	0.85		0.38		--		0.64	
CaO	18.76		19.11		16.91		14.92	
Na_2O	1.29		1.37		1.49		2.72	
K_2O	0.13		0.12		0.18		0.19	
P_2O_5	--		--		--		--	
Numbers of ions on the basis of 32(O)								
Si	8.397	15.67	8.385	15.645	8.688	16.00	9.099	15.92
Al	7.145		7.116		7.218		6.703	
Ti	0.010		0.010		--		0.012	
Fe^{3+}	0.113		0.138		0.089		0.106	
Mn	--	4.55	--	4.60	0.003	3.95	--	4.06
Mg	0.240		0.109		--		0.173	
Ca	3.807		3.949		3.371		2.892	
Na	0.474		0.512		0.538		0.954	
K	0.031		0.030		0.042		0.044	
P	--		--		--		--	
Ab	10.4		11.1		13.6		23.5	
An	88.9		88.2		85.4		75.4	
Or	0.7		0.7		1.0		1.1	
Name	Bytownite		Bytownite		Bytownite		Bytownite	
Trace Elements (ppm)								
Rb	1		2		--			
Sr	595		604		660			
Ba	47		27		23			

TABLE 1

Chemical analyses and structural formulae of plagioclases
Olivine-Pyroxene Gabbronorite Series

AMPHIBOLE GABBRO SERIES

	LP-184		CM-28		LP-117	LP-112	
SiO_2	49.59		49.68			54.11	
Al_2O_3	33.32		30.44			28.54	
TiO_2	0.02		0.03			0.03	
Fe_2O_3	0.79		1.27			0.90	
MnO	0.01		0.01			0.01	
MgO	0.20		0.16			0.31	
CaO	15.87		15.24			10.59	
Na_2O	1.93		2.73			4.79	
K_2O	0.08		0.13			0.12	
P_2O_5	0.09		0.01			--	
Numbers of ions on the basis of 32(O)							
Si	8.893	16.05	9.137	15.98		9.823	16.056
Al	7.044		6.600			6.106	
Ti	0.003		0.004			0.004	
Fe^{3+}	0.106		0.143			0.123	
Mn	0.002	3.81	0.002	4.06		0.002	3.856
Mg	0.053		C.044			0.084	
Ca	3.049		3.003			2.060	
Na	0.671		0.974			1.686	
K	0.018		0.031			0.028	
P	0.013		0.002			--	
Ab	17.8		24.0			44.7	
An	81.7		75.1		73.0	54.6	
Or	0.5		0.9			0.7	
Name	Bytownite		Bytownite		Bytownite	Labradorite	
Trace Elements (ppm)							
Rb	11		2			7	
Sr	579		795			1199	
Ba	104		90			59	

TABLE 1

Chemical analyses and structural formulae of plagioclases
Olivine-Pyroxene Gabbronorite Series

AMPHIBOLE GABBRO SERIES

	LP-200		LP-104		C-57	
SiO_2	45.28		46.57		44.24	
Al_2O_3	33.11		33.29		33.87	
TiO_2	0.05		0.02		0.01	
Fe_2O_3	0.68		0.99		0.59	
MnO	--		0.01		0.01	
MgO	0.12		0.19		0.20	
CaO	17.72		17.52		17.11	
Na_2O	1.11		1.47		1.48	
K_2O	0.11		0.10		0.18	
P_2O_5	0.01		0.06		--	
Numbers of ions on the basis of 32(0)						
Si	8.509	16.00	8.573	16.02	8.366	16.00
Al	7.335		7.224		7.548	
Ti	0.062		0.006		0.001	
Fe^{3+}	0.096		0.206		0.085	
Mn	--	4.03	0.002	4.07	0.002	4.11
Mg	0.034		0.052		0.056	
Ca	3.568		3.456		3.466	
Na	0.404		0.525		0.544	
K	0.026		0.023		0.043	
P	0.002		0.009		--	
Ab	10.0		12.9		13.2	
An	89.4		86.2		85.7	
Or	0.6		0.6		1.1	
Name	Bytownite		Bytownite		Bytownite	
Trace Elements (ppm)						
Rb	1		6		1	
Sr	1012		1001		695	
Ba	26		85		22	

TABLE 2

OLIVINES

Chemical analyses and structural formations of olivines

	LP113		MP-22		C-16		C-36		LP107		MP-18	
SiO_2	38.36		37.13		37.76		38.61		37.70		36.90	
Al_2O_3	--		1.61		2.92		1.74		0.93		0.98	
TiO_2	0.02		0.03		0.03		0.04		0.11		0.03	
FeO	20.53		19.53		19.50		20.48		20.71		21.76	
MnO	0.24		0.27		0.26		0.30		--		0.29	
MgO	41.46		39.28		37.61		36.51		36.44		36.95	
CaO	--		0.32		0.42		0.52		0.24		0.46	
Numbers of ions on the basis of 4(O)												
Si	0.984	0.98	0.973		0.982		1.011		1.011		0.987	
Al	--		0.050		0.090		0.054		0.029		0.031	
Ti	--		0.001		0.001		0.001		0.002		0.001	
Fe^{2+}	0.440	2.03	0.428	2.03	0.424	1.99	0.449	1.95	0.464	1.96	0.487	2.01
Mn	0.005		0.006		0.006		0.007		--		0.007	
Mg	1.585		1.535		1.458		1.425		1.456		1.473	
Ca	--		0.009		0.012		0.015		0.007		0.013	
Mg#	78.3		78.2		77.5		76.0		75.8		75.1	
Name	Chrysolite		Chrysolite		Chrysolite		Chrysolite		Chrysolite		Chrysolite	
Trace elements (ppm)												
Nb			4		4		6		1		6	
Ni			369		504		229		114		163	
Co			94		95		91		58		90	
Cr			73		93		42		--		63	
V			8		17		15		--		9	

TABLE 2

OLIVINES

Chemical analyses and structural formations of olivines

	MP-86		C-34		MP-59		MP-31		C-44		C-9-A	
SiO_2	41.24		38.82		37.85		40.32		39.51		37.80	
Al_2O_3	1.30		0.51		1.61		1.80		1.93		0.92	
TiO_2	0.07		0.07		0.04		2.03		0.10		0.03	
FeO	20.28		20.83		22.05		20.03		20.66		23.38	
MnO	0.31		0.33		0.32		0.57		0.33		0.37	
MgO	34.21		33.61		33.72		30.31		31.13		34.62	
CaO	1.88		3.01		1.33		3.49		4.29		1.20	
Numbers of ions on the basis of 4(O)												
Si	1.063		1.037		1.016		1.054		1.046		1.008	
Al	0.040		0.016		0.051		0.055		0.060		0.029	
Ti	0.001		0.001		0.001		0.040		0.002		0.001	
Fe^{2+}	0.437	1.85	0.466	1.92	0.495	1.94	0.438	1.82	0.457	1.88	0.521	1.97
Mn	0.007		0.007		0.007		0.013		0.007		0.008	
Mg	1.315		1.339		1.349		1.181		1.228		1.376	
Ca	0.052		0.086		0.038		0.098		0.122		0.034	
Mg#	75.0		74.2		73.1		72.9		72.9		72.5	
Name	Chrysolite		Chrysolite		Chrysolite		Chrysolite		Chrysolite		Chrysolite	
Trace elements (ppm)												
Nb	5		5		6		4		6		3	
Ni	301		92		193		23		195		104	
Co	87		197		90		145		82		211	
Cr	151		110		106		8		356		34	
V	51		41		22		61		82		14	

TABLE 3

Chemical analyses and structural formulae of clinopyroxenes
Olivine pyroxene Gabbro norite Series

OLIVINE - CLINOPYROXENE SUB-SERIES

	C-44		MP31		MP86		CM85		C-9-A		C-34		CM63A	
SiO_2	51.39		51.35		50.71		53.45		46.47		53.12		51.26	
Al_2O_3	4.17		4.34		4.30		3.02		6.79		4.24		5.60	
TiO_2	0.59		0.62		0.45		0.31		0.38		0.40		0.63	
FeO	7.70		7.24		7.28		7.37		7.42		8.42		7.80	
MnO	0.31		0.29		0.17		0.19		0.14		0.18		--	
MgO	18.55		17.21		17.26		15.60		15.46		16.78		13.54	
CaO	20.47		20.53		19.97		21.94		16.37		21.08		19.68	
Na_2O	--		0.28		--		--		--		0.18		0.58	
K_2O	--		0.02		--		0.100		--		--		0.11	
Numbers of ions on the basis of 6(0)														
Si	1.843	2.00	1.861	2.00	1.866	2.00	1.934	2.00	1.830	2.00	1.883	2.00	1.901	2.00
Al^{iv}	0.157		0.139		0.134		0.066		0.170		0.117		0.099	
Al^{vi}	0.019		0.047		0.053		0.062		0.145		0.060		0.146	
Ti	0.016		0.017		0.012		0.008		0.011		0.011		0.018	
Fe^{2+}	0.231		0.219		0.224		0.223		0.244		0.250		0.242	
Mn	0.009	2.05	0.009	2.04	0.005	2.03	0.006	2.00	0.005	2.00	0.005	2.02	--	1.98
Mg	0.991		0.930		0.947		0.841		0.907		0.886		0.748	
Ca	0.786		0.797		0.787		0.850		0.690		0.800		0.782	
Na	--		0.020		--		--		--		0.012		0.042	
K	--		0.001		--		0.005		--		--		0.005	
Mg	48.3		45.6		46.7		42.1		45.4		43.9		37.8	
Fe	13.4		14.3		14.5		15.1		20.1		15.9		20.3	
Ca	38.3		40.1		38.8		42.8		34.5		40.2		41.9	
Name	Augite		Augite		Augite		Augite		Augite		Augite		Augite	
Mg#	87.1		80.9		80.9		79.0		78.8		78.0		75.5	
Trace Elements (ppm)														
Rb	3		5				3		6		5		3	
Sr	71		102				24		116		29		52	
Ba	49		65				35		34		39		37	
Y	72		31				1		9		12		18	
Zr	29		--				3		5		11		--	
Nb	2		6				--		4		4		9	
Ni	32		14				215		87		72		6	
Co	22		31				64		31		30		27	
Cr	41		16				1494		905		478		--	
V	203		208				--		279		279		--	

TABLE 3

Chemical analyses and structural formulae of clinopyroxenes
Olivine pyroxene Gabbro norite Series

AMPHIBOLE - ORTHOPYROXENE SUB-SERIES

	CM68		C-53		CM165		CM88	
SiO_2	52.35		50.87		52.78		49.72	
Al_2O_3	4.51		2.98		5.75		4.97	
TiO_2	0.44		0.52		0.44		0.66	
FeO	7.90		10.62		10.84		16.51	
MnO	--		0.32		--		--	
MgO	13.03		14.31		11.94		16.98	
CaO	20.13		19.42		16.35		7.50	
Na_2O	0.51		0.53		1.03		0.29	
K_2O	0.12		--		0.16		0.12	
Numbers on ions on the basis of 6(O)								
Si	1.945	2.00	1.914	2.00	1.956	2.00	1.903	2.00
Al^{iv}	0.055		0.086		0.044		0.097	
Al^{vi}	0.142	1.97	0.046	2.02	0.208	1.95	0.128	1.98
Ti	0.012		0.015		0.012		0.019	
Fe^{2+}	0.245		0.301		0.336		0.529	
Mn	--		0.010		--		--	
Mg	0.722		0.831		0.660		0.969	
Ca	0.801		0.783		0.649		0.308	
Na	0.037		0.039		0.074		0.022	
K	0.006		--		0.008		0.006	
Mg	36.6		41.1		33.8		48.9	
Fe	20.6		18.2		28.7		34.1	
Ca	42.8		40.7		37.5		17.0	
Name	Augite		Augite		Augite		Sub Ca-Augite	
Mg#	74.7		73.4		66.3		64.7	
Trace elements (ppm)								
Rb	--		5		12		3	
Sr	51		73		93		50	
Ba	19		69		45		27	
Y	15		11		41		17	
Zr	27		--		15		--	
Nb	4		--		4		1	
Ni	2		23		15		39	
Co	32		54		35		59	
Cr	--		29		--		--	
V	--		202		--		--	

TABLE 4

Chemical analyses and structural formulae of orthopyroxenes
Olivine — pyroxene Series

OLIVINE - CLINOPYROXENE SUB-SERIES

	MP-31	LP107	CM92	C-44	MP-87
SiO_2	51.78	51.60	49.91	50.43	49.37
Al_2O_3	3.13	3.94	3.22	3.23	2.47
TiO_2	0.37	0.35	0.16	0.32	0.13
FeO	14.75	14.01	15.88	14.79	17.64
MnO	0.52	--	0.35	0.55	0.36
MgO	29.15	26.62	28.77	25.02	26.43
CaO	3.82	1.49	0.97	3.58	1.03
Na_2O	0.05	0.14	--	0.10	--
K_2O	0.02	0.07	0.12	--	0.01
Numbers of ions on the basis of 6(O)					
Si	1.827 ⎤ 2.00	1.889 ⎤ 2.00	1.833 ⎤ 2.00	1.878 ⎤ 2.00	1.865 ⎤ 2.00
Al^{iv}	0.130	0.111 ⎦	0.139	0.122 ⎦	0.110
Al^{vi}	--	0.059 ⎤ 2.02	--	0.020 ⎤ 2.05	--
Ti	0.010	0.010	0.004	0.009	0.004
Fe^{iii}	0.033 ⎦	--	0.024 ⎦	--	0.021 ⎦
Fe^{ii}	0.402 ⎤ 2.10	0.429	0.464 ⎤ 2.09	0.461	0.536 ⎤ 2.08
Mn	0.016	--	0.011	0.017	0.012
Mg	1.533	1.453	1.575	1.389	1.488
Ca	0.144	0.058	0.038	0.143	0.042
Na	0.003	0.010	--	0.007	--
K	0.001 ⎦	0.003 ⎦	0.006 ⎦	-- ⎦	-- ⎦
Mg	73.0	71.9	75.4	67.8	71.5
Fe	20.0	24.6	22.5	24.9	26.5
Ca	7.0	3.5	2.1	7.3	2.0
Name	inverted pigeonite	Bronzite	Bronzite	inverted pigeonite	Bronzite
Mg#	77.9	77.2	76.3	75.1	72.8
Trace elements (ppm)					
Rb	1	--	7	2	
Sr	14	20	13	16	
Ba	33	11	6	24	
Y	--	5	1	--	
Zr	--	--	--	--	
Nb	3	11	--	--	
Ni	19	29	1181	24	
Co	76	34	92	96	
Cr	17	--	264	17	
V	102	--	---	89	

TABLE 4

Chemical analyses and structural formulae of orthopyroxenes
Olivine — pyroxene Series

AMPHIBOLE - ORTHOPYROXENE SUB-SERIES

	CM152		CM165		CM60		CM109	
SiO_2	50.17		51.12		49.30		50.82	
Al_2O_3	3.19		2.15		4.93		2.26	
TiO_2	1.80		0.49		0.61		0.93	
FeO	23.09		24.28		23.37		24.61	
MnO	0.58		--		--		--	
MgO	17.49		16.66		14.41		15.02	
CaO	4.63		2.55		3.84		3.56	
Na_2O	0.08		0.03		0.33		0.28	
K_2O	0.39		0.08		0.11		0.14	
Numbers of ions on the basis of 6(0)								
Si	1.883	2.00	1.982	2.00	1.924	2.00	1.978	2.00
Al^{iv}	0.117		0.018		0.076		0.022	
Al^{vi}	0.024		0.081		0.151		0.081	
Ti	0.051		0.014		0.018		0.027	
Fe^{iii}	--		--		--		--	
Fe^{ii}	0.725		0.787		0.763		0.801	
Mn	0.018		--		--		--	
Mg	0.979		0.963		0.838		0.871	
Ca	0.186	2.01	0.106	1.96	0.161	1.96	0.148	1.96
Na	0.006		0.002		0.025		0.021	
K	0.019		0.004		0.005		0.007	
Mg	48.7		49.2		42.8		44.4	
Fe	40.8		45.1		47.5		46.6	
Ca	10.5		5.7		9.7		9.0	
Name	inverted pigeonite				inverted pigeonite		inverted pigeonite	
Mg#	57.4		55.0		52.3		52.1	
Trace elements (ppm)								
Rb	12		--		18		17	
Sr	18		25		112		19	
Ba	157		10		35		55	
Y	8		13		27		54	
Zr	51		--		3		47	
Nb	--		29		23		15	
Ni	20		77		48		47	
Co	119		59		55		67	
Cr	63		--		--		--	
V	--		--		--		--	

TABLE 4

Chemical analyses and structural formulae of orthopyroxenes
Olivine — pyroxene Series

AMPHIBOLE - ORTHOPYROXENE SUB-SERIES

	C-48		C-53		CM68		CM119	
SiO_2	50.46		51.34		51.10		51.11	
Al_2O_3	2.56		1.55		3.94		2.03	
TiO_2	1.31		0.25		0.60		0.29	
FeO	18.15		18.83		17.43		23.21	
MnO	0.57		0.62		--		--	
MgO	22.66		22.59		19.17		17.61	
CaO	3.67		2.28		4.54		2.11	
Na_2O	--		--		0.63		0.17	
K_2O	--		--		0.13		0.09	
Numbers of ions on the basis of 6(O)								
Si	1.996	2.00	1.947	2.00	1.931	2.00	1.986	2.00
Al^{iv}	0.004		0.053		0.069		0.014	
Al^{vi}	0.109		0.017		0.107		0.079	
Ti	0.037		0.007		0.017		0.008	
Fe^{iii}	--		--		--		--	
Fe^{ii}	0.566		0.597		0.551		0.754	
Mn	0.018	2.02	0.020	2.01	--	1.99	--	1.97
Mg	1.260		1.277		1.08		1.020	
Ca	0.147		0.093		0.184		0.088	
Na	--		--		0.046		0.013	
K	--		--		0.006		0.005	
Mg	62.4		63.5		54.3		51.8	
Fe	30.3		31.9		33.8		42.8	
Ca	7.3		4.6		11.9		5.4	
Name	inverted pigeonite		inverted pigeonite		inverted pigeonite		inverted pigeonite	
Mg#	69.0		68.1		66.2		57.5	
Trace elements (ppm)								
Rb	3		4		17		14	
Sr	4		9		35		13	
Ba	30		26		24		12	
Y	1		1		11		5	
Zr	3		--		--		7	
Nb	1		4		17		9	
Ni	1		16		50		34	
Co	45		46		49		58	
Cr	44		51		--		--	
V	130		105		--		--	

TABLE 5

Chemical analyses and structural formulae of the amphiboles
Olivine — Pyroxene Gabbronorite Series

	OLIVINE - CLINOPYROXENE SUB-SERIES											
	C-16		MP-22		MP-18		C-36		MP-86		MP-31	
SiO_2	43.25		43.87		41.37		41.13		44.00		40.21	
Al_2O_3	13.78		14.51		19.78		17.88		11.11		13.17	
TiO_2	0.45		0.13		0.09		0.28		0.73		2.25	
FeO	8.04		8.21		9.41		9.26		11.44		9.72	
MnO	0.15		0.15		0.14		0.14		0.17		0.20	
MgO	18.47		17.16		15.59		14.97		18.49		15.24	
CaO	10.51		10.94		10.46		10.93		10.72		12.16	
Na_2O	1.74		1.08		1.78		0.98		0.82		1.87	
K_2O	0.12		0.11		0.06		0.12		0.21		0.14	
Numbers of ions on the basis of 23(O)												
Si	6.259	8.00	6.345	8.00	5.872	8.00	6.018	8.00	6.388	8.00	6.043	8.00
Al^{iv}	1.741		1.655		2.128		1.982		1.612		1.957	
Al^{vi}	0.609		0.818		1.181		1.101		0.289		0.376	
Ti	0.049		0.016		0.010		0.031		0.080		0.254	
Fe^{2+}	0.973	5.63	0.993	5.55	1.117	5.62	1.133	5.55	1.389	5.78	1.222	5.29
Mn	0.018		0.018		0.017		0.017		0.021		0.025	
Mg	3.984		3.699		3.298		3.265		4.001		3.414	
Ca	1.629		1.695		1.591		1.713		1.667		1.958	
Na	0.488	2.14	0.303	2.02	0.490	2.09	0.278	2.01	0.231	1.94	0.545	2.53
K	0.022		0.020		0.011		0.022		0.039		0.027	
Name	Pargasitic Hornblende		Tschermakitic Hornblende		Pargasite		Tschermakite		Tschermakitic Hornblende		Pargasite	
Mg#	80.4		78.8		74.7		74.2		74.2		73.6	
Trace Elements (ppm)												
Rb	4				1				8		4	
Sr	86				114				70		129	
Ba	53				12				83		168	
Y	21				--				24		128	
Zr	31				--				32		12	
Nb	10				--				4		5	
Ni	170				34				142		19	
Co	29				74				41		29	
Cr	144				44				378		59	
V	121				31				262		414	

TABLE 5

Chemical analyses and structural formulae of the amphiboles
Olivine — Pyroxene Gabbronorite Series

OLIVINE - CLINOPYROXENE SUB-SERIES

	CM92		C-44		MP-87		LP179		CM47	
SiO_2	45.39		43.73		42.62		41.96		46.85	
Al_2O_3	14.02		13.38		15.66		13.87		8.46	
TiO_2	1.31		2.21		1.20		2.00		1.18	
FeO	11.48		10.22		10.28		11.52		14.07	
MnO	0.20		0.22		0.16		--		--	
MgO	17.83		15.26		14.41		14.66		13.18	
CaO	10.43		12.10		11.07		11.32		12.29	
Na_2O	2.02		2.04		3.30		1.80		0.56	
K_2O	0.31		0.13		0.35		0.24		0.30	
Numbers of ions on the basis of 23(0)										
Si	6.243	8.00	6.249	8.00	6.118	8.00	6.149	8.00	6.919	8.00
Al^{iv}	1.757		1.751		1.882		1.851		1.081	
Al^{vi}	0.516		0.503		0.768		0.544		0.391	
Ti	0.136		0.238		0.130		0.220		0.131	
Fe^{2+}	1.321	5.65	1.221	5.24	1.234	5.23	1.412	5.38	1.738	5.16
Mn	0.023		0.027		0.019		--		--	
Mg	3.655		3.251		3.083		3.202		2.901	
Ca	1.537		1.853		1.703		1.777		1.945	
Na	0.539	2.13	0.565	2.44	0.918	2.69	0.511	2.33	0.160	2.16
K	0.054		0.024		0.064		0.045		0.057	
Name	Pargasite		Pargasite		Magnesio-Taramite		Ferroan Pargasite		Magnesio Hornblende	
Mg#	73.4		72.7		71.4		69.4		62.5	
Trace Elements (ppm)										
Rb	5		4		5				4	
Sr	113		130		72				51	
Ba	90		154		141				59	
Y	10		121		67				17	
Zr	--		11		66				36	
Nb	241		5		6				9	
Ni	72		23		101				29	
Co	395		28		37				--	
Cr	--		59		106				490	
V	--		374		291				--	

TABLE 5

Chemical analyses and structural formulae of the amphiboles
Olivine — Pyroxene Gabbronorite Series

AMPHIBOLE - ORTHOPYROXENE SUB-SERIES

	CM152		C-48		CM68		CM88		CM119	
SiO	50.94		48.36		46.77		44.70		49.05	
Al_2O_3	9.14		10.94		9.19		10.98		7.15	
TiO_2	1.20		0.89		1.86		2.15		0.97	
FeO	14.03		9.48		12.18		12.26		13.51	
MnO	0.23		0.23		--		--		--	
MgO	11.93		13.95		13.39		12.58		13.27	
CaO	10.77		13.94		12.46		12.68		12.34	
Na_2O	0.99		1.01		1.62		1.98		0.72	
K_2O	0.91		0.05		0.38		0.31		0.49	
Numbers of ions on the basis of 23(0)										
Si	7.204	8.00	6.855	8.00	6.813	8.00	6.559	8.00	7.162	8.00
Al^{iv}	0.796		1.145		1.187		1.441		0.838	
Al^{vi}	0.728		0.683		0.391		0.458		0.393	
Ti	0.128		0.095		0.204		0.237		0.107	
Fe^{2+}	1.659	5.06	1.124	4.88	1.484	4.99	1.504	4.95	1.650	5.04
Mn	0.028		0.028		--		--		--	
Mg	2.515		2.947		2.907		2.751		2.888	
Ca	1.632		2.117		1.945		1.993		1.931	
Na	0.271	2.067	0.278	2.40	0.458	2.47	0.563	2.62	0.204	2.23
K	0.164		0.009		0.071		0.058		0.091	
Name	Magnesio Hornblende		Magnesio Hornblende		Edenite		Edenitic Hornblende		Magnesio Hornblende	
Mg#	73.4		72.4		66.2		64.6		63.6	
Trace Elements (ppm)										
Rb	7		5		15		--		19	
Sr	73		107		99		120		41	
Ba	277		181		44		52		61	
Y	37		25		39		57		51	
Zr	1323		--		71		45		70	
Nb	--		5		1		4		19	
Ni	63		12		62		41		20	
Co	88		60		48		45		47	
Cr	107		25		--		--		--	
V	--		182		--		--		--	

TABLE 5

Chemical analyses and structural formulae of the amphiboles
Olivine — Pyroxene Gabbronorite Series

AMPHIBOLE - GABBRO SERIES

	CM28		LP184		LP117		LP200		C-57	
SiO	50.04		42.48		42.68		39.90		43.39	
Al_2O_3	8.17		13.29		13.29		13.51		12.71	
TiO_2	0.60		2.38		2.12		2.75		1.97	
FeO	12.56		12.40		12.40		12.23		13.98	
MnO	--		--		--		--		0.21	
MgO	15.78		14.11		13.40		13.05		12.91	
CaO	11.38		11.14		11.25		10.88		11.55	
Na_2O	0.55		1.83		1.57		1.80		1.86	
K_2O	0.26		0.21		0.29		0.25		0.16	
Numbers of ions on the basis of 23(0)										
Si	7.089	8.00	6.212	8.00	6.287	8.00	6.075	8.00	6.380	8.00
Al^{iv}	0.911		1.788		1.713		1.925		1.620	
Al^{vi}	0.454		0.503		0.594		0.499		0.582	
Ti	0.064		0.262		0.235		0.315		0.218	
Fe^{2+}	1.488	5.34	1.516	5.36	1.527	5.30	1.557	5.33	1.547	5.20
Mn	--		--		--		--		0.026	
Mg	3.332		3.076		2.942		2.962		2.829	
Ca	1.727		1.745		1.775		1.775		1.819	
Na	0.151	1.93	0.519	2.30	0.448	2.28	0.531	2.36	0.530	2.38
K	0.047		0.039		0.054		0.049		0.030	
Name	Magnesio-Hornblende		Ferroan Pargasite		Ferroan Pargasitic Hornblende		Ferroan Pargasite		Ferroan Pargasitic Hornblende	
Mg#	69.1		67.0		65.8		65.5		64.6	
Trace Elements (ppm)										
Rb	4		3		3		13		6	
Sr	33		181		130		227		107	
Ba	59		79		101		93		167	
Y	8		27		25		28		64	
Zr	36		32		46		25		28	
Nb	6		8		8		8		4	
Ni	18		35		13		9		9	
Co	--		--		--		--		35	
Cr	466		338		507		56		62	
V	--		--		--		--		481	